3rd International Workshop on State of the Art in Nuclear Cluster Physics 2014

AA008926

Journal of Physics: Conference Series Volume 569

Yokohama, Japan
26-30 May 2014

Editors:

Taiichi Yamada **Yoshiko Kanada-En'yo**

ISBN: 978-1-63439-700-1
ISSN: 1742-6588

Printed from e-media with permission by:

Curran Associates, Inc.
57 Morehouse Lane
Red Hook, NY 12571

CURRAN ASSOCIATES INC. proceedings.com

Some format issues inherent in the e-media version may also appear in this print version.

Printed by Curran Associates, Inc. (2015)

For permission requests, please contact the Institute of Physics
at the address below.

Institute of Physics
Dirac House, Temple Back
Bristol BS1 6BE UK

Phone: 44 1 17 929 7481
Fax: 44 1 17 920 0979

techtracking@iop.org

Additional copies of this publication are available from:

Curran Associates, Inc.
57 Morehouse Lane
Red Hook, NY 12571 USA
Phone: 845-758-0400
Fax: 845-758-2634
Email: curran@proceedings.com
Web: www.proceedings.com

Table of contents

Volume 569

2014

3rd International Workshop on "State of the Art in Nuclear Cluster Physics"
26–30 May 2014, Yokohama, Japan

Accepted papers received: 11 November 2014
Published online: 8 December 2014

Preface

Papers

Journal of Physics: Conference Series **569** (2014) 012001 doi:10.1088/1742-6596/569/1/012001

Present status of nuclear cluster physics (Theory)

Hisashi Horiuchi

Research Center for Nuclear Physics, Osaka University, Osaka 567-0047, Japan

E-mail: horiuchi@rcnp.osaka-u.ac.jp

Abstract. Present status is discussed, from the theoretical viewpoint, for some selected topics of nuclear cluster physics. What is stressed here is that many of basic concepts of nuclear cluster physics since the days of K. Wildermuth are now examined and revived in new lights. An important example is the duality character of the ground-state wave function which was well known already in 1950 s and has been recently revived in order to explain observed large magnitudes of E0 transitions between cluster states and the ground state in many nuclei. The other topics we discuss are (1) spatial localization of clusters v.s. TH R wave function, (2) coexistence of cluster states with mean-field-type states, and (3) spatial localization of clusters and ab initio calculations of cluster states.

1. Introduction

According to the request of the organizers, I discuss present status of nuclear cluster physics from the theoretical side by selecting some topics. What is stressed in my talk is the fact that many of basic concepts of nuclear cluster physics since the days of K. Wildermuth are now examined and revived in new lights with fruitful results.

An important example is the concept of the duality of the shell-model wave function of the ground state which means that the ground-state wave function can be rewritten in the form of cluster-model wave function. The duality character of the ground-state wave function which was well known already in 1950's was recently revived in order to explain observed large magnitudes of E0 transitions between cluster states and the ground state in many nuclei.

Other topics we selected in addition to the duality of the shell model wave function of the ground state are (1) spatial localization of clusters v.s. THSR wave function, (2) coexistence of cluster states with mean-field-type states, and (3) spatial localization of clusters and ab initio calculations of cluster states.

2. The duality of the shell model wave function of the ground state

2.1. Recognition of the duality of the ground state wave function in 1950's

When Wildermuth started his microscopic cluster model study of nuclear structure in 1958 [1], the shell model had already been well established because it was proposed about 10 years ago by Mayer and Jensen. Wildermuth's cluster model was constructed based on the duality property of the shell-model wave function of the ground state which means that the ground-state wave function can be rewritten in the form of cluster-model wave function. A typical example is the duality of the doubly-closed-shell wave function of the ^{16}O ground state which is equivalent to

^{12}C + α cluster-model wave function;

$$\det |(0s)^4(0p)^{12}| = c_L\mathcal{A}\left[R_{(4-L)/2,L}(r_{C-\alpha},3\nu)[Y_L(\widehat{r}_{C-\alpha})\phi_L(^{12}\mathrm{C})]_{J=0}\phi(\alpha)\right] g(\boldsymbol{X}_G,16\nu), \quad (1)$$

where $R_{nL}(r)$ is radial H.O. function, $g(r,\gamma) = (2\gamma/\pi)^{3/4}\exp(-\gamma r^2)$ and $\boldsymbol{X}_G$ is the total center-of-mass coordinate. The duality character of ground state shell-model wave functions was known by many people who studied microscopic cluster-model wave functions in 1950's. For example, erring and Skyrme wrote at the beginning of the abstract of their paper [2] "It is shown that it is possible to write down α-particle wave functions for the ground states of ^{8}Be, ^{12}C, and ^{16}O, which become, when antisymmetrized, identical with shell-model wave functions."

Widermuth considered that the formation of cluster states is due to the activation of the clustering degree of freedom embedded in the ground state due to its duality nature. Namely he considered that, while the ground state has the heaviest overlap of clusters, excited cluster-states are formed by relaxing the overlap of clusters. He assigned cluster structures to many excited states in light nuclei [3]. For example, in ^{16}O he assigned ^{12}C(0^+) + α structure to 0^+(6.06 MeV), 2^+(6.91 MeV), 4^+(10.36 MeV), 1^-(9.58 MeV), and 3^-(11.62 MeV) and ^{12}C(2^+) + α structure to 2^+(9.84 MeV). Wildermuth's assignments are mostly supported by the present nuclear structure study. But, at the days around 1960, the assignments of cluster structure to excited states were not regarded as convincing. Therefore this situation even meant that the duality could lead to the unnecessity of the cluster model, because its wave function can be rewritten by the shell-model wave function.

A convincing signature of cluster structure which is not easily describable by shell model or by deformed mean-field model is "spatial localization of clusters". Spatial localization of clusters has been one of the most important concepts of nuclear clustering. Before discussing spatial localization of clusters in the next section, we discuss recent finding that the duality character of the ground state is well reflected in the observed large magnitudes of electric monopole ($E0$) transitions between cluster states and the ground state.

2.2. Experimental proof of the duality

In spite of large difference of structure between cluster states and the ground state, observed $E0$ transitions between them are recently pointed out to be strong and comparable with the single-nucleon strength [4]. Single-nucleon strength of $E0$ transition is roughly given by $M(E0) \sim \langle u_f|r^2|u_i\rangle \sim (3\ 5)R^2$ with R being nuclear radius which is about 5.4 fm^2 for $R = 3$ fm. $E0$ transition of the Hoyle state in ^{12}C and those of many cluster states in ^{16}O are good examples. The observed values of them are 5.4 fm^2 for $0_1^+ \leftrightarrow 0_2^+$ in ^{12}C, and 3.6 fm^2, 4.0 fm^2, 3.3 fm^2, for $0_1^+ \leftrightarrow 0_2^+$, $0_1^+ \leftrightarrow 0_3^+$, $0_1^+ \leftrightarrow 0_5^+$, respectively. From the shell-model viewpoint, this looks very contradictory because cluster states are described by superposition of many-particle many-hole configurations. However, the cluster model can explain easily and naturally why $E0$ transitions are large, by the use of the duality character of the ground-state wave function. This fact is explained as follows in the case of ^{16}O: Since 0_2^+ and 0_3^+ states are dominantly described by the wave functions $\mathcal{A}\{\chi_L(r)[Y_L(\hat{r})\phi_L(^{12}\mathrm{C})]_0\phi(\alpha)\}$ with $L = 0$ and $L = 2$ respectively, the expression of the ground state as $\mathcal{A}\{R_{(4-L)/2,L}(r)[Y_L(\hat{r})\phi_L(^{12}\mathrm{C})]_0\phi(\alpha)\}$ by its duality property teaches us that the $E0$ transitions between 0_2^+ and 0_3^+ and the ground state are nothing but the transitions of relative motion between $\chi_L(r)$ and $R_{(4-L)/2,L}(r)$. Namely, the $E0$ transitions are caused by a single degree of freedom (relative motion), which is the reason why observed $M(E0)$ are comparable to single-particle value. It is to be noticed that when the $E0$ transition operator $\mathcal{O}(^{16}\mathrm{O})$ is expressed as $\mathcal{O}(^{16}\mathrm{O}) = \mathcal{O}(^{12}\mathrm{C}) + \mathcal{O}(\alpha) + [(1/2)(12\times 4)/16]r^2$, only the relative-coordinate operator $[(1/2)(12\times 4)/16]r^2$ makes non-zero contribution in the matrix element $\langle 0_1^+|\mathcal{O}(^{16}\mathrm{O})|0_2^+ \text{or } 0_3^+\rangle$.

Journal of Physics: Conference Series **569** (2014) 012001 doi:10.1088/1742-6596/569/1/012001

2.3. Bayman-Bohr theorem of the duality

Bayman and Bohr gave a mathematically neat description of the duality by the aid of $SU(3)$ symmetry in 1958 59 [5]. With this theorem, Eq. (1) is expressed more precisely as

$$|(0s)^4(0p)^{12}(\lambda,\mu)=(0,0)\rangle = C_0\mathcal{A}\left\{[R_{(4,0)}(\boldsymbol{r}_{\mathrm{C}-\alpha},3\nu)\phi_{(0,4)}(^{12}\mathrm{C})]_{(0,0)}\phi(\alpha)\right\} g(\boldsymbol{X}_G,16\nu). \quad (2)$$

The H.O. function $R_{nL}(r,\gamma)Y_{LM}(\hat{r})$ has $SU(3)$ symmetry $(2n+L,0)$ and hence is written as $R_{(2n+L,0)LM}(\boldsymbol{r},\gamma)$, while $\phi_{LM}(^{12}\mathrm{C})$ has $SU(3)$ symmetry (0,4) and is written as $\phi_{(0,4)LM}(^{12}\mathrm{C})$. In the equality of Eq.(2), we note the conservation of the $SU(3)$ symmetry (λ,μ) = (0,0) in both left and right sides. Therefore we have $\mathcal{A}\{[R_{(4,0)}(\boldsymbol{r}_{\mathrm{C}-\alpha},3\nu)\phi_{(0,4)}(^{12}\mathrm{C})]_{(\lambda,\mu)}\phi(\alpha)\} = 0$ for $(\lambda,\mu) \neq$ (0,0). Ref. [5] was published soon after the proposal of the SU(3) shell model by Elliott in 1958 [6].

Thanks to the Bayman-Bohr theorem, we can judge easily whether the ground-state wave function ψ has duality character or not. When ψ is dominantly described by a single SU(3) shell-model wave function, it has duality character. In heavy nuclei, since the Elliott's SU(3) symmetry is largely lost in the ground state, we can say that ψ does not have duality character. But this does not mean the denial of the possible existence of excited states whose wave functions have duality character.

3. Spatial localization of clusters v.s. THSR wave function

An important convincing proof of the spatial localization of clusters is the phenomenon of the inversion doublet rotational bands with small energy gap between plus and minus parity bands. Since the inversion doublet comes from the parity-violating intrinsic state, the inversion doublet bands with non-small alpha-decay widths should have spatial localization of clusters. Observed typical examples of inversion-doublet bands are in ^{16}O and ^{20}Ne [7]. In ^{16}O, the plus-parity band is $K^\pi = 0^+$ band upon the first excited state with $J^\pi = 0^+$ at E_x = 6.05 MeV, and the minus-parity band is $K^\pi = 0^-$ band upon the 1^- state at E_x = 9.58 MeV which is slightly above the ^{12}C + α threshold. The $K^\pi = 0^-$ band states have large α-decay reduced widths comparable to the Wigner-limit value. Hence in accordance with Wildermuth, spatially-localized cluster structure of ^{12}C + α is assigned to this inversion-doublet. In the case of ^{20}Ne, the plus-parity band is the ground-state band and the minus-parity band is $K^\pi = 0^-$ band upon the 1^- state at E_x = 5.80 MeV which is slightly above the ^{16}O + α threshold. To this inversion-doublet, spatially-localized cluster structure of ^{16}O + α is assigned.

A very surprizing fact is that the R M CM wave functions of the inversion-doublet bands of ^{20}Ne were found, recently, to be almost the same as single THSR (Tohsaki-Horiuchi-Schuck-Roepke) wave functions [8]:

$$|\left\langle\Phi(\text{single THSR})|\Phi(^{16}\mathrm{O}+\alpha\ \mathrm{R\ \ M})\right\rangle|^2 \approx 100 \qquad \text{for inversion doublet band states.} \quad (3)$$

As is seen in the fact that the THSR wave function was originally introduced for describing α-codensate-like states [9], it expresses non-localized motion of clusters. Therefore this relation looks inconsistent with the idea of the parity-violating deformation due to the spatially localized ^{16}O+α clustering. This seeming inconsistency is resolved by considering the inter-cluster auli repulsion due to the antisymmetrization (Fermi statistics of nucleons) [10]. The inter-cluster auli repulsion prevents two clusters to come close to each other and hence gives rise to the spatial localization of ^{16}O and α clusters. In Fig. 1, we show the nucleon-density distribution of the ^{16}O + α THSR wave function before projection where we clearly see the spatial localization of ^{16}O and α clusters. We may say that dynamics prefers non-localized clustering while kinematics (auli repulsion) makes the system look like localized clustering.

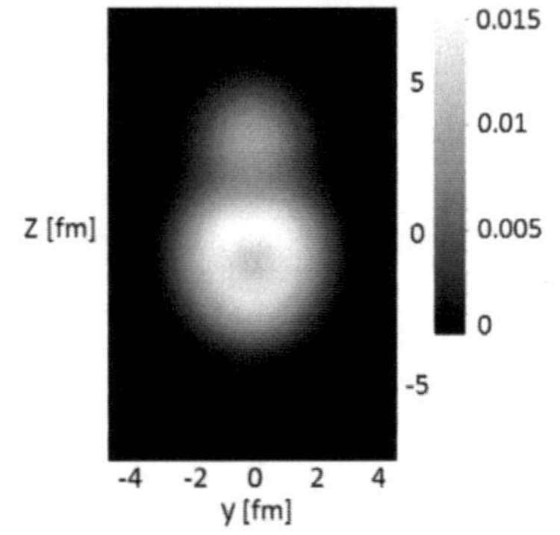

Figure 1. Nucleon-density distribution of the ^{16}O + α THSR wave function before projection.

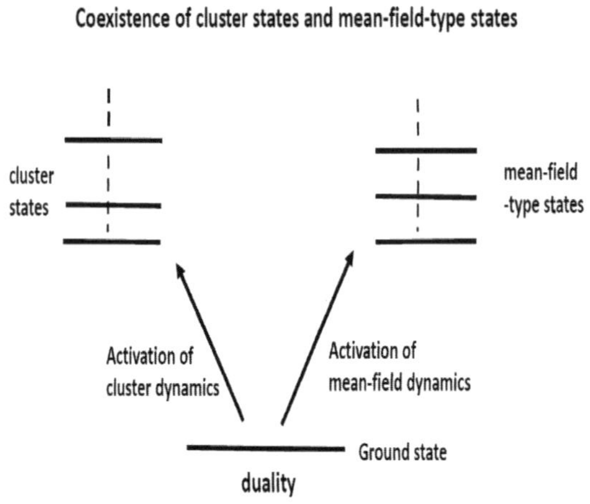

Figure 2. Coexistence of cluster states and mean-field-type states due to the duality of the ground state.

4. Coexistence of cluster states with mean-field-type states

Coexistence of cluster states with mean-field-type states is a natural consequence of the duality nature of the ground state, as is shown in Fig. 2. However, as mentiond before, at the days around 1960 when Wildermuth proposed his microscopic cluster model, the assignments of cluster structure to excited states were not regarded as convincing. Different from the days of Wildermuth, the assignments of cluster structure to many excited states are now regarded as convincing in many nuclei. Signatures or evidences of cluster structure have been pursued in many ways. We now see coexistence of cluster states with mean-field-type states in a wide region of nuclear chart including unstable nuclei.

In several nuclei, the number of cluster states is larger than that of mean-field-type states up to rather high excitation energy, say, 15 MeV. A typical example is the ^{16}O nucleus which is a double-closed-shell nucleus implying a fundamental nucleus for the shell model.

The coexistence feature is now seen up to pf-shell region. In this region, cluster states include molecular resonances in addition to α-cluster states and mean-field-type states include superdeformed states in addition to normal deformed states. In Fig. 3 AMD reproduction is shown of the coexistence feature of cluster states with mean-field-type states in ^{44}Ti [11].

Formation of molecular resonance states is due to the duality property of usually not the ground state but excited mean-field-type states, which is in contrast to alpha-cluster states formed through the duality of the ground state. For example, in ^{28}Si, ^{16}O + ^{12}C molecular-resonance states are considered to be formed on the basis of the duality property of the excited rotational band with prolate deformation upon the 0^+ state at 6.69 MeV [12]. This band has, as its dominant componet of the intrinsic state, $|(sd)^{12}[4^3](\lambda,\mu)=(12,0)\rangle$ which has the duality character that it is equivalent to ^{16}O + ^{12}C cluster-model wave function. Long time ago the ^{16}O + ^{12}C Brink- CM calculation reported that the calculation reproduced both ^{16}O + ^{12}C resonance states and the prolate band states upon the 0^+ state at 6.69 MeV [13]. Another example is the case of ^{16}O + ^{16}O molecular resonance states which are considered to be formed on the basis of the superdeformed states in ^{32}S [14, 15]. Ref. [15] reports that the AMD- CM calculation along the energy curve with respect to the quadrupole deformation gives rise to ^{16}O + ^{16}O molecular resonance states. The dominant component of the superdeformed-state wave

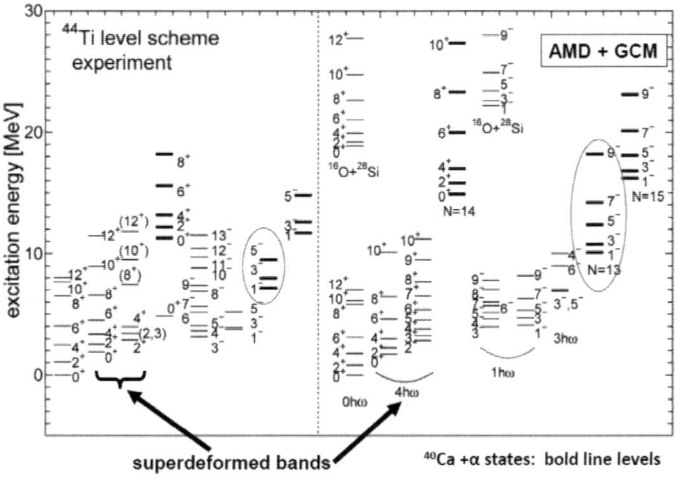

Figure 3. Reproduction by AMD of the coexistence of cluster states with mean-field-type states in ^{44}Ti. ^{40}Ca + α cluster states are shown by bold lines.

function is $|(sd)^{12}(pf)^4[4^4](\lambda,\mu) = (24,0)\rangle$ which has the duality character that it is equivalent to ^{16}O + ^{16}O cluster-model wave function. Fig. 4 gives a schematical picture of the two-step formation of molecular state on the basis of the duality of the mean-field-type excited state.

In neutron-rich nuclei, clustering is largely governed by valence-neutron structures around clustered core. We have molecular orbits spread in the whole nucleus, atomic orbits aroud single clusters, and others. Fig. 5 shows AMD study of such variety of nuclear structures in ^{22}Ne where the core part ($N = Z$=10) of the ground band states has dominantly (8,0) $SU(3)$ symmetry which has the duality of ^{16}O $+\alpha$ clustering [16].

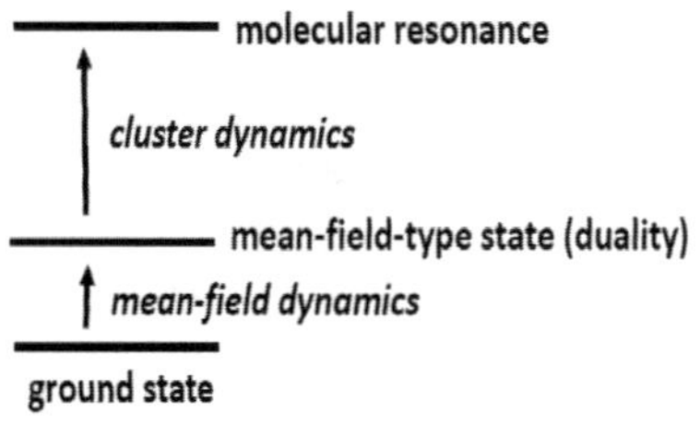

Figure 4. Two-step formation of molecular state due to the duality of the mean-field-type excited state.

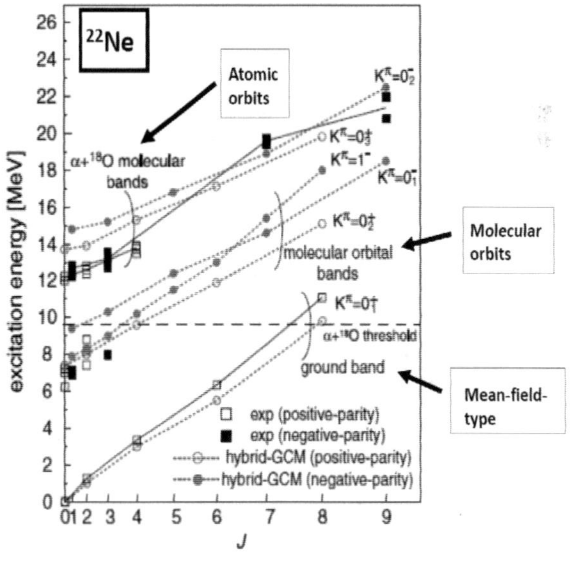

Figure 5. Coexistence of cluster states and mean-field-type states in neutron-rich ^{22}Ne studied with AMD.

Journal of Physics: Conference Series **569** (2014) 012001 doi:10.1088/1742-6596/569/1/012001

5. Spatial localization of clusters and ab initio calculations of cluster states

Convincing evidences of clustering are now abundantly given by *ab initio* calculations of nuclear structure. A well-known example is the quantum Monte Carlo calculation of ^{8}Be by Wiringer et al. [17] which showed spatially-localized 2α clusters which are separated from each other by about 2 fm.

Spatial localization of clusters requires very high $\hbar\omega$ for the reproduction of cluster states in shell model. If the $\hbar\omega$ space is not wide enough, cluster states can not be su ciently bound. This fact tells us that *an ab initio* shell model calculation gives us a good method to discriminate cluster states from mean-field-type states. For example, no-core shell-model calculations have not succeeded until now to reproduce the low excitation energy of the Hoyle state of ^{12}C [18]. Therefore, no-core shell-model calculations indicate very clearly that the Hoyle state is largely different from the mean-field-type state. This theoretical method of discriminating cluster states from mean-field-type states is very convincing and was not existent in the days of Wildermuth. Quite recently a similar study was reported in Ref. [19] where ^{8}Be was investigated by TOSM (tensor optimized shell model). Different from no-core shell model, TOSM includes very high $\hbar\omega$ space, but on the other hand it adopts only $0p-0h$ and $2p-2h$ configurations. Thus, for the well-developed 2α structure of the ^{8}Be ground band states, TOSM gives insu cient description.

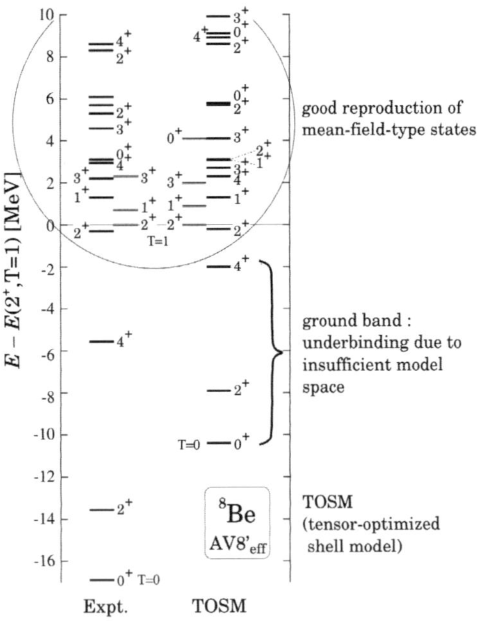

Figure 6. Reproduction of ^{8}Be levels by TOSM. The calculated energy position of the first $T = 1$ 2^+ state is shifted to be the same as the observed energy position. Although the mean-field-type states are well reproduced by TOSM, the calculated levels of the ground band are underbound compared with the observed levels, which is due to the insu ciency of TOSM for describing 2α cluster states.

Ab initio approaches have a powerful merit which is absent in other approaches to nuclear clustering. In *ab initio* approaches clustering can be judged not only by density distribution but also by tensor-force effects. A typical example is given in Ref. [20] where it is shown that the probabilities of S, P, and D states of the second 0^+ state of ^{4}He are very close to those of ^{3}He and t. This result strongly supports the $3N + N$ cluster structure of the second 0^+ state.

It is of course desirable that we have supports to the α-condensate model of the Hoyle state by *ab initio* approaches. However, at present there is no fully converged *ab initio* description of the Hoyle state. We can even say that the very frontier of *ab initio* study of nuclear structure is the study of the Hoyle state. Recent result by the quantum Monte Carlo approach given in Ref. [21] is maybe most advanced result. When we compare the density distribution $r^2\rho(r)$ of the Hoyle state between the quantum Monte Carlo calculation and the 3α THSR wave function [22], we see good correspondence between two density distributions, although the quantum Monte Carlo calculation has not yet fully converged for the Hoyle state [23].

6. Summary

(i) Many of basic concepts of nuclear cluster physics since the days of K. Wildermuth are now examined and revived in new lights.

(ii) The concept of the duality of the shell-model wave function of the ground state was the starting point of Wildermuth's cluster model.

(iii) Widermuth considered that the formation of cluster states is due to the activation of the clustering degree of freedom embedded in the ground state due to its duality nature. But, at the days around 1960, the assignments of cluster structure to excited states were not regarded as convincing.

(iv) Recently it has been discussed that the duality character of the ground state explains well and naturally observed large magnitudes of $E0$ transitions between cluster states and the ground state.

(v) Spatial localization of clusters which is a convincing signature of cluster structure has recently been discussed that it comes from the inter-cluster auli repulsion in nonlocalized cluster dynamics.

(vi) Abundance of cluster states is seen in a wide region of nuclear chart including unstable nuclei and up to high excitation energy. Molecular resonance states are discussed through the duality of excited mean-field states, which is in contrast to alpha-cluster states formed through the duality of the ground state.

(vii) Ab initio calculations of nuclear structure are presenting convincing evidences of clustering which include spatial localization, characteristic tensor-force contribution, and gas-like clustering.

References

[1] Wildermuth K and Kanellopoulos Th 1958 *Nucl. Phys.* **7** 150; 1958 59 *Nucl. Phys.***9** 449.
[2] Perring J K and kyrme T H R 1956 *Proc. Phys. Soc.* **69** 600.
[3] heline R K and Wildermuth K 1960 *Nucl. Phys* **21** 196.
[4] Kawabata T *et al.* 2007 *Phys. Lett.* B **646** 6.
[5] Bayman B F and Bohr A 1958 59 *Nucl. Phys.* **9** 596.
[6] Elliott J P 1958 *Proc. Roy. Soc.* (London) A **245** 128, 562.
[7] Horiuchi H and Ikeda K 1968 *Prog. Theor. Phys.* **40** 277.
[8] Zhou B *et al.* 2012 *Phys. Rev.* C **86** 014301; 2013 *Phys. Rev. Lett.* **110** 262501.
[9] Tohsaki A, Horiuchi H, chuck P and Röpke G 2001 *Phys. Rev. Lett.* **87** 192501.
[10] Zhou B *et al.* 2014 *Phys. Rev.* C **89** 034319.
[11] Kimura M and Horiuchi H 2006 *Nucl. Phys.* A **767**, 58.
[12] Horiuchi H, Ikeda K and Kato K 2012 *Prog. Theor. Phys.* upple. **192**, 1.
[13] Baye D 1976 *Nucl.Phys.* A **272** 445; Baye D and Heenen P H 1977 *Nucl.Phys.* A **283** 176.
[14] Ohkubo and Yamashita K 2002 *Phys. Rev.* C **66** (R) 021301.
[15] Kimura M and Horiuchi H 2004 *Phys. Rev.* C **69** (R) 051304.
[16] Kimura M 2007 *Phys. Rev.* C **75** 034312.
[17] Wiringa R B, Pieper C, Carlson J and Pandharipande V R 2000 *Phys. Rev.* C **62** 014001.
[18] Navrátil P, Vary J P and Barrett B R 2000 *Phys. Rev.* C **62** 054311; Roth R, Langhammer J, Calci A, Binder and Navrátil P 2011 *Phys. Rev. Lett.* **107** 072501.
[19] Myo T, Umeya A, Horii K, Toki H and Ikeda K 2014 *Prog. Theor. Exp. Phys.* 033D01.
[20] Hiyama E, Gibson B F and Kamimura M 2004 *Phys. Rev.* C **70** (R) 031001.
[21] Wiringa R B 2012 *501. Wilhelm und Else Heraeus-Seminar on Nuclear Ground-State Properties of the Lightest Nuclei: Status and Perspectives.*
[22] Funaki Y, Tohsaki A, Horiuchi H, chuck P and Röpke G 2005 *Eur. Phys. J.* A **24** 321.
[23] Horiuchi H 2013 *Journ. Phys. Conference Series* **436** 012003.

Journal of Physics: Conference Series **569** (2014) 012002 doi:10.1088/1742-6596/569/1/012002

Present status of nuclear cluster physics and experimental perspectives

C Beck

Département de Recherches Subatomiques, Institut Pluridisciplinaire Hubert Curien, IN_2P_3-CNRS and Université de Strasbourg - 23, rue du Loess BP 28, F-67037 Strasbourg Cedex 2, France

E-mail: `christian.beck@iphc.cnrs.fr`

Abstract.
Knowledge on nuclear cluster physics has increased considerably as nuclear clustering remains one of the most fruitful domains of nuclear physics, facing some of the greatest challenges and opportunities in the years ahead. The occurrence of "exotic" shapes in light $N{=}Z$ α-like nuclei and the evolution of clustering from stability to the drip-lines are being investigated more and more accurately both theoretically and experimentally. Experimental progresses in understanding these questions were recently examined and will be further revisited in this introductory talk: clustering aspects are, in particular, discussed for light exotic nuclei with a large neutron excess such as neutron-rich Oxygen isotopes with their complete spectrocopy.

1. Introduction

One of the greatest challenges in nuclear science is understanding the structure of light nuclei from both experimental and theoretical perspectives. Figure 1 was used by Catford to summarize the different types of clustering discussed during the last Cluster Conference in Debrecen [1]. Most of these structures were investigated in an experimental context by using either some new approaches or developments of older methods [2]. Starting in the 1960s the search for resonant structures in the excitation functions for various combinations of light α-cluster ($N{=}Z$) nuclei in the energy regime from the Coulomb barrier up to regions with excitation energies of E_x=20−50 MeV remains a subject of contemporary debate [3, 4]. These resonances [4] have been interpreted in terms of nuclear molecules [3]. The question of how quasimolecular resonances may reflect continuous transitions from scattering states in the ion-ion potential to true cluster states in the compound systems was still unresolved in the 1990s [3, 4]. In many cases, these resonant structures have been associated with strongly-deformed shapes and with α-clustering phenomena [5, 6], predicted from the Nilsson-Strutinsky approach, the cranked α-cluster model [5], or other mean-field calculations [6, 7]. In light α-like nuclei, clustering is observed as a general phenomenon at high excitation energy close to the α-decay thresholds [5, 8]. This exotic behavior has been perfectly illustrated by the famous "Ikeda-Diagram" for $N{=}Z$ nuclei in 1968 [9], which has been modified and extended by von Oertzen [10] for neutron-rich nuclei more than 10 years ago, as shown in the left panel of figure 2. Clustering is a general feature [11] not only observed in typical light neutron-rich nuclei [12], but also in less common systems such as the neutron halo ^{11}Li [13] and/or ^{14}Be nuclei, for instance [14]. The problem of cluster formation has also been treated extensively for very heavy systems by Gupta [7], by

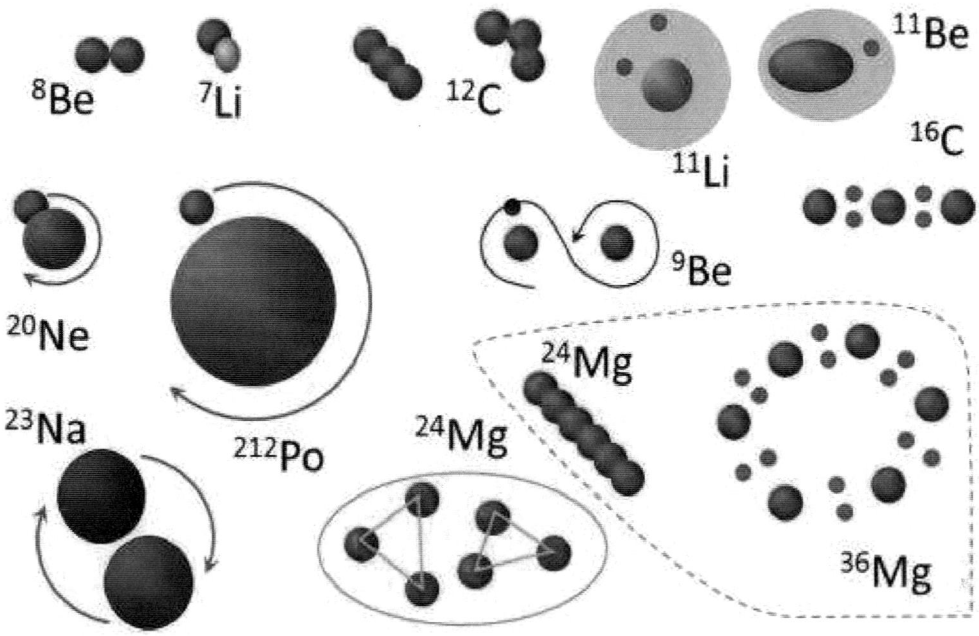

Figure 1. Different types of clustering in nuclei that have been discussed at this workshop. (Figure adapted from Ref.[1] courtesy from Catford).

Zagrebaev and Greiner [15] and by Simenel [16] where giant molecules and collinear ternary fission may co-exist [17]. Finally, signatures of α clustering have also been discovered in light nuclei undergoing from ultrarelativistic nuclear collisions [18, 19]. In this introductory talk, I will limit myself first to the light ^{12}C and ^{16}O α-like nuclei in Section 2, then to α clustering, deformations and α condensates in heavier α-like nuclei in Section 3, and, finally, to clustering in light neutron-rich nuclei in Section 4.

2. Renewed interest in the spectroscopy of ^{12}C and ^{16}O α-like nuclei

The renewed interest in ^{12}C was mainly focused to a better understanding of the nature of the so called "Hoyle" state [20] that can be described in terms of a bosonic condensate, a cluster state and/or a α-particle gas [21, 22, 23]. Much experimental progress has been achieved recently as far as the spectroscopy of ^{12}C near and above the α-decay threshold is concerned [24, 25, 26, 27, 28, 29]. More particularly, the 2_2^+ "Hoyle" rotational excitation in ^{12}C has been observed by several experimental groups [24, 26, 27]. The most convincing experimental result, displayed in figure 3, comes from measurements of the ^{12}C(γ,α)^{8}Be reaction performed at the HIGS facility [27]. The angular distributions of the alpha particles in the region of 9-10 MeV are consistent with an L=2 pattern, including a dominant 2^+ component. This 2_2^+ state that appears at around 10 MeV is considered to be the 2^+ excitation of the "Hoyle" state (in agreement with the previous experimental investigation of Itoh et al. [24]) according to the α cluster [30] and α condensation models [21, 23]. On the other hand, the experiment ^{12}C(α,α)^{12}C* carried out at the Birmingham cyclotron [29], UK, populates a new state compatible with an equilateral triangle configuration of three α particles. Still, the structure of the "Hoyle" state remains controversial as experimental results of its direct decay into three α particles are found to be in disagreement [31, 32, 33, 34, 35, 36].

Journal of Physics: Conference Series **569** (2014) 012002 doi:10.1088/1742-6596/569/1/012002

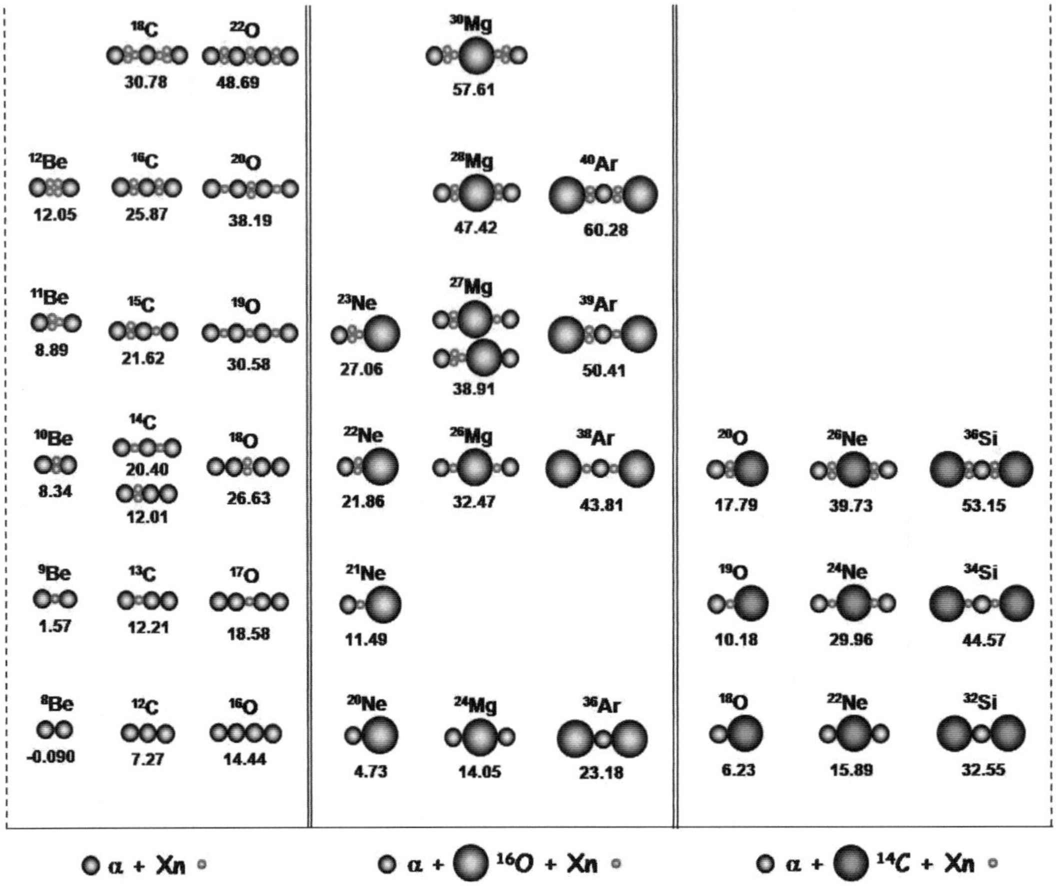

Figure 2. Schematic illustration of the structures of molecular shape isomers in light neutron-rich isotopes of nuclei consisting of α-particles, ^{16}O- and ^{14}C-clusters plus some covalently bound neutrons (Xn means X neutrons). The so called "Extended Ikeda-Diagram" [10] with α-particles (left panel) and ^{16}O-cores (middle panel) can be generalized to ^{14}C-cluster cores (right panel). The lowest line of each configuration corresponds to parts of the original "Ikeda-Diagram" [9]. However, because of its deformation, the ^{12}C nucleus is not included, as it was earlier [9]. Decay threshold energies (in MeV) are given for the relevant decompositions of clusters. (Figure adapted from Ref.[11] courtesy from von Oertzen).

In the study of Bose-Einstein Condensation (BEC), the α-particle states in light $N=Z$ nuclei [21, 22, 23], are of great importance. At present, the search for an experimental signature of BEC in ^{16}O is of highest priority. A state with the structure of the "Hoyle" state [20] in ^{12}C coupled to an α particle is predicted in ^{16}O at about 15.1 MeV (the 0_6^+ state), the energy of which is $\approx$ 700 keV above the 4α-particle breakup threshold [37]. However, any state in ^{16}O equivalent to the "Hoyle" state [20] in ^{12}C is most certainly going to decay exclusively by particle emission with very small γ-decay branches, thus, very efficient particle-γ coincidence techniques will have to be used in the near future to search for them. BEC states are expected to decay by alpha emission to the "Hoyle" state and could be found among the resonances in α-particle inelastic scattering on ^{12}C decaying to that state. In 1967 Chevallier et al. [38, 39] could excite these states in an α-particle transfer channel leading to the ^{8}Be–^{8}Be final state and proposed

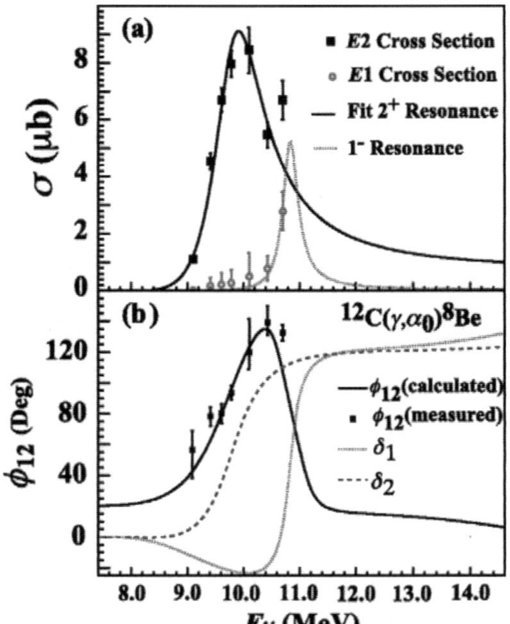

Figure 3. a) Measured E1 and E2 cross sections of $^{12}C(\gamma,\alpha_0)^8Be$. b) Measured E1 and E2 phase angles. (Figure taken from Ref. [27] courtesy from Gai).

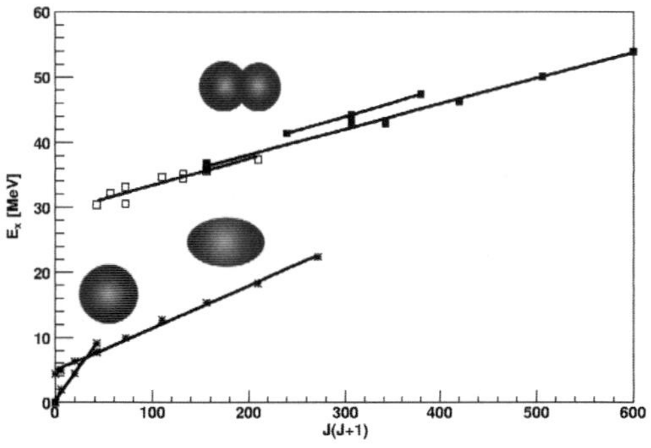

Figure 4. Experimental rotational bands and schematical deformed shapes in ^{36}Ar. Excitation energies of the ground state (spherical shape) and SD (ellipsoidal shape) bands [55], respectively, and the energies of HD (dinuclear shape) band from the quasimolecular resonances observed in the $^{12}C+^{24}Mg$ (open rectangles) and $^{16}O+^{20}Ne$ (full rectangles) reactions are plotted as a function of J(J+1). The relevant energies of the resonance can be found in Refs. [46, 58, 59]. (Figure reproduced from Refs. [46]).

that a structure corresponding to a rigidly rotating linear arrangement of four alpha particles may exist in ^{16}O. Very recently, a more sophisticated experimental setup was used at Notre Dame [40]: although the excitation function is generally in good agreement with the previous results [38, 39] a phase shift analysis of the angular distributions does not provide evidence to support the reported hypothesis of a 4α-chain state configuration. Experimental investigations are still underway to understand the nuclear structure of high spin states of both ^{16}O and ^{20}Ne nuclei for instance at Notre Dame [41] and/or iThemba Labs [42] facilities. Another possibility might be to perform Coulomb excitation measurements with intense ^{16}O and ^{20}Ne beams at intermediate energies.

3. Alpha clustering, deformations and alpha condensates in heavier nuclei

The real link between superdeformation (SD), nuclear molecules and alpha clustering [6, 43, 44, 45, 46] is of particular interest, since nuclear shapes with major-to-minor axis ratios of 2:1 are typical ellipsoidal elongations for light nuclei (corresponding to a quadrupole deformation parameter $\beta_2 \approx 0.6$). Furthermore, the structure of possible octupole-unstable 3:1 nuclear shapes (hyperdeformation (HD) with $\beta_2 \approx 1.0$) has also been discussed for actinide nuclei [45] in terms of clustering phenomena. Typical examples for the possible relationship between quasimolecular bands and extremely deformed (SD/HD) shapes have been widely discussed in the literature for $A = 20 - 60$ α-conjugate N=Z nuclei, such as ^{28}Si [47, 48, 49, 50, 51], ^{32}S [6, 48, 52, 53, 54], ^{36}Ar [45, 46, 55, 56, 57, 58, 59], ^{40}Ca [60, 61, 62, 63], ^{44}Ti [6, 64, 65], ^{48}Cr [66] and ^{56}Ni [67, 68, 69, 70, 71, 72].

Highly deformed shapes and SD rotational bands have been discovered in several light α-conjugate (N=Z) nuclei, such as ^{36}Ar and ^{40}Ca by using γ-ray spectroscopy techniques [55, 60].

In particular, the extremely deformed rotational bands in ^{36}Ar [55] (shown as crosses in figure 4) might be comparable in shape to the quasimolecular bands observed in both ^{12}C+^{24}Mg (shown as open rectangles in figure 4) and ^{16}O+^{20}Ne (shown as full rectangles in figure 4) reactions. Ternary clusterizations are also predicted theoretically, but were not found experimentally in ^{36}Ar so far [57]. On the other hand, ternary fission of ^{56}Ni – related to its hyperdeformed (HD) shapes – was claimed to be identified from out-of-plane angular correlations measured in the ^{32}S+^{24}Mg reaction with the Binary Reaction Spectrometer (BRS) at the VIVITRON Tandem facility of the IPHC, Strasbourg [71], though this remains to be confirmed [72]. This possibility [71] is not limited to light N=Z compound nuclei, true ternary fission [15, 17, 73] can also occur for very heavy [17, 73] and superheavy [74] nuclei.

The next natural question to be addressed is whether dilute-gas-like structures (i.e. BEC) [21, 22, 23] also exist in medium-mass α-conjugate nuclei as predicted by several theoretical investigations [75, 76, 77]. Several recent undergoing experiments indicate that it might be the case at least for ^{24}Mg [78, 79], ^{36}Ar [80] and ^{56}Ni [81] and much work is in progress in this field [22].

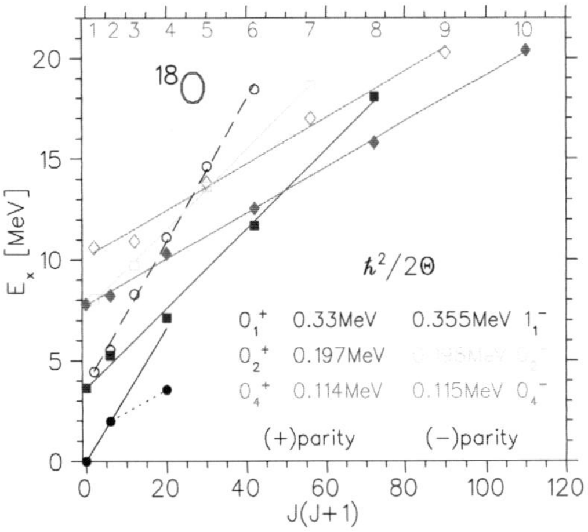

Figure 5. Overview of six rotational band structures observed in ^{18}O. Excitation energy systematics for the members of the rotational bands forming inversion doublets with K=0 are plotted as a function of $J(J$+1$)$. The curves are drawn to guide the eye for the slopes. The indicated slope parameters contain information on the moments of inertia. Square and open circle symbols correspond to cluster bands, whereas diamonds symbols correspond to molecular bands.(Figure adapted from [82] courtesy from von Oertzen).

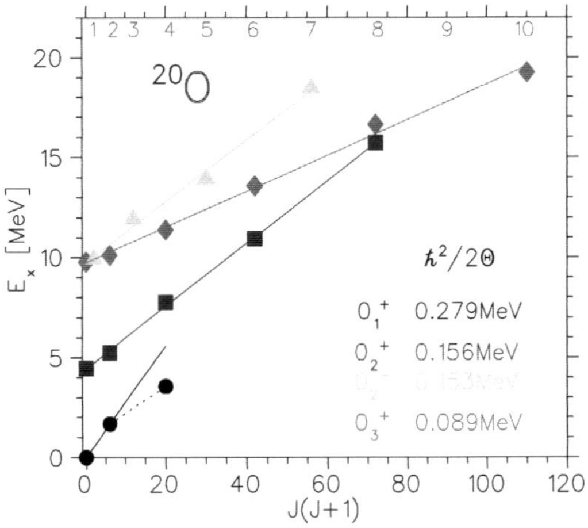

Figure 6. Overview of four rotational band structures observed in ^{20}O. Excitation energy systematics for the members of the rotational bands forming inversion doublets with K=0 are plotted as a function of $J(J$+1$)$. The curves are drawn to guide the eye for the slopes. The indicated slope parameters contain information on the moments of inertia. Square and triangle symbols correspond to cluster bands, whereas diamonds symbols correspond to molecular bands.(Figure adapted from [83] courtesy from von Oertzen).

4. Clustering in light neutron-rich nuclei

As discussed previously, clustering is a general phenomenon observed also in nuclei with extra neutrons as it is presented in an extended "Ikeda-Diagram" [9] proposed by von Oertzen [10] (see the left and middle panels of figure 2). With additional neutrons, specific molecular structures appear with binding effects based on covalent molecular neutron orbitals. In these diagrams α-clusters and ^{16}O-clusters are the main ingredients. Actually, the ^{14}C nucleus may play a similar role in clusterisation as the ^{16}O nucleus does. Both of them have similar properties as a cluster: i) closed neutron p-shells, ii) first excited states are well above $E^* = 6$ MeV, and iii) high binding energies for α-particles.

A general picture of clustering and molecular configurations in light nuclei can be drawn from a detailed investigation of the oxygen isotopes with $A \geq 17$. Here I will only present recent results on the even-even oxygen isotopes: ^{18}O [82] and ^{20}O [83]. But very striking cluster states have also been found in odd-even oxygen isotopes such as: ^{17}O [84] and ^{19}O [85].

Figure 5 gives an overview of all bands in ^{18}O as a plot of excitation energies as a function of $J(J+1)$ together with their respective moments of inertia. In the assignment of the bands both the dependence of excitation energies on $J(J+1)$ and the dependence of measured cross sections on $2J+1$ [82] were considered. Slope parameters obtained in a linear fit to the excitation energies [82] indicate the moment of inertia of the rotational bands given in figure 5. The intrinsic structure of the cluster bands is reflection asymmetric, the parity projection gives an energy splitting between the partner bands. The assignment of the experimental molecular bands are supported by either generator-coordinate-method [86] or Antisymmetrized Molecular Dynamics (AMD) calculations [87].

The bands of ^{20}O [83] shown in figure 6 can be compared with those of ^{18}O displayed in figure. 5. The first doublet (K=$0_2^{\pm}$) has a slightly larger moment of inertia (smaller slope parameter) in ^{20}O, which is consistent with its interpretation as ^{14}C–^{6}He or ^{16}C–^{4}He molecular structures (they start well below the thresholds of 16.8 MeV and 12.32 MeV, respectively). The second band, for which the negative parity partner has yet to be determined, has a slope parameter slightly smaller than in ^{18}O. This is consistent with the study of the bands in ^{20}O by Furutachi et al. [87], which clearly establishes parity inversion doublets predicted by AMD calculations for the ^{14}C–^{6}He quasimolecular (cluster) and ^{14}C-2n-α molecular structures. The corresponding moments of inertia illustrated in figure 4 and figure 5 are strongly suggesting large deformations for the cluster structures. It may be concluded that the reduction of the moments of inertia of the lowest bands of ^{18,20}O is consistent with the assumption that the strongly bound ^{14}C nucleus have equivalent properties to ^{16}O. It is interesting to note that the Quantum Mechanical Fragmentation Theory (QMFT) of Gupta [7] reaches to the same conclusion on the possibility of ^{14}C clustering [88]. Therefore, the "Ikeda-Diagram" [9] and the "Extended Ikeda-Diagram" consisting of ^{16}O cluster cores with covalently bound neutron [10] must be further extended to include also the ^{14}C cluster cores as illustrated in right panel of figure 2.

5. Summary and outlook

The connection of α-clustering, quasimolecular resonances, α condensates in very light nuclei and extreme deformations (SD, HD, ...) in heavier nuclei as investigated by more and more sophisticated experimental devices, has been discussed in this introductory talk. In particular, high-precision spectroscopy techniques help us to uncover important parts of the complete spectroscopy of the so called "Hoyle" state in ^{12}C. The origin of carbon for life is likely to be understood in the very near future with answer to the question of the "Hoyle" state structure. Similarly the quest for the 4α states of ^{16}O near the ^{8}Be+^{8}Be and ^{12}C+α decay thresholds, which correspond to the "Hoyle" state. Results have also been presented on neutron-rich oxygen isotopes displaying very well defined quasimolecular bands in agreement with AMD predictions. Consequently, the "Extended Ikeda-Diagram" has been further extended for light neutron-rich nuclei by inclusion of the ^{14}C cluster as a core, similarly to the ^{16}O one. The search for extremely elongated configurations (HD) in rapidly rotating medium-mass nuclei, which has been pursued by γ-ray spectroscopy measurements, will have to be performed in conjunction with charged-particle techniques in the near future since such states are most likely going to decay by particle emission (see [2, 71, 72]).

Acknowledgments

I would like to thank the Sotancp3 organizers for giving me the opportunity to introduce this workhop. Christian Caron (Springer) is acknowledged for initiating the new series of three volumes of *Lecture Notes in Physics* dedicated to "Clusters in Nuclei".

References

[1] Catford W 2012, *J. Phys.: Conferences Series* **436** 012095
[2] Papka P and Beck C 2012 Clusters in Nuclei - Vol.2 ed. Beck C, *Lecture Notes in Physics* **848** 299
[3] Greiner W, Park Y Jae, and Scheid W 1995 *Nuclear Molecules*, (Singapore: World Scientific)
[4] Beck C *et al.* 1994 *Phys. Rev. C* **49** 2618
[5] Freer M 2007 *Rep. Prog. Phys.* **70** 2149
[6] Horiuchi H 2010 Clusters in Nuclei - Vol.1, ed. Beck C, *Lecture Notes in Physics* **818** 57
[7] Gupta R K 2010 Clusters in Nuclei - Vol.1, ed. Beck C, *Lecture Notes in Physics* **818** 232
[8] von Oertzen W, Freer M, and Kanada-En'yo Y 2006 *Phys. Rep.* **432** 43
[9] Horiuchi H and Ikeda K 1968 *Prog. Theor. Phys.* **40** 277
[10] von Oerzten W 2001 *Eur. Phys. J. A* **11** 403
[11] von Oerzten W and Milin M, 2014 Clusters in Nuclei - Vol.3, ed. Beck C, *Lecture Notes in Physics* **875** 147
[12] Kanada-En'yo Y *et al.* 2010 Clusters in Nuclei - Vol.1, ed. Beck C, *Lecture Notes in Physics* **818** 129
[13] Ikeda K *et al.* 2010 Clusters in Nuclei - Vol.1, ed. C. Beck, *Lecture Notes in Physics* **818** 165
[14] Nakamura N and Kondo Y 2012 Clusters in Nuclei - Vol.2, ed. Beck C, *Lecture Notes in Physics* **848** 67
[15] Zagrebaev V and Greiner W 2010 Clusters in Nuclei - Vol.1, ed. Beck C, *Lecture Notes in Physics* **818** 267
[16] Simenel C 2014 Clusters in Nuclei - Vol.3, ed. Beck C, *Lecture Notes in Physics* **875** 95
[17] Kamanin D V and Pyatkov Y 2014 Clusters in Nuclei - Vol.3, ed. Beck C, *Lecture Notes in Physics* **875** 183
[18] Zarubin P L 2014 Clusters in Nuclei - Vol.3, ed. Beck C, *Lecture Notes in Physics* **875** 51
[19] Broniowski W and Ruiz Arriola E 2014 *Phys. Rev. Lett.* **112** 112501
[20] Hoyle F 1954 *Astrophys. J. Suppl. Ser.* **1** 121
[21] Tohsaki A, *et al.* 2001 *Phys. Rev. Lett.* **87** 192501
[22] von Oertzen W 2010 Clusters in Nuclei - Vol.1, ed. Beck C, *Lecture Notes in Physics* **818** 102
[23] Yamada T *et al.* 2012 Clusters in Nuclei - Vol.2, ed. Beck C, *Lecture Notes in Physics* **848** 229
[24] Itoh M *et al.* 2004 *Nucl. Phys. A* **738** 268
[25] Freer M *et al.* 2011 *Phys. Rev. C* **83** 034314
[26] Itoh M *et al.* 2011 *Phys. Rev. C* **84** 054308
[27] Zimmerman W R *et al.* 2014 *Phys. Rev. Lett.* **110** 152502
[28] Kokalova Tz *et al.* 2013 *Phys. Rev. C* **87** 057307
[29] Marin-Lambarri D J *et al.* 2014 *Phys. Rev. Lett.* **113** 012502
[30] Uegaki E *et al.* 1977 *Prog. Theor. Phys.* **57** 1262
[31] Freer M *et al.* 1994 *Phys. Rev. C* **49** R1751
[32] Raduta Ad R *et al.* 2011 *Phys. Lett. B* **705** 65
[33] Manfredi J *et al.* 2012 *Phys. Rev. C* **85** 037603
[34] Kirsebom O S *et al.* 2012 *Phys. Rev. Lett.* **108** 202501
[35] Rana T K *et al.* 2013 *Phys. Rev. C* **88** 021601(R)
[36] Itoh *et al.* 2014 *Phys. Rev. Lett.* submitted
[37] Funaki Y *et al.* 2008 *Phys. Rev. Lett.* **101** 082502
[38] Chevallier P *et al.* 1967 *Phys. Rev.* **160** 827
[39] Brochard F *et al.* 1976 *Phys. Rev. C* **13** 967
[40] Curtis N *et al.* 2013 *Phys. Rev. C* **88** 064309
[41] Kokalova Tz *et al.* 2013 *Phys. Rev. C* **87** 057309
[42] Papka P and Kokalova Tz 2014 (private communications)
[43] Beck C 2004 *Nucl. Phys.* **A 738** 24
[44] Beck C 2004 *Int. J. Mod. Phys.* **E13** 9
[45] Cseh J *et al.* 2009 *Phys. Rev. C* **80** 034320
[46] Beck C *et al.* 2011 *Acta Physica Polonica B* **42** 747
[47] Taniguchi Y *et al.* 2009 *Phys. Rev. C* **80**, 044316

[48] Ichikawa T, Kanada-En'yo Y, and Möller P, 2011 *Phys. Rev. C* **83** 054319
[49] Jenkins D G *et al.* 2012 *Phys. Rev. C* **86** 064308
[50] Jenkins D G 2013 Clusters in Nuclei - Vol.3, ed. Beck C, *Lecture Notes in Physics* **875** 25
[51] Darai J, Cseh J and Jenkins D G 2012 *Phys. Rev. C* **86** 064309
[52] Kimura M and Horiuchi H 2004 *Phys. Rev. C* **69** 051304
[53] Lönnroth T *et al.* 2010 *Eur. Phys. J. A* **46** 5
[54] Pandit D *et al.* 2010 *Phys. Rev. C* **81** 061302; 2005 *Phys. Rev. C* **72** 021601
[55] Svensson C E, *et al.* 2000 *Phys. Rev. Lett.* **85** 2693
[56] Beck C *et al.* 2008 *AIP Conf. Proc.* **1098** 207
[57] Beck C *et al.* 2009 *Phys. Rev. C* **80** 034604
[58] Beck C *et al.* 2013 *J. Phys.: Conferences Series* **436** 012014
[59] Beck C 2014 *IASEN Conf. Proc.* (to be published)
[60] Ideguchi E *et al.* 2001 *Phys. Rev. Lett.* **87** 222501
[61] Rousseau M *et al.* 2002 *Phys. Rev. C* **66** 034612
[62] Taniguchi Y *et al.* 2007 *Phys. Rev. C* **76** 044317
[63] Norrby M *et al.* (unpublished)
[64] O'Leary C D *et al.* 2000 *Phys. Rev. C* **61** 064314
[65] Fukada M *et al.* 2009 *Phys. Rev. C* **80** 064613
[66] Salsac M D *et al.* 2008 *Nucl. Phys.* **A 801** 1
[67] Nouicer R *et al.* 1999 *Phys. Rev. C* **60** 041303(R)
[68] Rudolph D *et al.* 1999 *Phys. Rev. Lett.* **82** 3763
[69] Beck C *et al.* 2001 *Phys. Rev. C* **63** 014607
[70] Bhattacharya C *et al.* 2002 *Phys. Rev. C* **65** 014611
[71] von Oertzen W *et al.* 2008 *Eur. Phys. J. A* **36** 279
[72] Wheldon C *et al.* 2008 *Nucl. Phys. A* **811** 276
[73] Pyatkov Y F *et al.* 2010 *Eur. Phys. J. A* **45** 29
[74] Zagrebaev V I *et al.* 2010 *Phys. Rev. C* **81** 044608
[75] Yamada T and Schuck P 2004 *Phys. Rev. C* **69** 024309
[76] Ebran J P *et al.* 2012 *Nature* **487** 341
[77] Girod M and Schuck P 2013 *Phys. Rev. Lett.* **111** 132503
[78] Kawabata T *et al.* 2013 *J. Phys.: Conferences Series* **436** 012009
[79] Morelli L *et al.* 2014 *J. Phys.: (London)* **G41** 075107
[80] Akimone H *et al.* 2014 presented at this Sotancp3 Workshop
[81] Akimone H *et al.* 2013 *J. Phys.: Conferences Series* **436** 012010
[82] von Oertzen W *et al.* 2010 *Eur. Phys. J. A* **43** 17
[83] von Oertzen W *et al.* 2009 *AIP Conf. Proc.* **1165** 19
[84] Milin M *et al.* 2009 *Eur. Phys. J. A* **41** 335
[85] Bohlen H G *et al.* 2011 *Eur. Phys. J. A* **47** 44
[86] Descouvemont P and Baye D 1985 *Phys. Rev. C* **31** 2274
[87] Furutachi N *et al.* 2008 *Prog. Theor. Phys.* **119** 403
[88] Bansai M *et al.* 2011 *J. Phys.: Conferences Series* **321** 012046

3rd International Workshop on "State of the Art in Nuclear Cluster Physics" IOP Publishing
Journal of Physics: Conference Series **569** (2014) 012003 doi:10.1088/1742-6596/569/1/012003

Alpha condensates and nonlocalized cluster structures

Yasuro Funaki[1]

[1] RIKEN, Nishina Center, Wako 351-0198, Japan

E-mail: funaki@riken.jp

Abstract. We discuss a container structure for non-gaslike cluster states, in which single Tohsaki-Horiuchi- chuck-Röpke (TH R) wave functions are shown to be almost 100 % equivalent to the full solutions of the corresponding RGM GCM equations, for the inversion doublet band states in ^{20}Ne, α-linear-chain states, and $\alpha + \alpha + \Lambda$ cluster states in $^{9}_{\Lambda}$Be. The recognition of the fact that the TH R wave function describes well not only gaslike cluster states but also non-gaslike cluster states is a recent remarkable development of nuclear cluster physics. This fact tells us that the cluster structure is composed of cluster-mean-field motion under the constraint of inter-cluster Pauli repulsion, in which we call the cluster-mean-field potential the container. We demonstrate that the evolution of the cluster structure of a nucleus is governed by the size parameter of the cluster-mean-field potential (container), for ^{16}O nucleus.

1. Introduction

The gaslike cluster structure is a novel type of nuclear clustering. The structure was first suggested in the study of ^{12}C, in which the Hoyle state (the second 0^+ state of ^{12}C at 7.65 MeV) was found to have a large root-mean-square (r.m.s.) radius and a loosely coupled S-wave component dominantly with respect to the relative α-α motions. The microscopic cluster model calculations such as the resonating group method (R M), generator coordinate method (CM), and orthogonality condition model (OCM) reproduced almost of all experimental data available of the Hoyle state and revealed its structure [1]. After that, about 20 years later, the Hoyle state was reinvestigated by using the α-condensation-like wave function, which we now refer to as the Tohsaki-Horiuchi-Schuck-Röpke (THSR) wave function [2], focusing on an analogy to the α condensation in nuclear matter [3]. After it was found in 2003 that the fully-microscopic wave functions of the 3α R M and CM obtained long time ago are almost 100 equivalent to single THSR wave functions, we have had a new understanding of the Hoyle state that is the 3α condensate, where the 3α particles loosely couple with each other like a gas and occupy the lowest energy S-orbit of mean-field-like single-α potential [4]. This equivalence relation of the R M CM wave function to single THSR wave function is also found in the case of ^{8}Be ground state [5].

In contrast, non-gaslike cluster states have been recognized as having localized structures of clusters, which are obviously quite different from the α condensate states mentioned above. A typical example is the inversion doublet band states of $\alpha + {}^{16}$O in ^{20}Ne [6], which have parity-violating intrinsic deformation. Another example is a linear-chain-state of α particles, where all α clusters are aligned on a line with a rigid-body configuration [7]. $^{9}_{\Lambda}$Be nucleus is also such an example, where the additional Λ particle shrinks the core ^{8}Be and compact and non-gaslike 2α cluster structure seems to be realized. Thus, the concept of the localized clustering is the important basis to understand these structure states. The very recent calculations, however, upset this common sense by revealing that all those structure states can also be represented by the single configurations of the THSR wave function with nearly 100 accuracy [8, 9, 10, 11, 12]. The THSR wave function has a container structure that the clusters are nonlocalized and move around in a whole nuclear volume (container) characterized by a size parameter "B". We will report in this contribution that the container structure is realized not only for the gaslike cluster states but also for the non-gaslike cluster states, by showing that the THSR-type wave functions describe well non-gaslike cluster states mentioned above.

2. THSR description of inversion-doublet bands in ^{20}Ne

In this section, we show that the ^{16}O + α Brink- CM wave functions of the states belonging to the inversion-doublet bands of ^{20}Ne have almost 100 square overlaps with single THSR wave functions. The THSR wave function was introduced in order to describe cluster-gaslike states, especially α-condensate-like states.

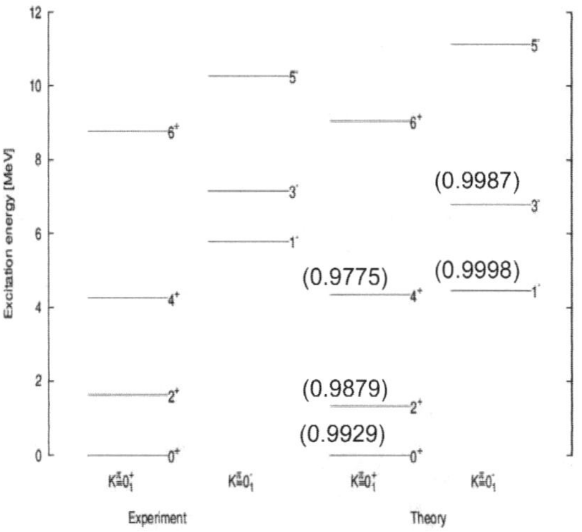

Figure 1. Comparison of the energy spectrum of inversion-doublet rotational bands in ^{20}Ne between experiment and theory. The squared overlap values between the single THSR wave functions and Brink- CM wave functions are shown in parentheses.

The THSR wave function of ^{16}O + α clustering is given by

$$\Psi^{\mathrm{THSR},J} = C_J P^J_{M0} \int d\boldsymbol{R}\, \exp\left\{-\frac{4(R_x^2+R_y^2)}{5\beta_x^2} - \frac{4R_z^2}{5\beta_z^2}\right\} \Psi^{\mathrm{B}}_{^{16}\mathrm{O}+\alpha}\Big(\frac{4}{5}\boldsymbol{R}, -\frac{1}{5}\boldsymbol{R}\Big), \tag{1}$$

$$= C'_J \exp(-\frac{10}{b^2}\boldsymbol{X}_G^2)\; P^J_{M0}\, \mathcal{A}\left[\exp\left\{-\frac{8(r_x^2+r_y^2)}{5B_x^2} - \frac{8r_z^2}{5B_z^2}\right\} \phi(^{16}\mathrm{O})\phi(\alpha)\right], \tag{2}$$

where $B_k^2 = b^2 + 2\beta_k^2$ $(k = x, y, z)$, $\boldsymbol{r} = \boldsymbol{X}_\alpha - \boldsymbol{X}_{\mathrm{O}}$, and $\boldsymbol{X}_G$ is total c.o.m. coordinate. $\Psi^{\mathrm{B}}_{^{16}\mathrm{O}+\alpha}(\frac{4}{5}\boldsymbol{R}, -\frac{1}{5}\boldsymbol{R})$ is the α + ^{16}O Brink wave function and C_J and C'_J are normalization constants. This THSR wave function Ψ_J^{THSR} has positive parity and therefore can have only even angular momentum J. The negative-parity THSR wave function with odd J is constructed by the following procedure

$$\Phi^{\mathrm{THSR},J^-} = \lim_{S_z\to 0} C_J^{(-)}\; P^J_{M0}\, \frac{1-P_\pi}{2}\, \mathcal{A}\left[\exp\left\{-\frac{8(r_x^2+r_y^2)}{5B_x^2} - \frac{8(r_z-S_z)^2}{5B_z^2}\right\} \phi(^{16}\mathrm{O})\phi(\alpha)\right], \tag{3}$$

where $C_J^{(-)}$ is normalization constant and $(1 - P_\pi)/2$ is the projection operator onto negative parity. If $B_x = B_z = B$, $\Phi^{\mathrm{THSR},(J^-)}$ has the following form

$$\Phi^{\mathrm{THSR},(J^-)} = D_J^{(-)} \mathcal{A}\left[r^J Y_{JM}(\hat{r}) \exp\left\{ -\frac{8r^2}{5B^2} \right\} \phi(^{16}\mathrm{O})\phi(\alpha) \right]. \tag{4}$$

In Fig. 1, we show the calculated energy spectrum in comparison with the experimental data. The squared overlap values of the single THSR wave function with the $\alpha + {}^{16}$O Brink- CM wave function are shown in parentheses. As mentioned above, we can see that both wave functions are almost 100 equivalent to each other. The inversion-doublet rotational bands in ^{20}Ne consist of the plus-parity band which is the ground-state band and the minus-parity band which is the $K^\pi = 0^-$ band upon the 1^- state at $E_x = 5.80$ MeV. The minus-parity band levels are all above the ^{16}O + α threshold and have large α-decay widths whose reduced widths are comparable with the Wigner limit value. In general the inversion-doublet rotational bands are generated from a parity-violating intrinsic deformed state. As the parity-violating intrinsic state of the inversion-doublet bands, spatially-localized cluster structure of ^{16}O + α is assigned [6]. Therefore the existence of the inversion-doublet rotational bands has been regarded as a convincing evidence of the spatial localization of clusters. On the other hand, the THSR wave function describes the nonlocalized motion of clusters. The almost 100 equivalence of the Brink- CM wave functions and single THSR wave functions for the ^{20}Ne inversion-doublet band states is thus really striking.

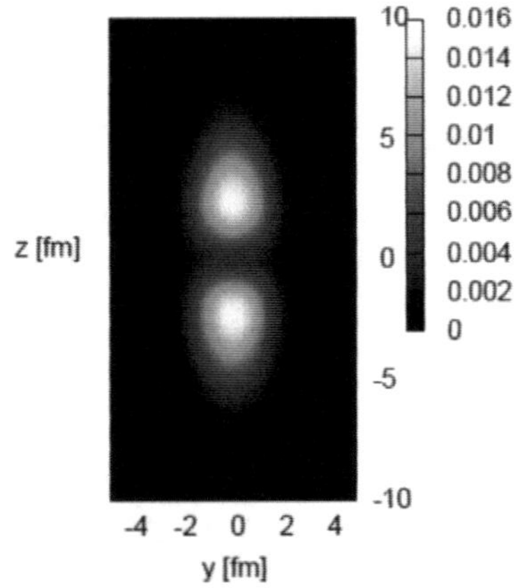

Figure 2. Nucleon-density distribution of the 2α prolate THSR wave function with $(\beta_x, \beta_y, \beta_z) = (1.78, 1.78, 8.75$ fm) [10].

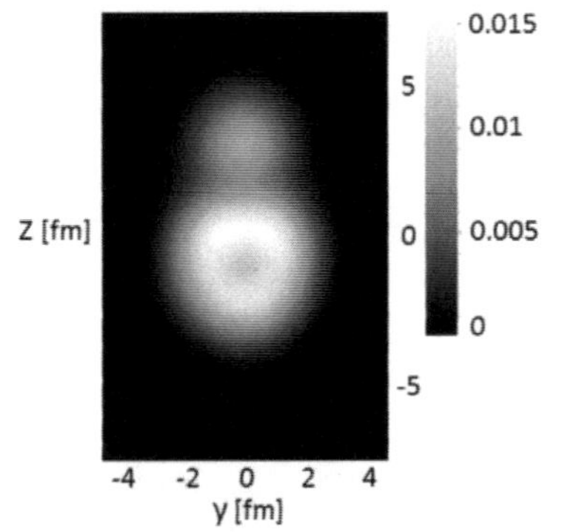

Figure 3. Nucleon-density distribution of the ^{16}O + α hybrid-Brink-THSR wave function with $S_z = 0.6$ fm and $(\beta_x, \beta_y, \beta_z) = (0.9, 0.9, 2.5$ fm) [10].

The problem of localized clustering vs nonlocalized clustering was already noticed in 2002 in the THSR study of α-α clustering in ^{8}Be. Ref. [5] reported that the Brink- CM wave function of the ground state of ^{8}Be is 100 equivalent to single THSR wave function. This equivalence looks like questioning the well-known dumbbell picture for the intrinsic shape of ^{8}Be. However it was shown recently [10] that the nucleon-density distribution of the intrinsic THSR wave function of 2α ground state displays clearly localized clustering of α-α as seen in Fig. 2. The reason why the THSR wave function which adopts the relative-motion wave function expressing the nonlocalized clustering yields the nucleon-density distribution showing the localized clustering is attributed to the inter-cluster auli repulsion which makes two clusters stay apart from each other avoiding close coming to each other. Figure 3 shows the nucleon intrinsic density distribution of the ^{16}O + α hybrid-Brink-THSR wave function Eq. (3), which is defined as the one-body density of the intrinsic wave function,

$$\Phi^{\rm hyb-B-THSR} = \mathcal{A}\left[\exp\left\{-\frac{8(r_x^2+r_y^2)}{5B_x^2} - \frac{8(r_z-S_z)^2}{5B_z^2}\right\}\phi(^{16}{\rm O})\phi(\alpha)\right], \qquad (5)$$

where S_z is set to be 0.6 fm and $(\beta_x, \beta_y, \beta_z) =$ (0.9,0.9,2.5 fm) [10]. We observe in this figure that, in spite of the small value of $S_z = 0.6$ fm, the inter-cluster distance between ^{16}O and α is about 3.6 fm. Clearly the large inter-cluster distance of about 3.6 fm cannot be attributed to the small value of $S_z = 0.6$ fm, but should be attributed to the effective spatial localization of ^{16}O and α clusters in the prolate THSR wave function with $(\beta_x, \beta_y, \beta_z) =$ (0.9,0.9,2.5 fm) with $S_z = 0$, which is nothing but the intrinsic THSR wave function giving the minimum energy for the 0^+ state.

3. THSR description of 3α and 4α linear-chain-states

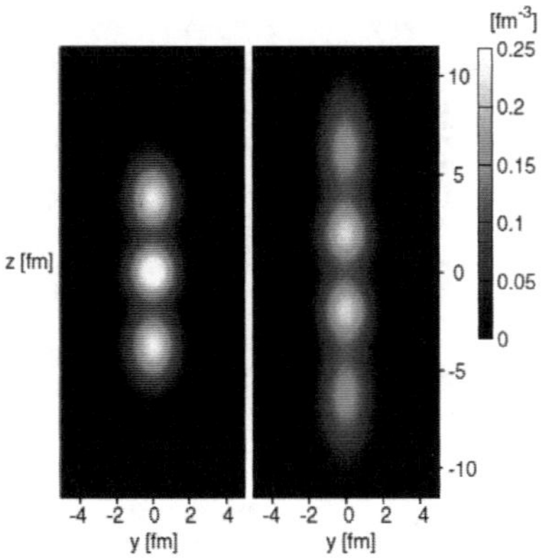

Figure 4. Intrinsic density profiles of the 3α- (Left) and 4α- (Right) linear-chain states from the THSR wave functions before angular-momentum projection at ($\beta_x = \beta_y$=0.1 fm,β_z=5.1 fm) and at ($\beta_x = \beta_y$=0.1 fm,β_z=8.2 fm), respectively. Figure taken from Ref. [11].

If we follow the conclusion of the study of the inversion doublet bands of ^{20}Ne, the basic feature of the cluster dynamics is nonlocalized clustering and the localization of clusters is due to the inter-cluster auli repulsion. A confirmation of this conclusion was given in Ref. [11] by the investigation of the linear-chain structure of α clusters which is a typical and well-known example of localized cluster structure. It was reported that the Brink- CM wave functions of the 3α and 4α linear-chain states have about 98 and 94 square overlaps with 3α and 4α single THSR wave functions, respectively. These large overlap values are to be compared with the maximum square overlap values with 3α and 4α single Brink wave functions which are 78 and 48 , respectively. These results urge us to regard that the Brink- CM wave functions of the 3α and 4α linear-chain states have one-dimensional α-condensate-like character rather than the conventional picture of localized α clusters on a line. However, just like the case of ^{16}O + α inversion doublet, the nucleon-density distribution of the intrinsic THSR wave functions of linear-chain states show the localized α clusters because of inter-α auli repulsion, which is displayed in Fig. 4.

4. THSR description of $^{9}_{\Lambda}$Be

The above-mentioned THSR descriptions of localized clustering states is not only for ordinary nuclei but also possible for hypernuclei. For example, the so-called Hyper-THSR wave function describing the $2\alpha + \Lambda$ hypernucleus with good angular momentum can be introduced as follows [12]:

$$\Psi_J^{\rm H}(\boldsymbol{\beta}) = \hat{P}^J_{MK}\Phi^{\rm H-THSR}_{2\alpha-\Lambda}(\boldsymbol{\beta}), \qquad \Phi^{\rm H-THSR}_{2\alpha-\Lambda}(\boldsymbol{\beta}) = \Phi^{\rm THSR}_{2\alpha}(\boldsymbol{\beta})\sum_{\kappa} f_\Lambda(\boldsymbol{\beta},\kappa)\varphi_\Lambda(\kappa), \tag{6}$$

where the Λ particle simply couples to the 2α core nucleus in an S wave. Its radial part is expanded in terms of aussian basis functions, $\varphi_\Lambda(\kappa) = (\pi/2\kappa)^{-3/4}\exp(-\kappa r^2_{2\alpha-\Lambda})$ with respect to the width parameter κ, where $\boldsymbol{r}_{2\alpha-\Lambda} = \boldsymbol{r}_\Lambda - (\boldsymbol{X}_1 + \boldsymbol{X}_2)/2$. Here we also assume an axially-symmetric deformation in Eq. (6), $\beta_x = \beta_y \equiv \beta_\perp \neq \beta_z$. The coe cients of the expansion $f_\Lambda(\beta_\perp, \beta_z, \kappa)$ are then determined by solving the following Hill-Wheeler-type equation of motion,

$$\sum_{\kappa'} \langle \hat{P}^J_{M0}\Phi^{\rm THSR}_{2\alpha}(\beta_\perp,\beta_z)\varphi_\Lambda(\kappa)|\hat{H} - E(\beta_\perp,\beta_z)|\hat{P}^J_{M0}\Phi^{\rm THSR}_{2\alpha}(\beta_\perp,\beta_z)\varphi_\Lambda(\kappa')\rangle f_\Lambda(\beta_\perp,\beta_z,\kappa') = 0. \tag{7}$$

This Hyper-THSR wave function is characterized only by the parameters $\beta_\perp$ and β_z, which correspond to a spatial extension of the whole nucleus. It is reported that the Λ particle invokes spatial core shrinkage in many hypernuclei. It should be mentioned that such core shrinkage effect is expected to be taken into account very naturally by this parametrization of $\boldsymbol{\beta}$, in general hypernuclei, since it specifies the dilatation of the whole nucleus. In the present work, only the S-wave component of the Λ particle is considered, for simplicity.

In order to compare the single Hyper-THSR wave function with the Brink- CM wave function, we calculate the following squared overlap:

$$\mathcal{O}_J(\beta_\perp,\beta_z) = \frac{|\langle\Psi^H_J(\beta_\perp,\beta_z)|\widetilde{\Psi}^B_J\rangle|^2}{\langle\Psi^H_J(\beta_\perp,\beta_z)|\Psi^H_J(\beta_\perp,\beta_z)\rangle\langle\widetilde{\Psi}^B_J|\widetilde{\Psi}^B_J\rangle}, \tag{8}$$

with

$$\widetilde{\Psi}^B_J = \sum_{R,S} f^{(J,\lambda=0)}(R,S)[u_J(\boldsymbol{R}),\psi_{\lambda=0}(\boldsymbol{S})]_J. \tag{9}$$

The above wave function $\widetilde{\Psi}^B_J$ is the Brink- CM wave function projected onto the model space with the angular-momentum channel $(L,\lambda) = (J,0)$.

The squared overlap between the single THSR wave functions and the Brink- CM wave functions defined in Eq. (9) are 99.5 , 99.4 , and 97.7 for the 0^+, 2^+, and 4^+ states, respectively. The calculated r.m.s. radii are 2.31 fm, 2.29 fm, and 2.20 fm, for the 0^+, 2^+, and 4^+ states, respectively. These values are all much smaller than the calculated r.m.s. value for ^{8}Be, 2.8 fm, and therefore in $^9_\Lambda$Be very localized $2\alpha + \Lambda$ cluster structure is realized, due to the shrinkage effect of the Λ particle. Thus this equivalence result means that THSR wave function works very well for localized cluster structure states in hypernuclei.

5. Evolution of cluster structures

We here demonstrate for the 4α system that the evolution of cluster states can be expressed by dilatation of a container. We first obtain the eigenstates of ^{16}O, $\Phi^{J=0}_{4\alpha,\lambda}$, by solving the Hill-Wheeler equation by taking $\boldsymbol{\beta}$ as generator coordinates with a constraint of $\beta_0 \equiv \beta_\perp = \beta_z$. These wave functions are expressed as the superposed ones of the THSR wave function with different β_0 values. It is interesting to see how these superposed wave functions are expressed by a single component THSR wave function $\Phi^{\rm THSR,(J=0)}_{4\alpha}(\beta_0)$, since this point is concerned with the container picture, in which the $\boldsymbol{\beta}$ parameter characterizes a shape of the whole nucleus. We have reported in Ref. [13] that the $(0^+_2)_{\rm THSR}$, $(0^+_3)_{\rm THSR}$, and $(0^+_4)_{\rm THSR}$ states, i.e. $\lambda = 2,3,4$ for the

Journal of Physics: Conference Series **569** (2014) 012003 doi:10.1088/1742-6596/569/1/012003

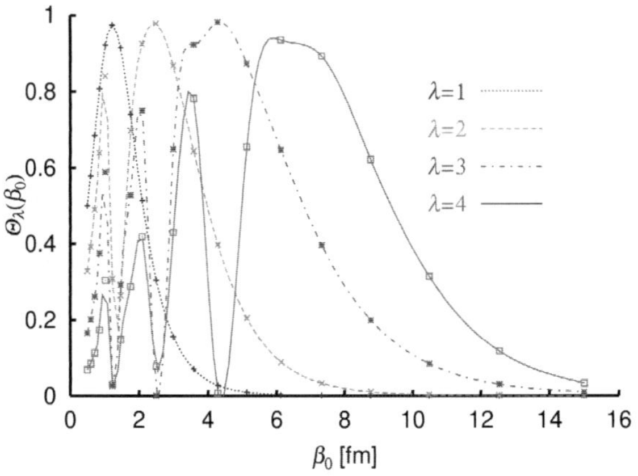

Figure 5. Overlap amplitudes $\Theta_\lambda(\beta_0)$ of Eq. (12) between the wave functions $\Phi^{J=0}_{4\alpha,\lambda}$ and $\widetilde{\Phi}_\lambda(\beta_0)$ with $\lambda = 1, \cdots, 4$, as a function of β_0. Figure taken from [13].

wave functions, $\Phi^{J=0}_{4\alpha,\lambda}$, respectively, in the THSR ansatz correspond to the $(0^+_2)_{\rm OCM}$, $(0^+_4)_{\rm OCM}$, and $(0^+_6)_{\rm OCM}$ states in the OCM calculation [14], which have the structures of ^{12}C(0^+) and α in an S-wave, of ^{12}C(0^+) and α in a higher nodal S-wave, and of the 4α condensate, respectively.

We first construct orthonormal wave functions from the THSR wave functions as follows:

$$\widetilde{\Phi}_\lambda(\beta_0) = N_\lambda P_{\lambda-1} \Phi^{\rm THSR,(J=0)}_{4\alpha}(\beta_0), \quad (\lambda = 1, \cdots, 4), \tag{10}$$

where N_λ is the normalization constant and $P_{\lambda-1}$ are projection operators defined by $P_0 = 1$ and for $\lambda \geq 2$ by

$$P_{\lambda-1} = 1 - \sum_{k=1}^{\lambda-1} |\Phi^{J=0}_{4\alpha,k}\rangle\langle\Phi^{J=0}_{4\alpha,k}|, \ (\lambda = 2, 3, 4). \tag{11}$$

The wave function $\widetilde{\Phi}_\lambda(\beta_0)$ depends on the parameter β_0 and is orthogonal to the wave functions $\Phi^{J=0}_{4\alpha,\lambda'}$ of the $(0^+_{\lambda'})_{\rm THSR}$ states with $\lambda' < \lambda$. We then calculate the following squared overlap amplitudes between the wave functions $\widetilde{\Phi}_\lambda(\beta_0)$ and $\Phi^{J=0}_{4\alpha,\lambda}$,

$$\Theta_\lambda(\beta_0) = \left| \left\langle \Phi^{J=0}_{4\alpha,\lambda} \middle| \widetilde{\Phi}_\lambda(\beta_0) \right\rangle \right|^2, \quad (\lambda = 1, \cdots, 4). \tag{12}$$

Table 1. The largest values of the squared overlaps in Fig. 5 are shown for the four THSR states, $(0^+_\lambda)_{\rm THSR}$ $(\lambda = 1, \cdots, 4)$, together with the corresponding β_0 values. Figure taken from Ref. [13].

	0^+_λ			
	$\lambda = 1$	$\lambda = 2$	$\lambda = 3$	$\lambda = 4$
$\Theta_\lambda(\beta_0)$	0.98	0.98	0.98	0.96
β_0 (fm)	1.2	2.5	4.0	6.5

In Fig. 5 we show $\Theta_\lambda(\beta_0)$ of Eq. (12) as a function of β_0. For any wave functions $\Phi^{J=0}_{4\alpha,\lambda}$ $(\lambda = 1, \cdots, 4)$, $\Theta_\lambda(\beta_0)$ $(\lambda = 1, \cdots, 4)$ takes values close to 100 at optimal values of β_0.

We also show in Table 1 the largest squared overlaps and the corresponding β_0 values for the four states. We can see that the optimal β_0 values increase from the $(0^+_1)_{\rm THSR}$ to the $(0^+_4)_{\rm THSR}$ states. The r.m.s. radii also increase with the increase of β_0. This means that the clustering excitations, which are usually described by the eigenfunctions of the Hill-Wheeler equation, are very accurately characterized by the variation of the β_0 value, i.e. dilatation of the container, from the most compact $(0_1^+)_{\rm THSR}$ state to the most dilute gaslike $(0_4^+)_{\rm THSR}$ state with the 4α condensate structure, through the intermediate density $(0_2^+)_{\rm THSR}$ and more dilute $(0_3^+)_{\rm THSR}$ states with the ^{12}C + α cluster structures. This means that the evolution of the cluster excitations, which are given as a result of solving the Hill-Wheeler equation, can be described by dilatation of the container, from a smaller value of β_0 to a larger one. This also gives a support of the correctness of the container picture. The more sophisticated description of ^{12}C + α configuration on the container picture is to use an extended version of the THSR wave function in Ref. [15].

6. Conclusion

We demonstrated that the single THSR wave functions is almost 100 equivalent to the full solutions of the R M CM equations for the inversion doublet band states in ^{20}Ne, α-linear-chain states, and $\alpha+\alpha+\Lambda$ cluster states in $^{9}_{\Lambda}$Be. The recognition of the fact that the THSR wave function describes well not only cluster-gaslike states but also non-gaslike cluster states is a recent remarkable development of nuclear cluster physics. This fact tells us that the cluster structure is composed of cluster-mean-field motion under the constraint of inter-cluster auli repulsion. The single-cluster wave function is the aussian wave packet around the c.o.m. coordinate which is the lowest orbit of the cluster-mean-field potential that we call the container. The evolution of the cluster structure of a nucleus is governed by the size parameter of the cluster-mean-field potential (container). We demonstrated that the ^{16}O nucleus, starting from the compact shell-model structure for the ground state, develops the cluster structure, from the $\alpha + {}^{12}$C structure to the 4α structure, with the increase of the size parameter $\boldsymbol{\beta}$.

References

[1] Fujiwara Y, Horiuchi H, Ikeda K, Kamimura M, Katō K, uzuki Y, and Uegaki E, *Prog. Theor. Phys. Supple.* 68 (1980) Chapt.2, p.29
[2] Tohsaki A, Horiuchi H, chuck P, and Röpke G, *Phys. Rev. Lett.* 2001 **87** 192501
[3] Röpke G, chenell A, chuck P, and Nozières P, 1998 *Phys. Rev. Lett.* **80** 3177
[4] Funaki Y, Tohsaki A, Horiuchi H, chuck P, and Röpke G, 2003 *Phys. Rev.* C **67** 051306(R)
[5] Funaki Y, Horiuchi H, Tohsaki A, chuck P, and Röpke G, 2002 *Prog. Theor. Phys.* **108** 297
[6] Horiuchi H and Ikeda K, 1968 *Prog. Theor. Phys.* **40** 277
[7] Morinaga H, 1966 *Phys. Lett.* **21** 78
[8] Zhou B, Funaki Y, Horiuchi H, Ren Z, Röpke G, chuck P, Tohsaki A, u C, and Yamada T, 2012 *Phys. Rev.* C **86** 014301
[9] Zhou B, Funaki Y, Horiuchi H, Ren Z, Röpke G, chuck P, Tohsaki A, u C, and Yamada T, 2013 *Phys. Rev. Lett.* **110** 262501
[10] Zhou B, Funaki Y, Horiuchi H, Ren Z, Röpke G, chuck P, Tohsaki A, u C, and Yamada T, 2014 *Phys. Rev.* C **89** 034319
[11] uhara T, Funaki Y, Zhou B, Horiuchi H, and Tohsaki A, 2014 *Phys. Rev. Lett.* **112** 062501
[12] Funaki Y, Yamada T, Hiyama E, Zhou B, and Ikeda K, ar iv: 1405.6067, to appear in 2014 *Prog. Theor. Ext. Phys.*
[13] Funaki Y, Yamada T, Tohsaki A, Horiuchi H, Röpke G, and chuck P, 2010 *Phys. Rev.* C **82** 024312
[14] Funaki Y, Yamada T, Horiuchi H, Röpke G, chuck P, and, Tohsaki A, 2008 *Phys. Rev. Lett.* **101** 082502
[15] Funaki Y, ar iv: 1408.5855

Clustering in non-self-conjugate nuclei ^{10}Be and ^{18}O

G.V. Rogachev[1,2], M.L. Avila[2,3], A.N. Kuchera[2,4], L.T. Baby[2], J. Belarge[2], J. Blackmon[5], V.Z. Goldberg[1], E.D. Johnson[2], K.W. Kemper[2], E. Koshchiy[1,2], L. Linhardt[5], K. Macon[5], D. Santiago-Gonzalez[2,5] and I. Wiedenhöver[2]

[1] Department of Physics & Astronomy and Cyclotron Institute, Texas A&M Univ., College Station, Texas, USA
[2] Department of Physics, Florida State University, Tallahassee, Florida, USA
[3] Physics Division, Argonne National Laboratory, Lemont, Illinois, USA
[4] National Superconducting Cyclotron Laboratory, Michigan State Univ., East Lansing, Michigan, USA
[5] Department of Physics & Astronomy, Louisiana State Univ., Baton Rouge, Louisiana, USA

E-mail: rogachev@tamu.edu

Abstract. Clustering phenomena in ^{10}Be and ^{18}O were studied by means of resonance elastic scattering of α-particles on ^{6}He and ^{14}C. Excitation functions for α+^{6}He and α+^{14}C were measured and detailed R-matrix analyses of the excitation functions was performed. We compare the experimental results with the predictions of modern theoretical approaches and discuss properties of cluster rotational bands.

1. Introduction.

Clustering is known to play an important role in the structure of light $N = Z$, even-even nuclei. A number of known structure peculiarities in nuclei, such as ^{8}Be, ^{12}C, ^{16}O, and ^{20}Ne, are associated with clustering (see recent review [1] and Refs. therein).

It has been proven to be more difficult to study clustering phenomena in non-self-conjugate, $N \neq Z$ nuclei. Clustering might manifest itself in a much more complex way in these nuclei. This is because the "extra" nucleons introduce additional degrees of freedom which may modify, create, enhance or destroy cluster structures. In addition to that there are difficulties with the experimental studies, which require a more complicated analysis due to the presence of low-lying nucleon decay channels and higher level density. On the other hand, the investigation of clustering phenomena in non-self conjugate, $N \neq Z$ nuclei seems very promising. "Valence" nucleons may be responsible for exotic, molecular-type structures [2]. Also, nucleon and α-decay thresholds are typically close in these nuclei, which allows more direct exploration of the interplay between the single nucleon and cluster degrees of freedom by measuring the corresponding partial widths and correlations. We studied the cluster structure of non-self conjugate nuclei ^{10}Be and ^{18}O by measuring the excitation functions for α+^{6}He and α+^{14}C resonance elastic scattering.

2. Cluster structures in ^{10}Be.

It has been suggested in several theoretical papers (see *e.g.* [3, 4, 5]) that highly clustered, double α- core structure, α:nn:α, may exist in ^{10}Be. This structure would form a positive-parity

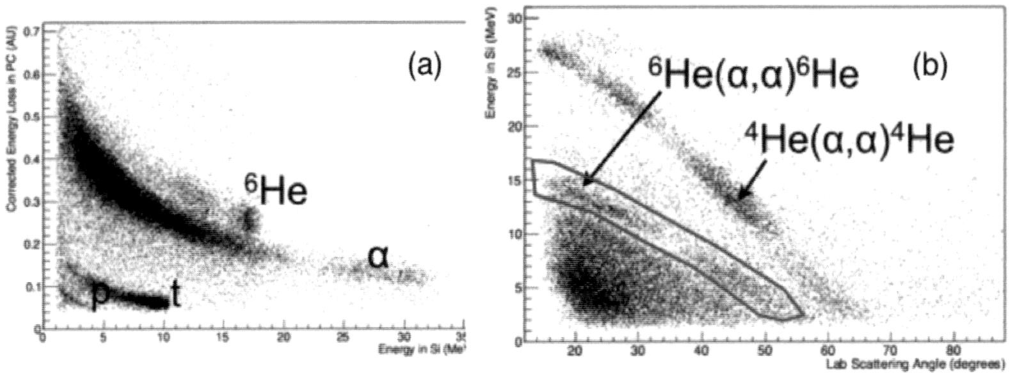

Figure 1. Identification of ^{6}He+α elastic scattering events in ANASEN. Part (a) shows the particle ID plot with energy deposited in the silicon array on the x-axis and energy loss measured by proportional counters in the y-axis. Part (b) is the energy - scattering angle correlation plot for α-particles. Bananas that correspond to the kinematics for ^{6}He+α and α+α elastic scattering are labeled.

rotational band based on the 0_2^+ state (0^+ at 6.17 MeV, 2^+ at 7.54 MeV and 4^+ at 10.15 MeV). The cluster structure of ^{10}Be has been studied in many experiments (see [7, 8, 9, 10, 11] and Refs. therein) but our knowledge is still incomplete. S_α factors are still unknown for the 0^+ and 2^+ states at 6.17 MeV and 7.54 MeV (although, results of Ref. [12] indicate unusually large α-strength that far exceeds a single particle limit for the 2^+ state while the α-strength was measured to be close to the single particle limit for the corresponding isobaric analog resonance in ^{10}B [14]). The spin-parity assignment for the state at 10.15 MeV was controversial at the time when this project started, with 3^- and 4^+ conflicting assignments [10, 11]. Existence of the next, 6^+ member of this band was predicted in [6] based on SU(3) algebraic model, but never observed experimentally. Confirmation of the 4^+ spin-parity assignment and search for the 6^+ member of the α:nn:α rotational band in the excitation function for ^{6}He+α elastic scattering were the main goals of the project. The excitation function for ^{6}He+α resonance elastic scattering in the excitation energy range 9.5 to 13.4 MeV was reported recently in [15]. We present new experimental data, which extends the previously measured ^{6}He+α excitation function to higher excitation energies, up to 15.4 MeV. These measurements were part of the commissioning run for new active target detector ANASEN.

Measurements were performed at the John D. Fox Superconducting Linear Accelerator Laboratory at Florida State University. The ^{6}He beam was produced by the in-flight rare isotope beam facility RESOLUT with intensity of about 10^5 pps. The new active target detector - Array for Nuclear Astrophysics and Structure with Exotic Nuclei (ANASEN) - was used to perform the measurements. ANASEN is a gas filled detector, which has an array of position sensitive silicon detectors that surrounds the beam axis and a cylindrical array of position sensitive proportional counters. The recoils that emerge from the place of interaction go through the proportional counter first. The location and energy loss in the active gas volume are measured. If the second hit is recorded by the silicon detector then the track of the recoil can be reconstructed using position information provided by the proportional counter and the silicon array. Particle identification is performed using energy loss in the active gas volume and the total energy measured by the silicon array (see Fig. 1(a)). The ^{6}He+α elastic scattering events can be

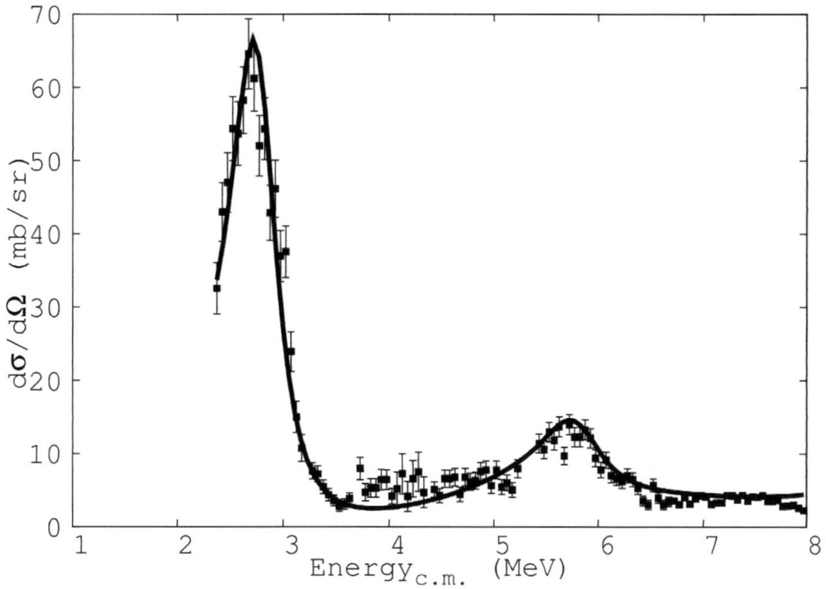

Figure 2. Excitation function for ^{6}He+α elastic scattering at 90±5° in c.m.. Solid line is the model R-matrix calculation with two resonances, 4^+ and 6^+ at 2.75 and 5.9 MeV in c.m. respectively. See text for details.

identified using the energy-angle kinematic dependence reconstructed from the tracks of the recoils, as shown in Fig. 1(b).

The excitation function for ^{6}He+α resonance elastic scattering in the energy range between 2.2 and 8 MeV in c.m.s. (9.7 and 15.5 MeV excitation energy in ^{10}Be) measured near 90° in c.m. is shown in Fig. 2. Obvious features of this excitation function is a peak at 2.75 MeV (10.16(2) MeV excitation energy in ^{10}Be) and a small peak near 5.9 MeV (13.4 MeV excitation energy). The first peak corresponds to the state that was first observed in [7] and [8]. The 3^- spin-parity was assigned to it in [10]. However, this was contradicted by [11] where the ^{6}He+α angular distribution at 2.75 MeV energy in c.m. showed a maximum at 90 degrees, indicating positive parity. The 4^+ spin-parity assignment was made for the state in [11]. This spin-parity assignment was confirmed recently in [15], where the excitation function for ^{6}He+α was measured in the energy range from 2.4 to 6 MeV in c.m. The excitation function shown in Fig. 2 is similar to that measured in [15], but the higher energy data obtained in this work allowed us to see that there is also another peak at 5.9 MeV, with the cross section sharply decreasing toward higher energies.

The angular distribution for ^{6}He+α elastic scattering at 2.72 MeV in c.m is shown in Fig. 3. The R-matrix fit with the 4^+ state at 2.79 MeV reproduces the corresponding angular distribution rather well. No other spin-parity assignment can produce a fit of comparable quality. We confirm the 4^+ spin-parity assignment for this state that was first made in [11] and later confirmed in [15]. The partial α width for the 10.16 MeV state determined from R-matrix fit exceeds the single-particle limit for the purely α-cluster state and we agree with the conclusions made in [11, 15] that this is indeed a highly clustered 4^+ state, that may belong to the α:nn:α rotational band. The peak at 5.9 MeV in the ^{6}He+α excitation function (Fig. 2) is very intriguing. It appears that this structure corresponds to a positive parity high spin state because

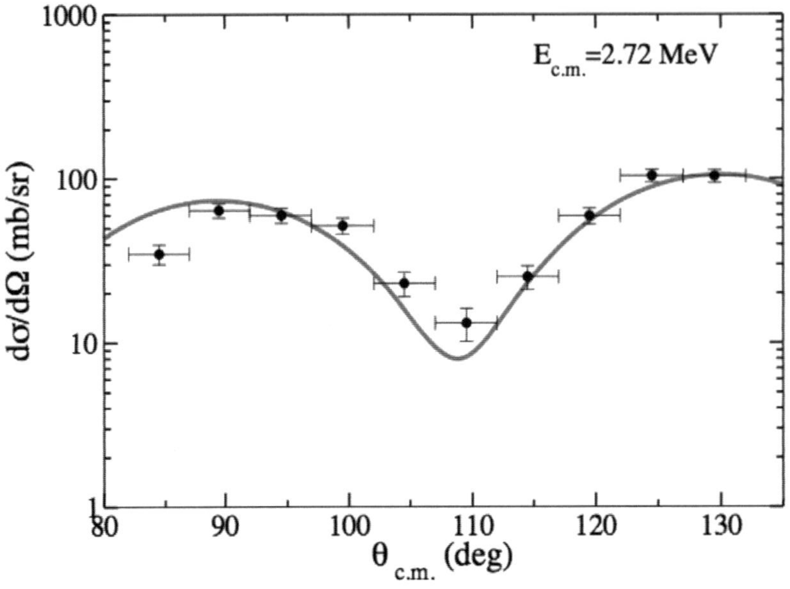

Figure 3. Angular distribution for ^{6}He+α elastic scattering at 2.72 MeV in c.m. The solid line is the R-matrix fit with 4^{+} state at 2.79 MeV.

it is most prominent near 90° in the limited angular range (70-130°) measured in this work. A definitive spin-parity assignment is handicapped by what appears to be direct ("potential") scattering that dominates angular distribution at this energy (around 6 MeV in c.m.). Model R-matrix calculations indicate that the shape and width of the resonance may correspond to a highly clustered 6^{+} state that predominantly decays by α emission to the ground and 2^{+} first excited states in ^{6}He. However, more analysis and/or new experimental data at angles close to 180° may be necessary to confirm or rule out the existence of this state. If confirmed, this state would be a strong candidate as the next, 6^{+} member of the α:nn:α rotational band in ^{10}Be.

3. Search for inversion doublet quasi-rotational bands in ^{18}O

One of the most striking manifestations of α-clustering in light nuclei is the appearance of a sequence of highly clustered states that form rotational bands of alternate parities. The positive parity α-cluster rotational band is found at a lower energy and the corresponding negative parity band is shifted up by several MeV toward higher excitation energies. These bands are called inversion doublets and have been conclusively identified in ^{16}O and ^{20}Ne (see [1] and references therein). There have been numerous attempts to find the members of the inversion doublet rotational bands in ^{18}O that correspond to an ^{14}C(g.s.)+α configuration ([16] e.g.), in analogy to the well known rotational bands in ^{16}O and ^{20}Ne. In spite of significant effort, the assignment of the band members remains controversial. The identification of states that can be considered as the members of the inversion doublet rotational bands in ^{18}O using excitation function for ^{14}C+α elastic scattering is the main focus of the discussion below.

The excitation function for ^{14}C+α elastic scattering was measured in the broad excitation energy and angular range using the Thick Target Inverse Kinematics (TTIK) technique [17]. The experiment was performed at Florida State University, John D. Fox Superconducting Linear

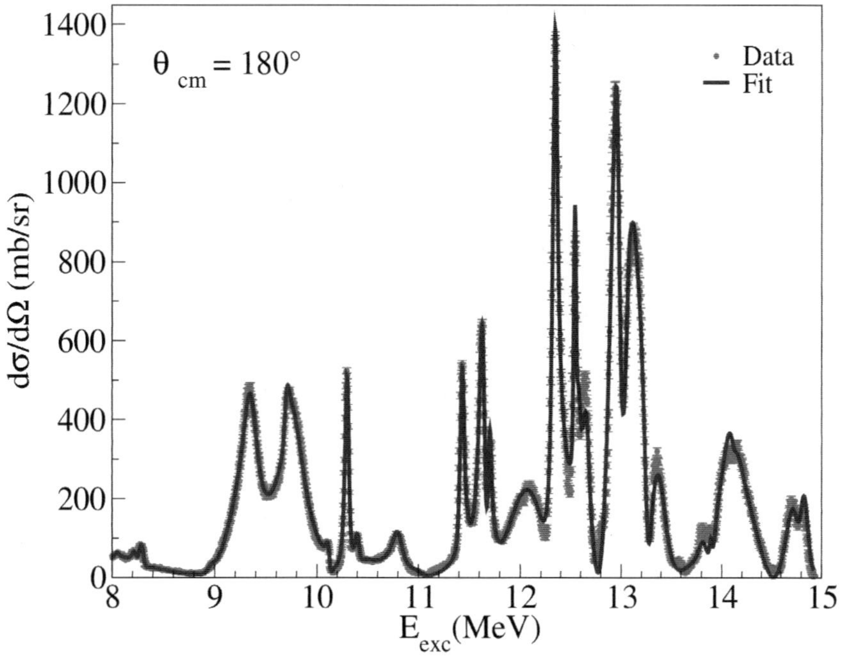

Figure 4. Excitation function for ^{14}C+α elastic scattering at 180° in c.m. The solid line is the best R-matrix fit.

Accelerator facility. The Thick Target Inverse Kinematics (TTIK) technique allows one to measure a large range of excitation energies without the need to change the initial energy of the beam, making the experiment more efficient and less time consuming. A 42 MeV ^{14}C beam and 99.9% pure helium gas (^{4}He) were used.

The 180° excitation function for ^{14}C+α elastic scattering measured in this experiment is shown in Fig. 4. A detailed R-matrix analysis was performed and spectroscopic information on 54 states in ^{18}O in the broad excitation energy range from 8 to 15 MeV were extracted. A complete description of the results of this experiment is given in [18]. Here we only consider those aspects of ^{18}O structure that are related to the inversion doublet α-cluster quasi-rotational bands. Selected states in ^{18}O that were observed in this experiment and are relevant for the discussion below are listed in Table 1.

Assignments of the ^{14}C$_{g.s.}$+α rotational bands have been suggested in many experimental works ([19, 20, 16] e.g.). The states 0^+(3.63 MeV), 2^+(5.24 MeV), 4^+(7.11 MeV) and 6^+(11.69 MeV) have been considered as members of the positive parity rotational band, which resembles the ^{20}Ne g.s. rotational band. (In [19] the 4^+ state at 10.29 MeV was suggested as a member of this rotational band instead of the 4^+ at 7.11 MeV, however our results indicate that this state has only moderate α-strength (Table 1).) An 8^+ state at 17.6 MeV and 18.06 MeV was suggested by [20] and [16] respectively as the fifth member of this rotational band. Only the 6^+ state at 11.7 MeV is within the energy range measured in this work and we confirm that this is a clustered state with α-strength θ_α^2=0.23. The α-strength is defined in a standard way as the dimensionless reduced α-width, θ_α^2, that is the ratio between the α reduced width and the single particle limit θ_α^2=$\gamma_\alpha^2/(\hbar^2/(\mu R^2))$, where R=5.2 fm and μ is a reduced mass. However, there exists the second strong 6^+ state at 12.58 MeV (θ_α^2 = 0.38). This forces us to conclude that clustering in ^{18}O is more complicated and cannot be described by a single pair of inversion doublet rotational bands. This is also supported by the data on the cluster states with negative parity.

Table 1. Parameters of the selected states in ^{18}O observed in this experiment. Only the most clustered states and states mentioned in the discussion are included. See Ref. [18] for complete results.

E^* (MeV)	J^π	Γ_t (keV)	Γ_α (keV)	Γ_n (keV)	θ^2_α
8.04(2)	1^-	2.0(7)	2.0(7)	-	0.02
8.290(6)	3^-	8.5(9)	2.9(2)	5.6(7)	0.18
9.19(2)	1^-	220(30)	200(10)	20(10)	0.20
9.35(2)	3^-	180(30)	110(30)	70(5)	0.48
9.70(1)	3^-	140(10)	15(2)	125(10)	0.04
9.76(2)	1^-	700(120)	630(10)	70(50)	0.46
10.290(4)	4^+	29(4)	19(2)	10(2)	0.09
10.80(3)	1^-	690(110)	630(90)	60(30)	0.29
11.627(4)	5^-	40(5)	30(3)	10(3)	0.13
11.699(5)	6^+	23(2)	12(1)	11(1)	0.23
12.339(4)	5^-	39(4)	26(2)	13(2)	0.06
12.576(9)	6^+	70(20)	50(10)	20(10)	0.38
13.08(1)	5^-	180(20)	120(10)	60(15)	0.17
14.1(1)	5^-	560(70)	260(20)	300(40)	0.23
14.7(1)	5^-	280(70)	230(20)	50(25)	0.16

The location of the band head and identification of the other members of the negative parity inversion doublet, α-cluster, quasi-rotational band in ^{18}O is a subject that has a rich history and has been discussed in many theoretical and experimental papers (see [16] and ref. therein). The 1^- state at 8.035 MeV was proposed as the band head for this band in recent work of W. von Oertzen, at el. [16] (the same suggestion was made in [21]). Our result rules out this state as a member of the 0^- band due to its small dimensionless reduced α-width ($\theta^2 = 0.02$). In fact, none of the states identified in [16] as members of the negative parity inversion doublet (1^- (8.04 MeV), 3^- (9.7 MeV), 5^- (13.6 MeV), 7^- (18.63 MeV)) can belong to this band (except maybe a 7^- state that lies beyond the energy region studied in this work). The 3^- state at 9.7 MeV has θ^2_α of only 0.04 (an order of magnitude less than the 3^- at 9.3 MeV) and the 5^- state at 13.6 MeV is not observed at all in this work, which rules out the assignment made in [16]. We observed the highly clustered 1^- and 3^- states at 9.19 MeV and 9.35 MeV with θ^2_α 0.2 and 0.48 respectively and it is tempting to identify them as the members of the inversion doublet band. However, as in the case of positive parity band, we observed a 1^- state at 9.76 MeV with θ^2_α=0.46, which makes the situation complicated. There is also another strong 3^- state ($\theta^2_\alpha = 0.18$) at excitation energy of 8.28 MeV. It is 1 MeV below the most clustered 3^- state at 9.35 MeV (θ^2=0.48). There is no dominant 5^- α-cluster state in the spectrum. Instead, there are several 5^- states with substantial α-strength spread out over a 3 MeV energy interval between 11.5 and 15 MeV. Their combined α-strengths add up to 0.8. The experimental data presented here reveal that there is a strong splitting of α- strength between several states of the same spin-parity and it is not possible to identify a single, dominant α-cluster rotational band of either positive or negative parity. This is contrary to some theoretical predictions [22, 23] and suggestions made in previous experimental studies ([19, 20, 21, 16] e.g.).

4. Summary

The cluster structure of non-self-conjugate nuclei ^{10}Be and ^{18}O were studied using the α-particle resonance elastic scattering approach. The excitation functions for ^{6}He+α and ^{14}C+α were measured and R-matrix analyses were performed to extract parameters for the observed resonances. We confirmed the 4^+ spin-parity assignment for the 10.16 MeV state in ^{10}Be and observed a peak in the ^{6}He+α excitation function at 15.9 MeV near 90° in c.m. This may be due to positive parity high spin state. It is possible that this is the next, 6^+ member of the α:nn:α rotational band but more data, especially at angle close to 180 ° in c.m. may be needed to confirm.

Detailed spectroscopic information on 54 states in ^{18}O in the energy range between 8 and 15 MeV have been obtained in this work. The main conclusion is that unlike for $N = Z$, ^{16}O and ^{20}Ne nuclei, the α-strength is spread between several states for each spin-parity and it is not possible to define an inversion doublet rotational bands in the same sense as for ^{16}O and ^{20}Ne nuclei. This conclusion is in contradiction to some theoretical predictions [22, 23] and suggestions made in previous experimental studies ([19, 20, 21, 16] e.g.).

The authors would like to acknowledge the financial support provided by the National Science Foundation under grants No. PHY-456463 and MRI grant No. 0820941.This material is also based upon work partially supported by the U.S. Department of Energy, Office of Science, under Award Number DE-FG03-93ER40773.

References

[1] M. Freer, Rep. Prog. Phys. **70**, 2149 (2007).
[2] W. von Oertzen, Z. Phys. A **354**, 37 (1996).
[3] A. Dote, et al., Phys. Rev. C **56**, 1877 (1997).
[4] N. Itagaki and S. Okabe, Phys. Rev. C **61**, 044306 (2000).
[5] M. Ito, K. Kato and K. Ikeda, Phys. Lett. B **636**, 293 (2006).
[6] R. Wolski, et al., Phys. of Atom. Nucl. **73**, 1405 (2010).
[7] S. Hamada, et al., Phys. Rev. C **49**, 3192 (1994).
[8] N. Soic, et al., Europhys. Lett. **34**, 7 (1996).
[9] M. Milin, et al., Europhys. Lett. **48**, 616 (1999).
[10] N. Curtis, et al., Phys. Rev. C **64**, 044604 (2001).
[11] M. Freer, et al., Phys. Rev. Lett. **96**, 042501 (2006).
[12] J.A. Liendo, Phys. Rev. C **65**, 034317 (2002).
[13] J.A. Liendo, et al., Phys. Rev. C **63**, 034301 (2001).
[14] A.N. Kuchera et al., Phys. Rev. C **84**, 054615 (2011).
[15] D. Suzuki, et al., Phys. Rev. C **87**, 054301 (2013).
[16] W. von Oertzen, et al., Eur. Phys. J. A **43**, 17 (2010).
[17] K.P. Artemov, et al., Sov. J. Nucl. Phys. **52**, 408 (1990).
[18] M.L. Avila, et al., arXiv:1406.6734v1 (2014), Phys. Rev. C. (accepted).
[19] G. Morgan, et al., Nucl. Phys. A **148**, 480 (1970).
[20] A. Cunsolo, et al., Phys. Let. B **112**, 121 (1982).
[21] N. Curtis, et al., Phys. Rev. C **66**, 024315 (2002).
[22] P. Descouvemont and D. Baye, Phys. Rev. C **31**, 2274 (1985).
[23] N. Furutachi, et al., Prog. of Theor. Phys. **119**, 403 (2008).

Clustering aspects in $N = Z$ nucleus ^{24}Mg studied by antisymmetized molecular dynamics

Masaaki Kimura and Yohei Chiba

Department of Physics, Hokkaido University, 060-0810 apporo, Japan

E-mail: masaaki@nucl.sci.hokudai.ac.jp

Abstract. Cluster structure of highly excited states of ^{24}Mg has been investigated by the antisymmetrized molecular dynamics. Imposing the constraints on the expectation values of harmonic oscillator quanta, α+^{20}Ne, ^{12}C+^{12}C and 6α cluster wave functions were generated without any *a priori* assumption. The isoscalar monopole excitation function is also calculated and reasonably agrees with the observation. It is found that α+^{20}Ne, ^{12}C+^{12}C and 6α clusters appear as the prominent peaks in the strength function.

1. Introduction

The recent development of nuclear structure models has enabled the description of a rich variety of nuclear structure changes and dynamics. Especially, the study of the $N = Z$ *sd*-shell nuclei such as ^{24}Mg, ^{24}Si and ^{32}S is of particular interest and importance because many exotic clusters will appear as the highly excited states and they will coexist with and couple to the low-lying collective states.

Among the *sd*-shell nuclei, ^{24}Mg is the nucleus that has been most intensively studied in detail. The triaxial deformed low-lying collective states have been discussed for a long time [1, 2], and in the highly excited region, many cluster states are expected to exist. According to the Ikeda's diagram [3], a couple of threshold energies that decompose the system into clusters such as ^{12}C+^{12}C, α+^{20}Ne and 2α+^{16}O are located at approximately 10 to 20 MeV in excitation energy. Cluster model studies [4, 5, 6, 7] have been devoted to reveal them, but to date, little is known and their existence has not been established well, except for the well-known ^{12}C+^{12}C molecular resonances [8, 9, 10]. It is also noted that very recently irod and Schuck [11] have pointed out the possible existence of 6α cluster state around 30 MeV based on the Hartree-Fock-Bogoliubov (HFB) calculation.

Thus, the excited states of ^{24}Mg must contain rich dynamics of nuclear many-body system, triaxial deformed collective states and many clusters. Furthermore, recent new experimental information obtained by the inelastic α scattering sheds new light on the highly excited states of ^{24}Mg. It is reported that several excited states exist around 10 to 20 MeV, and because their isoscalar (IS) monopole transition strengths are fairly large [12], they may be associated with the cluster states similar to the cases of ^{11}B [13, 14] ,^{12}C and ^{16}O [15, 16].

Our aim in this theoretical study is to show the existence of many clusters in ^{24}Mg and to investigate the relationship between those cluster states and the observed large IS monopole strengths. For this purpose, we employ the theoretical framework of antisymmetrized molecular dynamics (AMD) [17, 18], because it is capable of describing the various structures in the wide energy region without any *a priori* assumption. We show that AMD is quite successful in explaining the observed IS monopole strengths including the giant resonance and predicts the existence of the ^{12}C+^{12}C, α+^{20}Ne and 6α cluster states with large IS monopole transition strengths.

2. Theoretical Framework

2.1. Constrained energy minimization

We start from the microscopic Hamiltonian with the ogny D1S effective interaction [19] and use the parity-projected wave function as the variational wave function,

$$\hat{H} = \sum_i^A \hat{t}_i - \hat{t}_{cm} + \frac{1}{2}\sum_{ij}^A \hat{v}_{NN}(ij) + \frac{1}{2}\sum_{ij}^Z \hat{v}_C(ij), \tag{1}$$

$$\Phi^\pi = \frac{1+\pi\hat{P}_x}{2}\Phi_{int}, \quad \Phi_{int} = \mathcal{A}\{\varphi_1, \varphi_2, ..., \varphi_A\}, \tag{2}$$

$$\varphi_i = \exp\left\{-\sum_{\sigma=x,y,z} \nu_\sigma\left(r_\sigma - \frac{Z_{i\sigma}}{\sqrt{\nu_\sigma}}\right)^2\right\} \otimes (a_i\chi_\uparrow + b_i\chi_\downarrow) \otimes (\text{proton or neutron}), \tag{3}$$

where the single particle wave packet φ_i describing the motion of each nucleon is represented by the deformed aussian [20], and the parameters $\boldsymbol{Z}_i$, a_i, b_i and ν_σ are determined by the energy minimization with constraints. To deal with the low-lying quadrupole collectivity and highly excited cluster states simultaneously, we introduce two different constraints. The first is imposed on the nuclear matter quadrupole deformation parameters β and γ [21, 22] to describe the low-lying collectivity, and we denote the wave functions obtained with it as $\Phi_{\beta\gamma}$. As the second constraint, the method used in Refs. [23, 24] is extended, and we put the constraint on the following values,

$$\Delta N = N_x + N_y + N_z - N_0, \quad \lambda = N_z - N_y, \quad \mu = N_y - N_x, \quad (N_z \geq N_y \geq N_x), \tag{4}$$

where N_x, N_y, N_z represent the expectation values of the harmonic oscillator quanta in x, y, z directions and N_0 is the lowest auli-allowed value equal to 28 in the case of ^{24}Mg. We denote thus-obtained wave functions as Φ_N. As we see in the next section, N-basis successfully describes various cluster states as well as the single-particle excited states.

2.2. Generator coordinate method and isoscalar monopole strength

We further introduce the basis wave functions defined as,

$$\Phi_{r^2} = (1 - e^{-\mu\hat{O}})\Phi_{\beta\gamma} \simeq \mu\hat{O}\Phi_{\beta\gamma}, \quad \hat{O} = \sum_{i=1}^A r_i^2, \tag{5}$$

where μ is arbitrary small number and $\hat{O}$ is the IS monopole transition operator. By definition, the wave functions Φ_{r^2} describe $1p1h$ $(2\hbar\omega)$ configuration built on $\Phi_{\beta\gamma}$ by the IS monopole transition operator. Then, the wave functions generated by energy minimizations and by Eq. (5) are projected to the eignenstate of the angular momentum and superposed to perform the generator coordinate method (CM),

$$\Psi^{J\pi}_{M\alpha} = \frac{2J+1}{8\pi^2}\sum_{Ki} g^{J\pi}_{Ki\alpha}\int d\ \ D^{J*}_{MK}(\ \ )\hat{R}(\ \ )\Phi^\pi_i. \tag{6}$$

Solving the Hill-Wheeler equation, we obtain the eigenenergies $E^{J\pi}_\alpha$ and wave functions $g^{J\pi}_{Ki\alpha}$ for the ground and excited states of ^{24}Mg.

Using the wave function of the ground and excited 0^+ states, we calculate the IS monopole transition strength $S(E)$, its sums m_n and reduced transition probability $B(IS0)$,

$$S(E) = |\langle\Psi_n^{0+}(E')|\hat{O}|\Psi_{g.s.}^{0+}\rangle|^2 E\delta(E'-E), \quad B(IS0;0_n^+) = |\langle\Psi_n^{0+}|\hat{O}|\Psi_{g.s.}^{0+}\rangle|^2, \tag{7}$$

$$m_n = \int_0^\infty dE' E'^n |\langle\Psi^{0+}(E')|\hat{O}|\Psi_{g.s.}^{0+}\rangle|^2 \delta(E'-E). \tag{8}$$

We smear the strength function $S(E)$ by Lorentzian of 100 keV width for the sake of the presentation.

3. Results

3.1. Wave functions obtained by energy minimization

Figure 1 shows the density distributions of the various intrinsic wave functions obtained by the energy minimizations. The panels (a) and (b) show typical example of the wave functions $\Phi_{\beta\gamma}$, and they are the dominant components of the ground and the 0_2^+ states, respectively. Both of them are largely deformed with triaxility showing the importance of the quadrupole collectivity in the low-lying states. Indeed, the CM calculation with only the basis wave functions $\Phi_{\beta\gamma}$

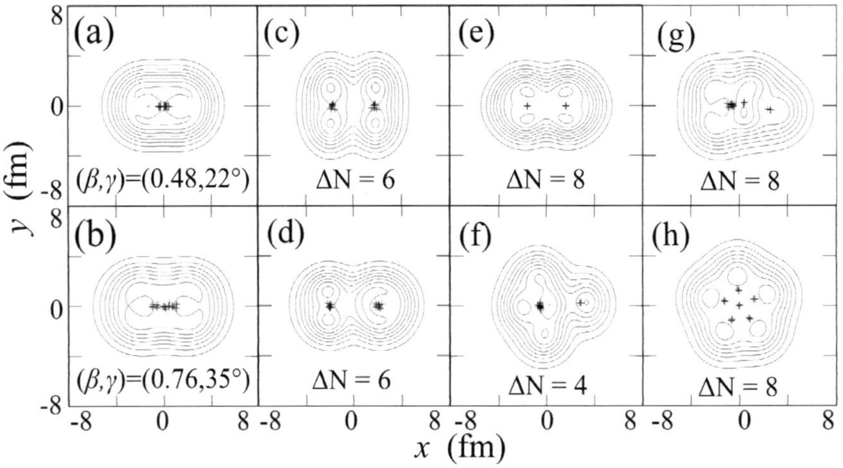

Figure 1. The density distributions of the intrinsic wave functions in the $z = 0$ plane. (a) and (b) show the example of $\Phi_{\beta\gamma}$ and (c)-(h) show those of Φ_N. Crosses in the figure show the centroid of the aussian wave packet describing nucleons.

[22] reasonably explains the low-lying spectrum of ^{24}Mg as shown in Fig. 2. The ground band, the γ and β bands built on the 2_2^+ and 0_2^+ states are nicely reproduced including their intra- and inter-band $B(E2)$ strengths. It is also noted that there is no cluster state in $\Phi_{\beta\gamma}$ as stated in Ref. [22]. Therefore, by a quick glance of this result, it looks as if there was no room for the clustering in this nucleus.

The use of the constraint on ΔN, λ and μ reveals the other facet of ^{24}Mg. It generates various cluster states shown in Fig. 1 (c)-(h) as well as the single-particle excited states that are not on the energy surface of $\Phi_{\beta\gamma}$. It is found that by increasing the value of ΔN, various cluster states start to emerge and their configurations such as inter-cluster distance and orientation of clusters vary depending on the values of $\Delta N, \lambda$ and μ. By applying the constraint of $\Delta N = 2$ ($2\hbar\omega$ excitation), the ^{12}C+^{12}C cluster states start to appear and they grow by increasing ΔN.

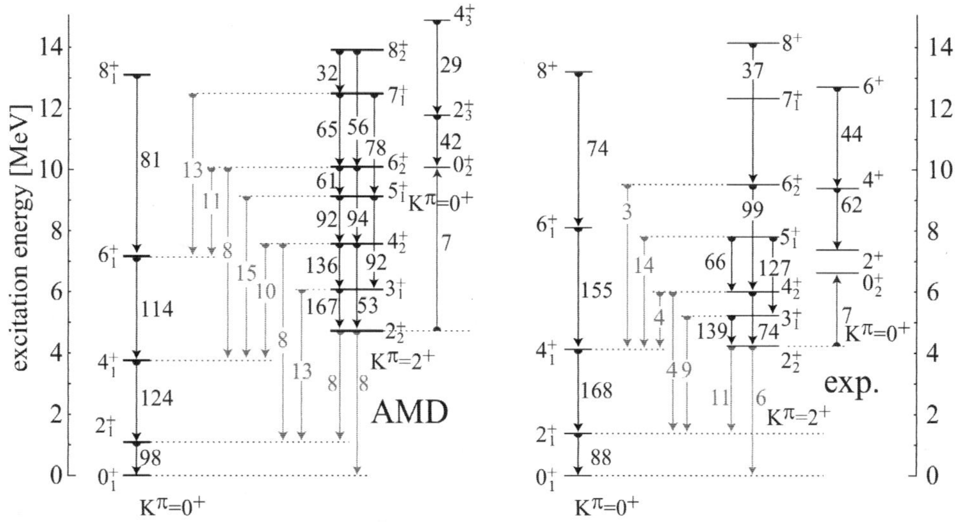

Figure 2. The calculated and observed excitation spectra of ^{24}Mg. The CM calculation employs only $\Phi_{\beta\gamma}$ as basis wave functions. Arrows in the figure show the intra and inter-band $E2$ transition strengths.

Figure. 1 (c)-(e) are examples of the ^{12}C+^{12}C cluster states having the excitation energies around 15 MeV obtained by $\Delta N = 6$ and 8. It is noted that the inter-cluster distances and orientations of ^{12}C clusters are different in Fig.1 (c)-(e) depending on the applied values of ΔN, λ and μ.

The ^{20}Ne+α cluster states also start to appear with $\Delta N = 2$, whose excitation energies are typically around 10 MeV. Similar to the case of the ^{12}C+^{12}C cluster states, the distance and orientation of the clusters depend on the applied constraint as shown in Fig. 1 (f) and (g). It is also noted that the ^{20}Ne cluster is strongly distorted as confirmed from the density distributions. This may owe to the fact that the ^{20}Ne cluster itself has the α+^{16}O cluster structure . Therefore, for example, the wave function shown in Fig. 1 (g) can be alternatively understood as 2α+^{16}O cluster structure. Indeed, if we increase the value of ΔN, the wave function (g) continuously evolves into the pronounced 2α+^{16}O cluster structure.

A very exotic and interesting cluster structure composed of six α particles emerges with $\Delta N \geq 8$ as shown in Fig. 1 (h). Five α clusters consist a pentagon and the last α cluster locates on top of it. The six alpha cluster state of ^{24}Mg is also reported by irod and Schuck based on the HFB calculation with the constraint on the matter radius [11]. However, those 6α states have different properties. While they reported a hexahedral configuration with larger radius and excitation energy, the present result has a pentagon configuration with smaller radius and excitation energy. The difference may originates in the difference of the applied constraint, and we expect that the hexahedral configuration will be also obtained in our calculation, if we increase the value of ΔN further.

Thus, without any *a priori* assumption on nuclear structure, AMD shows the existence of various clusters and demonstrates the evolution of the clustering as function of the excitation energy. It must be emphasised that we just imposed the constraint on the oscillator quanta, nevertheless various cluster structure including very exotic 6α state are obtained. This is an eloquent proof of the fact that the clustering is an important degree-of-freedom of nuclear excitation together with the collective and single-particle excitations.

3.2. Cluster correlations in the low-lying states and duality character of the ground state

We performed the CM calculation superposing the basis wave functions with and without cluster structure, and calculated the excitation spectrum up to $E_x \sim 40$ MeV. Here, we focus on the properties of the ground and 0_2^+ states to see how the cluster correlation persists in and affects the low-lying states. Table 1 summarizes the properties of the ground and 0_2^+ states

Table 1. The energies ($B.E.$), excitation energies (E_x) in MeV, root mean square radius r_{rms} (fm) and IS monopole transition probabilities $B(IS0)$ (fm^4) calculated by using different basis sets.

0_1^+	E	r_{rms}	0_2^+	E	E_x	r_{rms}	$B(IS0)$
(a) $\Phi_{\beta\gamma}$	-198.2	3.06	(a) $\Phi_{\beta\gamma}$	-188.1	10.1	3.08	140
(b) $\Phi_{\beta\gamma}+\Phi_{r^2}$	-198.3	3.06	(b) $\Phi_{\beta\gamma}+\Phi_{r^2}$	-188.1	10.2	3.11	153
(c) $\Phi_{\beta\gamma}+\Phi_{r^2}+\Phi_N$	-199.2	3.06	(c) $\Phi_{\beta\gamma}+\Phi_{r^2}+\Phi_N$	-189.9	9.30	3.11	122
exp.	-198.3	3.06	exp.	-192.9	6.43	–	180

calculated by using (a) $\Phi_{\beta\gamma}$, (b) $\Phi_{\beta\gamma}$ and Φ_{r^2} and (c) $\Phi_{\beta\gamma}$, Φ_{r^2} and Φ_N basis functions. As already mentioned, the excitation spectrum is reasonably described by $\Phi_{\beta\gamma}$ alone indicating the importance of the quadrupole collectivity in the low-lying states. Indeed, the addition of Φ_{r^2} as basis function rarely changes the properties of the ground and 0_2^+ states. However, it is found that the low-lying states gain non-negligible additional binding energies by the addition of Φ_N. The ground and 0_2^+ states gain 0.9 and 1.8 MeV, respectively, that shows the cluster correlations in these states. In particular, the 0_2^+ state has a large contribution from the $\alpha+^{20}$Ne cluster shown in Fig. 1 (g) and gains large correlation energy. As a result, the excitation energy of the 0_2^+ state that is overestimated by $\Phi_{\beta\gamma}$ is improved by the addition of Φ_N. Therefore, we see that the cluster correlation persists and has non-negligible contribution to the binding energy, even though the quadrupole collectivity dominates these states .

The other point that must be mentioned here is the duality character of the ground state that is known from the early days of the cluster studies [25]. Namely, the shell model (mean-field) wave function with SU(3) symmetry can be rewritten in the form of the cluster model wave function [26], which implies that the degree-of-freedom of cluster excitation is embedded in the ground state. Indeed, it is recently found that the IS monopole transition activates the degree-of-freedom of cluster excitation [15] giving rise to the large IS monopole transition probabilities between the ground and cluster states in many light-mass nuclei [13, 14, 16]. In the present study, it is also found that the ground state has duality character of mean-field and clusters, even though the deviation from SU(3) symmetry is large due to the strong spin-orbit splitting. The ground state has 66 and 52 of overlaps with ^{12}C+^{12}C and ^{20}Ne+α cluster wave functions shown in Fig. 1. This finding suggests a possibility to explore the cluster states by using IS monopole transition as a probe.

3.3. Isoscalar monopole strength and clusters

Here we demonstrate that the cluster states including 6α states can be experimentally explored by using IS monopole excitations. Figure 3 shows the strength functions of the IS monopole mode calculated by using (a) $\Phi_{\beta\gamma}$ only, (b) $\Phi_{\beta\gamma}$ and Φ_{r^2} (c) $\Phi_{\beta\gamma}$, Φ_{r^2} and Φ_N. We first verify the energy weighted sum rule (EWSR) and the position of the giant monopole resonance (MR) in our results. As clearly seen in Fig. 3 (a), the strength function calculated by only $\Phi_{\beta\gamma}$ is too weak and exhaust only 35 of the EWSR(Table. 2). The addition of Φ_{r^2} which describes $1p1h$ states built on $\Phi_{\beta\gamma}$ greatly increases the strengths in the region of $E_x > 20$ that corresponds to the IS MR. However, it does not largely change the strengths below $E_x < 15$ MeV, and the calculated centroid energy of IS MR overestimates the observed value. By including cluster

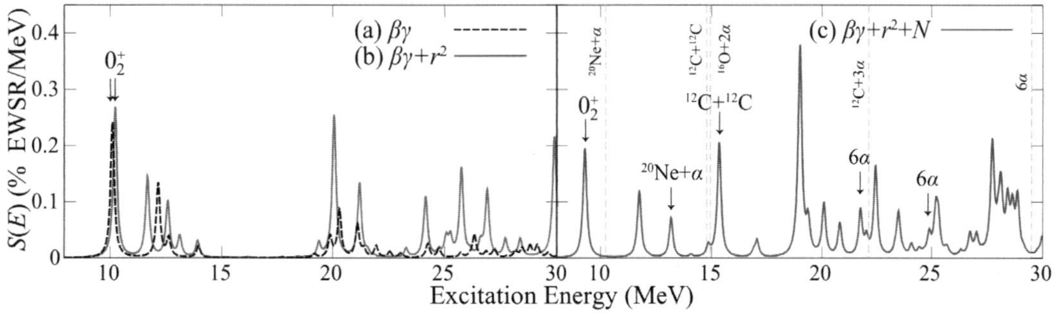

Figure 3. The IS monopole strength functions calculated with (a) $\Phi_{\beta\gamma}$ (b) $\Phi_{\beta\gamma}+\Phi_{r^2}$ (c) $\Phi_{\beta\gamma}+\Phi_{r^2}+\Phi_N$ basis sets. The vertical dashed lines indicate threshold energies of the cluster decays.

Table 2. Calculated and observed energy weighted sum m_1 in fraction of EWSR and centroid energy of IS MR evaluated by m_1/m_0 and $\sqrt{m_3/m_1}$ in MeV. (a)-(c) shows the present results obtained by using $\Phi_{\beta\gamma}$, Φ_{r^2} and Φ_N. Numbers in parenthesis are the energy weighted sum m_1 with all of obtained states, while other numbers are calculated by excluding the 0_2^+ state. These values are compared with the experimental data [29] and the results by the quasi particle random phase approximation [27].

Basis	m_1	m_1/m_0	$\sqrt{m_3/m_1}$
(a) $\Phi_{\beta\gamma}$	26 (35)	20.29	24.24
(b) $\Phi_{\beta\gamma}+\Phi_{r^2}$	101 (116)	25.61	29.26
(c) $\Phi_{\beta\gamma}+\Phi_{r^2}+\Phi_N$	90 (103)	22.24	25.18
Exp.	82 ± 9	$21.9^{+0.3}_{-0.2}$	$24.7^{+0.5}_{-0.3}$
eru et al.	94	20.57	

states described by Φ_N, the centroid energy of IS MR and EWSR are reasonably reproduced simultaneously. Thus, AMD successfully describes IS monopole strength function in a wide energy region. It is also noted that quasi particle random phase approximation (QR A) calculation [27] also reports very similar results except for the low-energy region of $E_x < 18$ MeV.

As clearly seen in Fig. 3, the addition of Φ_N changes the structure of strength function below $E_x < 18$ MeV. In particular, it generates prominent peaks at 13.7 and 15.3 MeV which are missing in Fig. 3 (a)(b) and QR A result [27]. The states at $E = 13.7$ MeV is dominated by the ^{20}Ne+α cluster state shown in Fig. 1 (f) and its $B(IS0)$ is 1.1 W.u. exhausting 2 of EWSR. The ^{12}C+^{12}C cluster state appears as the peak at 15.3 MeV that is just above the the ^{12}C+^{12}C threshold. Its $B(IS0)$ is 2.8 W.u. and exhaust 5 of EWSR which is as large as that of the 0_2^+ state. Very interesting and important point is that these peaks are missing in QR A calculation [27], but look existing in the observed data [28, 29]. Therefore, it is very tempting to conclude that the α+^{20}Ne and ^{12}C+^{12}C cluster states were successfully populated by the experiment, although further experimental data is needed to confirm it. Finally, we discuss 6α cluster state with pentagon shape. By the coupling with other cluster and $1p1h$ states, it is fragmented into two states at 22.4 and 25 MeV that are unfortunately embedded in the broad distribution of IS MR. However, they are still visible in Fig. 3 (c) and should be experimentally accessible, because they will selectively decay by α particle emission not by neutron emission. Thus, ^{24}Mg has a rich variety of cluster states in the energy range of 10~30 MeV, and many of them should be detectable by their enhanced IS monopole transition strengths.

4. Summary

We have investigated cluster structure of highly excited states of ^{24}Mg on the basis of the antisymmetrized molecular dynamics. By imposing the constraints on the expectation values of harmonic oscillator quanta, α+^{20}Ne, ^{12}C+^{12}C and 6α cluster structures were generated without any *a priori* assumption. We also studied the IS monopole transition strength. By including the cluster states as the basis function, the calculated strength function reasonably agrees with the observation. It is found that α+^{20}Ne, ^{12}C+^{12}C and 6α cluster structures appear as the prominent peaks in the energy range of 10~30 MeV.

Acknowledgments

The numerical calculations were performed on the HITACHI SR16000 at KEK and IT . One of the authors (M.K.) acknowledges the supports by the rants-in-Aid for Scientific Research on Innovative Areas from ME T (rant No. 2404:24105008) and JS S KAKENHI rant No. 25400240.

References

[1] Girod M and Grammaticos B 1983 *Phys. Rev.* C **27** 2317
[2] Bender M and Heenen P H 2008 *Phys. Rev.* C **78** 024309
[3] Ikeda K, Tagikawa N and Horiuchi H 1968 *Prog. Theor. Phys. Suppl.* Extra Number **464**
[4] Matsuse T, Kondō Y and Abe Y 1978 *Prog. Theor. Phys.* **59** 1009
[5] Katō K and Bando H 1979 *Prog. Theor. Phys.* **62** 644
[6] Baye D and Descouvemont P 1984 *Nucl. Phys.* A **419** 397
[7] Descouvemont P and Baye D 1987 *Nucl. Phys.* A **475** 219
[8] Almqvist E, Bromely D A and Kuehner J A 1960 *Phys. Rev. Lett.* **4** 515
[9] Cosman E R *et al.* 1975 *Phys. Rev. Lett.* **35** 265
[10] Erb K A *et al.* 1976 *Phys. Rev. Lett.* **37** 670
[11] Girod M and chuck P 2013 *Phys. Rev. Lett.* **111** 132503
[12] Kawabata T *et al.* 2012 *Prog. Theor. Phys. Suppl.* **196** 198
[13] Kanada-En yo Y 2007 *Phys. Rev.* C **75** 024302
[14] Kawabata T *et al.* 2007 *Phys. Lett.* B **646** 6
[15] Yamada T *et al.* 2008 *Prog. Theor. Phys.* **120** 1139
[16] Yamada T *et al.* 2012 *Phys. Rev.* C **85** 034315
[17] Kanada-En yo Y, Kimura M and Horiuchi H 2003 *C. R. Physique* **4**, 497
[18] Kanada-En yo Y, Kimura M and Ono A 2012 *PTEP* **2012** 01A202
[19] Berger J F, Girod M and Gogny D 1991 *Comput. Phys. Comm.* **63** 365
[20] Kimura M 2004 *Phys. Rev. C* **69** 044319
[21] uhara T and Kanada-En yo Y 2010 *Phys. Rev.* C **82** 044301
[22] Kimura M, Yoshida R and Isaka M 2012 *Prog. Theor. Phys.* **127** 287
[23] Kanada-En yo Y and Kimura M 2005 *Phys. Rev.* C **72** 064322
[24] Kimura M 2007 *Phys. Rev.* C **75** 034312
[25] Wildermuth K and Kanellopoulos Th 1958 *Nucl. Phys.* **7** 150
[26] Bayman B F and Bohr A 1958 59 *Nucl. Phys.* **9** 596
[27] Peru and Goutte H 2008 *Phys. Rev.* C **77** 044313
[28] Youngblood D H, Lui Y -W and Clark H L 1999 *Phys. Rev.* C **60** 014304
[29] Youngblood D H, Lui Y -W., Chen F and Clark H L 2009 *Phys. Rev.* C **80** 064318

Journal of Physics: Conference Series **569** (2014) 012006 doi:10.1088/1742-6596/569/1/012006

Crystal, cluster and quantum liquid nuclear states

J.-P. Ebran
CEA,DAM,DIF, F-91297 Arpajon, France

E. Khan
Institut de Physique Nucléaire, Université Paris-Sud, IN2P3-CNRS, F-91406 Orsay Cedex, France
E-mail: khan@ipno.in2p3.fr

T. Nikšić
Physics Department, Faculty of Science, University of Zagreb, 10000 Zagreb, Croatia

D. Vretenar
Physics Department, Faculty of Science, University of Zagreb, 10000 Zagreb, Croatia

Abstract. Nuclear states are often described as quantum liquid states. Using the localisation parameter, it is possible to understand cluster states in nuclei as hybrid states between quantum liquid and crystal states. Recent calculations discussing the effect of the depth of the confining potential, as well as various lengthscales ratios, are presented in order to provide a complementary view on the cluster phenomenon in nuclei.

1. Introduction: states of matter

It is well known that when temperature decreases and density increases, a system of A constituents interacting with a short range attractive interaction undergoes from a classical gaseous state to a liquid one. When further decreasing the temperature and increasing the density, the system becomes a solid, which is microscopically described as a crystal structure. Microscopically the system went from weakly interacting constituents (gas) towards more interacting ones (liquid) to bound states into crystal having constituents fixed at periodic nodes (solid). What happens if the density further increases, that is adding constituents between the nodes of the crystal ? In this case, the constituents at nodes start to overlap, forming a molecule of constituents, and the system becomes clusterised. When the density further increases, the dense system becomes homogeneous and this is the quantum liquid state [1]. If the density increases again the inner structure of the constituents starts to impact, and the system cannot be considered of only A interacting constituents anymore. Figure 1 summarizes the global picture of the above discussed states of matter.

Quantum effects start to have an impact from the solid to the quantum liquid states. This happens when the typical dispersion of the constituents is non-negligible compared to the inter-constituents distance. The wavelength of a constituent can be approximated by [1]:

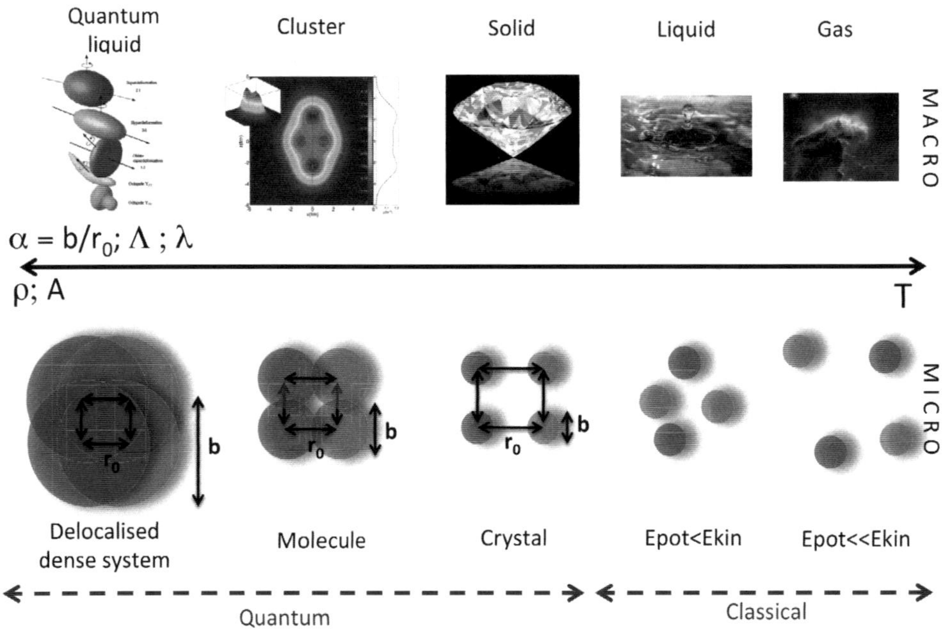

Figure 1. Sketch of the states of matter.

$$\lambda = \frac{h}{p} \simeq \frac{\hbar}{\sqrt{2mkT}} \tag{1}$$

where the kinetic energy of the constituent is approximated by kT. One therefore sees that decreasing temperature (or increasing density) generates a larger dispersion of the wavefunction of the constituent, which can lead to quantum effects when λ is non-negligible compared to the inter-constituents distance r_0. This means that in a dense system (quantum liquid) the filling of the space between the nodes of the crystal is done so to achieve a homogeneous density but with the quantum side-effect that the constituents get also delocalised.

Several concomitant quantities can be further used to analyse this effect:

- i) λ/r_0 as mentioned above. This is also discussed in Ref. [2].
- ii) b_0/r_0 where b_0 is a sort of interacting-constituent Compton wavelength:

$$b_0 \hat{=} \frac{\hbar}{\sqrt{2mV_0}} \tag{2}$$

where V_0 is the typical magnitude of the inter-constituent interaction. This ratio is similar to i) where the kinetic temperature has been approximated by the potential magnitude V_0. This ratio represents the quantality parameter used by Mottelson [3]:

Journal of Physics: Conference Series **569** (2014) 012006 doi:10.1088/1742-6596/569/1/012006

$$\Lambda \hat{=} 2 \left(\frac{b_0}{r_0}\right)^2 \tag{3}$$

- iii) b/r_0 where b is the typical dispersion of a constituent, taking into account the finite size of the system. This is the localisation parameter defined in Ref. [4]:

$$\alpha \hat{=} \frac{b}{r_0} = \frac{\sqrt{\hbar} A^{1/6}}{(2mV_0' r_0^2)^{1/4}} \tag{4}$$

  where V'_0 is the depth of the confining potential. It allows to study the evolution of the states with respect to the number of constituents A [5] and is well adapted to systems where finite-size effects are relevant ($A \lesssim 10^3$), such as nuclei. The quantality (Eq. (3)), on the contrary only depends on the interaction and would be the same in nuclei or in infinite nuclear matter. It does not consider finite size effects.

From this picture, cluster states in nuclei are hybrid states between quantum liquid and crystal states. They can be produced in light nuclei [6] and/or in dilute nuclear systems as discussed and noticed both experimentally and theoretically [7, 8]. In Ref. [5] it has been shown that a nuclear crystal cannot be produced in a nucleus, contrarily to in the crust of neutron stars. Another way to produce a nuclear cluster or a nuclear crystal might be by heating the nucleus. Using Eq. (1), the temperature to be reached to produce a cluster ($\lambda \sim r_0$) or a crystal ($\lambda \sim 0.1 r_0$) are 10 MeV and 1 GeV, respectively. These may be too high in order to keep the finite system bound.

The well-known large mean-free path of nucleons in nuclear matter [9] shows that nuclei usually behave as quantum liquid states. However, as a function of the number of nucleons A, the localisation parameter Eq. (4) shows that the delocalisation induced by the interactions is quenched in light nuclei due to the effect of the confining potential [5]. This effect is depicted on Fig. 1 and is due to the saturation of nuclear interactions, which allows the localisation parameter to depend on A while the depth of the confining potential remains rather constant.

The depth of the confining potential is therefore a key value and one can wonder about its actual magnitude. The answer is not straightforward since it is not an observable. In other fields of physics, such as in atomic traps, it is possible to experimentally tune the depth of the confining potential [10]. A deeper potential generates more localised structures. What happens in the case of nuclei ? How deep is the confining nuclear potential ? The question cannot be addressed experimentally, but it can be theoretically by using functionnals predicting different depth of the nuclear confining potential.

This is the case of relativistic functionnals compared to non-relativistic ones. In the relativistic approach, the depth of the potential is related to the so-called scalar and vector potentials generated by the mesons exchange. It has a deep connection with the spin-orbit potential, also related to these scalar and vector ones [4]. In the non-relativistic approach, the depth of the confining potential depends on the magnitude of the attraction of the nucleon-nucleon interaction, which remains a free parameter. As a result the depth of the relativistic potential is significantly larger than the non-relativistic one. This induces more localisation and it may be possible that it has sounder roots than the non-relativistic one. Fig. 3 shows the comparison between relativistic and non-relativistic potentials in ^{20}Ne.

It should be noted that studying the depth of the potential highlights this important quantity on clusterisation features and allows to connect the nuclear cluster phenomenology to various analogous previous approaches in other fields of physics such as in condensed matter [10, 11, 12], where the depth of the confining potential is known to impact clusterisation.

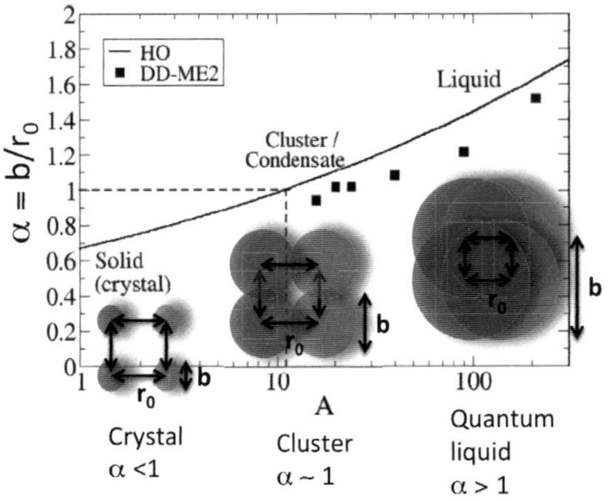

Figure 2. Localisation parameter as a function of the mass number in nuclei.

2. EDF approaches and cluster states

As a result, light nuclei densities are more localised in the relativistic approaches, as can be seen on numerous Figs. of Refs. [13, 14]. Therefore relativistic approaches predict clusterisation in light nuclear systems in agreement with cluster phenomenology. Moreover this means that relativistic energy density functionnals (EDFs) are able to describe both quantum liquid and clusters aspects of nuclei and hence encompasses interesting features about the origin of clusterisation in nuclei. Let us therefore underline the specificities of EDF approaches with respect to clusters, since this way of studying clusters is maybe not well known by the community because at the interface between two domains:

- EDF approaches allow for the most general antisymmetrized product of nucleonic wave functions to describe a nuclear state : there is no a priori assumption on the nucleon's wavefunction
- Correlations beyond the mean-field are effectively included by the EDF itself, its parameter being determined from nuclear observables including full correlations, allowing for saturation and α particle description
- Results are obtained in the intrinsic frame of the nucleus, allowing for cluster shapes
- EDFs are well suited to investigate nuclear structure on the whole nuclear chart including the quantum liquid vs cluster state transition

The contribution of J.-P. Ebran in the present proceedings [14] shows the large variety of clusters states which can by predicted with a microscopic relativistic EDF.

3. Towards a global picture

The fact that EDF approaches are able to describe clusterisation effects in nuclei may help to move towards a global picture of nuclear states. It should be stressed again that the present

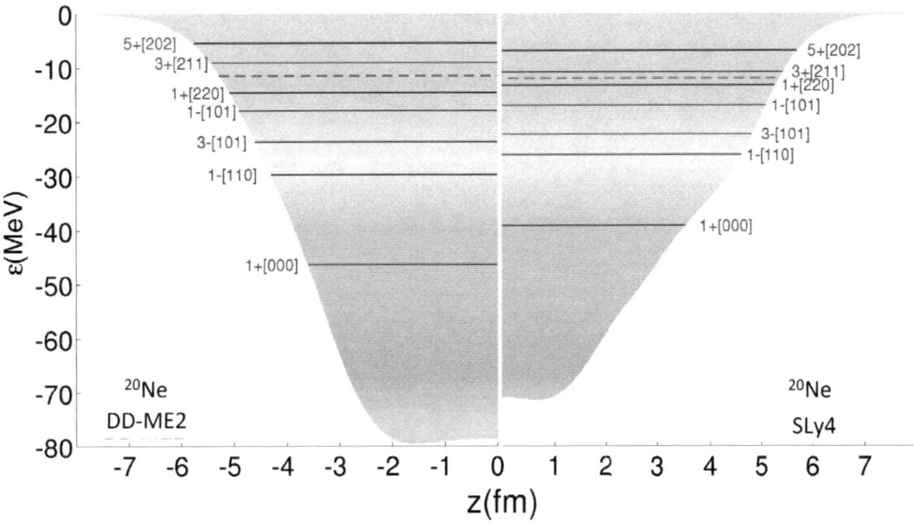

Figure 3. Comparison between the relativistic and non-relativistic potentials and single-neutrons energies in the mean-field framework.

approach is the regular relativistic one, used to describe usual quantum liquid states in nuclei without any further assumption [15]: clusters states naturally come out.

What is already known from decades of study on cluster states in nuclei ? i) Deformation and single particle levels degeneracy play an important role [6], ii) excited states are easier to clusterise, as we know from the Ikeda picture [16] and iii) neutron excess can favorize clusterisation, the excess neutron playing a role analogous to molecular bonding [6].

All these 3 features are included, predicted and understood by the present fully microscopic EDF approaches: i) level degeneracy and deformation studies shows that both large deformation and single particle level degeneracy raising favor clusterisation [13, 14], ii) excited states are more likely to clusterise [17, 18], and iii) molecular bonding are naturally predicted in neutron-rich nuclei using relativistic EDF in ^{10}Be for instance [13, 14]. This shows that EDFs are able to recover what is known from the cluster states phenomenology.

It should be also noted that this provides a global framework to understand haloes and cluster states [5]: in both cases it is the energy of the system which is related to the spatial localisation in nuclei: large delocalisation outside of the potential for weakly bound systems in the case of haloes, and localisation inside a deep potential in the case of clusters.

4. The origin of clusterisation

One can go one step further and try to investigate about the origin of clusterisation in the view of the present EDF tool. It is known that when diluting a nuclear system, at about 1/3 of the saturation density, it reaches a transition from a homogeneous to a cluster system [7, 8]:

at low density, the system gains energy to have separated clusters at saturation density rather than having a homogeneous diluted density. This points out saturation as a relevant quantity to understand the origin of clusterisation

Fig. 4 shows the competition between the localisation parameter and the quantality one [18]. In this framework cluster states are hybrid states where quantality (related to the nuclear interaction) is large enough to provide delocalised nucleons, but in light systems, the radius of the confining potential attenuates this delocalisation and provides some clusterisation effect. This is possible because the depth of the confining potential remains rather constant over the nuclear chart, due to saturation properties.

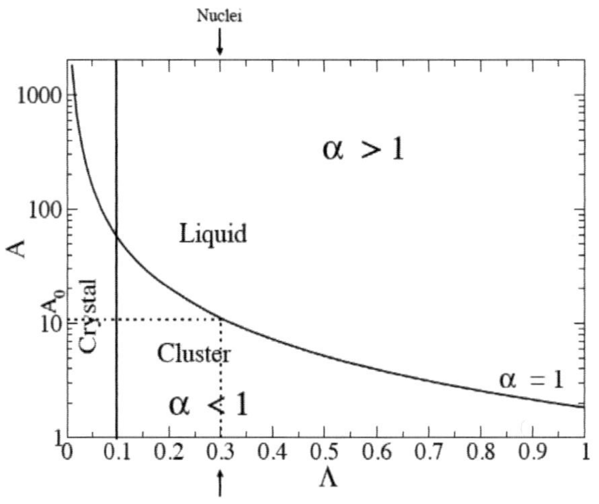

Figure 4. Effects of both quantality and localisation parameter on fermionic systems of A constituents [18].

5. Conclusion

The role of the depth of the confining potential and of the localisation parameter in nuclear clusterisation has been analysed, allowing for a connection with similar cluster studies in other fields of physics. Saturation plays a key role in the nuclear clusterisation. Relativistic EDF are able to describe clusterisation phenomena and to provide microscopic light on its phenomenology. These EDF approaches do not contain any further a priori assumption aimed to describe clusterisation. The present approach allows to better connect the EDF approach with previous nuclear cluster studies, and also the nuclear cluster field with clusters studies in other areas of physics.

References

[1] Pines D and Nozieres P 1966 *The theory of quantum liquids* (Benjamin Inc.)
[2] Zinner N T and Jensen A S 2013 *J. Phys. G: Nucl. Part. Phys.* **40** 053101
[3] Mottelson B *Nuclear Structure* 1996 (Les Houches, Session LXVI) 25

[4] Ebran J P, Khan E, Nikšić T and Vretenar D 2012 *Nature* **487** 341
[5] Ebran J P, Khan E, Nikšić T and Vretenar D 2013 *Phys. Rev. C* **87** 044307
[6] von Oertzen W, Freer M, Kanada-En'yo Y 2006 *Phys. Rep.* **432** 43
[7] Girod M and Schuck P, 2013 *Phys. Rev. Lett.* **111** 132503
[8] Typel S, Röpke G, Klähn T, Blaschke D, and Wolter H H 2010 *Phys. Rev. C* **81** 015803
[9] Rios A and Somà V 2012 *Phys. Rev. Lett.* **108** 012501
[10] Greiner M, Mandel O, Esslinger T, Hänsch T and Bloch I 2002 *Nature* **415** 39
[11] Falakshahi H and Waintal X 2005 *Phys. Rev. Lett.* **94** 046801
[12] Yannouleas C and Landman U 2007 *Rep. Prog. Phys.* **70** 2067
[13] Ebran J P, Khan E, Nikšić T and Vretenar D 2014 *Arxiv:1406.2473*
[14] Ebran J P, Khan E, Nikšić T and Vretenar D *see the present volume*
[15] Vretenar D, Afanasjev A V, Lalazissis G, Ring P 2005 *Phys. Rep.* **409** 101
[16] Ikeda K, Takigawa N and Horiuchi H, 1968 *Prog. Theor. Phys. Suppl.* **464**
[17] Ichikawa T, Maruhn J A, Itagaki N, Matsuyanagi K, Reinhard P G, and Ohkubo S 2012 *Phys. Rev. Lett.* **109** 232503
[18] Ebran J P, Khan E, Nikšić T and Vretenar D 2014 *Phys. Rev. C* **89** 031303(R)

Journal of Physics: Conference Series **569** (2014) 012007 doi:10.1088/1742-6596/569/1/012007

Nonlocalized cluster dynamics and container picture

Bo Zhou

Department of Physics, Nanjing University, Nanjing 210093, China
Meme Media Laboratory, Hokkaido University, Sapporo 060-0810, Japan

E-mail: bo@nucl.sci.hokudai.ac.jp

Abstract. A container picture is proposed for understanding cluster dynamics where the clusters make nonlocalized motion occupying the lowest orbit of the cluster mean-field potential characterized by the size parameter "B" in the THSR (Tohsaki-Horiuchi-Schuck-Röpke) wave function. The nonlocalized cluster aspects of the inversion-doublet bands in ^{20}Ne which have been considered as a typical manifestation of localized clustering are discussed. In the α+^{16}O system, localization is entirely of kinematical origin, that is, due to the inter-cluster Pauli repulsion. It is concluded that this feature is general for nuclear cluster states.

1. Introduction

The formation of clusters is one of the most important features in light nuclei together with the formation of the nucleon mean field [1–4]. The α-condensate-like state is a very novel cluster state in light nuclei which has attracted increasing interest in recent years (For example, see the reviews Ref. [5] and Ref. [6]). This state can be considered as a gas-like state of clusters in which the center-of-mass motion of each α cluster in nuclei occupies the same $0s$ orbit. The proposed THSR (Tohsaki-Horiuchi-Schuck-Röpke) wave function [7] has been proved to be very suitable for the realistic description of the dilute gas-like state of clusters. Actually in the case of the Hoyle state (the second 0^+ state) of ^{12}C, it was found that the full microscopic solutions [8] of the 3α RGM (resonating group method) [9] and that [10] of the GCM (generator coordinate method) with the Brink wave function [9] are almost 100% equivalent to single 3α THSR wave functions [11]. Also in the case of the ground-state band of ^{8}Be, the full microscopic solutions of the 2α RGM or its equivalent GCM with the Brink wave function were found to be practically 100% equivalent to single 2α THSR wave functions [12].

Although the THSR wave function was devised for describing the gas-like state of clusters, it was already found a decade ago in Ref. [11] that the wave functions of the ground state of ^{12}C with normal density obtained by 3α RGM and 3α Brink-GCM calculations have a large value of about 0.93 for the squared overlap with *single* 3α THSR wave functions. Recently the present authors found [13] that the ^{16}O + α Brink-GCM wave functions of the states of the ground-state rotational band of ^{20}Ne with normal density are almost 100% equivalent to single ^{16}O + α THSR wave functions. These results show that the THSR wave function has an ability which was not expected at first, namely it can be used to study not only the gas-like cluster states with low density but also (cluster) states with normal density.

In ^{20}Ne, the ground-state rotational band with $K^\pi = 0_1^+$ is known to constitute an inversion doublet together with the negative-parity rotational band with $K^\pi = 0_1^-$ built upon the 1^- state at the excitation energy $Ex = 5.79$ MeV. The existence of the inversion doublet bands has been regarded as being a clear manifestation of the existence of the parity-violating intrinsic state due to the ^{16}O + α localized clustering. In general, in non-identical two-cluster systems, the existence of inversion-doublet bands has been regarded as a clear indication of the existence of the localized cluster structure together with the observation of the large cluster decay widths. This argument implies that we have to regard the states of the ground-state rotational band of ^{20}Ne as having a ^{16}O + α localized clustering. However, it was found that, also in this negative-parity rotational band, the ^{16}O + α Brink-GCM wave functions are almost 100% equivalent to single ^{16}O + α THSR wave functions. Thus both rotational bands constituting the inversion doublet are found to have nonlocalized cluster structure of ^{16}O and α.

Thus, the typical cluster systems, 2α, 3α, and α+^{16}O, are all well described by a *single* THSR wave function. These facts urged us to introduce the container picture of cluster dynamics underlying the THSR wave function. Different from the traditional cluster models, in the container picture, the clusters make nonlocalized motion occupying the lowest orbit of the cluster mean-field potential characterized by the size parameter (See Ref. [14] for details).

2. Hybrid-Brink-THSR wave function

Recently, to extend and further clarify the concept of nonlocalized clustering even in non-gas-like cluster states with more compact density, we proposed a new type of microscopic cluster wave function [15], which we call hybrid-Brink-THSR wave function,

$$\Phi_{\text{cluster}}(\boldsymbol{\beta}_i, \boldsymbol{S}_i) = \int d^3R_1 \ldots d^3R_n \exp\{-\sum_{i=1}^{n} \sum_{k=x,y,z} \frac{R_{ik}^2}{\beta_{ik}^2}\} \Phi^B_{\text{cluster}}(\boldsymbol{R}_1 + \boldsymbol{S}_1, \ldots, \boldsymbol{R}_n + \boldsymbol{S}_n) \quad (1)$$

$$\propto \mathcal{A}[\prod_{i=1}^{n} \exp\{-A_i \sum_{k=x,y,z} \frac{(X_{ik} - S_{ik})^2}{2B_{ik}^2}\} \phi(C_i)], \quad (2)$$

$$\Phi^B_{\text{cluster}}(\boldsymbol{S}_1, \ldots, \boldsymbol{S}_n) = \mathcal{A}[\prod_{i=1}^{n} \exp\{-A_i \frac{(\boldsymbol{X}_i - \boldsymbol{S}_i)^2}{2b^2}\} \phi(C_i)]. \quad (3)$$

Here $\boldsymbol{\beta}_i \equiv (\beta_{ix}, \beta_{iy}, \beta_{iz})$, and $\boldsymbol{X}_i$ and $\phi(C_i)$ are the center-of-mass coordinate and the internal wave function of the cluster C_i, respectively. Different clusters C_i can have different mass numbers A_i and variational parameters $\boldsymbol{\beta}_i$. The oscillator parameter of the cluster C_i is called b, which also can be adopted so as to have different values for different clusters. Φ^B_{cluster} is the corresponding general Brink model wave function [16].

In Eq. (1), another generator coordinate $\boldsymbol{S}_i$ is introduced to the original THSR wave function. It can be seen from Eq. (2) that this hybrid wave function combines the important characters of the Brink model as in Eq. (3) and the THSR wave function in a very simple way. When $\boldsymbol{S}_i$=0, Eq. (1) corresponds to the THSR wave function and $\boldsymbol{\beta}_i$ or $\boldsymbol{B}_i$ becomes the size parameter. When $\beta_{ik} = 0$, i.e., $B_{ik} = b$ $(k = x, y, z)$, this equation is nothing more than the Brink wave function Eq. (3) and $\boldsymbol{S}_i$ is the position parameter of the cluster C_i.

As we know, the THSR model provides a nonlocalized clustering picture for the cluster structure rather than the localized clustering represented by the Brink model [17]. Since these two different kinds of pictures for clustering are both included in the hybrid-Brink-THSR wave function as the aforementioned two limits, this hybrid wave function provides a very nice way for verifying which picture is more adequate for understanding the relative motions of the cluster structures in nuclei.

Now, based on the above hybrid-Brink-THSR wave function, the following cluster wave function of ^{20}Ne can be obtained, as it was considered in Ref. [15],

$$\Phi_{\text{Ne}}(\boldsymbol{\beta}, \boldsymbol{S}) = \int d^3R \exp\{-(\frac{4R_x^2}{5\beta_x^2} + \frac{4R_y^2}{5\beta_y^2} + \frac{4R_z^2}{5\beta_z^2})\} \Phi^B_{\text{Ne}}(\frac{4}{5}(\boldsymbol{R} + \boldsymbol{S}), -\frac{1}{5}(\boldsymbol{R} + \boldsymbol{S})) \quad (4)$$

$$\propto \exp(-\frac{10X_G^2}{b^2})\widehat{\Phi}_{\rm Ne}(\boldsymbol{\beta}, \boldsymbol{S}),$$

$$\widehat{\Phi}_{\rm Ne}(\boldsymbol{\beta}, \boldsymbol{S}) = \mathcal{A}[\exp(-\sum_{k=x,y,z} \frac{8(r_k - S_k)^2}{5B_k^2})\phi(\alpha)\phi(^{16}{\rm O})], \tag{5}$$

where $B_k^2 = b^2 + 2\beta_k^2$, $(k = x, y, z)$, $\boldsymbol{r} = \boldsymbol{X}_1 - \boldsymbol{X}_2$, $\boldsymbol{X}_G = (4\boldsymbol{X}_1 + 16\boldsymbol{X}_2)/20$, and $\widehat{\Phi}_{\rm Ne}(\boldsymbol{\beta}, \boldsymbol{S})$ is the intrinsic wave function where the spurious center-of-mass motion is eliminated from $\Phi_{\rm Ne}(\boldsymbol{\beta}, \boldsymbol{S})$ in Eq. (4). $\boldsymbol{X}_1$ and $\boldsymbol{X}_2$ represent the center-of-mass coordinates of the α cluster and the ^{16}O cluster, respectively. All calculations are performed with restriction to axially symmetric deformation, that is, $\beta_x = \beta_y \neq \beta_z$ and $\boldsymbol{S} \equiv (0, 0, S_z)$. The spin and parity eigenfunctions can be obtained by the angular-momentum projection technique [13].

As the nuclear interaction, we adopt the same effective nuclear force, Volkov No.1 with the Majorana parameter $M = 0.59$, and the same oscillator parameter $b = 1.46$ fm, as were used in our previous papers [13, 15].

3. Nonlocalized cluster dynamics

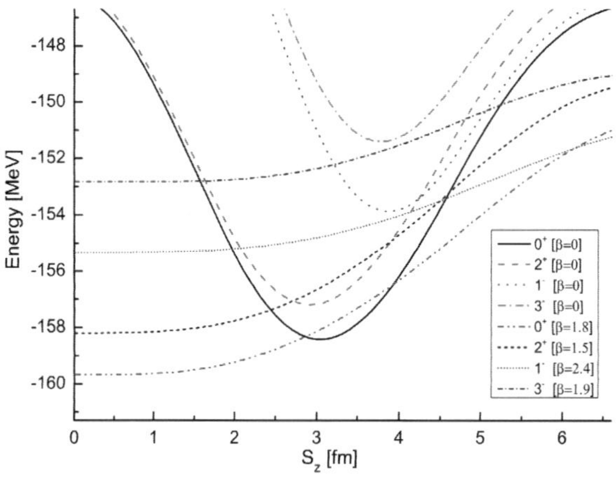

Figure 1. Energy curves of $J^\pi = 0^+$, 2^+, 1^-, and 3^- states with different widths of Gaussian relative wave functions in the hybrid model.

After variational calculations for the parameters S_z and $\boldsymbol{\beta}$ in the projected hybrid-Brink-THSR wave function, it is surprising to find that the inter-cluster distance parameter S_z becomes zero for the inversion doublet bands in ^{20}Ne. This means this hybrid-Brink-THSR wave function tends to a pure THSR wave function in describing the cluster states of the inversion doublet bands in ^{20}Ne. Thus, in spite of the fact that a pure Brink wave function gives a distinct energy minimum point with non-zero S_z, the localized clustering picture cannot be supported.

Fig. 1 shows the energy curves of the lower excited states of the inversion doublet bands with different widths of the Gaussian relative wave functions in the hybrid model. If $\boldsymbol{\beta}$ is fixed at 0, the hybrid-Brink-THSR wave function becomes the Brink wave function. In this case, S_z is the inter-cluster distance parameter and is usually regarded as a parameter describing the dynamics of the cluster system. For instance, the minimum energy of the ground state of ^{20}Ne appears at

$S_z = 3.0$ fm. For the $J^\pi = 1^-$ state, the optimum position appears at $S_z = 3.9$ fm. The non-zero values of S_z seem to indicate that the $\alpha+^{16}$O structure of ^{20}Ne favours localized clustering. This is just the traditional concept of localized clustering. Now, we believe that this argument is misleading [15]. The non-zero minimum point S_z simply occurs since the width of the Gaussian wave function of the relative motion in the Brink model is fixed to a narrow wave packet, characterized by the parameter b. If we take non-zero values for $\boldsymbol{\beta}$, namely, $\beta_x = \beta_y = \beta_z$=1.8 fm, 1.5 fm, 2.4 fm, and 1.9 fm for $J^\pi = 0^+$, 2^+, 1^-, and 3^- states, respectively, according to their minimum positions in the contour maps, then we find that the minimum points appear at $S_z = 0$ in Fig. 1. This indicates that the separation distance parameter S_z does not play any physical role in describing the $\alpha+^{16}$O cluster structure, even for the negative-parity states. Instead of that, the new parametrization by $\boldsymbol{\beta}$, which characterizes nonlocalized clustering, is more appropriate for describing the cluster structure in ^{20}Ne.

It should be noted that although S_z does not give any contribution to the energy gain, it still plays an important role in providing negative-parity states. We can prove that the negative-parity states can be constructed even in the limiting situation, $S_z \to 0$.

Table 1. $E^{J^\pi}_{\rm min}(\beta_x = \beta_y, \beta_z)$ are the minimum energies at the corresponding values of $\beta_x = \beta_y$ and β_z in the hybrid model. The squared overlaps between the single normalized projected THSR-type wave functions $\hat{\Phi}^{\rm THSR}_{\rm Min}$ corresponding to the minimum energies and the normalized Brink GCM wave functions are also listed. Units of energies are MeV.

State	$E^{J^\pi}_{\rm min}(\beta_x = \beta_y, \beta_z)$	$\lvert\langle\hat{\Phi}^{\rm THSR}_{\rm Min}\lvert\hat{\Phi}^{\rm Brink}_{\rm GCM}\rangle\rvert^2$
0^+	-159.85(0.9, 2.5)	0.9929
2^+	-158.53(0.0, 2.2)	0.9879
4^+	-155.50(0.0, 1.8)	0.9775
1^-	-155.38(3.7, 1.4)	0.9998
3^-	-153.07(3.7, 0.0)	0.9987

On the other hand, the exact solution of the $\alpha+^{16}$O cluster system can be obtained by superposing the single Brink wave functions, that is the Brink-GCM wave function.

$$\sum_j \langle \Phi^{J^\pi}_{\rm Brink}(R_i)|\widehat{H} - E|\Phi^{J^\pi}_{\rm Brink}(R_j)\rangle f(R_j) = 0. \tag{6}$$

Here, $\Phi^{J^\pi}_{\rm Brink}(R_i)$ can be obtained directly from the projected Brink wave function $\Phi^{J^\pi}_{\rm Brink}(\frac{4}{5}\boldsymbol{R}, -\frac{1}{5}\boldsymbol{R})$ with $\boldsymbol{R} = (0, 0, R_i)$. Thus, by solving the Hill-Wheeler equation Eq. (6), we can obtain the following Brink-GCM wave function,

$$\Phi^{J^\pi}_{\rm GCM} = \sum_i f(R_i)\Phi^{J^\pi}_{\rm Brink}(R_i). \tag{7}$$

Thus, we can compare the single THSR-type wave function with the exact Brink-GCM wave function [13] for the description of the $\alpha+^{16}$O cluster system by calculating their squared overlap $|\langle\hat{\Phi}^{\rm THSR}_{\rm min}|\hat{\Phi}^{\rm Brink}_{\rm GCM}\rangle|^2$. In Table 1, we find that the obtained single THSR-type wave functions have 99.29%, 98.79%, 97.75%, 99.98%, and 99.87% squared overlaps for J^π=0^+, 2^+, 4^+, 1^-, and 3^- states of ^{20}Ne, respectively, with the corresponding Brink-GCM solutions. These high squared overlaps mean that the single THSR-type wave functions are almost 100% equivalent to the corresponding RGM/GCM wave functions, thus, these obtained single angular-momentum projected THSR-type wave functions can accurately describe the states of the inversion doublet bands in ^{20}Ne [13,15]. Moreover, the concept of nonlocalized clustering proposed by the THSR-type wave function obtained from the hybrid-Brink-THSR wave function is essential to correctly understand the $\alpha+^{16}$O cluster structure in ^{20}Ne.

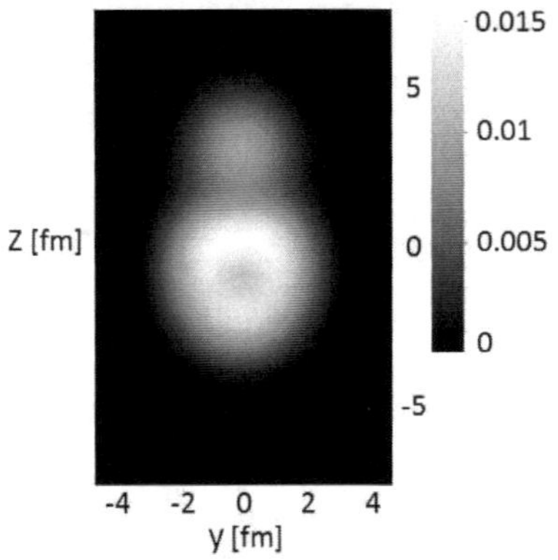

Figure 2. Density distribution of the ^{16}O + α hybrid-Brink-THSR wave function with S_z = 0.6 fm and $(\beta_x, \beta_y, \beta_z)$ = (0.9 fm, 0.9 fm, 2.5 fm).

But how to understand the rotational band based on the concept of nonlocalized clustering? Since the THSR wave function is a state of good parity, a pure THSR wave function is not suitable for expressing a parity-breaking density distribution of the ^{16}O-α clustering. However, as we will see below, if we use a hybrid-Brink-THSR wave function with small S_z parameter (for small values of S_z the energies in Fig. 1 are practically degenerate), the density distribution of this hybrid-Brink-THSR wave function which is quite close to a prolate THSR wave function shows clearly the effective spatial localization of ^{16}O and α clusters, which can lead to the rotational bands in ^{20}Ne. In Fig. 2 we show the density distribution of the hybrid-Brink-THSR wave function with S_z = 0.6 fm and $(\beta_x, \beta_y, \beta_z)$ = (0.9 fm, 0.9 fm, 2.5 fm). We observe in this figure that, in spite of the small value of S_z = 0.6 fm, the inter-cluster distance between ^{16}O and α is about 3.6 fm. Namely the large inter-cluster distance of about 3.6 fm between ^{16}O and α is not due to the parameter S_z but due to the effective spatial localization of ^{16}O and α clusters in the prolate THSR wave function with $(\beta_x, \beta_y, \beta_z)$ = (0.9 fm, 0.9 fm, 2.5 fm) which is just the wave function reduced from the hybrid-Brink-THSR wave function by letting go S_z to zero. Thus, the non-zero inter-cluster separation is a natural result of the Pauli repulsion in the THSR wave function while it is artificial to express this separation by the localized parameter in Brink wave function.

In two-cluster systems, cluster states generally have effective localization of clusters because of the inter-cluster Pauli repulsion. However, in three or more cluster systems, the spatial arrangement of clusters are not necessarily geometrical, namely clusters can be nonlocalized, although the inter-cluster separations are non-zero simultaneously because of the inter-cluster Pauli repulsion. However, as is discussed in Ref. [18], if a cluster state is forced to have strongly-prolate deformation, the state can have effective localization of clusters like in the case of 3α linear-chain structure. When the inter-cluster separations are large, the spatial arrangement of clusters can be non-rigid and gas-like.

4. Container picture

Clusters in the THSR wave function in low density systems make mutually independent nonlocalized motion occupying the lowest orbit of the harmonic-oscillator-like mean-field potential of clusters characterized by the size parameter B whose magnitude is similar to the radius of the system. In systems of 3α's [7, 11, 19] and 4α's [7, 20] we know that the excitation mode of the system is well described by the Hill-Wheeler equation of the size parameter B treated as the generator coordinate. Therefore we see that the excitation of the system is described firstly by the dynamics of the size parameter B which is adopted as the generator coordinate and secondly by the excitation of the single-particle motion of clusters in the cluster mean-field potential. We will call our new understanding of nuclear cluster dynamics the container picture of nuclear clustering, by which we aim to stress that the central quantity of cluster dynamics is the size parameter B of the self-consistent mean-field potential of clusters which we call the container. The name "container picture" may sound more appropriate for three (or more)-cluster systems because, for example, in the 3α system it describes the ground state (small B-parameter) and 3α-gas states (large B-parameter) on the same footing. In this container picture the existence of cluster-gas states is natural and the formation mechanism of cluster-gas states is just the spatial expansion of the container (B-parameter going from small to large). When we compare the container picture of cluster dynamics with the traditional description of cluster dynamics which uses explicitly wave functions with inter-cluster separation coordinates, the new understanding corresponds to a collective-motion picture characterized by the size parameter B. When B has obtained a large value the clusters become more or less independent. They have, however, to respect the excluded volume which is due to the Pauli principle, what leads to scattering processes among the clusters. It is in this way that, e.g., the α condensate is depleted by about 30% in the Hoyle state of ^{12}C [21].

Now we explain how the idea of the parity-violating deformation of localized ^{16}O + α clustering for the inversion-doublet bands of ^{20}Ne can be justified in this container picture of cluster dynamics which assumes nonlocalized clusters. The parity-violating deformation is a property of the intrinsic state which is the instantaneous (or adiabatic) quantum state of the rotation of the nucleus. Since the instantaneous configuration of two clusters is of prolate shape, the prolate THSR wave function is the intrinsic state of the system and the oblate THSR wave function is not the intrinsic state but rather a mathematical object which expresses the rotation-average of the intrinsic state. The spherical THSR wave function expresses the time average of the fully three-dimensional rotational motion, namely the angular-momentum projected state of the intrinsic state (the prolate THSR wave function). We, however, also need to notice the fact that two clusters can not come close to each other because, as just mentioned, of the inter-cluster Pauli repulsion, which implies that two clusters in the intrinsic state (the prolate THSR wave function) are effectively localized in space. Thus, the prolate THSR wave function has the parity-violating deformation of localized ^{16}O + α clustering. We can say that dynamics prefers nonlocalized clustering but kinematics makes the system look like localized clustering. Of course, this localization is most pronounced in the necessarily strongly prolate two cluster systems. In systems with low density α clusters in number more than two have more space to move independently and are, therefore, less localized in spherical containers.

In a word, the container picture of cluster dynamics has three important ingredients. The first is to regard the motion of clusters as being mutually independent and described by the nonlocalized lowest orbit of the self-consistent mean-field potential of clusters. The second is the collective excitation of the system which is described by the Hill-Wheeler equation with respect to the size parameter(s) B of the mean-field potential. The third is the inter-cluster Pauli repulsion which, in the case of two-cluster systems, is the origin of the molecular structure of clusters and which, in cases of more α clusters, like in the Hoyle state of ^{12}C, leads to $\alpha - \alpha$ scattering processes which somewhat depopulate the α condensate.

5. Summary

It is found that the single THSR wave functions for the inversion doublet bands of ^{20}Ne are nearly 100% equivalent to the corresponding exact RGM solution. We believe that the concept of nonlocalized clustering proposed by a THSR-type wave function is essential in correctly understanding the cluster structures in nuclei. We further proposed the container picture for understanding cluster dynamics where the clusters make nonlocalized motion occupying the lowest orbit of the cluster mean-field potential characterized by the size parameter.

Acknowledgments

The author is very grateful to his collaborators, Dr. Y. Funaki, Profs. H. Horiuchi, Z. Ren, G. Röpke, P. Schuck, A. Tohsaki, C. Xu, and T. Yamada for lots of helpful comments and discussions in this work. This work is supported by the National Natural Science Foundation of China (Nos. 11035001, 10975072, 10735010, 11375086, 11175085, 11235001, and 11120101005), by the 973 Program of China (No. 2010CB327803 and No. 2013CB834400), by CAS Knowledge Innovation Project No. KJCX2-SW-N02, by Research Fund of Doctoral Point (RFDP) (No. 20100091110028), and by the Project Funded by the Priority Academic Program Development of Jiangsu Higher Education Institutions (PAPD).

References

[1] Wildermuth K and Tang Y C 1977 *A Unified Theory of the Nucleus* (Braunschweig: Vieweg)
[2] von Oertzen W, Freer M and Kanada-En'yo Y 2006 *Phys. Rep.* **432** 43
[3] Freer M 2007 *Rep. Prog. Phys.* **70** 2149
[4] Ren Y and Ren Z 2012 *Phys. Rev.* C **85** 044608
[5] Horiuchi H, Ikeda K and Kato K 2012 *Prog. Theor. Phys. Supple.* **192** 1
[6] Yamada T, Funaki Y, Horiuchi H, Röpke G, Schuck P and Tohsaki A 2012 *Lecture Notes in Physics* **848** 229
[7] Tohsaki A, Horiuchi H, Schuck P and Röpke G 2001 *Phys. Rev. Lett.* **87** 192501
[8] Fukushima Y and Kamimura M 1978 *Supple. of J. Phys. Soc. Japan* **44** 225 ; Kamimura M 1981 *Nucl. Phys.* A **351** 456
[9] Horiuchi H, Ikeda K, Kamimura M, Saito S, Tamagaki R and Tohsaki A 1977 *Prog. Theor. Phys. Supple.* No. **62**
[10] Uegaki E, Okabe S, Abe Y and Tanaka H 1977 *Prog. Theor. Phys.* **57** 1262 ; Uegaki E, Abe Y, Okabe S and Tanaka H 1978 *Prog. Theor. Phys.* **59** 1031
[11] Funaki Y, Tohsaki A, Horiuchi H, Schuck P and Röpke G 2003 *Phys. Rev.* C **67** 051306(R)
[12] Funaki Y, Horiuchi H, Tohsaki A, Schuck P and Röpke G 2002 *Prog. Theor. Phys.* **108** 297
[13] Zhou B, Ren Z, Xu C, Funaki Y, Yamada T, Tohsaki A, Horiuchi H, Schuck P and Röpke G 2012 *Phys. Rev.* C **86** 014301
[14] Zhou B, Funaki Y, Horiuchi H, Ren Z, Röpke G, Schuck P, Tohsaki A, Xu C and Yamada T 2014 *Phys. Rev.* C **89** 034319
[15] Zhou B, Funaki Y, Horiuchi H, Ren Z, Röpke G, Schuck P, Tohsaki A, Xu C and Yamada T 2013 *Phys. Rev. Lett.* **110** 262501
[16] Brink D M 1966 *Proc. Intern. School of Physics 'Enrico Fermi'* course 36 ed Block C (Academic Press)
[17] Funaki Y, Horiuchi H, von Oertzen W, Röpke G, Schuck P, Tohsaki A and Yamada T 2009 *Phys. Rev.* C **80** 064326
[18] Suhara T, Funaki Y, Zhou B, Horiuchi H and Tohsaki A 2014 *Phys. Rev. Lett.* **112** 062501
[19] Funaki Y, Tohsaki A, Horiuchi H, Schuck P and Röpke G 2005 *Eur. Phys. J.* A **24** 321
[20] Funaki Y, Yamada T, Tohsaki A, Horiuchi H, Röpke G and Schuck P 2010 *Phys. Rev.* C **82** 024312
[21] Yamada T and Schuck P 2005 *Eur. Phys. J.* A **26** 185

Journal of Physics: Conference Series **569** (2014) 012008 doi:10.1088/1742-6596/569/1/012008

One-dimensional α condensation of α-linear-chain states

T. Suhara[1], Y. Funaki[2], B. Zhou[3], H. Horiuchi[4], and A. Tohsaki[4]

[1] Matsue College of Technology, Matsue 690-8518, Japan
[2] Nishina Center for Accelerator-Based cience, The Institute of Physical and Chemical Research (RIKEN), Wako 351-0198, Japan
[3] Department of Physics, Nanjing University, Nanjing 210093, China
[4] Research Center for Nuclear Physics (RCNP), Osaka University, Osaka 567-0047, Japan

E-mail: suhara@matsue-ct.ac.jp

Abstract. We present a new picture that the α-linear-chain structure for ^{12}C and ^{16}O has one-dimensional α condensate character. The wave functions of linear-chain states which are described by superposing a large number of Brink wave functions have extremely large overlaps of nearly 100% with single Tohsaki-Horiuchi- chuck-Röpke (TH R) wave functions, which were proposed to describe the α condensed "gas-like" states. Although this new picture is different from the conventional idea of the spatial localization of α clusters, the density distributions are shown to have localized α-clusters which is due to the inter-α Pauli repulsion.

1. Introduction

The cluster formation is one of the fundamental degrees of freedom in nuclear structure [1]. In particular, for the last decade, the physics of α-particle condensation has triggered much interest [2, 3]. The best achievement in this direction is to have revealed that the Hoyle state (the 0_2^+ state at 7.67 MeV in ^{12}C) does not only have the 3α-cluster structure but also has an α-condensate-like structure of the 3α clusters, which occupy an identical S orbit of mean-field potential [4]. In ^{16}O the analog to the Hoyle state was also proposed to exist theoretically as the 0_6^+ state, which might be identified as the observed 15.1 MeV state [5].

The above α condensate character was clarified with the use of the condensate-type microscopic cluster model wave function, so-called Tohsaki-Horiuchi-Schuck-Röpke (THSR) wave function [2, 6]. The wave function does not put α clusters on spatial positions any more like localized clustering but generates a quantum orbit occupied by the clusters, where they move in a non-localized way in a mean-field type potential of the α clusters. The THSR wave function has been used for the analysis of the "gas-like" state. For the "gas-like" states such as ^{8}Be and the Hoyle state, the solutions of 2α and 3α resonating group method (R M) equation under the bound state approximation were found to be almost 100 equivalent to single 2α and 3α THSR wave functions, respectively [3].

Quite recently, in Refs. [7, 8, 9], the THSR-type α + ^{16}O cluster model wave function was applied to the investigation of the inversion doublet bands of α + ^{16}O cluster structure in ^{20}Ne. In spite of the fact that the existence of the inversion doublet bands proves most convincingly the localized clustering of α + ^{16}O [10], they showed that the single THSR wave functions are almost 100 equivalent to the solutions of α+^{16}O R M equation. Since the THSR wave function expresses non-localized clustering, this result requires us to change the traditional understanding of clustering as the localized cluster structure.

Another typical example of the localized clustering is the α-linear-chain state, where α clusters are aligned on a line. Thus the above mentioned results urge us to examine the non-localized picture of the THSR ansatz for α-linear-chain states that α clusters occupy an identical lowest orbit of a prolate-deformed mean-field potential.

The 3α-linear-chain state, which was first proposed by Morinaga about 60 years ago, was used to explain a structure of the Hoyle state [11, 12]. Although this idea for the Hoyle state was later replaced by the interpretation as a "gas-like" state of 3α clusters, the higher 0^+ state observed at 10.3 MeV has still been the possible candidate of the chain state, for which antisymmetrized molecular dynamics (AMD) and fermionic molecular dynamics (FMD) calculations predict a 3α structure with slight bending from the linear-chain arrangement [13, 14].

As for the 4α-linear-chain states in ^{16}O, Chevallier *et al.* observed the resonant 2^+, 4^+, and 6^+ states which lie on a $J(J+1)$ rotational trajectory with a large moment of inertia, consistent with a linear-chain of 4α particles [15]. Theoretical calculation of α reduced widths supported this picture [16]. Freer *et al.* also observed those states [17, 18]. On the other hand, at higher angular-momentum region, recently Ichikawa *et al.* reported the formation and the stability of 4α-linear-chain state based on the cranked Hartree-Fock method [19].

In addition to these examples, various theoretical and experimental studies of $N\alpha$-linear-chain states have been made for almost half a century [20, 21]. Even now, the understanding is insu cient. To clarify the nature of linear-chain states is important from historical view. In this and previous papers [22], we compare the solutions of 3α- and 4α-linear-chain states obtained by the generator coordinate method (CM) by using Brink wave functions with the THSR wave functions. From this comparison, we present a new understanding of the linear-chain structure.

2. Formulation

First we explain the Brink wave function [23]:

$$\Phi_J^{(B)}(\boldsymbol{R}_1,\cdots,\boldsymbol{R}_n) = \hat{P}^J_{M0}\hat{P}^+\mathcal{A}[\varphi_1^\alpha\cdots\varphi_n^\alpha], \tag{1}$$

$$\varphi_i^\alpha = \phi_{\boldsymbol{R}_i}\phi_{\boldsymbol{R}_i}\phi_{\boldsymbol{R}_i}\phi_{\boldsymbol{R}_i}\chi_{p\uparrow}\chi_{p\downarrow}\chi_{n\uparrow}\chi_{n\downarrow}, \tag{2}$$

$$\phi_{\boldsymbol{R}_i} = (\pi b^2)^{-3/4}\exp\Big[-\frac{1}{2b^2}(\boldsymbol{r}-\boldsymbol{R}_i)^2\Big], \tag{3}$$

where $\boldsymbol{R}_i$ is the position parameter of the i-th α particle, and $\hat{P}^J_{M0}$ and $\hat{P}^+$ are the projection operators onto the angular momentum JM and positive parity, respectively.

We consider model space of the pure linear-chain state, by taking $R_{ix}=R_{iy}=0$. In this model space we perform the CM calculation by taking $\boldsymbol{R}_i=(0,0,R_{iz})$ as discretized generator coordinates. The CM wave function can thus be written as,

$$\Phi_J^{(\mathrm{B-GCM})} = \sum_{R_{1z},\cdots,R_{nz}} f(R_{1z},\cdots,R_{nz})\Phi_J^{(\mathrm{B})}(R_{1z},\cdots,R_{nz}). \tag{4}$$

The coe cients $f(R_{1z},\cdots,R_{nz})$ can then be determined by solving the Hill-Wheeler equation. In order to obtain the converged solution of the above equation, we adopt 100 and 300 parameter values for $R_{1z},\cdots,R_{nz}$. These are determined by randomly generated distances between the adjacent clusters, $d_i \equiv R_{(i+1)z}-R_{iz}$, within a range of 2 fm $\le d_i \le$ 10 fm.

Next, we show the THSR wave function as follows:

$$\Phi_J^{(\mathrm{THSR})} = \hat{P}^J_{M0}\Phi^{(\mathrm{THSR})}(\beta_x,\beta_y,\beta_z) = \hat{P}^J_{M0}\mathcal{A}\Big\{\prod_{i=1}^{n}\exp\Big[-\sum_{k=x,y,z}\frac{2}{B_k^2}(X_{ik}-X_{Gk})^2\Big]\phi(\alpha_i)\Big\} \tag{5}$$

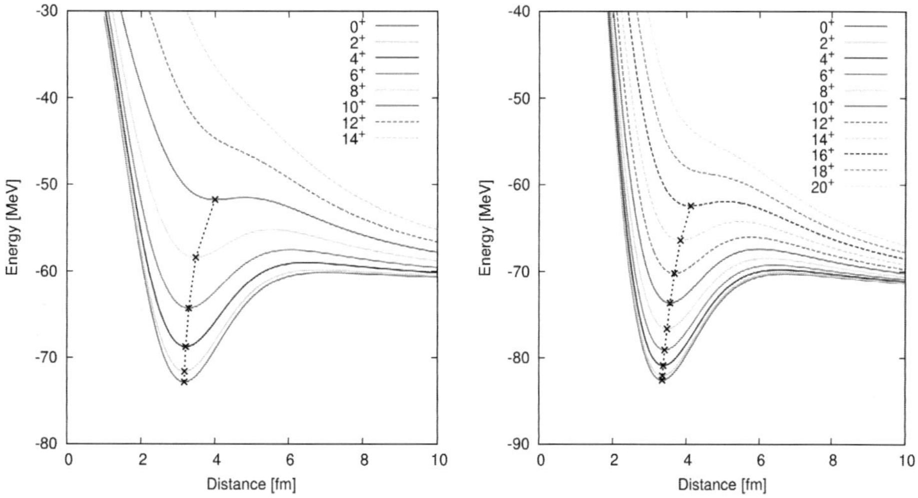

Figure 1. The energy curves for J^π states in equidistant linear-chain configuration, $d_1 = d_2$ for ^{12}C (Left) and $d_1 = d_2 = d_3$ for ^{16}O (Right). The minima are denoted by crosses.

with the antisymmetrizer $\mathcal{A}$ operating on all nucleons and $B_k^2 = b^2 + 2\beta_k^2$. $\boldsymbol{X}_i$ and $\boldsymbol{X}_G$ are the center-of-mass coordinate of the i-th α particle and the total center-of-mass coordinate, respectively. $\phi(\alpha_i)$ is the intrinsic wave function of the i-th α particle.

In the THSR model space, a linear-chain structure is realized by taking strongly prolate-deformed values for the parameter $(\beta_x, \beta_y, \beta_z)$, i.e. $\beta_x = \beta_y \sim 0$ and non-zero β_z value. This has the one-dimensional α condensate structure, in which the α clusters move independently inside a mean-field potential with strongly prolate-deformed shape. This THSR-type picture therefore is very different from the traditional one for the linear-chain structure, in which localized clustering in one dimension has been considered to occur. Our aim is to show that the solutions of the linear-chain Brink- CM equation are quite close to single THSR-type linear-chain wave functions.

For the NN interaction, the Volkov No.2 force [24] is used with Majorana parameter $M = 0.59$ and $M = 0.63$ for ^{12}C and ^{16}O, respectively. For these choices of the Majorana parameter, the binding energies of the ground states of ^{12}C and ^{16}O are calculated to be -91.8 MeV and -129.9 MeV, respectively, which are in good agreement with the corresponding experimental data, -92.2 MeV and -127.6 MeV. As the size parameter of the α particle, we adopt $b = 1.376$ fm in Eq. (3), which gives a minimum energy of the α particle, -28.0 MeV.

3. Results and discussion

First we show the energy curves for ^{12}C in the left panel of Fig. 1, where an equal distance between α particles is assumed, i.e. $d_1 = d_2$. For the states with $J^\pi = 0^+, \cdots, 10^+$, the minimum-energy states appear, which are denoted by crosses. The distances giving the minimum energies are pushed outside as the increase of the angular momentum. This indicates that the centrifugal force, which is strengthened by the increase of the angular momentum, is leading to a more elongated linear-chain structure. The minima tend to disappear at higher angular momentum. Similar calculation is performed for ^{16}O, which is shown in the right panel of Fig. 1, where $d_1 = d_2 = d_3$ is assumed. The minimum-energy states survive up to around $J^\pi = 16^+$.

In table 1, we list the energy minima for angular-momentum projected states (J^π) and the lowest eigenenergies by the CM calculation (CM). The energy of 0^+ state obtained by CM

Table 1. Minimum energies given by angular-momentum-projected and CM wave functions of Brink model, in a unit of MeV. The distance parameter values (d_1, d_2) for ^{12}C and (d_1, d_2, d_3) for ^{16}O, respectively, which give the minimum energies, are also shown, in a unit of fm. The threshold energies for ^{12}C and ^{16}O are calculated to be $E^{th}_{3\alpha} = -83.97$ MeV and $E^{th}_{4\alpha} = -111.96$ MeV, respectively.

	^{12}C (3α)			^{16}O (4α)		
	(d_1, d_2)	Energy	CM	(d_1, d_2, d_3)	Energy	CM
0^+	$(2.5, 4.1)$	-73.9	-76.8	$(4.2, 3.4, 2.8)$	-84.2	-90.1
2^+	$(2.5, 4.1)$	-72.7	-75.8	$(4.2, 3.4, 2.8)$	-83.7	-89.7
4^+	$(2.6, 4.2)$	-70.0	-73.6	$(4.2, 3.4, 2.9)$	-82.6	-88.9

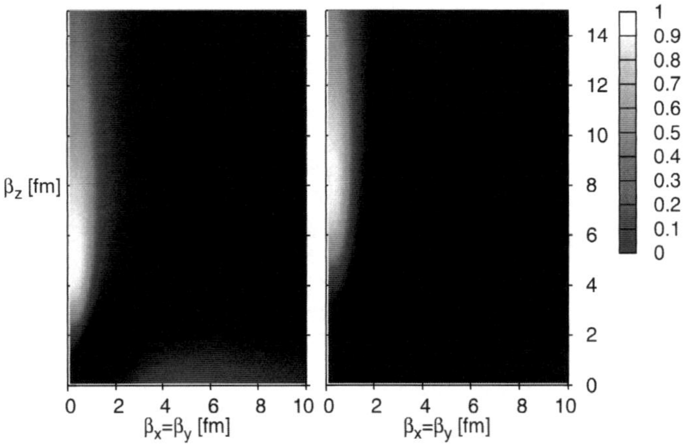

Figure 2. The squared overlap surfaces between the linear-chain wave functions obtained by CM and the single THSR wave functions in two parameter spaces, $\beta_x = \beta_y$ and β_z, for $J^\pi = 0^+$ states in ^{12}C (Left) and ^{16}O (Right).

is -76.8 MeV, 7.2 MeV above the 3α threshold energy, which is still about 4 MeV higher than the energy of 0^+ state observed at 10.3 MeV, for which AMD and FMD calculations predict a 3α structure with slight bending from the linear-chain arrangement. This difference may come from the lack of degree of freedom of bending motion of the α cluster from the linear-chain configuration in our present model space.

In Fig. 2, we show the squared overlap surfaces between the Brink- CM wave functions of Eq. (4) and the THSR wave functions Eq. (5) for $J^\pi = 0^+$ states of ^{12}C and ^{16}O, respectively, in two parameter spaces, $\beta_x = \beta_y$ and β_z. Surprisingly, we can see very large maximum values for both cases. The region of having large values is widely ranged in the extremely prolate-deformed region. We list the maximum values in table 2, for $J^\pi = 0^+, 2^+, 4^+$ states in ^{12}C and ^{16}O, together with the corresponding $(\beta_x = \beta_y, \beta_z)$ values to give the maxima. The energies of the THSR wave functions are also shown. The maximum squared overlaps are more than 0.98 and 0.93 in all the J^π states listed for ^{12}C and ^{16}O, respectively. The energies of the THSR wave functions are very close, with a few hundred keV, to those of the CM wave functions listed in table 1. These results of course mean that the THSR wave function very well describes the pure linear-chain structure. This obviously urges us to reconsider the conventional picture of linear-chain structure. The one-dimensional "gas" of the α particles with the condensate character

Table 2. Maximum squared overlaps between the CM wave functions obtained based on the Brink wave function and the single THSR wave functions, for $J^\pi = 0^+, 2^+, 4^+$ states for ^{12}C and ^{16}O. $(\beta_x = \beta_y, \beta_z)$ giving maxima and their energies are denoted, in units of fm and MeV, respectively.

	^{12}C (3α)			^{16}O (4α)		
	$(\beta_x = \beta_y, \beta_z)$	Max.	Energy	$(\beta_x = \beta_y, \beta_z)$	Max.	Energy
0^+	$(0.1, 5.1)$	0.987	-76.6	$(0.1, 8.2)$	0.944	-89.8
2^+	$(0.1, 5.4)$	0.989	-75.6	$(0.1, 8.4)$	0.942	-89.4
4^+	$(0.1, 6.6)$	0.981	-73.4	$(0.1, 9.0)$	0.931	-88.8

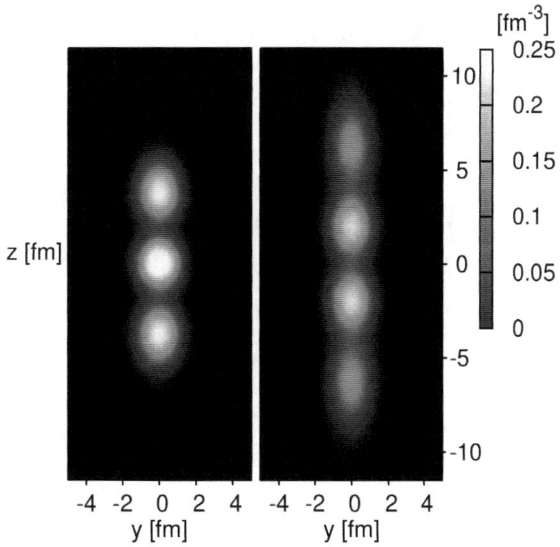

Figure 3. Intrinsic density profiles of the 3α- (Left) and 4α- (Right) linear-chain states generated from the THSR wave functions before angular-momentum projection at ($\beta_x = \beta_y =$ 0.1 fm, $\beta_z = 5.1$ fm) and ($\beta_x = \beta_y = 0.1$ fm, $\beta_z = 8.2$ fm), respectively.

is very different from the localized clustering described by a single Brink wave function. The maximum squared overlaps between the CM wave functions and single Brink wave functions are only 78 and 48 for the 0^+ states of ^{12}C and ^{16}O, respectively. These values are significantly smaller than the squared overlaps, 99 and 94 , between the CM wave functions and THSR wave fucntions.

We show in Fig. 3 the intrinsic density profiles of THSR wave functions at $x = 0$ for ^{12}C ($\beta_x = \beta_y = 0.1$ fm, $\beta_z = 5.1$ fm) and ^{16}O ($\beta_x = \beta_y = 0.1$ fm, $\beta_z = 8.2$ fm), which give the maximum squared overlaps for $J^\pi = 0^+$ states, respectively. We can clearly see the 3α and 4α clusters, which are aligned on a straight line and are all localized. This result gives us a new understanding of clustering. Even for the states which are described by non-localized-type wave function, localized nature of clustering can appear in density distribution due to the auli principle. We can say that dynamics prefers non-localized clustering but kinematics coming from the auli principle makes the system look like localized clustering.

4. Summary

We investigated the 3α- and 4α-linear-chain states with the use of Brink wave functions, where the α-particle motion is restricted to one dimension along z-axis. We obtained energy minima for a wide range of angular-momentum states for both cases. We obtained the Brink- CM wave functions with su cient energy convergence by superposing 100 and 300 Brink wave functions adopted as bases to solve the Hill-Wheeler equation, for ^{12}C and ^{16}O, respectively. Then we compared the Brink- CM wave functions with the single THSR wave functions, which provide the α condensate structure. The maximum squared overlaps between them amount to more than 98 for ^{12}C and 93 for ^{16}O, for $J^\pi = 0^+, 2^+, 4^+$ states, at the shapes with strongly prolate deformation. These large overlap values at the extremely deformed shapes mean that the α-linear-chain states have the one-dimensional α condensate character, where the α clusters are trapped into a one-dimensional potential in a non-localized manner, like a "gas". Although this new picture is different from the conventional idea of the spatial localization of α clusters, the density distributions are shown to have localized α-clusters which is due to the inter-α auli repulsion. To clarify the magnitude of the auli effect in linear-chain states, we are investigating, for example, momentum distribution of α clusters in the z-direction.

Acknowledgments

The authors wish to thank . Schuck, . Röpke, T. amada, . . Ren, and C. u for many helpful discussions concerning this work. This work was supported by JS S KAKENHI rant Nos. 25400256, 25400288, and 25887049.

References

[1] Wildermuth K and Tang Y C 1997 *A Unified Theory of the Nucleus* (Vieweg, Braunschweig)
[2] Tohsaki A, Horiuchi H, chuck P and Röpke G 2001 *Phys. Rev. Lett.* **87** 192501
[3] Funaki Y, Horiuchi H, von Oertzen W, Röpke G, chuck P, Tohsaki A and Yamada T 2009 *Phys. Rev.* C **80** 064326
[4] Funaki Y, Tohsaki A, Horiuchi H, chuck P and Röpke G 2003 *Phys. Rev.* C **67** 051306(R)
[5] Funaki Y, Yamada T, Horiuchi H, Röpke G, chuck P and Tohsaki A 2008 *Phys. Rev. Lett.* **101** 082502
[6] Funaki Y, Horiuchi H, Tohsaki A, chuck P and Röpke G 2002 *Prog. Theor. Phys.* **108** 297
[7] Zhou B, Ren Z, u C, Funaki Y, Yamada T, Tohsaki A, Horiuchi H, chuck P and Röpke G 2012 *Phys. Rev.* C **86** 014301
[8] Zhou B, Funaki Y, Horiuchi H, Ren Z, Röpke G, chuck P, Tohsaki A, u C and Yamada T 2013 *Phys. Rev. Lett.* **110** 262501
[9] Zhou B, Funaki Y, Horiuchi H, Ren Z, Röpke G, chuck P, Tohsaki A, u C and Yamada T 2014 *Phys. Rev.* C **89** 034319
[10] Horiuchi H and Ikeda K 1968 *Prog. Theor. Phys.* **40** 277
[11] Morinaga H 1956 *Phys. Rev.* **101** 254
[12] Morinaga H 1966 *Phys. Lett.* **21** 78
[13] Neff T and Feldmeier H 2004 *Nucl. Phys.* A **738** 357
[14] Kanada-En yo Y 2007 *Prog. Theor. Phys.* **117** 655
[15] Chevallier P, cheibling F, Goldring G, Plesser I and achs M W 1967 *Phys. Rev.* **160** 827
[16] uzuki Y, Horiuchi H and Ikeda K 1972 *Prog. Theor. Phys.* **47** 1517
[17] Freer M *et al* 1995 *Phys. Rev.* C **51** 1682
[18] Freer M *et al* 2004 *Phys. Rev.* C **70** 064311
[19] Ichikawa T, Maruhn J A, Itagaki N and Ohkubo 2011 *Phys. Rev. Lett.* **107** 112501
[20] Freer M and Merchant A C 1997 *J. Phys. G: Nucl. Part. Phys.* **23** 261
[21] Freer M 2007 *Rep. Prog. Phys.* **70** 2149
[22] uhara T, Funaki Y, Zhou B, Horiuchi H and Tohsaki A 2014 *Phys. Rev. Lett.* **112** 062501
[23] Brink D M 1966 *Proc. Int. School of Physics Enrico Ferm, Course 36 Varenna* ed C Bloch (Academic Press, New York)
[24] Volkov A B 1965 *Nucl. Phys.* A **74** 33

Journal of Physics: Conference Series **569** (2014) 012009 doi:10.1088/1742-6596/569/1/012009

Cluster structure of broad resonances near threshold in ^{12}C and ^{16}O

M Itoh[1], H Akimune[2], M Fujiwara[3], U Garg[4], H Hashimoto[3], T Kawabata[5], K Kawase[3], S Kishi[5], T Murakami[5], K Nakanishi[3], Y Nakatsugawa[5], B K Nayak[4], H Sakaguchi[5], S Terashima[5], M Uchida[7], Y Yasuda[3], M Yosoi[3], J Zenihiro[6]

[1] Cyclotron and Radioisotope Center (CYRIC), Tohoku University, Sendai, Miyagi 980-8578, Japan
[2] Department of Physics, Konan University, Kobe, Hyogo 658-8501, Japan,
[3] Research Center for Nuclear Physics (RCNP), Osaka University, Ibaraki, Osaka 567-0047
[4] Physics Department, University of Notre Dame, Notre Dame, Indiana 46556, USA
[5] Department of Physics, Kyoto University, Kyoto 606-8502, Japan
[6] RIKEN Nishina Center for Accelerator-Based Science, Wako, Saitama 351-0198, Japan
[7] Department of Physics, Tokyo Institute of Technology, Tokyo 152-8551, Japan

E-mail: `itoh@cyric.tohoku.ac.jp`

Abstract. We investigated decay properties of the broad resonance around 10 MeV in ^{12}C and measured inelastic scattering of 386 MeV α particles on ^{16}O at extremely forward angles including 0°. We have found that the lower component of the broad 0^+ resonance at 10 MeV in ^{12}C decayed to the ^{8}Be + α channel, and that the higher component decayed both to the ^{8}Be + α channel and to the ^{8}Be(2^+) + α channel. On the ^{16}O nucleus, we have found two 0^+ resonances at 16.7 MeV and 18.8 MeV, which are near the 4α breakup threshold. On the analogy to the broad resonances near the 3α threshold in ^{12}C, they are considered to have the ^{12}C(0_2^+) + α cluster and the ^{8}Be + ^{8}Be cluster structures, respectively. It is consistent with the result of the experiment for the ^{12}C(^{4}He,^{8}Be)^{8}Be and ^{12}C(^{4}He,^{12}C$^*\to{}^8$Be + α)α reactions.

1. Introduction

Cluster structure in the nucleus is considered to appear near the threshold energy of disassembling to the constituent clusters, as indicated by Ikeda diagram [1]. The 0_2^+ state at $E_x = 7.65$ MeV in ^{12}C, which is known as the Hoyle state, is a typical 3α cluster state, which is 0.38 MeV above the 3α breakup threshold. However, its configuration of α cluster is still controversial. In 1953, Morinaga proposed it had a linear 3α chain structure [2]. According to the semi-microscopic and microscopic α-cluster models, the linear 3α configuration was found to be energetically unstable and the Hoyle state was considered to have loosely coupled 3α gaslike structure [3, 4, 5]. In 2001, Tohsaki *et al.* presented that this loosely coupled 3α gaslike structure was similar to the Bose-Einstein condensate of the 3α clusters [6], in which the 3α clusters enter into the lowest S-orbit.

Considering the α condensate structure as a special case of the loosely coupled α gaslike structure, many analog states are expected such as the 2^+ and the broad 0^+ states around 10 MeV in ^{12}C [7, 8]. This 2^+ state is considered as the 2^+ excitation of the Hoyle state and also

Journal of Physics: Conference Series **569** (2014) 012009 doi:10.1088/1742-6596/569/1/012009

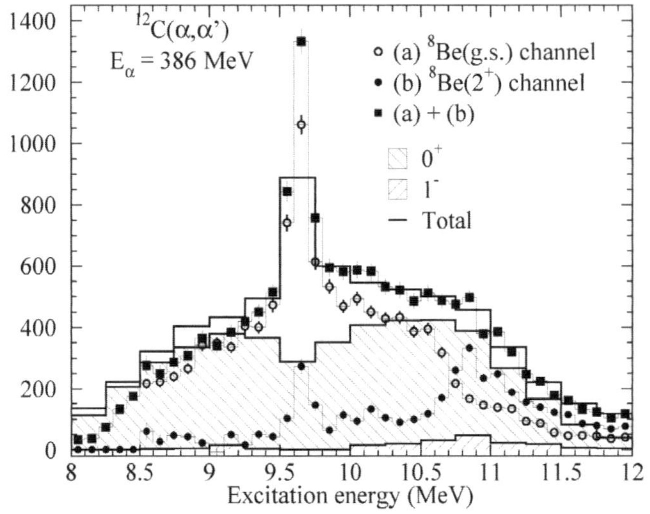

Figure 1. The coincidence energy spectra at 0° in the 10 MeV region is superimposed into the singles energy spectrum of inelastic α scattering of ^{12}C. (Figure adapted from Ref. [18].)

predicted in almost all α cluster models. In the α condensate model [7], it is considered that one of 3α clusters in the 3α condensate state is excited from the lowest S-orbit to the D-orbit. The broad 0 state around 10 MeV predicted in Ref. [8] is considered as the higher nodal state of the Hoyle state, in which one of 3α clusters is excited up to the second S-orbit. On the other hand, in the AMD and FMD calculations [9, 10], the linear-like 3α state is also predicted around 10 MeV, in which the dominant component is the 2^+ state of ^{8}Be (^{8}Be$_{2^+}$) and α cluster structure.

In the experimental side, recently, the 2^+ state at 10 MeV in ^{12}C has been established [11, 12, 13, 14]. Although the broad 0^+ states at 10 MeV was also observed in Ref [12], their $E0$ strength distribution around 10 MeV indicated the possibility of consisting of two components. These two components may correspond to the higher nodal state of the Hoyle state and the linear-like 3α state described above.

As for the ^{16}O nucleus, the 4α condensate state is predicted at 15.1 MeV 0^+ state, which is 0.66 MeV above the 4α breakup threshold [15]. The 4α linear chain structure is also proposed around 16 MeV from the measurement of the ^{12}C(^{4}He,^{8}Be)^{8}Be reaction [16]. Ohkubo and Hirabayashi predicted the band head 0^+ state of the Hoyle state (^{12}C[0_2^+]) + α cluster structure appeared at 16.9 MeV in ^{16}O by solving the coupled channel scattering equations [17]. Consequently, the situation of the 4α cluster configurations at above 4α breakup threshold in ^{16}O is very similar to that of the 3α cluster states in ^{12}C.

In this contribution, we report two studies on the α gaslike structure of broad resonance states near Nα breakup thresholds in ^{12}C and ^{16}O.

2. Experiments

The experiments were performed at the ring cyclotron facility in the Research Center for Nuclear Physics (RCNP), Osaka University. In the ^{12}C experiment, the coincidence measurement of a decay α particle via the ^{12}C(α,α')^{12}C*[α + X] reaction at E_α = 386 MeV was carried out. Details of the experimental setup and procedure are described in Refs. [18, 19]. Only a brief

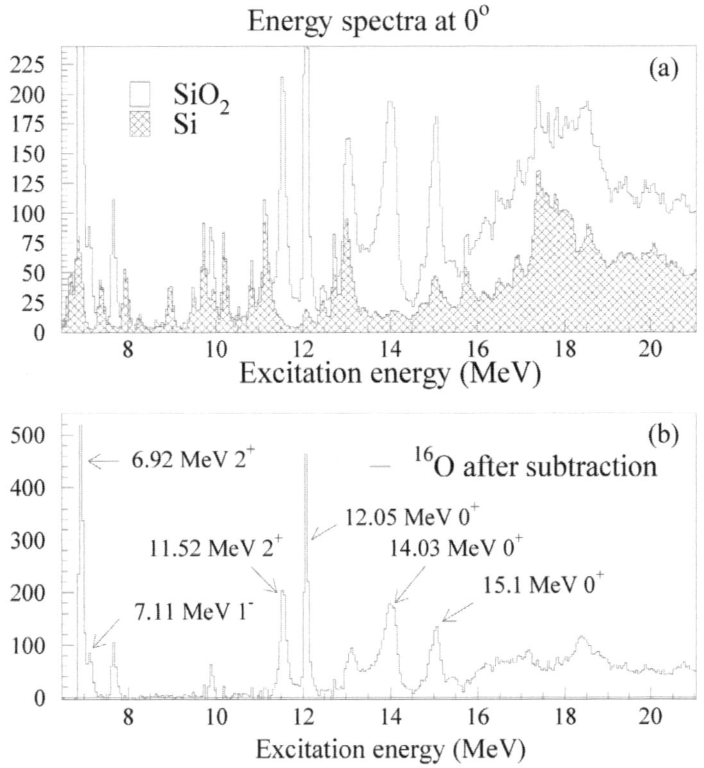

Figure 2. (a) The energy spectra of inelastic α scattering on SiO_2 and Si at 0° (b) The energy spectrum on ^{16}O obtained by subtracting the Si contaminant from the SiO_2 spectrum.

outline is provided below. The excitation energy of ^{12}C was determined from the momentum of a scattered α particle, which was analyzed by the GRAND RAIDEN spectrometer. The decay α particle from the excited ^{12}C was detected by the silicon detectors (SSD) array, which composed of fourteen SSDs installed in the scattering chamber at backward angles from 115° to 255°. The decay α particle was identified by the time of flight method.

From a missing mass spectrum in the $^{12}C(\alpha,\alpha')^{12}C^*[\alpha + X]$ reaction, the decay α particles were classified into two channels. One was the ground state of ^8Be ($^8Be_{gs}$) channel. The other was the $^8Be_{2^+}$ channel, although it included the direct 3α decay. Figure 1 shows the coincidence energy spectrum at 0° in the $^{12}C(\alpha,\alpha')^{12}C[\alpha + X]$ reaction. The singles energy spectrum (solid line) in the $^{12}C(\alpha,\alpha')$ and 0^+ (right hatched area) and 1^- (left hatched area) components obtained from the multipole decomposition analysis (MDA) in Ref. [12] are superimposed onto the coincidence energy spectra. Open and closed circles show the coincidence energy spectra with the $^8Be_{gs}$ and the $^8Be_{2^+}$ channels, respectively. The solid triangles show the sum of these two coincidence spectra. Since the scattering plane can not be defined in the 0° scattering, the ϕ-angle distribution of the α-decay is uniform. Furthermore, assuming the detectors for decay α particles covered the whole angles, this summed spectrum can reproduce a inclusive spectrum of the $^{12}C(\alpha,\alpha')$ reaction at 0°. This means the validity that the energy spectrum at 0° is considered to consist of these two components.

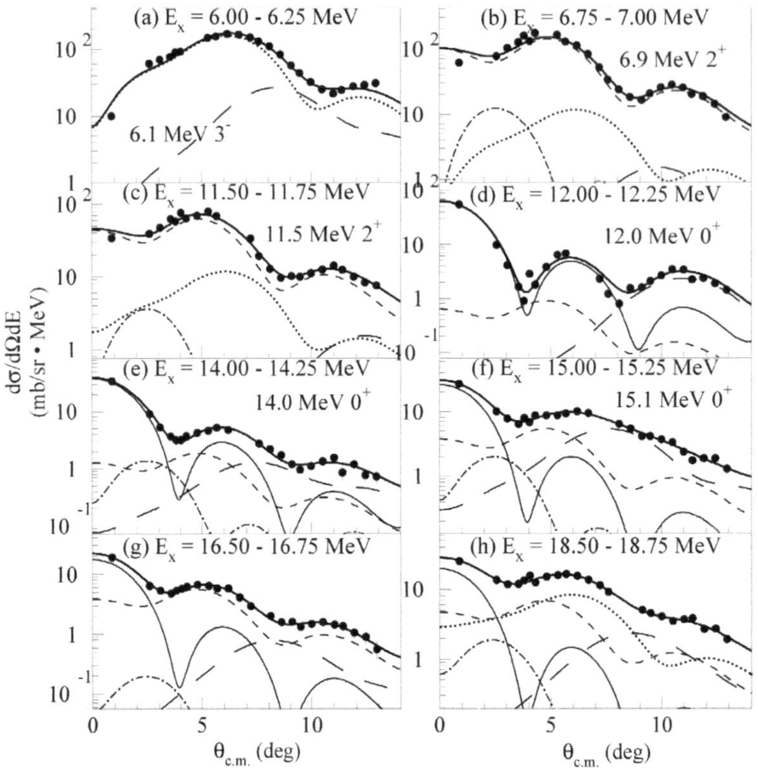

Figure 3. Typical angular distributions of the double-differential cross section for various excitation energy bins in ^{16}O. Thick solid lines show the fits to the data from multipole decomposition. In each panel, the contributions from $L = 0$ (thin sold), $L = 1$ (dot-dashed), $L = 2$ (dashed), $L = 3$ (dotted) and $L \geq 4$ (long-dashed) are also displayed. The excitation energy and J^π value displayed in each panel indicate a dominant component in each excitation energy bin.

In the experiment of ^{16}O, we measured energy spectra of inelastic α scattering at $E_\alpha = 386$ MeV on the SiO_2 and Si targets with thicknesses of 2.0 mg/cm^2 and 2.2 mg/cm^2, respectively, using the GRAND RAIDEN spectrometer. In order to obtain energy spectra of the ^{16}O(α,α') reaction, the Si contaminant was subtracted from the SiO_2 spectrum. Figure 2 shows the energy spectra of inelastic α scattering at 0° for SiO_2, Si, and ^{16}O after subtraction. The 2^+ states at 7.42 and 7.93 MeV, and the 1^- state at 8.90 MeV appeared in Fig. 2(a) are completely subtracted in the ^{16}O spectrum in Fig. 2(b). The candidate for the 4α condensate state at 15.1 MeV is clearly seen in Fig. 2(b).

3. Multipole decomposition analysis

In order to identify the multipolarities of the states in ^{16}O, we performed the MDA. In the MDA, the experimentally obtained cross sections are expressed as the sum of the contributions from the various multipole components as

$$\sigma^{exp}(\theta, E_x) = \sum_L a_L(E_x)\sigma_L^{calc}(\theta, E_x) \tag{1}$$

where θ, E_x are the scattering angle and the excitation energy, respectively. $\sigma_L^{calc}(\theta, E_x)$ is the DWBA cross section for the transferred angular momentum, L. In the present MDA, the energy spectrum at each angle was divided into 0.25-MeV bins. In the DWBA calculation, we adopted a single folding model with the Gaussian form N-α interaction. The interaction parameters were obtained from the fit of the angular distribution of elastic α scattering on ^{16}O, which were taken from Ref. [20].

Figure 3 shows typical results of the MDA. The 6.1 MeV 3^-, 6.9 MeV 2^+, 11.5 MeV 2^+, 12.0 MeV 0^+, 14.0 MeV 0^+, and 15.1 MeV 0^+ states in ^{16}O are clearly identified. We also found much 0^+ component above the 4α breakup threshold energy at 14.4 MeV as shown in Fig. 3(g) and 3(h). Figure 4(a) shows the obtained $E0$ strength distribution in ^{16}O, which is roughly in agreement with that obtained in Ref. [21] except the high excitation energy region above 25 MeV. In the calculation with the relativistic random-phase approximation, most of the $E0$ strength due to the collective monopole excitation of nucleons appeared at over 20 MeV [22]. On the other hand, according to the 4α-OCM [23] and the AMD + αGCM [24] calculations, the $E0$ strength lower than 16 MeV is dominated by the monopole excitation to cluster states. In our result, two 0^+ bumps are appeared at 16-20 MeV, which corresponds to the region between the cluster excitation and the collective monopole excitation of nucleons. In the next section, we discuss these two 0^+ bumps.

4. Discussion

4.1. Analogy to the broad resonance in ^{12}C

Figure 4(b) and 4(c) show the $E0$ strength distributions for ^{16}O and ^{12}C at above the 4α and the 3α breakup thresholds, respectively. The $E0$ strength distribution for ^{12}C is taken from Ref. [12]. In the case of the $E0$ strength distribution in ^{12}C, it is considered to have three resonances, a narrow peak of the Hoyle state at 7.65 MeV, the broad resonance of the higher nodal state of the Hoyle state, and the linear-like 3α cluster state, near the 3α breakup threshold as described in Sec. 2. In the case of ^{16}O, it indicates there are three resonances near the 4α breakup threshold. The lowest narrow resonance is the candidate for the 4α condensate state at 15.1 MeV in ^{16}O. The second broad resonance at 16.7 MeV, which is newly found in the present experiment, is considered to have the ^{12}C[0_2^+] + α cluster structure. Considering in the framework of the α condensate model, the higher nodal state appeared at about 1.5 MeV above the 3α condensate state of the Hoyle state. On the analogy to the higher nodal state in ^{12}C, one of the 4α clusters in the lowest S-orbit may be excited up to the second S-orbit by about 1.5 MeV. The third resonance at 18.8 MeV, which is also newly found in the present experiment, is considered to have the ^{8}Be + ^{8}Be cluster structure due to the following reason.

4.2. Comparison with the resonance scattering experiment

Figure 5 shows the comparison between the $E0$ strength distribution in ^{16}O (the left panel of Fig. 5) and the result of the recent experiment for the ^{12}C(^{4}He,^{8}Be)^{8}Be and ^{12}C(^{4}He,^{12}C*$\rightarrow$^{8}Be + α)α reactions [25] (the right panel of Fig. 5). In the excitation energy at 16.7 MeV in ^{16}O, the cross section for the ^{12}C(^{4}He,^{12}C*[0_2^+])α reaction channel is larger than that of other reaction channels. On the other hand, in the excitation energy at 18.8 MeV in ^{16}O, the cross section for the ^{12}C(^{4}He,^{8}Be)^{8}Be reaction channel is relatively large compared to other channels. Therefore, we conclude that the second resonance at 16.7 MeV is considered to have the ^{12}C[0_2^+] + α cluster structure, and that the third resonance at 18.8 MeV is considered to have the ^{8}Be + ^{8}Be cluster structure.

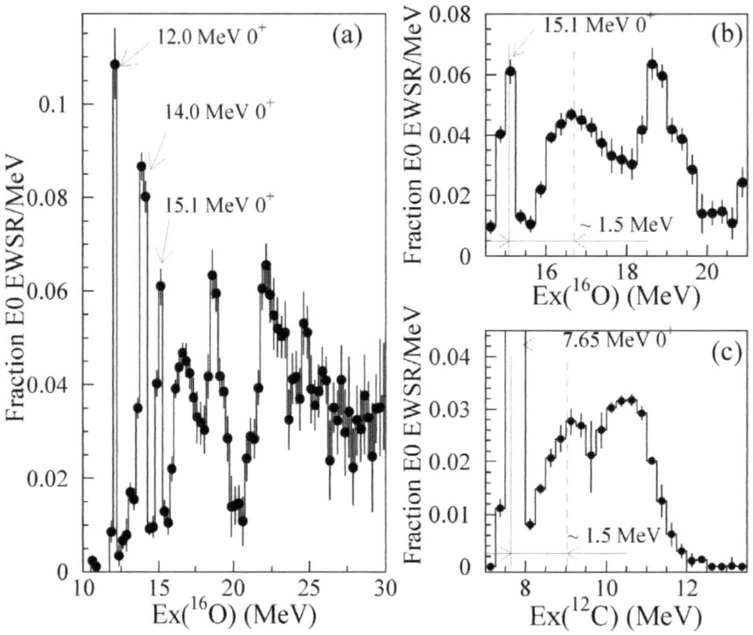

Figure 4. $E0$ strength distributions. (a) A distribution of the $E0$ EWSR fraction in ^{16}O obtained in present work. (b) Enlarged view of the $E0$ EWSR fraction distribution from 14.5 to 21.0 MeV in ^{16}O. (c) The $E0$ EWSR fraction distribution in ^{12}C from 7.0 to 13.5 MeV, which was obtained in Ref. [12], are shown.

5. Summary

In summary, we investigated the α gaslike structure of broad resonance states near Nα breakup threshold in ^{12}C and ^{16}O. From the coincidence measurement of a decay-α particle via the ^{12}C(α,α')^{12}C*[α + X] reaction, the decay property of the broad resonance at 10 MeV is different between the low excitation energy component and the high excitation energy component. They correspond to the higher nodal of the Hoyle state predicted by the CSM calculation and the linear-like 3α state predicted by the AMD calculation. From the measurement of inelastic scattering of 386 MeV α particles on ^{16}O and the multipole decomposition analysis, we have newly found two broad resonances at 16.7 MeV and 18.8 MeV, which are considered to correspond to the ^{12}C[0_2^+] + α cluster and the ^{8}Be + ^{8}Be cluster structures, respectively. It is consistent with the result of the recent experiment for the the ^{12}C(^{4}He,^{8}Be)^{8}Be and ^{12}C(^{4}He,^{12}C*$\rightarrow^8$Be + α)α reactions [25].

Acknowledgments

We are very much indebted to T. Wakasa for providing us with elastic α scattering data on ^{16}O. This work was supported in part by Aid for Scientific Research Grant No. 24740139 from the Japan Ministry of Education, Sports, Culture, Science, and Technology.

References

[1] Ikeda K, Takigawa N and Horiuchi H 1968 *Prog. Theor. Phys. Suppl.* Extra Number **464**
[2] Morinaga H 1956 *Phys. Rev.* **101** 254
[3] Horiuchi H 1974 *Prog. Theor. Phys.* **51** 1266, 1975 *Prog. Theor. Phys.* **53** 447

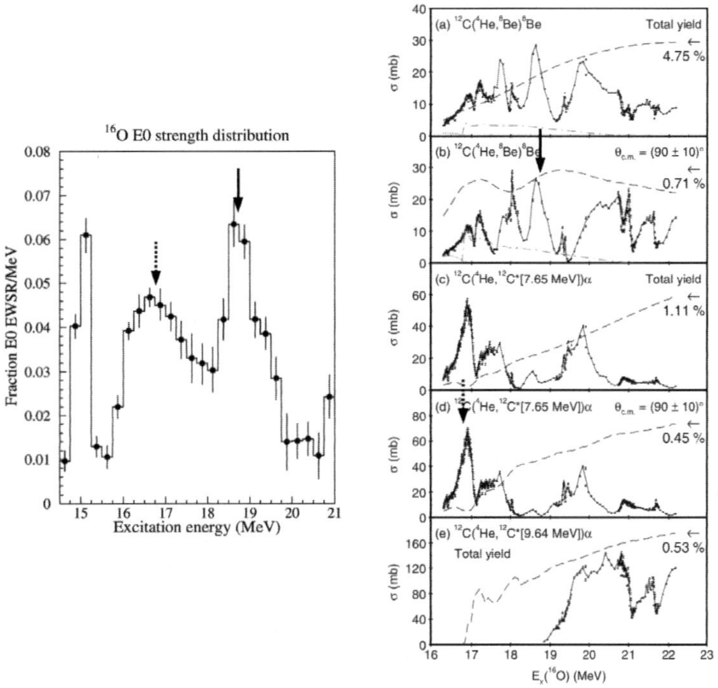

Figure 5. The comparison between the $E0$ strength distribution obtained in present study and the excitation function for the ^{12}C(^{4}He,^{8}Be)^{8}Be and ^{12}C(^{4}He,^{12}C$^{*}\rightarrow^{8}$Be + α)α reactions [25]. The solid and dotted arrows indicate peaks of the higher and the lower components of the broad 0^{+} resonance in ^{16}O, respectively. The right panel of the excitation functions is taken from Ref. [25].

[4] Uegaki E *et al.* 1977 *Prog. Theor. Phys.* **57** 1262
[5] Kamimura M 1981 Nucl. Phys. **A351** 456
[6] Tohsaki A, Horiuchi H, Shuck P and Röpke G 2001 *Phys. Rev. Lett.* **87** 192501
[7] Funaki Y *et al.* 2005 *Eur. Phys. J. A* **24** 321
[8] Kurokawa C and Kato K 2007 *Nucl. Phys. A* **792** 87
[9] Neff T and Feldmeier H 2004 *Nucl. Phys. A* **738** 357-361
[10] Kanada-En'yo Y 2007 *Prog. Theor. Phys.* **117** 655-680
[11] Itoh M *et al.* 2004 *Nucl. Phys. A* **738** 268
[12] Itoh M *et al.* 2011 *Phys. Rev. C* **84** 054308
[13] Freer M *et al.* 2009 *Phys. Rev. C* **80** 041303(R)
[14] Zimmerman W R *et al.* 2013 *Phys. Rev. Lett.* **110** 152502
[15] Funaki Y *et al.* 2008 *Phys. Rev. Lett.* **101** 082502
[16] Chevallier P *et al.* 1967 *Phys. Rev.* **160** 827
[17] Ohkubo S and Hirabayashi Y 2010 *Phys. Lett. B* **684** 127-131
[18] Itoh M *et al.* 2013 *J. Phys.: Conf. Series* **436** 012006
[19] Itoh M 2006 *Mod. Phys. Lett. A* **21** 2359
[20] Wakasa T *et al.* 2007 *Phys. Lett. B* **653** 173-177
[21] Lui Y -W, Clark H L and Youngblood D H 2001 *Phys. Rev. C* **64** 064308
[22] Ma Z, Giai N V, Toki H and L'Huillier M 1997 *Phys. Rev. C* **55** 2385-2388
[23] Yamada T *et al.* 2012 *Phys. Rev. C* **85** 034315
[24] Kanada-En'yo Y 2014 *Phys. Rev. C* **89** 024302
[25] Curtis N *et al.* 2013 *Phys. Rev. C* **88** 064309

Journal of Physics: Conference Series **569** (2014) 012010 doi:10.1088/1742-6596/569/1/012010

To condense, or not to condense, that is the question

Tzany Kokalova Wheldon

School of Physics and Astronomy, University of Birmingham, Birmingham, Edgbaston, Birmingham, B15 2TT, UK

E-mail: t.kokalova@bham.ac.uk

Abstract. Recent key experimental results on the spectroscopy of ^{12}C have been made, which, when taken together, provide much needed clarity on the structure of the ground- and Hoyle-state excitations and provided new avenues for future research. High excitation energies in light nuclei are being explored with evermore precision and recent examples of this in ^{12}C, ^{16}O and ^{20}Ne are presented. These steps forward are examined and the outstanding issue of clustering versus condensation explored.

1. Introduction

The α particle is a very robust nucleus – possessing no bound excited states despite the first threshold lying at 19.81 MeV – and has a total spin of zero. These properties have two consequences: one is the geometric arrangement of pre-formed α particles in the nucleus, *i.e.* cluster structure, and the other is the possibility of the existence of gas-like states, obeying Bose statistics, namely α condensation.

Alpha clustering in light nuclei has been of great interest for many years. Theories of clustering date from the 1930's and so far many varied theoretical models have been developed. Recently the idea of condensation of α particles has been intensely investigated and much theoretical progress has been made. Theoreticians and experimentalists from Japan have played key roles in this. One of the wonders of nature, the Hoyle state in ^{12}C, has been proposed as the best condensate candidate and recently significant experimental and theoretical effort has been invested to understand, in detail, its unorthodox structure. In particular, the identification of excitations of the Hoyle state has been a key goal.

This paper addresses the current experimental progress in understanding excitations in ^{12}C and also the search for analogues of the Hoyle-state in heavier α-conjugate nuclei. The Birmingham group has recently undertaken an extensive programme investigating pertinent states in several nuclei. New studies at high-excitation energies in ^{12}C, using inelastic scattering at the Birmingham cyclotron, and measurements of ^{20}Ne made at Notre Dame, will be presented. Additionally, results for ^{16}O relevant to the long-standing evidence for ^{8}Be – ^{8}Be cluster structures and four-α chain states will be discussed. Finally, the prospect of condensed bosonic-nuclear matter in heavier nuclei will be addressed.

2. Carbon-12 ground-state symmetries

Before understanding the crucial Hoyle state in ^{12}C, it is revealing to first examine structures associated with the ground state band. In 2012 the Birmingham group performed an inelastic scattering experiment using a 40 MeV ^{4}He beam impinging on a 100 $\mu g/cm^2$ $^{nat.}$C target. An array of four double-sided silicon strip detectors were placed symmetrically around the target position, to measure at least three of the four final-state particles; the scattered ^{4}He projectile and the ^{12}C* break-up α particles. More details on the set-up can be found in [1, 2]. The data were reconstructed using energy and momentum conservation and events were selected by placing gates on both the total Q-value, defining the reaction of interest, and on those in which two of the four alpha-particles originated from the ^{8}Be ground state. The resulting energy spectrum of ^{12}C is shown in figure 1 with the additional requirement that the scattered beam particle and two of the three break-up particles were detected. A state at 22.4 MeV is clearly visible.

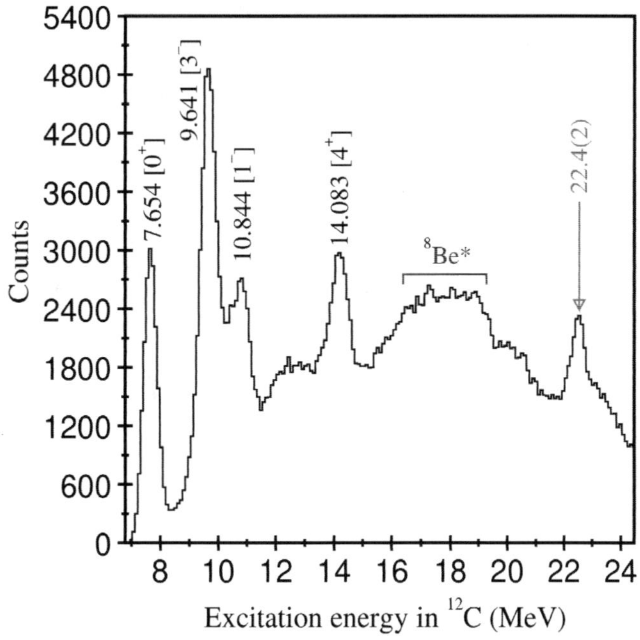

Figure 1. Carbon-12 excitation energy calculated from two detected α particles and the reconstructred third, undetected, break-up α particle. The newly observed state at 22.4(2) MeV can be clearly seen. The contamination by ^{8}Be states below 20 MeV is also observed. Based on figure 2 of [1].

By analysing the angular correlations of the initial break-up particles – α and ^{8}Be – for a subset of the data in figure 1, the periodicity of the oscillations for the 22.4 MeV resonance was found to be consistent only with $I = 5$. Given that the reaction exclusively involved spin-zero particles, the states populated are restricted to possessing natural parity leading to a spin-parity assignment of $I^\pi = 5^-$. This state has now also been observed in a second study exploiting ^{3}He inelastic scattering [3] (see later in this section). Calculations made using the algebraic cluster model, predict a characteristic sequence of states comprising, with increasing excitation energy, 0^+, 2^+, 3^-, $4^\pm$ and 5^-. In this model, the three α particles lie at the corners of an equilateral triangle; $\mathcal{D}_{3h}$ symmetry. The model allows for rotation about various axes as well as expansion and contraction. Other key features are softness (making nuclei more like liquids rather than rigid structures), a near equality of rotational and vibrational energies (such that it is difficult to distinguish between these two types of motion) and non-point-like identical constituents with a spatial extent comparable in size to the overall structure [4]. The energies predicted by the algebraic model for ^{12}C are shown in figure 2 together with the corresponding experimentally measured states. The 0^+ and 2^+ excitations correspond to rotations about an axis lying in the plane of the triangle with rotations about an axis perpendicular to the plane contributing to the higher lying spins and parities. The juxtaposed 4^- and 4^+ states together with the newly observed 5^- state is crucial evidence in favour of $\mathcal{D}_{3h}$ symmetry. See also [5].

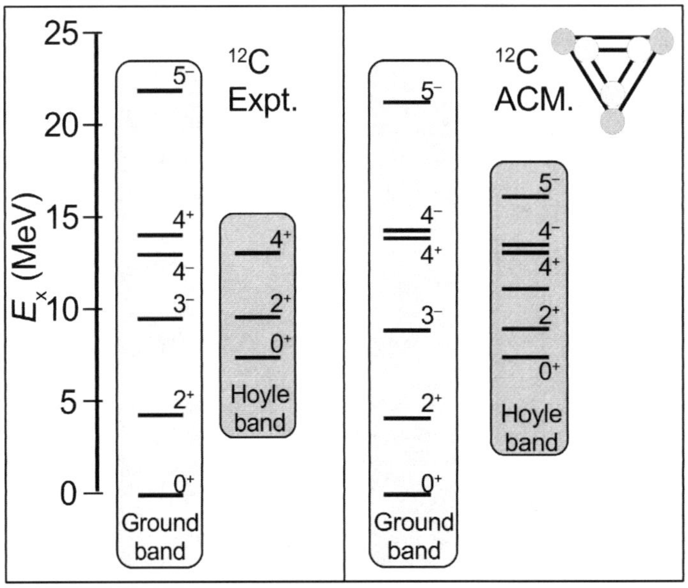

Figure 2. Left: experimentally observed states currently assigned to the ground- and Hoyle bands. Right panel: corresponding predicted energies in ^{12}C using the algebraic cluster model (see text for details) and performed by R. Bijker for [1].

In the process of obtaining evermore precise data on the higher-energy domain of ^{12}C it is not just the $I^\pi = 5^-$ state that has been discovered in recent experimental work. The study that independently confirmed the existence of the 22.4 MeV resonance via the ^{12}C(^{3}He,^{3}He)^{12}C* break-up reaction at 46 MeV [3], additionally found evidence for a series of states at 16.3, 17.2, 18.4, 19.7 and 25.1 MeV. These can be seen in figure 3. These states all decay via the ^{8}Be ground state and some of the higher-lying resonances may be related to excitations of the Hoyle state, beyond what is currently established (see section 3). Despite ^{12}C being studied extensively, such 5^- and other newly observed resonances demonstrate the importance of understanding structures with precision to the excitation-energy limit.

3. Hoyle-state excitations

Returning to the calculations presented in figure 2, it can been seen that the algebraic model not only makes predictions for the ground-state band described in section 2, but also describes the Hoyle state and its excitations as a symmetric stretching vibration of the equilateral triangle. This is at odds with results from state-of-the-art chiral effective field theory (EFT) calculations on the lattice which predict a bend-arm or open triangle configuration [6]. While both the EFT lattice and algebraic models agree on the triangular structure of the ground-state band, the calculations diverge for excited structures. It is here where precision experimental results can play a pivotal discriminating role.

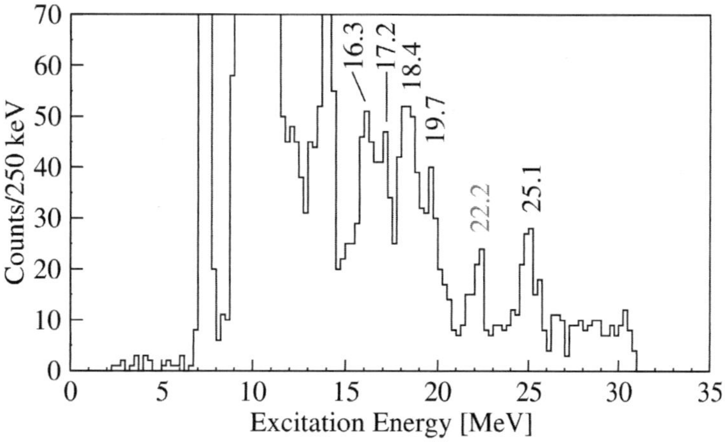

Figure 3. Carbon-12 high-excitation energy regime populated following break-up induced by ^{3}He inelastic scattering. The states shown decay via the ^{8}Be ground-state. The energies of the newly observed states are labelled in MeV in black and the 5^- in red. Intense, known excitations at lower energies are not labelled. Adapted from figure 5 of [3].

Recently several experimental advances have been made in identifying excited states built on the Hoyle state following a concerted international effort using a variety of techniques including break-up reactions and photo-disintegration via Compton backscattered photons. Beginning in 2004 by Itoh *et al.* [7] and later with Freer and coworkers [8] using inelastic proton scattering from carbon at the K600 spectrometer at iThemba Labs, South Africa, evidence for a possible 2^+ excitation of the Hoyle state began to emerge. This culminated in the confirmation by several separate studies [9, 10, 11] and convergence on the properties of the newly discovered resonance; $I^\pi = 2^+$, $E_x = 9.8 - 10.0$ MeV and Γ=800 keV [10, 11]. The proximity of the 2_2^+ to the well known 3_1^- state at 9.641 MeV that typically dominates this excitation energy region has led to a more precise characterisation of the properties of the 3^- state in order to fully understand the various contributions at each point. Making use of the proton inelastic scattering data with an experimental resolution of 23(1) keV, an R-matrix fit to the data yields a consistent width of 48(2) keV [12] at laboratory scattering angles of 10°, 16° and 28°. This width is significantly larger than the literature value of 34(5) keV [13].

Furthermore, in 2011, Freer *et al.* [14] obtained evidence using the two break-up reactions ^{12}C(^{4}He,^{4}He+^{4}He+^{4}He)^{4}He and ^{9}Be(^{4}He, ^{4}He+^{4}He+^{4}He)n for a possible 4^+ state – postulated to be the second excitation of the Hoyle state – lying at 13.3 MeV with a width of 1.7 MeV. This has been further elucidated by Marín-Lámbarri *et al.* [2], once again using angular correlations to confirm the $I^\pi = 4^+$ assignment.

With these new data, an early predictions for the underlying structure of the Hoyle state can be ruled out, specifically, a linear chain of three-α particles. This is by a simple calculation of the moment-of-inertia and relating this to the energy separation of the Hoyle state and the 2_2^+ state, which is now known to be $\approx$2.2 MeV. However, without further information, distinguishing between other predictions is challenging. For example, as mentioned earlier the chiral EFT lattice calculations predict a bent-arm structure and excitations of such a structure are not inconsistent with the present data. However, another theoretical approach treats the Hoyle state as a condensate of α particles all sharing a common s-orbit wavefunction [15]. In this model angular momentum is generated via exciting one of the α-particle bosons to a d-wave orbital [16, 17]. Again, the current data do not rule out such a scenario and are also consistent with the algebraic stretching mode discussed at the beginning of this section. What is missing to differentiate between the model predictions are electromagnetic transition rates linking the band members. Firstly, measuring enhanced γ decay rates would verify the common structural lineage and secondly, the transition rates constitute precision data to challenge the nuclear models. The difficulties lie in reliably measuring the anticipated 10^{-5} -10^{-7} branching ratios for such internal transitions. Such investigations are not without precedent having been performed by Datar and coworkers for the 4^+ to 2^+ transition in ^{8}Be [18] – a branching ratio of 1.5×10^{-7}.

With so much activity, both theoretical and experimental, the current status of our knowledge of the Hoyle state has recently undergone a timely review by Freer and Fynbo [19].

4. Heavier systems

The discussion so far has concentrated on the ^{12}C system, and the Hoyle state in particular, which although an important – and undoubtedly the most studied – multi-α state, it is expected to be just one example with similar levels predicted in the heavier α-conjugate nuclei ^{16}O, ^{20}Ne *etc.* Indeed, there is long-standing evidence for a linear-chain of four α-particles in ^{16}O as evidenced by ^{8}Be–^{8}Be structures reported by Chevallier *et al.* [20]. Using the ^{12}C(^{4}He,^{8}Be)^{8}Be reaction, evidence for a rigid linear chain emerged from angular distributions for the observed resonances indicating spin and parity assignments of $I^\pi = 2^+$, 2^+, 4^+ and 6^+ at $E_x = 16.95$, 17.15, 18.05, and 19.35 MeV, respectively. To search for further evidence, *e.g.* an associated 8^+ above 20 MeV, Curtis *et al.* [21] used the same reaction at Notre Dame, USA, measuring the excitation function from 12.2 to 20.0 MeV in steps as small as 10 keV. By reconstructing the kinematics of the break-up events, selections could be made for those originating from two ^{8}Be nuclei, or conversely, events that decayed via ^{12}C* + α. The resulting two excitation functions are shown in figure 4.

In order to extract spectroscopic information the ^{8}Be centre-of-mass angular distributions were fitted with

$$W(\theta) = \left| \sum_{L=0}^{L_{max}} \rho_L e^{i\phi_L} P_L(cos(\theta)) \right|^2 , \tag{1}$$

where ρ_L is the amplitude, ϕ_L the phase, P_L Legendre polynomials and fitting was performed for $L_{max} = 2$, 4, 6 and 8. However, the results of this phase-shift analysis not only found no evidence of a ^{8}Be–^{8}Be 8^+ structure, but also did not confirm the presence of the previously measured 6^+ state, effectively ruling out the existence of a four-α linear chain band in this region of the ^{16}O energy spectrum.

The search for a chain state in ^{16}O was conducted in the <24 MeV region as the four-α threshold lies at 14.44 MeV. For the next α conjugate nucleus, ^{20}Ne, the analogous five-α threshold lies at 19.17 MeV, delineating the region to look for Hoyle cognate states. An α-particle beam delivered by the Notre Dame tandem was once again used, but this time to explore the resonant ^{16}O(^{4}He,^{20}Ne*) break-up reaction cross-section as a function of excitation energy. The analysis probed both the ^{8}Be–^{8}Be–α and ^{12}C*–^{8}Be channels, the latter being a sub-set of the former in this case. The excitation functions are shown in figure 5 and for the ^{8}Be–^{8}Be–α channel (left panel) exhibit significant strength above 25 MeV. These results represent the first data on these channels in high excitation energies in ^{20}Ne and offer a tantalising glimpse into where to look for Hoyle counterparts in this nucleus.

5. Experimental challenges

These recent studies in ^{16}O and ^{20}Ne have focused on the emission of heavy clusters – *i.e.* ^{8}Be and ^{12}C* – as opposed to α-particle emission. However, the question remains, as for the ^{12}C system in section 3, about how to distinguish between a classically clustered state and a condensed gas-like state of α particles?

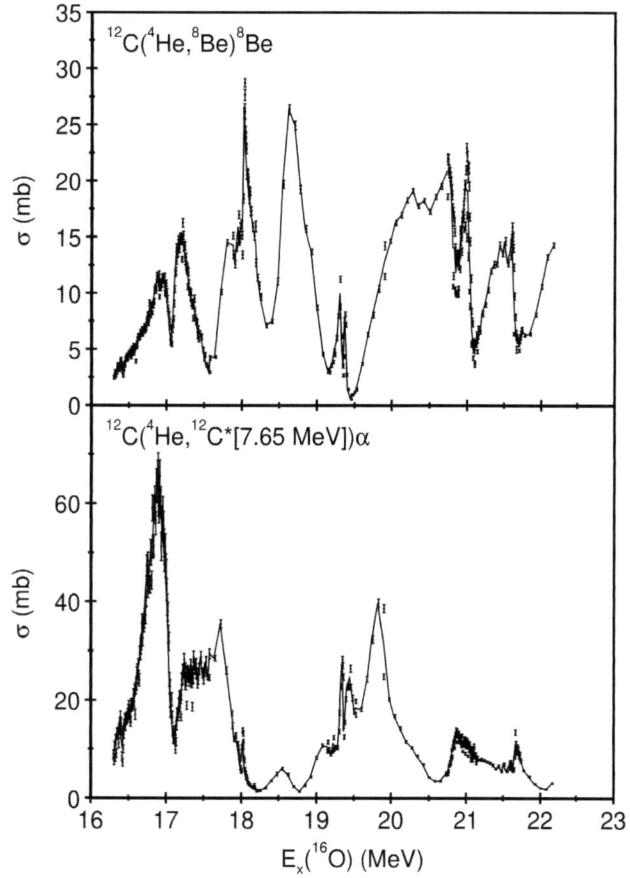

Figure 4. Excitation functions for ^{16}O [22]. (Top panel): for states decaying via the ^{8}Be–^{8}Be channel; (bottom panel): for states breaking up into ^{12}C* + α.

One such method, proposed in [24], is to search for a signature of these exotic gas-like states in reactions involving heavier nuclei [25] and compare the strengths of all the decay channel partitions. However, to do this requires a special experimental set-up the capabilities of which needs be: observation of the total kinetic energy distribution of the emitted particles (down to very low energies); comparison to known compact states and the ability to establish a shift due to Coulomb barrier differences with such a state; observation of high multiplicity α emission; extraction of decay probabilities for all particle partitions, *i.e.* a search for enhanced coherent multi-α-particle emission *e.g.* ^{8}Be, ^{12}C* .... Finally, to confirm the population of a condensate rather than a clustered state would require the analysis to show that the decay probabilities are equal for all partitions.

The continued advances in particle detection mean that such experiments, though challenging, can now be designed

6. Summary

Clustered states and their intriguing properties, still, after many decades of study, afford a unique insight into the manifestation of the nuclear force in many-body systems. The data on ^{12}C has reached a level of precision where many of the outstanding questions are on the cusp of being resolved, both from an experimental and theoretical outlook. To resolve the issue of Hoyle analogue states the focus of study will need to shift to heavier α conjugate nuclei and the

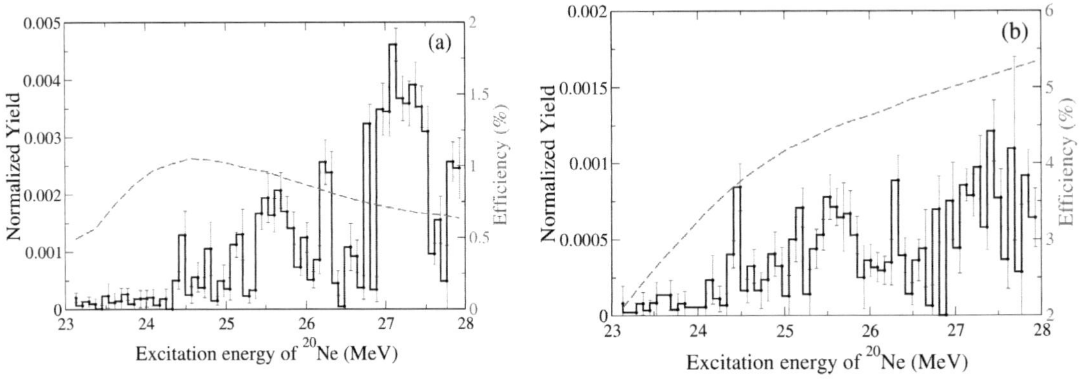

Figure 5. Excitation functions for ^{20}Ne for events breaking up into (a) ^{8}Be–^{8}Be–α and (b) ^{12}C*–^{8}Be. Adapted from figure 3 of [23]. The red dashed lines (corresponding to the right axes) demonstrate the change in detector efficiency for the two processes, obtained from Monte-Carlo simulations.

initial steps have been taken in ^{20}Ne, while activity on ^{16}O is considerably advanced. One of the most stimulating areas for the future looks set to be the possibilities of more exotic states of matter such as α-particle condensates or signatures for such states around heavy cores. The next SOTANCP workshop may well be the forum where matters related to condensation take centre stage.

References

[1] Marín-Lámbarri D J, Bijker R, Freer M, Gai M, Kokalova Tz, Parker D J and Wheldon C 2014 *Phys. Rev. Lett.* **113** 012502
[2] Marín-Lámbarri D J, Freer M, Kokalova Tz, Wheldon C and Parker D J 2014 *J. Phys. Conf. these proceedings*
[3] Wheldon C, Kokalova Tz, Freer M, Glenn A, Parker D J, Roberts T and Walmsley I, 2014 *Phys. Rev. C* **90** 014319
[4] Bijker R and Iachello F 2002 *Ann. Phys.* **298** 334 and 2000 *Phys. Rev.* C **61** 067305
[5] Gai M, Bijker R, Freer M, Kokalova Tz, Marín-Lámbarri D J and Wheldon C 2014 *J. Phys. Conf. these proceedings*
[6] Epelbaum E, Krebs H, Lähde T A, Lee D and Meißner U-G 2012 *Phys Rev. Lett.*
[7] Itoh M *et al.* 2004 *Nucl. Phys.* A **738** 268
[8] Freer M *et al.* 2009 *Phys. Rev.* C **80** 041303(R)
[9] Itoh M *et al.* 2011 *Phys. Rev.* C **84** 054308
[10] Freer M *et al.* 2012 *Phys. Rev.* C **86** 034320
[11] Zimmerman W R *et al.* 2013 *Phys. Rev. Lett.* **110** 152502
[12] Kokalova Tz *et al.* 2013 *Phys. Rev.* C **87** 057307
[13] Ajzenberg-Selove F 1990 *Nucl. Phys.* A **506** 1
[14] Freer M *et al.* 2011 *Phys. Rev.* C **83** 034314
[15] Tohsaki A, Horiuchi H, Schuck P, and Röpke G, 2001 *Phys. Rev. Lett.* **87** 192501
[16] Funaki Y, Tohsaki A, Horiuchi H, Schuck P and Röpke G 2005 *Eur. Phys. J.* A **24** 321
[17] Yamada T and Schuck P 2005 *Eur. Phys. J.* A **26** 185
[18] Datar V M, Kumar S, Chakrabarty D R, Nanal V, Mirgule E T, Mitra A and Oza H H 2005 *Phys. Rev. Lett.* **94** 122502
[19] Freer M and Fynbo H O U 2014 *Prog. in Part. and Nucl. Phys.* in press (doi:10.1016j.ppnp.2014.06.001)
[20] Chevallier P, Scheibling F, Goldring G, Plesser I and Sachs M W 1967 *Phys. Rev.* **160** 827
[21] Curtis N *et al.* 2013 *Phys. Rev.* C **88** 064309
[22] Curtis N private communication
[23] Kokalova Tz *et al.* 2013 *Phys. Rev.* C **87** 057309
[24] Kokalova Tz, Itagaki N, von Oertzen W and Wheldon C 2006 *Phys. Rev. Lett.* **96** 192502
[25] Kokalova Tz *et al.* 2005 *Eur. Phys. J.* A **23** 19

Journal of Physics: Conference Series **569** (2014) 012011 doi:10.1088/1742-6596/569/1/012011

The Rotation-Vibration Structure of ^{12}C

M Gai[1], R Bijker[2], M Freer[3], T Kokalova[3], D J Marin-Lambarri[3], C Wheldon[3]

1. LNS at Avery Point,University of Connecticut , Groton, CT 06340-6097, USA and WNSL, Dept of Physics, Yale University, New Haven, CT 06520-8124
2. Instituto de Ciencias Nucleares, Universidad Nacional Autonoma de Mexico, A.P. 70-543, 04510 Mexico, D.F., Mexico
3. School of Physics and Astronomy , University of Birmingham , Birmingham B15 2TT, UK

E-mail: moshe.gai@yale.edu

Abstract. The newly measured high spin $J^\pi = 5^-$ state at 22.4(2) MeV in ^{12}C reported in this conference, fits very well to the predicted (ground state) rotational band of an oblate equilateral triangular spinning top with a $\mathcal{D}_{3h}$ symmetry characterized by the sequence of states: $0^+, 2^+, 3^-, 4^\pm, 5^-$ with almost degenerate 4^+ and 4^- (parity doublet) states. Such a $\mathcal{D}_{3h}$ symmetry was observed in triatomic molecules, and it is observed here for the first time in nuclear physics. We discuss a classification of other rotation-vibration bands in ^{12}C such as the (0^+) Hoyle band and the (1^-) bending mode band and suggest measurements in search of the predicted ("missing") states that may shed new light on clustering in ^{12}C and light nuclei. In particular, the observation (or non observation) of the predicted ("missing") states in the Hoyle band will allow us to conclude the geometrical arrangement of the three alpha particles composing the Hoyle state at 7.654 MeV in ^{12}C.

1. Introduction

The recent observation of the 2^+ Hoyle rotational excitation in ^{12}C [1] allows for the first time the study of the Rotation-Vibration Structure of ^{12}C. These new data and the recently discovered 4^- [2, 3] and 4^+ [4] states in ^{12}C are in agreement with the predicted spectrum of an oblate spinning top with a $\mathcal{D}_{3h}$ symmetry [5, 6]. It was predicted [5, 6] that the three alpha-particle system of ^{12}C leads to the ground state rotational band including the most unusual sequence of states: $0^+, 2^+, 3^-, 4^\pm, 5^-$. The new high spin 5^- state reported by the Birmingham group in this conference [7] as well the previously published 4^- state [2, 3] lead to a J(J+1) trajectory as predicted by this U(7) model [8, 9] including the nearly degenerated 4^- and 4^+ states as shown in Fig. 1.

2. The Algebraic Cluster Model

The spectrum of ^{12}C predicted by the Algebraic Cluster Model (ACM) [5, 6] is shown in Fig. 2 where it is also compared to the measured spectrum of ^{12}C [8]. In addition to the ground state rotational band this U(7) model [5, 6] predicts the Hoyle state at 7.65 MeV in ^{12}C to be the first vibrational breathing mode of the three alpha-particle equilateral configuration leading to the same rotational structure albeit with a larger moment of inertia (by a factor of 2). Recent measurements revealed the 2^+ [1] and 4^+ [4] members of the predicted Hoyle rotational band and we are currently searching [10] for the 4^- predicted by the ACM to be nearly degenerated

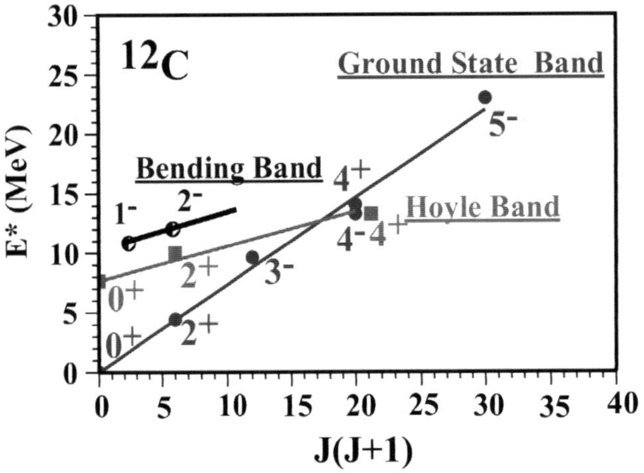

Figure 1. Rotational band structure of the: ground-state band, the Hoyle band and the bending vibration band in ^{12}C.

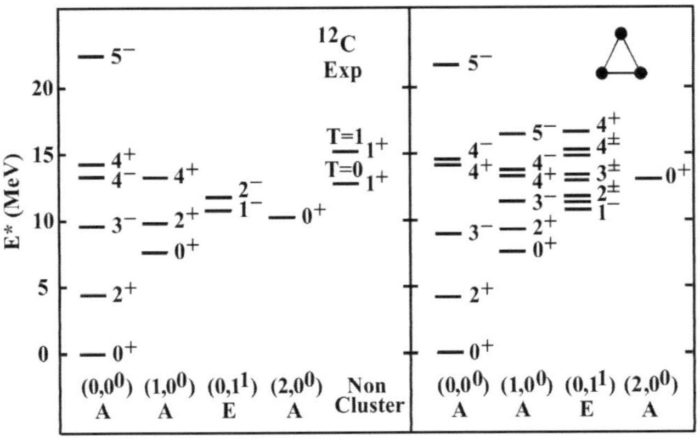

Figure 2. Comparison between the low-lying experimental spectrum of ^{12}C and the energies of the oblate symmetric top. The levels are organized in columns corresponding to the ground state band and the vibrational bands with A and E symmetry [5, 6] of an oblate top with triangular symmetry. The last column on the left-hand side, shows the lowest observed non-cluster (1^+) levels.

with the 4^+ state and the 3^- (broad) state that was suggested to lie between 11 and 14 MeV [2]. The observation (or lack there) of these "missing" states will allow us to determine wether the Hoyle state is composed of three alpha-particles in an equilateral triangle arrangement [8, 11, 12] or an obtuse triangle [13] or whether it is better described as vibrational excitation of a "diffuse gas" of three alpha-particles [14].

The U(7) model also predicts the 1^- state at 10.84 MeV to be the vibrational bending mode with a rotational band including the 1^- and a degenerate $2^{\pm}$ states. We are searching [10] for the third 2^+ of ^{12}C that is predicted by the U(7) model to lie near the observed 2^- state at 11.8 MeV.

In conclusion the ACM appears to open a new chapter in cluster physics of light nuclei and it presents an opportunity for further experimental investigation of light nuclei.

3. Acknowledgments
This material is based upon work supported by the U.S. Department of Energy, Office of Science, Office of Nuclear Physics, under Award Number DE-FG02-94ER40870.

References
[1] Itoh M *et al.* 2004 Nucl. Phys. **A738**, 268; Itoh M *et al.* 2011 Phys. Rev. C **84**, 054308; Freer M *et al.* 2009 Phys. Rev. C **80**, 041303(R); Zimmerman W R *et al.* 2011 Phys. Rev. C **84**, 027304(BR); Zimmerman W R *et al.* 2013 Phys. Rev. Lett. **110**, 152502
[2] Freer M *et al.* 2007 Phys. Rev. C **76**, 034320 (2007)
[3] Kirsebom O S *et al.* 2010 Phys. Rev. C **81**, 064313
[4] Freer M *et al.* 2011 Phys. Rev. C **83**, 034314
[5] Bijker B and Iachello F 2000 Phys. Rev. C **61**, 067305
[6] Bijker R and Iachello F 2002 Ann. Phys. **298**, 334
[7] Marin-Lambarri D J 2014 Invited talk, SOTANCP3, Yokohama, Japan, May 26-30
[8] Marin-Lambarri D J, Bijker R, Freer M, Gai M, Kokalova Tz, C. Wheldon, 2014 Phys. Rev. Lett. **113**, 012502.
[9] Popular press of Ref. [8]: Kevin Dusling, Physics, Synopsis:
http://physics.aps.org/synopsis-for/10.1103/PhysRevLett.113.012502,
Hamish Johnston, Physicsworld.com: http://physicsworld.com/cws/article/news/2014/jul/08/carbon-nu.
[10] A 12C(e,e') experiment in progress at Darmstadt, P. von Neumann-Cosel, private communication, 2013.
[11] Chernykh M, Feldmeier H, Neff T, von Neumann-Cosel P, and Richter A 2007 Phys. Rev. Lett. **98**, 032501
[12] Yoshiko Kanada-Enyo 2007 Prog. of Theo. Phys. **117**, 655
[13] Epelbaum E, Krebs H, Lee D, Meissner Ulf-G 2012 Phys. Rev. Lett. **106**, 192501; Epelbaum E, Krebs H, Lahde T, Lee D, and Meissner Ulf-G 2012 Phys. Rev. Lett. **109**, 252501
[14] Funaki Y, Horiuchi H, von Oertzen W, Ropke G, Schuck P, Tohsaki A, and Yamada T 2009 Phys. Rev. C **80**, 64326 and references therein

Journal of Physics: Conference Series **569** (2014) 012012 doi:10.1088/1742-6596/569/1/012012

Theory for Quartet Condensation in Fermi Systems with Applications to Nuclear Matter[1]

P Schuck[1 2 3]

[1]Institut de Physique Nucléaire, CNRS, UMR8608, Orsay, F-91406, France
[2]Université Paris-Sud, Orsay, F-91505, France
[3]Laboratoire de Physique et de Modélisation des Milieux Condensés, CNRS et, Université Joseph Fourier, 25 Av. des Martyrs, BP 166, F-38042 Grenoble Cedex 9, France

Abstract.
α clustering and α condensation in lighter nuclei is presently strongly and increasingly discussed in the literature both from the experimental side as from the theoretical one. In proto-neutron stars a macroscopic condensate of α particles may occur. A discussion of the present status of the theory for quartet condensation in general and for α particle condensation in nuclear matter in particular will be presented.

Introduction. Since 2001, when the idea of a possible existence of α condensate type of states in $n\alpha$ nuclei was advanced and formulated for the first time [1], many exciting new results, theoretical and experimental ones, have been produced. In this contribution, we would like to dwell on theoretical aspects of quartet condensation in general and in particular in nuclear matter.

Let us start with the reminder that nuclear clustering and in particular α clustering would not exist, if we did not have in nuclear physics *four* different types of fermions (proton/neutron spin up/down), all attracting one another. We should be aware of the fact that this is a rather singular situation in fermionic many body systems. However, the possibility of future trapping of four different kinds of cold fermionic atoms, may open a new field of cluster physics with similar features as in nuclei. Also the possibility of the condensation of bi-excitons in semiconductors may be a field of interest in this context. In a mean field description of an isolated α particle (what with, e.g., Skyrme forces gives reasonable results, if the c.o.m. motion is treated correctly) the four fermions can occupy the lowest 0S-level. Would there be only neutrons, only two of them can be in the 0S-level, the other two neutrons would have to be in the energetically very penalising 0P-state. That is why α particles exist, tetra-neutrons not. The ensuing fact is that α particles are very strongly bound ($E/A \sim 7$ MeV), almost as strong as the most bound nucleus which is ^{56}Fe ($E/A \sim 8$ MeV). In addition the first excited state of the α particle (~ 20 MeV) is by factors higher than that of any other nucleus. The α particle can, therefore, be considered as an almost inert ideal bosonic particle. As we will see in the discussion below, inspite of its strong binding, α particle condensation can only exist in the so-called BEC (Bose-Einstein

[1] This work is part of a larger collaboration on nuclear clusters with Y. Funaki, H. Horiuchi, Z. Ren, G. Roepke, A. Tohsaki, Chang Xu, T. Yamada, and Bo Zhou

Condensation) phase what implies low density. There is no analogue to the BCS phase of pairing where the Cooper pairs can have very large extensions, strongly overlapping with one another, still being fully antisymmetrised. This is the reason why α condensation only can be present at low densities where the α particles do not overlap strongly (this holds, if the system consists of protons and neutrons and α's. If other clusters as t, ^{3}He,d are around, the situation may change, see below). These considerations apply for nuclear matter as well as for finite nuclei. The Hoyle state in ^{12}C which can to a good approximation be described as a product of three α particles occupying all the lowest 0S state of their bosonic mean field has a density which is by a factor 3-4 lower than the one of the ground state of ^{12}C. In the ground state there exist α-type of correlations but there is no condensation phenomenon. Here, we want to focus our considerations on infinite matter.

α particle condensation in infinite matter. The in medium four-body equation can be written in the following form [2]

$$(E_{\alpha,\mathbf{K}} - \varepsilon_1 - \varepsilon_2 - \varepsilon_3 - \varepsilon_4)\Psi^{\alpha,\mathbf{K}}_{1234} = (1 - f_1 - f_2)v_{121'2'}\Psi^{\alpha,\mathbf{K}}_{1'2'34} + (1 - f_1 - f_3)v_{131'3'}\Psi^{\alpha,\mathbf{K}}_{1'23'4} + \ldots \quad (1)$$

In total, there are six terms coming from permutations. The ε_i are kinetic energies plus mean field corrections; v_{1234} are the matrixelements of the two body interaction, and f_i is a Fermi-Dirac distribution of the uncorrelated nucleons accounting for phase space blocking. Repeated indices are summed over and index numbers comprise momenta and spins. The above equation considers, therefore, *one* quartet in a gas of uncorrelated nucleons at temperature T. The analogous two body equation can be used to determine the critical temperature T_c for the onset of superfluidity or supraconductivity where T_c has to be determined so that the eigenvalue comes at two times the chemical potential μ. This is the famous Thouless criterion of BCS theory. In analogy with pairing, one has to find the critical temperature T_c^α so that the eigenvalue of the four body equation (1) comes at 4μ for a quartet at rest. The in medium four body equation is very difficult to solve. Nontheless, the solution has been found employing the Faddeev-Yakubovsky equations and using the Malfliet-Tjohn bare nucleon-nucleon interaction which yields realistic nucleon-nucleon phase shifts and properties of an isolated α particle [3]. To simplify the problem, we made in addition a very easy to handle variational ansatz of the four body wave function in (1). It consists of a mean field ansatz for the α particle projected on good total momentum. In momentum space this is (the singlet spin-isospin part is suppressed)

$$\Psi_{1234} \propto \delta(\mathbf{K} - \mathbf{k}_1 - \mathbf{k}_2 - \mathbf{k}_3 - \mathbf{k}_4)\varphi(\mathbf{k}_1)\varphi(\mathbf{k}_2)\varphi(\mathbf{k}_3)\varphi(\mathbf{k}_4) \quad (2)$$

Inserting this ansatz into (1), one obtains a nonlinear HF-type of equation for the S-wave function $\varphi(\mathbf{k})$. Of course, for quartet condensation, we choose $\mathbf{K} = 0$. With the mean field ansatz (2), one cannot use a bare force. We adjusted an effective separable force with two parameters which are chosen to reproduce binding energy and radius of the free α particle. The full Faddeev-Yakubovsky solution of (1) is shown for symmetric matter in Fig. 1 (crosses). We see that the ansatz (2) which very much eases the otherwise difficult solution of (1) works very well (continuous line). Also shown is the critical temperature for deuteron condensation. The striking feature is that α particle condensation abruptly breaks down already at very low density which approximately coincides with the point where the α's start to overlap appreciately (this fact was already found in [2] using a somewhat different variational ansatz for Ψ_{1234}). On the other hand deuteron condensation goes on up to very high densities and the limit is only triggered by the range of the effective force (which has been readjusted to reproduce the deuteron properties). This is so for symmetric nuclear matter. For strong asymmetry, deuteron condensation breaks down earlier than α condensation because the α particle according to its much stronger binding

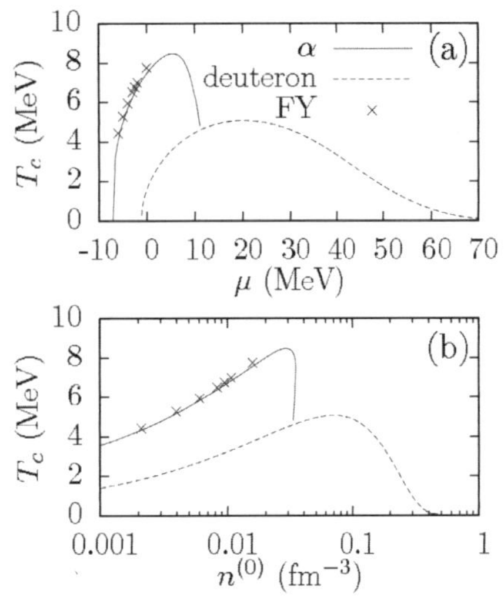

Figure 1. Critical temperatures for α particle and deuteron condensation in symmetric nuclear matter as a function of μ (a) and density $n^{(0)}$ (b).

is less sensitive to asymmetry [4]. The above features describe the afore mentioned phenomenon that α condensation only exists in the BEC phase, i.e., at low density, whereas deuteron continuously goes from negative to positive chemical potentials where for the latter the deuterons turn into large size Cooper pairs. More on this can be found in [2] [3]. We should mention that our calculation of T_c^α is only reliable rather close to the break down point, i.e., to the maximum. For lower densities, the T_c^α should join the one for condensation of ideal bosons (α's). To describe this feature, one should extend our theory to the so-called Nozières Schmitt-Rink (NSR) theory [5] for pairing, see also [6], to α particle condensation. This, however, has not been worked out so far and remains a task for the future.

At zero temperature, there are many α's which go into the condensate phase. For this, we have to set up an approach analogous to the nonlinear BCS theory. Equation(1) corresponds to the linearised version and only describes *one* α particle in an otherwise uncorrelated gas (at finite T) of fermions. In finite nuclei, there may exist such a situation even at zero temperature. This is the case of ^{212}Po which can to a certain extent be viewed as an α particle sitting on top of the doubly magic core of ^{208}Pb which can be well described by a HF-mean field approach, i.e. a Fermi gas in a container, see [7].

After having considered the linearised version of the equation for the quartet order parameter at the critical temperature, let us now try to write down, in analogy to the BCS case, the fully non-linear system of equations for the quartet order parameter. To clearly see the analogy to the BCS case, let us repeat the latter equations in a slightly unusual way. The pairing order parameter allowing for non-zero c.o.m. momentum of the pairs, obeys the equation

$$(\varepsilon_{k_1} + \varepsilon_{k_2})\kappa_{\mathbf{k}_1\mathbf{k}_2} + (1 - n_{k_1} - n_{k_2})\Delta_{\mathbf{k}_1\mathbf{k}_2} = 2\mu\kappa_{\mathbf{k}_1\mathbf{k}_2} \tag{3}$$

where $\Delta_{\mathbf{k}_1\mathbf{k}_2} = \sum v_{k_1k_2k_{1'}k_{2'}}\kappa_{\mathbf{k}_{1'}\mathbf{k}_{2'}}$ and $\kappa_{\mathbf{k}_1\mathbf{k}_2} = \langle \mathrm{BCS}|c_{\mathbf{k}_1}c_{\mathbf{k}_2}|\mathrm{BCS}\rangle = u_{\mathbf{k}_1}v_{\mathbf{k}_2}$ is the usual pairing tensor (with spin indices suppressed) and $n_k = v_k^2 = 1 - u_k^2$ are the BCS occupation numbers. The ε_k's are, as before, the kinetic energies, eventually including a mean field correction. The occupation numbers can be obtained from the Dyson equation for the single particle Green's function (the latter is derived from the Gorkov equation in eliminating the anomalous Green's function)

$$G_k^\omega = G_k^0 + G_k^0 M_k^\omega G_k^\omega \quad \text{with} \quad M_k = \frac{\Delta_k \Delta_k^*}{\omega + \varepsilon_k} \tag{4}$$

the BCS mass operator where Δ_k is the diagonal part of the gap for cases where the pairs are at rest. From the single particle Green's function, obviously we can calculate the occupation numbers closing, thus, the typical BCS selfconsistency cycle. Inspired by the BCS case, we then write for the quartet order parameter, see (1)

$$(4\mu - \varepsilon_1 - \varepsilon_2 - \varepsilon_3 - \varepsilon_4)\kappa_{1234} = (1 - n_1 - n_2)v_{121'2'}\kappa_{1'2'34} + (1 - n_1 - n_3)v_{131'3'}\kappa_{1'23'4} + ... \tag{5}$$

with $\kappa_{1234} = \langle c_1 c_2 c_3 c_4 \rangle$. Again, we have to close the equation with the Dyson equation for the single particle Green's function to obtain the occupation numbers. However, the single particle massoperator now contains the quartet order parameter

$$M_1^\alpha = \sum_{234} \frac{\Delta_{1234}^{(13)}[\bar{n}_2^0\bar{n}_3^0\bar{n}_4^0 + n_2^0 n_3^0 n_4^0]\Delta_{1234}^{(13)*}}{\omega + \varepsilon_2 + \varepsilon_3 + \varepsilon_4} \delta(\mathbf{k}_1 + \mathbf{k}_2 + \mathbf{k}_3 + \mathbf{k}_4) \tag{6}$$

with $\bar{n}^0 = 1 - n^0$ and n_i^0 being the uncorrelated Fermi function, i.e., the Fermi step and $\Delta_{1234}^{(13)} = \Delta_{1234} - [(k_3 \leftrightarrow k_4) - (k_2 \leftrightarrow k_3)]$ where $\Delta_{1234} = \sum v_{123'4'}\kappa_{3'4'34}$.
Let us give some hints how to derive such a mass operator. The first observation is that the following non-linear destruction operator

$$q_\nu = u_{k_1}^\nu c_{k_1} - \frac{1}{3!} \sum v_{k_2k_3k_4}^\nu c_{k_2}^+ c_{k_3}^+ c_{k_4}^+ \tag{7}$$

kills the following quartet coherent state

$$|Z\rangle = e^{\frac{1}{4!}\sum_{k_1k_2k_3k_4} Z_{k_1k_2k_3k_4} c_{k_1}^+ c_{k_2}^+ c_{k_3}^+ c_{k_4}^+}|\text{vac}\rangle \quad \text{with} \quad Z_{k_1k_2k_3k_4} = \sum_\nu (u^{-1})_{k_1}^\nu v_{k_2k_3k_4}^\nu \tag{8}$$

Minimising the average single particle energy $E_\nu = \langle Z|\{q_\nu, [H - \mu\hat{N}, q_\nu^+]\}|Z\rangle / \langle Z|\{q_\nu, q_\nu^+\}|Z\rangle$, where $\{.,.\}$ is the anti-commutator, with respect to the amplitudes u^ν, v^ν, leads to a two by two eigenvalue problem for those amplitudes. Eliminating the v^ν amplitudes leads to an effective single particle equation where the energy dependent potential can be identified with the above mass operator performing some lowest order approximation for the three hole propagator.

Before trying to solve the system of equations, let us discuss the differences between the pairing and the quartet case. The first thing which strikes is that the three 'holes' only have to have total momentum $\mathbf{k}_2 + \mathbf{k}_3 + \mathbf{k}_4 = -\mathbf{k}_1$ and, therefore, we have a remaining sum over momenta. In the pairing case with only one 'hole', there is no sum. Furthermore in the pairing case the hole propagator has no phase space factor because 'forward' and 'backward' going parts add up to one: $\bar{n}^0 + n^0 = 1$. In the quartet case there are three hole propagators and the corresponding sum of phase space factors does NOT add up to one, i.e. $\bar{n}_1^0\bar{n}_2^0\bar{n}_3^0 + n_1^0 n_2^0 n_3^0 \neq 1$!! This makes a dramatic difference with the pairing case. In order to understand this, let us compare the level density of a single hole with the one of three holes:

$$g^{1h}(\omega) = \sum_k [\bar{n}_k^0 + n_k^0]\delta(\omega + \varepsilon_k) = \sum_k \delta(\omega + \varepsilon_k) \tag{9}$$

$$g^{3h}(\omega) = \sum_{\mathbf{k}_1\mathbf{k}_2\mathbf{k}_3} [\bar{n}_{k_1}^0\bar{n}_{k_2}^0\bar{n}_{k_3}^0 + n_{k_1}^0 n_{k_2}^0 n_{k_3}^0]\delta(\omega + \varepsilon_{k_1} + \varepsilon_{k_2} + \varepsilon_{k_3}) \tag{10}$$

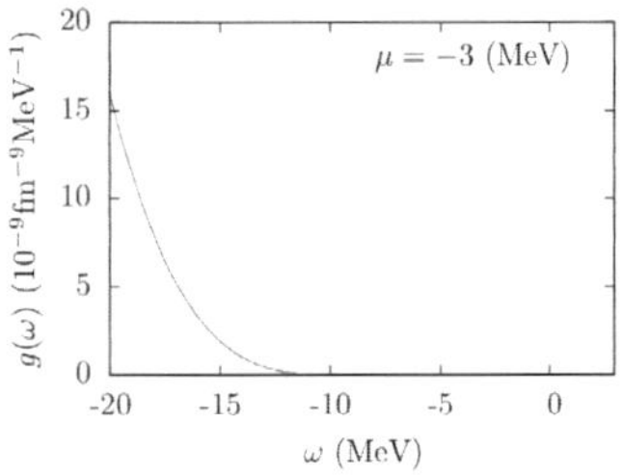

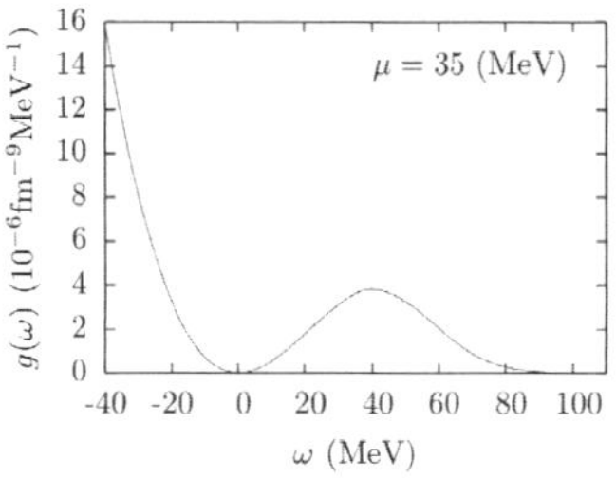

Figure 2. 3h level densities for negative and positive chemical potentials, respectively. Note that on the horizontal axis, the origine corresponds to $\mu = 0$.

The 3h level density is shown in Fig.2. We see that for positive μ there is a striking difference with the 1h level density. At positive μ, $g^{1h}(\omega = \mu)$ is obviously finite (not shown), whereas $g^{3h}(\omega = 3\mu)$ goes through zero (this remains true for any value of positive μ, even small). This is because phase space constraints and energy conservation cannot be fullfilled simultaneously at the Fermi surface in the latter case as is easily verified. That is, in the quartet case, exactly at the point where the correlations should be built up, namely at the Fermi-level, there is no level density! As a consequence, no quartet condensation is possible for positive μ. On the contrary for negative μ, $n^0 = 0$ and, thus, the phase space factor in the case of g^{3h} is also equal to one and then there is no qualitative difference with the $1h$ case. This explains in a natural way why quartet condensation is not possible at positive chemical potential.
In conclusion for positive μ only pairing survives whereas quartetting breaks down and only exists in the BEC phase with negative μ. The situation may be different when other light clusters are present, i.e. in a mixed gas of, e.g., nucleons, tritons (^{3}He), deuterons. Then a nucleon with momentum $\mathbf{k}$ may, eventually, directly pair with, e.g., a triton of momentum $-\mathbf{k}$ (or the other way round), rather similar to the standard pairing situation besides the fact that now two fermions with different masses have to pair up. Similar considerations hold for the pairing of two deuterons (pairing of 'bosons'). In compact star physics such situations may exist when the star is cooling down. The extension of our theory to this scenario is a task for the future. The full solution of the nonlinear set of equations (6) and (7) is again very much eased in taking for the order parameter the factorisation ansatz (2). The most interesting result is that the occupation numbers, e.g., for μ around zero is far from being close to saturation. At $k = 0$ it is only approximately $n_{k=0} \sim 0.30$. This scenario is analogous to pairing in the BEC regime. More results can be found in [8].

Remarks on Finite Nuclei. As we know from pairing, a direct observation of condensation phenomena only is possible in finite nuclei. Of course, in such small systems, there cannot exist a condensation in the macroscopic sense. Nevertheless, as we know very well, only a handfull of Cooper pairs suffices to show clear signatures of pairing in nuclei. For α particle condensation it is the same story. We also only can expect that there exists about a handfull of α particles, essentially in lighter $n\alpha$ nuclei, in a gaseous phase at low density. It is, indeed, surprising that such states at low density with about $\rho = \rho_0/3 - \rho_0/4$ with ρ_0 the density at saturation, indeed exist as excited quite long lived states in those nuclei. The most fameous example is the Hoyle state in ^{12}C at 7.65 MeV, just about 300 keV above the 3α threshold. We will not dwell much on the successful theoretical description of this state (and others, e.g., in ^{16}O) with the THSR wave function [1], since this has been presented in the recent literature a great number of times [9]. Let us only make a couple of remarks. The THSR wave function is schematically written

Journal of Physics: Conference Series **569** (2014) 012012 doi:10.1088/1742-6596/569/1/012012

for a finite number of quartets as

$$\Psi_{n\alpha}^{\mathrm{THSR}} = \mathcal{A}[\Phi_\alpha\Phi_\alpha....\Phi_\alpha] \tag{11}$$

where the single α wave function Φ_α depends on four spatial coordinates, the spin-isospin part being suppressed. This wave function is fully antisymmetric due to the antisymmetriser $\mathcal{A}$ and is analogous to the number projected BCS wave function $\Psi^{\mathrm{BCS}} = \mathcal{A}[\phi_{\mathrm{pair}}\phi_{\mathrm{pair}}....\phi_{\mathrm{pair}}]$ where ϕ_{pair} is the Cooper pair wave function depending on two spatial coordinates. The calculus with the α condensate wave function is very much facilitated with a variational ansatz where Φ_α is split into a product of a c.o.m. gaussian with a large width parameter B times another, intrinsic, gaussian depending only on the relative coordinates of the α particle and having a width parameter b which corresponds to the size of an isolated α particle. The first important remark to be made is that this THSR wave function contains two important limits: if $B = b$ then it corresponds to a pure harmonic oscillator Slater determinant. If $B \gg b$, then the α's are so distant from one another that the Pauli principle among the different α's can be neglected and, thus, the antisymmetriser be dropped. The THSR wave function is then a pure product state of α particles, i.e., a condensate state of ideal bosons. Reality is, of course, in between those limits and one important task is to find out whether reality is closer to a Slater determinant or to a Bose condensate. That this question must be carefully investigated, can also be deduced from the fact that a number projected BCS wave function always leads to a non trivial pairing solution. For example for ^{208}Pb, one obtains a non trivial BCS solution inspite of the fact that ^{208}Pb is certainly not superfluid. Only when the original particle number breaking BCS theory with $|\mathrm{BCS}\rangle = e^{\sum z_{kk'}c_k^+c_{-k}^+}|\mathrm{vac}\rangle$ has a non-trivial solution, we can speak of a superfluid nucleus. For ^{208}Pb BCS has no solution. One way to analyse whether the THSR approach leads primarily to an α condensate or to a Slater determinant, is to investigate the bosonic occupation numbers. It has been found that in the Hoyle state the α's are condensed to 70% [9], the remaining part being scattered out of the condensate due to the action of the Pauli principle.

Conclusions. As we have seen, the existence of α gas and α condensate states in nuclear systems where the α's play practically the role of elementary bosons, is fascinating. Nuclear physics is at the forefront of this kind of physics. In the future, experiments with cold atoms trapping four (or more) different kinds of fermions may also open wide perspectives in the field of cluster physics. For more reading on α cluster states, we recommend to consult our review article [9].

References

[1] Tohsaki A, Horiuchi H, Schuck P, and Röpke G 2001 *Phys. Rev. Lett.* **87**, 192501
[2] Röpke G, Schnell A, Schuck P, Nozières P 1998 ıPhys. Rev. Lett. **80**, 3177
[3] Sogo T, Lazauskas R, Roepke G, Schuck P 2009 *Phys. Rev.* C **79**, 051301
[4] Sogo T, Roepke G, Schuck P 2010 *Phys. Rev.* C **82**, 034322
[5] Nozières P, Schmitt-Rink S 1985 *J. Low Temp. Phys.* **59**, 195
[6] Jin M, Urban M, Schuck P 2010 *Phys. Rev. C* **82**, 024911
[7] Roepke G *et al.*, arXiv:1407.0510.
[8] Sogo T, Roepke G, Schuck P 2010 *Phys. Rev.* C **81**, 064310
[9] Yamada T, Funaki Y, Horiuchi H, Roepke G, Schuck P, Tohsaki A 2011 arXiv:1103.3940, "Clusters in Nuclei (Lecture Notes in Physics)-Vol.2-", ed. by C. Beck, Springer Verlag, Berlin

Applying new tools to old problems—experimental studies of resonances in ^{12}C

O S Kirsebom, H O U Fynbo, A M Howard and K L Laursen

Department of Physics and Astronomy, Aarhus University, 8000 Aarhus C, Denmark

E-mail: oliskir@phys.au.dk

Abstract. We report preliminary results from an experimental study of the $p + {}^{11}$B reaction at beam energies of 2.00 MeV, 2.63 MeV and 3.12 MeV, corresponding to three known 0^+, 2^+ and 3^- resonances in ^{12}C at excitation energies of 17.79 MeV, 18.38 MeV and 18.81 MeV. The resonances have small γ-decay branches to lower-lying resonances of the order of 10^{-6}–10^{-5}. By detecting the three outgoing α particles in coincidence and measuring their momenta, we obtain complete kinematics information. From the combined energy of the α particles we determine the energy of the γ transition. In this way, we identify two previously observed transitions, $(0^+, 17.79) \rightarrow (1^+, 12.71)$ and $(3^-, 18.38) \rightarrow (3^-, 9.64)$, and one new transition, $(2^+, 18.81) \rightarrow (1^+, 12.71)$. The results demonstrate the usefulness of γ decay as a probe of the low-lying resonance spectrum of ^{12}C in the search for new broad (cluster) resonances.

1. Introduction

1.1. Turning a disadvantage into an advantage

The famous Dutch footballer Johan Cruyff supposedly coined the phrase "every disadvantage has its advantage". In nuclear-reaction experiments, high multiplicity used to be considered a disadvantage because it was very difficult to detect more than two particles in coincidence. Consequently, only *partial* kinematic information was obtained, which made it difficult to single out the reaction channels of interest because a clear signature was missing (such as a peak in the sum-energy spectrum). Within the last two decades, however, multi-particle coincidence measurements have become feasible thanks to the advent of compact arrays of segmented silicon detectors, and with *complete* kinematic information now available, the disadvantage of having many particles in the final state turns into an advantage: Not only can we make a clean selection, but one that is *cleaner* than for low-multiplicity channels because high-multiplicity channels are comparatively rarer. High-multiplicity studies are, of course, closely connected to the topic of this conference, as clustering phenomena in nuclei manifest themselves most strongly at excitation energies close to or somewhat above particle-decay thresholds.

1.2. Putting two old accelerators back into use

Aarhus University has two Van de Graaff accelerators at its disposal, which have been in operation since the 1960s. The accelerators have been used primarily for material analysis and student exercises, but recently we have started using them for nuclear-physics experiments. The smaller accelerator goes up to 400 kV, the larger to about 4 MV (nominally 5 MV). The results presented here were obtained at the 5 MV accelerator. Results obtained at the 400 kV accelerator will be presented by K. L. Laursen in a separate contribution.

We use a compact array of double-sided silicon strip detectors (DSSDs) to revisit "well-known" nuclear reactions induced by few-MeV beams of ^{1}H, ^{3}He or ^{4}He on light-element targets such as Li, Be, B, C and N. We focus on reactions with high Q value leading to final states with high multiplicity, examples of which being ^{6}Li(^{3}He, $p2\alpha$) and ^{10}B(^{3}He, $p3\alpha$). The combination of high granularity and large solid-angle coverage allows us to measure multi-particle coincidences with high resolution and high efficiency. Having complete kinematic information, we are able to improve on previous studies both quantitatively and qualitatively. In particular, we are able to extract new and valuable information on resonances in light nuclei of relevance both to nuclear astrophysics and nuclear structure. Since the demand for the two Van de Graaff accelerators is low, we are able to run for extended periods of time (several weeks), making it possible to accumulate enough statistics to study rare decay modes.

2. First experiment: $p + {}^{11}$B

2.1. Motivation

In recent years, the topic of α clustering in ^{12}C has generated numerous papers, both theoretical and experimental, and at the present conference, a considerable number of contributions have been devoted to the topic. Arguably the most significant step forward on the experimental side has been the identification of a new 2^+ resonance at an excitation energy of 10.1 MeV [1]. The properties of the resonance remain to be pinned down. Whether it can be viewed as a rotational excitation of the Hoyle state is at present not clear[1]. Recent experiments have uncovered evidence of additional new resonances[2], but the search is challenging. The "easy" (relatively narrow) resonances have all been found, leaving only the "tough" (broad) ones. A more complete experimental picture of the resonance spectrum of ^{12}C is necessary to advance the theoretical understanding of α clustering in ^{12}C. It is also a necessary prerequisite for an accurate determination of the stellar triple-α rate at elevated temperatures [2]. Clearly, a complementary approach is necessary. No single reaction or decay will give us the full answer.

2.2. Experimental approach

The reaction $p + {}^{11}$B has been used by many experimenters to study resonances in ^{12}C above the proton-decay threshold. Many experimenters have studied the breakup into three α particles. Hanna *et al.* have studied the γ decay to lower-lying states [3]. We have collected complete-kinematics data on both decay mechanisms. Here, we present preliminary results on the γ-decay mechanism. The full results (on both decay mechanisms) will be presented in forthcoming papers. Hanna *et al.* were mainly interested in the structure of the initially populated state in ^{12}C. In contrast, we are mainly interested in the state populated in the subsequent γ decay. A schematic illustration is given in Figure 1.

Like the β decays of ^{12}B and ^{12}N, γ decay is a highly *selective* probe of the low-lying resonance spectrum in ^{12}C in the sense that the decay is governed by selection rules that strongly favour the population of states with certain spin-parities [4]. In β decay the spin and parity of the initial state, and hence the selection rules, are fixed. In γ decay there is no similar restriction as we can use the $p + {}^{11}$B reaction (or another suitable reaction) to populate states with spin-parities of our preference (to the extent that such states exist). Thus, γ decay constitutes not only a highly selective but also an *adjustable* probe. Its main disadvantage is the low branching ratio for γ decay, typically of the order of 10^{-6}–10^{-5}. The spin-parities that can be reached via E1,

[1] See the contribution of Garrido.

[2] See the contributions of Itoh, Kokalova, Gai, Laursen and Marín-Lámbarri.

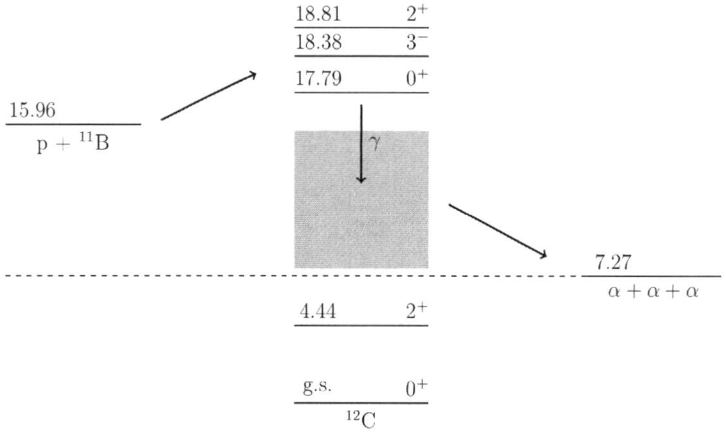

Figure 1. Schematic illustration of experimental approach: The reaction $p + {}^{11}$B is used to populate selected resonances in ^{12}C that γ decay to lower-lying resonances, which subsequently break up into three α particles. The α particles are detected and the occurrence of a γ transition is deduced from the missing energy.

Table 1. Spin-parities that can be reached via E1, M1 and E2 transitions. The initial states are labeled by their excitation energy in MeV and spin-parity.

Eℓ/Mℓ	17.79, 0^+	18.38, 3^-	18.81, 2^+
E1	1^-	$2^+, 3^+, 4^+$	$1^-, 2^-, 3^-$
M1	1^+	$2^-, 3^-, 4^-$	$1^+, 2^+, 3^+$
E2	$1^+, 2^+$	$1^-, \ldots, 5^-$	$0^+, \ldots, 4^+$

M1 and E2 transitions in the present experiment are listed in Table 1. Isospin selection rules provide additional selectivity.

2.3. Setup and data analysis

The detector setup is shown in Figure 2. It consists of two 24 × 32 annular DSSDs, which cover the forward-most angles (16°–42°) and the backward-most angles (142°–166°), and two 16 × 16 square DSSDs, which cover the intermediate angles (53°–127°). The annular DSSDs are 1 mm and 64 μm thick, respectively, and the square DSSDs are both 60 μm thick. The detectors view the target from a distance of 4 cm and provide a solid-angle coverage of 40% of 4π, giving in a triple-α coincidence detection efficiency of the order of 6%. Both ADC (energy) and TDC (time) information is recorded. For the present experiment, the target was 7 μg/cm^2 natural boron on a 4 μg/cm^2 carbon backing. The typical beam intensity was 0.3 nA as measured in a Faraday cup downstream of the detector setup. The beam diameter was less than 3 mm at the target site. Measurements were performed at beam energies of 2.00 MeV, 2.63 MeV and 3.12 MeV, corresponding to three known $T = 1$ resonances with spin-parities of 0^+, 2^+ and 3^- and excitation energies of 17.79 MeV, 18.38 MeV and 18.81 MeV. The duration of the measurements were 41 h, 30 h and 32 h, respectively.

The occurrence of a γ transition can be inferred from the missing energy carried away by the

Journal of Physics: Conference Series **569** (2014) 012013 doi:10.1088/1742-6596/569/1/012013

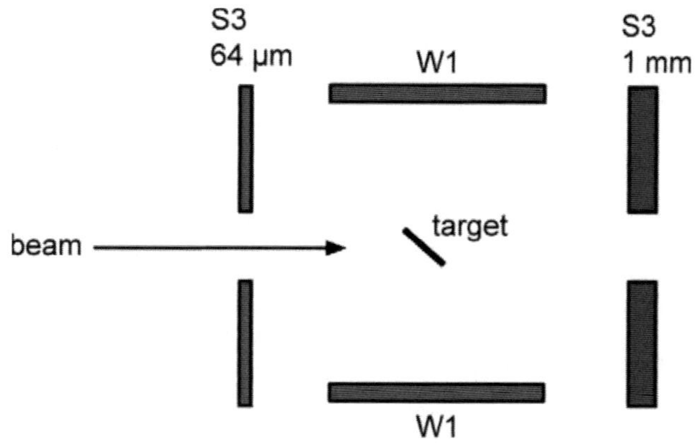

Figure 2. Schematic illustration of the detector setup (top view). The beam comes in from the left. The target is surrounded by four DSSDs in a rectangular configuration. Two annular DSSDs (S3) cover forward and backward angles and two square DSSDs (W1) cover intermediate angles.

γ ray. The γ ray itself is not detected. As demonstrated in a previous study [5], the technique is particularly useful to identify γ transitions to broad states. The present setup does not allow us to distinguish between different types of ions. At the beam energies used in the present study, the only energetically allowed three-body final state in the $p + {}^{11}\mathrm{B}$ reaction is $\alpha + \alpha + \alpha$. However, reactions on ${}^{10}\mathrm{B}$ can give $\alpha + \alpha + {}^{3}\mathrm{He}$. Random coincidences between, say, an elastically scattered proton and two α particles can also give a triple coincidence that "mimics" a triple-α event. The background due to such false triple-α events is reduced by imposing a number of cuts in the data analysis: TDC information is used to reduce the background due to random coincidences by a factor of ~ 40. Further reduction is achieved by placing an anti-gate on the kinematic curve of elastically scattered protons and the kinematic curve of α particles from the reaction ${}^{10}\mathrm{B}(p, \alpha){}^{7}\mathrm{Be}$. Finally, we require the total momentum to be conserved (the momentum carried away by the γ ray is negligible).

3. Preliminary results

In our data, we find clear evidence of the following γ-ray transitions,

(i) 0^+, 17.79 $\rightarrow$ 1^+, 12.71
(ii) 3^-, 18.38 $\rightarrow$ 3^-, 9.64
(iii) 2^+, 18.81 $\rightarrow$ 1^+, 12.71

We also find tentative evidence of other transitions, but a careful analysis of the background is necessary before firm conclusions will be drawn. Transitions (i) and (ii) were observed previously by Hanna *et al.* [3] whereas transition (iii) appears not to have been observed previously. Monte Carlo simulations will be performed in the near future to determine the detection efficiency of the setup. This will enable us to extract branching ratios.

The complete-kinematics data collected on the 2^+ resonance are shown in Figures 3 (a) and (b). Figure (a) shows the momentum deficit, defined as

$$\Delta p = |\,\vec{p}_p - \sum_{i=1,2,3} \vec{p}_{\alpha_i}| \tag{1}$$

where $\vec{p}_p$ is the momentum of the beam and $\vec{p}_{\alpha_i}$ are the α-particle momenta, versus the excitation energy in ^{12}C, determined as

$$E_x = S_{3\alpha} - \frac{1}{12}E_p + \sum_{i=1,2,3} E_{\alpha_i} \tag{2}$$

where $S_{3\alpha} = 7.27$ MeV is the triple-α threshold energy in ^{12}C, E_p is the beam energy and E_{α_i} are the α-particle energies. The momenta and the energies are all measured in the laboratory frame. TDC information has been used to reduce the background due to random coincidences. Further reduction has been achieved by placing an anti-gate on the kinematic curve of elastically scattered protons and the kinematic curve of α particles from the reaction $^{10}\mathrm{B}(p,\alpha)^7\mathrm{Be}$. By imposing the additional requirement $\Delta p < 55$ MeV/c (dashed line), we obtained the excitation-energy spectrum shown in Figure (b). The main peak (E) corresponds to the ordinary decay mechanism, *i.e.*, breakup into three α particles without a preceding γ transition; (C) is the previously unobserved γ transition to the 12.71 MeV state; (B) and (D) are potential candidates for new γ transitions; finally, (A) corresponds to $\alpha + \alpha + {}^3$He events from reactions on ^{10}B. By gating on the narrow ground state of ^{8}Be in the α-α relative-energy spectrum, it is possible to single out the breakups that proceed via the ground state of ^{8}Be. By considering only these breakups, one obtains the gray-shaded spectrum. Note that peak (C) is absent in the gray-shaded spectrum. This is consistent with the proposed interpretation, as parity conservation forbids the breakup of unnatural-parity states via the 0^+ ground state of ^{8}Be.

4. Summary and outlook

At Aarhus University in Denmark we have built a new facility to study nuclear reactions induced by few-MeV beams of ^{1}H, ^{3}He or ^{4}He on light-element targets such as Li, Be, B, C and N. The detection system consists of a compact arrangement of segmented silicon detectors, allowing many-particle final states to be studied in complete kinematics. Here, we have reported preliminary results from the first experiment at the new facility. We have studied the decay of three selected resonances in ^{12}C populated via $p + {}^{11}$B. The resonances have small γ-decay branches (10^{-6}–10^{-5}) to lower-lying resonances which subsequently break up into three α particles. We have determined the energy of the γ transition from the combined energy of the α particles. In this way, we have identified two previously observed transitions, $(0^+, 17.79) \rightarrow (1^+, 12.71)$ and $(3^-, 18.38) \rightarrow (3^-, 9.64)$, and one new transition, $(2^+, 18.81) \rightarrow (1^+, 12.71)$. We find tentative evidence for additional transitions, but further analysis is required to fully understand the background. Simulations will be performed in the near future to determine the detection efficiency of the setup. This will allow us to extract branching ratios. We are in the process of improving the setup. Important goals include improving the time resolution and acquiring an isotope enriched ^{11}B target.

Acknowledgments

We acknowledge financial support from the European Research Council under the ERC starting grant LOBENA, No. 307447.

References

[1] Zimmerman W R, *et al.* 2013 *Phys. Rev. Lett.* **110** 152502
[2] Fynbo H O U, *et al.* 2005 *Nature* **433** 136–9
[3] Hanna S S, Feldman W, Suffert M and Kurath D 1982 *Phys. Rev.* C **25** 1179–86
[4] Hyldegaard S, *et al.* 2010 *Phys. Rev.* C **81** 024303
[5] Kirsebom O S, *et al.* 2009 *Phys. Lett.* B **680** 44–9

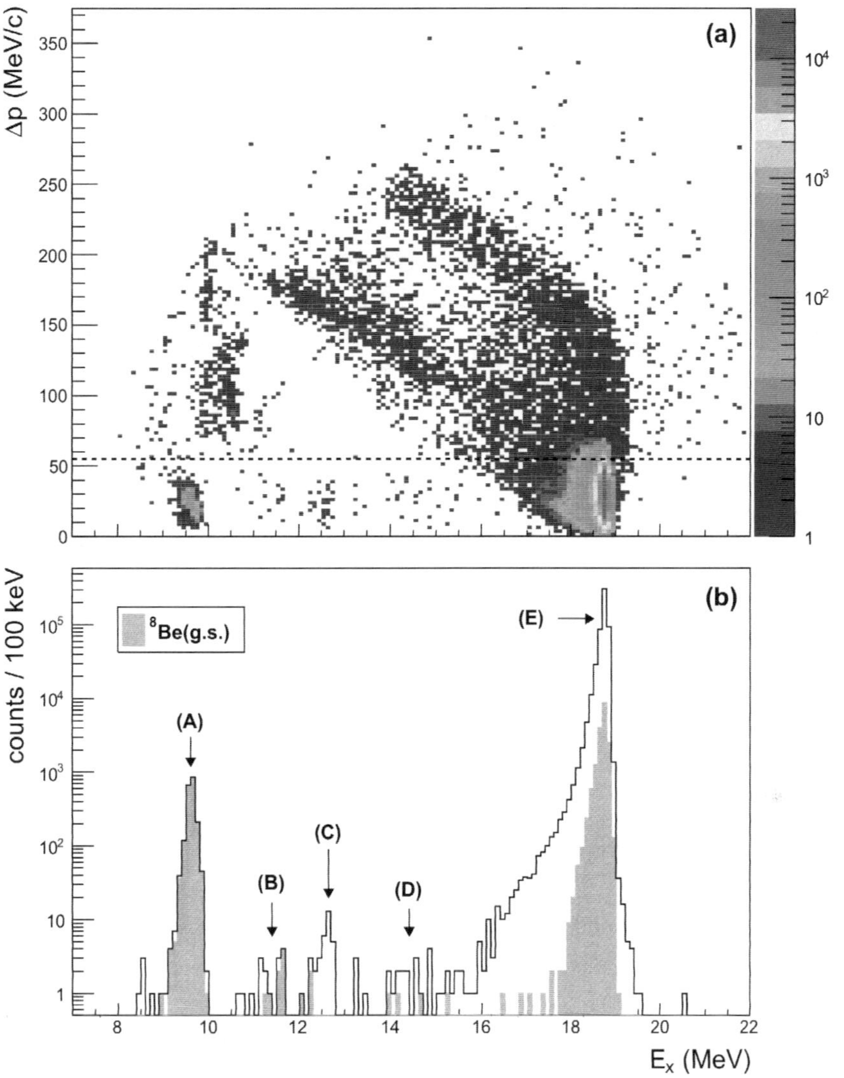

Figure 3. Complete-kinematics data collected for the 2^+, 18.81 MeV state. Figure (a) shows the momentum deficit, cf. Eq. (1), versus the excitation energy, cf. Eq. (2). TDC information and kinematic cuts have been used to reduce the background due to random coincidences. Events below the dashed line fulfil momentum conservation. Figure (b) shows the excitation-energy spectrum obtained by considering only events below the dashed line. The interpretation of the five peaks (A)–(E) is discussed in the text. The gray-shaded area is the spectrum obtained by including only decays proceeding via the ground state of ^{8}Be.

Journal of Physics: Conference Series **569** (2014) 012014 doi:10.1088/1742-6596/569/1/012014

Missing monopole strength of the Hoyle state in the alpha inelastic scattering

T Kawabata[1], T Kadoya[1], N Yokota[1], S Adachi[1], T Baba[1], T Furuno[1], H Fujimura[2], M Fujiwara[3], K Hatanaka[3], Y Ishii[1], M Itoh[4], T Ito[3], Y Maeda[5], Y Matsuda[3], M Murata[1], T Sato[4], A Tamii[3], M Tsumura[1], H Watanabe[1], and J Zenihiro[6]

[1]Department of Physics, Kyoto University, Sakyo, Kyoto 606-8502, Japan.
[2]Department of Physics, Wakayama Medical University, Wakayama 641-8509, Japan.
[3]Research Center for Nuclear Physics, Osaka University, Ibaraki, Osaka 567-0047, Japan.
[4]Cyclotron and Radioisotope Center, Tohoku University, Sendai, Miyagi 980-8578, Japan.
[5]Faculty of Engineering, University of Miyazaki, Miyazaki 889-2155, Japan.
[6]RIKEN (The Institute for Physical and Chemical Research), Wako, Saitama 351-0198, Japan.

E-mail: kawabata@scphys.kyoto-u.ac.jp

Abstract. Cross sections for the alpha inelastic scattering exciting the low-lying monopole states in ^{12}C, ^{16}O, ^{24}Mg, ^{28}Si, and ^{40}Ca were measured and compared with the distorted-wave Born-approximation (DWBA) calculation to examine the puzzle of the missing monopole strength of the Hoyle state. It was found the DWBA calculation using the density-dependent αN interaction systematically overestimates the cross sections for the 0^+ transitions, and the puzzle is a universal problem in light nuclei but not special in the Hoyle state. Since the DWBA calculation using the density-independent interaction reasonably well reproduces the experiment, this puzzle might be related to the density dependence of the effective interaction. The coupled-channel effect for the alpha inelastic scattering is also examined. The coupled-channel effect reduces the calculated cross section, and solve the puzzle of the missing monopole strengths in part, but the improvement of the density dependence of the effective interaction is still necessary to solve the puzzle.

1. Introduction

The second 0_2^+ state in ^{12}C, known as the Hoyle state, is the most famous 3α cluster state, and locates at an excitation energy higher than the 3α-decay threshold by 0.4 MeV. The Hoyle state plays a very important role in the nucleosynthesis in the universe. This state is of great interest in the last decade because this state is theoretically described by introducing a novel concept of the nuclear structure, *i.e.*, this state has a dilute-gas-like structure where three α clusters are weakly interacting and are condensed into the lowest s-orbit [1]. Those condensed α clusters have a very sharp delta-function-like momentum distribution, thus, their density distribution expands into large volume and becomes dilute [2].

The structure of the Hoyle state was extensively studied from both the theoretical and experimental sides. The electron scattering data were precisely compared with the model calculations, and these calculations reasonably well reproduce the experiment. On the other hand, the cross sections for the alpha inelastic scattering exciting the Hoyle state were significantly overestimated by the theoretical calculation, thus, the monopole strength for the Hoyle state had to be scaled down by a factor of 2–4 to explain the experimental data. D. T. Khoa *et al.* claimed this puzzle of the missing monopole strength should be attributed to the exotic nature of the Hoyle state. Since the loosely bound dilute structure of this state enhances absorption in the exit $\alpha+^{12}\mathrm{C}^*(0_2^+)$ channel, the strong imaginary part of the transition potential must be introduced to solve the puzzle [3].

Recently, it was demonstrated that the large monopole strength is a signature of the spatially well-developed cluster state in light nuclei [4]. The alpha inelastic scattering was extensively utilized to measure monopole strengths and to examine the cluster structures in light nuclei. It is, therefore, an important question whether the puzzle of the missing monopole strength is a special phenomena to the Hoyle state only as suggested in Ref. [3] or is a universal phenomena observed in various nuclei.

In the present work, a systematic measurement of the alpha inelastic scattering exciting the low-lying monopole states in ^{12}C, ^{16}O, ^{24}Mg, ^{28}Si, and ^{40}Ca was carried out and the measured cross sections were compared with the model calculations.

2. Experiment and results

The experiment was performed at the Research Center for Nuclear Physics, Osaka University, using a 130-MeV α beam. The α beam extracted from the AVF cyclotron was achromatically transported to the target. Scattered α particles were momentum analyzed by the high-resolution spectrometer Grand Raiden [5]. The ^{nat}C, ^{24}Mg, ^{nat}SiO$_2$, ^{nat}Si, and ^{nat}Ca targets with a thickness of 1–5 mg/cm^2 were used.

The cross sections for the alpha elastic scattering from ^{12}C, ^{16}O, ^{24}Mg, ^{28}Si, and ^{40}Ca were measured at $\theta_{lab} = 3.7°$–$55°$, whereas the cross sections for the alpha inelastic scattering were measured at $\theta_{lab} = 0°$–$13.4°$. The measured cross sections for the elastic scattering are shown in Fig. 1, and the cross sections for the inelastic scattering exciting some of the low-lying 0^+ and 2^+ states are shown in Fig. 2.

3. Discussion

3.1. Elastic scattering

The measured cross sections for the alpha elastic scattering from ^{12}C, ^{16}O, ^{24}Mg, ^{28}Si, and ^{40}Ca are compared with the optical-model calculation in Fig. 1. The optical potentials $U(r)$ were obtained by the single folding prescription where the ground-state density distribution $\rho_0(r)$ is folded by the phenomenological αN interaction $V(|\boldsymbol{r}-\boldsymbol{r}'|, \rho_0(r'))$.

$$U(r) = \int d\boldsymbol{r}' \rho_0(r') V(|\boldsymbol{r}-\boldsymbol{r}'|, \rho_0(r')) \, .$$

The ground-state density distributions are taken from the electron scattering data [6]. The αN interaction is parametrized by

$$\begin{aligned} V(|\boldsymbol{r}-\boldsymbol{r}'|, \rho_0(r')) &= -V\left(1+\beta\rho_0^{2/3}(r')\right)\exp\left(-\frac{|\boldsymbol{r}-\boldsymbol{r}'|^2}{\alpha_V}\right) \\ & \quad -iW\left(1+\beta\rho_0^{2/3}(r')\right)\exp\left(-\frac{|\boldsymbol{r}-\boldsymbol{r}'|^2}{\alpha_W}\right), \end{aligned}$$

where β determines the density dependence of the interaction. In the present work, the two different values of β were used. $\beta = -1.9$ taken from Ref. [7] gives the density-dependent (DD)

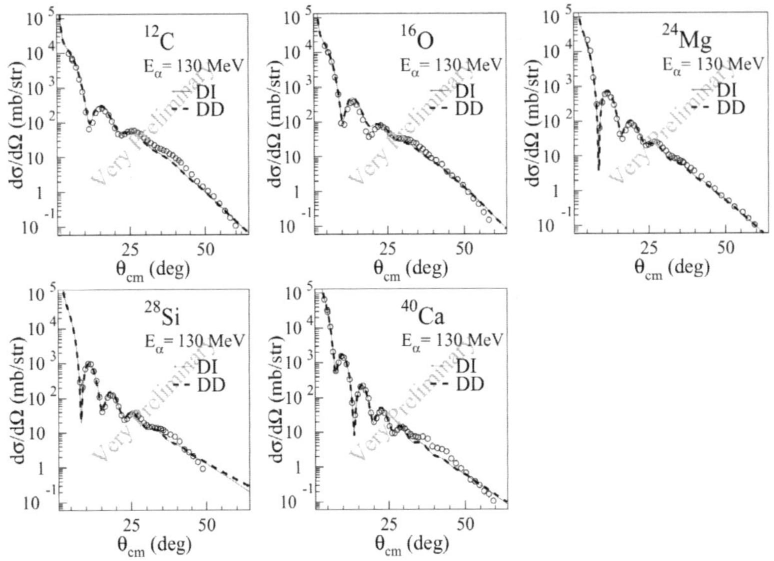

Figure 1. Measured cross sections for the alpha elastic scattering from ^{12}C, ^{16}O, ^{24}Mg, ^{28}Si, and ^{40}Ca compared with the optical-model calculation. The optical-model calculation with the DI and DD interactions are shown by the thin solid and thick dashed lines, respectively.

interaction, and $\beta = 0$ gives the density-independent (DI) interaction. The four parameters α_V, α_W, V, and W are determined to fit the elastic scattering data. The optical-model calculation with the DI and DD interactions are shown by the thin solid and thick dashed lines in Fig. 1, respectively. Both the DI and DD interactions give the almost same results and reasonably well describe the experimental data.

3.2. Inelastic scattering

The measured cross sections for the alpha inelastic scattering exciting some of the low-lying 0^+ and 2^+ states in ^{12}C, ^{16}O, ^{24}Mg, ^{28}Si, and ^{40}Ca are compared with the distorted-wave Born-approximation (DWBA) calculation in Fig. 2. The transition potentials $\delta U_\lambda(r, E_x)$ used in the DWBA calculation were obtained by folding the macroscopic transition densities $\delta\rho^{(\lambda)}(r, E_x)$ with the phenomenological αN interaction which is determined to fit the elastic scattering.

$$\delta U_\lambda(r, E_x) = \int d\mathbf{r}'\delta\rho^{(\lambda)}(r, E_x)\left[V\left(|\mathbf{r}-\mathbf{r}'|, \rho_0(r')\right) + \rho_0(r')\frac{\partial V\left(|\mathbf{r}-\mathbf{r}'|, \rho_0(r')\right)}{\partial\rho_0(r')}\right].$$

The macroscopic transition densities for the 0^+ and 2^+ states are given as following.

$$\delta\rho^{(0)}(r') = -\delta_0\left(3 + r'\frac{d}{dr'}\right)\rho_0(r'), \qquad \delta\rho^{(2)}(r') = -\delta_2\frac{d}{dr'}\rho_0(r').$$

The cross sections calculated by the DI and DD interactions are shown by the thin solid and thick dashed lines, respectively. The normalization factors for the macroscopic transition densities were determined to reproduce the known electromagnetic transition strengths. The shaded bands in Fig. 2 show uncertainties of the calculated cross sections due to the errors in the electromagnetic transition strengths. It should be noted that all the parameters in the present DWBA calculation are determined from the previous experimental data and no parameters are tuned to fit the present results in Fig. 2.

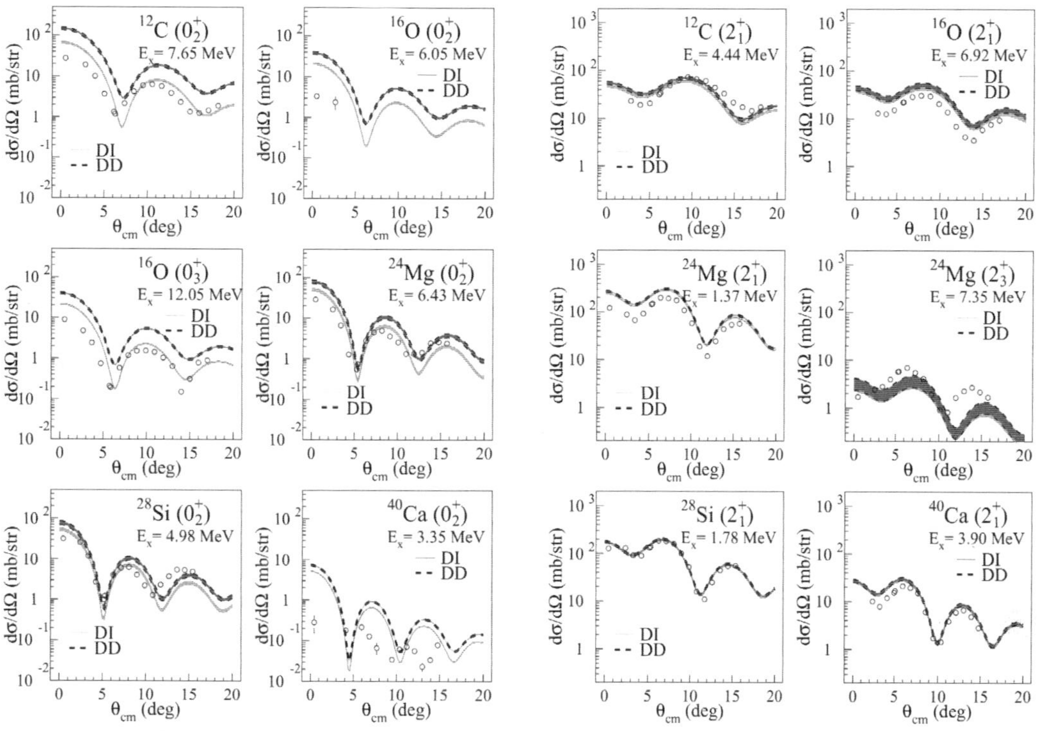

Figure 2. Measured cross sections for the alpha inelastic scattering exciting several low-lying 0^+ and 2^+ states in ^{12}C, ^{16}O, ^{24}Mg, ^{28}Si, and ^{40}Ca compared with the distorted-wave Born-approximation (DWBA) calculation. The DWBA calculation with the DI and DD interactions is shown by the thin solid and thick dashed lines, respectively. The shaded bands show uncertainties due to the errors in the electromagnetic transition strengths.

For the 2^+ states, the DWBA calculations using the DI and DD interactions predict the almost same cross sections and both the calculations reasonably reproduces the experimental results. On the other hand, the DWBA calculations using the DI and DD interactions give quite different results for the 0^+ states. The DD interaction gives much larger cross sections than the DI interaction, especially for the lighter nuclei such as ^{12}C and ^{16}O. The DWBA calculation using the DD interaction systematically overestimates the cross sections for the 0^+ states, and this result is quite similar to the previous result in Ref. [3]. It is worth to mention that the previous calculation in Ref. [3] was carried out using the density-dependent NN interaction. The preset results suggest the puzzle of the missing monopole strengths is not a special problem in the Hoyle state, but a universal problem in light nuclei.

For further clarification of the problem, the transition potentials for the 0_2^+ and 2_1^+ states are shown in Figs. 3(a) and (b). The transition potentials calculated by the DI and DD interactions are shown by the thin and thick lines, and the real and imaginary parts of the transition potentials are given by the solid and dashed lines, respectively. It is remarkable that the transition potentials with the DD and DI interactions are completely different for the 0_2^+ state although they are quite similar to each other for the 2_1^+ state. This difference comes from the radial dependence of the transition densities as seen in Fig. 3(c). The transition density for the 0^+ transition has a finite value at the origin where the matter density is high, whereas the transition density for the 2^+ transition is zero at the origin. Because the Pauli blocking effect

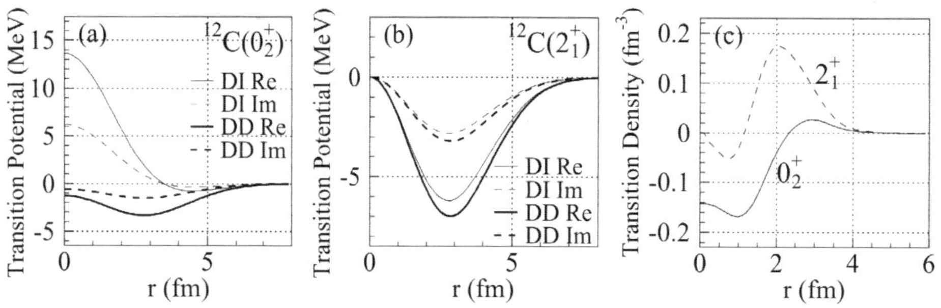

Figure 3. (a) Transition potentials for the 0_2^+ state in ^{12}C. The thin and thick lines show the transition potentials calculated by using the DI and DD interactions, and the solid and dashed lines show the real and imaginary parts of the transition potentials, respectively. (b) Same with (a) but for the 2_1^+ state. (c) Macroscopic transition densities for the 0^+ (solid) and 2^+ (dashed) transitions.

attenuates the effective interaction around the origin, the DD transition potentials for the 0^+ transitions become shallower.

Since the imaginary part of the DD transition potential for the 0^+ transition is much shallower than that of the DI transition potential, the DD interaction suppresses the absorption to enhance the cross section. This is the reason that the DD interaction overestimates the cross sections for the 0^+ transitions and causes the puzzle of the missing monopole strengths. It should be noted the DI interaction gives relatively better results for the 0^+ transitions than the DD interaction in Fig. 2. It should be considered that this puzzle is related to the density dependence of the effective interaction, and the density dependence of the effective interaction must be improved to solve the puzzle.

3.3. Uncertainties in the DWBA calculation

Uncertainties in the DWBA calculation mainly come from (a) the effective αN interaction, (b) transition densities, and (c) coupled-channel effects. Since the effective interaction has been discussed in the previous subsection, the uncertainties in the transition densities and coupled-channel effects are discussed in this subsection.

Since the transition densities for the DWBA calculation are obtained by the macroscopic model in the present work, the radial distribution of the transition densities is determined by the spin and parity only. Therefore, the state-by-state variation of the nuclear structure is ignored. To examine the uncertainties of the DWBA calculation come from the transition densities, Fig. 4 shows the calculated cross sections for the 0_2^+ and 2_1^+ states in ^{12}C using the macroscopic transition densities compared with those using the microscopic transition densities, which are obtained from the THSR model [8]. The solid and dashed lines show the calculated cross sections using the macroscopic and microscopic transition densities, respectively, whereas the thin and thick lines present the calculated cross section using the DI and DD interactions. The macroscopic and microscopic transition densities give similar cross sections, therefore, the uncertainties due to the transition densities are quite small in the present DWBA calculation.

The coupled-channel effects are examined in Fig. 5. The solid lines show the cross sections for the 0_2^+ and 2_1^+ states in ^{12}C calculated by the DWBA, and the dashed lines present those by the coupled-channel calculation taking the couplings between the 0_1^+, 2_1^+, 4_1^+, 2_2^+, and 3_1^- states into account. The transition densities concerning the 3_1^- states are taken from the RGM calculation [9] and the other transition densities are taken from the THSR model. The cross

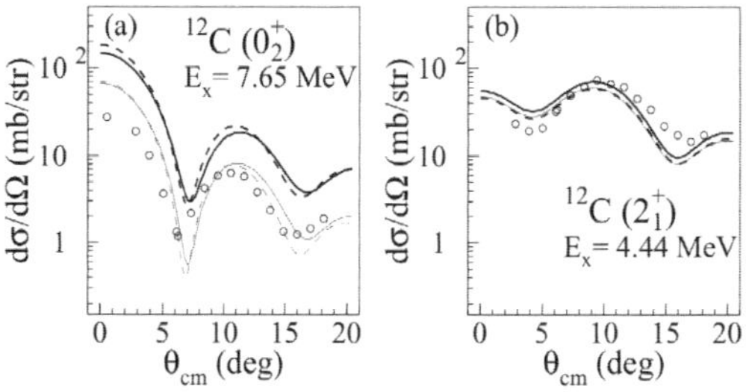

Figure 4. Calculated cross sections for the 0_2^+ and 2_1^+ states in ^{12}C using the macroscopic transition densities are compared with those using the microscopic transition densities.

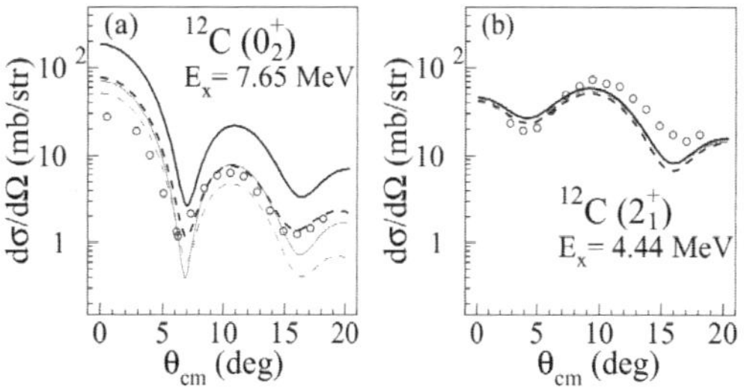

Figure 5. Cross sections for the 0_2^+ and 2_1^+ states in ^{12}C obtained by the DWBA (solid lines) and coupled-channel (dashed lines) calculations.

sections shown by the thin and thick lines are calculated by using the DI and DD interactions.

The coupled-channel effects significantly reduce the cross section for the 0_2^+ state, but do not change the cross section for the 2_1^+ state. According to the cluster model calculations [8, 9], the coupling between the 0_2^+ and 2_2^+ states is much stronger than that between the 0_1^+ and 2_1^+ states because the 2_2^+ state is considered to be a single α excited state of the 0_2^+ state where one α cluster is excited into the d-orbit from the lowest s-orbit. The strong absorption in the $\alpha+^{12}\mathrm{C}^*(0_2^+)$ channel pointed out in Ref. [3] should be attributed to this strong coupling between the 0_2^+ and 2_2^+ states in part, but the improvement of the density dependence of the effective interaction is still necessary to solve the puzzle of missing monopole strengths.

4. Summary

In the present work, a systematic measurement of the alpha inelastic scattering exciting the low-lying monopole states in ^{12}C, ^{16}O, ^{24}Mg, ^{28}Si, and ^{40}Ca was carried out and the measured cross sections were compared with the model calculations. It was found that the DWBA calculation using the DD interaction systematically overestimates the cross sections for the 0^+ transitions and the puzzle of the missing monopole strengths is a universal problem in light nuclei not special in the Hoyle state. Since the DWBA calculation using the DI interaction reasonably well reproduces the experiment, this puzzle might be related to the density dependence of the effective interaction.

Uncertainties in the DWBA calculation are also discussed. It was found that the uncertainties due to the transition densities are quite small, but the coupled-channel effects between the 0_2^+ and 2_2^+ states are sizable. The strong absorption in the $\alpha+^{12}\mathrm{C}^*(0_2^+)$ channel pointed out in Ref. [3] should be attributed to this strong coupling between the 0_2^+ and 2_2^+ states in part, but the improvement of the density dependence of the effective interaction is still necessary to solve the puzzle of missing monopole strengths.

References

[1] Tohsaki A, Horiuchi H, Schuck P, and Röpke G 2001 *Phys. Rev. Lett.* **87** 192501
[2] Yamada T and P. Schuck 2005 *Eur. Phys. J. A* **26** 185
[3] Khoa Dao T and Cuong Do Cong 2008 *Phys. Lett. B* **660** 331
[4] Kawabata T *et al.* 2007 *Phys. Lett. B* **646** 6
[5] Fujiwara M *et al.* 1999 *Nucl. Instrum. Meths. A* **422** 484
[6] De Vries H, De Jager C W, and De Vries C 1987 *Atomic Data and Nuclear Data Tables* **36** 495
[7] Satchler G R and Khoa Dao T 1997 *Phys. Rev. C* **55** 285
[8] Funaki Y, Tohsaki A, Horiuchi H, Schuck P, and Röpke G 2003 *Phys. Rev. C* **67** 051306
[9] Kamimura M 1981 *Nucl. Phys. A* **351** 456

Journal of Physics: Conference Series **569** (2014) 012015 doi:10.1088/1742-6596/569/1/012015

The isoscalar transition strengths of the cluster states of ^{12}C

Dao T. Khoa[1], Do Cong Cuong[1] and Yoshiko Kanada-En'yo[2]

[1]Institute for Nuclear cience and Technology, VINATOM
179 Hoang Quoc Viet Rd., Hanoi, Vietnam.
[2] Department of Physics, Kyoto University, Kyoto 606-8502, Japan.

Abstract. An elaborate folding model + coupled channel analysis of the inelastic $\alpha+^{12}$C scattering has been done using the inelastic form factors based on the nuclear transition densities predicted by the antisymmetrized molecular dynamics. The form factors for the observed isoscalar states of ^{12}C were fine-tuned against the measured (α,α') cross sections to estimate the corresponding $E\lambda$ transition strengths. A significant $E2$ transition strength has been allocated for the second 2^+ state of ^{12}C from our consistent coupled channel analysis of the (α,α') data measured at $E_\alpha = 240$ and 386 MeV. The $E\lambda$ transition strengths of the 0_2^+, 3_1^-, 0_3^+, and 1_1^- states were also carefully deduced from the same folding model + coupled channel analysis.

The excited states of ^{12}C at energies near the α-decay threshold are extensively studied recently [1, 2] because of the interesting α-cluster structure established in several cases, like the isoscalar 0_2^+ state at 7.65 MeV in ^{12}C, known as the Hoyle state that is vital for the carbon synthesis in cosmos. Along with a three α-cluster structure of the Hoyle state shown about three decades ago by the Resonating roup Method (R M) calculations [3, 4, 5], an interesting α-condensate scenario [2] for this state has been suggested [6, 7], where three α clusters are shown to condense into the lowest S state of their potential.

iven a nonspherical shape of the Hoyle state, an excited rotational band with the angular momentum $J^\pi = 2^+, 4^+, \ldots$ built upon this state was suggested long ago by Morinaga [8]. In the α-condensate scenario, it is also natural that the next level in the potential containing three α-particles should be a 2^+ state formed by promoting an α-particle from the S to D level. The second 2^+ state of ^{12}C has been predicted by the antisymmetrized molecular dynamics (AMD) [9, 10] and some other structure studies at the excitation energy around 10 MeV, i.e., about 2 MeV above the α-decay threshold, with a pronounced ^{8}Be+α structure [1]. Although some evidence for a broad 2^+ resonance was found in several experiments that might be assigned to the 2_2^+ state of ^{12}C, a clear identification of this state could be made just recently in the high-precision (α,α') experiment at $E_\alpha = 386$ MeV [11] and the photodissociation of carbon [12, 13].

To validate conclusion made in the structure calculation, the wave functions must be carefully tested in the study of nuclear reactions. Since the spin- and isospin zero α-particle is a very good projectile to excite the isoscalar (IS) states, the nuclear transition densities predicted by AMD [9] for the known IS excitations of ^{12}C have been used in the present work for a detailed folding model + coupled channel analysis of the (α,α') data taken at $E_\alpha = 240$ [14] and 386 MeV [11]. The generalized folding model of Ref. [15] was used to evaluate the complex optical potential (O) and inelastic scattering form factor (FF) from the (complex) effective nucleon-nucleon

(NN) interaction between the projectile nucleon i and target nucleon j as

$$U_{A\to A^*} = \sum_{i\in\alpha;j\in A,j'\in A^*} [\langle ij'|v_{\rm D}|ij\rangle + \langle ij'|v_{\rm EX}|ji\rangle], \quad (1)$$

where A and A^* denote the target in the entrance- and exit channel of the (α, α') scattering, respectively. The direct ($v_{\rm D}$) and exchange ($v_{\rm EX}$) parts of the central NN interaction are those of the density dependent CDM3 6 interaction [16], supplied by an imaginary part [17] determined from the JLM results for the complex nucleon O in the nuclear matter limit [18]. Eq. (1) gives the O if $A^* = A$ and inelastic scattering FF if otherwise. The AMD nuclear transition densities enter the folding calculation (1) such that the IS transition strength for a 2^λ-pole nuclear transition $|J_i\rangle \to |J_f\rangle$ is given by the reduced nuclear transition rate as $B({\rm IS}\lambda; J_i \to J_f) = |M({\rm IS}\lambda; J_i \to J_f)|^2$, with the 2^λ-pole transition moment determined from the corresponding nuclear transition density (see explicit expressions in Ref. [19]). Because the

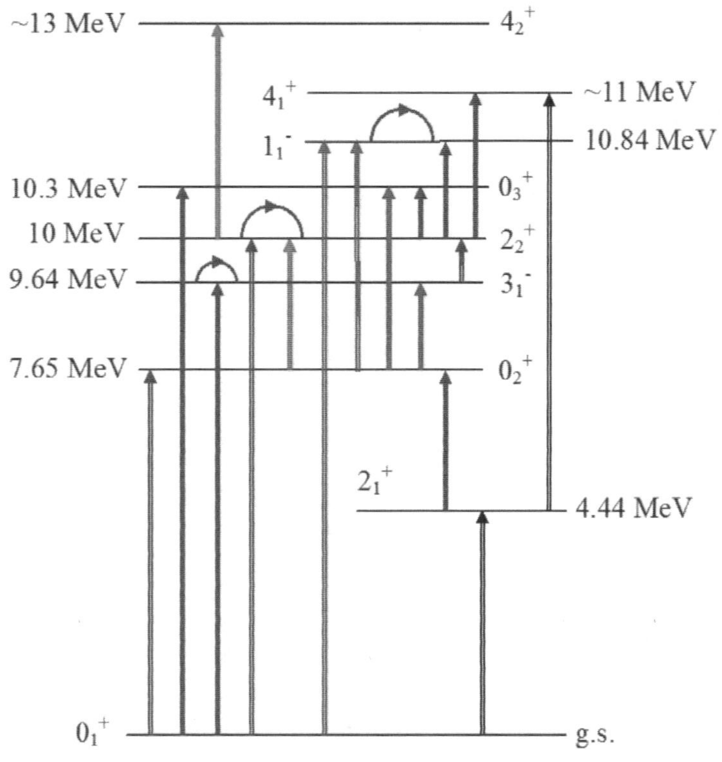

Figure 1. The coupling scheme used in the present CC analysis of the elastic and inelastic $\alpha+^{12}$C scattering data measured at $E_\alpha = 240$ and 386 MeV.

exit channel contains ^{12}C* being in an excited state that is generally more dilute, with the predicted $\langle r^2\rangle^{1/2}$ radius larger than that of the ground state (see Table 1), the O of each exit channel ($U_{A^*\to A^*}$) has been computed separately using the diagonal ($\lambda = 0$) density of ^{12}C* given by the AMD. Such a treatment of the exit O lead to a better agreement of the calculated cross sections with the (α,α') data, which helped to deduce accurately the $B(E\lambda)$ values for the excited states under study [19].

The AMD has been proven to give a realistic description of the shell-model like 0^+_1, 2^+_1 and 3^-_1 states as well as the cluster 0^+_2, 0^+_3 and 2^+_2 states. iven very strong $E\lambda$ strengths predicted for IS transitions between the cluster states, we have calculated all the corresponding FF's for the coupled channel (CC) calculation. In total, 39 diagonal O and inelastic FF's were obtained in the folding calculation using the AMD transition densities.

Table 1. Excitation energies and $E\lambda$ transition strengths of the IS states of ^{12}C under study. The calculated values are predicted by the AMD [9], and the best-fit values are given by the present folding model + CC analysis. $M(E\lambda)$ is given in e fm$^{\lambda+2}$ for the 0^+ and 1^- states.

J^π	$\langle r^2\rangle^{1/2}_{\rm calc}$ (fm)	$E_{\rm calc}$ (MeV)	$E_{\rm exp}$ (MeV)	Transition	calc. (e^2fm$^{2\lambda}$)	best-fit (e^2fm$^{2\lambda}$)	exp. (e^2fm$^{2\lambda}$)	Ref.
2^+_1	2.668	4.5	4.44	$B(E2;2^+_1\to0^+_1)$	8.4	8.4 ± 1.5	7.4 ± 0.2	[11]
							7.7 ± 1.0	[14]
							8.0 ± 0.8	[21]
				$B(E2;2^+_1\to4^+_1)$	28.5			
0^+_2	3.277	8.1	7.65	$M(E0;0^+_2\to0^+_1)$	6.6	4.5 ± 0.5	3.7 ± 0.2	[14]
							5.4 ± 0.2	[22]
				$B(E2;0^+_2\to2^+_1)$	25.5		13.0 ± 2.0	[23]
				$B(E3;0^+_2\to3^-_1)$	3122			
				$M(E0;0^+_2\to0^+_3)$	16.7			
				$M(E1;0^+_2\to1^-_1)$	12.5			
3^-_1	3.139	10.8	9.64	$B(E3;3^-_1\to0^+_1)$	74.4	59.5 ± 3.2	35.9 ± 1.4	[11]
							34.3 ± 5.7	[14]
							87.1 ± 1.3	[24]
				$B(E2;3^-_1\to3^-_1)$	70.1			
				$B(E3;3^-_1\to2^+_2)$	136.7			
				$M(E1;3^-_1\to2^+_2)$	3.71			
0^+_3	3.985	10.7	10.3	$M(E0;0^+_3\to0^+_1)$	2.3	2.9 ± 0.3	3.0 ± 0.2	[14]
				$B(E2;0^+_3\to2^+_2)$	1553			
2^+_2	3.993	10.6	9.84	$B(E2;2^+_2\to0^+_1)$	0.4	0.6 ± 0.1	0.37 ± 0.02	[11]
			10.03				0.73 ± 0.14	[12]
			10.13				1.57 ± 0.14	[13]
				$B(E2;2^+_2\to0^+_2)$	102			
				$B(E2;2^+_2\to2^+_2)$	518			
				$B(E2;2^+_2\to4^+_1)$	13.5			
				$B(E2;2^+_2\to4^+_2)$	1071			
1^-_1	3.424	12.6	10.84	$M(E1;1^-_1\to0^+_1)$	1.58	0.34 ± 0.04	0.31 ± 0.04	[14]
				$M(E1;1^-_1\to2^+_2)$	3.73			
				$B(E3;1^-_1\to2^+_2)$	1679			
				$B(E2;1^-_1\to1^-_1)$	20.1			

The present CC analysis (see the coupling scheme shown in Fig. 1) has included also the $E2$ reorientation of the 2^+, 3^- and 1^- states, with the reorientation FF$_{A^*\to A^*}$ given by the diagonal transition ($\lambda=2$) densities of ^{12}C* (see the corresponding $B(E2)$ values in Table 1). Thus, the present CC scheme contains three more couplings compared to that of our previous CC analysis [19]. One can see in Fig. 1 that the indirect routes to the 2_2^+ state via coupling to other cluster states are properly taken into account. Although many (α, α') experiments with ^{12}C target were performed in the last century, the accurate measurement of (α, α') spectrum for the cluster states around 10 MeV has been done only recently at the α energies of 240 and 386 MeV [11, 14]. In this work, we discuss mainly results of our folding model + CC analysis obtained for the Hoyle and 2_2^+ states. The IS transition strengths obtained for other states of ^{12}C are presented in Table 1, with the CC results being discussed in details in Ref. [19].

All the DWBA and CC calculations were done using the code ECIS97 written by Raynal [20]. To account for the higher-order dynamic contributions to the folded O [15] and to fine tune the strength of the CDM3 6 interaction for each energy, the real and imaginary folded O 's were renormalized by the coe cients $N_{\rm R}$ and $N_{\rm I}$, respectively, which were adjusted to the best optical model fit to the elastic data. These factors were then used to renormalize the complex folded FF's for the DWBA calculation [14, 15]. In the CC calculation, $N_{\rm R(I)}$ were re-adjusted again to account only for higher-order effects caused by the nonelastic channels not included into the CC scheme shown in Fig. 1, and then used to renormalize the complex folded FF's. Finally, the AMD transition densities were scaled to the best CC fit to the (α, α') data, giving the best-fit $E\lambda$ transition strengths presented in Table 1.

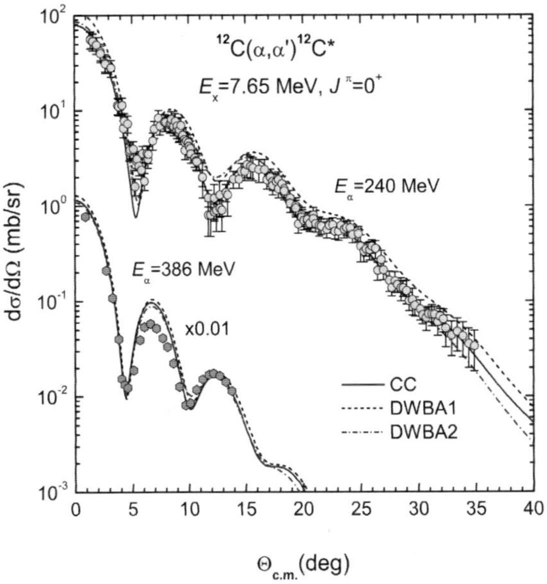

Figure 2. DWBA and CC descriptions of the (α, α') data for the Hoyle state, measured at E_α = 240 MeV [14] and 386 MeV [11]. The DWBA1 results were obtained using the same O for both the entrance and exit channels, and the DWBA2 and CC results were obtained with the O of the exit channel computed separately at the energy $E_\alpha - Q$, using the diagonal ($\lambda = 0$) density of ^{12}C*.

The CC analysis of the (α, α') data for the Hoyle state (see Fig. 2) has revealed important coupling effects that are best seen in the results obtained at the α energy of 240 MeV. As discussed in Ref. [25], the original multipole decomposition analyses (MDA) of the (α, α') data at different energies consistently gave a much weaker $E0$ transition strength of the Hoyle state, with $M(E0; 0_2^+ \to 0_1^+) \approx 3.6 \sim 3.8$ e fm^2 [14] that is about 30 weaker than the experimental value $M(E0)_{\rm exp} \approx 5.4$ e fm^2 deduced from the (e, e') data [22]. The DWBA1 calculation using the (rescaled) AMD transition density would give the best-fit $M(E0) \approx 3.65$ e fm^2, about the same as that given by the R M transition density rescaled to fit the (α, α') data in the DWBA [25]. The present CC calculation included all strong $E\lambda$ transitions from the Hoyle state to the neighboring cluster states (see Fig. 1), and the best-fit $E0$ strength given by the folding model + CC analysis of the 240 MeV data is $M(E0) \approx 4.5$ e fm^2 which is about 20 stronger than

Journal of Physics: Conference Series **569** (2014) 012015 doi:10.1088/1742-6596/569/1/012015

that given by the standard DWBA analysis. We expect that a full coupled reaction channel analysis of the (α, α') data including different breakup channels would yield the $M(E0)$ value closer to the (e, e') data, and thus explain the missing monopole strength of the Hoyle state in (α, α') scattering that can only be accounted for in the DWBA by an enhanced absorption in the exit channel [25, 26].

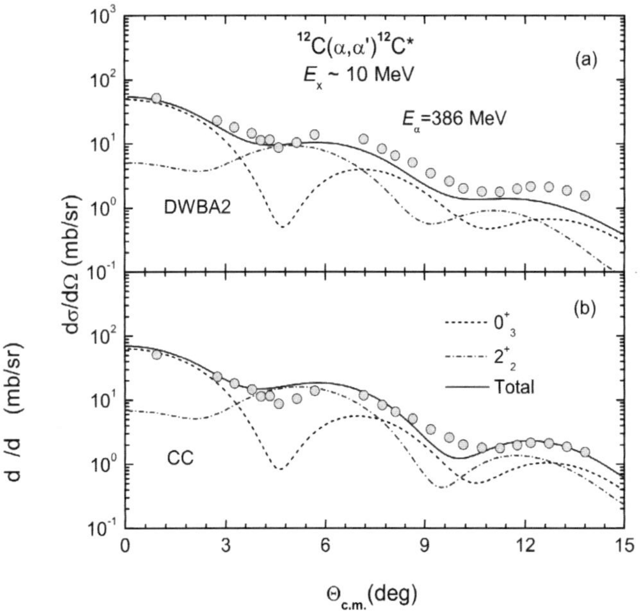

Figure 3. DWBA (a) and CC (b) descriptions of the (α, α') data measured at $E_\alpha = 386$ MeV for the 0_3^+ and 2_2^+ states [11]. The DWBA2 and CC results were obtained in the same way as described in the caption of Fig. 2. The FF for the 2_2^+ state has been scaled to the best CC fit to the data, giving $B(E2\downarrow) \approx 0.6\ e^2\text{fm}^4$.

The MDA of the (α, α') data measured at $E_\alpha = 386$ MeV has shown a broad 0_3^+ resonance and a narrower 2_2^+ state centered at the excitation energies $E_x \approx 9.93$ and 9.84 MeV, respectively. The (α, α') angular distribution deduced for the wide bump centered at $E_x \approx 10$ MeV [11] contains mainly the contributions from the 2_2^+ and 0_3^+ states (see Fig. 3). iven the $E0$ strength of the 0_3^+ state accurately determined in the CC analysis [19] of the 240 MeV data, the $E2$ strength of the 2_2^+ state remains the only parameter in the CC analysis of the 386 MeV (α, α') data. After the FF for the 2_2^+ state is scaled to the best CC fit to the (α, α') data, we obtained $B(E2; 2_2^+ \to 0_1^+) \approx 0.6\ e^2\text{fm}^4$, in a good agreement with $B(E2\downarrow)_{\text{exp}} \approx 0.73\ e^2\text{fm}^4$ deduced from the photodissociation data [12]. Note that this value is lower than $B(E2\downarrow) \approx 1.57\ e^2\text{fm}^4$, deduced from the revised analysis of the $^{12}\text{C}(\gamma, \alpha)^8\text{Be}$ data [13].

A question now is why the 2_2^+ state has not been found at $E_x \approx 10$ MeV by the MDA of the (α, α') data taken at $E_\alpha = 240$ MeV [14]. iven the realistic $E\lambda$ strengths of the IS states found in our folding model + CC analysis of both data sets, we have analysed again the 240 MeV (α, α') data measured for several energy bins around 10 MeV. The best-fit $E\lambda$ transition strengths of the 2_2^+, 0_3^+ and 1_1^- states (Table 1) were distributed over the considered energy bins by an averaging procedure [19] to estimate the corresponding inelastic FF. From the CC results obtained with and without the contribution from the 2_2^+ state one can see in Fig. 4 that this state is indeed present in the energy bins around $E_x \sim 10$ MeV. Such a subtle effect could not be resolved by the original MDA of the 240 MeV (α, α') data.

We are indebted to M. Itoh, B. John, T. Kawabata, and W.R. immerman for their helpful communications. This work has been supported, in part, by VINATOM and the LIA project of

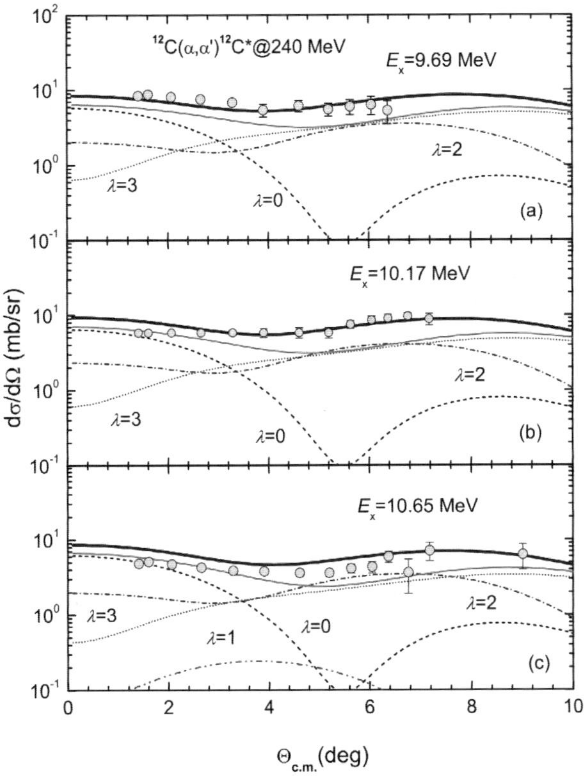

Figure 4. Differential (α, α') cross sections measured at $E_\alpha = 240$ MeV [14] for the 475 keV-wide energy bins centered at $E_x = 9.69$ MeV (a), 10.17 MeV (b), and 10.65 MeV (c), in comparison with the CC results given by different 2^λ-pole transition strengths. The total cross sections obtained with and without the contribution from the 2_2^+ state are shown as the thick (blue) and thin (red) solid lines, respectively.

the Ministry of Science and Technology of Vietnam.

[1] M. Freer 2007 *Rep. Prog. Phys.* **70** 2149
[2] Y. Funaki, H. Horiuchi, W. von Oertzen, G. Röpke, P. chuck, A. Tohsaki and T. Yamada 2009 *Phys. Rev.* C **80** 064326
[3] E. Uegaki, . Okabe, Y. Abe and H. Tanaka 1977 *Prog. Theor. Phys.* **57** 1262
[4] M. Kamimura 1981 *Nucl. Phys.* A **351** 456
[5] R. Pichler, H. Oberhummer, A. Csoto and .A. Moszkowski 1997 *Nucl. Phys.* A **618** 55
[6] A. Tohsaki, H. Horiuchi, P. chuck and G. Röpke 2001 *Phys. Rev. Lett.* **87** 192501
[7] Y. Funaki, A. Tohsaki, H. Horiuchi, P. chuck and G. Röpke 2003 *Phys. Rev.* C **67** 051306(R)
[8] H. Morinaga 1956 *Phys. Rev.* **101** 254
[9] Y. Kanada-En yo 2007 *Prog. Theor. Phys.* **117** 655
[10] D.T. Khoa, D.C. Cuong and Y. Kanada-En yo 2011 *Phys. Lett.* B **695** 469
[11] M. Itoh et al. 2011 *Phys. Rev.* C **84** 054308
[12] W.R. Zimmerman et al. 2013 *Phys. Rev. Lett.* **110** 152502
[13] W.R. Zimmerman, PhD Dissertation, http: digitalcommons.uconn.edu dissertations 230
[14] B. John, Y. Tokimoto, Y.W. Lui, H.L. Clark, . Chen and D.H. Youngblood 2003 *Phys. Rev.* C **68** 014305
[15] D.T. Khoa and G.R. atchler 2000 *Nucl. Phys.* A **668** 3
[16] D.T. Khoa, G.R. atchler and W. von Oertzen 1997 *Phys. Rev.* C **56** 954
[17] D.C. Cuong, D.T. Khoa and G. Colò 2010 *Nucl. Phys.* A **836** 11
[18] J.P. Jeukenne, A. Lejeune and C. Mahaux 1977 *Phys. Rev.* C **16** 80
[19] D.C. Cuong, D.T. Khoa and Y. Kanada-En yo 2013 *Phys. Rev.* C **88** 064317
[20] J. Raynal, in *Computing as a Language of Physics* (IAEA, Vienna, 1972) p.75; J. Raynal, coupled-channel code ECI 97 (unpublished)
[21] . Raman, C.W. Nestor Jr. and P. Tikkanen 2001 *At. Data and Nucl. Data Tables* **78** 1
[22] P. trehl 1970 *Z. Phys.* **234** 416
[23] P.M. Endt 1979 *At. Data and Nucl. Data Tables* **23** 3; F. Ajenberg- elove 1990 *Nucl. Phys.*A **506** 1
[24] T. Kibedi and R.H. pear 2002 *At. Data and Nucl. Data Tables* **80** 35
[25] D.T. Khoa and D.C. Cuong 2008 *Phys. Lett.* B **660** 331
[26] D.T. Khoa 2008 *Intern. J. Modern Phys.* E **17** 2055

Journal of Physics: Conference Series **569** (2014) 012016 doi:10.1088/1742-6596/569/1/012016

Three-α model calculations of low-lying ^{12}C states and the triple-α reaction

Souichi Ishikawa

cience Research Center, Hosei University, 2-17-1 Fujimi, Chiyoda, Tokyo 102-8160, Japan

E-mail: ishikawa@hosei.ac.jp

Abstract. Properties of some ^{12}C resonant states at low energies are studied in three-α-particle model. Calculations for breakup reactions of ^{12}C bound state leading to three-body continuum states are performed at energies up to a few MeV above the breakup threshold by applying the Faddeev three-body formalism in coordinate space. Calculated cross sections reveal some resonant peaks. Decay modes and the structure at interior region of these resonant states as well as their effects on the triple-alpha reaction rate will be reported.

1. Introduction

Low-lying resonant states of ^{12}C nucleus are interesting objects to study, because some curious structures having a particular configuration of α-particles, such as a linear chain, an equilateral triangle, or a bent-arm, have been predicted to exist [1, 2, 3, 4, 5]. Furthermore these states, especially the second 0^+ state $^{12}\mathrm{C}(0_2^+)$ (the Hoyle state) [6], are known to play an important role in the triple-alpha (3α) process,

$$\alpha + \alpha + \alpha \to {}^{12}\mathrm{C} + \gamma, \tag{1}$$

by which three α-particles (3-α) are fused into a ^{12}C nucleus in stars.

In a recent paper [7], the present author studied an inverse reaction of (1), namely the E2-photodisintegration of $^{12}\mathrm{C}(2_1^+)$ bound state leading to final 3-α continuum states with total angular momentum $J = 0$,

$$^{12}\mathrm{C}(2_1^+) + \gamma(\mathrm{E2}) \to \alpha + \alpha + \alpha \ (J = 0), \tag{2}$$

and then calculated the 3α reaction rate for stellar temperatures of $T = 10^7$ K to 10^9 K. It turned out that the calculated reaction rate for $T > 10^8$ K agrees with a standard rate by the Nuclear Astrophysics Compilation of Reaction Rates (NACRE) [8], and is about one order of magnitude larger than the NACRE rate at $T = 10^7$ K.

In Ref. [7], a wave function for the process (2) was defined and solved by applying the Faddeev three-body formalism in coordinate space. Calculations were performed for an energy range of the 3-α system from 10 keV to 600 keV, which includes the energy of the two-α resonant state $^8\mathrm{Be}(0_1^+)$ (92 keV) and that of the Hoyle state (380 keV). In the present work, aiming to calculate the 3α reaction rate at higher temperatures as 10^{10} K, the calculations are extended by increasing energies of the 3-α system up to several MeV, where some additional resonant states with higher angular momenta exist. In the following, after shortly reviewing a formalism to calculate the disintegration process in Sec. 2, interaction models among α-particles used in this work, and then results for resonant energies, energy spectrum of decaying α-particles, the structure of resonant states at interior region, and their effects on the thermal 3α reaction rate will be presented in Sec. 3. Summary will be given in Sec. 4.

2. Formalism

Let us consider the photodisintegration process of the $^{12}\mathrm{C}(J_A^\pi)$ bound state by the Eλ photon of the energy E_γ, leading to 3-α continuum states of the angular momentum J. Once the cross section for the process $\sigma_\gamma(E_\gamma)$ is obtained, the 3α reaction rate $\langle\alpha\alpha\alpha\rangle$ is calculated by

$$\langle\alpha\alpha\alpha\rangle = \frac{(3)^{3/2}3!\ 8\pi c}{(m_\alpha c^2)^3}\left(\frac{\hbar c}{k_{\mathrm{B}}T}\right)^3 \sum_{J_A,J,\lambda}(2J_A+1)\int_0^\infty dE_\gamma E_\gamma^2 \sigma_\gamma(E_\gamma) e^{-(E_\gamma-|E_A|)/k_{\mathrm{B}}T}, \tag{3}$$

where m_α is the α particle mass and E_A is the energy of the initial bound state.

Let us define a wave function for the disintegration process of a bound state $|\Psi_A\rangle$ by the electromagnetic interaction H_γ:

$$\Psi(\boldsymbol{x},\boldsymbol{y}) = \langle\boldsymbol{x},\boldsymbol{y}|\frac{1}{E+\imath\epsilon-H}H_\gamma|\Psi_A\rangle, \tag{4}$$

where $\boldsymbol{x}$ is the relative coordinate of an α-pair, $\boldsymbol{y}$ is the relative coordinate of the third (spectator) α-particle with respect to the center of mass of the pair, E is the 3-α energy in the center of mass system, $E = E_\gamma - |E_A|$, and H is the Hamiltonian of the 3-α system. The wave function Ψ will be obtained by applying the Faddeev three-body formalism [9] to solve integral equations in coordinate space with accommodating the long range Coulomb force effects. (See Refs. [7, 10] for the details of the calculations.) This procedure leads to a three-body breakup amplitude $F^{(B)}(\hat{\boldsymbol{q}},\hat{\boldsymbol{p}},E_q)$, where $\boldsymbol{q}$ is relative momentum of a α-α pair in the final state, $\boldsymbol{p}$ is the momentum of the spectator α particle, and E_q is the relative α-α energy. Here, the amplitude $F^{(B)}(\hat{\boldsymbol{q}},\hat{\boldsymbol{p}},E_q)$ is the complex conjugate of an amplitude for the capture process (1), and from which the photodisintegration cross section is given by

$$\sigma_\gamma(E_\gamma) = \frac{1}{8\pi}\frac{1}{2J_A+1}\frac{1}{\hbar c K^3}\left(\frac{3}{4}\right)^2\int d\hat{\boldsymbol{q}}d\hat{\boldsymbol{p}}dE_q qp|F^{(\mathrm{B})}(\hat{\boldsymbol{q}},\hat{\boldsymbol{p}},E_q)|^2, \tag{5}$$

where $K=\sqrt{\frac{m_\alpha}{\hbar^2}E}$.

3. Results

3.1. Interactions

The mode D of the Ali-Bodmer potential [11] with the (point) Coulomb potential is used for the α-α interaction. In addition, 3-α potentials (3α) are introduced to reproduce the properties of ^{12}C, namely binding energies and resonance energies. We used two different functional forms: one is taken from Ref. [12]:

$$W_\Delta = \sum_{J=0,2}\hat{P}_J W_J \exp\left(-\frac{r_{12}^2+r_{23}^2+r_{31}^2}{a^2}\right) + [\text{c.p.}], \tag{6}$$

and the other is from Ref. [13]:

$$W_V = \sum_{J=0,2}\hat{P}_J W_J \exp\left(-\frac{r_{12}^2+r_{23}^2}{a^2}\right) + [\text{c.p.}], \tag{7}$$

Table 1. The range and strength parameters of the 3α models used in this work, and results for the reaction (8) (See the text).

Model	Eq.	a (fm)	$W_0^{(b)}$ (MeV)	$W_0^{(c)}$ (MeV)	W_2 (MeV)	$E_{peak}[^{12}\mathrm{C}(2_2^+)]$ (MeV)	$B(\mathrm{E2},2_2^+ \to 0_1^+)$ ($\mathrm{e^2fm^4}$)
Δ_L	(6)	3.39	-29.2	-30.95	-15.3	2.4	2.1
Δ_S	(6)	2.61	-99.0	-101.22	-52.7	2.5	0.92
V_L	(7)	3.33	-14.3	-12.21	-7.7	1.9	3.6
V_S	(7)	2.50	-42.8	-39.64	-22.4	2.2	1.8

where [c.p.] denotes the cyclic permutations and $\hat{P}_J$ is the projection operator on a 3-α state with the total angular momentum J. We examined two different values for the range parameter a both for W_Δ (6) and W_V (7). Calculations with these 3α models will be denoted as Δ_L, Δ_S, V_L, and V_S, in which the subscript L (S) indicates the choice of a long-range (short-range) parameter. Chosen values of the range and strength parameters will be described below (Table 1).

3.2. Photodisintegration cross section (Transition strength function)

We studied, in addition to the E2 photodisintegration of the $^{12}\mathrm{C}(2_1^+)$ state (2), that of the $^{12}\mathrm{C}(0_1^+)$ state,

$$^{12}\mathrm{C}(0_1^+) + \gamma(\mathrm{E2}) \to \alpha + \alpha + \alpha \ (J = 2). \tag{8}$$

In solving Ψ in (4), the maximum angular momentum of the α-α subsystem in 3-α partial wave components is chosen to be 4, while was 2 in Ref. [7], which causes little change compared to the results in Ref. [7].

In calculating the reaction (2), the 3α strength parameter for $J = 2$ state, W_2, is determined to reproduce the binding energy of $^{12}\mathrm{C}(2_1^+)$, and that for $J = 0$ state, $W_0^{(c)}$, is determined to reproduce the Hoyle state energy, as shown in Table 1.

In the following, the E2-strength function defined by

$$\frac{dB(E\lambda)}{dE} = \frac{\lambda\,[(2\lambda+1)!!]^2}{(2\pi)^3(\lambda+1)} \frac{\sigma_\gamma(E_\gamma)}{k_\gamma^{\lambda-1/2}}, \tag{9}$$

where $\lambda = 2$ in the present case and k_γ is the momentum of the photon, will be presented instead of the photodisintegration cross section $\sigma_\gamma(E_\gamma)$ to accent peak structures corresponding to resonances.

Fig. 1 shows the calculated E2 strength functions for the reaction (2), in which we observe, in addition to sharp peak corresponding to $^{12}\mathrm{C}(0_2^+)$ at $E = 0.38$ MeV, two peaks: one is around $E = 1$ MeV and the other is around $E = 4$ MeV. The lower-energy peak corresponds to the resonant state $^{12}\mathrm{C}(0_3^+)$ whose existence was predicted by complex scaling calculations [14, 15]. Such a two-peak structure of 0^+ ^{12}C state is indicated experimentally in Ref. [16]. For higher-energy peak, which is assigned as $^{12}\mathrm{C}(0_4^+)$, there is a rather large amount of model dependence in position, width, and height of the strength-peak. Therefore further experimental information about $^{12}\mathrm{C}(0_4^+)$ state is useful for the improvement of 3-α model interactions.

For the reaction (8), the 3α strength parameter for $J = 0$ state had to be changed from $W_0^{(c)}$ by at most 10 to reproduce the $^{12}\mathrm{C}(0_1^+)$ bound state energy (shown as $W_0^{(b)}$ in Table 1). The construction of a 3α that reproduces both of bound and continuum states is left for a future work. For the $J = 2$ strength of 3α , the same values are used as in previous ones.

Journal of Physics: Conference Series **569** (2014) 012016 doi:10.1088/1742-6596/569/1/012016

Calculations of the E2-strength function for the process (8) are shown in Fig. 2. The results show a resonant structure corresponding to ^{12}C(2_2^+) state, which appeared in some theoretical calculations, (*e.g.*, [2, 3, 4, 5, 14, 15]), and experimentally observed [16, 17, 18, 19]. Calculated peak energy, $E_{peak}[^{12}\mathrm{C}(2_2^+)]$, and the E2 transition strength $B(\mathrm{E2},2_2^+ \to 0_1^+)$ are summarized in Table 1. The calculated peak energies are subequal or small by a few hundred keV compared to the experimental results: 2.48(15) MeV [18] and 2.76(11) MeV [19]. Calculated values of the strength $B(\mathrm{E2},2_2^+ \to 0_1^+)$ are larger than the recent result of $^{12}\mathrm{C}(\gamma,\alpha)$ reaction, 0.73(13) $e^2\mathrm{fm}^4$ [19], but not so much different from values obtained from $^{12}\mathrm{C}(\alpha,\alpha')$ reaction, 1.83(9) $e^2\mathrm{fm}^4$ or 1.6(2) $e^2\mathrm{fm}^4$ [16]. These might suggest that we should examine to change the strength parameter W_2 as well as for W_0.

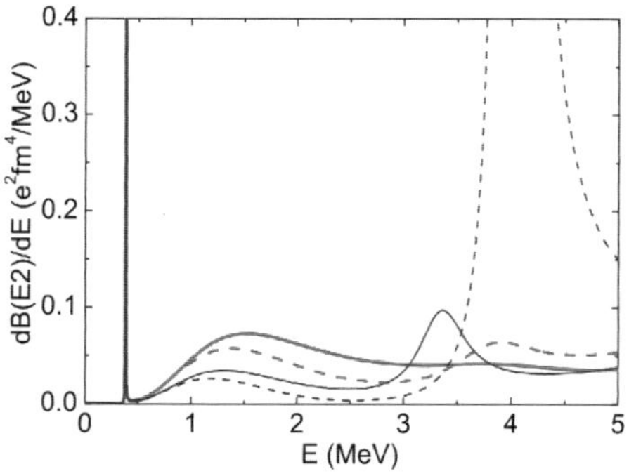

Figure 1. E2 strength function for the process (2) as a function of the 3-α energy E. The (red) thick dashed and solid lines indicate results of the Δ_L and Δ_S, respectively. The (blue) thin dashed and solid lines indicate results of the V_L and V_S, respectively.

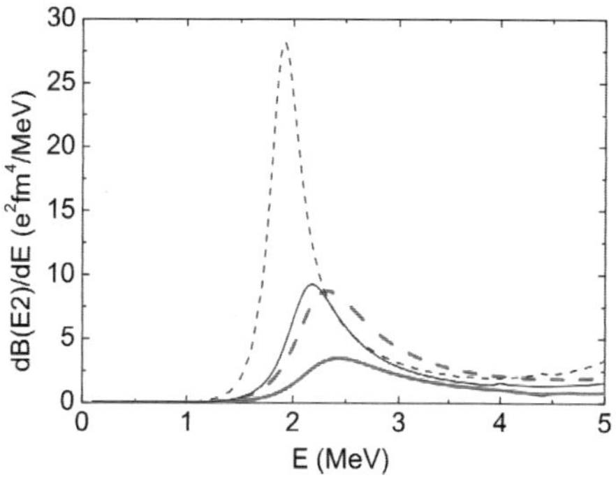

Figure 2. E2 strength function for the process (8) as a function of the 3-α energy E. The (red) thick dashed and solid lines indicate results of the Δ_L and Δ_S, respectively. The (blue) thin dashed and solid lines indicate results of the V_L and V_S, respectively.

3.3. Spectrum of decaying α particles

Fig. 3 shows the spectrum of decaying α-particles from the Hoyle state for the V_S model, which is obtained by integrating $|F^{(B)}(\hat{\boldsymbol{q}},\hat{\boldsymbol{p}},E_q)|^2$ in Eq. (5) over $\hat{\boldsymbol{q}}$ and $\hat{\boldsymbol{p}}$. The spectrum has a peak at $E_q = E_r[^8\mathrm{Be}(0_1^+)]$ corresponding to a sequential process, in which the Hoyle state first decays to ^{8}Be(0_1^+) and α-particle and then ^{8}Be(0_1^+) to two α-particles. A contribution of such sequential decay (SD) process is calculated by integrating the spectrum just around $E_q = E_r[^8\mathrm{Be}(0_1^+)]$. More than 99 of the total decay is coming from the SD process for the Hoyle state, which is consistent with experimental results of Refs. [20, 21, 22, 23], but not with that of Ref. [24]. While such large contributions of the SD channel are calculated for the decay processes of ^{12}C(0_3^+) and ^{12}C(2_2^+) states, a rather small (about 50 of the total) SD contribution is calculated for ^{12}C(0_4^+).

3.4. Structure of the Hoyle state

Although the function $\Psi(\boldsymbol{x},\boldsymbol{y})$ (4) is not the eigen-state of the 3-α Hamiltonian, its interior structure is expected to reflect that of the resonance state because of the 3-α reen's function. Actually, $\Psi(\boldsymbol{x},\boldsymbol{y})$ at the resonance energy has a concentration of the amplitude at interior region.

Journal of Physics: Conference Series **569** (2014) 012016 doi:10.1088/1742-6596/569/1/012016

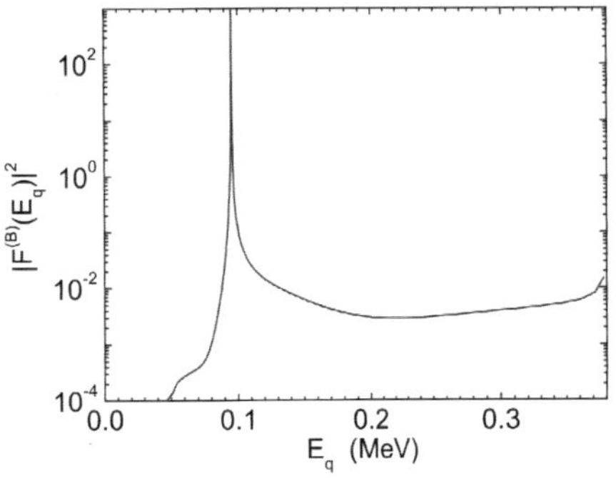

Figure 3. Energy distribution of decaying α-particles from the Hoyle state for the V_S model.

In Fig. 4, the density distribution $\rho(x, y)$ of the disintegration wave functions,

$$\rho(x, y) = x^2 y^2 \int d\hat{\boldsymbol{x}} d\hat{\boldsymbol{y}} |\Psi(\boldsymbol{x}, \boldsymbol{y})|^2, \tag{10}$$

calculated at the Hoyle resonance energy, is plotted. Here, the wave function $\Psi(\boldsymbol{x}, \boldsymbol{y})$, which is not normalizable since it is continuum state function, is artificially normalized within the region: $\{x < 12 \text{ fm}, y \leq 12 \text{ fm}\}$.

The density has three distinct local peaks denoted by A, B, and C in the figure, which are located at $(x, y) \sim (2.6 \text{ fm}, 2.3 \text{ fm})$, $(3.3 \text{ fm}, 4.4 \text{ fm})$, and $(5.6 \text{ fm}, 1.8 \text{ fm})$, respectively. Similar peak structure is appeared in calculations of a 3-α eigen-state performed in Refs. [25, 26].

The peak A corresponds to the configuration of an equilateral triangle of side length 2.6 fm. The peaks B and C correspond to a bent-arm configuration, in which three α-particles compose an isosceles triangle with two equal sides of length 3.3 fm and the third side length being 5.6 fm. Because of the symmetric property of the wave function, a bent-arm configuration appears at two points for the density distribution.

Since each peak is associated with wide slopes, three α-particles may take each triangle configuration rather loosely. The probability to find 3-α taking the equilateral triangle configuration is estimated by integrating the density $\rho(x, y)$ for the domain of a square, 1.5 fm on a side, around the peak A. This gives about 10 . Similar procedure for the peak B (C) gives about 20 (10). Therefore, probabilities for the equilateral triangle and the bent-arm configurations are roughly about 10 and 30 , respectively. The density distribution shows that the Hoyle state has a mixed configuration of the equilateral triangle and the bent-arm.

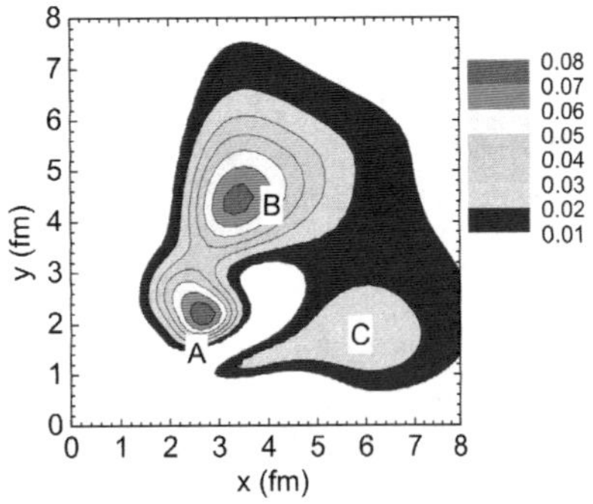

Figure 4. Contour plot of the density distribution $\rho(x, y)$ of the Hoyle state for the V_S model. See the text for the meaning of the characters, 'A', 'B', and 'C' in the figure.

3.5. 3α reaction rate

The 3α reaction rates calculated by Eq. (3) including both of the processes (2) and (8) are shown as the ratio to the NACRE 3α rate [8] in Fig. 5. The dashed-dotted line demonstrates that effects of 3-α states with $J = 0$ are small on the 3α rate for $T > 2 \times 10^9$ K, and the difference among the other four lines indicates an importance of detailed information on the ^{12}C(2_2^+) resonant state.

The calculated rates at lower temperatures, as $T \sim 10^7$ K, are large by an order at most than the NACRE rate, which is consistent with recent calculation by the imaginary time method [27], but not with those of Refs. [28, 25] that give a huge enhancement by many orders.

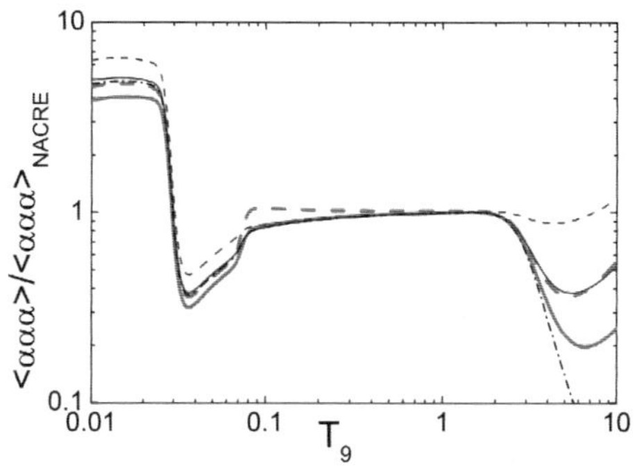

Figure 5. The ratio of the calculated 3α reaction rates to the NACRE 3α rate as a function of the temperature $T_9 = T/10^9$ K. The (red) thick dashed and solid lines indicate results of the Δ_L and Δ_S, respectively. The (blue) thin dashed and solid lines indicate results of the V_L and V_S, respectively. The (black) dashed-dotted line denotes the rate calculated with only the $J = 0$ 3-α state (2) for the V_S model.

4. Summary

Low energy 3-α continuum states are studied via the breakup reactions of the ground 0_1^+ state and the first 2_1^+ excited state of ^{12}C by the E2 photon leading to 2^+ and 0^+ final states, respectively. The Ali-Bodmer α-α potential together with some forms of 3-α potential are used for the 3-α calculations. The calculated cross sections of the processes show some peaks corresponding to ^{12}C(0_2^+), ^{12}C(0_3^+), ^{12}C(0_4^+), and ^{12}C(2_2^+) resonant states. The sequential decay process dominates for 3-α decay processes of ^{12}C(0_2^+), ^{12}C(0_3^+), and ^{12}C(2_2^+), but not for ^{12}C(0_4^+). The density distribution of the wave function for the process with the Hoyle state, ^{12}C(0_2^+), has some peaks corresponding to the equilateral triangle configuration (10) and the bent-arm configuration (30). Contribution of the ^{12}C(2_2^+) resonant state to the 3α reaction rate is significant for temperatures of 2×10^9 K $< T < 10^{10}$ K, although a rather large dependence on the choice of 3-α potential requires further studies.

References

[1] Morinaga H 1956 *Phys. Rev.* **101** 254
[2] Uegaki E, Okabe , Abe Y and Tanaka H 1977 *Prog. Theor. Phys.* **57** 1262
[3] Kanada-En yo Y 2007 *Prog. Theor. Phys.* **117** 655
[4] Chernykh M, Feldmeier H, Neff T, von Neumann-Cosel P and Richter A 2007 *Phys. Rev. Lett.* **98** 032501
[5] Epelbaum E, Krebs H, Lahde T A, Lee D and Mei ner U-G 2012 *Phys. Rev. Lett.* **109** 252501
[6] Hoyle F 1954 *Astrophys. J. Suppl.* **1** 121
[7] Ishikawa 2013 *Phys. Rev.* C **87** 055804
[8] Angulo C *et al* 1999 *Nucl. Phys.* A **656** 3
[9] Faddeev L D 1961 *Sov. Phys.-JETP* **12** 1041
[10] Ishikawa 2009 *Phys. Rev.* C **80** 054002
[11] Ali and Bodmer A R 1966 *Nucl. Phys.* **80** 99
[12] Fedorov D V and Jensen A 1996 *Phys. Lett.* B **389** 631
[13] Portilho O, Agrello D A and Coon A 1983 *Phys. Rev.* C **27** 2923
[14] Kurokawa C and Kato K 2007 *Nucl. Phys.* A **792** 87

[15] Ohtsubo -I, Fukushima Y, Kamimura M and Hiyama E 2013 *Prog. Theor. Exp. Phys.* **2013** 73D02
[16] Itoh M *et al* 2011 *Phys. Rev.* C **84** 054308
[17] Freer M *et al* 2009 *Phys. Rev.* C **80** 041303
[18] Freer M *et al* 2012 *Phys. Rev.* C **86** 034320
[19] Zimmerman W R *et al* 2013 *Phys. Rev. Lett.* **110** 152502
[20] Freer M *et al* 1994 *Phys. Rev.* C **49** R1751
[21] Manfredi J *et al* 2012 *Phys. Rev.* C **85** 037603
[22] Kirsebom O *et al* 2012 *Phys. Rev. Lett.* **108** 202501
[23] Rana T K *et al* 2013 *Phys. Rev.* C **88** 021601
[24] Raduta A R *et al* 2011 *Phys. Lett.* B **705** 65
[25] Nguyen N B, Nunes F M and Thompson I J 2013 *Phys. Rev.* C **87** 054615
[26] Vasilevsky V, Arickx F, Vanroose W and Broeckhove J 2012 *Phys. Rev.* C **85** 034318
[27] Akahori T, Funaki Y and Yabana K 2014 *Preprint* nucl-th 1401.4390
[28] Ogata K, Kan M and Kamimura M 2009 *Prog. Theor. Phys.* **122** 1055

Journal of Physics: Conference Series **569** (2014) 012017 doi:10.1088/1742-6596/569/1/012017

Application of the complex scaling method for few-body scattering problems including the breakup

Rimantas Lazauskas

Université de trasbourg, IPHC, CNR , 23 rue du Loess, 67037 trasbourg, France.

E-mail: `rimantas.lazauskas@iphc.cnrs.fr`

Abstract. A formalism based on the complex-scaling method is used to solve a few-particle scattering problem in configuration space. This method allows to use trivial boundary conditions and is compatible with most of the bound state techniques. In this contribution calculations of neutron scattering on triton is presented using realistic nuclear Hamiltonians for neutron energies above four-nucleon breakup threshold.

1. Introduction

Rigorous description of the quantum-mechanical collisions is one of the most complex problems in theoretical physics. During the last two decades exact numerical solutions for bound states containing dozens of particles became available, however the full solution of the scattering problem (containing elastic, rearrangement and breakup channels) remained limited to the three-body case. Therefore there is a natural interest in adapting bound-state techniques to solve the scattering problems. In particular, some recent efforts have been successful [1, 2, 3, 4, 5, 6].

The main di culty to solve the scattering problem in configuration space is related to the fact that, unlike the bound state wave functions, scattering wave functions are not localized. One is therefore obliged to find solutions for multidimensional differential equations, which satisfy extremely complex boundary conditions. Number of the open channels increases rapidly with number of particles and or total energy of the system, strongly complicating asymptotic structure of the wave function of the system. In particular the asymptotic structure of the systems wave function becomes complex if breakup channels in three or more particle clusters are open. Notably breakup asymptotic for charged clusters are not known analytically. Therefore it is rather obvious that one will not be able to handle such a complicate boundary behavior directly for the systems containing more than three or four particles [1].

Therefore, finding a method which enables us to solve the scattering problem without an explicit use of the asymptotic form (or singularities in momentum space) of the wave function is of great importance. In this contribution I will present a formalism based on the complex-scaling method [7, 6]. Several applications have been already provided proving e ciency of the method in describing elastic and three-body breakup reactions for Hamiltonians which may combine short-range, Coulomb as well as optical potentials [8, 10]. In this presentation this formalism

[1] One should notice that the scattering wave functions are localized in momentum space, however they contain complex structure of singularities which are as difficult to treat numerically as a boundary conditions in configuration space [2, 4]

Journal of Physics: Conference Series **569** (2014) 012017 doi:10.1088/1742-6596/569/1/012017

is applied to solve four-nucleon scattering problem above the breakup threshold using realistic interactions.

2. Guidelines of the formalism

In our recently published review article [8] we have explained in detail how to implement the complex scaling method to solve the general scattering problem for a few-particle systems. Here, without loosing generality, we will briefly summarize the main ideas of this formalism and in particular its application in conjunction with the Faddeev- akubovski (F) for a system of four identical particles.

The method of complex scaling (or coordinates) for two- and three-particle scattering was proposed by Nuttall and Cohen [7]. The main idea of the method is that, the configuration is composed of only single incoming wave Ψ^{in}, determined by the incident state and thus known *a priori*, and as many outgoing waves as there are open channels. Regardless their complexity the outgoing wave functions are some analytic functions of the coordinates which will fall off exponentially for rotated $r = |r|e^{i\theta}$, where $0 < \theta < \pi$. Thus one may set $\Psi^{out} = \Psi - \Psi^{in}$ with Ψ^{out} representing the difference between the complete scattering wave function Ψ and the wave function Ψ^{in} describing the incident state. Ψ^{out} should contain only outgoing waves and thus fall off exponentially for the complex scaled coordinates. The physical observables are straightforwardly related with the scattering amplitudes, which on their turn might be expressed as integrals involving Ψ^{out}, regardless the fact that only transform of this function is available for the complex scaled coordinates. Therefore the whole calculation can be performed using a wave functions part Ψ^{out} which decreases exponentially and thus using standard bound-state techniques to approximate it.

When considering four-nucleon scattering problem we start with Faddeev- akubovski (F) equations for a system of four identical particles [2]. In this particular case, we start by separating the incoming plane wave from the F components (see figure 1):

$$K(\vec{x},\vec{y},\vec{z}) = K^{out}(\vec{x},\vec{y},\vec{z}) + K^{in}(\vec{x},\vec{y},\vec{z}) \tag{1}$$

$$H(\vec{x},\vec{y},\vec{z}) = H^{out}(\vec{x},\vec{y},\vec{z}) + H^{in}(\vec{x},\vec{y},\vec{z}). \tag{2}$$

Note, that the four-body F equations involve components of two types, K and H. These components in their $z \to \infty$ asymptotes separate respectively elastic 3+1 and 2+2 particle channels. Therefore, in the last equation, one has $H^{in} \equiv 0$ when considering collision of 3+1 particle clusters, whereas $K^{in} \equiv 0$ when considering 2+2 particle collisions. Asymptotes of the F components K^{out} and H^{out} will contain only various combinations of the outgoing waves. These components are solutions of the driven F equations, which have the form:

$$\begin{aligned}(E - H_0 - V_{12})\,K^{out} \quad &- \quad V_{12}(P^+ + P^-)\left[(1+Q)K^{out} + H^{out}\right] \\ &= V_{12}(P^+ + P^-)\left[(1+Q)H^{in} + QK^{in}\right], \end{aligned} \tag{3}$$

$$(E - H_0 - V_{12})\,H^{out} \quad - \quad V_{12}\tilde{P}\left[(1+Q)K^{out} + H^{out}\right] = V_{12}\tilde{P}\left[(1+Q)K^{in}\right], \tag{4}$$

where $P^- = P_{23}P_{12}$, $P^+ = P_{12}P_{23}$, $Q = P_{34}$ and $\tilde{P} = P_{13}P_{24}$ are particle permutation operators.

It is practical to use smooth complex scaling operator (CSO), which acts only on hyperradial variable $r^2 = x^2 + y^2 + z^2$ and thus does not affect coordinate transformation operators. The smooth CSO is expressed as

$$\widehat{S} = e^{i\theta r\frac{\partial}{\partial r}} = e^{i\theta(x\frac{\partial}{\partial x} + y\frac{\partial}{\partial y} + z\frac{\partial}{\partial z})}, \tag{5}$$

[2] Equivalent formulation is possible for the chrödinger equation. We prefer to use FY equations simply because we dispose some earlier developed tools, which are proper to these equations.

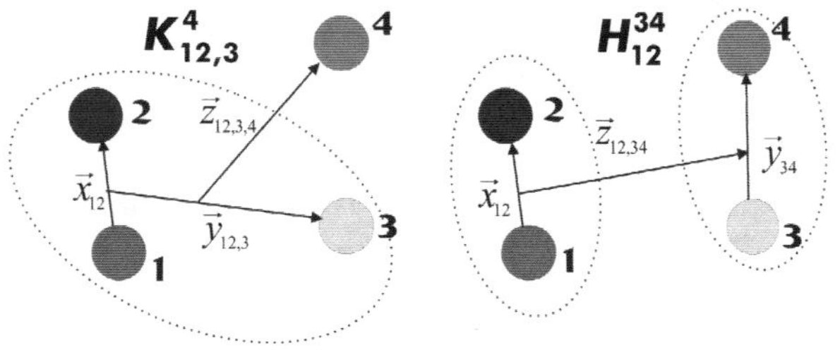

Figure 1. The F components $K^4_{12,3}$ and H^{34}_{12} for a given particle ordering. As $z \to \infty$, the K components describe 3+1 particle channels, while the H components contain asymptotic states of 2+2 channels.

where θ is the complex scaling angle. The action of the complex scaling operator on outgoing wave gives

$$\widehat{S}\exp(ikr) = \exp(-kr\sin\theta)\exp(ikr\cos\theta). \tag{6}$$

As aforementioned one may get exponentially decreasing functions K^{out} (or H^{out}) by acting on them with the complex scaling operator (5). On contrary through the same operation functions K^{in} (or H^{in}) gets exponentially divergent. Nevertheless these functions appear in the F equations (3-4) only together with potential energy operator terms, and thus if interactions are exponentially bound the kernel of the complex scaled equations becomes compact.

The scattering amplitudes are extracted from the solutions of eqs. (3-4) by using integral relations, obtained through the reen's theorem [6, 8]. In particular, scattering amplitude for 3+1 particle channels is provided by:

$$A_{31}(\vec{k}) = \frac{m}{\hbar^2}\int\int\int_V d\mathcal{V}_{xyz}\left\{6\mathcal{K}_K^\dagger(H_0 - E)(K^{out} + K^{in}) + 3\mathcal{K}_H^\dagger(H_0 - E)(H^{out} + H^{in})\right\} \tag{7}$$

with $\mathcal{K}_K = \frac{1}{2}(P^+ + P^-)Q(1 + P^+ + P^-)K^{in}$ and $\mathcal{K}_H = \tilde{P}(1 + P^+ + P^-)K^{in}$. The scattering amplitudes are scalars and do not depend on r, therefore they are not affected by the complex scaling operator $\hat{S}A(\vec{k}) = A(\vec{k})$.

Equations (3-4) are solved by expanding spin, isospin and angular dependence of the F components in partial-waves. The radial dependence of the F components, on variables (x, y, z), is expanded by using Lagrange-Laguerre mesh technique [9].

3. Results

Already in early seventies McDonald and Nuttall applied the complex scaling method to solve neutron-deuteron scattering problem above the breakup threshold [7]. At that time very primitive nucleon-nucleon interaction model was used, which has not been compatible with two-nucleon scattering data, therefore their pioneering results has been di cult to analyze. Lately interest in complex scaling method have been revived by Japanese scientists in relation with various nuclear reactions due to external probe perturbation [11, 12, 13, 14, 15, 16]. During the few last years I have also pursued the pioneering studies of Nuttall in applying complex scaling method to solve different 3-body scattering problems. In particular accurate and reliable results for phaseshifts and inelasticity parameters in 3 and 4-nucleon systems using phenomenological Malfliet and Tjon potential have been obtained [6, 17]. Also it has been demonstrated that it is possible to extract accurately the breakup amplitudes for neutron-deuteron scattering [6]. Validity of the complex scaling method have been also demonstrated for three-body Hamiltonians which may include optical potentials in conjunction with one repulsive Coulomb interaction [10].

Lately we have developed computer code which is capable to handle four-nucleon problems for the realistic Hamiltonians. In particular we have considered $E_{lab} = 22.1$ MeV neutron's scattering on triton, i.e. energy above the four-nucleon breakup threshold. Calculations have been performed for total angular momenta $J \leq 4$ (but excluding $J = 4^+$ case). To reach the convergence one had to add all the partial waves with partial angular momenta satisfying $max(l_x, l_y, l_z) \leq 4$ condition. In table 1 we compare our calculated phaseshifts δ and inelasticity parameters η using INO 04 potential. Results of this work are compared with ones of Deltuva *et al.* [4] obtained using momentum space calculations based on complex energy method. All the phaseshifts and inelasticity parameters agree within three significant digits. Single exception is for inelasticity parameter of $^1S_0$ wave; this inelasticity is very close to unity (demonstrating that this channel is strongly dominated by the elastic scattering) and is very di cult to determine accurately. In figure 2 different model predictions for the elastic differential cross section and neutron analyzing power A_y are presented. Once again we should remark very nice agreement for INO 04 model with results of Ref. [4]. One may remark small model dependence for the elastic cross section, which is probably related with the fact that AV.18 model is not able to reproduce properly three and four-nucleon breakup thresholds. There is also small discrepancy with respect to experimental data, which might be also due to the underestimated experimental error bars near the cross section minima. On the other hand agreement for the calculated neutron analyzing power A_y is quite nice.

Finally, in the figure 3 total elastic and breakup cross section dependence on the binding energy of the triton, provided by three different realistic nucleon-nucleon interaction models (namely INO 04, CD-Bonn and AV18), is presented. One may observe rather simple linear correlation pattern for both cross sections. Thus suggesting that at this energy proper reproduction of the breakup thresholds is very important, whereas sensibility to the short-range details of the nuclear Hamiltonian is rather weak.

Table 1. Some calculated phaseshifts δ and inelasticity parameters η for 22.1 MeV neutron scattering on triton using INO 04 potential. This work results are compared with the ones of [4].

	η		δ (deg.)	
W	This work	Ref. [4]	This work	Ref. [4]
$^1S_0$	0.985	0.990	62.74	62.63
$^3P_0$	0.959	0.959	43.07	43.04
$^3P_2$	0.949	0.950	65.25	65.27

3.1. Acknowledgments

This work was granted access to the H C resources of IDRIS under the allocation 2009-i2009056006 made by ENCI (rand Equipement National de Calcul Intensif). I thank the staff members of the IDRIS for their constant help. I would also like to acknowledge the financial assistance provided by the French-Japanese cooperation agreement EDC25719.

References

[1] Efros V D, Leidemann W and Orlandini G 1997 *Phys. Rev. Lett.* **78** 4015
[2] Uzu E, Kamada H and Koike Y 2003 *Phys. Rev.* C **68** 061001
[3] Rubtsova O A, Pomerantsev V N and Kukulin V I 2009 *Phys. Rev.* C **79** 064602

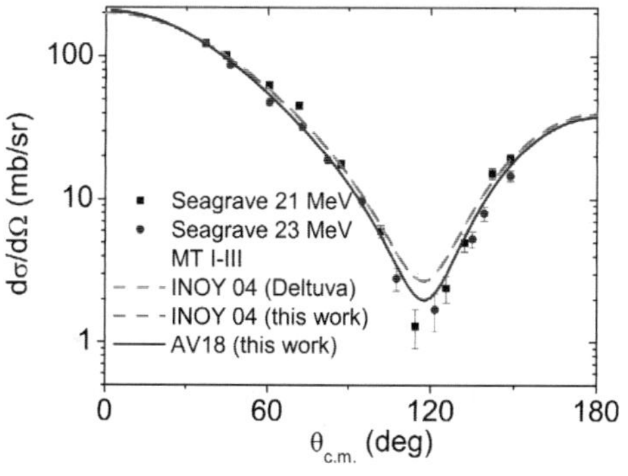

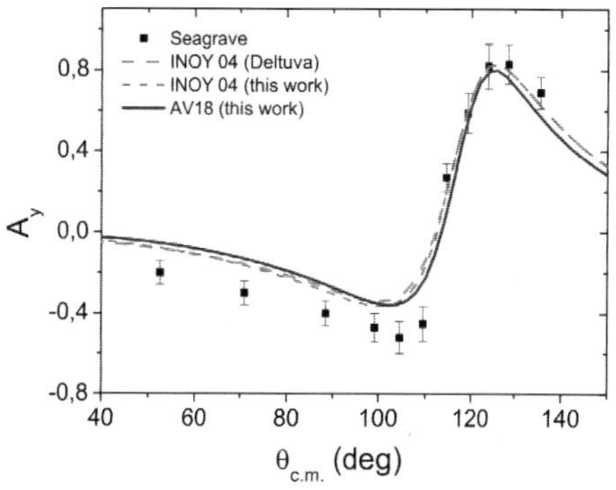

Figure 2. Calculated $n-^3H$ elastic differential cross sections (left panel) and neutron analyzing power A_y (right panel) for neutrons of laboratory energy 22.1 MeV and 22.1 MeV compared with experimental results of Seagrave et al. [18].

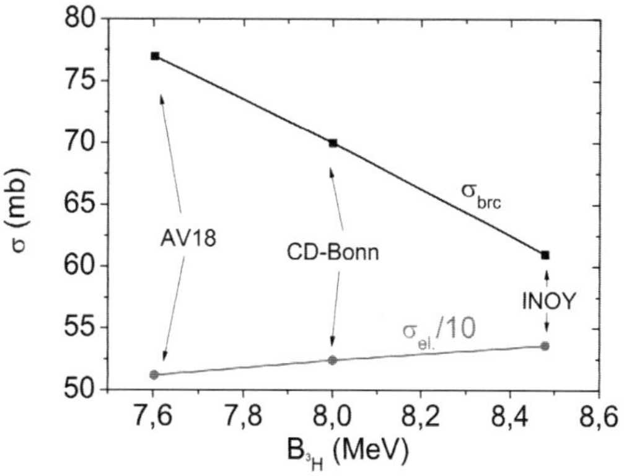

Figure 3. Dependence of the calculated $n -^3 H$ total elastic and inelastic (breakup) cross sections on the triton binding energy.

[4] Deltuva A and Fonseca A C 2012 *Phys. Rev.* C **86** 011001
[5] Kievsky A, Viviani M and Marcucci L E 2012 *Phys. Rev.* C **85** 014001
[6] Lazauskas R and Carbonell J 2011 *Phys. Rev.* C **84** 034002
[7] Nuttal J and Cohen H L 1969 *Phys. Rev.* **188** p 1542; McDonald F A and Nuttall J *Phys. Rev.* C **6** 121
[8] Deltuva A *et al.* 2014 *PPNP* **74** 55
[9] Baye D 2006 *Phys. Status Solidi* B **243** 1095; Baye D, Heenen P H 1986 *J. Phys.* A **19** 2041
[10] Deltuva A *et al.* 2013 *Clusters in Nuclei"* Vol.3 *LNP* **875** p 1
[11] Myo T *et al.* 2001 *Phys. Rev.* C **63** 054313.
[12] uzuki R *et al.* 2005 *Prog. Theor. Phys.* **113** 1273
[13] Aoyama *et al.* 2006 *Prog. Theor. Phys.* **116** 1
[14] Kikuchi Y 2010 *Phys. Rev.* C **81** 044308
[15] Kikuchi Y 2011 *Phys. Rev.* C **84** 064610
[16] Kikuchi Y 2013 *Phys. Rev.* C **87** 034606
[17] Lazauskas R 2012 *Phys. Rev.* C **86** 044002
[18] eagrave J *et al.* 1972 *Annals of Physics* **74** 250

Journal of Physics: Conference Series **569** (2014) 012018 doi:10.1088/1742-6596/569/1/012018

Unscreened cross-sections for nuclear astrophysics via the Trojan Horse Method

A. Tumino[1,2], C. Spitaleri[1,3], M. La Cognata[1,3], L. Lamia[1,3], R.G. Pizzone[1,3], M.L. Sergi[1,3]

[1]Laboratori Nazionali del Sud - INFN, Catania, Italy
[2]Universitá degli Studi di Enna "Kore", Enna, Italy
[3]Dipartimento di Metodologie Fisiche e Chimiche per l'Ingegneria - Università di Catania, Catania, Italy

E-mail: tumino@lns.infn.it

Abstract. The bare nucleus astrophysical S(E) factor is the Nuclear Physics parameter to determine the reaction rates in stellar plasmas. Whilst not being accessed in direct measurements, it can be easily determined using the Trojan Horse Method, successful indirect technique for nuclear astrophysics. The basic features of the method will be discussed and some recent results will be presented.

1. Introduction

The cross sections of nuclear reactions relevant for astrophysics are key parameters to better understand the stellar evolution. For this reason they should be measured at the Gamow energy. There are many different types of stars and each has its own unique set of nuclear reactions that change in character as the star evolves, in particular towards the end of its life. In spite of the great efforts to measure the rates of the most relevant reactions, there is still considerable uncertainty about their values. This is due to the difficulty to measure the rates at the Gamow energy. Although temperatures involved are high, on the order of hundred million degrees, the corresponding reaction rates are extremely small, making it difficult for them to be measured directly in the laboratory. Indeed, for charged particle reactions the Coulomb barrier causes a strong suppression of the cross-section, which drops exponentially with decreasing energy. The standard way to determine the behavior of $\sigma(E)$ at low energy is the extrapolation from higher energies (usually E>100 keV) by means of the astrophysical S(E)-factor

$$S(E) = E\sigma(E)\exp(2\pi\eta), \quad (1)$$

with η the Coulomb parameter of the colliding nuclei, and $\exp(2\pi\eta)$ the inverse of the Gamow factor that removes the energy dependence of $\sigma(E)$ due to the barrier tunneling. However, even an easier extrapolation can be source of additional uncertainties for $\sigma(E)$ due, for instance, to the presence of unexpected resonances.

Another critical issue in the laboratory measurement of nucleosynthesis processes is represented by the electron screening effect. Indeed both target and projectile are usually embedded in neutral/ionized atoms or molecules and their electron clouds are responsible for a reduction of the Coulomb barrier. This, in turn, leads to an increased cross section for screened nuclei, $\sigma_s(E)$, compared to the cross section for bare nuclei $\sigma_b(E)$ [1, 2]. Therefore, the so called screening factor, defined as

$$f_{\rm lab}(E) = \sigma_s(E)/\sigma_b(E) \approx \exp(\pi\eta U_e/E)\,, \tag{2}$$

where U_e is the so-called "electron screening potential" [1, 2], has to be taken into account to determine the bare nucleus cross section.
In the stellar plasma, the cross section σ_{pl}(E) is related to the bare nucleus cross section by a similar enhancement factor:

$$f_{pl}(E) = \sigma_{pl}(E)/\sigma_b(E) \approx \exp(\pi\eta U_{pl}/E) \tag{3}$$

that can be calculated once the plasma screening potential U_{pl} is known, depending on important properties of the plasma such as the Debye-Hückel radius. A measurement of U_e, which is needed to calculate σ_s(E) from Eq. 2, would also help to better understand U_{pl}.
Low-energy fusion reactions measured so far have indeed shown the exponential enhancement according to equation (2)[2]. However, the deduced U_e values are larger than the adiabatic limit, provided by the atomic models as the difference between the electron binding energies of the separate atoms in the entrance channel and that of the composite atom [2, 3]. This disagreement in laboratory experiments is yet to be justified, and prevents the effects under astrophysical conditions to be fully understood.
A week point in the laboratory approach - and thus in the deduced U_e value - is the necessity to make an assumption for the energy dependence of σ_b(E) at ultra-low energies. Thus, indirect techniques ([4] and references therein) such as the Trojan Horse Method (THM) ([5] and references therein) , have been introduced to overcome all these difficulties. In particular, the THM has been successfully applied to determine σ_b(E) for reactions between charged particles. Here we recall the basic ideas of the THM and we show some recent results.

2. Short Description of Trojan Horse Method

The THM ([5, 6, 7] and references therein) makes use of an appropriate $A + a \to c + C + s$ two-to-three body process to measure the astrophysical $A + x \to c + C$ two-body reaction of interest by establishing a relation between the two reactions based on the nuclear reaction theories. The three-body process is chosen in such a way that target a (or equivalently the projectile) is described in terms of the $x \oplus s$ cluster structure with x the target (or equivalently the projectile) of the two-body reaction. The selected part of the three-body phase space is the one where the quasi free kinematics applies: the other cluster s remains spectator to the process, and $A + x \to c + C$ can be regarded as an half-off-energy-shell (HOES) two-body reaction, usually referred to as a quasi-free reaction. Since the three-body process occurs at an energy above the Coulomb barrier, the main feature is the real suppression of both Coulomb barrier and screening effects in the HOES two-body cross section. Nevertheless, the quasi-free $A+x$ process can occur even at very low sub-Coulomb energies because the $A + a$ relative motion is compensated for by the $x - s$ binding energy. This is indeed a different approach to the THM [5] compared to the original idea of Baur [8], where the initial velocity of the projectile A is compensated for by the $x - s$ intercluster motion. In that framework a quite large momentum of the order of 200 MeV/c or more is needed. But the relative yield of the experimental momentum distribution at such momenta can be very small, in particular for a l=0 inter-cluster motion (for example p-n motion inside ^{2}H or α-d motion inside ^{6}Li). This would make very critical the separation from other competitive reaction mechanisms. Moreover, the theoretical description of the tails of the momentum distribution is a hard task, their shape being very sensitive to it. In our approach to the THM, the intercluster motion is only needed to determine the accessible astrophysical

Journal of Physics: Conference Series **569** (2014) 012018 doi:10.1088/1742-6596/569/1/012018

energy region. It corresponds to a cutoff in momentum distribution of s of few tens of MeV/c. In this framework, the so called "quasi-free two-body energy" is given by:

$$E_{QF} = \frac{m_x}{m_x + m_A} E_A - B_{x-s}. \tag{4}$$

where E_A represents the beam energy, m_x and m_A are the masses of x and A particles respectively, and B_{x-s} is the binding energy for the $x - s$ system.
In the Impulse Approximation, based essentially on the assumption that the interaction of the spectator with particles C and c is neglected, the three body-cross cross section can be factorized as:

$$\frac{d^3\sigma}{dE_c d\Omega_c d\Omega_C} \propto [KF\ |\varphi_a(\mathbf{p}_{sx})|^2] \left(\frac{d\sigma}{d\Omega_{c.m.}}\right)^{\mathrm{HOES}} \tag{5}$$

where KF is a kinematical factor containing the final state phase-space factor. It is a function of the masses, momenta and angles of the outgoing particles [30]; $\varphi_a(\mathbf{p}_{sx})$ is proportional to the Fourier transform of the radial wave function $\chi(\mathbf{r})$ for the $x - s$ inter-cluster relative motion; $(d\sigma/d\Omega_{c.m.})^{\mathrm{HOES}}$ is the HOES differential cross section for the binary reaction at the center of mass energy $E_{c.m.}$ given in post-collision prescription by

$$E_{c.m.} = E_{cC} - Q_{2b}\,. \tag{6}$$

Here, Q_{2b} is the Q-value of the binary reaction and E_{cC} is the relative energy of the outgoing particles c and C, which spans the accessible astrophysical region defined above.

The factorization of Eq.5 is strictly valid in Plane Wave Impulse Approximation, which changes only the absolute magnitude but not the energy dependence of the two-body cross section.

3. Experimental Details

In THM experiments, the decay products (c and C) of the virtual two-body reaction of interest are usually detected and identified by means of telescopes (silicon detector or ionization chamber as ΔE detector and PSD as E detector) placed in a phase space region where quasi free kinematics is dominant. After the selection of the reaction channel, a critical point is the separation of the quasi free mechanism from other reaction mechanisms feeding the same particles in the final state, e.g. sequential decay and direct break-up. An observable which is very sensitive to the reaction mechanism is the shape of the experimental momentum distribution $\Phi(p_s)$ of the spectator. To reconstruct the experimental $\Phi(p_s)$ distribution, the energy sharing method [9] can be applied for each pair of coincidence QF angles: the quasi-free coincidence yield with a cutoff in relative energy of few hundreds of keV at most is divided by the kinematic factor, providing a quantity which is proportional to the product of the momentum distribution for the spectator with the differential two-body cross section (see Eq.5). The window is chosen in such a way that the differential two-body cross section in this range can be considered almost constant. Thus, the quantity defined above represents essentially the momentum distribution for the spectator that in PWIA can be compared with the Fourier transform of the radial $x - s$ bound state wave function.
Data analysis is limited to the region where the agreement between the two distributions exists. Usually a window of few tens of MeV/c is chosen not to move too far from E_{QF}, according to the prescriptions reported in [15]. Therefore, $((d\sigma/d\Omega_{cm})^{\mathrm{HOES}}$ can be extracted from the three-body coincidence yield by simply inverting Eq.5. The Coulomb barrier in $((d\sigma/d\Omega_{cm})^{\mathrm{HOES}}$ is suppressed [10] and this is due to the virtuality of particle x. This seems to be the only consequence of off-energy-shell effects as suggested by the agreement between HOES and on-energy-shell (OES) cross-sections for the ^{6}Li(n,α)^{3}H reaction [11].

The suppression of the Coulomb barrier has been tested also in scattering processes addressing the study of the p+p elastic scattering through the ^{2}H(p, pp)n reaction as reported in [12, 13]. All experimental works provided so far provide conclusive evidence of the suppression of Coulomb effects in the THM method.
Thus, in a final step, to relate the HOES excitation function to the relevant OES one, Coulomb corrections have to be considered. If one looks at the angular distributions no correction is needed because once the energy is fixed, it would mean to introduce simply a scaling factor. Thus, the OES data are directly comparable with HOES ones projected onto the emission angle of C (or c) in the $C - c$ center of mass system, $\theta_{c.m.}$, as given by [16]:

$$\theta_{c.m.} = \arccos \frac{(\mathbf{v}_{\mathrm{A}} - \mathbf{v}_{\mathbf{x}}) \cdot (\mathbf{v}_{\mathbf{c}} - \mathbf{v}_{\mathrm{C}})}{|\mathbf{v}_{\mathrm{A}} - \mathbf{v}_{\mathbf{x}}||\mathbf{v}_{\mathbf{c}} - \mathbf{v}_{\mathrm{C}}|} \tag{7}$$

where the vectors $\mathbf{v}_{\mathrm{A}}$, $\mathbf{v}_{\mathbf{x}}$, $\mathbf{v}_{\mathbf{c}}$, $\mathbf{v}_{\mathrm{C}}$ are the velocities of projectile, transferred particle and outgoing nuclei respectively. These quantities are calculated from their corresponding momenta in the lab-system, where the momentum of the transferred particle is equal and opposite to that of s when the quasi-free break-up takes place in the target, otherwise a little different formula has to be used [16]. If HOES data are projected onto the E_{Cc} axis, Coulomb suppression has to be introduced before comparison with OES data. In a simple approach this is done by means of the penetrability factor:

$$P_l(k_{Ax}R) = \frac{1}{G_l^2(k_{Ax}R) + F_l^2(k_{Ax}R)} \tag{8}$$

with F_l and G_l regular and irregular Coulomb wave functions and R the so called cutoff radius [22], which is usually chosen as the sum of the radii of nuclei A and x. This procedure does not provide the absolute value of the two-body cross section that can be easily derived from a scaling to the direct data available at higher energies.

4. Applications

The THM has been applied already many times to charged particle reactions connected with fundamental astrophysical problems [17], from Big-Bang to stellar nucleosynthesis. A list of reactions studied by means of the THM is given in Table 1 together with the relevant references. In particular, low-energy cross sections for reactions producing or destroying Lithium isotopes are fundamental for several astrophysical problems yet to be solved, e.g. understanding Big Bang nucleosynthesis and the "Lithium depletion" in the Sun or in other galactic stars. In addition they offer very clear examples on the capabilities of the method in measuring the bare nucleus cross section. An experimental program was undertaken to study p-capture reactions on ^{6,7}Li, main responsible for their destruction [18, 19, 22, 23, 24, 25, 26, 27, 28, 29, 50]. The extracted astrophysical S(E) factors were compared with those from direct measurements and found in fair agreement in the energy range where screening effects are negligible. Recently, an updated reaction rate for the ^{6}Li(p, α)^{3}He reaction has been determined using direct data of [53, 54] for a more accurate normalization in the energy region below 300 keV (see [29] for the detailed description). The new normalization gives a S(0)=3.44 $\pm$ 0.39 MeV b, with an S(E) factor shown in Fig.1 as red solid dots from [23, 29] and black solid dots from [22]. Direct data are shown as open symbols as described in [29]. The blue and red ones refer to direct data of [53, 54]. The solid line represents the polynomial fit to the THM bare nucleus S(E) factor as reported in [29]. Fitting the low-energy data below E_{cm}=70 keV with the screening function (Eq.2) and leaving U_e as free parameter, this provides U_e=350$\pm$100 eV. The new reaction rate deviates from 5% to 15% with respect to the value reported in the NACRE compilation [52] as the temperature decreases from 1 down to 10^{-3} T9 (see [29] for details). Astrophysical implications of these results have been investigated performing calculations on PMS models for ^{6}Li and on the solar

Table 1. Two-body reactions studied via the THM with measured two-to-three body TH reactions and relevant references for each process

Direct reaction	TH reaction	ref
^{7}Li(p,α)^{4}He	^{7}Li(d,α α)n	[19]
^{7}Li(p,α)^{4}He	^{7}Li(^{3}He,α α)^{2}H	[20]
^{6}Li(d,α)^{4}He	^{6}Li(^{6}Li,α α)^{4}He	[21]
^{6}Li(p,α)^{3}He	^{6}Li(d,α ^{3}He)n	[22, 23]
^{11}B(p,α)^{8}Be	^{11}B(d ,^{8}Be α)n	[30]
^{10}B(p,α)^{7}Be	^{10}B(d,^{7}Be α)n	[31, 32]
^{9}Be(p,α)^{6}Li	^{9}Be(d,^{6}Li α)n	[33, 34]
^{2}H(^{3}He,p)^{4}He	^{6}Li(^{3}He,p α)^{4}He	[35]
^{2}H(d,p)^{3}H	^{2}H(^{6}Li,t p)^{4}He	[36]
^{15}N(p,α)^{12}C	^{15}N(p,α^{12}C)n	[37]
^{18}O(p,α)^{15}N	^{18}O(p,α^{15}N)n	[38, 39]
^{1}H(p,p)^{1}H	^{2}H(p,pp)n	[12, 13]
^{2}H(d,p)^{3}H	^{2}H(^{3}He,t p)^{1}H	[40, 41, 42]
^{2}H(d,n)^{3}He	^{2}H(^{3}He,^{3}He n)^{1}H	[40, 41, 42]
^{19}F(p,α)^{16}O	^{2}H(^{19}F,α^{16}O)n	[43]
^{17}O(p,α)^{14}N	^{2}H(^{17}O,α^{14}N)n	[44]
^{4}He(^{12}C,^{12}C)^{4}He	^{6}Li(^{12}C,α^{12}C)^{2}H	[45]
n(^{6}Li,t)^{4}He	^{2}H(^{6}Li,t^{4}He)^{1}H	[11, 46]
^{11}B(p,α)^{8}Be	^{2}H(^{11}B,α^{8}Be)n	[48]
^{13}C(α,n)^{16}O	^{6}Li(^{13}C,n^{16}O)^{2}H	[49]

"Lithium problem" for ^{7}Li. A full account is reported in [29] and [25] respectively. In particular for the ^{7}Li(p,α)^{3}He case, it appears that the THM cross section does not significantly change the surface Lithium abundance for the present Sun, with respect to previous results reported in the NACRE compilation [52] because of the small discrepancy between the two rates. The Lithium problem for the Sun is not solved and the observed Lithium surface abundance is not reproduced by the model.

However, uncertainties on these nuclear reaction rate are now so small to move the "Lithium problem" to astrophysics. Other input parameters such as original metallicity, helium abundance, convection efficiency and stellar mass have much larger uncertainties that seriously affect any comparison between theory and observations and need to be better constrained. Recently, the new direct data by [53, 54] have been used for a better normalization also in the ^{7}Li(p,α)^{4}He channel [27]. The new normalization of the THM S(E) factor is shown in Fig.2 as black solid dots. Red and blue open circles represent data from [53] and [54] respectively. The other open symbols refer to direct data of [57]. The solid line represents the polynomial fit to the THM bare nucleus S(E) factor as reported in [27]. The new S(0) turns out to be S(0)=53 $\pm$ 5 keV b, while the fit of the low energy region including the screening function (Eq.2) returns a U_e value of U_e=425$\pm$60eV. The updated S(0) value deviates by 4% when compared with the previous normalization in [19], while U_e deviates by about 23%. The new reaction rate experiences a variation with respect to the NACRE one [52] from about 13% at

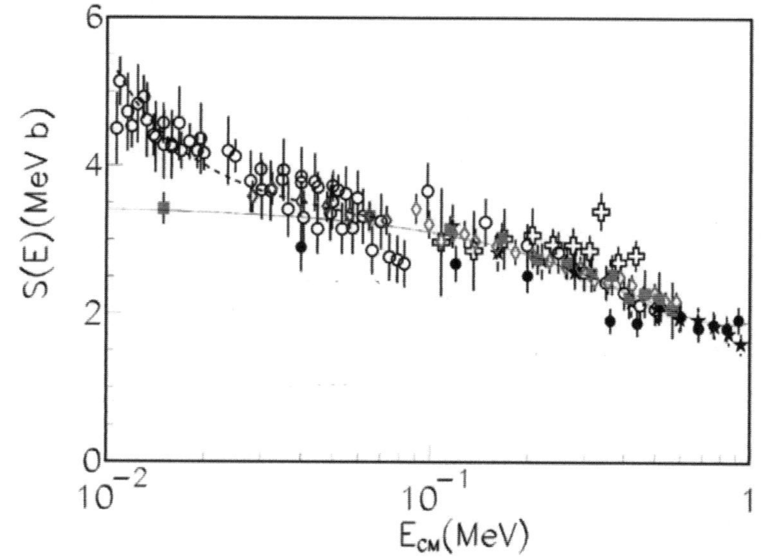

Figure 1. THM bare nucleus S(E) factor for the $^{6}Li(p,\alpha)^{3}He$ reaction (green solid dots from [29] and black solid dots for [22]. Direct data represented with open symbols as described in [29]. The blue and green ones refer to the most recent direct experiment reported in [53, 54]. The solid line represents the polynomial fit to the THM bare nucleus S(E) factor as reported in [29].

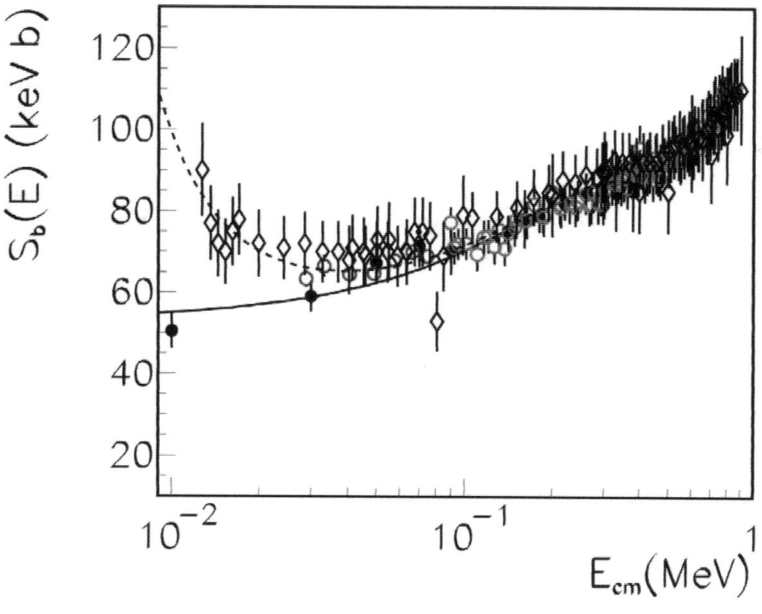

Figure 2. THM bare nucleus S(E) factor for the $^{7}Li(p,\alpha)^{4}He$ reaction (black solid dots from [19]. Direct data represented with open symbols as described in [27]. The red and blue open circles refer to the most recent direct experiment reported in [53] and [54] respectively. The solid line represents the polynomial fit to the THM bare nucleus S(E) factor as reported in [27].

around T9=10^{-3} to about 5% at higher temperatures and astrophysical calculations need to be readdressed for several scenarios.

5. Bibliography

[1] Assenbaum HJ *et al* 1987 *Z . Phys.* A **327** 461
[2] Strieder F *et al* 2001 *Naturwissenschaften* **88** 461
[3] Fiorentini G *et al* 1995 *Z. Phys.* A **350** 289
[4] Tumino A *et al* 2013 *Few Body Syst.* **54**, Issue 7-10 869
[5] Spitaleri C *et al* 2011 *Phys. At. Nucl.* **74** 1763
[6] Tumino A *et al* 2013 *Few Body Syst.*, **54**,Issue 5-6 745
[7] Tumino A *et al* 2014 *Acta Physica Polonica B* **45** 181
[8] Baur G 1986 *Phys. Lett.* B **178** 135
[9] Kasagi J *et al* 1975 *Nucl. Phys.* A **239** 233
[10] Mukhamedzhanov A M et al., *Eur. Phys. J.* A **27** (2006) 205
[11] Tumino A *et al* 2005 *Eur. Phys. J.* A **25** 649
[12] Tumino A *et al* 2007 *Phys. Rev. Lett.* **98** 252502
[13] Tumino A *et al* 2008 *Phys. Rev.* C **78** 064001
[14] Jackson J D and Blatt J M 1950 *Rev. Mod. Phys.* **22** 77
[15] Shapiro I S *et al* 1965 *Nucl. Phys.* A **61** 353
[16] Jain M *et al* 1970 it Nucl. Phys. A **153** 49
[17] Copi C J, Schramm D N, Turner M S 1995 Science **627** 192
[18] Spitaleri C *et al* 1999 *Phys. Rev.* C **60** 055802
[19] Lattuada M *et al* 2001 Astrophysical Journal **562** 1076
[20] Tumino A *et al* 2006 *Eur. Phys. J.* A direct 1 DOI: 10.1140/epja/i2006-08-038-1
[21] Spitaleri C *et al* 2001 *Phys. Rev.* C **63** 005801
[22] Tumino A *et al* 2003 *Phys. Rev.* C **67** 065803
[23] Tumino A *et al* 2004 *Prog. Theor. Phys. Suppl.* **154** 341
[24] Aliotta M *et al* 2000 *Eur. Phys. J.* A **9** 435
[25] Pizzone RG *et al.*, A& A **398**, (2003) 423
[26] Pizzone RG *et al.*, *Phys. Rev.* C **83**, (2011) 045801
[27] Lamia L *et al* 2012 *A&A* **A185** 541
[28] Pizzone RG *et al.* 2005 A& A **438** 779
[29] Lamia L *et al* 2013 *Astrophysical J.* **768** 65
[30] Spitaleri C *et al* 2004 *Phys. Rev.* C **63** 055806
[31] Lamia L *et al* 2007 *Nucl. Phys. A* **787** 309
[32] Lamia L *et al* 2008 *Il Nuovo Cimento* **31** N.4 423
[33] Qun-Gang Wen *et al* 2008 *Phys. Rev.* C **78** 035805
[34] Qun-Gang Wen *et al.* 2011 *J.Phys. G: Nucl. Part. Phys.*, **38** 085103
[35] La Cognata M *et al* 2006 *Eur. Phys. J.* A **27** 249
[36] Rinollo A *et al* 2005 *Nucl. Phys.* A **758** 146
[37] La Cognata M *et al* 2007 *Phys. Rev.* C **76** 065804
[38] La Cognata M *et al* 2008 *Phys. Rev. Lett.* **101** 52501
[39] La Cognata M *et al* 2010 *Astrophysical J.* **708** 796
[40] Tumino A *et al* 2011 *Phys. Lett.* B **700** 111
[41] Tumino A *et al* 2011 *Phys. Lett.* B **705** 546
[42] Tumino A *et al* 2014 *Astrophysical Journ.* **785** 96
[43] La Cognata M *et al* 2011 *Astrophysical J. Lett.* **739** L54
[44] Sergi ML *et al* 2010 *Phys.Rev.* C **R82** 032801
[45] Spitaleri C *et al* 2000 *Eur. Phys. Journ.* A7 181
[46] Gulino M *et al* 2010 *J.Phys. G: Nucl. Part. Phys.* **37** 125105
[47] Lamia L *et al* 2009 *Il Nuovo Cimento*, **31** 423
[48] Lamia L *et al* 2012 *Journ. of Phys. G: Nucl. Part. Phys.* **39** 015106
[49] La Cognata M *et al* 2013 *Phys. Rev. Lett.* **49** 015106
[50] Lamia L *et al* 2012 *Phys. Rev. C* **85** 025805
[51] Grevesse N & Sauval AJ 1998, *Space Sci. Rev.*, **85** 161
[52] Angulo C *et al* 1999 *Nucl. Phys.* A **656** 3
[53] Cruz J *et al* 2005 *Phys. Lett. B* **624** 181
[54] Cruz J *et al.* *2008 Journ. of Phys. G: Nucl. Part. Phys.* **635** *192*
[55] Degl'Innocenti S *et al* 2008 *Ap&SS* **316** 25
[56] Ciacio F, Degl'Innocenti S & Ricci B 1997 *A&A* **123** 449
[57] Engstler S *et al* 1992 *Z. Phys.* A **342** **471**

Nuclear clusters studied with alpha resonant scatterings using RI beams at CRIB

H. Yamaguchi[1], D. Kahl[1], T. Nakao[1,2], Y. Wakabayashi[3,1], T. Hashimoto[4,1], S. Hayakawa[5,1], T. Kawabata[6], T. Teranishi[7], Y.K. Kwon[8], D.N. Binh[9,1], L.H. Khiem[10], N.N. Duy[10], S. Kubono[2,1], T. Suhara[11], Y. Kanada-En'yo[7], J.Y. Moon[8], A. Kim[12], N. Iwasa[13], P.S. Lee[14], K.Y. Chae[12], S.M. Cha[12], M.S. Gwak[12], D.H. Kim[15], E. Milman[2]

E-mail: yamag@cns.s.u-tokyo.ac.jp
[1] Center for Nuclear Study (CNS), University of Tokyo.
[2] Advanced Science Research Center, Japan Atomic Energy Agency
[3] RIKEN Nishina Center.
[4] Research Center for Nuclear Physics, Osaka University.
[5] Laboratori Nazionali del Sud, Istituto Nazionale di Fisica Nucleare.
[6] Department of Physics, Kyoto University.
[7] Department of Physics, Kyushu University.
[8] Institute for Basic Science.
[9] 30 MeV Cyclotron Center, Tran Hung Dao Hospital.
[10] Institute of Physics, Vietnam Academy of Science and Technology.
[11] Matsue College of Technology.
[12] Department of Physics, Sungkyunkwan University.
[13] Department of Physics, Tohoku University.
[14] Department of Physics, Chung-Ang University.
[15] Department of Science Education, Ewha Womans University.

Abstract. Alpha resonant scattering is a simple and promising method to study α-cluster structure in nuclei. It has several good features which enable us to perform measurements with short-lived and relatively low-intense RI beams.

Several measurements on alpha resonant scattering have been carried out at CRIB (CNS Radioactive Ion Beam separator), which is a low-energy RI beam separator at Center for Nuclear Study (CNS) of the University of Tokyo. Recent α resonant scattering studies at CRIB, using ^{7}Li, ^{7}Be and ^{10}Be beams with a helium gas target, are discussed.

1. Introduction

CRIB [1,2] is a radio-isotope (RI) beam separator operated by Center for Nuclear Study (CNS), the University of Tokyo, installed in the RIBF facility of RIKEN Nishina Center. CRIB can produce low-energy ($<$ 10 MeV/u) RI beams in flight, using primary heavy-ion beams from the AVF cyclotron of RIKEN (K=70). Most of the RI beams are produced via direct reactions such as (p, n), (d, p) and (^{3}He, n), taking place at an 8-cm-long gas target with a maximum pressure of 760 Torr. A cryogenic target system, in which the target gas can be cooled down to about 90 K, is currently available, and an intense ^{7}Be beam of 2×10^{8} pps was produced using the system [3]. as the production target. The secondary beam is purified with an magnetic analysis using dipole magnets, and with a Wien filter, which can separate the beams according to their velocities. For relatively light RI beams such as ^{7}Be, we can obtain a good purity of almost 100% after the Wien filter.

The low-energy RI beams at CRIB are suitable for studies of astrophysical reactions and nuclear structure. An experimental method extensively used is the thick-target in inverse kinematics method [4]. In that method, the beam energy is degraded in a thick target, and thus reactions can occur at various center-of-mass energies in the same target. We detect light particles emitted after reactions, and reconstruct the kinematics. This method has several advantages, namely, (a) using inverse kinematics, we can study reactions with short-lived RI which cannot be used as the target, (b) we can perform simultaneous measurements of cross section of various excitation energies without varying the incoming RI beam energy, and (c) we can perform measurements at 180° in center-of-mass angle, where the Coulomb scattering is minimal. Many fruitful results have been obtained at CRIB with this method [5–10].

A recent major topic of our interest is the measurement of alpha-induced reactions. Several (α, p) reactions, such as ^{14}O(α, p), ^{11}C(α, p), ^{21}Na(α, p), ^{18}Ne(α, p), ^{30}S(α, p), and ^{22}Mg(α, p) have been measured at CRIB. The elastic resonant scatterings of α particle using ^{14}O, ^{21}Na, ^{7}Li [10] and ^{7}Be [11] beams have also been measured. These elastic-scattering measurements are relevant both for the astrophysical (α, γ) reaction rates, and α cluster structure in nuclei. Using an α particle as the probe, we can selectively study α-cluster like states as resonances. Recent results of α resonant scattering measurements are presented below.

2. ^{7}Li+α elastic resonant scattering

The exotic cluster structures in the ^{11}B and ^{11}C nuclei are attracting much attention in recent years [12]. The $3/2_3^-$ state in ^{11}B at the excitation energy $E_{\rm ex}$=8.56 MeV is regarded as a dilute cluster state [13], where two α particles and ^{3}He are weakly interacting. In particular, the alpha cluster structure in ^{11}B was studied by measuring its isoscalar monopole and quadrupole strengths in the ^{11}B(d,d') reaction, and the 8.56-MeV state was suggested to have a dilute cluster structure [14, 15]. A natural method to study the α cluster structure in ^{11}B is to form such a cluster state by the elastic scattering of ^{7}Li and α.

The ^{7}Li+α system is also related to the astrophysical reaction, ^{7}Li(α, γ)^{11}B. Studies on the resonance parameters should contribute to the precise determination of the ^{7}Li(α, γ)^{11}B reaction rate at high temperature ($T_9 > 1$). In the low temperature stellar environment which ignites the p-p chains, ^{7}Li is destroyed via the ^{7}Li(p, α)^{4}He reaction, and the ^{7}Li(α, γ)^{11}B reaction is much weaker in comparison. However, in some high temperature phenomena, the ^{7}Li(α, γ)^{11}B reaction may play important roles. For example, in the ν-process in core-collapse supernovae [16] ^{11}B is mainly produced via ^{4}He$(\nu,\nu'p)^{3}$H$(\alpha,\gamma)^{7}$Li$(\alpha,\gamma)^{11}$B, but the production can be enhanced with the ^{12}C$(\overline{\nu_e},e^+n)^{11}$B reaction. A precise comparison between the observed abundance ratio of ^{11}B/^{7}Li in stars and a calculation using the experimental ^{7}Li(α, γ)^{11}B reaction rate may provide a constraint on the neutrino mixing parameter θ_{13}, as suggested in Ref. [16].

Figure 1 schematically shows the experimental setup for the ^{7}Li+α scattering measurement. A ^{7}Li beam was accelerated at the AVF cyclotron and transported to the final focal plane (F3) of CRIB. The beam had a well-defined energy of 13.7 MeV, and collimated by a 3 × 3-mm square aperture at F3. An MCP (Micro-Channel Plate) was used to measure the position and timing of the beam. The helium gas was housed in a 50-mm diameter duct and a small target chamber. The helium gas was at 920 Torr and sealed with a thin (2.5 μm) Havar foil as a beam entrance window. Alpha particles recoiling to the forward angles in the laboratory frame were detected by the "ΔE-E detector", which consisted of 20-μm- and 480-μm-thick silicon-strip detectors, placed directly in the gas chamber. The measurement was performed for 2.9 $\times 10^{10}$ ^{7}Li particles injected into the gas target over 60 hours.

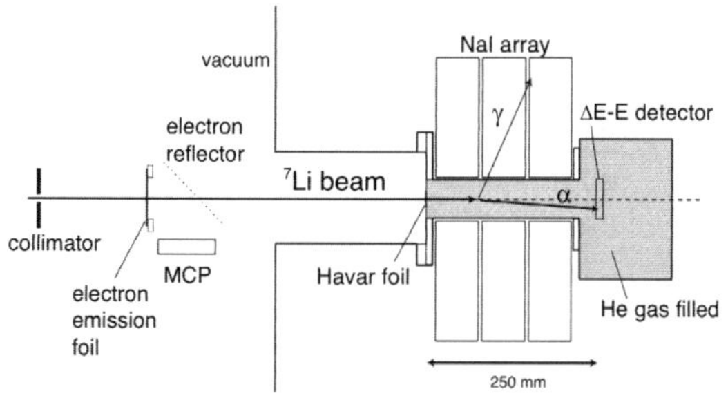

Figure 1. Arrangement of the detectors and targets in ^{7}Li+α measurement.

Our measurement corresponds to $E_{\rm ex}$=10–13 MeV and the center-of-mass angles of 160–180°. The excited states of ^{11}B in this energy region have been studied by ^{7}Li+α elastic scattering [17, 18] or using other methods [19]; however, some of the resonance parameters are still uncertain. Especially, the alpha widths for most of the resonances were not determined with a good precision. Using inverse kinematics, the excitation function around 180° in the center-of-mass system was measured for the first time, and we observed strong resonant structure as shown in the left panel in Figure 2.

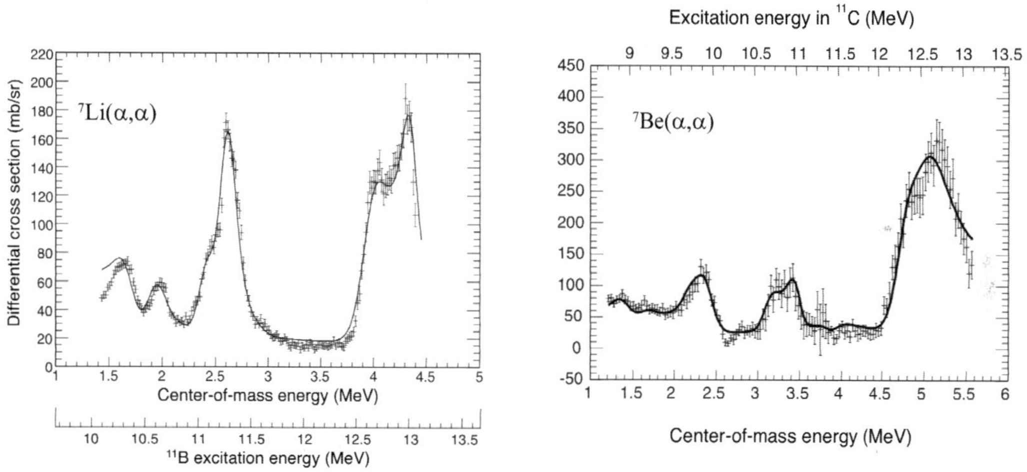

Figure 2. Excitation functions of ^{7}Li+α (left) and ^{7}Be+α (right) elastic scattering around $\theta_{\rm c.m.}$=170°. The curves are by R-matrix calculation.

Rotational bands in ^{11}B and ^{11}C, which might be related to the cluster structure, had been discussed in [20, 21]. Based on the experimental result, we proposed a negative-parity cluster band which was later found to be consistent with a theoretical calculation [22]. The theoretical interpretation of these cluster states in the band was that they have a 3-body (2α+t) cluster configuration.

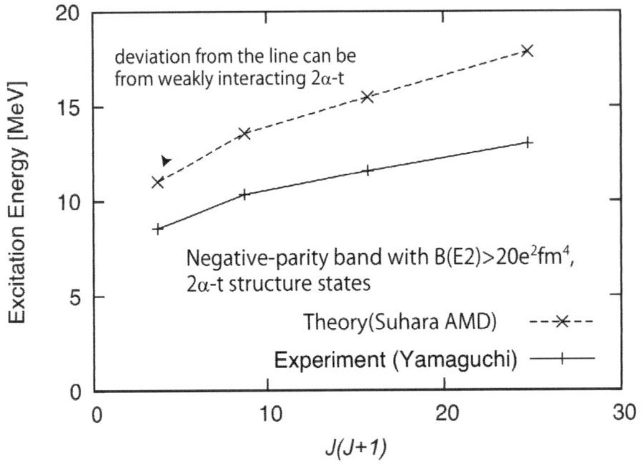

Figure 3. The band structure in ^{11}B proposed by the measurement [10] and AMD calculation [22].

3. ^{7}Be+α elastic resonant scattering

The measurement of ^{7}Be+α scattering can be associated with the ^{7}Be(α, γ) reaction. The ^{7}Be(α, γ) reaction is considered to play an important role in the hot p-p chain and related reaction sequences [23]. Several reaction sequences including the ^{7}Be(α, γ) reaction should take place in some high-temperature environments at $T_9 > 0.2$, where T_9 is the temperature in GK. In the νp-process in core-collapse supernovae [24], the ^{7}Be(α, γ) reaction may contribute as much as the triple-α process to the synthesis of elements heavier than boron at the relevant temperature of T_9 = 1.5–3, according to a theoretical calculation [25]. The Gamow energy window at the highest temperature T_9 = 3 corresponds to the excitation energy $E_{\rm ex}$=8.2–9.6 MeV in ^{11}C. By our study, the resonant reaction rate should be evaluated more precisely by determining α widths for the resonances at high temperatures.

In 2010, we performed the measurement of ^{7}Be+α resonant elastic and inelastic scatterings with the thick-target method in inverse kinematics at CRIB [11]. The setup was almost identical to the one used in the above-described ^{7}Li+α measurement. A low-energy ^{7}Be beam at 14.7 MeV was produced using a 2.3-mg/cm^2-thick hydrogen gas target and a primary ^{7}Li beam at 5.0 MeV/u. The typical ^{7}Be beam intensity used in the measurement was 1–2 $\times$ 10^5 per second at the secondary target, and the main measurement using a thick helium-gas target was performed for 4 days, injecting 2.9 $\times 10^{10}$ ^{7}Be particles into the target. We obtained an excitation function with several peaks, which correspond to the resonance structure in ^{11}C.

The obtained excitation function of the elastic scattering is shown in the right panel in Figure 2. An R-matrix analysis was performed to deduce the parameters of the observed resonance structure. A similar measurement was independently carried out at other facilities [26], but our measurement was with γ-ray detection to identify inelastic scatterings, and some differences in the energy and cross section were found in the obtained spectra [11].

As for the cluster levels, a similar discussion as ^{11}B can be applied to ^{11}C. Two positive-parity rotational bands, $K = 3/2^+$ and $5/2^+$, had been suggested in [21]. We observed a strong resonance with $J^\pi = 9/2^+$ at 12.4 MeV in ^{11}C, and it can be the missing member of the $K = 3/2^+$ rotational band. We propose a new negative-parity band also in ^{11}C, as shown in Fig. 4. The members of the band could have a 2α-^{3}He cluster structure. The head is the 8.10-MeV ($J^\pi = 3/2^-$) state, the mirror of the 8.56-MeV state in ^{11}B. The second member is the 9.78-MeV state, assigned as $J^\pi = 5/2^-$ previously. The third member can be the state at 11.03 MeV. Our assignment was either $J^\pi = 5/2^-$ or $7/2^-$, and the latter assignment agrees

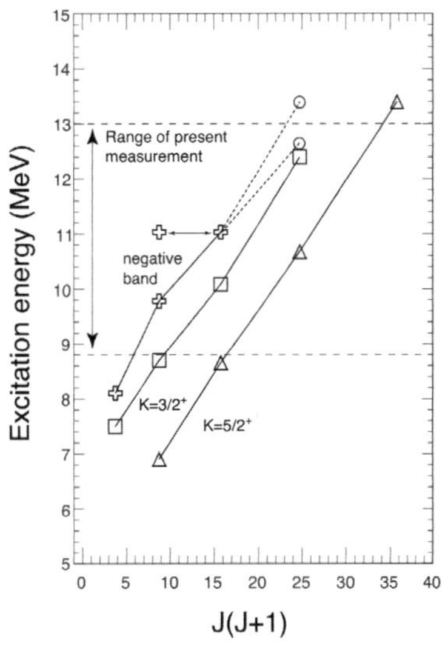

Figure 4. Two positive-parity rotational bands in ^{11}C suggested in [21], and a newly proposed negative-parity band.

with the systematics of this negative band. The systematics predicts there can be a $J^\pi = 9/2^-$ state around 13 MeV. A candidate is the 12.65-MeV state. In the present work, this state was regarded to form a doublet together with the 12.4-MeV state. A similar doublet was observed in the mirror nucleus [10], and the higher state was considered to have $J^\pi = 9/2^-$. If the tentative assignment in another work of $J^\pi = 7/2^+$ was wrong, the 12.65-MeV state may have $J^\pi = 9/2^-$, as in the mirror nucleus. Another candidate is the resonance at 13.4 MeV [27], which is known to have a certain α width, but its J^π has not been determined. In Fig. 4, these states are shown as circles and connected as dotted lines under the assumption that they have $J^\pi = 9/2^-$. The energy of the band head (8.10 MeV) appears as lower than the systematics expected from the higher state. This lowering is in agreement with the mirror state [10, 22].

4. ^{10}Be+α

Another interest is on the clustering in the ^{14}C nucleus. Suhara and En'yo [28, 29] obtained a band (0_5^+, 2_6^+, 4_6^+) in ^{14}C which could be explained as linear-chain cluster levels. It was predicted that these levels appear a few MeV or more above the ^{10}Be+α threshold. The investigation in [29] shows that the AMD wave function has a configuration in which two α particles and two neutrons are closely located, while the remaining α particle is more distant. This implies states having such a linear-chain configuration could be accessed from the ^{10}Be+α channel. Therefore we are planning a measurement of ^{10}Be+α resonant scattering at CRIB. In Jan. 2014, we carried out the first ^{10}Be beam production test at CRIB and successfully produced a ^{10}Be beam with 2 $\times$ 10^4 pps, at an energy of 2.0 MeV/u. Using the beam, we are going to perform a measurement of resonant scattering and we expect to observe α-cluster states in ^{14}C.

Acknowledgments

This work was partly supported by JSPS KAKENHI (No. 21340053,25800125) from the Ministry of Education, Culture, Sports, Science and Technology (MEXT) of Japan, and the Priority Centers Research Program in Korea (2009-0093817).

References

[1] Kubono S, Yanagisawa Y, Teranishi T, Kato S, Kishida T, Michimasa S, Ohshiro Y, Shimoura S, Ue K, Watanabe S and Yamazaki N 2002 *Eur. Phys. J.* **A13** 217

[2] Yanagisawa Y, Kubono S, Teranishi T, Ue K, Michimasa S, Notani M, He J J, Ohshiro Y, Shimoura S, Watanabe S, Yamazaki N, Iwasaki H, Kato S, Kishida T, Morikawa T and Mizoi Y 2005 *Nucl. Instrum. Meth. Phys. Res., Sect. A* **539** 74–83

[3] Yamaguchi H, Wakabayashi Y, Amadio G, Hayakawa S, Fujikawa H, Kubono S, He J, Kim A and Binh D 2008 *Nucl. Instrum. Meth. Phys. Res., Sect. A* **589** 150–156

[4] Artemov K P, Belyanin O P, Vetoshkin A L, Wolskj R, Golovkov M S, Gol'dberg V Z, Madeja M, Pankratov V V, Serikov I N, Timofeev V A, Shadrin V N and Szmider J 1990 *Sov. J. Nucl. Phys* **52** 408

[5] Teranishi T, Kubono S, Shimoura S, Notani M, Yanagisawa Y, Michimasa S, Ue K, Iwasaki H, Kurokawa M, Satou Y, Morikawa T, Saito A, Baba H, Lee J H, Lee C S, Fülöp Z and Kato S 2003 *Phys. Lett. B* **556** 27–32

[6] Teranishi T, Kubono S, Yamaguchi H, He J J, Saito A, Fujikawa H, Amadio G, Niikura M, Shimoura S, Wakabayashi Y, Nishimura S, Nishimura M, Moon J Y, Lee C S, Odahara A, Sohler D, Khiem L H, Li Z H, Lian G and Liu W P 2007 *Phys. Lett. B* **650** 129–134

[7] He J J, Kubono S, Teranishi T, Notani M, Baba H, Nishimura S, Moon J Y, Nishimura M, Iwasaki H, Yanagisawa Y, Hokoiwa N, Kibe M, Lee J H, Kato S, Gono Y and Lee C S 2007 *Phys. Rev. C* **76** 055802

[8] Yamaguchi H, Wakabayashi Y, Kubono S, Amadio G, Fujikawa H, Teranishi T, Saito A, He J, Nishimura S, Togano Y, Kwon Y, Niikura M, Iwasa N, Inafuku K and Khiem L 2009 *Phys. Lett. B* **672** 230–234

[9] He J J, Kubono S, Teranishi T, Hu J, Notani M, Baba H, Nishimura S, Moon J Y, Nishimura M, Iwasaki H, Yanagisawa Y, Hokoiwa N, Kibe M, Lee J H, Kato S, Gono Y and Lee C S 2009 *Phys. Rev. C* **80** 015801

[10] Yamaguchi H, Hashimoto T, Hayakawa S, Binh D N, Kahl D, Kubono S, Wakabayashi Y, Kawabata T and Teranishi T 2011 *Phys. Rev. C* **83** 034306

[11] Yamaguchi H, Kahl D, Wakabayashi Y, Kubono S, Hashimoto T, Hayakawa S, Kawabata T, Iwasa N, Teranishi T, Kwon Y, Binh D N, Khiem L and Duy N 2013 *Phys. Rev. C* **87** 034303

[12] Kanada-En'yo Y 2007 *Phys. Rev. C* **75** 024302

[13] Tohsaki A, Horiuchi H, Schuck P and Röpke G 2001 *Phys. Rev. Lett.* **87** 192501

[14] Kawabata T, Akimune H, Fujimura H, Fujita H, Fujita Y, Fujiwara M, Hara K, Hara K Y, Hatanaka K, Ishikawa T, Itoh M, Kamiya J, Kishi S, Nakamura M, Nakanishi K, Noro T, Sakaguchi H, Shimbara Y, Takeda H, Tamii A, Terashima S, Toyokawa H, Uchida M, Ueno H, Wakasa T, Yasuda Y, Yoshida H P and Yosoi M 2004 *Phys. Rev. C* **70** 034318

[15] Kawabata T, Akimune H, Fujita H, Fujita Y, Fujiwara M, Hara K, Hatanaka K, Itoh M, Kanada-En'yo Y, Kishi S, Nakanishi K, amd Y Shimbara H S, Tamii A, Terashima S, Uchida M, Wakasa T, Yasuda Y, Yoshida H P and Yosoi M 2007 *Phys. Lett. B* **646** 6–11

[16] Yoshida T, Kajino T, Yokomakura H, Kimura K, Takamura A and Hartmann D H 2006 *Phys. Rev. Lett.* **96** 091101

[17] Cusson R 1966 *Nucl. Phys.* **86** 481–508

[18] Bohlen H, Marquardt N, von Oertzen W and Gorodetzky P 1972 *Nucl. Phys. A* **179** 504–512

[19] Ajzenberg-Selove F 1990 *Nucl. Phys. A* **506** 1

[20] Ragnarsson I, Åberg S, Håkansson H B and Sheline R 1981 *Nucl. Phys. A* **361** 1–22

[21] Soić N, Freer M, Donadille L, Clarke N, Leask P, Catford W, Jones K, Mahboub D, Fulton B, Greenhalgh B, Watson D and Weisser D 2004 *Nucl. Phys. A* **742** 271–290

[22] T Suhara and Y Kanada-En'yo 2012 *Phys. Rev. C* **85** 054320

[23] Wiescher M, Görres J, Graff S, Buchmann L and Thielemann F K 1989 *Astrophys. J.* **343** 352–364

[24] Fröhlich C, Hauser P, Liebendörfer M, Martínez-Pinedo G, Thielemann F K, Bravo E, Zinner N, Hix W, Langanke K, Mezzacappa A and Nomoto K 2006 *Astrophys. J.* **637** 415

[25] Wanajo S, Janka H T and Kubono S 2011 *Astrophys. J.* **729** 46

[26] Freer M, Achouri N L, Angulo C, Ashwood N I, Bardayan D W, Brown S, Catford W N, Chipps K A, Curtis N, Demaret P, Harlin C, Laurent B, Malcolm J D, Milin M, Munoz-Britton T, Orr N A, Pain S D, Price D, Raabe R, Soić N, Thomas J S, Wheldon C, Wilson G and Ziman V A 2012 *Phys. Rev. C* **85**(1) 014304

[27] Tilley D, Kelley J, Godwin J, Millener D J, Purcell J E, Sheu C G and Weller H R 2004 *Nucl. Phys. A* **745** 155

[28] Suhara T and Kanada-En'yo Y 2010 *Phys. Rev. C* **82** 044301

[29] Suhara T and Kanada-En'yo Y 2011 *Phys. Rev. C* **84**(2) 024328

Journal of Physics: Conference Series **569** (2014) 012020 doi:10.1088/1742-6596/569/1/012020

Treatment of the Mirror ^{3}H(α,γ)^{7}Li and ^{3}He(α,γ)^{7}Be Reactions in the Algebraic Version of the Resonating Group Model

A S Solovyev[1], S Yu Igashov[1], and Yu M Tchuvil'sky[2]

[1] All-Russia Research Institute of Automatics (VNIIA), 127055 Moscow, Russia
[2] Scobeltsyn Institute of Nuclear Physics, Lomonosov Moscow State University, 119991 Moscow, Russia

E-mail: alexander.solovyev@mail.ru

Abstract. A unified microscopic approach based on the algebraic version of the resonating group model has been realized for description of the radiative capture reactions ^{3}H(α, γ)^{7}Li and ^{3}He(α, γ)^{7}Be, which play an important role for modern nuclear astrophysics. The astrophysical *S*-factors of the reactions and branching ratios between capture to the ground and first excited states of the ^{7}Li and ^{7}Be nuclei have been calculated. The comparison with the most recent experimental data demonstrates a good agreement.

1. Introduction

The mirror ^{3}H(α, γ)^{7}Li and ^{3}He(α, γ)^{7}Be reactions of the radiative capture are rather significant for the nuclear astrophysics [1, 2]. Both of these reactions are important for the primordial nucleosynthesis investigation, in particular, for solution of the so-called ^{7}Li problem [3]. Moreover, the ^{3}He(α, γ)^{7}Be reaction is very interesting for study of the stellar nucleosynthesis including kinetics of processes at the Sun since it is the starting point for the second and third chains of the pp-cycle of hydrogen burning. For that reasons, the experimental interest, which arose since the middle of the last century [4–6], does not decrease up till now [7–19].

The radiative capture cross section is exponentially small at low sub-barrier energies because of Coulomb repulsion between colliding nuclei. That is why the experimental measurement of the cross section turns out to be extremely hard (or even impossible in some cases) task at the energies which are important for astrophysics. Thus microscopic theoretical calculation tuned by comparison with the experimental results obtained at higher energies occurs the most justified and promising approach for determination of the cross section or astrophysical *S*-factor at low energies.

The resonating group model (RGM) [20] is one of the most consistent realizations of the cluster model. Pioneering RGM-calculation of the ^{3}He(α, γ)^{7}Be and ^{3}H(α, γ)^{7}Li reactions was performed in 1981 [21]. More recent RGM-calculation is presented, for example, in [22]. Most part of calculations of the ^{3}He(α, γ)^{7}Be reaction was made before appearance of the modern experimental data [16–18] covering the middle energies. The data [5] were the sole ones presented in the last century which cover the middle energies. Therefore, the calculations were focused on the description of experiment [5] and extrapolation the data to the low energy domain. Modern data [16–18] occurred in quantitative and qualitative discrepancy with [5] in the middle energy region. As a result, the questions concerning energy dependence of the *S*-factor of the ^{3}He(α, γ)^{7}Be reaction appeared again.

The present work is focused on the microscopic treatment of the ^{3}H(α, γ)^{7}Li and ^{3}He(α, γ)^{7}Be reactions in the framework of the algebraic version (AV) [23] of the RGM which looks promising for these purposes. The astrophysical *S*-factors of the ^{3}H(α, γ)^{7}Li and ^{3}He(α, γ)^{7}Be reactions and branching ratios between capture to the ground and first excited states of the ^{7}Li and ^{7}Be nuclei have been calculated. Parameters of the dynamic model are tuned to reproduce the results of new experiments.

2. Ingredients of the theoretical model

2.1. The AVRGM basis wave functions

The main idea of the AVRGM consists in expansion of the relative motion wave function of colliding nuclei (clusters) over the oscillator basis

$$f_{\nu lm}(\mathbf{q}) = (-1)^{(\nu-l)/2}\left(\frac{2\Gamma((\nu-l+2)/2)}{r_0^3\,\Gamma((\nu+l+3)/2)}\right)^{1/2}\left(\frac{q}{r_0}\right)^l L_{(\nu-l)/2}^{(l+1/2)}\left(\frac{q^2}{r_0^2}\right)\exp\left(-\frac{q^2}{2r_0^2}\right)Y_{lm}(\mathbf{n}_{\mathbf{q}}), \qquad (1)$$

where Γ, $L_n^{(\beta)}$, Y_{lm} are gamma-function, generalized Laguerre polynomial, and spherical harmonic respectively, ν is number of oscillator quanta, *l*, *m* are relative orbital momentum and its projection, **q** is Jacobi vector of the relative motion of clusters, r_0 is oscillator radius. As a result, the total wave function of the system is sought in the form of expansion over the AVRGM basis

$$\Psi_{J^\pi Mls\nu} = N_{J^\pi ls\nu}\,\mathrm{A}\left\{\sum_{m+\sigma=M} C_{lm\,s\sigma}^{JM}\left[\phi_{s_1}^{(1)}\phi_{s_2}^{(2)}\right]_{s\sigma} f_{\nu lm}(\mathbf{q})\right\}. \qquad (2)$$

Here $N_{J^\pi ls\nu}$ is the norm, A is antisymmetrization operator, $\phi_{s_1}^{(1)}$, $\phi_{s_2}^{(2)}$ are internal wave functions of clusters with spins s_1, s_2 coupled to the channel spin *s*, σ – its projection, *J*, *M* are total angular momentum and its projection, π is parity of system, $C_{lm\,s\sigma}^{JM}$ is Clebsch–Gordan coefficient. In this study the functions $\phi_{s_1}^{(1)}$ and $\phi_{s_2}^{(2)}$ are chosen in the form of the lowest compatible with the Pauli exclusion principle wave functions of the translationally invariant oscillator shell model.

2.2. The AVRGM equations sets

Substituting the total wave function expansion over the basis (2) into the projected many-particle Schrödinger equation we can obtain the set of the AVRGM equations for unknown expansion coefficients. As a result, for the continuum states we arrive the nonhomogeneous set

$$\sum_{\nu=\nu_0}^{\nu_{\mathrm{as}}-2}\left(\langle J^\pi Mls\mu|H|J^\pi Mls\nu\rangle - E\delta_{\mu\nu}\right)C_{J^\pi Mls\nu}^{(\mathrm{C})} = F_{J^\pi Mls\mu}, \quad \mu=\nu_0,\nu_0+2,\ldots,\nu_{\mathrm{as}}, \qquad (3)$$

$$F_{J^\pi Mls\mu} = -\sum_{\nu=\nu_{\mathrm{as}}}^{\nu_{\max}}\langle J^\pi Mls\mu|H|J^\pi Mls\nu\rangle C_{J^\pi Mls\nu}^{(\mathrm{as})}. \qquad (4)$$

Here ν_0 is the minimal compatible with the Pauli exclusion principle number of oscillator quanta, and the expansion coefficients $C_{J^\pi Mls\nu}^{(\mathrm{C})}$ are replaced by their asymptotic values $C_{J^\pi Mls\nu}^{(\mathrm{as})}$ [24] starting from the sufficiently large ν_{as}. The value $\nu_{\max}$ is large enough to neglect in (4) the terms with $\nu > \nu_{\max}$. For the bound states equations set is similar but homogeneous (with zero right side of the equations). The modified Hasegawa–Nagata potential [25] was applied as the strong NN-potential in the calculations.

2.3. Astrophysical S-factor and branching ratio of the radiative capture

Astrophysical *S*-factor $S(E_{\mathrm{c.m.}}) = E_{\mathrm{c.m.}}\,\sigma(E_{\mathrm{c.m.}})\exp(2\pi\eta)$ has more smooth energy dependence than cross section since the exponential smallness caused by Coulomb barrier penetrability is explicitly extracted from it.

Assuming predominance of the E1-transition the partial cross section of the discussed reactions in the AVRGM basis (2) can be written as

$$\sigma_{i\to f}(E_{\text{c.m.}}) = \frac{8\pi}{9\hbar(2l_i+1)}\left(\frac{E_\gamma}{\hbar c}\right)^3 \left| \sum_{\nu_i,\nu_f} C^{(\text{D})}_{J_f^{\pi_f} l_f s \nu_f} \left\langle J_f^{\pi_f} l_f s \nu_f \left\| M_1^{\text{E}} \right\| J_i^{\pi_i} l_i s \nu_i \right\rangle C^{(\text{C})}_{J_i^{\pi_i} l_i s \nu_i} \right|^2, \quad (5)$$

where the indices i, f denote initial and final states of the system respectively, E_γ is energy of the emitted photon, M_1^{E} is electric dipole operator (its matrix elements have been calculated in our work [26]). In the expression (5) the initial wave function is assumed to be a partial wave of a unit-flux scattering wave function and the channel spin s equals 1/2.

Branching ratio of the ^{3}H(α, γ)^{7}Li (^{3}He(α, γ)^{7}Be) reaction is determined by the expression $R(E_{\text{c.m.}}) = \sigma_1(E_{\text{c.m.}})/\sigma_0(E_{\text{c.m.}})$, where σ_0, σ_1 are the capture cross sections of the ^{3}H + ^{4}He (^{3}He + ^{4}He) system to the ground and first excited states of the ^{7}Li (^{7}Be) nucleus. Cross section σ_0 is the sum of three partial ones (5) with the following values of the quantum numbers (J, l): $(J_f, l_f) = (3/2, 1)$; $(J_i, l_i) = (1/2, 0), (3/2, 2), (5/2, 2)$. In turn, cross section σ_1 is the sum of two partial ones (5) with (J, l) possessing the following values: $(J_f, l_f) = (1/2, 1)$; $(J_i, l_i) = (1/2, 0), (3/2, 2)$. The total cross section is the sum of σ_0 and σ_1.

3. Results of calculations

The comparison of the calculated astrophysical S-factor and branching ratio of the ^{3}H(α, γ)^{7}Li reaction with experimental data [6–9] is presented in figure 1.

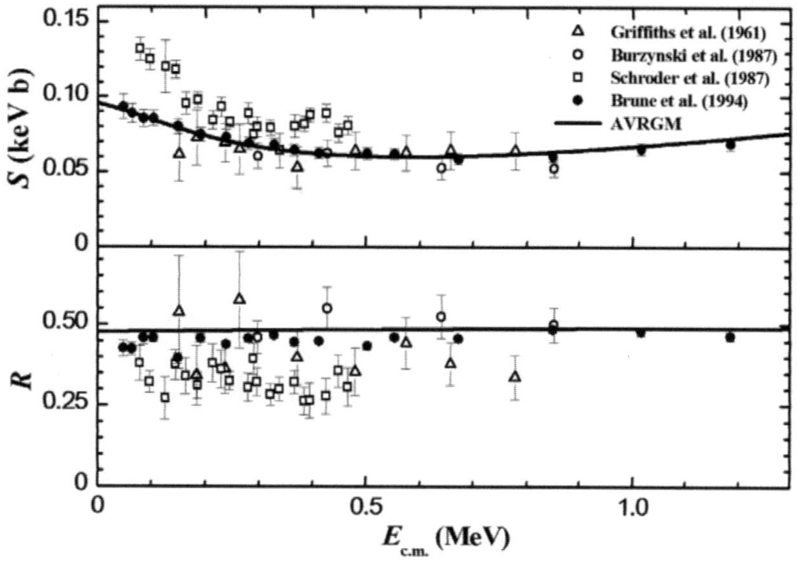

Figure 1. Astrophysical S-factor and branching ratio of the ^{3}H(α, γ)^{7}Li reaction.

Figure 1 demonstrates rather good agreement with the data [6, 7, 9]. It seems to be essential that the most recent data [9] with relatively small errors covering sufficiently wide energy range are well-described by the theoretical curves. There are only two adjustable parameters in this study: r_0 and the Majorana forces intensity g_c, which is introduced into the central part

$$V_{ij}=\sum_{n=1}^{3}V_n\left(w_n+(1-g_c)m_n-g_c m_n P_{ij}^{\sigma}P_{ij}^{\tau}+b_n P_{ij}^{\sigma}-h_n P_{ij}^{\tau}\right)\exp(-\mu_n r_{ij}^2) \qquad (6)$$

of the nucleon-nucleon potential (see [25]). The following values $r_0 = 1.386$ fm and $g_c = 1.021$ were adopted here. The calculated astrophysical *S*-factor and branching ratio of the ^{3}He(α, γ)^{7}Be reaction and the experimental data from [10–19] are presented in figure 2.

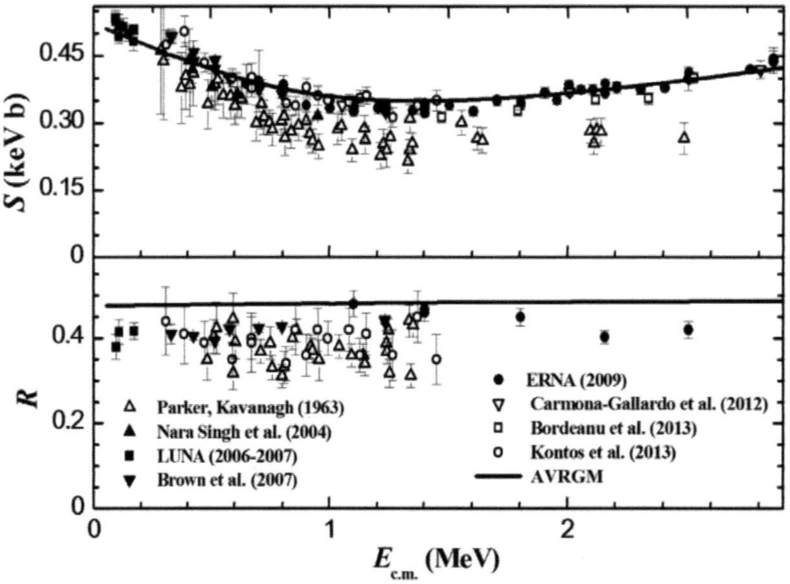

Figure 2. Astrophysical *S*-factor and branching ratio of the ^{3}He(α, γ)^{7}Be reaction.

The sole 'old' data [5] covering the middle energies (references to other data obtained in the last century can be found, for example, in [2]) are also marked in figure 2. It is very important that many early *S*-factor calculations were aimed at description of data [5] which, as it is seen from figure 2, do not explicitly agree with the newest data [16–18] in the middle energies region. The calculated *S*-factor of the ^{3}He(α, γ)^{7}Be reaction is in very good agreement with modern data [10–19]. The calculated branching ratio also demonstrates reasonable fit of the data. In the calculations we chose the following values of the parameters: $r_0 = 1.2$ fm, $g_c = 1.035$. It should be noted, that the use of these r_0 and g_c values for calculation of ^{3}H(α, γ)^{7}Li reaction provides satisfactory but worse results.

Table 1. Energies of the ^{4}He, ^{3}H, ^{7}Li nuclei.

	$E(^4\text{He})$	$E(^3\text{H})$	$E(^7\text{Li})$	$E(^7\text{Li}^*)$
Experiment	–28.296	–8.482	–39.244	–38.766
Calculation	–28.296	–6.467	–35.997	–35.930

Table 2. Energies of the ^{4}He, ^{3}He, ^{7}Be nuclei.

	$E(^4\text{He})$	$E(^3\text{He})$	$E(^7\text{Be})$	$E(^7\text{Be}^*)$
Experiment	–28.296	–7.718	–37.600	–37.171
Calculation	–26.703	–2.073	–29.746	–29.638

The calculated with the corresponding set of adjustable parameters and experimental [27, 28] values of the binding energies of the ^{4}He, ^{3}He, and ^{3}H nuclei, as well as the ground and first excited states energies of the ^{7}Li and ^{7}Be nuclei are presented in the tables. As it can be seen from the tables, the calculated energies are smaller than the corresponding experimental ones in magnitude. The model demonstrates reasonable sensitivity to the adjustable parameters: the variation of the r_0 within 1% influences upon S and R within 3% and 0.1% respectively. The similar variation of g_c results in variation S and R which not exceeds 18% and 1%.

4. Conclusion

In the present work the investigation of the isospin mirror ^{3}H(α, γ)^{7}Li and ^{3}He(α, γ)^{7}Be reactions was performed. Fully antisymmetrized microscopic wave functions of the single-channel AVRGM were used for the calculations of astrophysical S-factors and branching ratios of these reactions. Comparison with the most recent data shows well enough agreement. At the same time the calculated S-factors for the ^{3}H(α, γ)^{7}Li and ^{3}He(α, γ)^{7}Be reactions differ quantitatively and qualitatively from 'old' data [5] and [8] respectively. In fact, it turns out to be impossible to reproduce these data using reasonable values of the adjustable parameters. Obtained values of the S-factors show more smooth energy dependence than those data. On the whole, taking into account remaining underbinding of the clusters and the compound nuclei, one can conclude that further more comprehensive theoretical investigations of these reactions keep actuality.

References

[1] Adelberger E G *et al* 2011 *Rev. Mod. Phys.* **83** 195
[2] Xu Y *et al* 2013 *Nucl. Phys. A* **918** 61
[3] Iocco F *et al* 2009 *Phys. Rep.* **472** 1
[4] Holmgren H D and Johnston R L 1959 *Phys. Rev.* **113** 1556
[5] Parker P D and Kavanagh R W 1963 *Phys. Rev.* **131** 2578
[6] Griffiths G M, Morrow R A, Riley P J and Warren J B 1961 *Can. J. Phys.* **39** 1397
[7] Burzyński S, Czerski K, Marcinkowski A and Zupranski P 1987 *Nucl. Phys. A* **473** 179
[8] Schröder U *et al* 1987 *Phys. Lett. B* **192** 55
[9] Brune C R, Kavanagh R W and Rolfs C 1994 *Phys. Rev. C* **50** 2205
[10] Nara Singh B S, Hass M, Nir-El Y and Haquin G 2004 *Phys. Rev. Lett.* **93** 262503
[11] Bemmerer D *et al* 2006 *Phys. Rev. Lett.* **97** 122502
[12] Gyürky Gy *et al* 2007 *Phys. Rev. C* 75 035805
[13] Confortola F *et al* 2007 *Phys. Rev. C* **75** 065803
[14] Costantini H *et al* 2008 *Nucl. Phys. A* **814** 144
[15] Brown T A D *et al* 2007 *Phys. Rev. C* **76** 055801
[16] Di Leva A *et al* 2009 *Phys. Rev. Lett.* **102** 232502; **103** 159903(E)
[17] Carmona-Gallardo M *et al* 2012 *Phys. Rev. C* **86** 032801(R)
[18] Bordeanu C *et al* 2013 *Nucl. Phys. A* **908** 1
[19] Kontos A *et al* 2013 *Phys. Rev. C* **87** 065804
[20] Wildermuth K and Tang Y C 1977 *A Unified Theory of the Nucleus* (Braunschweig: Vieweg)
[21] Liu Q K K, Kanada H, Tang Y C 1981 *Phys. Rev. C* **23** 645
[22] Arai K, Baye D and Descouvemont P 2002 *Nucl. Phys. A* **699** 963
[23] Filippov G F and Okhrimenko I P 1980 *Phys. Atom. Nucl.* **32** 480
[24] Igashov S Yu 2008 *The J-Matrix Method. Developments and Applications* ed A D Alhaidari, E J Heller, H A Yamani and M S Abdelmonem (Berlin: Springer) part 2 pp 49–66
[25] Kanada H, Kaneko T, Nagata S and Nomoto M 1979 *Progr. Theor. Phys.* **61** 1327
[26] Solovyev A S, Igashov S Yu and Tchuvil'sky Yu M 2014 *Bull. Russ. Acad. Sci. Phys.* **78** 433
[27] Tilley D R *et al* 2002 *Nucl. Phys. A* **708** 3
[28] Audi G, Wapstra A H and Thibault C 2003 *Nucl. Phys. A* **729** 337

Journal of Physics: Conference Series **569** (2014) 012021 doi:10.1088/1742-6596/569/1/012021

Clustering structure of nuclei in deep inelastic processes

M. Hirai[a], **S. Kumano**[b,c,d], **K. Saito**[d,e], **T. Watanabe**[e]

[a] Nippon Institute of Technology, Saitama 345-8501, Japan
[b] KEK Theory Center, Institute of Particle and Nuclear Studies, KEK and Department of Particle and Nuclear Studies, Graduate University for Advanced Studies 1-1, Ooho, Tsukuba, Ibaraki, 305-0801, Japan
[c] Theory Group, Particle and Nuclear Physics Division, J-PARC Center 203-1, Shirakata, Tokai, Ibaraki, 319-1106, Japan
[d] J-PARC Branch, KEK Theory Center, Institute of Particle and Nuclear Studies, KEK 203-1, Shirakata, Tokai, Ibaraki, 319-1106, Japan
[e] Department of Physics, Faculty of Science and Technology, Tokyo University of Science 2641, Yamazaki, Noda, Chiba, 278-8510, Japan

E-mail: shunzo.kumano@kek.jp

Abstract. A clustering aspect is explained for the ^{9}Be nucleus in charged-lepton deep inelastic scattering. Nuclear modifications of the structure function F_2 are studied by the ratio $R_{\rm EMC} = F_2^A/F_2^D$, where A and D are a nucleus and the deuteron, respectively. In a JLab experiment, an unexpectedly large nuclear modification slope $|dR_{\rm EMC}/dx|$ was found for ^{9}Be, which could be related to its clustering structure. We investigated a mean conventional part of a nuclear structure function F_2^A by a convolution description with nucleon momentum distributions calculated by antisymmetrized (or fermionic) molecular dynamics (AMD) and also by a simple shell model. We found that clustering effects are small in the conventional part, so that the JLab result could be associated with an internal nucleon modification or a short-range nuclear correlation which is caused by high densities due to cluster formation.

1. Introduction

Nuclear modifications of the structure function F_2 are known as the European Muon Collaboration (EMC) effect [1], and such effects are shown by the ratio $R_{\rm EMC}(x) = F_2^A(x)/F_2^D(x)$, where A and D indicate a nucleus and the deuteron, respectively, and x is the Bjorken scaling variable. Various mechanisms contribute to the modifications. At large $x > 0.7$, nucleon Fermi motion in a nucleus gives rise to increase of the ratio $R_{\rm EMC}(x)$. At medium x $(0.3 < x < 0.7)$, nuclear binding is the major source of nuclear effects. In addition, modifications of internal nucleon structure could contribute. At small $x < 0.05$, the nuclear shadowing effects become prominent. The nuclear structure functions have been measured from small x to large x for many nuclei, and the nuclear modifications have been determined by global analyses of experimental data [2].

Nuclei are basically described by a shell model, where nucleons are assumed to move in an average central potential created by interactions of all the nucleons in a nucleus. However, cluster structure in a nucleus is not easily described by the shell model with a limited number of shells. We know that some nuclei in the region of the mass number $A \sim 10$ have clustering configurations. It is interesting to study whether such clustering features appear in the nuclear structure functions F_2^A.

In 2009, an anomalously large nuclear effect was reported by the JLab measurement on the beryllium-9 structure function F_2 as shown in Fig. 1 [3], where the slope is shown as a functions of the scaled average nuclear density. The scaled means that the factor $(A-1)/A$ is multiplied for removing the struck nucleon. Although other data are on the smooth curve, the beryllium-9 data is much larger than the one expected from its average nuclear density.

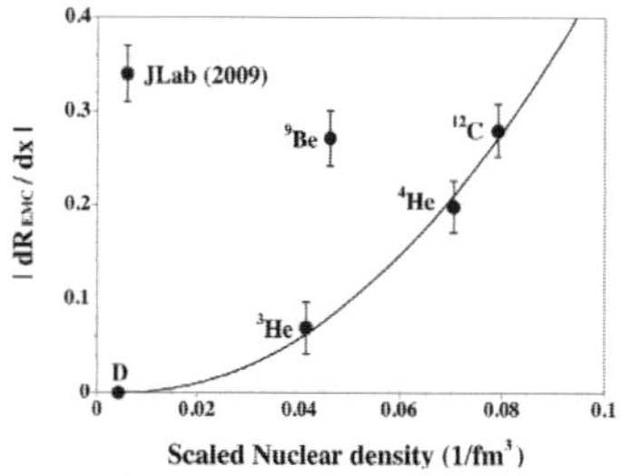

Figure 1. JLab data are shown for the nuclear modification slope $|d(F_2^A/F_2^D)/dx|$ as a function of the scaled nuclear density. The curve indicates a smooth function of the density to fit the data.

It is known that the beryllium-9 has a clustering configuration with two α nuclei and surrounding neutron clouds. Therefore, such an unexpected nuclear modification could come from the cluster formation in the nucleus. We investigated such a possibility by using a convolution description for the structure function F_2^A with nucleon momentum distributions calculated by antisymmetrized molecular dynamics (AMD) and also by a simple shell model for comparison [4]. Then, possible reasons are suggested for the large slope of ^{9}Be due to the cluster formation in the nucleus.

2. Clustering effects in nuclear structure functions

The standard method for describing the nuclear structure functions F_2^A is to use the convolution description [5] for describing a mean conventional part. The structure function F_2^A is given by the nucleonic one F_2^N convoluted with a nucleon momentum distribution in the nucleus $f(y)$ [1, 5] as shown in Fig. 2:

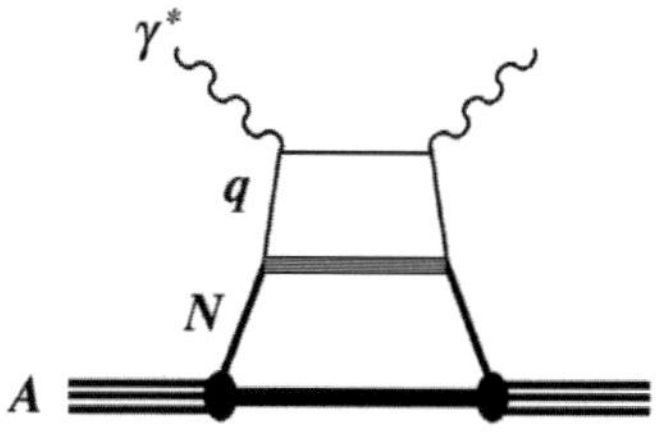

Figure 2. Convolution model for F_2^A. A quark momentum distribution in a nucleus is calculated by the integral of a corresponding quark distribution convoluted with a nucleon momentum distribution.

$$F_2^A(x,Q^2) = \int_x^A dy\, f(y)\, F_2^N(x/y, Q^2),$$

$$f(y) = \frac{1}{A}\int d^3p_N\, y\, \delta\left(y - \frac{p_N\cdot q}{M_N\nu}\right)|\phi(\vec{p}_N)|^2, \quad (1)$$

where y is the momentum fraction $y = M_A\, p_N\cdot q/(M_N\, p_A\cdot q) \simeq A\, p_N^+/(p_A^+)$ with a light-cone momentum p^+.

The momentum density $|\phi(\vec{p}_N)|^2$ is calculated by two methods, antisymmetrized molecular dynamics (AMD) [6] or fermionic molecular dynamics (FMD) [7], and a simple shell model for comparison. The AMD and FMD are essentially the same, so that we use the notation AMD hereafter. The AMD is a variational method, and its advantage is that there is no a priori assumption on nuclear structure whether it is a shell or cluster-like configuration. A nuclear wave function is given by the Slater determinant of single-particle wave functions

$$|\Phi(\vec{r}_1,\vec{r}_2,\cdots,\vec{r}_A)\rangle = \frac{1}{\sqrt{A!}}\det[\varphi_1(\vec{r}_1),\varphi_2(\vec{r}_2),\cdots,\varphi_A(\vec{r}_A)]. \quad (2)$$

Here, a nucleon is described by the single-particle wave function $\varphi_i(\vec{r}_j) = \phi_i(\vec{r}_j)\,\chi_i\,\tau_i$, where χ_i and τ_i indicate spin and isospin states, respectively. The function $\phi_i(\vec{r}_j)$ is the space part of the wave function, and it is assumed to be given by the Gaussian functional form:

$$\phi_i(\vec{r}_j) = \left(\frac{2\nu}{\pi}\right)^{3/4}\exp\left[-\nu\left(\vec{r}_j - \frac{\vec{Z}_i}{\sqrt{\nu}}\right)^2\right], \quad (3)$$

where ν is a parameter to express the extent of the wave packet. The center of the wave packet is given by $\vec{Z}_i/\sqrt{\nu}$, where $\vec{Z}_i$ is a complex variational parameter. The variational parameters are then determined by minimizing the system energy with a frictional-cooling method. We use effective NN interactions which could describe gross properties of nuclei:

$$\begin{aligned}
\text{2-body:}\quad & V_2 = (1 - m - mP_\sigma P_\tau)\left[v_{21}e^{-(r/r_{21})^2} + v_{22}e^{-(r/r_{22})^2}\right], \\
\text{3-body:}\quad & V_3 = v_3\,\delta^3(\vec{r}_1 - \vec{r}_2)\,\delta^3(\vec{r}_2 - \vec{r}_3), \\
\text{LS:}\quad & V_{LS} = v_{LS}\left[e^{-(r/r_{LS1})^2} - e^{-(r/r_{LS2})^2}\right] P(^3O)\,\vec{L}\cdot\vec{S},
\end{aligned} \tag{4}$$

where m, v_{21}, v_{22}, r_{21}, r_{22}, v_3, v_{LS}, r_{LS1}, and r_{LS2} are constants.

The momentum distributions of Eq. (1) are also calculated in a simple shell model, where the potential is the harmonic-oscillator type ($M_N\omega^2r^2/2$), for clarifying cluster effects by comparing two results. Its wave function is separated into radial- and angular-dependent parts: $\psi_{n\ell m}(r,\theta,\phi) = R_{n\ell}(r)Y_{\ell m}(\theta,\phi)$ where r, θ, and ϕ are spherical coordinates, and n, ℓ, and m are radial, azimuthal, and magnetic quantum numbers, respectively. The function $Y_{\ell m}(\theta,\phi)$ is the spherical harmonics, and the radial wave function is given by

$$R_{n\ell}(r) = \sqrt{\frac{2\kappa^{2\ell+3}(n-1)!}{[\Gamma(n+\ell+1/2)]^3}}\, r^\ell e^{-\frac{1}{2}\kappa^2r^2} L_{n-1}^{\ell+1/2}(\kappa^2r^2), \tag{5}$$

where $L_{n-1}^{\ell+1/2}(x)$ is the Laguerre polynomial, and κ is defined by $\kappa \equiv \sqrt{M_N\omega}$.

As a result, we obtain the two-dimensional spacial density of the ^{9}Be nucleus in Fig. 3. It obviously indicates the existence of two α clusters, together with surrounding neutron clouds. In a simple shell model, such clusters do not appear, and the density is a monotonic distribution. If the averages are taken over polar and azimuthal angles θ and ϕ, the distributions becomes the ones in Fig. 4. After the averages, the cluster structure is no longer apparent. Because the nucleons are confined mainly in two separate regions, the radial distribution is shifted toward large-r region for ^{9}Be in Fig. 4 as shown by the solid curve in comparison with the dashed one.

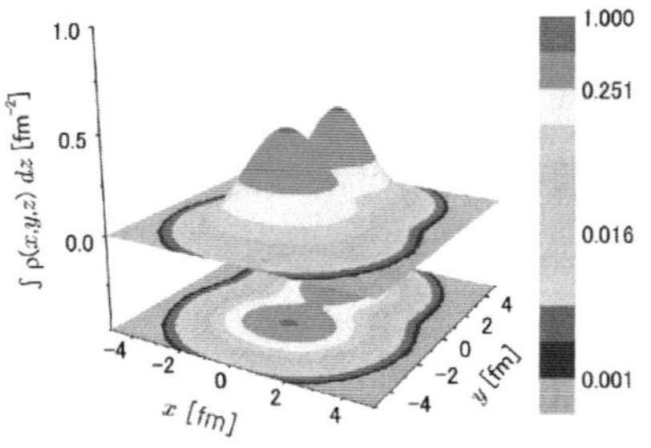

Figure 3. Coordinate-space density of ^{9}Be is shown in the AMD by taking the integral over the coordinate z: $\int dz\rho(x,y,z)$.

The momentum distributions in Fig. 5 are calculated from the spacial distributions. Both AMD and shell-model results are shown for ^{4}He and ^{9}Be. The momentum distribution of the AMD is shifted

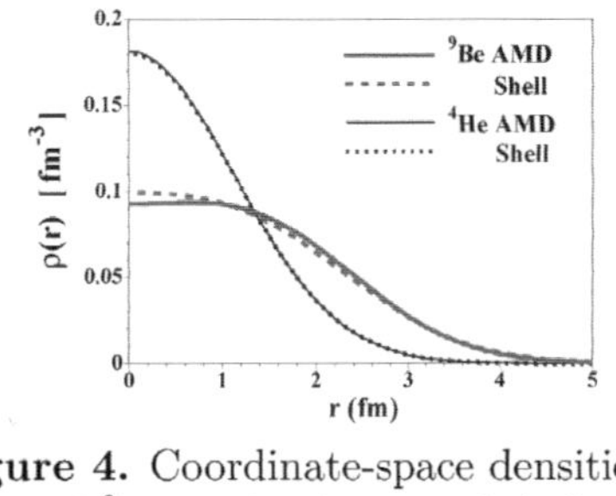

Figure 4. Coordinate-space densities for ^{4}He and ^{9}Be in the AMD and shell models.

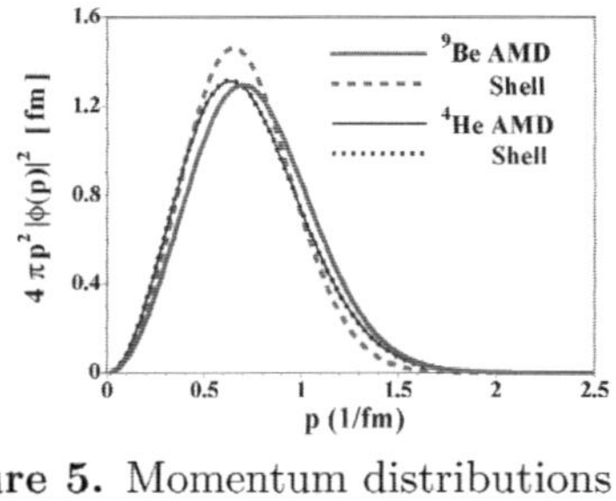

Figure 5. Momentum distributions for ^{4}He and ^{9}Be in the AMD and shell models.

toward the high-momentum region in ^{9}Be, which is caused by the fact that the dense clusters are formed within the ^{9}Be nucleus. If nucleons are confined in the small space regions of the clusters as indicated by the AMD, it leads to an increase of high momentum components.

The modifications of the momentum distributions should affect the nuclear structure function in Eq. (1), and the results are shown for the structure function F_2 in Fig. 6. There are two theoretical curves by the AMD and shell model. In general, the theoretical ratios are consistent with the data in the sense that the ratio decreases at medium x and it increases at large x. These decrease and increase are caused by the nuclear binding and the nucleon's Fermi motion, respectively. However, the simple convolution model is not sufficient to explain the experimental data. The differences should be attributed to the effects of short-range nucleon-nucleon correlations and modifications of internal nucleon structure. The correlation effects change the theoretical ratios toward the experimental data at $x = 0.6 - 0.8$. Since we are interested in an order-of-magnitude estimate of clustering effects, we did not step into such details in this work. In any case, the differences of the two curves, namely the clustering effects, are not large and they are of the same order of experimental errors. In Fig. 6, we found that the clustering effects are rather small in the mean conventional part of F_2^A. Therefore, there should be other sources for the large slope of ^{9}Be. We consider that the nuclear structure functions consist of the mean conventional part and the remaining one depending on the maximum local density:

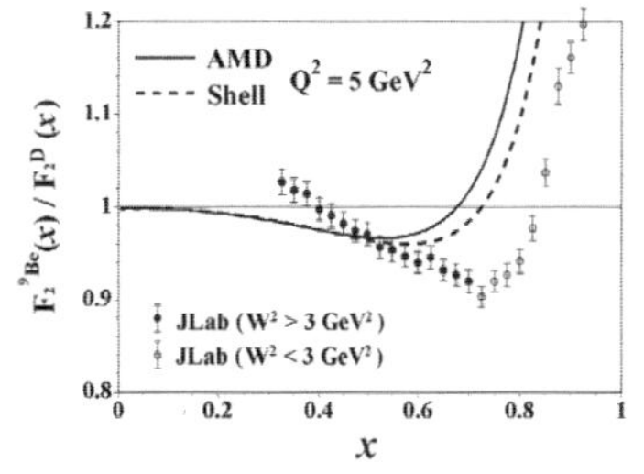

Figure 6. Comparison with JLab experimental data of F_2^{9Be}/F_2^D. The solid and dashed curves indicate the theoretical ratios by the AMD and shell models.

$$F_2^A = (\text{mean part}) + (\text{part created by large densities due to cluster formation}). \quad (6)$$

The first part was calculated by the convolution model as shown in Fig. 6, and the second part seems to be important for explaining the JLab ^{9}Be result. The second term is associated with the modification of internal nucleon structure and/or a short-range correlation caused by the cluster formation.

Although we do not step into detailed calculations in this work, we would like to find whether such an idea works. If the physics mechanism behind the JLab ^{9}Be result is related to the cluster formation, observables could be interpreted by the cluster densities. For such estimates, we calculated maximum local densities by both AMD and shell models, rather than the average nuclear densities. In Fig. 7, the JLab data are plotted by the maximum local densities calculated in the AMD and shell model. The maximum densities are almost the same in the nuclei, ^{3}He, ^{4}He, and ^{12}C, whereas the AMD density is much different from the shell-model one for ^{9}Be. According to the shell model with a limited number of shells, the cluster formation, namely the large densities, cannot be described, and it leads to a small maximum density of ^{9}Be in Fig. 7. It is also interesting to find that there is also a small cluster effect in the ^{12}C. From Fig. 7, we find that all the data are on the smooth curve if the density is calculated by the AMD for ^{9}Be; however, it is far away from the curve if it is evaluated by the shell model. The figure indicates that the large nuclear modification slope $|dR_{\rm EMC}/dx|$ of ^{9}Be should be related to the high densities created by the cluster formation.

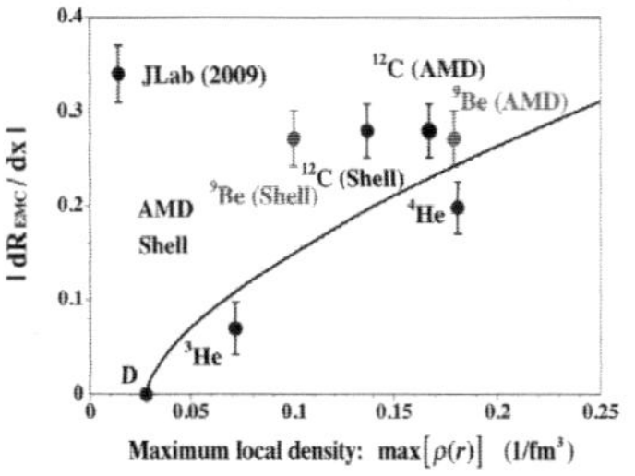

Figure 7. JLab data of the nuclear modification slopes are shown by the maximum local densities. The curve indicates a smooth function $|dR_{EMC}/dx| = a(\rho - \rho_D)^b$ to fit the JLab experimental data except for ^{9}Be.

The JLab measurement could be the first result on a clustering aspect of a nucleus in high-energy nuclear reactions. It could be associated with internal nucleon modifications and/or short-range nuclear correlations, which are caused by the high densities due to the cluster configuration. In future, there will be measurement for other nuclei, such as ^{6}Li, ^{7}Li, ^{10}B, and ^{11}B [8], so that much detailed information will become available.

3. Summary

The large nuclear modification slope of the structure function F_2 was found for ^{9}Be, which has a typical cluster configuration. We showed that the clustering effects are rather small in the mean conventional part by using the convolution model for the structure functions F_2^A with the momentum distributions calculated in the AMD and also in the shell model. The JLab result should be associated with the high-density formation due to the clusters in the ^{9}Be nucleus. We showed it by plotting the JLab data as a function of the maximum local density of a nucleus. We have not clarified the physics origin behind its result. However, it is likely to be modifications of internal nucleon structure in the nuclear medium and/or short-range nucleon-nucleon correlations in the dense clusters. There is an approved experiment at JLab to measure the EMC effects in the mass region $A \sim 10$, so that much details will be investigated in the near future for possible clustering effects in deep inelastic scattering (DIS). The cluster physics has been investigated mainly in low energy nuclear reactions; however, DIS processes could shed light on new aspects in both cluster physics and nuclear structure functions. Cluster physicists have an opportunity to play a major role in high-energy hadron physics.

Acknowledgements

This work was supported by the MEXT KAKENHI Grant Number 25105010.

References

[1] Geesaman D F, Saito K and Thomas A W 1995 *Ann. Rev. Nucl. Part. Sci.* **45** 337
[2] Hirai H, Kumano S and Miyama M 2001 *Phys. Rev.* D **64** 034003; Hirai M, Kumano S and Nagai T-H 2004 *Phys. Rev.* C **70** 044905; 2007 *Phys. Rev.* C **76** 065207
[3] Seely J *et al.* 2009 *Phys. Rev. Lett.* **103** 202301
[4] Hirai M, Kumano S, Saito K and Watanabe T 2011 *Phys. Rev.* C **83** 035202
[5] Ericson M and Kumano S 2003 *Phys. Rev.* C **67** 022201
[6] Kanada-En'yo Y, Kimura M and Horiuchi H 2003 *C. R. Physique* **4** 497
[7] Feldmeier H and Schnack J 2000 *Rev. Mod. Phys.* **72** 655
[8] Jefferson Lab PAC-35 proposal, PR12-10-008, Arrington J *et al.* 2009

Spatial measure of reaction size in proton scattering

M Tomita[1], M Iwasaki[1], R Otani[1], and M Ito[1,2]
[1]Department of Pure and Applied Physics,
Kansai University, Yamatecho, 3-3-35, Suita, Japan
[2]Research center for Nuclear Phtsics (RCNP),
Osaka University, Mihogaoka 10-1, Suita 567-0047, Japan

E-mail:k977550@kansai-u.ac.jp

Abstract. We introduce a scattering radius, which characterizes a spatial size of the scattering area, from partial wave decompositions of a cross section for a nucleon-nucleus scattering. The coupled-channel calculations of the p + ^{12}C scattering are performed in the range of E_{lab} = 29.95 MeV to 65 MeV, and the scattering radii for the elastic scattering and the various inelastic channels, which involve the rotational or vibrational excitations and the 3α excitations in ^{12}C, are derived from the partial wave decomposition. We found that the scattering radii for the inelastic channels with a well developed 3α structure are strongly enhanced in comparison to the scattering radius for the elastic and collective channels. This enhancement of the scattering radius for the 3α channel strongly suggests that the scattering radius is sensitive to a size of the intrinsic structure of the finally excited state in the scattering process.

1. Introduction

Clustering phenomena, in which a nucleus is decomposed into some subunits, are well known to appear in the excited states of the lighter mass system [1]. In ^{12}C, for instance, three α particles are strongly overlapped to each other in its ground 0_1^+ state, while they are weakly coupled in the excited 0_2^+ state, called Hoyle state, at E_x = 7.65 MeV [2,3]. One of characteristic features in cluster states is an extension of a matter radius. Previous theoretical calculations suggest that a radius of a cluster state is larger by about 50% than a radius of a ground state [1-3].

Unfortunately, a size of a cluster state is difficult to measure directly by experiments because a cluster state has a very short life-time. In a naive expectation, however, such a spatial extension may be reflected in observables of scattering phenomena. In recent studies, some enhancement factors originated from a spatial extension of a cluster structure are studied especially in the 3α state in ^{12}C [4, 5]. In Ref. [5], for instance, a diffraction radius of the Hoyle state in ^{12}C is speculated from the diffraction pattern observed in the differential cross section in the high energy proton scattering. In fact, a sign of an enhanced radius is confirmed for the Hoyle state of ^{12}C.

Recently, we have proposed the method of the "scattering radius" on the basis of a partial wave decomposition of a scattering cross section [6]. The scattering radius is considered to characterize a spatial size of the reaction area for exclusive reactions. We can always derive the scattering radius just by doing the partial wave analysis of a scattering cross section. For the first time this method has been applied to a simple two channel problem in the p + ^{12}C scattering going to the Hoyle state [6], where the prominent enhancement of the scattering radius in the Hoyle channel has been confirmed.

In this report, the method of the scattering radius is extended to the multi-channel problem in the p + ^{12}C scattering, going to the various inelastic channels, which contain the collective excitations of the 2_1^+ and 3_1^- states and the 3α excitations of the 0_2^+ and 2_2^+ states. The matter radius of the 2_2^+ state is expected to be strongly enhanced because the 2_2^+ state corresponds to the rotational excited states of the Hoyle state [3]. Thus, the scattering radius for the 2_2^+ state is very interesting in connection to the enhanced matter radius of this state.

2. Theoretical framework

2.1. Coupled-channel equations and nuclear interactions

We solve a set of the coupled-channel equation for the proton plus ^{12}C system, which is given in a symbolic form

$$\left(T_f(\boldsymbol{R}) + V_{f,f}(\boldsymbol{R}) - E_f\right)\chi_f(\boldsymbol{R}) = -\sum_{i\neq f} V_{f,i}(\boldsymbol{R})\,\chi_i(\boldsymbol{R}) \quad . \tag{1}$$

Here the subscripts of f and i design a channel. $T_f(\boldsymbol{R})$ represents the kinetic energy of the relative motion of the p + ^{12}C system with a relative coordinate $\boldsymbol{R}$, while $V_{f,i}(\boldsymbol{R})$ denotes the coupling potential for the transition from the channel i to the channel f. The total energy in the channel f, E_f, is given by the relation of $E_f = E - \varepsilon_f$ with a proton's incident energy E and an internal energy of ^{12}C, ε_f. $\chi_f(\boldsymbol{R})$ is the proton–^{12}C relative wave function for the channel f, which should be solved in the coupled-channel equations. In the present calculation, we include all the discrete states of ^{12}C; the rotational state of 2_1^+, the vibrational 3_1^- state, the 3α cluster states of 0_2^+ and 2_2^+.

The coupling potential ($V_{f,i}(\boldsymbol{R})$) has a component of the real central ($V_{f,i}^{CE}(\boldsymbol{R})$) and the phenomenological imaginary potential ($-\mathrm{i}U_{f,i}(\boldsymbol{R})$),

$$V_{f,i}(\boldsymbol{R}) = V_{f,i}^{CE}(\boldsymbol{R}) - \mathrm{i}U_{f,i}(\boldsymbol{R})\delta_{f,i} \ . \tag{2}$$

The central potential, $V_{f,i}(\boldsymbol{R})$, is calculated by the folding model [7, 8]. We employ the DDM3Y (Density-Dependent Michigan 3-range Yukawa) [9, 10] effective nucleon-nucleon force and the internal wave function of ^{12}C calculated by the microscopic cluster model [3].

In addition to the folding potential, we introduce an imaginary potential $-iU_{f,i}(\boldsymbol{R})$ in the diagonal transition, which has a Woods-Saxon form factor [6], in order to reproduce the observed differential cross sections as much as possible. In the present calculation, we neglect the imaginary part of the off-diagonal potential and the spin-orbit potential although the latter potential was included in the analysis in Ref. [6].

Definition of the scattering radius

In the usual treatment, the coupled-channel equations (1) are solved by the partial wave decomposition. In the present calculation, the proton is treated as the spin-less particle, and the initial state of the target nucleus is also in the spin-less states. Thus, the differential cross section can be decomposed into the partial wave component, σ($JLL'S$), which is designed by the total spin J, the initial orbital spin L (L=J), the final spin of a target S, and the final orbital spin L' for the scattering two particles.

In order to characterize a size of scattering area, we simply define the effective orbital spin $\bar{L}$ from the partial cross section σ($JLL'S$) like

Journal of Physics: Conference Series **569** (2014) 012022 doi:10.1088/1742-6596/569/1/012022

$$\bar{L} = \sqrt{\frac{\sum_{JL} \tilde{L}^4 \sigma(JLL'S)}{\sum_{JL} \tilde{L}^2 \sigma(JLL'S)}} \qquad (3)$$

with $\tilde{L} = \sqrt{L(L+1)}$ which is calculated from the initial orbital spin L. The effective orbital spin is obtained by the fourth and second powers of the magnitude of the orbital spin, $\tilde{L} = \sqrt{L(L+1)}$. In Eq. (3), we simply assume the same functional form as the root-mean-squared radius for the matter density distribution. The scattering radius R_{SC} corresponding to $\bar{L}$ can be simply obtained from the relation of $\bar{L} = kR_{SC}$, where k denotes the wave number of the incident channel, measured in the laboratory system.

3. Results

We calculate the differential cross sections of the elastic and inelastic scattering in the energy range of E_{lab}=29.95 MeV to 65 MeV, and optimize the parameters in the imaginary potential for the diagonal transition so as to reproduce the observed cross sections. Figure 1 shows the comparison of the theoretical calculation with the experimental data at E_{lab} = 65 MeV. As can be clearly confirmed from Fig.1, the observed cross sections are nicely reproduced by the present coupled-channel calculation although the reproduction of the detailed oscillation is still insufficient in the 0_2^+ channel.

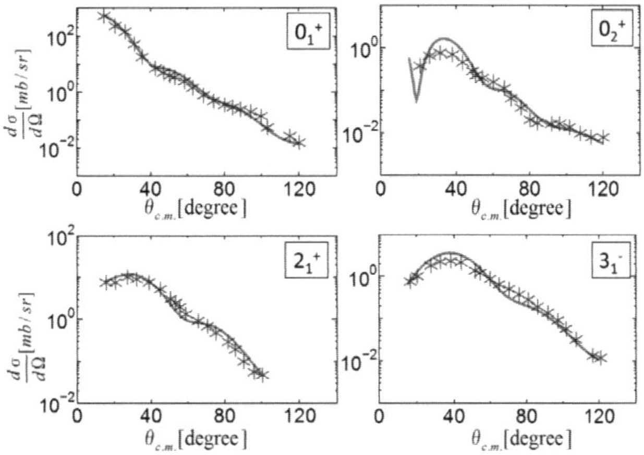

FIG. 1: Differential cross sections at E_{lab} = 65MeV in the 0_1^+ (g.s.), 0_2^+ (3α state), 2_1^+ and 3_1^- channels. The stars and curves represent the experimental data and the theoretical calculation, respectively.

From the scattering calculation in Fig.1, we can obtain the partial cross section. In Fig.2, we show the distributions of the partial cross section σ(*JL*) for the spin-less channels of the 0_1^+ (asterisks) and 0_2^+ (squares). In this figure, only the nuclear part of the partial cross sections is plotted, and the Coulomb part is subtracted in the elastic 0_1^+ channel. Here σ(*JL*) is originally labeled by *J* and *L*, but the relation of *J*=*L* is hold because the proton is treated as the spin-less particle. The values of σ(*J*=*L*) are plotted in Fig.2, which are normalized by the total cross section of $\sigma = \sum_J \sigma(J)$. In this figure, we can clearly observe the partial cross section of the 0_2^+ channel has the extended distribution. This extension strongly suggests that a size of the scattering area for the inelastic 0_2^+ channel is

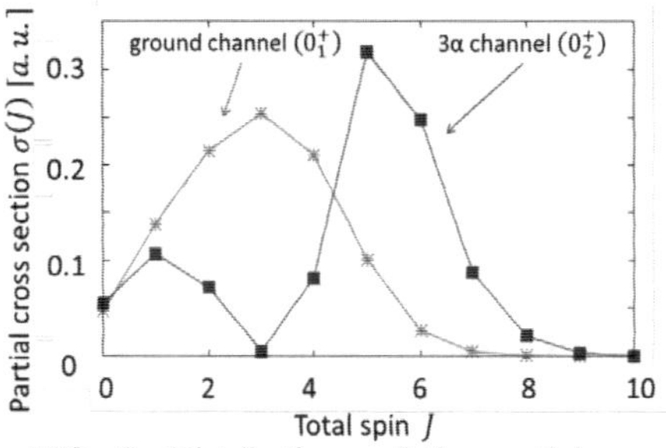

FIG. 2: Distributions of the partial cross sections at E_{lab} = 65 MeV. Both of the partial cross sections are divided by the total cross section.

larger than that for the elastic channel.

We have also observed the distributions of the partial cross sections of other inelastic channels. The distributions of the partial cross sections for the 3_1^- and 2_2^+ channels especially spread over a wide range of the total spin J. In order to characterize the spatial size of each reaction channel more clearly, the effective orbital spins $\bar{L}$ and the scattering radii, R_{SC} are calculated for all the channels from the partial cross section. The obtained effective orbital spins $\bar{L}$ and the scattering radii R_{SC} are summarized in table 1. In this table, the matter radii of all the states ($\bar{r}$) [2] are also shown in the bottom row for comparison.

	0_1^+	0_2^+	2_1^+	2_2^+	3_1^-
$\bar{L}$	4.64	6.37	5.65	6.82	5.96
R_{SC} [fm]	2.73	3.74	3.32	4.01	3.51
$\bar{r}$ [fm]	2.40	3.47	2.38	4.00	2.76

Table 1: The effective orbital spin $\bar{L}$ and the scattering radius R_{SC} at E_{lab} = 65 MeV. The matter radius $\bar{r}$ of the 0_1^+, 0_2^+, 2_1^+, 2_2^+ and 3_1^- states are also shown in the bottom column [2]. R_{SC} and $\bar{r}$ are shown in units of fm.

From table 1, we can see the enhancement of R_{SC} for the inelastic scattering going to the 0_2^+, 2_2^+ and 3_1^- states, which are the resonance states just above the 3α threshold, in comparison to the scattering radius of the 0_1^+ and 2_1^+ states. The enhancement of the scattering radii in the 0_2^+ and 2_2^+ channels can be attributed to the spatial extension of the 3α structures of these states. On the contrary, there is an enhancement of the scattering radius in the 3_1^- channel although this state does not have the 3α-cluster structure but the spatially compact structure [1, 2]. The 3_1^- state is a resonant state existing above the 3α threshold of $E_{3\alpha}$=7.27 MeV (E_x=9.6 MeV). Therefore, the 3_1^- state has an extended tails in its radial wave function. This extension is considered to be one of the origins for the enhanced scattering radius of the 3_1^- state. In particular, the R_{SC} of the 2_2^+ is strongly enhanced. This is very interesting result, because the 2_2^+ state is 3α rotational state, which has much more extended matter radius, and this large radius has already been pointed out by Funaki et al. [3].

The scattering radius does not necessarily mean the matter radius of a target nucleus itself but our scattering radius is consistent to the matter radius of the 2_2^+ state. Thus, the scattering radius derived from the partial cross section can be used for the measure of the spatial size of the individual reaction channels.

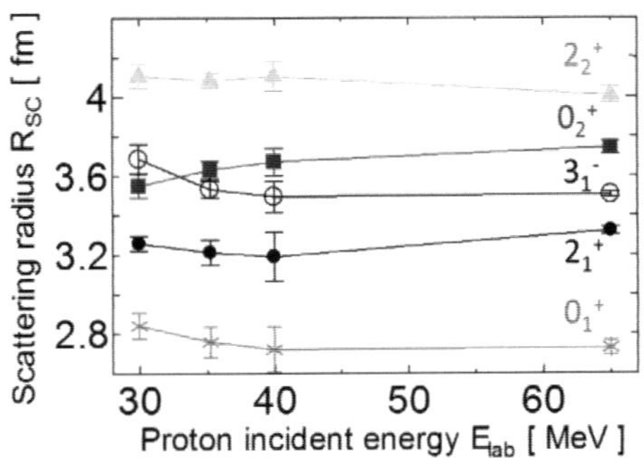

FIG. 3: The incident dependence of the scattering radius. Asterisk, black circle, open circle, square and delta are 0_1^+, 2_1^+, 3_1^-, 0_2^+ and 2_2^+ states respectively.

Finally, we discuss the incident energy dependence of the scattering radii R_{SC}, which are shown in Fig.3. The results in Fig.3 represent that the scattering radius doesn't depend so much on the incident energy. Therefore, the qualitative features of the scattering radius are almost invariant with respective to the variation of the incident energy. In all the energy range considered here, the scattering radius of the 2_2^+ state (triangles) is the largest of all the channels, while the enhancements are commonly observed in the 0_2^+ (squares) and 3_1^- (open circles) channels. In contrast to these enhanced channels, the scattering radii of the ground 0_1^+ (asterisk) and its rotational state of 2_1^+ (black circles) are clearly suppressed. The magnitude correlation in the scattering radii observed in Fig.3 is the consistent to the matter radii predicted in the 3α cluster model shown in table 1 although the order of 0_2^+ and 3_1^- is inversed only at E_{lab}=29.95 MeV .

The error bars attached to individual symbols mean the theoretical ambiguity of the potential parameters. In the present coupled-channel calculation, a simultaneous reproduction of all the observed cross sections is a little difficult in a common parameter set of the imaginary potential and the strength of the folding potential. Thus, we have searched the several parameter sets of the potential, which can reproduce each of exit channels. For example, one parameter set is optimized for the reproduction of the elastic channel, while the other parameter set can reproduce only the Hoyle state, and so on. From the searched potential parameter sets, we can also evaluate the various scattering radii. The error bars mean the standard deviation of a set of the scattering radii, calculated from a set of the potential parameters. The resultant magnitude of the error reaches about ±10%, at most, in all the exit channels although the error bars may be changed in the future parameter search. Thus, the qualitative features of the scattering radius are not changed so much even if much more tuned potentials in comparison to the present calculation are employed in the coupled-channels calculation.

Summery and discussion

In summary, we have introduced a scattering radius from the partial wave decomposition of the differential cross sections. The method of the scattering radius has been applied to the multi coupled-channel problem for the p + ^{12}C scattering, and the scattering radii of the elastic and inelastic scattering to 0_2^+, 2_1^+, 2_2^+ and 3_1^- states have been derived. We have found a clear enhancement of the scattering radius for the final inelastic channels in comparison to the elastic channel. In particular, we confirmed the enhanced scattering radius in the cluster channels (0_2^+, 2_2^+). This enhanced scattering radius for the cluster channels is attributed to the spatial extension of the 3α structure.

The scattering radius is considered to give size information on a spatial area, where the elastic scattering or the transition to a final channel occurs. In principle, the scattering radius can always be defined when the partial wave analysis is possible for the differential cross section. In a recent study, effective radii in the deuteron induced reaction are derived from the continuum-discretized coupled-channel, and radii for the inclusive reactions are clearly defined [11]. In contrast to this approach, the present method is useful for the definition of the radius in an exclusive reaction, going to a specific channel.

In the present coupled-channel calculation, we have treated only the p + ^{12}C scattering system. The systematic studies are very important to establish the validity of the method of the scattering radius. For example, the analysis of the alpha and ^{3}He scatterings is also important because the previous studies have pointed out the evolution of the Airy structures in the angular distribution in connection to the enhanced radius of the Hoyle state [12, 13]. Furthermore, the ^{16}O nucleus is another typical example that the α + ^{12}C [14, 15] and 4α [15] structures appear in the excited states. In particular, a recent study in Ref. [15] suggests that the 0_6^+ state corresponds to the 4α dilute state, and it is expected to have the extended matter radius. Thus, the analysis of the inelastic scattering of proton and alpha particle by the ^{16}O nucleus is very interesting. Systematic studies are now underway.

References

[1] Ikeda K et al., *Prog. Theor. Phys.* **68**, 1 (1980), and references therein.
[2] Fukushima Y and Kamimura M, *Proceedings of the International Conference Nuclear Structure, J. Phys. Soc. Jpn.* **44**, 225 (1977); Kamimura M, *Nucl. Phys.* A **351**,456 (1981).
[3] Funaki Y et al., *Phys. Rev.* C **67**, 051306(R) (2003), and references therein.
[4] Danilov A N et al., *Phys. Rev.* C **80**, 054603 (2009).
[5] Iida K, Koide S, Kohama A, Oyamatsu K, Mod. *Phys. Lett.* A **27**, 1250020 (2012).
[6] Tomita M, Iwasaki M, Otani R and Ito M, *Phys. Rev.* C **89**,034619 (2014).
[7] Ito M, Hirabayashi Y, and Sakuragi Y, *Phys. Rev.* C **66**,034307 (2002); Ito M, Sakuragi Y, and Hirabayashi Y, *Phys. Rev.* C **63** 064303(2001); ibid, *Eur. Phys. Jour.* A **5**, 373 (1999).
[8] Ito M, Sakuragi Y, and Hirabayashi Y, *Eur. Phys. Jour.* A **5**, 373 (1999).
[9] Satchler G R and Love W G, *Phys. Rep.* **55**, 183 (1979).
[10] Bertsch G, Borysowicz J, Macmanus H and Love W G, *Nucl. Phys.* A **284**, 399 (1977).
[11] Hashimoto S et al., *Phys. Rev.* C **83**, 054617 (2011).
[12] Ohkubo S, and Hirabayashi Y, *Phys. Rev.* C **70**,041602 (2004).
[13] Hamada Sh et al., *Phys. Rev.* C **87**,024311 (2013).
[14] Suzuki Y, *Prog. Theor. Phys.* **55**, 1751 (1976); ibid, *Prog. Theor. Phys.* **56**, 111 (1976).
[15] Funaki Y et al., *Phys. Rev. Lett.* 101, 082502 (2008).

Journal of Physics: Conference Series **569** (2014) 012023 doi:10.1088/1742-6596/569/1/012023

Probing α-cluster distribution via α-transfer reaction

Tokuro Fukui[1], Yasutaka Taniguchi[2], Tadahiro Suhara[3], Yoshiko Kanada-En'yo[4], and Kazuyuki Ogata[1].

[1] *Research Center for Nuclear Physics, Osaka University, Ibaraki, Osaka 567-0047, Japan.*
[2] *Department of Medical and General Sciences, Nihon Institute of Medical Science, Moroyama, Saitama 350-0435, Japan.*
[3] *Matsue College of Technology, Matsue, Shimane 690-8518, Japan.*
[4] *Department of Physics, Kyoto University, Kyoto 606-8502, Japan.*

E-mail: `tokuro@rcnp.osaka-u.ac.jp`

Abstract. The α-transfer reaction ^{16}O(^{6}Li,d)^{20}Ne is investigated in order to extract the α-clustering probability of ^{20}Ne. The relative wave function for the α-^{16}O system is calculated by a microscopic cluster model. We show the angular distribution of the transfer cross section is a good probe to see the radial dependence of the α-clustering probability in the surface region.

1. Introduction

The clustering phenomena have been predicted by theoretical studies that several states of unstable nuclei or of sd-shell nuclei have a cluster structure [1]. However, there is no direct measurement of the cluster structure except for resonance states decaying into constituent clusters. Therefore, it is desirable to establish how to extract the quantitative information on the clustering from observables.

In theoretical studies it is known that the α-cluster state develops in the surface region of nuclei. We take here ^{20}Ne as a typical nucleus having an α-^{16}O cluster structure. The purpose of the present study is to extract the probability of the α-clustering in the surface region from an α-transfer reaction, ^{16}O(^{6}Li,d)^{20}Ne. The transfer reaction is analyzed by means of the conventional distorted wave Born approximation (DWBA) with the overlap between ^{20}Ne and α-^{16}O, which is calculated by the microscopic cluster model.

2. Theoretical framework

2.1. Microscopic description of cluster wave function

As for the relative wave function between α and ^{16}O, we adopt a microscopic cluster model. The total wave function of ^{20}Ne with the resonating group method (R M) [2, 3, 4, 5] for the α-^{16}O configuration is given by

$$|\Psi_{\rm Ne}\rangle = \frac{1}{\sqrt{20!}}\mathcal{A}\left[\chi_l(r)Y_{l0}(\hat{\boldsymbol{r}})\phi(\alpha)\phi({\rm Ne})\right], \tag{1}$$

where r is the relative coordinate between α and ^{16}O, $\mathcal{A}$ stands for the antisymmetrization operator, and $\phi(C)$ is the intrinsic wave function of the nucleus C. χ_l can be expanded by the orthonormal set R_{nl} of the radial wave function of the harmonic oscillator (HO) as

$$\chi_l(r) = \sum_n a_n R_{nl}(r), \tag{2}$$

$$a_n = \int r^2 dr R_{nl}(r)\chi_l(r). \tag{3}$$

Here, n and l correspond to the principal quantum number and the orbital angular momentum of the HO, respectively. The relative wave function is defined by

$$u_l(r) = \sum_n a_n \sqrt{\mu_{nl}} R_{nl}(r) \tag{4}$$

with the eigenvalue μ_{nl} of the R M norm kernel [6]. For a normalized cluster wave function satisfying $\langle\Psi|\Psi\rangle = 1$, the relative wave function u_l is normalized to unity. Details of the formulation of the microscopic cluster model are given in Ref. [7]

2.2. Distorted wave Born Approximation (DWBA) formalism

In this paper the α-transfer reaction ^{16}O(^{6}Li,d)^{12}C is described with the post form distorted wave Born Approximation (DWBA) approach. The coordinates for the reaction system are illustrated in Fig. 1. The transition matrix for the α-transfer reaction is given by

$$T^{(\text{post})}_{\text{DWBA}} = \left\langle \Psi_f^{(-)} \middle| V_\alpha \middle| \Psi_i^{(+)} \right\rangle, \tag{5}$$

where the α-d interaction V_α in the final channel is adopted as the transition interaction, which causes the transition from the initial channel i to the final channel f. The total wave functions $\Psi_i^{(+)}$ and $\Psi_f^{(-)}$ for the initial and final channels, respectively, are written as

$$\Psi_i^{(+)}(\boldsymbol{r}_\alpha, \boldsymbol{r}_i) = \psi_\alpha(\boldsymbol{r}_\alpha)\chi_i^{(+)}(\boldsymbol{r}_i), \tag{6}$$

$$\Psi_f^{(-)}(\boldsymbol{r}_{\alpha\text{O}}, \boldsymbol{r}_f) = \psi_{\alpha\text{O}}(\boldsymbol{r}_{\alpha\text{O}})\chi_f^{(-)}(\boldsymbol{r}_f), \tag{7}$$

where ψ_α ($\psi_{\alpha\text{O}}$) is the relative wave function of the α-d (α-^{16}O) system and the distorted wave between ^{6}Li and ^{16}O (d and ^{20}Ne) is represented by $\chi_i^{(+)}$ ($\chi_f^{(-)}$). The superscript (+) and (−) represents the outgoing and incoming boundary conditions, respectively, on the scattering wave function. We adopt the cluster wave function defined by Eq. (4) for the radial part of $\psi_{\alpha\text{O}}$.

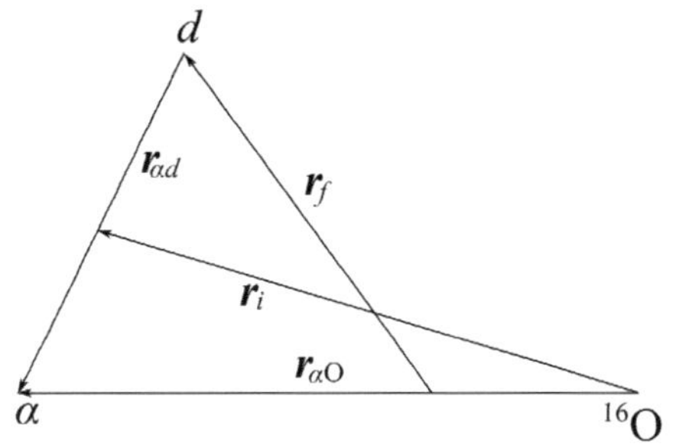

Figure 1. Illustration of the three-body system.

3. Result

We adopt the Volkov No. 2 effective interaction with the Majorana parameter $m = 0.62$ [8] to calculate the α-^{16}O relative wave function u_l. The width parameter $\nu = 0.16$ fm^{-2} is used for both α and ^{16}O. ψ_α is calculated with a two-range aussian interaction V_α [9].

We consider the ^{16}O(^{6}Li,d)^{20}Ne reaction at four incident energies: 20, 38, 42, and 75 MeV. At 20 and 38 MeV, we adopt phenomenological distorting potentials of a Woods-Saxon form given in Ref. [10]. At 42 (75) MeV, potential parameters are taken from Ref. [11] (Ref. [12]) and Ref. [13] (set 2 of Ref. [14]) for the initial and final channels, respectively.

To investigate the role of the α-cluster distribution in the transfer reaction ^{16}O(^{6}Li,d)^{20}Ne, the cross sections are calculated with the ^{20}Ne wave functions of the cluster model (CM), Eq. (4), and of the potential model (M). In M the α-^{16}O relative wave function is simply calculated with the Woods-Saxon potential $V_{\alpha\mathrm{O}}$ between α and ^{16}O: $V_{\alpha\mathrm{O}} = -V_0/[1+\exp\{(r_{\alpha\mathrm{O}} - r_0)/a\}]$. The parameters of $V_{\alpha\mathrm{O}}$ are listed in Table 1. Figures 2(a) and 2(b) show the α-^{16}O relative wave functions of the 0_1^+ (ground state) and the 1_1^- state (5.79 MeV), respectively. For the 1_1^- state we use a bound state approximation to calculate the relative wave function, taking the binding energy to be 0.2 MeV. By changing the parameter a, the M wave function (M2) can reproduce the behavior of the CM wave function in the surface region ($r_{\alpha\mathrm{O}} \gtrsim 5$ fm).

The transfer cross sections of ^{16}O(^{6}Li,d)^{20}Ne(0_1^+) as a function of the neutron emitting angle θ in the center-of-mass frame are compared with the experimental data [10, 12, 15] in Fig. 3. One sees the result with CM (solid line) agrees well with that with M2 (dotted line) up to the third maximum at all energies. On the other hand, the result with M1 (dashed line) deviates from the other two significantly. As shown in Fig. 2, CM and M1 gives the same distribution

Table 1. otential parameters for $V_{\alpha\mathrm{O}}$ in fm. The depth V_0 of $V_{\alpha\mathrm{O}}$ is determined so as to reproduce the binding energy 4.73 MeV and 0.20 MeV for 0_1^+ and 1_1^- states, respectively.

	0_1^+		1_1^-	
	r_0	a	r_0	a
M1	$1.25\times16^{1/3}$	0.65	$1.25\times16^{1/3}$	0.65
M2	$1.25\times16^{1/3}$	0.76	$1.25\times16^{1/3}$	0.83

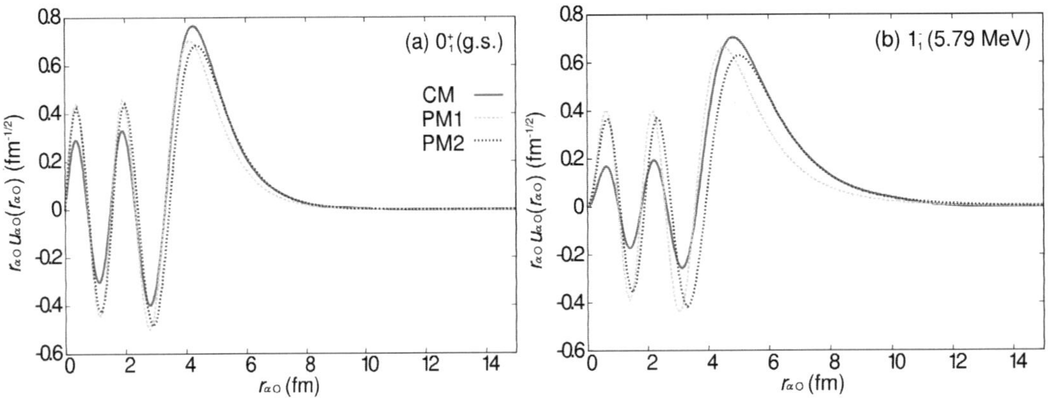

Figure 2. (a) The α-^{16}O relative wave functions for the 0_1^+ state calculated with CM (solid line) and two parameter sets of M: M1 (dashed line) and M2 (dotted line). (b) Same as in (a) but for the 1_1^- state.

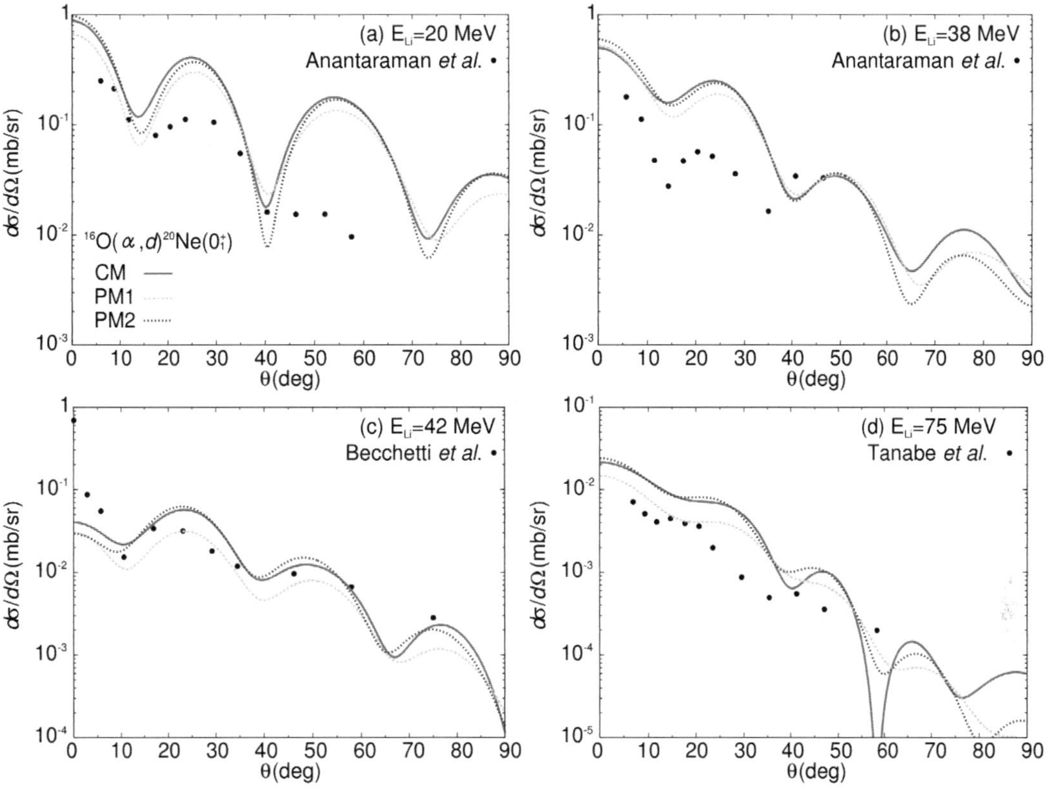

Figure 3. Transfer cross sections of ^{16}O(^{6}Li,d)^{20}Ne(0_1^+) at (a) 20 MeV, (b) 38 MeV, (c) 42 MeV, and (d) 75 MeV. In each panel, the solid line shows the calculation with the CM wave function. Results with M1 and M2 wave function are shown by the dashed and dotted lines, respectively. Experimental data are taken from Ref. [10, 12, 15].

in the surface region but are different from each other in the inner region. Whereas the two sets of M show a difference only in the surface region. Thus, the results of Fig. 3 suggest that the transfer cross section is not sensitive to the inner part of the structure of ^{20}Ne but it probes the α-^{16}O radial wave function in the surface region.

Note that the radial wave functions used here are normalized to unity. Nevertheless, M1 gives a significant difference from other two models on the cross sections. This indicates the di culty of a precise determination of the spectroscopic factor (SF) from transfer reactions. It is also important that the surface region in this study means about 5–8 fm in the α-^{16}O relative distance, i.e., still within a range of the nuclear interaction between the two clusters. Thus, the transfer process considered here is not governed by the asymptotic normalization coe cient (ANC). The α-clustering probability in the surface region will be a third alternative to the SF and the ANC for nuclear structural information to be extracted from reaction observables.

Unfortunately, however, agreement of the calculations with CM and M with the experimental data is not satisfactorily well. This may be due to ambiguity of the distorting potentials. We will fix this possible problem by adopting an $\alpha + d + {}^{16}$O three-body model in describing the transfer reaction. In this case we need the α-^{16}O and d-^{16}O distorting potentials, for which some global parameterizations can be used.

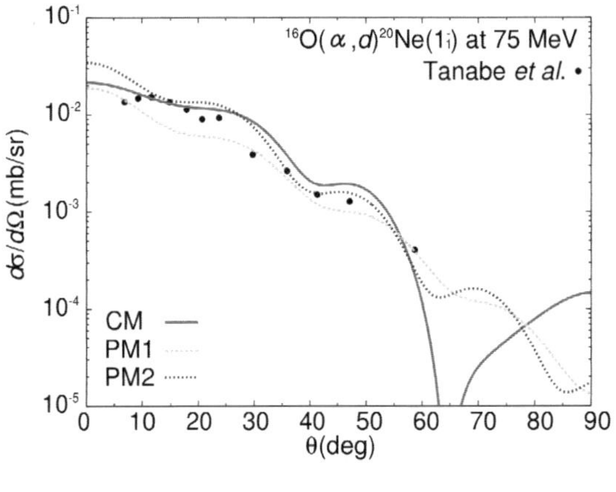

Figure 4. Same as in Fig. 3 but for the transfer cross section to the 1_1^- state of ^{20}Ne.

Figure 4 shows the transfer cross section populating the 1_1^- state of ^{20}Ne. One may draw a similar conclusion on this result to that for the transfer to the 0_1^+ state. It implies a possibility to probe a cluster structure also in a resonance state of a nucleus. This will be one of the advantages to use transfer reactions for the study of clustering phenomena.

4. Summary

We have analyzed the transfer reaction ^{16}O(^{6}Li,d)^{20}Ne to investigate the radial dependence of the α-cluster probability. The α-^{16}O relative wave function calculated microscopically are adopted in the DWBA analysis. We have found that the angular distribution of the transfer cross section is a good probe to see the radial dependence of the α-clustering probability. The procedure proposed in the present study can be useful and applicable to probe the cluster structure via observables in general systems such as unstable nuclei and sd-shell nuclei. As future work, to take into account the breakup channels of ^{6}Li with an $\alpha + d +{}^{16}$O three-body model by means of the continuum-discretized coupled-channels method (CDCC) [16, 17, 18] will be important. This will also minimize ambiguity of distorting potentials required in reaction calculations.

References

[1] Horiuchi H, Ikeda K and Katō K 2012 *Prog. Theor. Phys. Suppl.* **192** 1
[2] Wheeler J A 1937 *Phys. Rev.* **52** 1083
[3] Wheeler J A 1937 *Phys. Rev.* **52** 1107
[4] Wildermuth K and Kanellopoulos T 1958 *Nucl. Phys.* **7** 150
[5] Wildermuth K and Kanellopoulos T 1958 1959 *Nucl. Phys.* **9** 449
[6] Ikeda K and Tamagaki R 1977 *Prog. Theor. Phys. Suppl.* **62** 1
[7] Kanada-En yo Y, uhara T and Taniguchi Y 2014 *Prog. Theor. Exp. Phys.* **2014** 073D02
[8] Volkov A B 1965 *Nucl. Phys.* **74** 33
[9] akuragi Y, Yahiro M and Kamimura M 1986 *Prog. Theor. Phys. Suppl.* **89** 136
[10] Anantaraman N *et al.* 1979 *Nuclear Physics A* **313** 445
[11] Chua L, Becchetti F, Jënecke J and Milder F 1976 *Nucl. Phys.* **A273** 243
[12] Tanabe T *et al.* 1981 *Phys. Rev. C* **24** 2556
[13] Newman E, Becker L, Preedom B and Hiebert J 1967 *Nucl. Phys.* **A100** 225
[14] Hinterberger F, Mairle G, chmidt-Rohr U, Wagner G and Turek P 1968 *Nucl. Phys.* **A111** 265
[15] Becchetti F, Jënecke J and Thorn C 1978 *Nucl. Phys.* **A305** 313
[16] Kamimura M *et al.* 1986 *Prog. Theor. Phys. Suppl.* **89** 1
[17] Austern N, Iseri Y, Kamimura M, Kawai M, Rawitscher G and Yahiro M 1987 *Phys. Rep.* **154** 125
[18] Yahiro M, Ogata K, Matsumoto T and Minomo K 2012 *Prog. Theor. Exp. Phys.* **2012** 01A206

Journal of Physics: Conference Series **569** (2014) 012024 doi:10.1088/1742-6596/569/1/012024

^{8}Li+α resonant elastic scattering: a tool to study cluster states in ^{12}B

D Torresi[1,2], L Cosentino[1], P Descouveont[3], A Di Pietro[1], C Ducoin[4], P Figuera[1], M Fisichella[1,5], M Lattuada[1,2], C Maiolino[1], A Musumarra[1,2], M Papa[4], M G Pellegriti[1,4], M Rovituso[2], D Santonocito[1], G Scalia[2], V Scuderi[1] E Strano[1,2] and M Zadro[6]

[1]INFN Laboratori Nazionali del Sud, Catania, Italy
[2]Dipartimento di Fisica ed Astronomia, Università di Catania, Catania, Italy
[3]Physique Nucleaire Theorique et Physique Mathematique, Universite Libre de Bruxelles, Brussels, Belgium
[4]INFN Sezione di Catania, Catania, Italy
[5]Dipartimento di Fisica, Università di Messina, Messina, Italy
[6]Ruder Boškoviċ Institute, Zagreb, Croatia

E-mail: domenico.torresi@lns.infn.it

Abstract. The ^{8}Li+^{4}He elastic scattering excitation function was measured by using the inverse kinematic thick target method. The ^{8}Li beam was provided by the LNS radioactive facility EXCYT at an energy of 30 MeV and it was delivered into a large scattering chamber filled with ^{4}He gas. The detection system was made by three silicon telescopes and one MCP. This last detector was used to measure the number of incident particles as well as the time of flight allowing for the discrimination between elastic and inelastic scattering. The setup and the experimental technique will be described in details and the results, as well as the comparison of the data with an R-matrix calculation, will be shown and discussed.

1. Introduction

One of the interesting topics that can be explored by radioactive beams is the exotic clustering. In this type of clustering nuclei are described as formed by two or more clusters where, at least one of them, is an exotic nucleus itself.

Studying the structure of neutron rich nuclei, Kanada En'yo and collaborators [1] performed calculations in the Antisymmetrized Molecular Dynamics (AMD) framework in order to describe the structure of odd Boron isotopes. Calculations show how the cluster structure becomes more evident and more exotic with the increasing of neutron number in the case of boron isotopes. For instance, the structure of the ^{19}B ground state can be described with a very exotic di-nuclear configuration ^{11}Li-^{8}He, while the less neutron rich ^{13}B shows a single particle behavior. Analogous predictions concern also the even isotope ^{12}B. A calculation using the Generator Coordinate Method (GCM) has been performed by Descouvemont [2]. In this study the presence of a positive parity band with a large reduced width for α-decay, located just above the α-decay threshold, supports the picture of a α-^{8}Li cluster configuration for ^{12}B.

In this paper it will be described an experiment aimed to the study of the resonant elastic scattering of a ^{8}Li beam on a ^{4}He target by using the Thick Target Inverse Kinematic elastic

Journal of Physics: Conference Series **569** (2014) 012024 doi:10.1088/1742-6596/569/1/012024

scattering method (TTIK) [3, 4, 5].

2. The technique

The TTIK method has been used in order to investigate the α-^{8}Li structure of ^{12}B in the excitation energy range between 13.5 MeV and 19.6 MeV. In this technique the thick ^{4}He gas target has a threefold role. It acts at the same time as a degrader, as a target and as a shield for the detector that can be placed along the beam direction. Passing through the ^{4}He gas, the beam decreases its energy down to zero, hence the elastic scattering is induced in a continuous range of energies between the initial beam energy (30.6 MeV) and zero. While the heavier ^{8}Li beam is stopped inside the target the recoiling α particles, owing to their lower stopping power, reach the detectors placed at forward angles (backward angles in center of mass (CM) system).

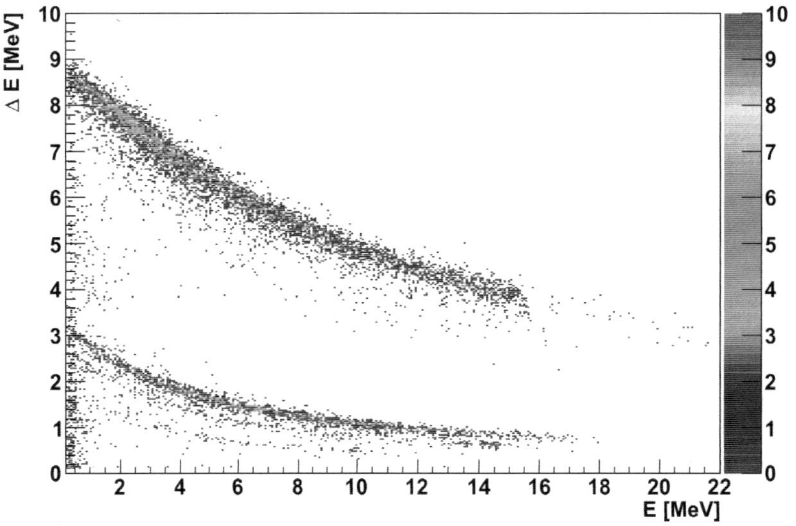

Figure 1. E versus ΔE plot for the telescope at 0° showing the separation of α-particles from hydrogen isotopes.

The ^{8}Li beam delivered at an energy of 30.6 MeV was produced by the radioactive beam facility EXCYT of INFN-LNS [6]. EXCYT uses the ISOL technique taking advantage of a K-800 superconducting cyclotron as a driver, and of a 13 MV Tandem for post-acceleration of the radioactive beams. Before entering the scattering chamber, the beam passes-through the aluminized Mylar foil (0.8 μm thick) of a Micro Channel Plate (MCP) detector. This last provides a reference signal for time measurements and, at the same time, allows for counting the beam particles ($\sim$5$\cdot$10^{4} pps). After the MCP detector, the ^{8}Li beam passes through a kapton foil (13μm thick) that separates the beam line from the CT2000 scattering chamber filled with ^{4}He enriched gas at a pressure P$\sim$700 mbar and temperature T$\sim$294 K. The recoiling α particles are detected by an array of three Silicon ΔE-E telescope detectors that are placed at forward angles θ=-5°, 0°, +5° with respect to the center of the chamber, $\sim$180 cm far from the entrance of the chamber. The first stage of the telescope is made by four-quadrant-Silicon-Pad detectors (50-60 μm thick) while the second stage is made by Double-Sided-Silicon-Strip-Detectors (DSSSD) 500-1000 μm thick. An experimental limitation of the TTIK method is represented by the contribution of other possible reaction mechanisms to the detected α spectrum. For example inelastic scattering if the center-of-mass energy is high enough to open the inelastic channel. Two α-particles can be detected with the same energy if one comes from an elastic event occurred in a given position of the chamber whilst the second one comes from an inelastic event occurred in a different position of the chamber. Thus, it is not possible to disentangle α coming from elastic and inelastic scattering, looking just at the detected energy of the α-particle.

This difficulty was overcome with by improving the technique with the time of flight measurement. The time elapsed between the ^{8}Li passing through the foil of the MCP and the detection of α particles was measured. In this case α-particles detected with the same energy but originated from different processes have different time of flight (ToF). Thus, in this way it is possible to discriminate elastic from inelastic events.

3. Analysis

In the following the results obtained from the most forward telescope ΔE-E, placed at θ_{lab}=0° with respect to the beam direction will be described. For this telescope, the ΔE nominal thickness is 47 μm. The ΔE-E plot is shown in Fig. 1. In this plot it is possible to notice that the α-particles are well separated from hydrogen isotopes. We can also see the presence of a large yield of tritons, coming most probably from the break-up channel ^{8}Li → ^{4}He+t+n+4.5 MeV.

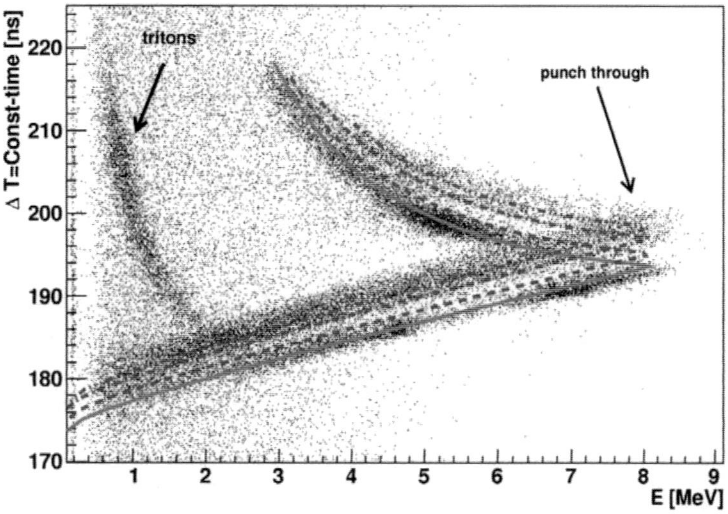

Figure 2. ToF versus ΔE plot for the detector at 0°. Continuous line is a calculation for the elastically scattered α-particles. Dotted lines correspond from bottom to the top respectively to the inelastic scattering for the three first excited states of ^{8}Li.

In Fig. 2, the ToF is shown as a function of the energy deposited in the ΔE detector. The lines are calculations of the ToF as a function of the detected energy both for the elastic and inelastic scattering corresponding to ^{8}Li first three excited levels. The calculation includes the kinematics, energy losses of ^{8}Li and α-particles in ^{4}He gas and energy losses of α-particles in the entrance dead layer of the detector. It can be seen that the calculated locus of the elastic events is in agreement with the data. As regards the inelastic events, we can notice that the inelastic scattering corresponding to the first excited state of ^{8}Li is scarcely populated. As one can observe, the elastic events are clearly distinguished from inelastic events.

A source of background comes from the ^{8}Li decay in two α (^{8}Li(β^-)$\rightarrow$ ^{8}Be* $\rightarrow 2\alpha$). The life time of ^{8}Li is 838 msec, for this reason the α particles are uncorrelated in time and hits the detectors at random time. The background of uncorrelated in time α-particles that ranges up to 5 MeV was taken into account and subtracted from the elastically scattered α particles.

For the events which stop in the ΔE detector the elastically scattered α-particles have been selected by a graphical cut on the plot ToF-ΔE. While for particle stopping in the E stage of the telescope it is possible to make a selection of the α particle locus from the ΔE-E plot and then select the elastically scattered α-particles from the ToF-E plot.

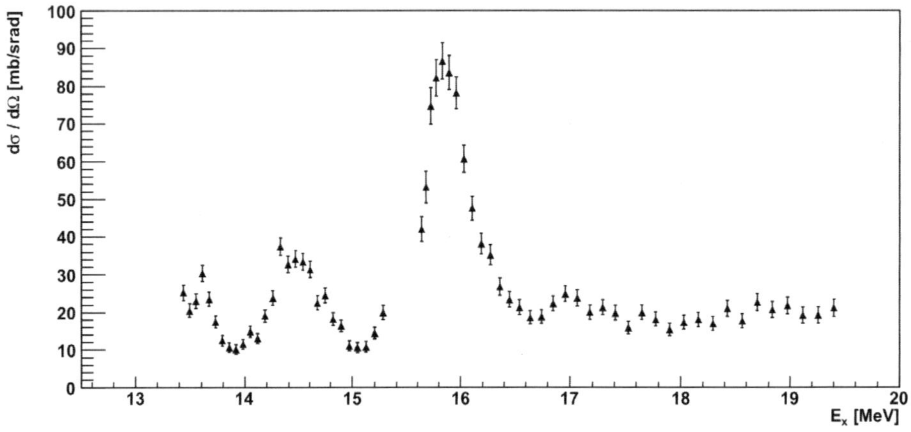

Figure 3. Elastic scattering differential cross section for most forward angles (~180°).

In order to reconstruct the elastic scattering excitation function from the spectrum of the detected α-particles it is required a precise knowledge of the stopping power of ^{4}He for ^{8}Li and α. For this reason during the same experiment a measurement of the stopping power was also performed.

The elastic scattering excitation function extracted for the telescope placed at 0° degree is shown in Fig. 3 and it manifests different structures. The analysis of the excitation functions was performed using an R-matrix approach. In order to simplify the problem the approximation of spin zero for ^{8}Li was considered. In this case, the spin of each level J is equal to the relative angular momentum l. In order to fit the data seven resonances were used. An additional background resonance at higher energy was introduced to fit the long tail in the high energy region. For the data fitting, the R-matrix calculation was convoluted with the experimental resolution function. The ^{8}Li+α excitation function is reproduced rather well by the R-matrix calculations as shown in Fig. 4. Some of the resonances was previously known [7, 8].

The reduced α-widths of the seven resonances used to perform the fit were calculate and their low values (less than 5%) indicate that there is no clear evidence of α-cluster configurations associated to the observed resonances.

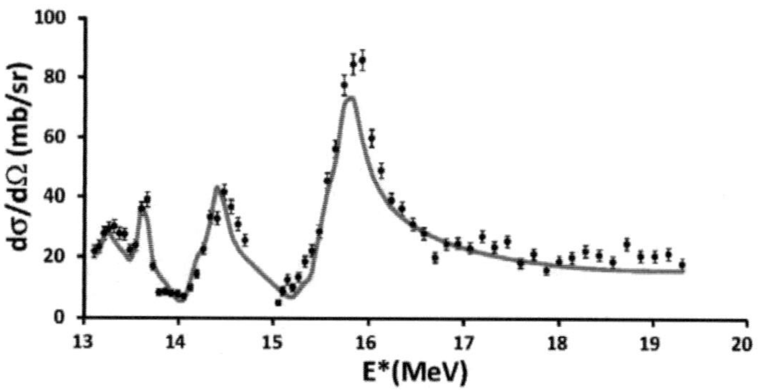

Figure 4. The elastic scattering cross section experimental data are shown as full circles, the solid red line is the R-matrix fit.

4. Conclusions

We have reported the measurements of the ^{8}Li-α elastic scattering excitation function by using the TTIK method. The technique included a time of flight measurement that allows the discrimination of elastic scattering from other reaction mechanisms. The excitation function was measured for excitation energy in the range of 13.5 and 19.5 MeV and it was analyzed using an R-matrix approach. The R-matrix calculation, that reproduces rather well the experimental data, includes seven resonances plus a background term. The reduced α-widths do not show evidence of strong α cluster configurations

References

[1] Y. Kanada En'yo et al., *Prog. Theor. Phys.* **93** (1995) 115.
[2] P. Descouvemont, *Nucl. Phys. A* **596** (1996) 285.
[3] K.P. Artemov et al., *Sov. J. Nucl. Phys.* **52** (1990) 408.
[4] K.M. Kallman et al., *Nucl. Instr. and Meth. in Phys. Res. A* **338** (1994) 413.
[5] V.Z. Goldberg et al., *Phys. Rev. C* **69** (2004) 024602.
[6] G. Ciavola et al., *Nucl. Phys. A* **701** (2002) 54c.
[7] F. Ajzenberg-Selove et al., *Nucl. Phys. A* **506** (1990) 1.
[8] N. Soić et al., *Nucl. Phys. A* **738** (2004) 347c.

Journal of Physics: Conference Series **569** (2014) 012025 doi:10.1088/1742-6596/569/1/012025

Many-body resonances and continuum states above many-body decay thresholds

K. Katō[1], T. Myo[2,3] and Y. Kikuchi[4]

[1] Nuclear Reaction Data Centre, Faculty of cience, Hokkaido University, apporo 060-0810, Japan

[2] General Education, Faculty of Engineering, Osaka Institute of Technology, Osaka, Osaka 535-8585, Japan

[3] Research Center for Nuclear Physics (RCNP), Osaka University, Ibaraki, Osaka 567-0047, Japan

[4] RIKEN Nishina Center, Wako, aitama 351-0198, Japan

E-mail: kato@nucl.sci.hokudai.ac.jp

Abstract. We study the resonance spectroscopy of the He isotopes and their mirror, proton-rich nuclei with an $\alpha + N + N + N + N$ cluster model. Many-body resonances are treated with the correct boundary condition as the Gamow states using the complex scaling method. We obtain the resonances up to the five-body decaying states in ^{8}He and ^{8}C. The spectrum agrees with the recent experiment systematically for energies and decay widths and we predict several resonances. We also discuss the mirror symmetry breaking of He isotopes and their mirror nuclei for the configurations and the spatial distributions of the valence nucleons.

1. Introduction

Unstable nuclei have often been observed in unbound states beyond the particle thresholds due to the weak binding nature of valence nucleons. The resonance spectroscopy of unbound states in unstable nuclei has been developed using radioactive-beam experiments [1, 2]. In addition to the energies and decay widths of resonances, information on their configurations and spatial properties is important to understand the structures of unstable nuclei. In proton- and neutron-rich nuclei, the correlations between valence nucleons and those between a valence nucleon and a core nucleus lead to the exotic nuclear properties in resonances as well as in the weakly bound states. The comparison of the structures between proton- and neutron-rich nuclei is important to examine the mirror symmetry in unstable nuclei with a large isospin.

In this report, we show the results of the recent applications of the complex scaling method (CSM) to the spectroscopy of unstable nuclei. We focus on the structures of the neutron-rich He isotopes and their mirror proton-rich nuclei, most of the states of which are unbound owing to the weak binding of valence nucleons to the α particle. We investigate the many-body resonances observed in these nuclei, with the correct boundary condition for the multi-particle emission using the CSM beyond the two-body case.

2. Complex-scaled cluster orbital shell model

We briefly explain the method of describing the resonance and non-resonant continuum states of He isotopes and their mirror nuclei on the basis of the cluster model with an α core. To

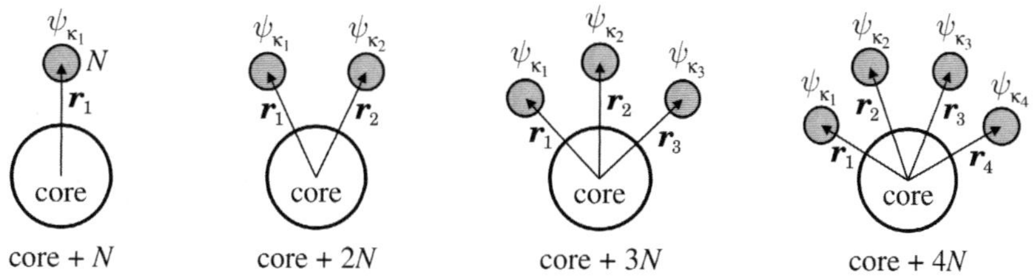

Figure 1. Sets of the spatial coordinates in CS-COSM.

describe the unbound states, it is essential to treat the boundary condition of the multi-particle emission and to solve the relative motion of each particle or cluster. For this purpose, we use the cluster orbital shell model with the complex scaling, called as CS-COSM, hereafter. We use the following Hamiltonian for the system consisting of α and valence nucleons [3, 4, 5, 6, 7]:

$$H_{\text{CS-COSM}} = \sum_{i=1}^{N_v}\left[\frac{\vec{p}_i^{\,2}}{2\mu}+V_i^{\alpha N}\right]+\sum_{i<j}^{N_v}\left[\frac{\vec{p}_i\cdot\vec{p}_j}{(A_c+1)\mu}+V_{ij}^{NN}\right], \tag{1}$$

where the operator $\vec{p}_i$ represents the relative momentum between a valence nucleon and α. The reduced mass μ is $mA_c/(A_c+1)$, where m is the nucleon mass and $A_c = 4$ of the α core. The number N_v is a valence nucleon number around the α core.

The total wave function $\Psi^{JT}_{\text{CS-COSM}}$ of the nucleus with a mass number A $(= A_c + N_v)$, a spin J, and an isospin T, is represented by the superposition of the various configurations Φ^{JT}_c as

$$\Psi^{JT}_{\text{CS-COSM}} = \sum_c C^J_c\,\Phi^{JT}_c(A), \qquad \Phi^{JT}_c(A) = \prod_{i=1}^{N_v} a^\dagger_{\kappa_i}|0\rangle, \tag{2}$$

where the vacuum $|0\rangle$ represents the α particle. The creation operator $a^\dagger_\kappa$ denotes the single-particle state of a valence nucleon above ^{4}He, with the quantum number $\kappa = \{n, \ell, j, t_z\}$ in a jj coupling scheme. Here, the index n is used to distinguish the different radial component of the single-particle state. The index c represents the set of κ as $c = \{\kappa_1, \cdots, \kappa_{N_v}\}$. The expansion coe cients $\{C^J_c\}$ in Eq. (2) are determined by the diagonalization of the Hamiltonian matrix elements. The schematic illustration of the coordinate set is shown in Fig. 1 for a core + N_vN systems with $N_v = 1, \cdots, 4$. We describe the radial component of the single nucleon wave function using the aussian expansion method [3, 8]. The antisymmetrization between a valence nucleon and α is treated on the orthogonality condition model [9], in which the single-particle state is imposed to be orthogonal to the $0s$ state occupied by an α core.

In the CS-COSM, the relative coordinates $\boldsymbol{r}_i$ between between a valence nucleon and α shown in Fig. 1, are transformed into $\boldsymbol{r}_i e^{i\theta}$ for $i = 1, \cdots, N_v$ to describe the many-body resonances and continuum states; here, θ is a scaling angle. Using the complex-scaled Hamiltonian $H^\theta_{\text{CS-COSM}}$, the complex-scaled Schrödinger equation is given as

$$H^\theta_{\text{CS-COSM}}\Psi^{JT,\theta}_{\text{CS-COSM}} = E^\theta_{JT}\Psi^{JT,\theta}_{\text{CS-COSM}}, \qquad \Psi^{JT,\theta}_{\text{CS-COSM}} = \sum_c C^{JT,\theta}_c\Phi^{JT}_c(A). \tag{3}$$

The energy eigenvalues E^θ_{JT} are obtained on a complex energy plane. In the CS-COSM, we adopt a finite number of the radial basis states for valence nucleons, which provides a discretized representation of the continuum states in the CSM.

In Eq. (1), the α-N interaction $V^{\alpha N}$ consists of the KKNN potential [3, 10] for the nuclear part and the folded Coulomb part. We use the Minnesota potential [11] as the nucleon-nucleon

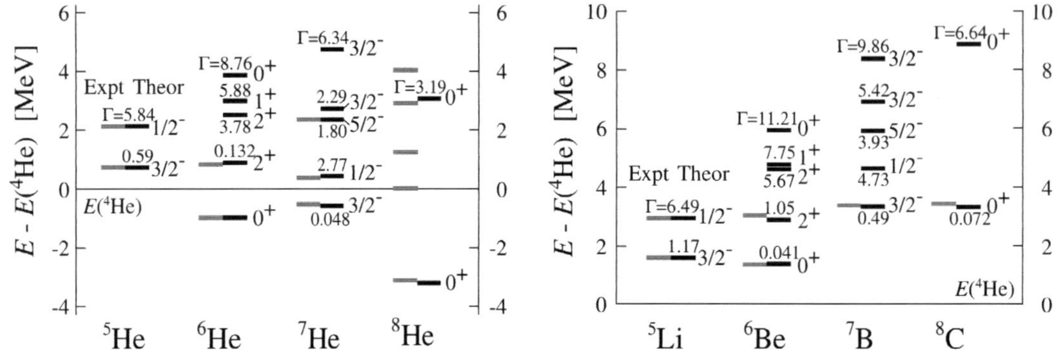

Figure 2. Energy levels of He isotopes (left) and their mirror nuclei (right) measured using the α energy. Units are in MeV. The black and gray lines correspond to the theoretical and experimental results, respectively. Small numbers denote theoretical decay widths.

interaction V^{NN} in addition to the point Coulomb interaction. For the single-particle states, we take the angular momenta $\ell \leq 2$ and adjust the two-neutron separation energy of ^{6}He to the experimental value of 0.975 MeV.

2.1. Results of He isotopes and their mirror nuclei

We show a systematics of energy levels observed experimentally and calculated by CS-COSM for He isotopes and their mirror nuclei in Fig. 2, measured from on the basis of the α particle. The small numbers near the levels represent the decay widths of the states. A good agreement is observed between the theoretically and experimentally obtained energy positions up to a five-body case of ^{8}He and ^{8}C. Further, there are several theoretical predictions of the excited states. The matter and charge radii of the ground states of ^{6}He and ^{8}He reproduce the recent experiments, as shown in Table 1. Hence, the CS-COSM wave functions describe the specially extended distributions of neutrons in the halo and skin structures observed in the He isotopes.

Firstly, we focus on the structures of ^{7}He and its mirror nucleus ^{7}B [6]. They are both unbound nuclei. Our CS-COSM predicts five resonances for each nucleus. For ^{7}He, the $3/2^-$ ground state is obtained as the two-body resonance located in the energy range between the thresholds of ^{6}He+n and α+$3n$. The other four states are four-body resonances above the α+$3n$ threshold energy as shown in Fig. 2. For ^{7}B, the resonances are all located above the α+$3p$ threshold energy and are interpreted as four-body resonances. There is no experimental data for the excited states of ^{7}B so far.

Table 1. Matter ($R_{\rm m}$) and charge ($R_{\rm ch}$) radii of ^{6}He and ^{8}He in comparison with the experimental values; a[12], b[13], c[14], d[15], and e[16]. Units are in fm.

		resent	Experiments
^{6}He	$R_{\rm m}$	2.37	$2.33(4)^{\rm a}$ $2.30(7)^{\rm b}$ $2.37(5)^{\rm c}$
	$R_{\rm ch}$	2.01	$2.068(11)^{\rm d}$, $2.060(8)^{\rm e}$
^{8}He	$R_{\rm m}$	2.52	$2.49(4)^{\rm a}$ $2.53(8)^{\rm b}$ $2.49(4)^{\rm c}$
	$R_{\rm ch}$	1.92	$1.929(26)^{\rm d}$, $1.959(16)^{\rm e}$

Table 2. *S*-factors of the ^{6}He-n components in ^{7}He.

	$^6\mathrm{He}(0^+_1)$-n	$^6\mathrm{He}(2^+_1)$-n
$3/2^-_1$	$0.63 + i0.08$	$1.60 - i0.49$
$3/2^-_2$	$0.00 - i0.01$	$0.97 + i0.01$
$3/2^-_3$	$0.01 + i0.00$	$0.04 - i0.01$
$1/2^-$	$0.95 + i0.03$	$0.07 - i0.02$
$5/2^-$	$0.00 + i0.00$	$1.00 + i0.01$

Table 3. *S*-factors of the ^{6}Be-p components in ^{7}B.

	$^6\mathrm{Be}(0^+_1)$-p	$^6\mathrm{Be}(2^+_1)$-p
$3/2^-_1$	$0.51 + i0.02$	$2.35 - i0.15$
$3/2^-_2$	$0.02 - i0.01$	$0.96 - i0.01$
$3/2^-_3$	$0.00 + i0.01$	$-0.01 - i0.06$
$1/2^-$	$0.93 - i0.02$	$0.10 - i0.01$
$5/2^-$	$0.00 + i0.00$	$1.04 - i0.01$

It is meaningful to discuss the mirror symmetry between ^{7}He and ^{7}B consisting of the α core and three valence neutrons or protons. We show the the spectroscopic factors (*S*-factors) of one-nucleon removal from each nucleus, the ^{6}He-n components of ^{7}He and the ^{6}Be-p components of ^{7}B. These quantities are important to examine the coupling behaviors between the $A = 6$ daughter nuclei and the last nucleon.

We list the results of the *S*-factors of ^{7}He and ^{7}B in Table 2 and 3, respectively. For the amow states, the *S*-factors can be complex numbers. In the results, most of the components are found to show almost the real values. A sizable difference is observed between the ground states of ^{7}He and ^{7}B in terms of the components including the $A = 6$ (2^+_1) states. The $^6\mathrm{Be}(2^+_1)$-p component in ^{7}B obtained as 2.35 (Table 3), is larger than the $^6\mathrm{He}(2^+_1)$-n component in ^{7}He as 1.60 (Table 2), by 47 for the real part. This difference indicates the breaking of the mirror symmetry in their ground states. The reason for the different values of the 2^+ couplings is that the ^{7}B ground state is located close to the $^6\mathrm{Be}(2^+_1)$ state by 0.45 MeV in the energy as shown in Fig. 2. This situation does not occur in ^{7}He, in which the energy difference between $^7\mathrm{He}(3/2^-_1)$ and $^6\mathrm{He}(2^+_1)$ is 1.46 MeV. The small energy difference between ^{7}B and $^6\mathrm{Be}(2^+_1)$ enhances the $^6\mathrm{Be}(2^+_1)$-p component in ^{7}B because of the increase of the coupling to the open channel of the $^6\mathrm{Be}(2^+_1)$+p threshold. Contrastingly, the $^6\mathrm{Be}(0^+_1)$-p component in ^{7}B becomes smaller than that of ^{7}He, because the energy difference between the ground states of ^{7}B and ^{6}Be is 1.97 MeV, larger than that in the case of ^{7}He as 0.40 MeV. The difference between the *S*-factors in ^{7}He and ^{7}B originates from the Coulomb repulsion, which acts to shift the entire energy of the ^{7}B states upward with respect to the ^{7}He energy.

In conclusion, the mirror symmetry is largely broken only in the ground states of ^{7}He and ^{7}B, while the excited states of the two nuclei keep the symmetry approximately. In relation to the *S*-factors, we have calculated the one-neutron removal strength of ^{7}He to form ^{6}He as a function of the observed excitation energy of ^{6}He [4]. The dominant component comes from the 2^+ state , the strength of which peaks at the resonance energy of 0.84 MeV of $^6\mathrm{He}(2^+_1)$.

Secondly, we compare the structures of ^{8}He and ^{8}C for their mirror 0^+ states as the five-body decaying states with an α core [7]. We investigate the role of the Coulomb interaction on the spatial properties of ^{8}C in comparison with ^{8}He. For this purpose, we calculate the various radii sizes of ^{8}He and ^{8}C, which reflect the spatial motion of four valence nucleons. We noted that the radius of amow states is obtained as a finite complex number, similar to the *S*-factors as was discussed. In the present analyses, most of the radii of resonances exhibit imaginary parts that are relatively smaller than the real ones, similar to case of the squared amplitudes [5, 7]. Hence, we discuss the spatial size of resonances using the real part of the complex radii.

The results of the radii of the $0^+_{1,2}$ states in ^{8}He and ^{8}C are listed in Table 4. We calculate the matter (R_m), proton (R_p), neutron (R_n), and charge (R_ch) radii, and the mean relative distances between α and a single valence nucleon ($r_{\mathrm{c}\text{-}N}$) and between α and the center of mass of the four valence nucleons ($r_{\mathrm{c}\text{-}4N}$).

Table 4. Various radii of the $0^+_{1,2}$ states of ^{8}He and ^{8}C in units of fm.

	$^8\mathrm{He}(0^+_1)$	$^8\mathrm{C}(0^+_1)$	$^8\mathrm{He}(0^+_2)$	$^8\mathrm{C}(0^+_2)$
R_m	2.52	$2.81 - 0.08i$	$7.56 + 2.04i$	$4.87 + 0.13i$
R_p	1.80	$3.06 - 0.10i$	$3.15 + 0.69i$	$5.46 + 0.15i$
R_n	2.72	$1.90 - 0.01i$	$8.53 + 2.32i$	$2.36 + 0.05i$
R_ch	1.92	$3.18 - 0.09i$	$3.22 + 0.67i$	$5.53 + 0.15i$
$r_{\mathrm{c}\text{-}N}$	3.55	$4.05 - 0.12i$	$11.03 + 3.11i$	$7.21 + 0.21i$
$r_{\mathrm{c}\text{-}4N}$	2.05	$2.36 - 0.03i$	$5.60 + 1.55i$	$3.68 + 0.13i$

For the 0^+_1 states, the matter radius of ^{8}C is larger than that of ^{8}He by about 12 in the real part. For 0^+_2 states, all the values of radii are complex numbers because of the resonances in the two nuclei. The imaginary parts are fairly large for $^8\mathrm{He}(0^+_2)$, but still smaller than the real ones. It is interesting that the matter radius of ^{8}C is smaller than that of ^{8}He in the 0^+_2 states. This relation is opposite to that observed for the ground states of ^{8}He and ^{8}C. For $^8\mathrm{He}(0^+_2)$, the observed large matter radius originates from the large neutron radius. For $^8\mathrm{C}(0^+_2)$, the large matter radius is due to the large proton radius, which is smaller than the neutron radius of ^{8}He.

We conclude that the relation of the spatial properties between ^{8}He and ^{8}C depends on the states, which can be explained in terms of the Coulomb interaction. The Coulomb interaction acts repulsively, thereby shifting the entire energy of ^{8}C upward with respect to the ^{8}He energy. In the ground state of ^{8}C, this repulsion extends the distances between α and a valence proton and between valence protons. On the other hand, the Coulomb interaction makes the barrier above the particle threshold in ^{8}C and the 0^+_2 resonance is affected by this barrier, the effect of which prevents the wave function of valence protons of ^{8}C from extending spatially. In ^{8}He, there is no Coulomb barrier for the four valence neutrons and the neutrons can extend to a large distance in the resonance. This role of the Coulomb interaction leads to the radius of $^8\mathrm{C}(0^+_2)$ being smaller than that of $^8\mathrm{He}(0^+_2)$.

Acknowledgments

This work was supported by JS S KAKENHI rant Numbers 24740175 and 25400241. Numerical calculations were performed on a supercomputer (NEC S 9) at RCN , Osaka University.

References

[1] Tanihata I, avajols H and Kanungo R 2013 *Prog. Part. Nucl. Phys.* **68** 215
[2] Charity R J *et al.* 2011 *Phys. Rev.* C **84** 014320
[3] Aoyama , Myo T, Katō K and Ikeda K, 2006 *Prog. Theor. Phys.* **116** 1
[4] Myo T, Ando R and Katō K 2009 *Phys. Rev.* C **80** 014315
[5] Myo T, Ando R and Katō K 2010 *Phys. Lett.* B **691** 150
[6] Myo T, Kikuchi Y and Katō K 2011 *Phys. Rev.* C **84** 064306
[7] Myo T, Kikuchi Y and Katō K 2012 *Phys. Rev.* C **85** 034338, 2013 *Phys. Rev.* C **87** 049902
[8] Hiyama E, Kino Y and Kamimura M 2003 *Prog. Part. Nucl. Phys.* **51** 223
[9] aito 1977 *Prog. Theor. Phys. Suppl.* **62** 11
[10] Kanada H, Kaneko T, Nagata and Nomoto M 1979 *Prog. Theor. Phys.* **61** 1327
[11] Tang Y C, LeMere M and Thompson D R 1978 *Phys. Rep.* **47** 167
[12] Tanihata I *et al.* 1992 *Phys. Lett.* B **289** 261
[13] Alkhazov G D *et al.* 1997 *Phys. Rev. Lett.* **78** 2313
[14] Kiselev O A *et al.* 2005 *Eur. Phys. J.* A **25**, *Suppl. 1* 215
[15] Mueller P *et al.* 2007 *Phys. Rev. Lett.* **99** 252501
[16] Brodeur M *et al.* 2012 *Phys. Rev. Lett.* **108** 052504

Journal of Physics: Conference Series **569** (2014) 012026 doi:10.1088/1742-6596/569/1/012026

Two-dineutron condensation in the excited state of ^{8}He

Fumiharu Kobayashi[1] and Yoshiko Kanada-En'yo[2]

[1] Department of Physics, Niigata University, Niigata 950-2102, Japan
[2] Department of Physics, Kyoto University, Kyoto 606-8502, Japan

E-mail: f-kobayashi@nt.sc.niigata-u.ac.jp

Abstract. One of the aims of this work is to investigate a two-dineutron condensate state in ^{8}He and to discuss the properties of two dineutrons in that state. We suggest that the excited $^8\mathrm{He}(0^+)$ state that is the candidate for the two-dineutron condensation would exist above the $\alpha + 4n$ threshold energy by a few MeV and this state contains two compact dineutrons in the lowest S-wave around an α core. Different from α clusters, dineutron clusters in that state is soft in size changing, because two neutrons are not bound in free space and dineutron correlation becomes weak in the low-density structure of the dineutron condensation.

1. Introduction

Recently, neutron-rich nuclei have been investigated eagerly and many exotic phenomena have been suggested in them. Dineutron correlation is one of the most interesting topics of the physics of neutron-rich nuclei. A dineutron is a spatially compact spin-singlet pair of two neutrons, which can be regarded as a kind of cluster (quasi-boson). Although two neutrons are not bound in free space, it is theoretically suggested that a compact dineutron can be formed in the low-density region of nuclear matter [1] and in the neutron-halo or neutron-skin region of neutron-rich nuclei [2]. Such compact dineutrons whose size is comparable to that of an α is expected to behave as bosonic clusters individually and, therefore, various dineutron cluster excitation such as dineutron cluster condensation, in which some dineutrons behave as bosons and occupy the lowest-S orbit simultaneously, may be realized. Cluster condensation, especially α condensation, attracts great interests as a novel and exotic structure in finite nuclei and has been investigated eagerly [3]. However, dineutron condensation has not been discussed so well, and the multi-dineutron dynamics and the systematics of formation of dineutron condensation are very interesting issues. Moreover, a dineutron is not naturally as rigid as an α cluster because of the unbound feature of two neutrons in free space, so the dineutron properties, that is, the degree of formation and the size, would change significantly depending on the circumstances such as the nuclear structure. We would like to clarify the diversity of the universal dineutron properties which has not been known at all.

In order to investigate dineutron correlation and condensation systematically, we have constructed a framework "a dineutron condensate (DC) wave function" [4]. This framework describes a nuclear system as a core plus one or a few dineutrons under explicitly considering the core structure such as deformation and excitation and the dineutron-size change. We apply the DC wave function to ^{8}He, which is well described with a system of an α core plus four valence neutrons [5]. One of the aims of this work is to investigate a two-dineutron condensate state in ^{8}He and to discuss the properties of two dineutrons in that state. We suggest that the excited ^{8}He(0^+) state that is the candidate for the two-dineutron condensation would exist above the $\alpha + 4n$ threshold energy by a few MeV and would have a gaslike structure where one α and two dineutron clusters are weakly interacting with each other and expanded greatly, similar to an α condensate state. We show that two dineutrons in that state are surely rather compact and behave as a kind of boson to occupy the lowest-S orbit simultaneously but these dineutrons are soft and change their size readily. Dineutron correlation would become weak to some extent in a very low-density structure such as a gaslike one of two-dineutron condensation in ^{8}He.

2. Framework

We superpose two kinds of wave functions to describe ^{8}He(0^+) states, the extended ^{6}He+$2n$ cluster wave function and the ^{8}He DC wave function. We project the those wave functions to $J^\pi = 0^+$, and we diagonalize the Hamiltonian by using those wave functions to obtain the 0^+ states. In the following, we explain these frameworks briefly.

2.1. Extended ^{6}He+$2n$ cluster wave function

The extended ^{6}He+$2n$ cluster wave function describes ^{8}He as a system of ^{6}He plus extended $2n$ (denoted as $2n^*$ hereafter) clusters. The ^{6}He cluster is composed of an α cluster and two valence neutrons in $(0p)^2$ orbits around the α and we describe these valence neutrons with a shifted aussian from the α particle. We locate the extended $2n$ cluster around ^{6}He described in such a way. By "extended", we mean that we take into account the dissociation of the $2n$ cluster due to the spin-orbit interaction. In the ordinary cluster wave function, the single-particle wave function is a aussian which have a real center parameter. A real part of the aussian center corresponds to the mean position of the nucleon. In the present one, we add an imaginary part, λ, to each aussian center as done in Ref. [6]. An imaginary part corresponds to the mean momentum of the nucleon. When $\lambda = 0$, the $2n^*$ cluster is a spin-singlet cluster, that is, a dineutron. When $\lambda \neq 0$, the spin-up and spin-down neutrons in the $2n^*$ cluster have the opposite momentum to gain the spin-orbit attraction. Then, the $2n^*$ cluster is not a spin-singlet cluster any longer exactly. In this way, we take into account the dissociation of spin-singlet pairs and describe the shell-model configurations such as $(0p_{3/2})^4$ which is expected to be significant in the $N = 6$ nuclei such as ^{8}He. The form of the ^{6}He+$2n^*$ cluster wave function is as follows.

$$\Phi_{^6\mathrm{He}+2n^*} = \mathcal{A}\Big\{\Phi_{^6\mathrm{He}}(\boldsymbol{R}_{^6\mathrm{He}}, b_\alpha)\ \phi_{n\uparrow}(\boldsymbol{R}_\uparrow, b_\alpha)\ \phi_{n\downarrow}(\boldsymbol{R}_\downarrow, b_\alpha)\Big\}. \tag{1}$$

$\Phi_{^6\mathrm{He}}$ and $\phi_{n\uparrow/\downarrow}$ are the wave functions of the ^{6}He cluster and the spin-up -down valence neutrons. $\boldsymbol{R}$ is the aussian centers, $\boldsymbol{R}_{^6\mathrm{He}} = (-1/4d_\lambda, 0, 0)$ and $\boldsymbol{R}_{\uparrow/\downarrow} = (3/4d_\lambda, \pm i\lambda, 0)$. $b_\alpha = 1.46$ fm is the aussian width and common to all the nucleons in this wave function. In this work, we superpose the wave functions with $\lambda = 0.0, 0.4$ fm and $d_\lambda = 1, \ldots, 4$ fm.

2.2. ^{8}He DC wave function

The DC wave function describes a system of a core plus one or a few dineutrons in the S wave around the core. In the present work, the core is $\alpha + 2n^*$ and one dineutron is distributed around the core. The form of the wave function is

$$\Phi_{\mathrm{DC}} = \mathcal{A}\Big\{\Phi_\alpha(\boldsymbol{R}_\alpha, b_\alpha)\ \phi_{n\uparrow}(\boldsymbol{R}_\uparrow, b_\lambda)\ \phi_{n\downarrow}(\boldsymbol{R}_\downarrow, b_\lambda) \times \psi_r(b_n)\ \psi_G(\beta)\Big\}, \tag{2}$$

$$\psi_r \propto \exp\left[-\frac{r^2}{4b_n^2}\right], \qquad \psi_G \propto \exp\left[-\frac{r_G^2}{\beta^2}\right], \tag{3}$$

where $\Phi_\alpha \phi_{n\uparrow} \phi_{n\downarrow}$ is the core wave function of $\alpha + 2n^*$ and is described with the same manner as the ^{6}He+$2n^*$ cluster wave function. ψ_r and ψ_G are the relative and center-of-mass parts of the dineutron wave function composed of two neutrons coupled to a spin singlet. They have the relative coordinate between the two neutrons, $\boldsymbol{r}$, and the center-of-mass coordinate of the two neutrons, $\boldsymbol{r}_G$, respectively. In the relative part, the aussian width is b_n and this parameter characterizes the relative distance between two neutrons, i.e., the dineutron size. In the center-of-mass part, the aussian width is β and characterizes the distance between the center of mass of the two neutrons and that of the core, i.e., the spatial expansion of the dineutron from the core. In the DC wave function, we vary these parameters to describe a state containing a dineutron with various sizes and distributions. We superpose the DC wave functions with $\beta = 2, \ldots, 8$ fm and five b_n values for each β. We prepare two kinds of parameter sets characterizing the $\alpha + 2n^*$ core, that is, $\lambda = 0.0$ fm and $d_\lambda = 1, \ldots, 8$ fm and $b_\lambda = b_n$, and $\lambda = 0.4$ fm and $d_\lambda = 1, \ldots, 4$ fm and $b_\lambda = b_\alpha$. In the former set, the $2n^*$ cluster is also a dineutron and has the same size as the other dineutron and they can be expanded widely. In the latter, the $2n^*$ cluster is dissociated at the surface of the α. Superposing these DC wave functions, we describe the configurations such as $\alpha + 2n[(0p_{3/2})^2]$ plus one developed dineutron and two-dineutron condensation where two dineutrons have the same compact size and both are in the lowest S wave.

By superposing these wave functions, we describe the shell-model configurations and one- or two-dineutron configurations effectively.

3. Results

We discuss the 0^+ states described by superposing the wave functions explained in Sec. 2. For the Hamiltonian in the present calculation, we use the effective two-body interactions, namely the Volkov No.2 force as the central force and the spin-orbit part of the 3RS force as the spin-orbit force. The parameters in the central force are chosen as $m = 0.55, b = h = 0.125$ and the strength of the spin-orbit force is $v_{\mathrm{LS}} = 2000$ MeV, chosen to reproduce the ^{8}He ground state energy. In the present framework, we cannot exactly remove the total center-of-mass motion since all the aussian widths of the single-particle wave functions for the DC wave functions are not the same. Thus, we extract the expectation value of the total center-of-mass kinetic energy from the Hamiltonian to treat the center-of-mass motion in an approximated way.

3.1. Energies and radii of $0^+_{1,2}$

We calculate the energy and the matter, proton and neutron radii of the $0^+_{1,2}$ states shown in Table 1. The 0^+_1 energy agrees with the experimental value. Although the neutron radius of the 0^+_1 state is underestimated compared with the experimental value, the large difference between the proton and neutron radii, which means a neutron skin, is well reproduced in the present calculation. We skip the details of the structure of the 0^+_1 state, but the $(0p_{3/2})^4$ and one-dineutron components are important in this state. In the 0^+_2 state, whose excitation energy

Table 1. Energies and matter, proton and neutron radii (r_m, r_p and r_n) in the $0^+_{1,2}$ states. As for 0^+_1, we show the experimental values in the second line.

	Energy (MeV)	r_m (fm)	r_p (fm)	r_n (fm)
0^+_1	−31.13	2.37	1.87	2.51
0^+_1(Expt.)	−31.40	2.49 − 2.52	1.76 − 2.15	2.64 − 2.69
0^+_2	−23.24	4.85	2.26	5.44

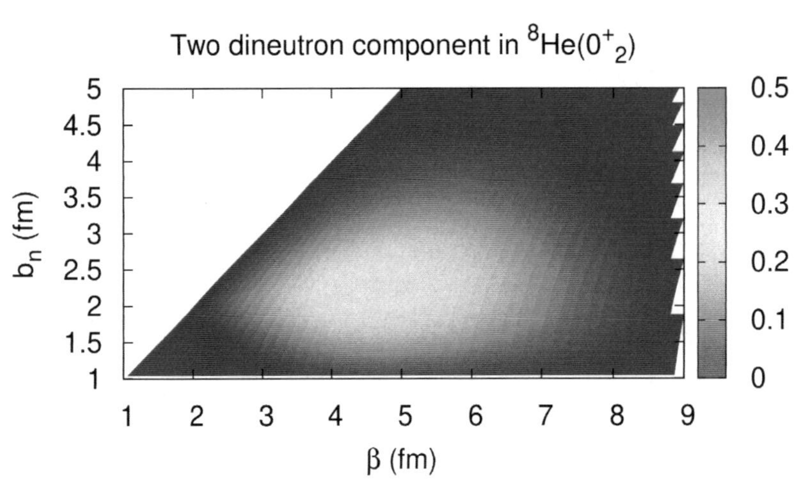

Figure 1. The two-dineutron component of the 0_2^+ state on the β-b_n plane. This figure is referred from Ref. [5].

is $\sim$ 8 MeV, the neutron radius is much larger than that of the 0_1^+ state. This means that the valence neutrons extend to large distances in this state. In the next subsection, we suggest that this state has a gaslike structure of one α and two compact dineutrons and that it is the two-dineutron condensate state.

3.2. Dineutron component in the 0_2^+ *state*

To see the two-dineutron component in the 0_2^+ state, we calculate the overlap with the DC wave function containing one α and two spin-singlet pairs of two neutrons.

$$\mathcal{N}_{\rm dineutron}(\beta, b_n) = |\langle \Phi_{\rm 2DC}(\beta, b_n) | \Psi_{^8{\rm He}(0_2^+)} \rangle|^2, \tag{4}$$

where $\Psi_{^8{\rm He}(0_2^+)}$ is the wave function of the 0_2^+ state described by superposing the ^{6}He+$2n^*$ wave functions and ^{8}He DC wave functions. $\Phi_{\rm 2DC}$ is the test DC wave function describing the system containing two dineutrons which have the same size, b_n, and expansion from the core, β. These two dineutrons are distributed around the core in the S wave as Eq. (2), so this wave function with small b_n and large β corresponds to the component of two-dineutron condensation, which means that two compact dineutrons considered as a kind of cluster are in the lowest S wave and expanded widely. We plot the overlap with $\Phi_{\rm 2DC}$ as a function of the dineutron size, b_n, and its spatial expansion from the core, β, in Fig. 1. A peak exists at $(\beta, b_n) \sim (5, 2)$ with a broad width in both the direction of β and b_n. The peak value of $b_n \sim 2$ fm is much smaller than the size of the total system, and, therefore, the valence neutrons in this state form two compact dineutrons. However, this peak has the broad width in the b_n direction (the peak spreads from $\sim$ 1.5 fm to $\sim$ 3 fm), which means that these dineutrons are certainly compact but their size changes readily. The broad width in the β direction (the peak spreads from $\sim$ 3 fm to $\sim$ 7 fm) means that the interaction between the α and two dineutron clusters is weak and that these clusters are expanded widely. This corresponds to the gaslike structure of one α and two dineutrons. It should be noticed that two dineutrons in this state are certainly compact but very soft in size changing unlike an α cluster which is rigid and hardly changes its size. The strength of dineutron correlation can greatly depend on the circumstances such as the density, and the dineutron correlation in a dilute condensate state becomes somewhat weaker than, for example, that in a ground state. Such a diverse property is an interesting point in dineutron correlation.

4. Summary

In this paper, we have mainly reported the possibility of the existence of the two-dineutron condensation in the excited state of ^{8}He. The candidate has a much larger neutron radius than that of the ground state and the four valence neutrons are very spatially extended. They form two spin-singlet pairs and they become compact to be a dineutron. We have shown that, in that state, two dineutrons are rather compact but soft in size changing. These two dineutrons are expanded spatially from the core keeping a compact size, and one α and two dineutron clusters have little correlation with each other to form a gaslike structure of α+two dineutrons.

Two neutrons are not bound in free space, so a dineutron is naturally soft and changes its size readily, especially in such low-density structure as the gas-like one of the 0_2^+ state. However, it is sure that a dineutron becomes so small that it can behave as a kind of cluster, and it is expected that a dineutron condensation can be realized in neutron-rich nuclei. It is challenging to investigate neutron-rich nuclei systematically in view of the dineutron condensation and to make clear the universal properties of the dineutron in such a system.

Acknowledgments

This work was supported by rant-in-Aid for Scientific Research from Japan Society for the romotion of Science (JS S). It was also supported by the rant-in-Aid for the lobal COE rogram "The Next eneration of hysics, Spun from Universality and Emergence" from the Ministry of Education, Culture, Sports, Science and Technology (ME T) of Japan. A part of the computational calculations of this work was performed by using the supercomputers at IT .

References

[1] Matsuo M 2006 *Phys. Rev.* C **73** 044309
[2] Hagino K, agawa H, Carbonell J and chuck P 2007 *Phys. Rev. Lett.* **99** 022506
[3] Tohsaki A, Horiuchi H, chuck P and Röpke G 2001 *Phys. Rev. Lett.* **87** 192501
[4] Kobayashi F and Kanada-En yo Y 2001 *Prog. Theor. Phys.* **126** 457
[5] Kobayashi F and Kanada-En yo Y 2013 *Phys. Rev.* C **88** 034321
[6] Itagaki N, Masui H, Ito M and Aoyama 2005 *Phys. Rev.* C **71** 064307

3rd International Workshop on "State of the Art in Nuclear Cluster Physics" IOP Publishing
Journal of Physics: Conference Series **569** (2014) 012027 doi:10.1088/1742-6596/569/1/012027

^{16}O + ^{16}O + valence neutrons in molecular orbitals structures of positive- and negative-parity superdeformed bands in ^{34}S

Yasutaka Taniguchi

Department of Medical and General ciences, Nihon Institute of Medical cience, himogawara 1276, Moroyama-machi, Iruma-gun, aitama 350-0435, Japan

E-mail: `yasutaka@nims.ac.jp`

Abstract. The structures of superdeformed (D) states in ^{34} are investigated using the antisymmetrized molecular dynamics and generator coordinate method (GCM). The GCM basis wave functions are calculated via energy variation with a constraint on the quadrupole deformation parameter β. By applying the GCM after parity and angular momentum projections, the coexistence of two positive- and one negative-parity D bands are predicted, and low-lying states and other deformed bands are obtained. The D bands have structures of ^{16}O + ^{16}O + two valence neutrons in molecular orbitals around the two ^{16}O cores in a cluster picture. The configurations of the two valence neutrons are δ^2 and π^2 for the positive-parity D bands and $\pi^1\delta^1$ for the negative-parity D band.

1. Introduction

Dynamic structural changes under excitation are significant properties of nuclei. Superdeformation and clustering are typical changes. Clustering is a typical structure in the light mass region. To clarify the dynamic structural changes of nuclei, it is necessary to study the nuclear structure in terms of both deformation and clustering. However, those studies have been insu cient.

S isotopes ($Z = 16$) are suitable nuclei for studying deformation and clustering caused by excitation. S isotopes are expected to be favorable for the formation of superdeformed (SD) bands because $Z = 16$ is considered a magic number of superdeformation. The existence of SD bands in S isotopes has been discussed in the frameworks of a mean-field model[1], a cluster model[2], and antisymmetrized molecular dynamics (AMD)[3]. In terms of clustering, S isotopes are key nuclei in the *sd* shell. S isotopes are analogs of Be isotopes because those isotopes can form systems consisting of two doubly closed shell nuclei (^{16}O and α for S and Be isotopes, respectively) and valence neutrons. In Be isotopes, structures consisting of α + α + valence neutrons in molecular orbitals are thought to develop in low-lying states, with the valence neutrons in molecular orbitals around two α cores[4–11]. The SD states in ^{32}S are predicted to contain a large amount of ^{16}O + ^{16}O cluster structure components[2, 3]. They suggest the existence of SD states that have ^{16}O + ^{16}O + valence neutrons in the molecular orbital structure in S isotopes.

By a γ spectroscopy experiment, the structures in ^{34}S are investigated up to the $J^\pi = 10^+$ and (9^-) states for positive- and negative-parity states, respectively, mainly for the yrast states[12].

In contrast to the yrast states, the structures of the non-yrast states with mp-mh configurations have never been clarified.

This paper aims to clarify the structures of SD states in ^{34}S using AMD and the generator coordinate method (CM). The coexistence of positive- and negative-parity SD bands and their structures are discussed, focusing on ^{16}O + ^{16}O + valence neutrons in molecular orbitals around the two ^{16}O cores.

2. Framework

The wave functions in low-lying states are obtained using parity projection and angular momentum projection (AM) and the CM with deformed-basis AMD wave functions. A deformed-basis AMD wave function $|\Phi\rangle$ is a Slater determinant of aussian wave packets.

The basis wave functions of the CM are obtained by energy variation with a constraint potential V_{cnst} after projection onto eigenstates of parity. The ogny D1S force is used as the effective interaction.

To obtain the deformed wave functions, the constraint potential V_{cnst} for the matter quadrupole deformation parameter β of the total system is used. The optimized wave functions are superposed after parity projection and AM by employing the quadrupole deformation parameter β,

$$\left|\Phi_M^{J\pi}\right\rangle = \sum_i f_i \hat{P}_{MK_i}^{J\pi} \left|\Phi(\beta_i)\right\rangle, \tag{1}$$

where $\hat{P}_{MK}^{J\pi}$ is the parity and total angular momentum projection operator, and $|\Phi(\beta_i)\rangle$ are optimized wave functions with a constraint for $\beta = \beta_i$. The coe cients f_i are determined by the Hill–Wheeler equation,

$$\delta\left(\left\langle\Phi_M^{J\pi}\middle|\hat{H}\middle|\Phi_M^{J\pi}\right\rangle - \epsilon\left\langle\Phi_M^{J\pi}\middle|\Phi_M^{J\pi}\right\rangle\right) = 0. \tag{2}$$

Then we obtain the energy spectra and the corresponding wave functions, which are expressed by the superposition of the optimum wave functions, $\{|\Phi(\beta_i)\rangle\}$.

3. Results

3.1. Energy curves

Figures 1 and 2 show the energy surfaces as functions of the quadrupole deformation parameter β for the positive- and negative-parity states, respectively, obtained by energy variation with a constraint on β after parity projection. The energies projected onto the $J^\pi = 0^+$, 3^-, and 4^- states are also shown.

In the positive-parity energy surface (Fig. 1), three excited local minima or shoulders exist around $\beta = 0.4$, 0.6, and 0.8 as well as the minimum at $\beta = 0.25$, which suggest the existence of three excited deformed bands in the positive-parity states. After AM onto the $J^\pi = 0^+$ states, more deformed states gain more binding energy, and the β values of the local minima become larger. In the negative-parity energy surface (Fig. 2), three local minima exist around $\beta = 0.2$, 0.5, and 0.7. In the slightly deformed region, $\beta < 0.6$, the $J^\pi = 3^-$ components have the lowest energies after AM . Highly deformed wave functions, $\beta > 0.6$, have few $J^\pi = 3^-$ components, and the $J^\pi = 4^-$ components have the lowest energies.

3.2. Level scheme and particle-hole configurations

Figure 3 shows the level scheme of the positive- and negative-parity states. Various rotational bands, called $K^\pi = 0^+_{\text{ND}}$, 0^+_{SD1}, and 0^+_{SD2} for positive parity and $K^\pi = 2^-_{\text{ND}}$, 4^-_{SD}, and 3^-_{SD} for negative parity, are obtained, as well as low-lying $0\hbar\omega$ and $1\hbar\omega$ states. The energies of the $0\hbar\omega$ and $1\hbar\omega$ states are consistent with experimental data.

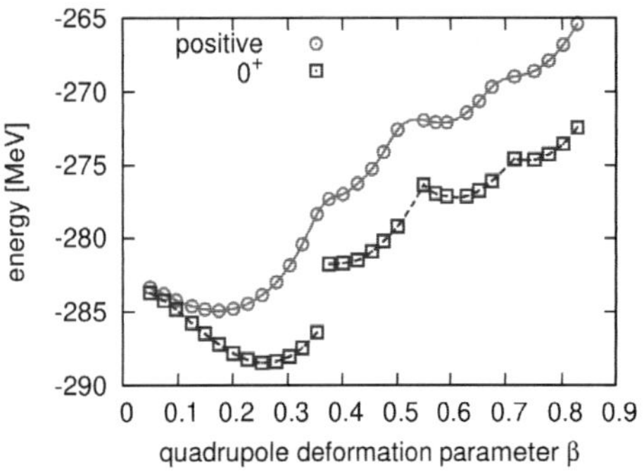

Figure 1. Energy surfaces as functions of quadrupole deformation parameter β for positive-parity states. Circles and squares show energies projected onto positive-parity and $J^\pi = 0^+$ states, respectively.

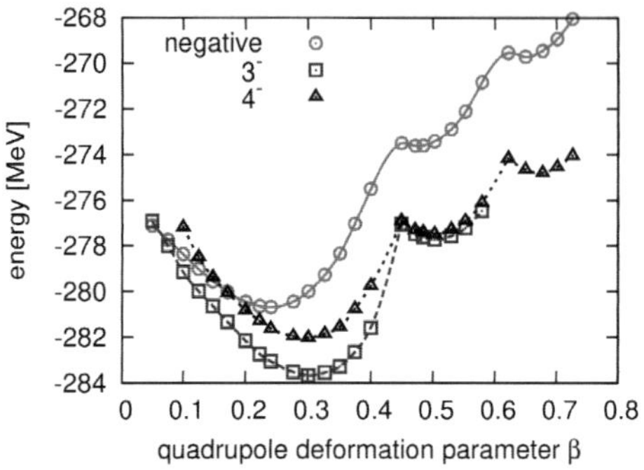

Figure 2. Energy surfaces as functions of quadrupole deformation parameter β for negative-parity states. Circles, squares, and triangles show energies projected onto positive-parity, $J^\pi = 3^-$, and $J^\pi = 4^-$ states, respectively.

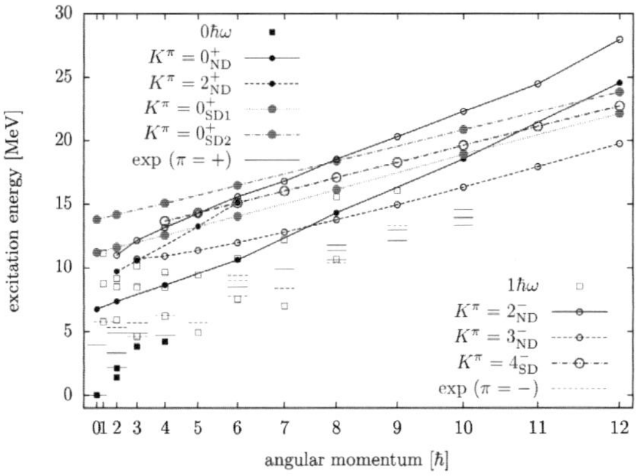

Figure 3. Excitation energies as functions of $J(J+1)$, where J is the total angular momentum. States with closed and open squares have dominant $0\hbar\omega$ and $1\hbar\omega$ configurations, respectively. Solid, dashed, dotted, and dot-dashed lines with closed circles show members of $K^\pi = 0^+_{\rm ND}$, $2^+_{\rm ND}$, $0^+_{\rm SD1}$, and $0^+_{\rm SD2}$ bands, respectively. Solid, dashed, and dot-dashed lines with open circles show members of $2^-_{\rm ND}$, $3^-_{\rm ND}$, and $4^-_{\rm SD}$ bands, respectively. Solid and dashed lines with no symbols show experimental values.

The members of the $K^\pi = 0^+_{\rm SD1}$, $0^+_{\rm SD2}$, and $4^-_{\rm SD}$ bands have mp-mh configurations for both the proton and neutron components; $(n_\pi, n_\nu) = (2,2)$, (2, 4), and (2, 3) for the $K^\pi = 0^+_{\rm SD1}$, $0^+_{\rm SD2}$, and $4^-_{\rm SD}$ bands, respectively, where n_π and n_ν are the numbers of protons and neutrons, respectively, excited from the sd-shell to the pf-shell. The values of the quadrupole deformation parameter β of dominant components of the SD bands are greater than 0.6. Their energies are within a few MeV, although the particle-hole configurations of the neutron components differ; they are $n_\nu = 2$, 3, and 4 for the $K^\pi = 0^+_{\rm SD1}$, $4^-_{\rm SD}$, and $0^+_{\rm SD2}$ bands, respectively. This shows the coexistence of two positive- and one negative-parity SD bands. The energy of the $K^\pi = 4^-_{\rm SD}$

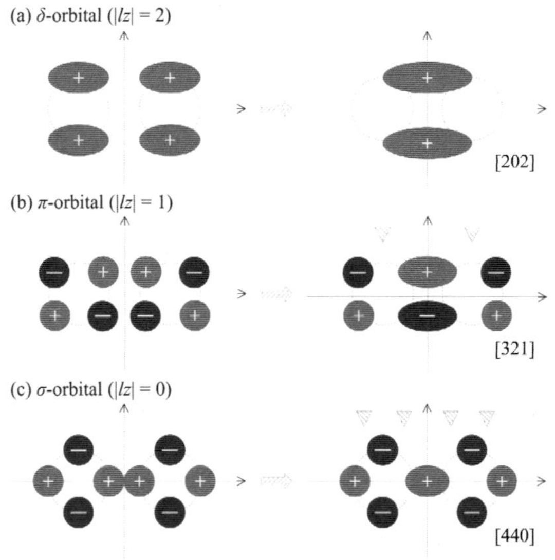

Figure 4. Schematic pictures of molecular orbitals generated from $0d$ orbits around two ^{16}O cores for (a) δ, (b) π, and (c) σ orbitals. Left and right columns show phases of $0d$ orbits around two ^{16}O cores and molecular orbitals, respectively. Dotted circles show two ^{16}O cores located on the z axis. Inverse triangles show locations of nodes in molecular orbitals in the z direction. Numbers in brackets show Nilsson quanta (see text).

band is intermediate between those of the $K^\pi = 0^+_{\rm SD1}$ and $0^+_{\rm SD2}$ bands.

The two highest single-particle orbits have $[202]^2$, $[321]^2$, and $[202]^1[321]^1$ configurations for the $K^\pi = 0^+_{\rm SD1}$, $0^+_{\rm SD2}$, and $4^-_{\rm SD}$ bands, respectively, in the Nilsson picture.

4. Discussions

4.1. Similarity of molecular orbitals around ^{16}O + ^{16}O cores and Nilsson orbits

Before discussing the structures of the SD states in ^{34}S, the quanta of molecular orbitals around the ^{16}O + ^{16}O cores are discussed in the Nilsson picture.

The lowest orbit around an ^{16}O core is a $0d_{5/2}$ orbit. By linear combination of $0d_{5/2}$ orbits around two ^{16}O cores, molecular orbitals around the two ^{16}O cores are formed. Figure 4 shows schematic illustrations of the formation of molecular orbitals around two spherical ^{16}O cores. Here, the two ^{16}O cores are located on the z axis (horizontal axis), and a cylindrical coordinate system is used in this section because of axial symmetry around the z axis.

The $0d_{5/2}$ orbits around the ^{16}O cores form δ, π, and σ orbitals, which are formed from two $(l, |l_z|) = (2, 2)$, $(2, 1)$, and $(2, 0)$ orbits, respectively, around the left and right ^{16}O cores. For the δ orbital, the $(l, |l_z|) = (2, 2)$ orbits have no node in the z direction, as shown in the left column of Fig. 4(a). Therefore, the δ orbital also has no node in the z direction, as shown in the right column of Fig. 4(a). Similarly, the numbers of nodes of the π and σ orbitals are two and four in the direction of the z axis, as shown in Figs. 4(b) and (c), respectively. In the radial direction, they have no node because a $0d$ orbit has no node in the radial direction. This shows that the quanta of the δ, π, and σ orbitals are [202], [321], and [440], respectively, in the Nilsson picture. When a system has a two-^{16}O core structure, the [202], [321], and [440] orbits correspond to the δ, π, and σorbitals, respectively. The parities of the δ, π, and σ orbitals are positive, negative, and positive, respectively.

4.2. Configurations of valence neutrons in the SD bands

The CM calculation yielded three SD bands, called the $K^\pi = 0^+_{\rm SD1}$, $0^+_{\rm SD2}$, and $4^-_{\rm SD}$ bands, the structures of which are interpreted as ^{16}O + ^{16}O + two valence neutrons that have δ^2, π^2, and $\delta^1\pi^1$ configurations, respectively, in a cluster picture.

Their proton components have $2\hbar\omega$ excited configurations and neck structures, and the neutrons, except for the two highest-energy orbits, have the same configuration. This configuration is the same as that of the SD band in ^{32}S, which contains a large amount of ^{16}O + ^{16}O cluster structure components[3]. Therefore, the three SD bands have structures of ^{16}O + ^{16}O + valence neutrons in a cluster picture.

The configurations of the valence neutrons of the $K^\pi = 0^+_{\rm SD1}$, $0^+_{\rm SD2}$, and $4^-_{\rm SD}$ bands are $[202]^2$, $[321]^2$, and $[202]^1[321]^1$, respectively. As shown in the previous section, the [202] and [321] orbits correspond to the δ and π molecular orbitals, respectively, around the two ^{16}O cores. Therefore, the configurations of the valence neutrons of the $K^\pi = 0^+_{\rm SD1}$, $0^+_{\rm SD2}$, and $4^-_{\rm SD}$ bands are interpreted as δ^2, π^2, and $\delta^1\pi^1$, respectively.

5. Conclusions

The structure of the SD states in ^{34}S were investigated using the AMD and CM. By superposing the AMD wave functions calculated via energy variation with a constraint on the quadrupole deformation parameter β, the coexistence of two positive- and one negative-parity SD bands is predicted. The SD states have mp-mh configurations, and they are interpreted as a structure consisting of ^{16}O + ^{16}O + valence neutrons in molecular orbitals around ^{16}O + ^{16}O cores in a cluster picture. Highly e cient γ spectroscopy experiments on high-spin states may reveal the structures of those deformed states.

Acknowledgments

This work was supported by JS S KAKENHI rant Number 25800124. The numerical calculations for this work were conducted under the Interdisciplinary Computational Science rogram of the Center for Computational Sciences, University of Tsukuba.

References

[1] Inakura T, Mizutori , Yamagami M and Matsuyanagi K 2003 *Nucl. Phys. A* **728** 52
[2] Ohkubo and Yamashita K 2002 *Phys. Rev. C* **66** 021301
[3] Kimura M and Horiuchi H 2004 *Phys. Rev. C* **69** 051304
[4] eya M, Kohno M and Nagata 1981 *Prog. Theor. Phys.* **65** 204
[5] Kanada-En yo Y, Horiuchi H and Doté A 1999 *Phys. Rev. C* **60** 064304
[6] Itagaki N and Okabe 2000 *Phys. Rev. C* **61** 044306
[7] Ito M, Kato K and Ikeda K 2004 *Phys. Lett. B* **588** 43
[8] Ito M 2006 *Phys. Lett. B* **636** 293
[9] Ito M, Itagaki N, akurai H and Ikeda K 2008 *Phys. Rev. Lett.* **100** 182502
[10] Ito M, Itagaki N and Ikeda K 2012 *Phys. Rev. C* **85** 014302
[11] Kanada-En yo Y, Kimura M and Ono A 2012 *Prog. Theor. Exp. Phys.* **2012** 01A202
[12] Mason P, Mărginean N, Lenzi M, Ionescu-Bujor M, Della Vedova F, Napoli D R, Otsuka T, Utsuno Y, Nowacki F, Axiotis M, Bazzacco D, Bizzeti P G, Bizzeti- ona A, Brandolini F, Cardona M, de Angelis G, Farnea E, Gadea A, Hojman D, Iordachescu A, Kalfas C, Kröll T, Lunardi , Martínez T, Petrache C M, Quintana B, Ribas R V, Rossi Alvarez C, Ur C A, Vlastou R and Zilio 2005 *Phys. Rev. C* **71** 014316

Nuclear clustering in the Energy Density Functional Approach

J.-P. Ebran
CEA,DAM,DIF, F-91297 Arpajon, France
E-mail: jean-paul.ebran@cea.fr

E. Khan
Institut de Physique Nucléaire, Université Paris-Sud CEA, IN2P3 CNRS, F-91406 Orsay Cedex, France

T. Nikšić
Physics Department, Faculty of Science, University of Zagreb, 10000 Zagreb, Croatia

D. Vretenar
Physics Department, Faculty of Science, University of Zagreb, 10000 Zagreb, Croatia

Abstract. Nuclear Energy Density Functionals (EDFs) are a microscopic tool of choice extensively used over the whole chart to successfully describe the properties of atomic nuclei ensuing from their quantum liquid nature. In the last decade, they also have proved their ability to deal with the cluster phenomenon, shedding a new light on its fundamental understanding by treating on an equal footing both quantum liquid and cluster aspects of nuclei. Such a unified microscopic description based on nucleonic degrees of freedom enables to tackle the question pertaining to the origin of the cluster phenomenon and emphasizes intrinsic mechanisms leading to the emergence of clusters in nuclei.

1. Introduction

Correlations can lead profound changes in a system's bulk properties. Already present in ideal quantum gases, where, at low temperature, quantum statistics manifest in entirely different behaviour of bosons and fermions, pending the irreducible representation of the permutation group they belong to, the situation is even more involved when correlations stem from the constituents' interactions: they dictate the phase of matter preferred by the system. This is particularly true in nuclear systems whose rich behaviour lies in the quantum arrangement of four kinds of fermions (spin-isospin different combinations) interacting, *inter alia*, by the complicated reminiscent strong force acting between their underlying quarks and gluons. Atomic nuclei usually display a quantum liquid like nature even in their ground state, in other words a phase with delocalized structure and elementary excitations with long mean free path whose eigenvalues therefore reflect the shape and space dependence of the confining potential over the whole nuclear volume [1]. The corollary of this feature is a quasi homogeneous repartition of the nucleons in the nuclear volume. In some cases, this picture breaks down. Nuclear clustering, *i.e.* the arrangement of nucleons in additional bound sub-structures in the nucleus itself, provides a typical example.

Clustering is an essential feature of many-nucleon dynamics that coexists with the nuclear mean-field. Therefore, although in most cluster models the existence of such structures is assumed a priori and the corresponding effective interactions are adjusted to the binding energies and scattering phase shifts of these configurations, a fully microscopic understanding of cluster formation necessitates a more general description that encompasses both cluster and quantum liquid aspects in light as well as in heavier nuclei. At present the only comprehensive approach to nuclear structure is based on the framework of energy density functionals (EDFs). Nuclear EDFs enable a complete and accurate description of ground-state properties and collective excitations over the whole nuclide chart [2, 3, 4] and therefore define an appropriate tool to describe the coexistence of cluster and quantum-liquid aspects of light nuclei (Fig 1). EDFs do not assume

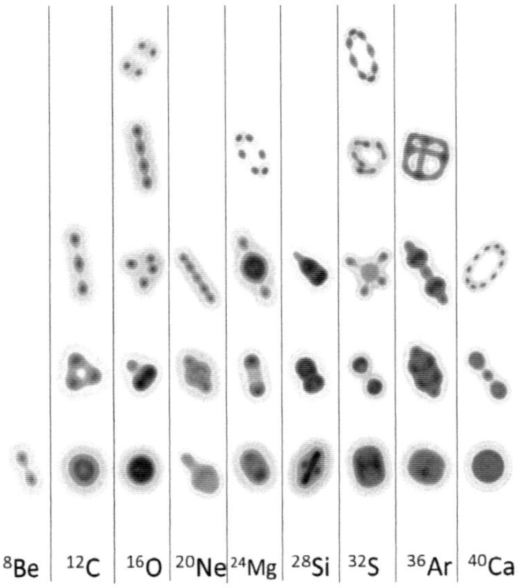

Figure 1: Total positive-parity projected density plots obtained for a number of excited states in $N = Z$ nuclei. For each nucleus the first density on the bottom corresponds to the equilibrium configuration.

any specific form for the nuclear A-body wave function apart to be a product wave function. Within this framework, we can identify three main conditions for α-clusterisation :

i) the effect of the depth of the mean potential
ii) the effect of nuclear deformation
iii) the effect of baryonic density

2. Effect of the depth of the confining potential

As studied in Refs. [5, 6], the depth of the confining potential impacts α-clusterisation through the localisation properties of the nucleonic single-particle orbitals. If in solid state physics, one is able to experimentally fine tune the depth of a potential trapping fermionic atoms to study its effect on the formation of dimers [7], any experimental study on the depth of the nuclear

Journal of Physics: Conference Series **569** (2014) 012028 doi:10.1088/1742-6596/569/1/012028

confining potential turns out in a delicate issue. Equivalent informations can be obtained in the theoretical side, by comparing various realisations of nuclear EDFs yielding the same observables but distinguishable by different depths of their mean potentials. Relativistic and the non-relativistic Gogny and Skyrme functionnals are such realisations. For example, ^{36}Ar observables are equivalenty described by the three aforementioned functionals (Fig. 2). On the other hand the relativistic functionnal involves a deeper confining potential than the non-relativistic ones. As explained in Refs. [5, 6], and also by E. Khan in the present volume, a deeper confining potential causes the nucleonic single-orbitals to be more localised leading to a clustering phase. Consequently, a more and more prononced degree of localisation of nucleons can be noticed in the total density inserts of Fig. 2 going from Skyrme SLy4, then Gogny D1S, to DD-ME2.

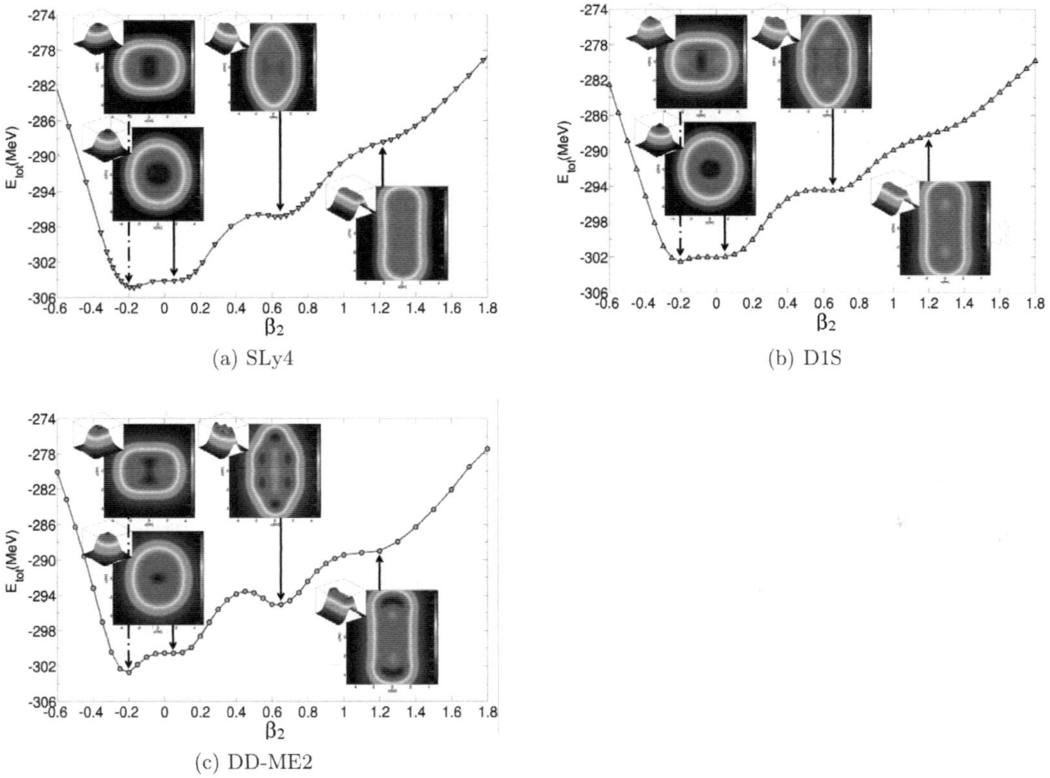

(a) SLy4

(b) D1S

(c) DD-ME2

Figure 2: Binding energy of ^{36}Ar versus the quadrupole deformation parameter with inserts displaying the total density at corresponding deformations.

3. Effect of the deformation

The relationship between α clusters and single particle states in deformed nuclei is a well known effect [8, 9, 10]. A harmonic oscillator confining potential, when deformed, loses at first the degeneracy of the single particle states presented at sphericity, however recreates it again when the ratio of its frequencies becomes the ratio of integer numbers. The new magic numbers generated can be expressed as combinations of the spherical ones, so that the deformed harmonic oscillator potential admits an interpretation in terms of a series of shifted overlapping spherical potentials. Clustering can then be viewed as a mass exchange process between these spherical potentials. We shall focus here on α clusterisation. As noted by Aberg [10], an isolated single-particle state in a deformed N=Z nuclei corresponds to an alpha clusterisation, because of both the Kramers (time-invariance) degeneracy and the isospin symmetry: 2 protons and 2 neutrons shall have similar wave functions and therefore drive their common localisation leading to alpha clusterisation. Hence, the degeneracy raising among levels seems to be the master condition for alpha clusterisation. Let us quantatively analyse this idea using the microscopic EDF framework on the example of ^{12}C. Increasing its quadrupole deformation on the prolate side make it pass through different cluserised states untill eventually reach the linear α-chain configuration. We then define the average energy difference between consecutive single particle levels in those systems as a function of the deformation parameter. When the consecutive level spacing is larger than 5 MeV, the value is kept at 5 MeV which seems sufficient to consider a degeneracy raising among levels. Fig. 3 shows a correlation with a maximum average degeneracy raising and the arrising of clusters : the more raised the single-particle degeneracy, the more alpha-localised the density.

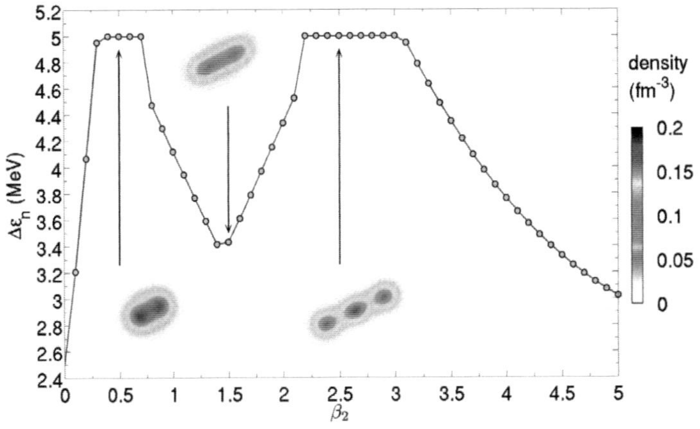

Figure 3: Mean gap value between consecutive occupied neutron levels versus the quadrupole deformation parameter. The inserts display the total density at corresponding deformations

4. Effect of the density

Infinite nuclear matter at subsaturation density finds energetically favorable to undergoe a Mott-like phase transition breaking its homogeneity by forming sub-uits at saturation density. The occurence of such a phenomenom in finite nuclei can be studied in the framework of EDF [11, 12]. In order to isolate the influence of density, *i.e.* to suppress the impact of deformation on clusterisation, calculations are performed imposing a vanishing quadrupole moment. Constraining on the total r.m.s. radius of the finite nucleus allows to reach low density phases. For example, inflating ^{16}O makes it go from a homogeneous spherical shape to a 4 α clusterized phase with tetragonal shape. The effect of the phase transition on the total binding energy is displayed in Fig. 4.

5. Conclusion

EDFs are a powerful tool to tackle both quantum liquid and cluster properties from fundamental grounds. They allow to investigate on the origin of the clustering phenomenon, linked not only to

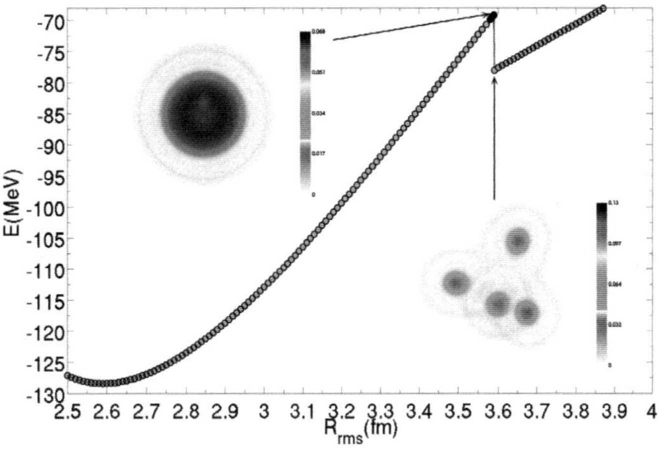

Figure 4: Total binding energy versus the r.m.s radius of ^{16}O with DD-ME2.

the depth of the confining potential as well as the baryon density, but also, in a finite size system, to the deformation. The study of cluster structures in neutron rich systems is straightforward [13](for instance Fig. 5), and the inclusion of beyond mean field correlations is in progress.

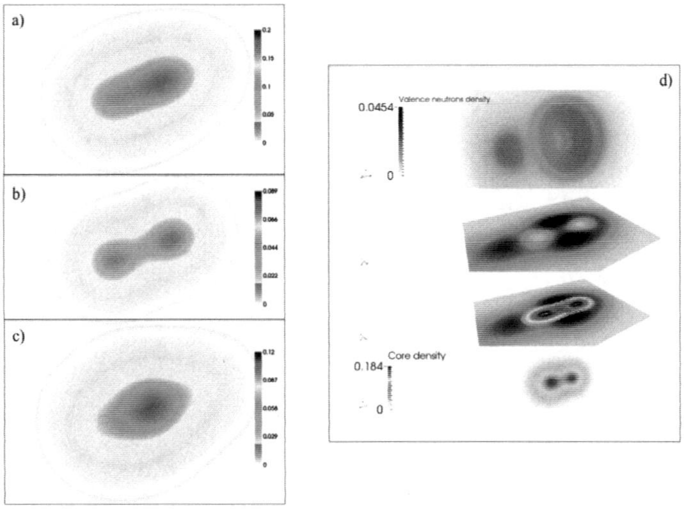

Figure 5: Left panel : total (a), proton (b) and neutron (c) intrinsic densities of ^{14}Be at equilibrium deformation. Right panel, from bottom to top: 3D density of the $\alpha + \alpha$ core; contour plots of the core density and the density of the valence neutrons in the (Oxz) plane; 3D density of the valence neutrons..

References

[1] Mottelson B 1996 *Nuclear Structure, Les Houches, Session LXVI, 25*
[2] Ichikawa T, Marhun J, Itagaki N and Ohkubo S 2011 *Phys. Rev. Lett.* **107** 112501

[3] Robledo L and Bertsch G 2011 *Phys. Rev. C* **84** 054302
[4] Lalazissis G, Ring P and Vretenar D 2004 *Lecture Notes in Physics* **641**
[5] Ebran J P, Khan E, Niksic T and Vretenar D 2012 *Nature* **487** 341
[6] Ebran J P, Khan E, Niksic T and Vretenar D 2013 *Phys. Rev. C* **87** 044307
[7] Greiner M *et al.* 2002 *Nature* **415** 39
[8] Kanada En'yo Y and Horiuchi H 2001 *Prog. Theor. Phys. Suppl.* **142** 205–263
[9] Horiuchi H 2011 *Lecture Notes in Physics* **818** 57
[10] Aberg S and Jönsson L O 1994 *Z. Phys. A* **349** 205
[11] Girod M and Schuck P 2013 *Phys. Rev. Lett.* **111** 132503
[12] Ebran J P, Khan E, Niksic T and Vretenar D 2014 *Phys. Rev. C* **89** 031303(R)
[13] Ebran J P, Khan E, Niksic T and Vretenar D 2014 *arXiv:1406.2473*

Journal of Physics: Conference Series **569** (2014) 012029 doi:10.1088/1742-6596/569/1/012029

Co-existence of states in quantum systems

Yoritaka Iwata

School of Science, The University of Tokyo, Hongo 7-3-1, 113-0033 Tokyo

E-mail: iwata@cns.s.u-tokyo.ac.jp

Abstract. Co-existence of different states is a profound concept, which possibly underlies the phase transition and the symmetry breaking. Because of a property inherent to quantum mechanics (cf. uncertainty), the co-existence is expected to appear more naturally in quantum-microscopic systems than in macroscopic systems. In this paper a mathematical theory describing co-existence of states in quantum systems is presented, and the co-existence is classified into 9 types.

1. Introduction

The boundary-value problem of nonlinear partial differential equation of elliptic-type:

$$\begin{cases} -\nabla^2 u - mu - V(u)u = 0 \quad \text{in } \Omega, \\ u = 0 \quad \text{on } \partial\Omega, \end{cases} \tag{1}$$

is studied, where m is a real number, and $\Omega \in R^3$ is a closed domain with a sufficiently smooth boundary. The unknown complex function u consists of the unknown state ψ and the reference state $\bar{\psi}$:

$$u = \psi - \bar{\psi},$$

where $\bar{\psi}$ (corresponding to a generalized concept of the vacuum) is not necessarily a solution of Eq. (1), although the most simplest case $\bar{\psi} = 0$ (the simplest vacuum) satisfies Eq. (1). Let a part of the inhomogeneous term $V(u)$, whose spectral set is assumed to be included in a real axis, satisfy

$$\partial_u (V(u)u)|_{u=0} = V_L. \tag{2}$$

For the simplicity V_L, which corresponds to the signed strength of linearized interaction being independent of u, is assumed to be a real number. As is readily seen, the function $u = \psi - \bar{\psi} = 0$ is always a solution of this problem (refer to the trivial solution). In this sense let us imagine a simple case when $\bar{\psi} = 0$, and then the emergence of a solution ψ from another solution $\bar{\psi} = 0$ is true if $u \neq 0$ is the solution of Eq. (1). Here we seek the non-trivial solution $u \neq 0$ ($\psi \neq \bar{\psi}$) to Eq. (1). The corresponding situation is nothing but the co-existence of different states ψ and $\bar{\psi}$.

Equation (1) is associated with the stationary problem of nonlinear Schrödinger equations as well as nonlinear Klein-Gordon equations. In the context of Klein-Gordon equations, it is possible to associate $\sqrt{-m}$ with the mass (if $m < 0$). Note that the statistical property inherent to many-body system, which might bring about rather interesting physical properties, is not taken into account in order to see the most fundamental properties associated with the co-existence in both nonlinear Schrödinger equations and nonlinear Klein-Gordon equations.

2. Theory describing the co-existence

2.1. Mathematical settings

Let X and Y be functional spaces

$$X = W_0^{1,2}(\Omega) \cap W^{2,2}(\Omega),$$
$$Y = L^2(\Omega)$$

respectively (for mathematical notation, see [1]). An inclusion relation $X \subset Y$ is true. For $u \in X$, a mapping $f : R^1 \times X \to Y$ is defined by

$$f(\lambda, u) := -\nabla^2 u - mu - V(u)u.$$

The original master equation is written by $f(\lambda, u) = 0$. Since the trivial solution $u = 0$ always exists, $f(\lambda, 0) = 0$ is satisfied. According to the Sobolev embedding theorem $-\nabla^2$ is a C^2-mapping from X to Y, where the detail setting of $V(u)$ is necessary to know the regularity of the mapping f. The space $W_0^{1,2}(\Omega)$ denotes all the functions included in $W^{1,2}(\Omega)$ satisfying $u|_{\partial\Omega} = 0$.

2.2. Linearized analysis

Linearized problem is derived. The Fréchet derivative of $f(\lambda, u)$ is calculated as

$$f_u(\lambda, 0)[u] = -\nabla^2 u - \lambda u = 0, \tag{3}$$

where $\lambda = m + V_L$. This corresponds to the master equation for the linearized eigen-value problem. It is well known that the linearized problem (with the Dirichlet boundary condition) is solvable. Furthermore it is known that a infinite set of eigen-values $\{\lambda_i\}_{i=0}^{\infty}$ of $-\nabla^2$ satisfy

- $0 < \lambda_0 < \lambda_1 \le \lambda_2 \le \cdots$;
- λ_0 is a simple eigen-value.

Let the eigen-function corresponding to the eigen-value λ_0 be u_0 (i.e., $-\nabla^2 u_0 = \lambda_0 u_0$). First, according to the simple property of the eigen-value λ_0, it is clear that

$$\mathrm{Ker}(f_u(\lambda_0, 0)) = \{tu_0;\ t \in R^1\},$$

so that the dimension of $\mathrm{Ker}(f_u(\lambda_0, 0))$ is equal to 1. Second, if there exists a solution $v \in X$ for $\nabla^2 v - \lambda_0 v = h$ with $h \in Y$, then

$$R(f_u(\lambda_0, 0)) = \left\{ h \in Y; \int_\Omega h(x)u(x)dx = 0 \right\},$$

so that $R(f_u(\lambda_0, 0))$ is a closed subset of Y with its co-dimension 1 (cf. the Riesz-Schauder theory [1]). Third, it is valid that

$$f_{u\lambda}(\lambda_0, 0)[u] = -\lambda_0 u \notin R(f_u(\lambda_0, 0)). \tag{4}$$

Consequently, according to the bifurcation theory [2, 3], $\lambda = \lambda_0$ has been clarified to be a bifurcation point (corresponding to $(\lambda_0, 0)$ in Fig. 1). Note that only sufficient conditions for the existence of the bifurcation point is presented in the bifurcation theory.

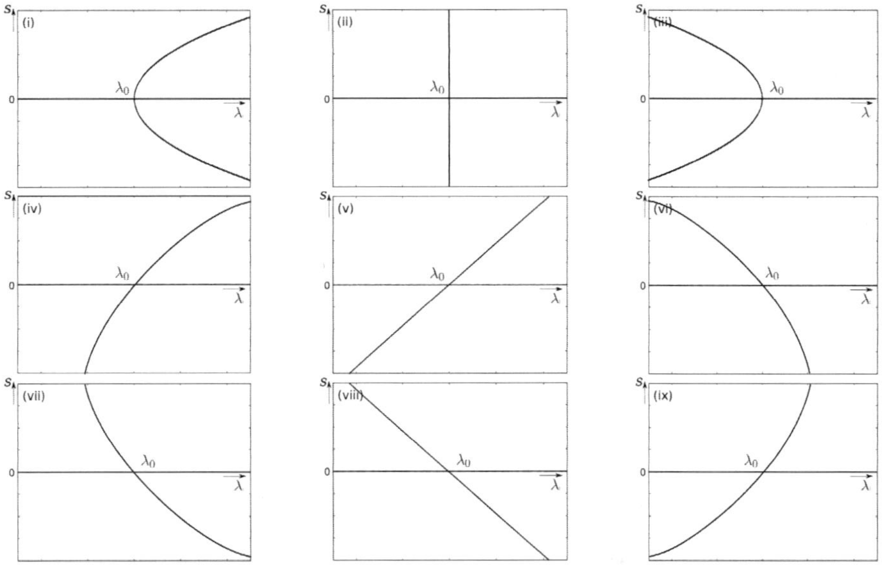

Figure 1. 9 types of co-existence based on Eqs. (5) and (6): cases (i), (ii), and (iii) appear if $\mu_s(0) = 0$, cases (iv), (v), and (vi)) appear if $\mu_s(0) > 0$, cases (vii), (viii), and (ix) appear if $\mu_s(0) < 0$; cases (i), (iv), and (vii) appear if $\mu_{ss}(0) > 0$, cases (ii), (v), and (viii) appear if $\mu_{ss} = 0$, cases (iii), (vi), and (ix) appear if $\mu_{ss} < 0$.

2.3. Nonlinear analysis

Co-existence of different states (i.e., existence of non-trivial solution $u \neq 0$) is shown. We set a closed interval $[-\epsilon_0, \epsilon_0]$ and a C^1-function $\lambda(s)$ satisfying $\lambda(0) = \lambda_0$, where s parametrizes the functional space X. Under the three conditions confirmed in Sec. 2.2, let the corresponding solution u be represented by

$$u(\lambda, s, x) = su_0(\lambda, x) + sz(\lambda, s, x),$$

where s is defined on the interval, and $z(\lambda, s, x)$ is a sufficiently smooth function of s defined on $R^1 \times R^1 \times X$. The function $z(\lambda, s, x)$ satisfies $z(\lambda, 0, x) = 0$ and

$$\int_\Omega z(x)u_0(x)dx = 0.$$

The function $u(\lambda, s, x)$ satisfies the condition $u(\lambda, 0, x) = 0$, which means the existence of the trivial solution. It is useful to define a linear operator

$$A := -\nabla^2 - \lambda_0,$$

with its domain X, and then it is readily seen that A is a self-adjoint operator in Y. The original equation is written by $Au = \mu(s)u + (V(u) - V_L)u$ with $\mu(s) = \lambda(s) - \lambda_0$, and the linearized problem is written by $Au_0 = 0$. By differentiating the original equation with respect to s, step by step

$$\begin{aligned}
(Au)_s &= \mu_s u + \mu u_s + \partial_s(V(u)u) - V_L u_s \\
&= \mu_s u + \mu u_s + (\partial_s V(u))u + V(u)u_s - V_L u_s \\
(Au)_{ss} &= \mu_{ss} u + 2\mu_s u_s + \mu u_{ss} + \partial_s^2(V(u)u) - V_L u_{ss} \\
&= \mu_{ss} u + 2\mu_s u_s + \mu u_{ss} + (\partial_s^2 V(u))u + 2(\partial_s V(u))u_s + V(u)u_{ss} - V_L u_{ss} \\
(Au)_{sss} &= \mu_{sss} u + 3\mu_{ss} u_s + 3\mu_s u_{ss} + \mu u_{sss} + \partial_s^3(V(u)u) - V_L u_{sss} \\
&= \mu_{sss} u + 3\mu_{ss} u_s + 3\mu_s u_{ss} + \mu u_{sss} + (\partial_s^3 V(u))u + 3(\partial_s^2 V(u))u_s + 3(\partial_s V(u))u_{ss} \\
&\quad + V(u)u_{sss} - V_L u_{sss}
\end{aligned}$$

where the functions are represented by $u_s = u_0 + sz_s + z$, $u_{ss} = sz_{ss} + 2z_s$, and $u_{sss} = sz_{sss} + 3z_{ss}$ respectively. The derivatives of the inhomogeneous terms become

$$\begin{aligned}
\partial_s V(u) &= (\partial_u V(u))\ u_s \\
\partial_s^2 V(u) &= (\partial_u^2 V(u))\ u_s^2 + (\partial_u V(u))\ u_{ss} \\
\partial_s^3 V(u) &= (\partial_u^3 V(u))\ u_s^3 + 3(\partial_u^2 V(u))\ u_s u_{ss} + (\partial_u V(u))\ u_{sss}.
\end{aligned}$$

By taking $s = 0$, the bi-linear forms become

$$\begin{aligned}
&(Au)_s|_{s=0} = \mu_s(0)u|_{s=0} + \mu(0)u_s|_{s=0} + \partial_u(V(u)u)u_s|_{s=0} - V_L u_s|_{s=0} \\
&\quad = V_L u_s|_{s=0} - V_L u_s|_{s=0} = 0, \\
&((Au)_s|_{s=0}, u_0) = 0,
\end{aligned}$$

$$\begin{aligned}
&(Au)_{ss}|_{s=0} = \mu_{ss}(0)u|_{s=0} + 2\mu_s(0)(u_0 + z|_{s=0}) + 2\mu(0)z_s|_{s=0} + \partial_s^2(V(u)u)|_{s=0} - 2V_L z_s|_{s=0} \\
&\quad = 2\mu_s(0)u_0 + \partial_s^2(V(u)u)|_{s=0} - 2V_L z_s|_{s=0}, \\
&((Au)_{ss}|_{s=0}, u_0) = 2\mu_s(0)(u_0, u_0) + (\partial_s^2(V(u)u)|_{s=0}, u_0) - (2V_L\ z_s|_{s=0}, u_0),
\end{aligned}$$

$$\begin{aligned}
&(Au)_{sss}|_{s=0} = \mu_{sss}(0)u|_{s=0} + 3\mu_{ss}(0)(u_0 + z|_{s=0} + 6\mu_s(0)z_s|_{s=0} + 3\mu(0)z_{ss}|_{s=0} \\
&\quad + \partial_s^3(V(u)u)|_{s=0} - 3V_L\ z_{ss}|_{s=0} \\
&\quad = 3\mu_{ss}(0)u_0 + 6\mu_s(0)z_s|_{s=0} + \partial_s^3(V(u)u)|_{s=0} - 3V_L\ z_{ss}|_{s=0}, \\
&((Au)_{sss}|_{s=0}, u_0) = 3\mu_{ss}(0)(u_0, u_0) + (6\mu_s(0)z_s|_{s=0}, u_0) + (\partial_s^3(V(u)u)|_{s=0}, u_0) - 3(V_L\ z_{ss}|_{s=0}, u_0),
\end{aligned}$$

where $u|_{s=0} = 0$, $z|_{s=0} = 0$, and $\mu(0) = 0$ are utilized, as well as Eq. (2). $(Au_{ss}|_{s=0}, u_0) = (u_{ss}|_{s=0}, Au_0) = 0$ due to $Au_0 = 0$. Consequently

$$2\mu_s(0) = -(\partial_s^2(V(u)u)|_{s=0}, u_0) + 2(V_L z_s|_{s=0}, u_0), \tag{5}$$

and the sign of $\lambda_s(0) = \mu_s(0)$ is determined by $-(\partial_s^2(V(u)u)|_{s=0}, u_0) + 2(V_L z_s|_{s=0}, u_0)$. In the same manner $(Au_{sss}|_{s=0}, u_0) = (u_{sss}|_{s=0}, Au_0) = 0$. It leads to

$$3\mu_{ss}(0) = -(\partial_s^3(V(u)u)|_{s=0}, u_0) - (6\mu_s(0)z_s|_{s=0}, u_0) + 3(V_L z_{ss}|_{s=0}, u_0), \tag{6}$$

and the sign of $\lambda_{ss}(0) = \mu_{ss}(0)$ is determined by $-(\partial_s^3(V(u)u)|_{s=0}, u_0) - (6\mu_s(0)z_s|_{s=0}, u_0) + 3(V_L z_{ss}|_{s=0}, u_0)$. In particular, if $\lambda_s(0) = \mu_s(0) = 0$ is true, the sign of $\lambda_{ss}(0)$ is determined by $-(\partial_s^3(V(u)u)|_{s=0}, u_0) + 3(V_L z_{ss}|_{s=0}, u_0)$. According to Eqs. (5) and (6), the co-existence of states is classified into 9 types (Fig. 1). In Figure 1, around the neighbour of the bifurcation point $(\lambda_0, 0)$, two solutions co-exist in types (iv) to (ix), while the transition from single-existence to co-existence is described in types (i) and (iii).

Table 1. Systematic analysis for ψ^k-interaction theory. Possible classification of co-existence is shown in the column "Type", where $\sigma = 4\eta(u_0 z_s|_{s=0}, u_0) - 2\eta(u_0^2, u_0)(z_s|_{s=0}, u_0)$.

k	$\partial_s V(u)$	$\partial_s^2 V(u)$	$\partial_s^2(V(u)u)\|_{s=0}$	$\partial_s^3(V(u)u)\|_{s=0}$	$\mu_s(0)$	$\mu_{ss}(0)$	Type
$=3$	$-\eta u_s$	$-\eta u_{ss}$	$-2\eta u_0^2$	$-12\eta u_0 z_s\|_{s=0}$	$\eta(u_0^2, u_0)$	σ	all
$=4$	0	$-2\eta u_s^2$	0	$-6\eta u_0^3$	0	$2\eta(u_0^3, u_0)$	(i),(ii),(iii)
≥ 5	0	0	0	0	0	0	(ii)

3. Application to ψ^k-interaction theory

If the Lagrangian includes the kth-order nonlinearity in its interaction part (for example, see textbooks of particle physics), the inhomogeneous term of the master equation becomes

$$V(u)u = -\eta u^{k-1},$$

for integers $k \geq 1$, where η is assumed to be a real number. Here $V_L = 0$ and $V(u)|_{s=0} = V(u)|_{u=0} = 0$ are true. The first derivative is

$$\partial_u V(u)|_{u=0} = -(k-2)\eta u^{k-3}|_{u=0}$$

for $k \geq 3$, so that it is equal to $-\eta$ for $k = 3$, and zero for $k \geq 4$. The second derivative is

$$\partial_u^2 V(u)|_{u=0} = -(k-2)(k-3)\eta u^{k-4}|_{u=0}$$

for $k \geq 4$, so that it is equal to zero for $k = 3$, -2η for $k = 4$, and zero for $k \geq 5$. The third derivative is

$$\partial_u^3 V(u)|_{u=0} = -(k-2)(k-3)(k-4)\eta u^{k-5}|_{u=0}$$

for $k \geq 5$, so that it is equal to zero for $k \leq 4$, -6η for $k = 5$, and zero for $k \geq 6$. Results are summarized in Table 1. In case of $k = 4$ (ψ^4-interaction theory), the non-trivial solution corresponds to type (i) of Fig. 1 if $\eta > 0$, to type (ii) if $\eta = 0$, and to type (iii) if $\eta < 0$. In particular when $\eta > 0$, the co-existence emerges only if $m > \lambda_0$ (cf. spontaneous symmetry breaking).

If there is no interaction (free particle condition: $\eta = 0$), $\mu_s(0) = \mu_{ss}(0) = 0$ is true, and the co-existence is classified into type (ii). If the interaction is linear ($V(u) = V_L \neq 0$; ψ^2-interaction theory), the derivatives are

$$\partial_u V(u)|_{u=0} = \partial_u^2 V(u)|_{u=0} = \partial_u^3 V(u)|_{u=0} = 0,$$

so that $\partial_s V(u) = \partial_s^2 V(u) = \partial_s^3 V(u) = 0$. It leads to $\partial_s^2(V(u)u)|_{s=0} = V_L u_{ss}|_{s=0}$ and $\partial_s^3(V(u)u)|_{s=0} = V_L u_{sss}|_{s=0}$ so that $\mu_s(0) = -(V_L z_s|_{s=0}, u_0) + (V_L z_s|_{s=0}, u_0) = 0$ and $\mu_{ss}(0) = -(V_L z_{ss}|_{s=0}, u_0) + (V_L z_{ss}|_{s=0}, u_0) = 0$ follows. The co-existence is classified into type (ii). As a result the nonlinearity can be identified by the classification other than type (ii).

Acknowledgments

This work was supported by HPCI Strategic Programs for Innovative Research Field 5 "The origin of matter and the universe". The author is grateful to Prof. Emeritus Dr. Hiroki Tanabe for reading the manuscript.

References

[1] Yosida K. 1980 Functional Analysis Sixth Edition (Berlin Heidelberg New York, Springer-Verlag).
[2] Crandall M. G. and Rabinowitz P. H. 1971 *J. Funct. Anal.* **8** 321.
[3] Crandall M. G. and Rabinowitz P. H. 1973 *Arc. Rational Mech. Anal.* 21 XI **52** 2 161.

Journal of Physics: Conference Series **569** (2014) 012030 doi:10.1088/1742-6596/569/1/012030

Dynamical cluster decay model applied to very light mass compound systems of mass A∼30 formed in heavy ion reactions

BirBikram Singh[1], Mandeep Kaur[1], Manoj K. Sharma[2] and Raj K. Gupta[3]

[1]Department of Physics, Sri Guru Granth Sahib World University, Fatehgarh Sahib-140406, India
[2]School of Physics and Materials Science, Thapar University, Patiala - 147004, India
[3]Department of Physics, Panjab University, Chandigarh-160014, India

E-mail: birbikram.singh@gmail.com

Abstract.
The study of the decay of $^{32}S^*$ and $^{31}P^*$ compound systems formed in $^{20}Ne+^{12}C$ and $^{19}F+^{12}C$ reactions, respectively, is further extended on the basis of collective clusterization process within the dynamical cluster model (DCM) of Gupta and collaborators, with the effects of deformations and orientations included, at an excitation energy E^*_{CN}=60 MeV. In the present study, we have investigated the effects of deformations and orientations on the target, i.e., ^{12}C like yield, denoted C-yield (σ_C), which contains fusion-fission (FF) decay cross-section, σ_{FF}, from compound nucleus process and deep inelastic orbiting (DIO) cross-section, σ_{DIO}, from non-compound nucleus process. As observed in one of our earlier study for $^{32}S^*$ system there is a competition between FF and DIO, while, for $^{31}P^*$ there is a contribution of FF cross-section only, in the total C-yield. The comparative analysis of C-Yield for the considerations of spherical and oriented nuclei, shows similar results with the only difference of the values of neck length parameter (ΔR), which are more for the later case. The calculated cross-sections σ_C show good agreement with experimental data for both the considerations.

1. INTRODUCTION

Within last couple of decades, light heavy ion collisions have been studied over a wide range of low bombarding energies per nucleons (E_{lab} ≤10 MeV) for various target + projectile combinations [1, 2, 3]. The fission process in very light systems has distinctive features *w.r.t.* heavy systems. In contrast to heavy systems, where symmetric break up is favoured, the dependence of macroscopic potential energy surface due to nuclear deformations and shape asymmetry favours the break up of lighter systems into two unequal fragments. There is a competition between fusion-fission and deep inelastic orbiting processes for theses systems [4, 5]. In the fusion-fission (FF) process, a completely equilibrated compound nucleus (CN) is formed, which decays in various exit channels, independent of the entrance channel, whereas deep inelastic orbiting (DIO) is the process of formation of long lived dinuclear molecular complex which acts as a 'door way to fusion' resulting into non compound nucleus (nCN) formation with strong memory of entrance channel.

The dynamics of very light compound systems $^{32}S^*$ and $^{31}P^*$ formed in $^{20}Ne+^{12}C$ and $^{19}F+^{12}C$ reactions, respectively, is studied using the dynamical cluster decay Model (DCM) of Gupta and collaborators [6, 7] with quadrupole deformations (β_2) and hot, compact optimal orientations of nuclei included. The structure information of CN enters into the model via preformation probability P_0 of the fragments. It is relevant to mention here that the decay of these systems was also studied [8] within the DCM with spherical consideration of the decaying fragments. However, the deformation effects were also included within the description of Süsmann central radii $C_t = C_1+C_2$, where $C_i = R_i - b^2/R_i$ (in fm) for the decaying fragments, $R_i = [1.28A_i^{1/3} - 0.76 + 0.8A_i^{1/3}][1 + 0.0007T^2]$ fm and surface thickness parameter b=0.99 fm. The study clearly points out the competition between FF and DIO in the C-yield of $^{32}S^*$ whereas it counts only from the CN process for the $^{31}P^*$ decay.

In the present work we intend to study the deformation and orientation effects in the C-yied of $^{32}S^*$ and $^{31}P^*$. For that we have also made calculations for the purely spherical considerations, i.e., with $R_t = R_1 + R_2$, in order to make comparisons. The quadruple deformation (β_2) and compact optimum orientations lowers the barrier height [9] which effects the FF cross-section σ_{FF} of CN. It has been given in the experimental data that the degree of competition between two classes FF and DIO is co-related with the number of open channels for the decay of light dinuclear system. Large number of open channels favours the occurrence of FF process with regard to mechanism which retain the memory of entrance channel [10].

Within the DCM, an empirical estimate of nCN decay, i.e., σ_{DIO} can be obtained from FF cross-section,

$$\sigma_{DIO} = \sigma_C^{Expt.} - \sigma_{FF}^{DCM}. \tag{1}$$

For nCN process we take P_0=1 for the incoming channel, since the target and projectile is considered to have not yet lost their identity.

In this paper, using the DCM for pure spherical consideration as well as deformations and orientations effects included for nuclei, we calculate the C-yield for the decay of $^{32}S^*$ and $^{31}P^*$ compound systems. The contribution of competing mechanisms is calculated for explaining the experimental data.

The DCM is briefly described in section 2. The calculations and discussions are given in section 3. The work is summarized and concluded in section 4.

2. The dynamical cluster-decay model (DCM)

The DCM [6, 7] is based on the collective coordinate of mass (and charge) asymmetry $\eta = (A_1 - A_2)/(A_1 + A_2)$ ($\eta_Z = (Z_1 - Z_2)/(Z_1 + Z_2)$), relative separation R, the quadrupole deformations $\beta_{\lambda i}$ ($\lambda = 2$) and orientations of two nuclei and fragments.

Using the decoupled approximation to R- and η-motions, the DCM defines the decay cross section, in terms of partial waves, as;

$$\sigma = \frac{\pi}{k^2}\sum_{\ell=0}^{\ell_c}(2\ell+1)P_0P; \qquad k = \sqrt{\frac{2\mu E_{c.m.}}{\hbar^2}} \tag{2}$$

where, P_0, the preformation probability, refers to η-motion and P, the penetrability, to R-motion, both depend upon angular momentum ℓ, temperature T, deformations $\beta_{\lambda i}$ and compact optimum orientations θ_i [9]. The tunnelling probability P, for each η, is calculated as the WKB integral,

$$P = exp(-\frac{2}{\hbar}\int_{R_a}^{R_b}\{2\mu[V(R,\beta_{\lambda_i},\theta_i,T) - Q_{eff}]\}^{1/2}dR), \tag{3}$$

with first and second turning points R_a and R_b. Q_{eff} is the Q-value of the decay process (Fig.1(a) and (b)) and the first turning point is given as: $R_a(T) = R_1(\alpha_1, T) + R_2(\alpha_2, T) + \Delta R(T)$,where $R_i(\alpha_i)$=$R_{0i}(T)[1 + \sum_\lambda \beta_{\lambda_i} Y_\lambda^{(0)}(\alpha_i)]$ and $R_{0i}(T)$=$[1.28A_i^{1/3} - 0.76 + 0.8A_i^{1/3}][1 + 0.0007T^2]$.

The P_0 is given by the solution of Schrödinger eqn. in η, at a fixed R=R_a, the first turning point for the penetration path for different ℓ-values,

$$\{-\frac{\hbar^2}{2\sqrt{B_{\eta\eta}}}\frac{\partial}{\partial\eta}\frac{1}{\sqrt{B_{\eta\eta}}}\frac{\partial}{\partial\eta} + V(\eta,\beta_{\lambda_i},\theta_i,T)\}\psi^\nu(\eta) = E^\nu\psi^\nu(\eta) \tag{4}$$

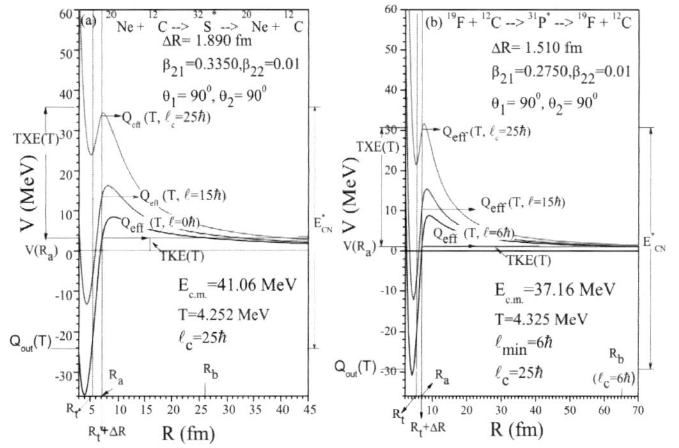

Figure 1. Scattering Potential for (a) $^{32}S^* \rightarrow ^{20}Ne + ^{12}C$, and (b) $^{31}P^* \rightarrow ^{19}Ne + ^{12}C$, with effects of deformations and optimum orientations included, at different ℓ-values.

with ν=0,1,2,3... referring to ground-state ($\nu = 0$) and excited-states solutions. Then the preformation probability is given by

$$P_0(A_i) = \mid \psi(\eta(A_i)) \mid^2 \sqrt{B_{\eta\eta}} \frac{2}{A}. \quad (5)$$

For fixed R=R_a, the fragmentation potential is given by

$$\begin{aligned} V(R, \eta, \beta_{\lambda_i}, \theta_i, T) &= \sum_{i=1}^{2} \Big[V_{LDM}(A_i, Z_i, \beta_{\lambda_i}, T) \Big] + \sum_{i=1}^{2} \Big[\delta U_i \Big] exp(-\frac{T^2}{T_0{}^2}) \\ &+ V_c(Z_i, \beta_{\lambda_i}, \theta_i, T) + V_P(A_i, Z_i, \beta_{\lambda_i}, \theta_i, T) + V_\ell(A_i, \beta_{\lambda_i}, \theta_i, T). \quad (6) \end{aligned}$$

The DCM treats all the decay processes as the dynamical collective mass motion of the preformed clusters through the interaction barrier.

3. Calculations and discussions

Figs. 1(a) and 1(b) give the scattering potentials (with the effects of deformations and compact orientations included) at different $\ell(\hbar)$ values for the decay of $^{32}S^*$ and $^{31}P^*$, respectively. We notice here that the value of ℓ_{min} (minimum ℓ value at which WKB integral starts contributing)=6 $\hbar$ and 0 $\hbar$ for the decay of $^{31}P^*$ and $^{32}S^*$, respectively, in the respective exit channels. Note that, in an earlier study [8], the values for the same were ℓ_{min}= 9 $\hbar$ and 0 $\hbar$ for the decay of $^{31}P^*$ and $^{32}S^*$, respectively. Interestingly, $\ell = \ell_c (= 25\ \hbar$ for both the cases) and the barrier does not vanish because these values of angular momentum, due to the general instabilities of the composite system against fission, lies well below the LDM angular momentum limit for which the fission barrier vanishes [11]. It is pointed out here that in an earlier study [8] barriers almost vanished at ℓ=ℓ_{max}(=23 $\hbar$ and 22 $\hbar$ for $^{32}S^*$ and $^{31}P^*$, respectively). It is relevant to mention here (as also given in the introduction) that this study was carried with the spherical considerations of the nuclei but within the description of Süsmann central radii, which incorporates some deformation effects.

Fig. 2 shows a comparative analysis of preformation probability P_0 plotted at two extreme values at $\ell = 0$ and ℓ_c for the decay of $^{32}S^*$ and $^{31}P^*$ formed in entrance channel reactions $^{20}Ne + ^{12}C$ and $^{19}F + ^{12}C$, having same excitation energy E^*_{CN}=60 MeV with deformations taken

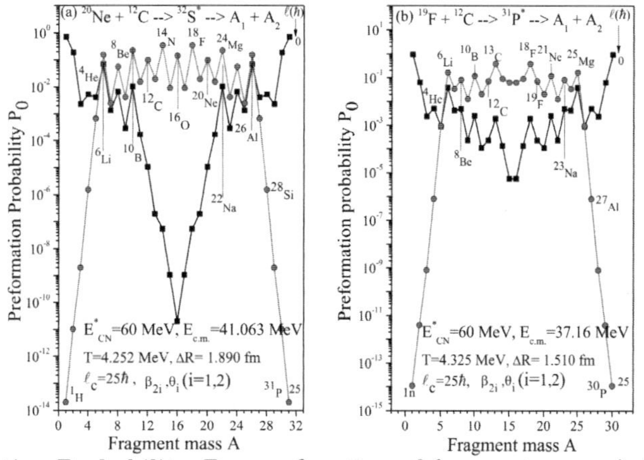

Figure 2. Preformation Probability P_0 as a function of fragment mass A for the decay of (a) ^{32}S*, and (b) ^{31}P*, at different ℓ-values.

upto β_2 and the hot optimum orientations. The preformation probability P_0 of the fragments (before tunnelling through the barrier) accounts for the structure effects in the decay process of the nuclear system. Here, we again observe, like an earlier study [8], that for ^{31}P* C-yield is very strongly preformed with A=13 (closely followed by A=12), whereas for ^{32}S* C-yield with A=12 is preformed with very strong competition from other fragments like ^{10}B and ^{14}N. In the calculation we have counted the contributions of σ_{FF} from masses A=11,12,13 and A=12,13 for the C-yield in the decay of ^{32}S* and ^{31}P*, respectively. Fig. 3 shows that the P starts

Table 1. The contributions of σ_{FF} and σ_{DIO} summed up to $l_c(\hbar)$ for the $C-yield$ (σ_C) from ^{32}S*. Whereas the same from ^{31}P* have only contribution from σ_{FF}.

					σ_C (mb)			
					DCM			
System	*Parameter*	ℓ_c ($\hbar$)	ΔR_{FF}(fm)	ΔR_{DIO}(fm)	FF	DIO	total	Expt.[5]
^{32}S*	$C_t + \Delta R$	22	1.35	0.949	49.6	72.1	121.7	122.6
^{31}P*		22	1.30	-	48	-	48	47.92±4.37
^{32}S*	$R_t + \Delta R$	27	1.685	1.34	72.8	49.1	121.9	122.6
^{31}P*	(Sph.)	26	1.460	-	50.4	-	50.4	47.92±4.37
^{32}S*	$R_t + \Delta R$	25	1.890	1.529	73.2	50.4	123.2	122.6
^{31}P*	(Ori.)	25	1.510	-	48.4	-	48.4	47.92±4.37

contributing at higher ℓ-values (it increases very fast upto higher ℓ-values). At ℓ_c it approaches to the maximum value near to one. But P_0 contribute at all ℓ-values, i.e., almost constant for all the ℓ-values.

Table 1 presents the contribution of σ_{FF} and σ_{DIO} for the C-yield, with the different spherical considerations along with that of oriented nuclei, from the compound system ^{32}S* and the only contribution of σ_{FF} from ^{31}P*. It is very important to point out here that the earlier calculations [8] for the σ_C, with the spherical considerations having Süsmann central

Journal of Physics: Conference Series **569** (2014) 012030 doi:10.1088/1742-6596/569/1/012030

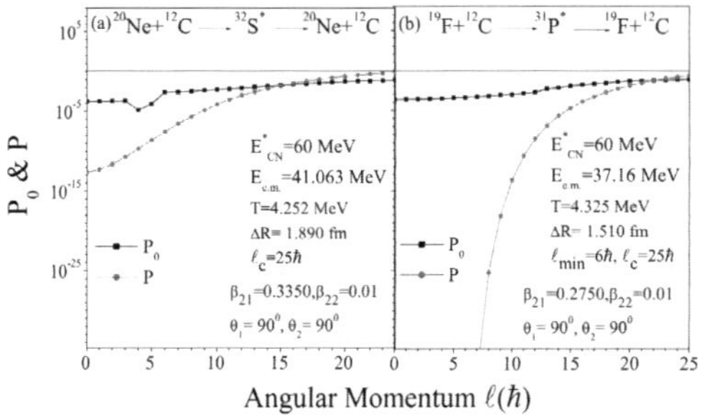

Figure 3. Preformation probability P_0 and Penetration probability P as a function ℓ for the ^{12}C decay of (a) ^{32}S* and (b) ^{31}P*.

radii $C_t + \Delta R$ incorporating deformation effects, have been improved further while adding the contributions (from σ_{FF} and σ_{DIO}) up to the ℓ=ℓ_c for the decay of ^{32}S*. Note that the calculated σ_C for the considerations of spherical and oriented nuclei, shows almost similar results with the only difference of the values of neck length parameter (ΔR) having highest values for the consideration of oriented nuclei. The calculated cross-sections show good agreement with experimental data for the different considerations. It will be highly interesting to further extend the the present study to investigate the effect of different excitation energies on the comparative contributions of σ_{FF} and σ_{DIO} in the C-yield of ^{32}S*.

4. Summary and Conclusions

Within the DCM, we have studied the decay of ^{32}S* and ^{31}P* at E^*_{CN}=60 MeV, having both the considerations of spherical as well oriented nuclei. The comparison of contributions of σ_{FF} and σ_{DIO} in the σ_C from ^{32}S* have been presented explicitly for these different considerations and the same for ^{31}P* system as well which is having purely CN decay, i.e., σ_{FF} contributions only. We further plan to extend the study to investigate the contributions of σ_{FF} and σ_{DIO} for C-yield from ^{32}S* at different excitation energies, as the experiment data for the same is also available. It is relevant to mention here that the calculations based on DCM not only successfully explains the dynamics of reactions under study, but also provide the very significant information about the quantitative contribution of the various processes in the decay of these very light mass compound systems having A$\sim$30.

Acknowledgement

One of us (B.B.S.) acknowledges the support by the Department of Science and Technology (DST), New Delhi, in the form of a Young Scientist Award under the SERC Fast Track Scheme, vide letter No. SR/FTP/PS-013/2011.

5. References

[1] Sanders SJ, et al., 1986 *Phys.Rev.* C **34** 1746.
[2] Sanders SJ, et al., 1991 *Phys.Rev.* C **44** 2676.
[3] Beck C, et al., 1992 *Z.Phys.* A **343** 309.
[4] Anjos RM, et al., 1994 *Phys.Rev.* C **49** 2018.
[5] Dey A, et al., 2007 *Phys.Rev.* C **76** 034.
[6] Gupta RK, et al., 2003 *Phys.Rev.* C **68** 014610; 2010 *Lecture Notes in Physics* 818, 223.
[7] Singh BB et al., 2006 *IJMP* E **15 No. 3** 699; 2008 *Phys.Rev.* C **77**, 054613; Proc. DAE Symp. on Nuc. Phys. 2011 **56** 474, 2012 **57** 550; 2013 **58** 380.
[8] Singh BB et al., **Fusion**14 Conf. Proc. EPJ web of Conferences (to be published).
[9] Gupta RK, et al., 2005 *Phys.Rev.* C **71** 014601.
[10] Beck C, et al., 1994 *Phys.Rev.* C **49** 2618; 1995 *Nucl.Phys.* A **583** 269.
[11] Beck C, et al., 1996 *Phys.Rev.* C **53** 1989.

Clustering in nuclear environment

G. Röpke
University of Rostock, Institut für Physik, 18051 Rostock, Germany
E-mail: gerd.roepke@uni-rostock.de

Abstract. The properties of few-body clusters (mass number $A \leq 4$) are modified if they are immersed in a nuclear medium. In particular, Pauli blocking that reflects the antisymmetrization of the many-body wave function is responsible for the medium modification of light clusters and the dissolution with increasing density. A more consistent description is given with takes also the contribution of correlations in the continuum into account. The relation between cluster formation in warm dense matter and in nuclear structure is discussed.

1. In-medium few-nucleon Schrödinger equation

We consider the propagation of A nucleons $[(A-Z)$ neutrons (n) and Z protons $(p)]$ in a nuclear medium described by the particle densities n_n and n_p as well as the temperature T. Using a thermodynamic Green function approach, the following in-medium wave equation can be derived (see [1])

$$[E_1^{\rm qu}(1) + \ldots + E_1^{\rm qu}(A) - E_{A\nu}^{\rm qu}(P)]\psi_{A\nu P}(1 \ldots A) \\ + \sum_{1' \ldots A'} \sum_{i<j} [1 - f(i) - f(j)] V(ij, i'j') \prod_{k \neq i,j} \delta_{kk'} \psi_{A\nu P}(1' \ldots A') = 0 \,. \tag{1}$$

This equation contains the effects of the medium in the quasiparticle shift $E_1^{\rm qu}(1)$ as well as in the Pauli blocking terms $[1 - f(i) - f(j)]$. The single-nucleon state $\{\vec{p}_1, \sigma_1, \tau_1\}$ is abbreviated by $i = 1$ etc. The single-nucleon quasiparticle shift $\Delta E^{\rm SE}(1) = E_1^{\rm qu}(1) - \hbar^2 p_1^2/2m$ is obtained from the single-nucleon self-energy which can be taken, for instance, in mean-field approximation. Standard expressions give a rigid shift in the energy and, if the momentum dependence is taken into account, the replacement of the nucleon mass m by an effective mass. The treatment of the single-nucleon shift in solving Eq. (1) is trivial [1] and will not be investigated here. In particular, the rigid energy shifts can be absorbed by a shift $\Delta E_{A\nu}^{\rm SE}(P) = \Delta E^{\rm SE}(1) + \ldots + \Delta E^{\rm SE}(A)$ of the cluster quasiparticle energy

$$E_{A\nu}^{\rm qu}(P) = \hbar^2 P^2/2Am + \Delta E_{A\nu}^{\rm SE}(P) + \Delta E_{A\nu}^{\rm Pauli}(P) + \Delta E_{A\nu}^{\rm Coul}(P) \tag{2}$$

and will not be considered here any more, the same also with the shift of the Coulomb energy. Note that the in-medium few-nucleon Schrödinger equation (1) describes the medium modification not only of possible bound states that may be considered as quasi-particles, but also the modification of the scattering states.

For exploratory calculations we use a simple separable interaction [2] with Gaussian form factor.

$$V(\mathbf{p}_1, \mathbf{p}_2; \mathbf{p}_1', \mathbf{p}_2') = \lambda e^{-(\mathbf{p}_1 - \mathbf{p}_2)^2/4\gamma^2} e^{-(\mathbf{p}_1' - \mathbf{p}_2')^2/4\gamma^2} \delta(\mathbf{p}_1 + \mathbf{p}_2 - \mathbf{p}_1' - \mathbf{p}_2') \,. \tag{3}$$

The parameter values can be fixed by experimental data. In the case of the deuteron ($A = 2$), with potential parameters $\lambda = -1287.4$ MeV fm^3 and $\gamma = 1.474$ fm^{-1} in Eq. (3), the binding energy and rms radius of the free deuteron are reproduced.

Our main concern are the effects of the Pauli blocking terms where the phase space occupation is described by the Fermi distribution function $f(i) = [\exp(\hbar^2 p_i^2/2mT - \mu_i/T) + 1]^{-1}$. Here, the chemical potential μ_τ is related to the nucleon density $n_{\tau_1} = (1/\text{volume}) \sum_{p_1,\sigma_1} f(1)$. In the case of zero temperature ($T = 0$), the Fermi function becomes the step function $\Theta(E_{\tau_1}^{\text{Fermi}} - \hbar^2 p_1^2/2m)$ where the chemical potential coincides with the Fermi energy $E_\tau^{\text{Fermi}} = (\hbar^2/2m)(3\pi^2 n_\tau)^{2/3}$. The Pauli blocking term leads to a shift in the energy $\Delta E_{A\nu}^{\text{Pauli}}(P)$ [2] and the in-medium modification of the wave function or the scattering phase shifts. The properties of bound states have been investigated elsewhere [1]. In particular we investigate the phase shifts and the consequences for the second virial coefficient. Different approaches are discussed.

2. In-medium phase shifts for $A = 2$

We first consider the (deuteron like) case $A = 2$. Furthermore we will use the Tamm-Dancoff (TD) expression $[1 - f(i) - f(j)] \to [1 - f(i)][1 - f(j)]$ which neglects the hole-hole contributions that are of relevance in deriving the gap equation if pairing is considered.

For the separable potential, the T-matrix can be solved. With the integral

$$I(E) = -\frac{\lambda}{2\pi^2} \int_0^\infty \mathrm{d}p \, \frac{p^2}{\hbar^2 p^2/m - E} \, \mathrm{e}^{-2p^2/\gamma^2} \, [1 - f(p)]^2 \tag{4}$$

we can now calculate the scattering phase-shift depending on the energy E of relative motion (zero angular momentum):

$$\tan[\delta_0(E)] = \frac{\operatorname{Im} T(12, 1'2', E)}{\operatorname{Re} T(12, 1'2', E)} = \frac{-\operatorname{Im} I(E)}{1 + \operatorname{Re} I(E)} \tag{5}$$

Note that for negative values of E, the solution of $1 + \operatorname{Re} I(-E_d) = 0$ gives the in-medium binding energy, for instance, of the deuteron (isospin singlet channel).

Obviously the bound state wave functions and energy eigenvalues $E_d(P, T, n_n, n_p)$ as well as the scattering phase shifts become dependent on temperature and density because of the Fermi distribution functions in the Pauli blocking term. We focus on the scattering phase shifts $\delta_0(E; P, T, n_n, n_p)$. For $P = 0$ and $n_n = n_p = n/2$, some results are shown in Fig. 1.

Having the scattering phase shifts to our disposal, we can calculate the second virial coefficient according to the generalized Beth-Uhlenbeck formula [4]. In addition to the bound state contribution $b_{^3S_1}^{\text{bound}}(0, T, n_n, n_p) = 3[\exp(E_d(0, T, n_n, n_p)/T) - 1]$ (P dependence of the in-medium binding energy neglected) we have the contribution of scattering states,

$$b_{^3S_1}^{\text{sc}}(P, T, n_n, n_p) = \frac{3}{\pi T} \int_0^\infty \mathrm{d}E \, \mathrm{e}^{-E/T} \left[\delta_0(E; P, T, n_n, n_p) - \frac{1}{2} \sin[2\delta_0(E; P, T, n_n, n_p)] \right] \tag{6}$$

If there exist a bound state (below the critical density $n_{\text{cr}}(T)$ where, e.g., $n_{\text{cr}}(0.1\,\text{MeV}) = 0.0016167$ fm^{-3}, $n_{\text{cr}}(5\,\text{MeV}) = 0.005472$ fm^{-3}), it is possible to combine the bound state and the scattering state contribution to the second virial coefficient extending the scattering phase shift to the bound state energy, see Fig. 1. There is a smooth behavior near the Mott point where the bound state is dissolved because of Pauli blocking. We give also a simple interpolation formula that can be used to estimate the contribution of the continuum correlations to the second virial coefficient and, furthermore, to the density fraction of correlated states:

$$b_{^3S_1}^{\text{sc}} \approx \frac{3[0.008709 \exp(-0.15403T)[0.3007 + 0.6662 \exp(-0.1014T)] + n^2]}{0.008709 \exp(-0.15403T) + n^2/[-0.1012 + 0.08611 \exp(0.09432T)] \exp[(26.3 + 483.8/T)n]} \tag{7}$$

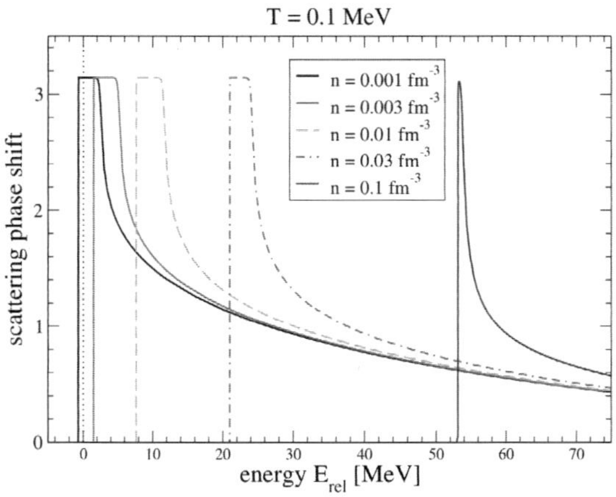

Figure 1. $T = 0.1$ MeV scattering phase shifts for different densities, TD blocking.

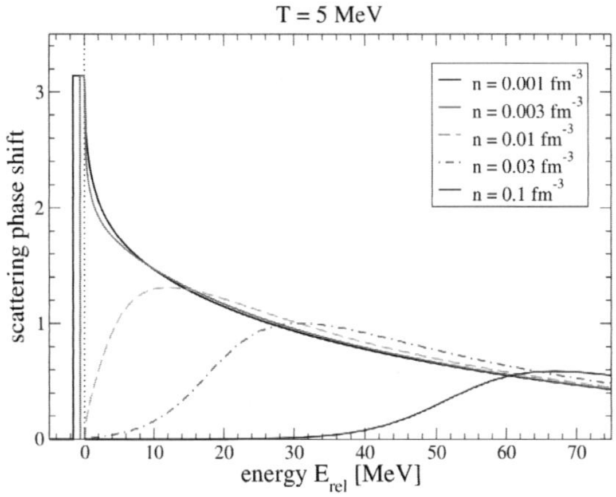

Figure 2. $T = 5$ MeV scattering phase shifts for different densities, TD blocking.

for $T > 1$ MeV. At lower temperatures, $b^{\rm sc}_{^3S_1} \propto \exp[-(\hbar^2/mT)(3\pi^2 n/2)^{2/3}]$.

An important feature is that at very low temperatures the phase shifts are constant, $\delta = \pi$, not only up to the edge of the continuum of scattering states but up to the Fermi energy (at $T = 0$, energy conserving nucleon-nucleon scattering for energies below $E_\tau^{\rm Fermi}$ is not possible because of Pauli blocking). Thus, the bound states merge with the continuum of scattering states at zero temperature only if the bound state energy per nucleon becomes larger than the Fermi energy.

At higher temperatures these sharp structures in the phase shifts shown in Fig. 1 are washed out, see Fig. 2. The bound state will be dissolved already when it merges with the continuum of scattering states ($E = 0$ when the self-energy rigid shifts are absorbed by the chemical potential).

Journal of Physics: Conference Series **569** (2014) 012031 doi:10.1088/1742-6596/569/1/012031

3. Particle-particle RPA and pairing for $A = 2$

Coming back to the so-called particle-particle Random-Phase Approximation (ppRPA) [3], the Pauli blocking term reads $[1 - f(i) - f(j)]$ which takes the hole-hole contributions into account. The in-medium phase shifts are shown in Fig. 3 for moderate temperatures of warm dense matter. At very low temperatures, sharp structures arise similar to the Tamm-Dancoff case, see Fig. 4. As in the Tamm-Dancoff case, we obtain a jump of the phase shifts at energies that are

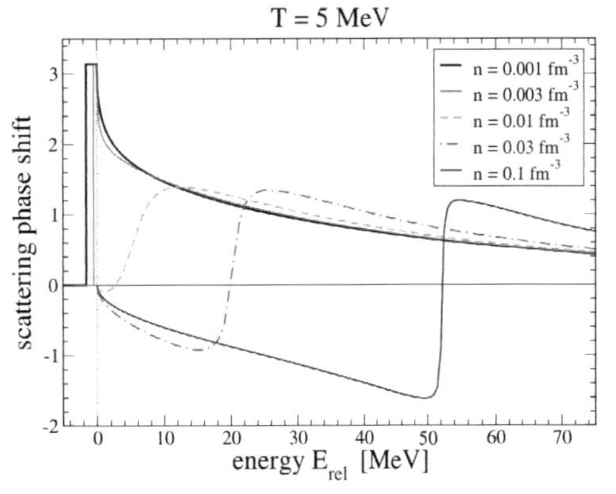

Figure 3. $T = 5$ MeV; ppRPA blocking.

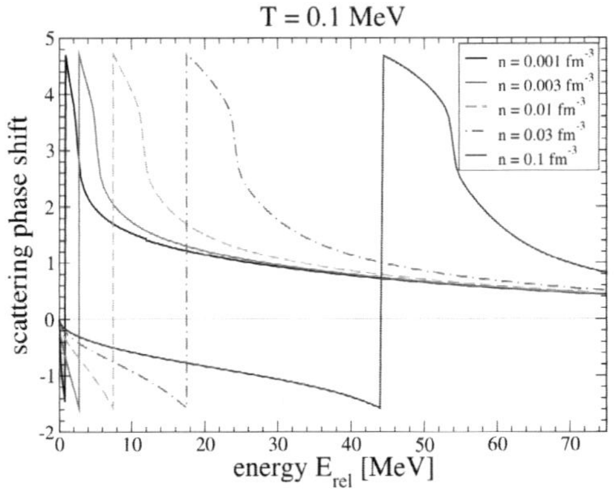

Figure 4. $T = 0.1$ MeV; ppRPA blocking.

near to the energies of bound state formation, but there is also a general negative shift what resembles the effect of repulsion.

We will not discuss here the consequences due to the use of different expressions for the Pauli blocking term (the critical densities are now $n_{\rm cr}(0.1\,{\rm MeV}) = 0.000235476$ fm^{-3}, $n_{\rm cr}(5\,{\rm MeV}) = 0.0045349$ fm^{-3}). The difference is the term $f(i)f(j) \propto n^2$ that is of higher order in the density, and one may argue that there are also further terms missing if we consider this order of density. However, there is a more fundamental difference between both expressions for the Pauli blocking. The Tamm-Dancoff expression is used, for instance, in the Brueckner theory to calculate the quasiparticle energy shifts of the nucleons. On the other hand, pairing may occur in nuclear matter at low temperatures. The corresponding Gorkov equation follows only if the ppRPA expression for the Pauli blocking is used. A detailed discussion is given in Ref. [4].

4. α-particle case $(A = 4)$

Because of the relatively high binding energy (28.3 MeV), the α-particle is of special interest in describing correlations and cluster formation in warm dense matter. The wave function of the free α-particle can be approximated solving the four-particle Schrödinger equation with a pair potential (3), where the potential parameters $\lambda = -1449.6$ MeV fm^3 and $\gamma = 1.152$ fm^{-1} are determined by the binding energy and the rms radius of the α-particle.

The medium modifications of the bound state energy and of the continuum states are of interest. Within a variational approach, results are obtained as described elsewhere [5]. As example, we give here the shift of the binding energy of the α-particle, see Fig. 5.

Using a variational approach, we considered the shift of the α particle in zero temperature matter [5]. Whereas at finite temperatures the bound state becomes dissolved already when the binding energy becomes zero, at zero temperature the bound state can disintegrate only when the free particle states are not Pauli blocked. Above the Fermi energy the decay is possible. This

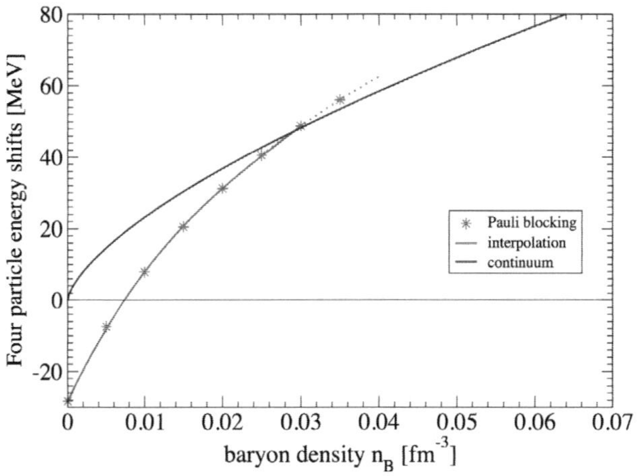

Figure 5. Internal four-nucleon energy (no c.o.m. motion) in a medium with nucleon density $n = n_n + n_p$. The continuum edge of free single-particle states is given by $4E_{\rm Fermi}$. At zero baryon density, the four-nucleon energy is given by the binding energy of the α particle, $B_\alpha^0 = 28.3$ MeV. With increasing density, the binding energy B_α^0 is decreasing due to the Pauli blocking. The four-nucleon bound state disappears at $n_B \approx 0.03$ fm^{-3}. A fit to the calculated values is given in Ref. [5].

way we can understand why at finite temperatures the Mott point is given by the zero binding energy condition, whereas at zero temperature the Fermi energy is of relevance. We conclude that at zero temperature the α-like bound state is not dissolved at $n \approx 0.01$ fm^{-3} where it merges in the continuum of unbound scattering states but at $n \approx 0.03$ fm^{-3} where it can decay into unbound, free states above the Fermi energy because they are not occupied. This value $n \approx 0.03$ fm^{-3} is in agreement with other approaches, in particular the fully antisymmetrized THSR formalism, see [6].

It should be mentioned that we may calculate the virial coefficient for the four-nucleon case by rescaling the two-nucleon result. Another important issue is the treatment of correlations in the medium. Already the ppRPA expression for the Pauli blocking leads to the formation of Cooper pairs in the medium. The THSR ansatz allows to treat four-nucleon correlations also in the medium, taking into account full antisymmetrization.

Acknowledgments
The author acknowledges the contributions of P. Schuck, Y. Funaki, H. Horiuchi, Zhongzhou Ren, A. Tohsaki, Chang Xu, T. Yamada, and Bo Zhou to this work.

References
[1] Röpke G 2009 *Phys. Rev.* C **79** 014002
[2] Röpke G 2011 *Nucl. Phys.* A **867** 66
[3] Ring P and Schuck P 1980 *The Nuclear Many-Body Problem* (Berlin: Springer)
[4] Schmidt M *et al.* 1990 *Ann. Phys.* **202** 57
[5] Röpke G *et al.* 2014 Bound clusters on top of doubly magic nuclei *Preprint* arXiv:1407.0510 [nucl.theory]
[6] Hiroki Takemoto *et al.* 2004 *Phys. Rev.* C **69** 035802

α clustering and flow in ultra-relativistic heavy-ion collisions

W Broniowski[1,2] and E Ruiz Arriola[2]

[1] Institute of Physics, Jan Kochanowski University, 25-406 Kielce, Poland
[2] H. Niewodniczański Institute of Nuclear Physics PAN, 31-342 Cracow, Poland
[3] Departamento de Física Atómica, Molecular y Nuclear and Instituto Carlos I de Física Teórica y Computacional, Universidad de Granada, E-18071 Granada, Spain

E-mail: Wojciech.Broniowski@ifj.edu.pl, earriola@ugr.es

Abstract. We show how ultra-relativistic collisions of light nuclei with heavy targets may be used to record snap-shots of the ground-state configurations and reveal information on cluster correlations. The development of collective flow in the formed fireball, which reflects the geometric correlations in the initial state, is essential for the method. As an illustration we analyze the ^{12}C-^{208}Pb collisions.

This talk is based on [1], where the relevant details can be found. We propose a novel way of investigating the ground-state correlations in light nuclei, which provides a surprising bridge between the lowest-energy nuclear structure and highest-energy nuclear reactions, where collective flow of the fireball develops. This flow transmutes the correlations in the initial state (such as those due to the α clusters) into specific measurable transverse-momentum asymmetries in the spectra of the produced hadrons.

While the concept of the α clustering is more than 80 years old [2] (for reviews see, e.g., [3, 4, 5, 6, 7, 8]), numerous issues still remain open. Even the ground state structure of light nuclei, such as ^{12}C, is a topic of active research (see various contributions to these proceedings). With this in mind, our method, inherently investigating multiparticle correlations leading to collective effects (flow) may provide a novel insight. We stress that since the clusterization phenomenon concerns multi-particle correlations (see, e.g., [9]), it is accessible directly only through observables which are many-body. Thus the typically studied one-body quantities, such as excitation spectra of the EM form factors, by definition cannot "prove" clusterization in a direct manner.

To model the collision process, we specifically need the distribution of centers of nucleons in ^{12}C, which is nothing but the ground state nuclear wave function squared, $|\Psi_A(\vec{x}_1, \ldots, \vec{x}_A)|^2$. While it would be best to incorporate realistic calculations (see, e.g., [13, 14]), in Ref. [1] we have applied a simple and practical procedure with α-clustered (or unclustered for comparison) random distributions. In the α-clustered case we randomly generate positions of the 12 nucleons, 4 in each cluster of a Gaussian shape and size r_α. The centers of the clusters are placed in an equilateral triangle of side length l. The short-distance NN repulsion is incorporated by precluding the centers of each pair of nucleons to be closer than the expulsion distance of 0.9 fm [15]. The parameters l and r_α are optimized such that the one particle density $\rho(r)$ of BEC [11] or FMD [12] calculations are accurately reproduced (standard unfolding the proton

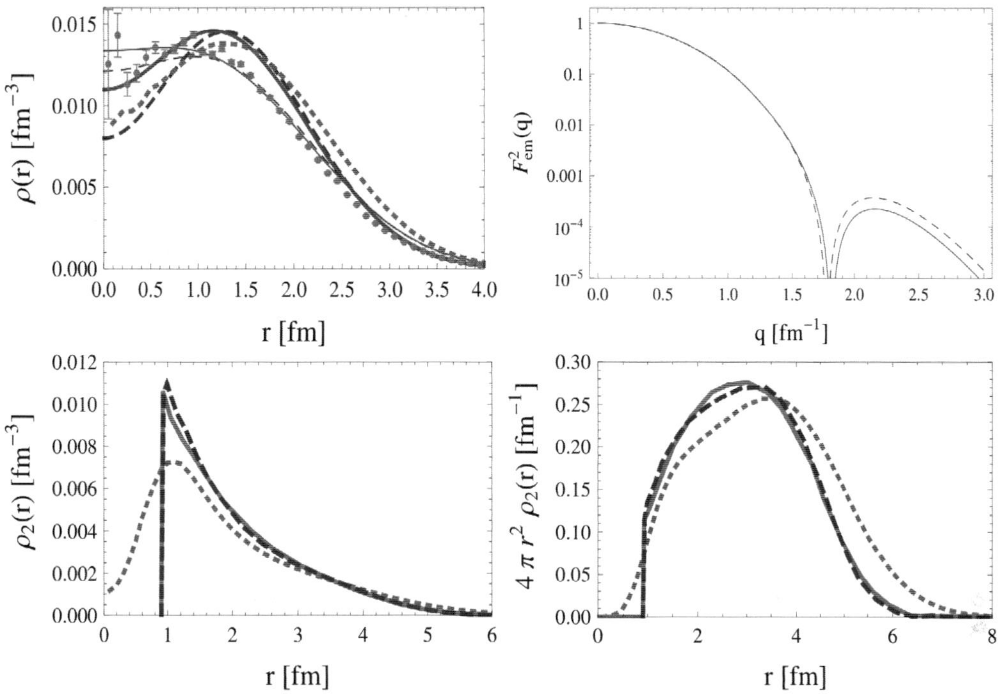

Figure 1. Upper left: Normalized one-particle distributions in ^{12}C. The electric charge density $\rho^{\rm em}(r)/Z$ (thin lines) and the corresponding distribution of the centers of nucleons $\rho_1(r)$ (thick lines) in ^{12}C for the data [10] and BEC [11] calculations (dashed lines), FMD [12] calculations (solid lines), and Jastrow correlated wave function [13] (dotted line). Upper right: Charge Form factor $F^{\rm em}(q)$ in ^{12}C as a function of the momentum transfer q (in fm^{-1}) for BEC (dashed line) and FMD calculations (solid line). Lower left: Normalized two particle distribution $\rho_2(r)$ in ^{12}C. We show our results for the fitted $\rho(r)$ in the FMD (thick line) and BEC cases (dashed line), and compare to the Jastrow correlated wave function [13] (dotted line). Lower right: The same as in lower left panel for the radial distribution $4\pi r^2 \rho_2(r)$.

charge density from the charge distribution $\rho^{\rm em}(r)$ is necessary), see Fig. 1. Note a large central depletion in the distributions, originating from the separation of the α clusters arranged in the triangular configuration. Besides, a fair reproduction of two particle densities $\rho_2(r)$ from multiclustered Jastrow correlated calculations [13] is observed. The radial distribution $4\pi r^2 \rho_2(r)$ peaks at the size of the triangle, $l \simeq 3$ fm. Thus, we deal with realistic nuclear distributions.

The essence of our approach is as follows: α clusters lead to substantial *intrinsic* deformation of nucleon distributions in light nuclei. Since the time of the reaction in ultrarelativistic collisions is much shorter from any characteristic time of the nuclear structure (the Lorentz contraction factor for the colliding nuclei reaches 100 at RHIC!), an instant snapshot of the frozen light nucleus configuration is made, revealing the lumpy structure when present. Consequently, when a geometrically deformed nucleus hits a large target at almost the speed of light, the created fireball in the transverse plane inherits the shape of the light nucleus (cf. Fig. 2-4). With a large target, the created fireball is abundant enough to evolve collectively, in full analogy to ultrarelativistic collisions of two heavy nuclei as studied recently in colliders (RHIC, LHC) or fixed-target experiments (SPS). Due to the initial fireball deformation, a deformed flow pattern develops, leading to azimuthal asymmetry in the transverse momentum distributions of the hadrons (mostly pions) produced in the collision. This asymmetry can be analyzed and measured event-by-event through well-established methods [16, 17, 18]. Here we focus on the promising $^{12}C+^{208}Pb$ system, as the ^{12}C has a large intrinsic triangularity [19, 12], resulting in large triangular flow, increasing strongly with the multiplicity of the produced hadrons [1].

In our study we apply GLISSANDO [20, 21] to model the early phase of the collision within the Glauber Monte Carlo approach. The eccentricity parameters ϵ_n are convenient measures of the harmonic components of the intrinsic deformation. They are defined for each collision (event) as the coefficients of the Fourier decomposition of the distribution in the transverse plane,

$$\epsilon_n e^{in\Phi_n} = \frac{\sum_j w_j \rho_j^n e^{in\phi_j}}{\sum_j w_j \rho_j^n}, \tag{1}$$

where j labels the *sources* in the event, ρ_j is the transverse position of the source, w_j its weight, n indicates the rank, and, finally, Φ_n is the principal axis angle in the event. The $n = 2$ deformation is referred to as *ellipticity*, and $n = 3$ as *triangularity*. The notion of the *source* is used in the Glauber-model sense, and indicates the wounded nucleons [22] or binary collisions.

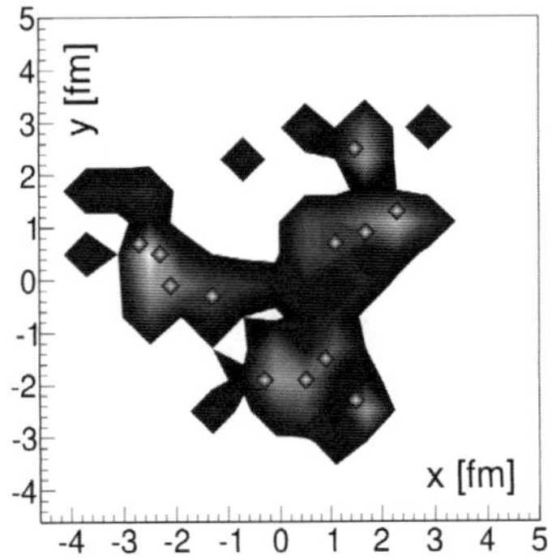

Figure 2. Fireball created in a ^{12}C-^{208}Pb collision, viewed in the transverse plane. The original location of the nucleons in ^{12}C is indicated by small diamonds. The fireball is formed from the collisions of the projectile and target nucleons within the Glauber model. The triangular arrangement of the three α clusters is reflected in the triangular shape of the fireball, warped to some extent with fluctuations.

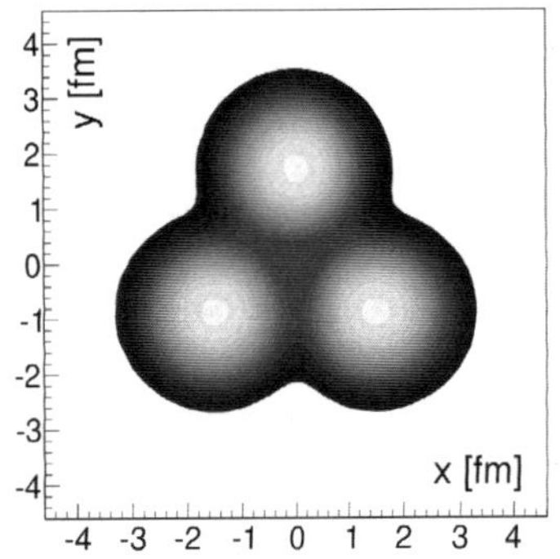

Figure 3. Average intrinsic triangular distribution of ^{12}C projected on the transverse plane.

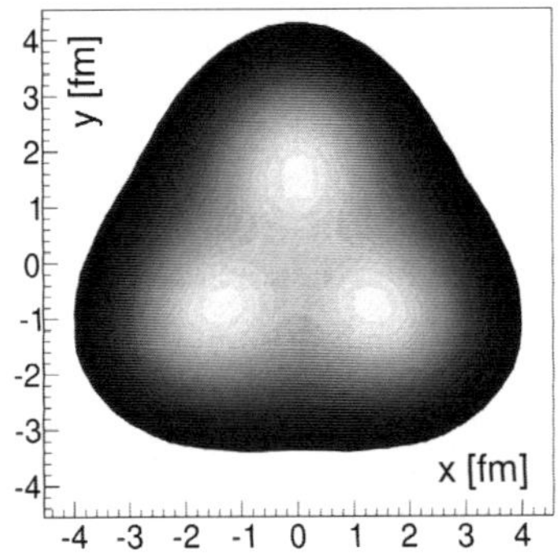

Figure 4. Corresponding average intrinsic transverse density of the fireball created in a flat-on ^{12}C-^{208}Pb collision.

Figure 5. Flat-on (left) and side-wise (right) orientations of the ^{12}C nucleus in collisions with ^{208}Pb. The former leads to a larger multiplicity, larger triangularity, and smaller ellipticity, while the latter, on the opposite, has smaller multiplicity, smaller triangularity, and larger ellipticity.

The collective dynamics of the intermediate evolution of an ultra-relativistic nuclear reactions is expected to be properly modeled with hydrodynamics (for recent reviews see, e.g., [23, 24] and references therein) or transport models [25]. These methods lead to an event-by-event transmutation of the initial-state anisotropies, quantified with the ϵ_n coefficients, into the harmonic flow coefficients of the transverse momentum distributions of the produced particles, known as v_n [16, 17, 18]. Since it is well established that the proportionality $v_n \sim \epsilon_n$ holds (see, e.g., a recent analysis [26]), one may gain the relevant information on the physically relevant v_n coefficients by studying the eccentricities ϵ_n, which is way simpler, as no costly hydrodynamic or transport studies are needed (such full-fledged studies are underway).

Before showing our key results (here shown for the distributions reproducing the BEC case), let us present a very specific and crucial for understanding correlation between the "geometry" and the multiplicity of particles produced in the collision (which is proportional to the number of the wounded nucleons, N_w). The orientation of the "little triangle" describing the ^{12}C nucleus with respect to the transverse plane is random in each event; sometimes the collision is flat-on, sometimes side-wise, or assumes any intermediate angle (cf. Fig. 5). In the flat-on case the damage created by the projectile in the target is largest, as the geometric cross section is highest (at ultra-relativistic energies the projectile nucleon wounds everything in its straight path!). At the same time, triangularity is highest and the ellipticity is lowest. For the side-wise orientation, the effects are opposite. Thus we find positive correlation between multiplicity and triangularity, and negative correlations between multiplicity and ellipticity.

This simple quantitative effect is clearly seen in actual simulations with GLISSANDO, presented in the left panel of Fig. 6. The displayed growth of the event-averaged ϵ_3 and the falloff of the event-averaged ϵ_2 are the advocated signatures of the α clusterization in ^{12}C. The case of the unclustered ^{12}C (i.e., with the nucleons distributed uniformly but with exactly the same one-body density as in the clustered case) is shown in the right panel of Fig. 6. We note a very similar behavior of $\langle\epsilon_2\rangle$ and $\langle\epsilon_3\rangle$ as functions of N_w. The fact that these are non-zero is due entirely to fluctuations [26] of the uniform system with a finite number of sources. In addition, we show the scaled standard deviations of the event-by-event values of ϵ_3 and ϵ_2. These measures are also sensitive to clusterization.

We note that the presented analysis is similar in spirit to the studies of the d-A collisions [27] and ^{3}H/^{3}He-A collisions [28] (to be investigated experimentally at RHIC), however, the collective effects in a larger system, such as in ^{12}C-^{208}Pb presented here, are expected to be much stronger, hence the shape-flow transmutation should be more visible. In particular, the separation of the geometric component of the flow from the non-flow effects due to fluctuations should be easier.

In summary, our method, which links the cluster features of lowest-energy nuclear structure physics with flow phenomena known from the highest-energy nuclear collisions, offers a novel possibility to study geometric correlations in the ground state. Viewed from the opposite direction, a detailed knowledge of the clustered nuclear wave functions (obtained, e.g., from the ab initio numerical analyses) will help to place important constraints on the models of the fireball evolution and thus gain information on the properties of the quark-gluon plasma.

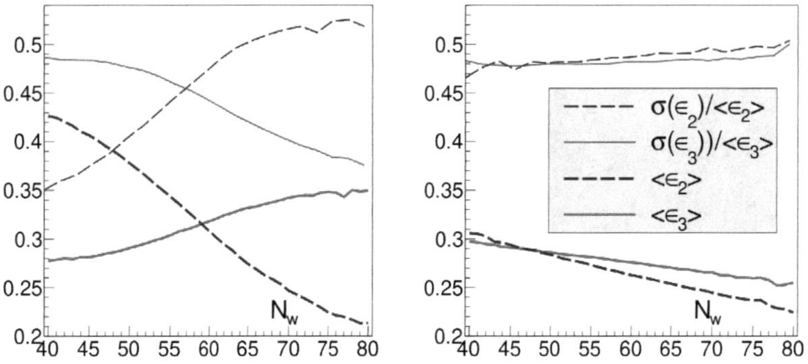

Figure 6. Event-by-event average ellipticity and triangularity of the fireball, as well as their scaled standard deviations, plotted as functions of the number of wounded nucleons [22], N_w. Left: the case where the ^{12}C nucleus is clustered according to the BEC model. Right: uniform distribution (no clusters). SPS energies ($\sigma_{NN}^{\mathrm{inel}} = 32$ mb), mixed Glauber model [1].

Acknowledgments
This research was supported by the Polish National Science Centre, grants DEC-2012/05/B/-ST2/02528 and DEC-2012/06/A/ST2/00390, Spanish DGI (grant FIS2011-24149) and Junta de Andalucía (grant FQM225).

References
[1] Broniowski W and Arriola E R 2014 *Phys.Rev.Lett.* **112** 112501 (*Preprint* 1312.0289)
[2] Gamow G 1931 *Constitution of atomic nuclei and radioactivity* (Oxford University Press)
[3] Brink D 1965 *Proc. Int. School Enrico Fermi, Course* **36**
[4] Freer M 2007 *Reports on Progress in Physics* **70** 2149
[5] Ikeda K, Myo T, Kato K and Toki H 2010 *Clusters in Nuclei-Vol.1* (Lecture Notes in Physics **818**, Springer)
[6] Beck C 2012 *Clusters in Nuclei-Vol.2* (Lecture Notes in Physics **848**, Springer)
[7] Okołowicz J, Nazarewicz W and Płoszajczak M 2013 *Fortsch.Phys.* **61** 66–79 (*Preprint* 1207.6225)
[8] Zarubin P I 2014 *Clusters in Nuclei-Vol.3* (Springer)
[9] Viollier R and Walecka J 1977 *Acta Phys.Polon.* **B8** 25
[10] De Vries H, De Jager C and De Vries C 1987 *Atom.Data Nucl.Data Tabl.* **36** 495–536
[11] Funaki Y, Tohsaki A, Horiuchi H, Schuck P and Ropke G 2006 *Eur.Phys.J.* **A28** 259–263
[12] Chernykh M, Feldmeier H, Neff T, von Neumann-Cosel P and Richter A 2007 *Phys.Rev.Lett.* **98** 032501
[13] Buendia E, Galvez F and Sarsa A 2004 *Phys.Rev.* **C70** 054315 (*Preprint* nucl-th/0405027)
[14] Wiringa R, Schiavilla R, Pieper S C and Carlson J 2014 *Physical Review C* **89** 024305 (*Preprint* 1309.3794)
[15] Broniowski W and Rybczyński M 2010 *Phys. Rev.* **C81** 064909 (*Preprint* 1003.1088)
[16] Ollitrault J Y 1992 *Phys. Rev.* **D46** 229–245
[17] Borghini N, Dinh P M and Ollitrault J Y 2001 *Phys.Rev.* **C64** 054901 (*Preprint* nucl-th/0105040)
[18] Voloshin S A, Poskanzer A M and Snellings R 2010 *Landolt-Boernstein,* Relativistic Heavy Ion Physics, *vol. 1/23* (Springer-Verlag) p 5
[19] Kanada-En'yo Y 2007 *Prog.Theor.Phys.* **117** 655–680 (*Preprint* nucl-th/0605047)
[20] Broniowski W, Rybczyński M and Bożek P 2009 *Comput. Phys. Commun.* **180** 69 (*Preprint* 0710.5731)
[21] Rybczyński M, Stefanek G, Broniowski W and Bożek P 2013 (*Preprint* 1310.5475)
[22] Białas A, Błeszyński M and Czyż W 1976 *Nucl. Phys.* **B111** 461
[23] Heinz U and Snellings R 2013 *Ann.Rev.Nucl.Part.Sci.* **63** 123–151 (*Preprint* 1301.2826)
[24] Gale C, Jeon S and Schenke B 2013 *Int.J.Mod.Phys.* **A28** 1340011 (*Preprint* 1301.5893)
[25] Lin Z W, Ko C M, Li B A, Zhang B and Pal S 2005 *Phys.Rev.* **C72** 064901 (*Preprint* nucl-th/0411110)
[26] Bzdak A, Bożek P and McLerran L 2014 *Nucl.Phys.* **A927** 15–23 (*Preprint* 1311.7325)
[27] Bożek P 2012 *Phys.Rev.* **C85** 014911 (*Preprint* 1112.0915)
[28] Sickles A M (PHENIX Collaboration) 2013 (*Preprint* 1310.4388)

Structure and decay correlations of two-neutron systems beyond the dripline

Z. Kohley[1,2,*], T. Baumann[1], D. Bazin[1], G. Christian[1,3], P. A. DeYoung[4], J. E. Finck[5], R.A. Haring-Kaye[6], J. Hinnefeld[7], N. Frank[8], E. Lunderberg[4], B. Luther[9], S. Mosby[1,3], W. A. Peters[1,3], J. K. Smith[1,3], J. Snyder[1,3], S.L. Stephenson[10], M. J. Strongman[1,3], A. Spyrou[1,3], M. Thoennessen[1,3], and A. Volya[11]

[1]National Superconducting Cyclotron Laboratory, Michigan State University, East Lansing, Michigan 48824, USA
[2]Department of Chemistry, Michigan State University, East Lansing, Michigan 48824, USA
[3]Department of Physics & Astronomy, Michigan State University, East Lansing, Michigan 48824, USA
[4]Department of Physics, Hope College, Holland, Michigan 49423, USA
[5]Department of Physics, Central Michigan University, Mt. Pleasant, Michigan 48859, USA
[6]Department of Physics and Astronomy, Ohio Wesleyan University, Delaware, Ohio 43015, USA
[7]Department of Physics and Astronomy, Indiana University at South Bend, South Bend, Indiana 46634, USA
[8]Department of Physics & Astronomy, Augustana College, Rock Island, Illinois 61201, USA
[9]Department of Physics, Concordia College, Moorhead, Minnesota 56562, USA
[10]Department of Physics, Gettysburg College, Gettysburg, Pennsylvania 17325, USA
[11]Department of Physics, Florida State University, Tallahasee, Florida 32306, USA

E-mail: *kohley@nscl.msu.edu

Abstract. The two-neutron unbound systems of ^{16}Be, ^{13}Li, ^{10}He, and ^{26}O have been measured using the Modular Neutron Array (MoNA) and 4 Tm Sweeper magnet setup. The correlations of the 3-body decay for the ^{16}Be and ^{13}Li were extracted and demonstrated a strong correlated enhancement between the two neutrons. The measurement of the ^{10}He ground state resonance from a ^{14}Be$(-2p2n)$ reaction provided insight into previous predictions that wavefunction of the entrance channel, projectile, can influence the observed decay energy spectrum for the unbound system. Lastly, the decay-in-target (DiT) technique was utilized to extract the lifetime of the ^{26}O ground state. The measured lifetime of $4.5^{+1.1}_{-1.5}$ $(stat.) \pm 3(sys.)$ ps provides the first indication of two-neutron radioactivity.

1. Introduction

The addition or removal of neutrons from stable isotopes has been shown in many cases to drastically alter the structure and properties of the nucleus [1–6]. Radioactive-ion beam facilities have made it possible to produce and study these exotic nuclei existing far from stability. Furthermore, techniques such as invariant-mass spectroscopy can allow for the study of nuclei that exist beyond the driplines providing a view into systems with the most extreme neutron-to-proton ratios (N/Z) possible.

These exotic dripline nuclei can exhibit unique types of decay which can offer additional insight into the structure of the nucleus. For example, Goldansky was the first to propose that the simultaneous emission of two protons could be observed given a scenario in which the intermediate state was positioned above the initial state [7]. Studies of two-proton decays have confirmed the presence of the direct two-proton decay mechanism and also presented more complex mechanisms, in some cases termed democratic decays, in which the intermediate state is very broad [8]. Therefore, the decay is not truly sequential or direct. Insight into the decay mode and wavefunction of the two-proton unbound system can be accessed through comparison of the measured and theoretical 3-body correlations [6, 9]. This provides a powerful tool for exploring nuclei beyond the proton dripline.

The discovery of the two-neutron halo system, ^{11}Li, initiated significant interest in the correlations within the 3-body system ($n+n+core$) [10]. Theoretical calculations have indicated that there is a strong dineutron component to the wavefunction of the two-neutron halo nuclei. Experimental measurements of the low-lying dipole strength distribution of ^{11}Li [11] and the ^{9}Li momentum distribution following a two-neutron knockout from ^{11}Li [12] provide support for a dineutron-like configuration. Two-neutron unbound systems provide a unique system in which the correlations of the neutrons from the ground state decay can be measured. The MoNA collaboration has developed a program to explore nuclei existing two neutrons beyond the dripline. Much like the unbound two-proton systems, measurements of the 3-body decay correlations should provide a connection to the wavefunction of the two-neutron unbound nucleus. In the following proceedings, the results from the MoNA collaboration measurements of the ^{16}Be [13], ^{13}Li [14], ^{10}He [15], and ^{26}O [16, 17] two-neutron unbound systems are presented.

2. Experiment

The Coupled Cyclotron Facility at the National Superconducting Cyclotron Laboratory at Michigan State University was used to produce radioactive ion beams at energies between 50-80 MeV/u. The two-neutron unbound nuclei of interest were then populated from reactions with the radioactive beams. The details of each experiment discussed in the proceedings are presented in Table 1.

Table 1. Details of the experiments for the measurements of the two-neutron unbound nuclei.

Unbound nucleus	Primary beam	Secondary beam	Reaction target	Secondary beam rate	Ref.
^{16}Be	120 MeV/u ^{22}Ne	53 MeV/u ^{17}B	470 mg/cm^2 ^{9}Be	250 pps	[13]
^{13}Li	120 MeV/u ^{18}O	53.6 MeV/u ^{14}Be	477 mg/cm^2 ^{9}Be	500 pps	[14]
^{10}He	120 MeV/u ^{18}O	59 MeV/u ^{14}Be	435 mg/cm^2 CD_2	1000 pps	[15]
^{26}O	140 MeV/u ^{48}Ca	82 MeV/u ^{27}F	705 mg/cm^2 ^{9}Be	14 pps	[16, 17]

Coincident detection of the two neutrons and the residual fragment from the decay of the two-neutron unbound system was accomplished using the modular neutron array (MoNA) [18, 19] and the large gap 4 Tm sweeper magnet [20]. The large area neutron detector, MoNA, is composed of 144 plastic scintillator bars. Light guides and photomultiplier tubes are attached to the ends of each bar for detection of the light produced from the interaction of the neutron(s).

The detector bars measure 200 cm × 10 cm × 10 cm each and the array is typically configured with 9 walls, each 16 bars high. The time-of-flight (ToF) of the neutrons to MoNA is measured with respect to a scintillator placed in front of the target. The angle and energy of a neutron can be determined from the interaction point in the detector bar and the ToF, respectively. The sweeper magnet bends the charged particles ∼43° into a suite of position sensitive charged-particle detectors. Both the particle identification and kinematical properties of the charged fragments can be determined from the sweeper magnet detectors (see Ref. [21] for more details). From the measured energy and angle of the neutrons and charged particle, the invariant mass of the 3-body system can be calculated. A detailed Monte Carlo simulation has been developed to simulate the production and decay of the unbound nuclei. While a general discussion of the Monte Carlo simulation is provided in Refs. [21, 22], the simulation of the neutron interactions in MoNA is of particular importance. The MoNA simulation is built upon the GEANT4 framework [23, 24], which allows for the tracking of each neutron interaction throughout the array. A custom neutron interaction model, referred to as MENATE_R [25, 26], was incorporated into the GEANT4 framework and allowed for discrete inelastic neutron-carbon interactions to be simulated based on experimental cross sections [22]. The inclusion of MENATE_R demonstrated a drastic improvement in the ability of the simulation to reproduce experimental neutron scattering [22]. In the following, all the experimental results have had a "causality cut" applied to the data to isolate events in which the first two interactions in MoNA represent true two neutron events. The causality cuts are based on the relative distance and velocity of the first two hits in MoNA (additional details about the cuts for each experiment can be found in Refs. [13–17, 22]).

3. 3-body correlations in the decay of ^{16}Be and ^{13}Li

Measurements of the ground state resonances for ^{16}Be and ^{13}Li indicated that both systems should exhibit a direct two-neutron decay since a sequential decay would be energetically forbidden due to the location of the intermediate states [14, 27, 28]. This offers a unique opportunity to examine the correlations of the two neutrons that are emitted directly from the ground state of the nucleus. The 3-body correlations are described within the Jacobi coordinate system defined in Fig. 1. A complete description of the three-body correlations can be obtained from the relative energy (E_x/E_T) and the angle (θ_k) calculated within the **T** and **Y** Jacobi systems [29, 30]. The relative energy is defined as the energy of the two-body system (frag+n or n+n), E_x, relative to the total three-body energy, E_T.

In Fig. 2 the relative energy and angle for both the **T** and **Y** systems are shown from the experimental data of the ^{13}Li and ^{16}Be ground state decays. The 3-body correlations of the

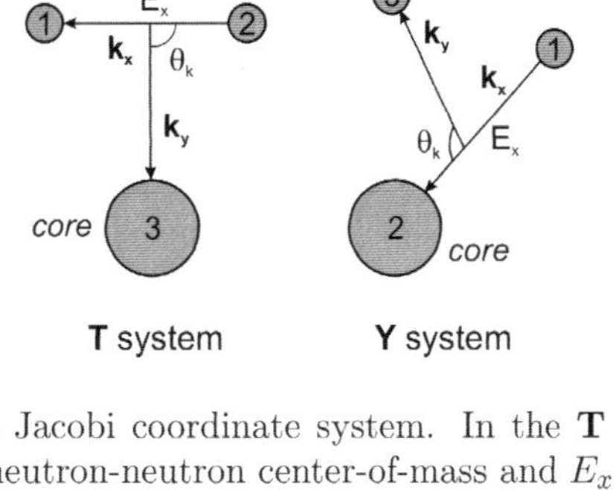

Figure 1. Depiction of the Jacobi coordinate system. In the **T** system θ_k is defined as the angle between the core and neutron-neutron center-of-mass and E_x is the relative energy of the neutron-neutron system. In the **Y** system θ_k is defined as the angle between the neutron-core and the other neutron and E_x is the relative energy of the neutron-core system.

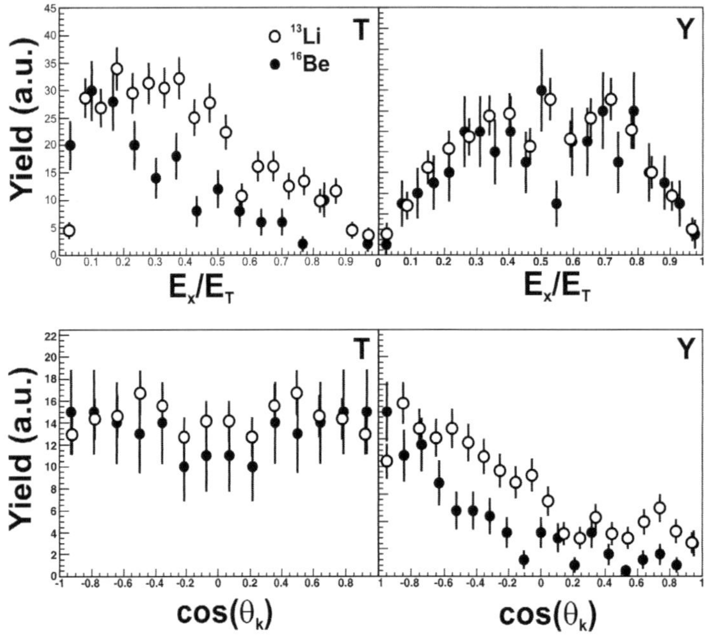

Figure 2. The relative energy (E_x/E_T) and angle ($\cos(\theta_k)$) as defined in the **T** and **Y** systems are shown for the experimentally measured decay of ^{13}Li (open circles) and ^{16}Be (closed circles).

two systems show a striking similarity with nearly identical features throughout the Jacobi coordinate plots. Of particular interest is the relative energy in the **T** system and $\cos(\theta_k)$ in the **Y** system which define the energy of the neutron-neutron system and the angle of each neutron relative to the neutron-core system, respectively. Therefore, these two observables are sensitive to the correlations between the neutrons. For both the ^{13}Li and ^{16}Be there is an enhancement in events with low E_x/E_T in the **T** system and for events with $\cos(\theta_k)$ near -1. These results indicate that the emitted neutrons are correlated in both energy and angle.

Further insight into the measured correlations can be garnered from the comparison with simulations of different decay modes. In Fig. 3, the experimental relative energy distribution from the **T** system, corresponding to the neutron-neutron energy, is shown in comparison to simulations for a 3-body phase-space [31, 32] and dineutron decay [14]. While the two simulations represent extreme cases, where the neutrons have no interaction (phase-space) and where the neutrons are emitted as a pair (dineutron), they can provide a general indication as to how the emission occurs. Both the ^{16}Be and ^{13}Li data are well reproduced by the dineutron decay, while the 3-body phase-space simulation significantly underpredicts the low energy portion of the distributions. The neutron-neutron energy from the dineutron simulation is governed by the nn scattering length (a_s). Simulations with the typical nn-scattering length of -18.7 fm and an increased value of -100 fm are shown. Particularly in the ^{16}Be case, the larger scattering length (increased neutron-neutron correlation) provides a better description of the data.

While the results demonstrate a strongly correlated emission of the neutrons, a full 3-body theoretical calculation is required to make the connection between the wavefunction of the unbound system and the correlation observed in the decay. It cannot be assumed that the observation of the dineutron-like decay is evidence of the existence of a dineutron in the ground state of the unbound system. However, theoretical calculation for dripline nuclei, such as ^{11}Li and ^{18}C [33, 34], as well as unbound nuclei, such as ^{26}O [35], do predict that the ground state wavefunction would contain a strong dineutron component.

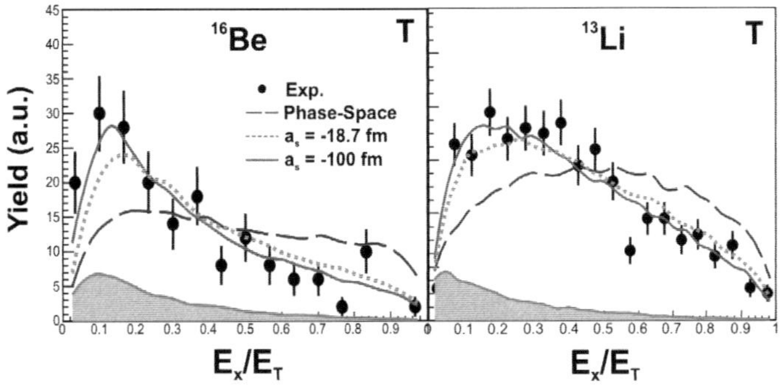

Figure 3. (Color online) Relative energy distribution, E_x/E_T, in the **T** system from the measured ^{16}Be and ^{13}Li decays. Simulations for the 3-body phase space and dineutron decay (represented by the nn-scattering length, a_s) are shown as the dashed and solid lines. The grey region represents the component of the distribution, determined from the dineutron simulation, that is from false two neutron hits.

4. Entrance-channel effects in population of ^{10}He

Recently, Grigorenko *et al.* suggested that the measured energy of a neutron unbound state energy can be "shifted" relative to the true energy of the state in cases where the state is populated from halo nuclei [36–39]. The argument is that the time-scale for the population and decay of a broad unbound state is so short that a component of the initial extended wavefunction of the halo projectile is observed in the outgoing channel. Therefore, the measured decay energy spectrum is modified due to the entrance channel wavefunction. If this theory can be verified, it would have profound impact on previous and future measurements of unbound neutron-rich nuclei populated from halo nuclei.

The two-neutron unbound ^{10}He nucleus has become the "test case" for this theory since it has been populated and measured using different reaction mechanisms, as shown in Fig. 4, and has a large resonance width ($\Gamma \sim$1.8 MeV). The determination of the ground-state energy of ^{10}He has sparked some controversy due to inconsistent experimental measurements. At the JINR in Dubna, the missing mass spectrum of ^{10}He was measured using a ^{8}He(t,p) transfer reaction by Golovkov in 2009 [37] and an updated measurement was completed by Sidorchuk in 2012 [38]. The results of these experiments (shown as solid stars in Fig. 4) indicate a ^{10}He ground state resonance above 2 MeV. In contrast, the GSI-LAND group measured the ^{10}He invariant mass using a 1-proton knockout from ^{11}Li and reported a g.s. resonance energy of 1.54 MeV [40]. This measurement along with other ^{11}Li$(-p)$ experiments are shown in Fig. 4 as open circles. An extended discussion of the different measurements can be found in Ref. [15]. A large discrepancy of about 500 keV remains between the transfer reaction measurements and the one-proton knockout reactions from ^{11}Li. This discrepancy was reconciled through the theoretical calculations of Grigorenko and Zhukov, which showed that the observed peak in the ^{10}He spectrum depends on the reaction mechanism and source size or wavefunction of the projectile [36]. Thus, the observed energy spectrum from the ^{11}Li$(-p)$ is shifted down in energy, while the mechanism of the transfer reaction does not perturb the observed energy spectrum.

The MoNA collaboration measured the ^{10}He ground state resonance using a ^{14}Be$(-2p2n)$ reaction and proposed that this would not exhibit a shifted energy spectrum due to the more complex $2p2n$-removal reaction [15]. Therefore, the ^{10}He g.s. measured from the ^{14}Be($-2p2n$) reaction should be around 2 MeV if the predictions of Grigorenko and transfer reaction measurements were correct. Instead, the ^{10}He g.s. measured from the ^{14}Be($-2p2n$) was observed at 1.6 MeV (see Fig. 4). This suggests that the measured unbound resonance was not dependent on the reaction mechanism or incoming wavefunction. However, new theoretical calculations have shown that the halo component of the ^{14}Be nucleus, even in a $2p2n$-removal, could also create a shift in the ^{10}He energy spectrum [39]. Thus, the situation is not resolved as to whether the observed decay energy spectra of neutron unbound nuclei, specifically ^{10}He, can be modified by unique reaction mechanisms involving halo nuclei. It is worth noting that Fortune proposed another explanation for differences observed in the ^{10}He g.s. measurements based on the prediction that ^{10}He has two low-energy 0^+ states [41]. Fortune suggests that the differences in the measured g.s. energies could be due to different relative populations of the ground and excited 0^+ states which would be dependent on the p and sd components of the projectile wavefunctions. Further investigation of this open question is needed.

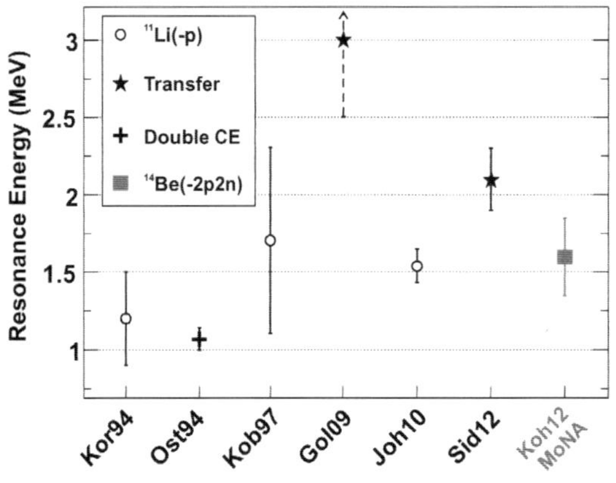

Figure 4. (Color online) Compilation of experimental measurements of the ^{10}He ground state resonance. The reaction mechanism used to populate ^{10}He is described in the legend. The associated references for each entry are: Kor94 [42], Ost94 [43], Kob97 [44], Gol09 [37], Joh10 [40, 45], Sid12 [38], and Koh12 [15].

5. Evidence of two-neutron radioactivity in ^{26}O

The ground state of the two-neutron unbound ^{26}O was measured for the first time at the NSCL using a one-proton knockout reaction from a ^{27}F beam. The ground state decay energy was constrained to be less than 200 keV [16]. Further experimental results from GSI-LAND have constrained this value to <120 keV [46]. Theoretical predictions had indicated that if ^{26}O had a purely $[d^2]$ configuration that the lifetime for such a low decay energy could be on the order of picoseconds [47]. This suggested that ^{26}O could provide the first case for the discovery of two-neutron radioactivity.

A new technique to determine the half-life was developed, termed Decay in Target (DiT), in which the relative velocity between the ^{24}O fragment and two neutrons was used as a probe for the distance the ^{26}O fragment traveled within the target [48]. Assuming a more typical unbound system, the decay should occur immediately ($\sim 10^{-21}$s), as shown in the top of Fig. 5(a). If the ^{26}O has a finite lifetime then the ^{26}O would traverse through part of the target before decaying (bottom of Fig. 5(a)). Therefore, the relative velocity between the ^{24}O and two neutrons should be dependent on the lifetime of ^{26}O. The relative velocity spectrum (V_{rel}) is shown in Fig. 5(b) compared with simulations for how the shape and location of the distribution would change as a function of lifetime. The results indicate that a 0 ps decay time is not the best fit. The distribution is shifted away from $V_{rel} = 0$ and is better reproduced by a lifetime around 4 ps. An unbinned maximum likelihood technique was employed to determine the statistical significance of the results [21, 49] and the reported lifetime of ^{26}O was $4.5^{+1.1}_{-1.5}$ $(stat.) \pm 3(sys.)$ ps. This represents the first evidence for new mode of radioactive decay: two-neutron radioactivity.

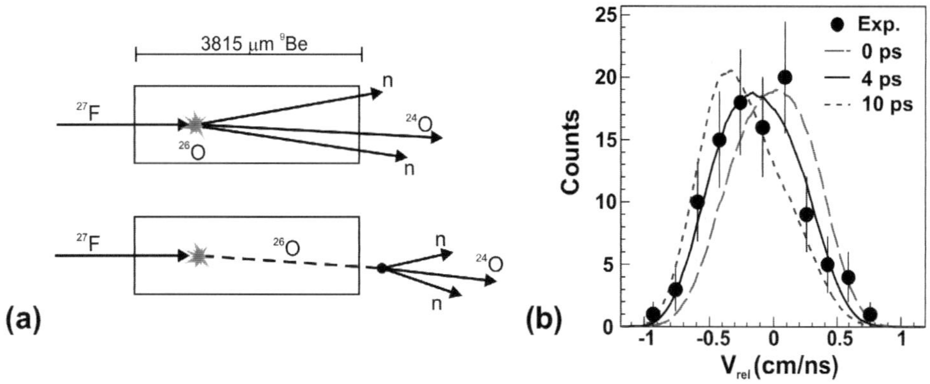

Figure 5. (Color online) (a) Illustration of the DiT technique. (b) Relative velocity spectrum for the ^{24}O fragment and two-neutrons from the decay of the ^{26}O ground state.

6. Conclusions

The exploration of neutron unbound systems beyond the dripline provides access to nuclear matter with neutron excess and, therefore, new phenomena can occur and be studied. The recent two-neutron decay studies of the MoNA collaboration have examined the 3-body decay correlations resulting from the two-neutron emission from the ground states of ^{16}Be and ^{13}Li, the predicted entrance-channel effects of populating neutron unbound systems from halo nuclei in population of ^{10}He from ^{14}Be, and the possibility for two-neutron radioactivity within the ^{26}O system. In all cases further investigation is required to gain a more complete understanding of the results. For example, full 3-body decay calculations are strongly encouraged which would allow for improved understanding of the strong neutron-neutron correlations observed in the decays of ^{16}Be and ^{13}Li. In particular, this would offer the opportunity to possibly connect the observed correlations to information about the ground state wavefunction of these systems. It will also be extremely important in the future to remeasure the ^{26}O lifetime with higher statistics to confirm the existence of two-neutron radioactivity.

Acknowledgements

The authors gratefully acknowledge the support of the NSCL operations staff for providing a high quality beam. This work was supported by the National Science Foundation under Grant No. PHY11-02511.

References

[1] Zeldes N 1956 *Nucl. Phys.* **2** 1
[2] Talmi I and Unna I 1960 *Phys. Rev. Lett.* **4** 469
[3] Baumann T, Spyrou A and Thoennessen M 2012 *Rep. Prog. Phys.* **75** 036301

[4] Brown B A 2001 *Prog. Part. Nucl. Phys.* **47** 517
[5] Thoennessen M 2004 *Rep. Prog. Phys.* **67** 1187
[6] Pfutzner M, Karny M, Grigorenko L V and Riisager K 2012 *Rev. Mod. Phys.* **84** 567
[7] Goldansky V I 1960 *Nucl. Phys.* **19** 482
[8] Egorova I A *et al.* 2012 *Phys. Rev. Lett.* **109** 202502
[9] Grigorenko L V and Zhukov M V 2003 *Phys. Rev. C* **68** 054005
[10] Tanihata I *et al.* 1985 *Phys. Rev. Lett.* **55** 2676
[11] Nakamura T *et al.* 2006 *Phys. Rev. Lett.* **96** 252502
[12] Ogawa Y, Suzuki Y and Yabana K 1994 *Nucl. Phys. A* **571** 784
[13] Spyrou A, Kohley Z, Baumann T, Bazin D, Brown B A, Chrstian G, DeYoung P A, Finck J E, Frank N, Lunderberg E, Mosby S, Peters W A, Schiller A, Smith J K, Synder J, Strongman M J, Thoennessen M and Volya A 2012 *Phys. Rev. Lett.* **108** 102501
[14] Kohley Z, Lunderberg E, DeYoung P A, Volya A, Baumann T, Bazin D, Christian G, Cooper N L, Frank N, Gade A, Hall C, Hinnefeld J, Luther B, Mosby S, Peters W A, Smith J K, Snyder J, Spyrou A and Thoennessen M 2013 *Phys. Rev. C* **87** 011304(R)
[15] Kohley Z *et al.* 2012 *Phys. Rev. Lett.* **109** 232501
[16] Lunderberg E, DeYoung P A, Kohley Z, Attanayake H, Baumann T, Bazin D, Christian G, Divaratne D, Grimes S M, Haagsma A, Finck J E, Frank N *et al.* 2012 *Phys. Rev. Lett.* **108** 142503
[17] Kohley Z *et al.* 2013 *Phys. Rev. Lett.* **110** 152501
[18] Luther B, Baumann T, Thoennessen M, Brown J, DeYoung P, Finck J, Hinnefeld J, Howes R, Kemper K, Pancella P, Peaslee G, Rogers W and Tabor S 2003 *Nucl. Instrum. Meth. A* **505** 33
[19] Baumann T *et al.* 2005 *Nucl. Instrum. Meth. A* **543** 517
[20] Bird M D, Kenney S J, Toth J, Weijers H W, DeKamp J C, Thoennessen M and Zeller A F 2005 *IEEE Trans. Appl. Supercond.* **15** 1252
[21] Christian G *et al.* 2012 *Phys. Rev. C* **85** 034327
[22] Kohley Z, Lunderberg E, DeYoung P A, Roeder B T, Baumann T, Christian G, Mosby S, Smith J K, Snyder J, Spyrou A and Thoennessen M 2012 *Nucl. Instrum. Meth. Phys. Res. A* **682** 59
[23] Agostinelli S, Allision J, Amako K, Apostolakis J, Araujo H, Arce P, Asai M, Axen D, Banerjee S, Barrand G, Behner F, Bellagamba L, Boudreau J *et al.* 2003 *Nucl. Instrum. Meth. A* **506** 250
[24] Allision J, Amako K, Apostolakis J, Araujo H, Dubios P A, Asai M, Barrand G, Capra R, Chauvie S, Chytracek R, Cirrone G A P, Cooperman G *et al.* 2006 *IEEE T. Nucl. Sci.* **53** 270
[25] Roeder B Development and validation of neutron detection simulations for EURISOL EURISOL Design Study, Report: [10-25-2008-006-In-beamvalidations.pdf, pp 31-44] (2008), www.eurisol.org/site02/physics and instrumentation/
[26] Desesquelles P, Cole A J, Dauchy A, Giorni A, Heuer D, Lleres A, Morand C, Sain-Martin J, Stassi P, Viano J B, Chambon B, Cheynis B, Drain D and Pastor C 1991 *Nucl. Instrum. Meth. A* **307** 366
[27] Spyrou A *et al.* 2011 *Phys. Rev. C* **84** 044309
[28] Snyder J *et al.* 2013 *Phys. Rev. C* **88** 031303(R)
[29] Ershov S N *et al.* 2010 *J. Phys. G: Nucl. Part. Phys.* **37** 064026
[30] Grigorenko L V *et al.* 2009 *Phys. Rev. C* **80** 034602
[31] James F CERN, Yellow Report No. 68-15 (1968).
[32] Brun R and Rademakers F 1997 *Nucl. Instrum. Meth. A* **389** 81 see also http://root.cern.ch/html/TGenPhaseSpace.html
[33] Hagino K, an dJ Carbonell H S and Schuck P 2007 *Phys. Rev. Lett.* **99** 022506
[34] Hagino K, Takahashi N and Sagawa H 2008 *Phys. Rev. C* **77** 054317
[35] Hagino K and Sagawa H 2014 *Phys. Rev. C* **89** 014331
[36] Grigorenko L V and Zhukov M V 2008 *Phys. Rev. C* **77** 034611
[37] Golovkov M S *et al.* 2009 *Phys. Lett. B* **672** 22
[38] Sidorchuk S I *et al.* 2012 *Phys. Rev. Lett.* **108** 202502
[39] Sharov P G, Egorova I A and Grigorenko L V 2014 *ArXiV* **1403.1748v1**
[40] Johansson H T *et al.* 2010 *Nucl. Phys. A* **842** 15
[41] Fortune H T 2013 *Phys. Rev. C* **88** 054623
[42] Korsheninnikov A A *et al.* 1994 *Phys. Lett. B* **326** 31
[43] Ostrowski A N *et al.* 1994 *Phys. Lett. B* **338** 13
[44] Kobayashi T, Yoshida K, Ozwaw A, Tanihata I, Korshninnikov A, Nikolski E and Nakamura T 1997 *Nucl. Phys. A* **616** 223c
[45] Johansson H T *et al.* 2010 *Nucl. Phys. A* **847** 66
[46] Caesar C *et al.* 2013 *Phys. Rev. C* **88** 034313
[47] Grigorenko L V, Mukha I G, Scheidenberger C and Zhukov M V 2011 *Phys. Rev. C* **84** 021303(R)

[48] Thoennessen M *et al.* 2013 *Nucl. Instrum. Meth. A* **729** 207
[49] Christian G *et al.* 2012 *Phys. Rev. Lett.* **108** 032501

Mechanism of Coulomb breakup reactions of two-neutron halo nuclei ^{6}He and ^{11}Li

Yuma Kikuchi[1], Takayuki Myo[2,3], Kiyoshi Katō[4], and Kiyomi Ikeda[1]

[1] Nishina Center for Accelerator-Based Science, RIKEN, Wako 351-0198, Japan.
[2] General Education, Faculty of Engineering, Osaka Institute of Technology, Osaka 535-8585, Japan.
[3] Research Center for Nuclear Physics (RCNP), Osaka University, Ibaraki 567-0047, Japan.
[4] Graduate School of Science, Hokkaido University, Sapporo 060-0810, Japan.

E-mail: yuma.kikuchi@riken.jp

Abstract. We investigate the three-body Coulomb breakup reactions of two-neutron halo nuclei and discuss the correlations of every binary subsystem such as of core-n and n-n by showing the invariant mass spectra. It is found that the final-state interactions of core-n and n-n binary subsystems dominantly determine the observed energy distributions of the breakup cross sections, such as the low-lying enhancements. Furthermore, we investigate the effects of the ^{9}Li core excitations on the Coulomb breakup cross section of ^{11}Li. It is shown that the integrated $E1$ strength at low energy region is reduced by $\sim$ 15 % by the inclusion of the ^{9}Li core excitations.

1. Introduction

Since the discovery of neutron halo nuclei such as ^{11}Be, ^{6}He, and ^{11}Li, extensive studies have been performed to understand their exotic structure caused by the weakly-bound nature of valence neutrons [1]. Theoretically, using the models based on the core + valence neutrons picture, the halo structure of the ground states and the excitations of halo nuclei have been investigated [2–5]. In two-neutron halo nuclei, it has been pointed out the correlation between two halo neutrons is important to explain the observed small two-neutron separation energies and large matter radii [3].

Experimentally, the Coulomb breakup reactions have been performed to expose the role of the n-n correlation in the two-neutron halo nucle. The observed Coulomb breakup cross section provides with interesting information on the properties of weak binding and the excitations of halo nuclei. It is also expected to gain a deeper understanding of the n-n correlation in the two-neutron halo nuclei through the Coulomb breakup reactions [6]. To understand the n-n correlation in the two-neutron halo nuclei, there are at least two problems to be solved: (i) clarifying the dominant breakup process such as the direct breakup to a noninteracting three-body continuum or sequential decay via the resonance of the binary subsystem, (ii) evaluating the influence of the final-state interactions (FSI) on the cross section.

It has been observed in ^{11}Li that the amount of the $(1s_{1/2})^2$ component (45±10%) of the halo neutrons is comparable to that of $(0p_{1/2})^2$ [7]. This fact indicates the breaking of the $N = 8$ magic number in the ground state of ^{11}Li. To reproduce this situation in ^{11}Li, most of the theoretical studies based on the ^{9}Li + n + n three-body models assume the $1s_{1/2}$ orbit is degenerated with the $0p_{1/2}$ one energetically by using the different ^{9}Li-n interactions for even and odd parity states [6; 8; 9]. However, Esbensen *et al.* [8] have mentioned that the ^{9}Li + n + n three-body model assuming the inert ^{9}Li core fails to explain the observed charge radius and the $E1$ strength in ^{11}Li consistently.

In Ref. [10], Myo *et al.* have shown that various physical quantities such as the matter and charge radii and the $E1$ strength of ^{11}Li are reproduced by using the coupled-channel ^{9}Li + n + n model, in which the 2p-2h excitations due to the tensor and pairing correlations in the ^{9}Li core are taken into account. It is important to investigate the effect of the ^{9}Li core excitations on the Coulomb breakup reaction of ^{11}Li.

In this study, we analyze the Coulomb breakup reaction of two-neutron halo nuclei, ^{6}He and ^{11}Li, and clarify the relation between the subsystem correlations of core-n and n-n and the Coulomb breakup cross section. To describe the three-body breakup states, we use the complex-scaled solutions of the Lippmann-Schwinger equation (CSLS) [11; 12]. We calculate not only the breakup cross section with respect to the total energy of the system but also invariant mass spectra of core-n and n-n by using CSLS, and discuss the effects of the binary subsystem correlations on the cross section. Also, we investigate the effect of the ^{9}Li excitation due to the tensor and pairing correlations on the Coulomb breakup cross section of ^{11}Li.

2. Method

We calculate the Coulomb breakup cross section by using the equivalent photon method and the $E1$ transition strength of halo nuclei. The Coulomb breakup cross section as function of the relative momenta is given as

$$\frac{d^6\sigma}{d\mathbf{k}d\mathbf{K}} = \frac{16\pi^3}{9\hbar c} N_{E1}(E_\gamma) \frac{d^6B(E1)}{d\mathbf{k}d\mathbf{K}}, \tag{1}$$

where $N_{E1}(E_\gamma)$ is the virtual photon number with the photon energy E_γ. Two momenta, $\mathbf{k}$ and $\mathbf{K}$, represent relative momenta in Jacobi coordinates for the three-body system. The $E1$ strength distribution is given as

$$\frac{d^6B(E1)}{d\mathbf{k}d\mathbf{K}} = \frac{1}{2J_{\rm gs}+1} \left| \langle \Psi^{(-)}(\mathbf{k},\mathbf{K}) || \hat{O}(E1) || \Phi_{\rm gs} \rangle \right|^2, \tag{2}$$

where $\Phi_{\rm gs}$ and $\Psi^{(-)}(\mathbf{k},\mathbf{K})$ are the wave functions for the ground and excited states, respectively, and the ground-state wave function for ^{6}He and ^{11}Li are the same ones as in Ref. [4] and [5], respectively. The total spin of the ground state is given as $J_{\rm gs}$. Using Eq. (2), we obtain the Coulomb breakup cross section as function of the total energy E as

$$\frac{d\sigma}{dE} = \iint d\mathbf{k}d\mathbf{K} \frac{d^6\sigma}{d\mathbf{k}d\mathbf{K}} \delta\left(E - \frac{\hbar^2k^2}{2\mu} - \frac{\hbar^2K^2}{2M}\right), \tag{3}$$

where μ and M are reduced masses corresponding to $\mathbf{k}$ and $\mathbf{K}$, respectively. Similarly, we can calculate the invariant mass spectra as

$$\frac{d\sigma}{d\varepsilon_1} = \iint d\mathbf{k}d\mathbf{K} \frac{d^6\sigma}{d\mathbf{k}d\mathbf{K}} \delta\left(\varepsilon_1 - \frac{\hbar^2k^2}{2\mu}\right), \tag{4}$$

where ε_1 is the energy of the binary subsystem. Using Eq. (4), we can discuss the effects of binary subsystem correlations, such as core-n and n-n, on the Coulomb breakup reaction.

Journal of Physics: Conference Series **569** (2014) 012034 doi:10.1088/1742-6596/569/1/012034

To evaluate the distributions given in Eqs. (3) and (4), it is necessary to obtain the scattering states, $\Psi^{(-)}(\mathbf{k},\mathbf{K})$. Here, we obtain $\Psi^{(-)}(\mathbf{k},\mathbf{K})$ using CSLS[11; 12]. The incoming scattering state $\Psi^{(-)}$ in the bra-representation is described as

$$\langle\Psi^{(-)}(\mathbf{k},\mathbf{K})| = \langle\Phi_0(\mathbf{k},\mathbf{K})| + \langle\Phi_0(\mathbf{k},\mathbf{K})|\hat{V}\lim_{\varepsilon\to 0}\frac{1}{E-\hat{H}+i\varepsilon}, \tag{5}$$

where Φ_0 is a solution of an asymptotic Hamiltonian $\hat{H}_0$. The interaction $\hat{V}$ is defined by subtracting $\hat{H}_0$ from the total hamiltonian $\hat{H}$ and becomes the source of FSI.

The Green's function in Eq. (5) is replaced with the complex-scaled Green's function in CSLS, and the relation between the Green's function in Eq. (5) and the complex-scaled one is given as

$$\lim_{\varepsilon\to 0}\frac{1}{E-\hat{H}+i\varepsilon} = U^{(-1)}(\theta)\frac{1}{E-\hat{H}^\theta}U(\theta) = \sum\!\!\!\!\!\!\int_n U^{(-1)}(\theta)|\chi_n^\theta\rangle\frac{1}{E-E_n^\theta}\langle\tilde{\chi}_n^\theta|U(\theta). \tag{6}$$

In derivation of the right-hand-side of Eq. (6), we insert the complete set constructed with $\{\chi_n^\theta\}$, being the eigenstates of the complex-scaled Hamiltonian $\hat{H}^\theta$. We here calculate the eigenstates and eigenvalues of $\hat{H}^\theta$, $\{\chi_n^\theta\}$ and $\{E_n^\theta\}$ by using the core + n + n three-body models [4; 5]. We obtain $\Psi^{(-)}(\mathbf{k},\mathbf{K})$ as

$$\langle\Psi^{(-)}(\mathbf{k},\mathbf{K})| = \langle\Phi_0(\mathbf{k},\mathbf{K})| + \sum\!\!\!\!\!\!\int_n\langle\Phi_0(\mathbf{k},\mathbf{K})|\hat{V}U^{-1}(\theta)|\chi_n^\theta\rangle\frac{1}{E-E_n^\theta}\langle\tilde{\chi}_n^\theta|U(\theta). \tag{7}$$

The scattering state in CSLS consists of two terms: The first term represents the noninteracting continuum states, and the second is the effects of FSI. This description of the scattering states is useful to investigate the effects of correlations in the final states after the breakup. The correlations in the ground state can also be seen in the transition to the first term.

3. Coulomb breakup cross sections of ^{6}He and ^{11}Li

We first show the Coulomb breakup cross section with respect to the total energy of ^{6}He and ^{11}Li, in Fig. 1, in comparison with the observed ones [13; 14]. In both cases, the calculated cross sections show the low-lying enhancements, and well reproduces the observed data. We also estimate the effect of FSI on the cross sections by taking only the first term in Eq. (7) in

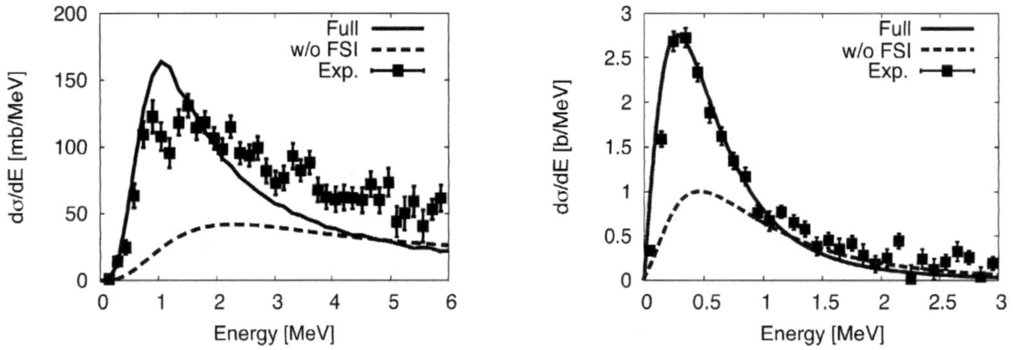

Figure 1. Obtained Coulomb breakup cross sections of ^{6}He and ^{11}Li, measured from the threshold energy of core + n + n. The left and right panels represent the cross sections for ^{6}He and ^{11}Li, respectively. The observed data, shown as solid squares, are taken from Ref. [13] for ^{6}He and from Ref. [14] for ^{11}Li.

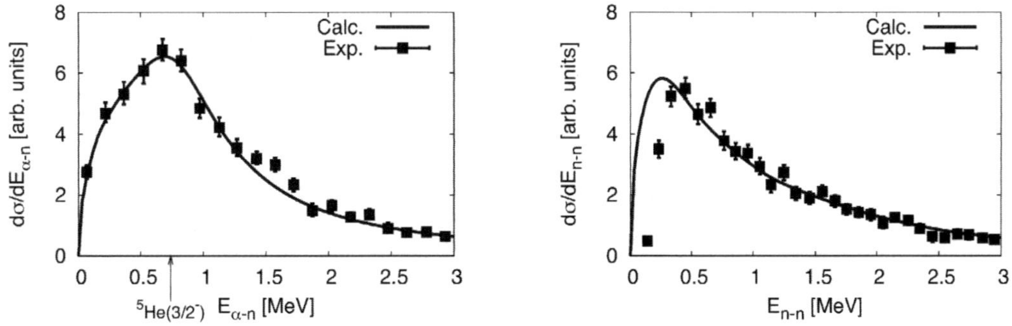

Figure 2. Invariant mass spectra of the Coulomb breakup reaction of ^{6}He. The left and right panels present the spectra for α-n and n-n, respectively. The observed data, shown as solid squares, are taken from Ref. [13]. An arrow in the left panel indicates the position of the ^{5}He$(3/2^-)$ resonances in our calculation.

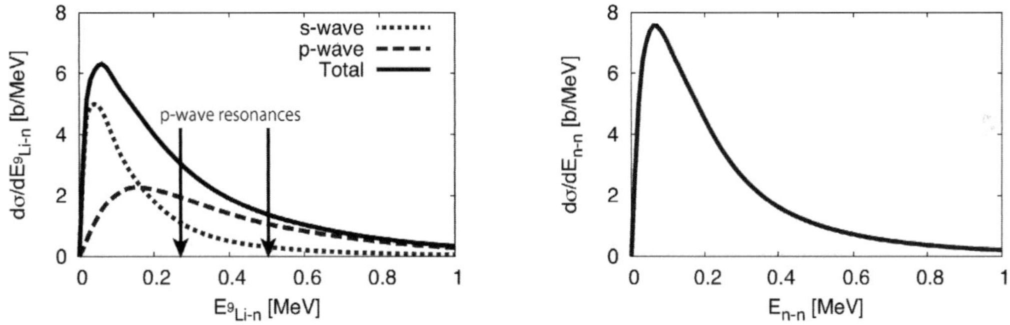

Figure 3. Invariant mass spectra of the Coulomb breakup reaction of ^{11}Li. The left and right panels present the spectra for ^{9}Li-n and n-n, respectively. In the left panel, three types of spectra are presented: The total distribution (solid), s-wave component (dotted), and p-wave component (dashed). Two arrows in the left panel indicate the position of the p-wave resonances in our calculation.

the calculation, and it is found that correlations in final states play decisive roles in reproducing the Coulomb breakup cross sections.

To discuss what kinds of FSI are important in the Coulomb breakup reactions, we calculate the invariant mass spectra for binary subsystems. We show the spectra for ^{6}He in Fig. 2. From the results, the invariant mass spectra for the α-n subsystem shows the peak at around 0.7 MeV corresponding to the ^{5}He$(3/2^-)$ resonance. The α-n correlation is clearly confirmed in the invariant mass spectra of the Coulomb breakup cross section. For the n-n subsystem, the low-lying enhancement is seen near the zero-energy region, which indicates the importance of the n-n virtual state in the final states.

In Fig. 3, we also show the calculated invariant mass spectra for ^{11}Li. From the results, we confirm that both invariant mass spectra for ^{9}Li-n and n-n show the low-lying enhancements. For ^{9}Li-n, it is seen that the peak comes from the s-wave component as shown In the left panel in Fig. 3. This result indicates the virtual-state correlation of the ^{9}Li-n subsystem. For n-n, we find the similar behavior to that for ^{9}Li-n, and the virtual-state correlation of the n-n subsystem is confirmed in the spectra. One the other hand, from Fig. 3, the p-wave resonances in ^{10}Li does not contribute to the invariant mass spectra for the ^{9}Li-n subsystem since the energies of the

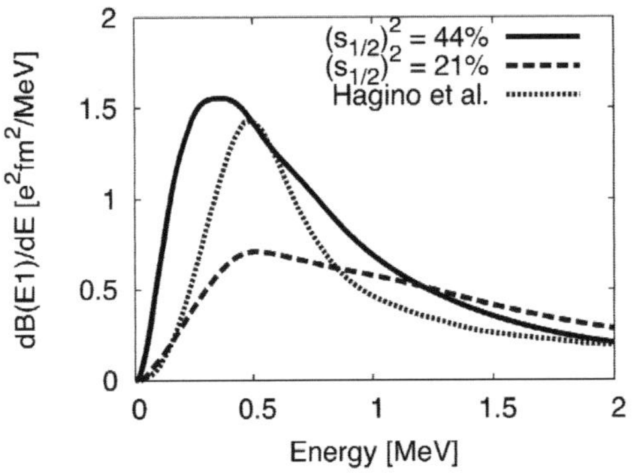

Figure 4. Comparison between the $E1$ strength distributions with different $(s_{1/2})^2$ components. The solid line represent the result of using the ground-state wave function with $(s_{1/2})^2 = 44\%$, which is used in the right panel of Fig. 1. The dashed line is the result with $(s_{1/2})^2 = 21\%$. The dotted line is taken from Ref. [9].

p-wave resonances, which are obtained at 275 keV and 506 keV in our calculation, are higher than the peak position of the total energy of the breakup cross section shown in the right panel of Fig. 1.

From the results for ^{6}He and ^{11}Li, it is found that the invariant mass spectra for core-n subsystems reflect the characters of the core-n correlations in two-neutron halo nuclei, while the behaviors of the spectra for n-n subsystem are commonly dominated by the n-n virtual-state correlation in two nuclei.

4. Effects of ^{9}Li core excitations on the Coulomb breakup reaction of ^{11}Li

To investigate the effects of the ^{9}Li core excitations on the Coulomb breakup reaction of ^{11}Li, we compare the $E1$ transition strength calculated in our coupled-channel ^{9}Li + n + n model [5] with the result in Ref. [9], in which the three-body model assuming the inert ^{9}Li core is employed.

We here calculate the $E1$ strength distribution using the different ^{11}Li wave function, in which the coupling to the 2p-2h configuration involved only by the pairing correlation in ^{9}Li core is taken into account. This restriction leads to the small $(s_{1/2})^2$ component as 21.0% in the ground state, which gives a similar value to that in the three-body model assuming the inert ^{9}Li core. The calculated distribution is shown in Fig.4 as dashed line in comparison with the result in the model assuming the inert core (dotted line). The wave function used in the results shown as dashed and dotted lines contain almost the same amount of the s-wave components in the ^{11}Li ground state. In two kinds of results, the $E1$ strength distributions commonly have peaks at around 0.5 MeV; however, there exists a large difference of strength around the peak energy. This is due to the fact that about 15% of the integrated strength in our calculation escapes to the higher excited states having the excited components of the ^{9}Li core.

We confirm that the observed Coulomb breakup cross section and the $E1$ strength distribution cannot be reproduced only by taking into account the coupling to the 2p-2h excitation due to the pairing correlation in ^{9}Li. To reproduce the observed cross section, it is essential to include the 2p-2h excitation coming from not only the pairing correlation but also the tensor one in the ^{9}Li core. By taking into account both correlations in ^{9}Li, we obtain the result shown as solid line in Fig. 4, which is used in the calculation of the cross section in the right panel in Fig. 2.

Journal of Physics: Conference Series **569** (2014) 012034 doi:10.1088/1742-6596/569/1/012034

5. Summary

In this study, we calculate the three-body Coulomb breakup cross sections and invariant mass spectra for binary subsystems of ^{6}He and ^{11}Li. Our calculations reproduce the Coulomb breakup cross sections well. From the result of cross sections, it is found that the FSI has a dominant contribution to explain the low-lying enhancement seen in the cross sections. It is confirmed that the subsystem correlations of the ^{5}He($3/2^-$) resonance and the ^{10}Li virtual state dominantly determine the distributions of the invariant mass spectra of the breakup reactions of ^{6}He and ^{11}Li, respectively, while the n-n virtual-state correlations are found in both cases. Furthermore, we investigated the effects of the ^{9}Li core excitations on the Coulomb breakup reaction of ^{11}Li. It is shown that the integrated $E1$ strength is reduced by $\sim$ 15 % by inclusion of the ^{9}Li core excitations. It is essential to take into account 2p-2h excitations due to the tensor and pairing correlations in the ^{9}Li core in reproducing the observed Coulomb breakup cross section of ^{11}Li.

References

[1] Tanihata I, Savajols H and Kanungo R 2013 *Prog. Part. Nucl. Phys.* **68** 215
[2] Zhukov M V, Danilin B V, Fedorov D V, Bang J M, Thompson I J and Vaagen J S 1993 *Phys. Rep.* **231** 151
[3] Aoyama S, Mukai S, Katō K and Ikeda K 1995 *Prog. Theor. Phys.* **93** 99
[4] Myo T, Katō K, Aoyama S and Ikeda K 2001 *Phys. Rev. C* **63** 054313
[5] Myo T, Kikuchi Y, Katō K, Toki H and Ikeda K 2008 *Prog. Theor. Phys.* **119** 561
[6] Esbensen H and Bertsch G F 1992 *Nucl. Phys. A* **542** 310
[7] Simon H *et al.* 1999 *Phys. Rev. Lett.* **83** 496
[8] Esbensen H, Hagino K, Mueller P and Sagawa H 2007 *Phys. Rev. C* **76** 024302
[9] Hagino K, Sagawa H, Nakamura T and Shimoura S 2009 *Phys. Rev. C* **80** 031301
[10] Myo T, Katō K, Toki H and Ikeda K 2007 *Phys. Rev. C* **76** 024305
[11] Kikuchi Y, Myo T, Takashina M, Katō K and Ikeda K 2009 *Prog. Theor. Phys.* **122** 499
[12] Kikuchi Y, Katō K, Myo T, Takashina M and Ikeda K 2010 *Phys. Rev. C* **81** 044308
[13] Aumann T *et al.* 1999 *Phys. Rev. C* **59** 1252
[14] Nakamura T 2007 *Nucl. Phys. A* **788** 243c

Journal of Physics: Conference Series **569** (2014) 012035 doi:10.1088/1742-6596/569/1/012035

Study of clustering structures through breakup reactions

Pierre Capel

Physique Nucléaire et Physique Quantique (CP229), Université Libre de Bruxelles (ULB)
50 av. F. D. Roosevelt, B-1050 Brussels, Belgium

E-mail: pierre.capel@ulb.ac.be

Abstract. Models for the description of breakup reactions used to study the structure of exotic cluster structures like halos are reviewed. The sensitivity of these models to the projectile description is presented. Calculations are sensitive to the projectile ground state mostly through its asymptotic normalisation coefficient (ANC). They also probe the continuum of the projectile. This enables studying not only resonant states of the projectile but also its non-resonant continuum both resonant and non-resonant. This opens the possibility to study correlations between both halo neutrons in two-neutron halo nuclei.

1. Introduction

The developments of radioactive-ion beams in the mid-80s, has permitted the exploration of the nuclear chart far from stability. Various unexpected features have been uncovered thanks to this technical breakthrough. Halo nuclei are of particular interest here as they exhibit a strongly clusterised structure that explains there very large radius compared to their isobars. Indeed these neutron-rich nuclei have a small binding energy for one or two neutrons. Thanks to this lose binding, the valence neutron(s) can tunnel far into the classically forbidden region to form a dilute halo around a more compact core [1]. ^{11}Be and ^{15}C are examples of one-neutron halo nuclei, while ^{6}He and ^{11}Li are archetypical two-neutron halo nuclei. Though less probable, proton halos are also possible. For example, ^{8}B is thought to exhibit a one-proton halo.

Being located close to the dripline, halo nuclei are very short lived and cannot therefore be studied through usual spectroscopic techniques. To study their structure, one must resort to indirect techniques such as reactions [1]. Breakup reaction is probably among the best tools to study exotic clusterised structures such as halo nuclei. During this reaction, the nucleus of interest is broken up into its constituents—the halo nucleon(s) and the core in this case—through its interaction with a target [2]. To extract valuable information about the cluster structure of the projectile from breakup observables, a good understanding of the reaction mechanism is needed and a clear knowledge of the influence of the projectile structure upon the measured cross sections is required. In this contribution, I will present the nonperturbative reaction models mostly used to analyse breakup data: the Continuum Discretised Coupled Channel model (CDCC), the Time-Dependent approach (TD) and the Dynamical Eikonal Approximation (DEA) (see also Ref. [3] for a recent review). I will then discuss recent results that show the influence of the projectile description upon breakup data. An outlook concludes these proceedings.

2. Breakup modelling

2.1. Theoretical framework

In most breakup models, the projectile P is described as a two-body system: an inert core c to which the halo nucleon f is loosely bound. The c-f interaction is described by a phenomenological potential V_{cf} that is adjusted to reproduce the energy and the quantum numbers of the bound states and maybe some resonant states of the nucleus. The Hamiltonian that describes this cluster structure hence reads

$$H_0 = T_r + V_{cf}(\boldsymbol{r}), \tag{1}$$

where $\boldsymbol{r}$ is the c-f relative coordinate. The eigenstates of H_0 are the wave functions describing the projectile internal structure

$$H_0\Phi_{lm}(\epsilon, \boldsymbol{r}) = \epsilon\Phi_{lm}(\epsilon, \boldsymbol{r}), \tag{2}$$

where l is the c-f relative angular momentum and m is its projection (the spins of the clusters are neglected here for simplicity). The negative-energy eigenstates correspond to the bound states of the system. In the following the ground state of energy ϵ_0 is denoted by Φ_0. The positive-energy states describe the c-f continuum, i.e. the broken up projectile.

The target T is usually described as a structureless particle and its interaction with the projectile constituents c and f is simulated by the optical potentials V_{cT} and V_{fT}, respectively. Within this framework, studying the P-T collision reduces to solving the following three-body Schrödinger equation

$$\left[T_R + H_0 + V_{cT}(\boldsymbol{R}_{cT}) + V_{fT}(\boldsymbol{R}_{fT})\right]\Psi(\boldsymbol{r}, \boldsymbol{R}) = E_T\Psi(\boldsymbol{r}, \boldsymbol{R}), \tag{3}$$

where $\boldsymbol{R}$ is the coordinate of the projectile centre of mass relative to the target, while $\boldsymbol{R}_{cT}$ and $\boldsymbol{R}_{fT}$ are the c-T and f-T relative coordinates, respectively. Eq.(3) is solved with the condition that the projectile, initially in its ground state, is impinging on the target:

$$\Psi(\boldsymbol{r}, \boldsymbol{R}) \underset{Z\to-\infty}{\longrightarrow} e^{iKZ+\cdots}\Phi_0(\boldsymbol{r}), \tag{4}$$

where K is the initial P-T wave number, which is related to the total energy $E_T = \hbar^2K^2/2\mu_{PT} + \epsilon_0$, with μ_{PT} the projectile-target reduced mass.

2.2. CDCC

In the Continuum Discretised Coupled Channel method (CDCC), Eq.(3) is solved expanding the three-body wave function Ψ upon the projectile eigenstates $\Phi_{lm}(\epsilon)$ [4, 5]

$$\Psi(\boldsymbol{r}, \boldsymbol{R}) = \sum_i \chi_i(\boldsymbol{R})\Phi_i(\boldsymbol{r}), \tag{5}$$

where i stands for l, m and ϵ. This leads to a set of coupled equations

$$\left[T_R + \epsilon + V_{ii}\right]\chi_i + \sum_{j\neq i} V_{ij}\chi_j = E_T\chi_i, \tag{6}$$

where $V_{ij} = \langle\Phi_i|V_{cT} + V_{fT}|\Phi_j\rangle$ are coupling the various channels.

Since the model aims at describing the breakup of the projectile, the expansion (5) must include a reliable description of the continuum of the projectile. To be tractable, this description is obtained by discretisation of the continuum [6].

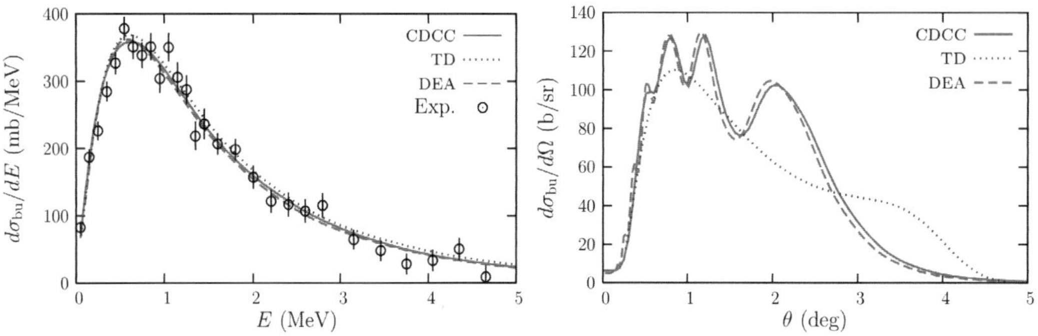

Figure 1. Comparison of CDCC, DEA and TD, three nonperturbative breakup models used to analyse experiments involving halo nuclei [9]. Energy distribution (left) and angular distribution (right) for ^{15}C impinging on Pb at 68AMeV, data from Ref. [2].

2.3. Time-dependent approach

In the Time-Dependent approach (TD) the three-body Schrödinger equation (3) is simplified using a semiclassical approximation, viz. approaching the P-T relative motion by a classical trajectory [?, ?]. Along that trajectory, the projectile experiences a time-dependent potential that simulates its interaction with the target. This leads to the resolution of a time-dependent Schrödinger equation

$$i\hbar\frac{\partial}{\partial t}\Psi(\boldsymbol{r},\boldsymbol{b},t) = [H_0 + V_{cT}(t) + V_{fT}(t)]\Psi(\boldsymbol{r},\boldsymbol{b},t), \tag{7}$$

which is solved with the initial condition $\Psi(\boldsymbol{r},\boldsymbol{b},t) \underset{t\to-\infty}{\longrightarrow} \Phi_0(\boldsymbol{r})$ in agreement with Eq.(4).

2.4. Dynamical Eikonal Approximation

The Dynamical Eikonal Approximation (DEA) is based on the eikonal approximation [7], which assumes that at sufficiently high beam energy the P-T relative motion does not deviate much from the incoming plane wave (4). The idea is thus to factorise that plane wave out of Ψ

$$\Psi(\boldsymbol{r},\boldsymbol{R}) = e^{iKZ}\widehat{\Psi}(\boldsymbol{r},\boldsymbol{R}) \tag{8}$$

to obtain a function $\widehat{\Psi}$ that varies smoothly with $\boldsymbol{R}$. Neglecting its second-order derivatives with respect to its first-order derivatives leads to the DEA equation [8]

$$i\hbar v\frac{\partial}{\partial Z}\widehat{\Psi}(\boldsymbol{r},\boldsymbol{b},Z) = [H_0 - \epsilon_0 + V_{cT} + V_{fT}]\widehat{\Psi}(\boldsymbol{r},\boldsymbol{b},Z), \tag{9}$$

which is solved with the boundary condition $\widehat{\Psi}(\boldsymbol{r},\boldsymbol{R}) \underset{Z\to-\infty}{\longrightarrow} \Phi_0(\boldsymbol{r})$ according to Eqs. (4) and(8).

In Ref. [9], these three models have been confronted to one another for ^{15}C impinging on Pb at 68AMeV. All three models agree very well with each other in the energy distribution (see Fig. 1 left) indicating that they describe the same reaction process. They also agree very well with the data of Ref. [10], confirming the halo structure of ^{15}C. In the angular distribution (see Fig. 1 right), CDCC and DEA agree very well with each other. TD however, due to its underlying semiclassical approximation, does not exhibit the oscillatory pattern, that corresponds to quantal interferences. Nevertheless, it gives the general trend of the distribution, which explains why it reproduces correctly observables integrated over the angles such as the energy distribution.

Journal of Physics: Conference Series **569** (2014) 012035 doi:10.1088/1742-6596/569/1/012035

This example illustrates the reliablity of reaction modelling. However it is mostly confined to a simple description of the projectile. Various groups have tried to go beyond this simple two-body description, e.g. by extending the aforementioned models to three-body projectiles. A 4-body CDCC reaction models, meaning a three-body projectile plus a target, has been developed by two groups [11, 12]. Unfortunately, due to the heavy computational requirement of CDCC, this extension was at first limited to the description of the elastic scattering of two-neutron halo nuclei, in which a breakup channel was included in a crude way. This was nevertheless interesting as it showed that both the three-body structure of the projectile and the inclusion of the breakup channel are necessary to reproduce experimental data.

Another group suggested instead to use a simpler reaction model within which a full description of the three-body continuum could be included [13]. They used the Coulomb-Corrected Eikonal model, which corresponds to an approximate solution of the DEA equation (9). First applied to ^{6}He (see Ref. [13] and Fig. 4), this model showed excellent agreement with experiment for the Coulomb breakup of ^{11}Li on Pb at 70AMeV [14], which confirmed the two-neutron halo in ^{11}Li and enabled to study the reaction mechanism for such kind of nuclei. This study suggests in particular that double-differential cross sections like Dalitz plots provide information about correlations between both halo neutrons (see also Sec. 3.3).

3. Sensitivity of breakup reactions to the projectile structure

3.1. Peripherality of breakup reactions

Breakup measurements are often used to infer a spectroscopic factor (SF) for one particular configuration of the projectile [2]. However, this is valid only if the reaction probes the whole range of the projectile wave function. To check whether or not breakup reactions are sensitive to the internal part of the projectile, calculations have been performed with descriptions of the projectile that differ significantly in the interior but exhibit identical tails [15]. Fig. 2 illustrates the results obtain in the case of ^{8}B on Ni at 25 MeV. The left panel shows the two bound-state wave functions considered in the test, while the right panel shows the corresponding angular distributions obtained within CDCC. As is clearly shown both curves are superimposed indicating that, albeit significant, changes in the interior of the projectile wave function do not affect breakup calculations. Similar results have been obtained at other energies, for different projectiles and targets, indicating the generality of this conclusion [15]. Breakup is thus a peripheral reaction in the sense that it probes only the tail of the projectile wave function, i.e. its Asymptotic Normalisation Coefficient (ANC).

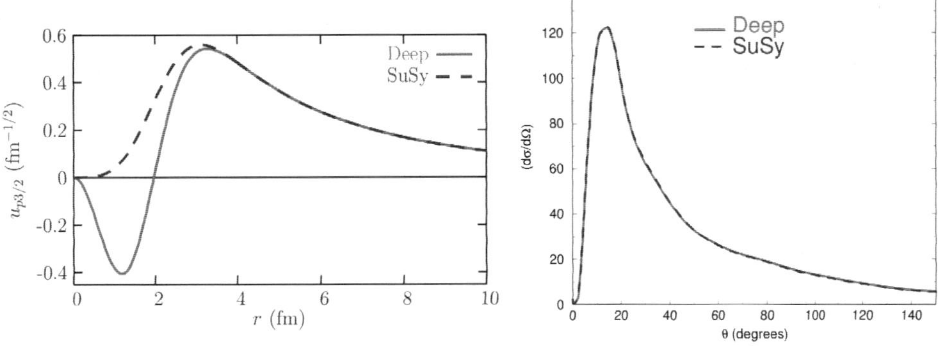

Figure 2. Peripherality of breakup reactions: bound-state wave functions that differ significantly in the interior but have identical asymptotics (left) lead to identical breakup cross section (right). CDCC calculation performed for the breakup of ^{8}B on Ni at 25 MeV [15].

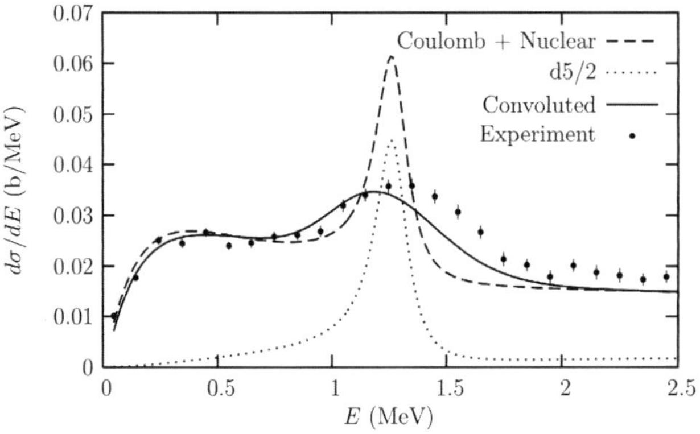

Figure 3. Nuclear breakup of ^{11}Be on C at 67AMeV. The calculation confirms the significant influence of the $d_{5/2}$ resonance upon the energy distribution [17]. Data from Ref. [2].

3.2. Sensitivity to resonant continuum

Along the Coulomb breakup of ^{11}Be, Fukuda *et al.* have measured its dissociation on a light target (C) (see Fig. 3) [2]. This energy distribution exhibits a very different behaviour than Coulomb-breakup ones (compare to Fig. 1 left): it exhibits a peak at the energy of a known $5/2^+$ resonance. This was interpreted as the signature of the ^{10}Be-n resonant state in nuclear-dominated breakup [2]. This was confirmed by a TD calculation in which the two-body description of ^{11}Be included this resonance within the $d5/2$ partial wave [17]. The calculated cross section exhibits a narrow peak (dashed line in Fig. 3), which, once folded with the experimental energy resolution, agrees very well with the measurements (solid line). This peak is observed only in the $d5/2$ partial wave, confirming that nuclear breakup can be used reliably to study the presence of cluster resonant states in nuclei far from stability.

The influence of resonant states upon breakup has also been shown for three-body projectiles like ^{6}He (see below and Refs. [13, 22]). This influence thus appears in more complicate structures than the simple two-body model presented here and one can expect that interesting information about the projectile continuum can be inferred from such measurements (see also Ref. [18]).

3.3. Influence of the non-resonant continuum

In the previous section, the influence of resonant states on breakup observables has been shown. It is therefore necessary to properly include these states in reaction modelling. The question addressed in this section is whether the description of the non-resonant continuum of the projectile influences breakup calculations or not. A study has shown that for two-body projectiles, breakup cross sections can be affected by up to 40% depending on the choice of V_{cf} in the continuum [16]. This is significant, in particular if one is interested in extracting spectroscopic information such as ANCs from breakup data.

The problem worsens for three-body projectiles. This is illustrated in Fig. 4, which shows the energy distribution for the breakup of ^{6}He on Pb at 240AMeV, i.e. the breakup cross section as a function of the total energy between the α and the two halo neutrons after dissociation. The calculation is made within the CCE [13] with a complete description of the continuum of ^{6}He. The total cross section is the top solid line, which includes the contribution of the non-resonant continuum and that of the narrow 2^+ resonance at about 1 MeV above the α-n-n threshold. As already discussed in Sec. 3.2, the latter is responsible for the narrow peak at that energy. The

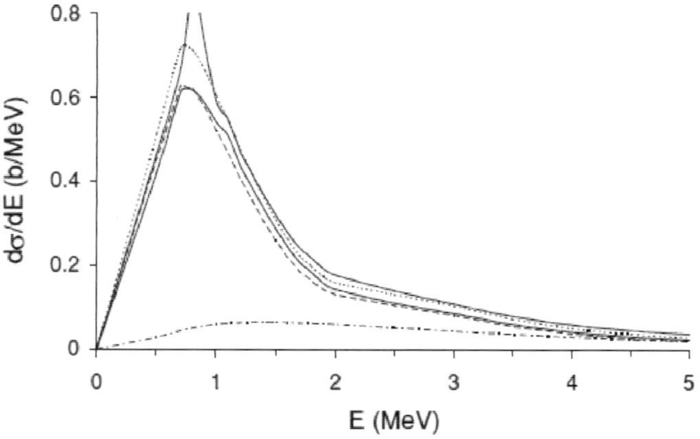

Figure 4. Influence of the non-resonant continuum upon breakup calculations. The Coulomb-breakup cross section of ^{6}He on Pb at $240A$MeV is shown to exhibit a significant $E1$ contribution (lower solid line) when final-state interactions are considered. This broad peak disappear when the continuum is described by plane waves (dashed-dotted line) [13].

non-resonant contribution is mostly the $E1$ strength shown by the lower solid line. This non-resonant continuum is obtained assuming proper interactions between the clusters [13]. When these interactions are switched off, i.e. when the continuum is described by mere plane waves, the cross section exhibits a much different behaviour (see the dash-dotted line in Fig. 4).

The analysis of these results indicates that these final-state interactions lead to a breakup that populates a slightly dominant α-dineutron structure in the ^{6}He continuum. Other theoretical works have confirmed this hypothesis of a significant effect of the final-state interactions in breakup calculations [19]. In all cases, complex structures are observed in the double-differential cross sections, which suggests that interesting information about the three-body continuum can be inferred from such observables. In particular one might hope to study the correlation between both halo neutrons through such cross sections. Experimental data will help us confirming these hopes. In any case, they would enable us to discriminate the different models of ^{6}He that have been used in these calculations.

4. Outlook

Breakup reactions are a powerful tool to study the structure of nuclei far from stability and in particular exotic cluster structures like halos [1]. However, the reaction mechanism is more complicated than initially thought and an accurate reaction model is needed to analyse reliably experimental data. Various such models exist; CDCC [4, 5], TD [?, ?] and DEA [8] have been reviewed in this contribution (see Ref. [3] for a more complete review). They compare well to each other at intermediate energy [9]. They also reproduce experimental data, which suggests that the reaction mechanism is now rather well understood, at least for two-body projectiles.

The extension of reaction models to more complicated structures, like two-neutron halo nuclei, is computationally challenging [20, 21, 13] but recent developments have shown it possible. This has opened the path to an extensive study of breakup observables for three-body projectiles.

Within this contribution, the analyses of the sensitivity of breakup calculations to the projectile description have also been reviewed. First it has been reminded that breakup reactions are peripheral in the sense that they are sensitive only to the tail of the projectile wave function and not to its whole range, i.e. they are better suited to extract ANCs than SFs [15]. Second, they are not only sensitive to the projectile ground state, but they also probe its continuum. In particular core-halo resonances can be studied through nuclear-dominated reactions [17, 22, 2]. Interestingly, the non-resonant continuum also influences the calculations [16]. This should be kept in mind when analysing experimental data, as it can affect values such as ANCs inferred from measurements. This sensitivity is an opportunity for three-body projectiles as recent calculations have shown that their continuum has a significant influence upon breakup calculations [13, 19]. In particular, double-differential cross sections (Dalitz plots) could be used to study the correlation between both halo neutrons in projectiles like ^{11}Li and ^{6}He.

Breakup reactions therefore exhibit great qualities to analyse exotic cluster structures. Experimental and theoretical efforts should be made to refine the information that can be obtained through breakup to expand our understanding of nuclear structure far from stability.

Acknowledgments

This research was supported in part by the Fonds de la Recherche Fondamentale Collective (grant number 2.4604.07F). This text presents research results of the Belgian Research Initiative on eXotic nuclei (BRIX), Program No. P7/12, on interuniversity attraction poles of the Belgian Federal Science Policy Office.

References

[1] Tanihata I *J. Phys. G* **22** (1996) 157
[2] Fukuda N, Nakamura T, Aoi N, Imai N, Ishihara M, Kobayashi T, Iwasaki H, Kubo T, Mengoni A, Notani M, Otsu H, Sakurai H, Shimoura S, Teranishi T, Watanabe Y X and Yoneda K *Phys. Rev. C* **70** (2004) 054606
[3] Baye D and Capel P *Clusters in Nuclei, Vol.2* (*Lecture Notes in Physics* vol **848**, 2012) Ed. Beck C (Springer Berlin Heidelberg) pp 121–163
[4] Kamimura M, Yahiro M, Iseri Y, Sakuragi Y, Kameyama H and Kawai M *Prog. Theor. Phys. Suppl.* **89** (1986) 1
[5] Tostevin J A, Nunes F M and Thompson I J *Phys. Rev. C* **63** (2001) 024617
[6] Yahiro M, Iseri Y, Kameyama H, Kamimura M and Kawai M *Prog. Theor. Phys. Suppl.* **89** (1986) 32
[7] Glauber R J 1959 *Lecture in Theoretical Physics* vol 1 ed Brittin W E and Dunham L G (New York: Interscience) p 315
[8] Baye D, Capel P and Goldstein G 2005 *Phys. Rev. Lett.* **95** 082502
[9] Capel P, Esbensen H and Nunes F M 2012 *Phys. Rev. C* **85** 044604
[10] Nakamura T, Fukuda N, Aoi N, Imai N, Ishihara M, Iwasaki H, Kobayashi T, Kubo T, Mengoni A, Motobayashi T, Notani M, Otsu H, Sakurai H, Shimoura S, Teranishi T, Watanabe Y X and Yoneda K 2009 *Phys. Rev. C* **79** 035805
[11] Matsumoto T, Hiyama E, Ogata K, Iseri Y, Kamimura M, Chiba S and Yahiro M 2004 *Phys. Rev. C* **70** 061601(R)
[12] Rodríguez-Gallardo M, Arias J M, Gómez-Camacho J, Johnson R C, Moro A M, Thompson I J and Tostevin J A 2008 *Phys. Rev. C* **77** 064609
[13] Baye D, Capel P, Descouvemont P and Suzuki Y 2009 *Phys. Rev. C* **79** 024607
[14] Pinilla E C, Descouvemont P and Baye D 2012 *Phys. Rev. C* **85** 054610
[15] Capel P and Nunes F M 2007 *Phys. Rev. C* **75** 054609 (pages 6)
[16] Capel P and Nunes F M 2006 *Phys. Rev. C* **73** 014615
[17] Capel P, Goldstein G and Baye D 2004 *Phys. Rev. C* **70** 064605
[18] Moro A M and Lay J A 2012 *Phys. Rev. Lett.* **109** 232502
[19] Kikuchi Y, Katō K, Myo T, Takashina M and Ikeda K 2010 *Phys. Rev. C* **81** 044308
[20] Matsumoto T, Egami T, Ogata K, Iseri Y, Kamimura M and Yahiro M 2006 *Phys. Rev. C* **73** 051602
[21] Rodríguez-Gallardo M, Arias J M, Gómez-Camacho J, Moro A M, Thompson I J and Tostevin J A 2009 *Phys. Rev. C* **80** 051601
[22] Kikuchi Y, Matsumoto T, Minomo K and Ogata K 2013 *Phys. Rev. C* **88** 021602

Journal of Physics: Conference Series **569** (2014) 012036 doi:10.1088/1742-6596/569/1/012036

CDCC studies on clustering physics

Takuma Matsumoto[1], Kazuyuki Ogata[2], Kosho Minomo[2], and Masanobu Yahiro[1]

[1] Department of Physics, Kyushu University, Fukuoka 812-8581, Japan.
[2] Research Center of Nuclear Physics (RCNP), Osaka University, Ibaraki 567-0047, Japan.

E-mail: matsumoto@phys.kyushu-u.ac.jp

Abstract. The continuum-discretized coupled-channels method (CDCC) has successfully been applied to studies on various reactions involving unstable nuclei. The four-body CDCC method is a new method to calculate three-body breakup continuum in a projectile, which gives an accurate analysis for four-body breakup systems. The eikonal reaction theory is a new approach to calculate inclusive breakup cross sections such as neutron removal cross sections. Moreover, CDCC is a useful tool for evaluation of nuclear data, which are important for nuclear engineering.

1. Introduction

Study on unstable nuclei is one of the main subject in nuclear physics. Breakup reactions involving such nuclei have played a key role in investigating the exotic properties. The observables such as breakup and neutron removal cross sections reflect information of the ground and resonance continuum states. Furthermore breakup reactions are also useful for searching the cluster structure of nuclei. Therefore in order to understand the exotic properties from the observables, an accurate analysis for describing the breakup mechanism is highly desirable.

The continuum-discretized coupled-channels method (CDCC) [1, 2, 3] is one of the most reliable methods for treating breakup processes in exclusive reactions such as elastic scattering, elastic-breakup reactions, and transfer reactions. CDCC has been proposed at first to solve three-body scattering problems, in which a projectile breaks up into two constituents. Recently, there are two important extensions for CDCC. One is a new method to calculate three-body breakup continuum in a projectile, which gives an accurate analysis for four-body breakup systems. This new method is called four-body CDCC [4, 5, 6, 7]. The other is a new approach to calculate inclusive breakup cross sections such as neutron removal cross sections. In the new approach called the eikonal reaction theory (ERT) [8], both nuclear and Coulomb breakup processes are consistently treated by CDCC without making the adiabatic approximation.

In SOTANCP3, we reviewed recent studies of CDCC. One is analyses of ^{6}He breakup reactions and another is an application of CDCC to nuclear engineering. These works have been already reported in the CDCC review paper [3].

2. Analysis of ^{6}He breakup reactions

^{6}He is known as a two-neutron halo nucleus, and the structure can be described by an n + n + ^{4}He three-body model. In the scattering of ^{6}He, the projectile easily breaks up into three constitutes. This reaction is treated as a four-body system with three-body projectile and a

Journal of Physics: Conference Series **569** (2014) 012036 doi:10.1088/1742-6596/569/1/012036

target nucleus. To describe the four-body breakup system, we adopt the four-body CDCC method.

For the ^{6}He breakup reaction, we set a four-body model Hamiltonian as

$$H = T_R + U_{nT} + U_{nT} + U_{\alpha T} + h_P, \tag{1}$$

where T_R is a kinetic energy for the relative coordinate R between the center-of-mass of ^{6}He and a target, U_{xT} is an optical potential between constituents x in ^{6}He and a target, and h_P is a internal model Hamiltonian of ^{6}He. Here, $U_{\alpha T}$ includes the Coulomb interaction between α and a target.

In the four-body CDCC method, the total wave function is expanded in terms of bound and discretized continuum states of ^{6}He,

$$\Psi^{(+)}_{\mathrm{CDCC}}(\boldsymbol{\xi}, \boldsymbol{R}) = \Phi_0(\boldsymbol{\xi})\chi_0^{(+)}(\boldsymbol{R}) + \sum_\nu \Phi_\nu(\boldsymbol{\xi})\chi_\nu^{(+)}(\boldsymbol{R}). \tag{2}$$

Here Φ_0 and Φ_ν represent bound and discretized continuum wave functions with the internal coordinate $\boldsymbol{\xi}$, which is represented by a set of Jacobi coordinates, $\boldsymbol{\xi} = (\boldsymbol{r}, \boldsymbol{y})$. Relative wave functions between ^{6}He and target, $\chi_0^{(+)}$ and $\chi_\nu^{(+)}$, are solved with the outgoing boundary condition. Momenta conjugate to coordinates $\boldsymbol{R}$ and $(\boldsymbol{y}, \boldsymbol{r})$ are defined by $\boldsymbol{P}$ and $(\boldsymbol{p}, \boldsymbol{k})$, respectively. To obtain the discretized continuum wave functions, the pseudo-state discretization method [9, 10, 11, 12] is applied. In the pseudo-state discretization, the internal Hamiltonian of ^{6}He is diagonalized with the L^2-type basis functions. In the present analysis, we adopt the Gaussian basis functions [13]. The details of the four-body CDCC method are shown in Ref. [6].

The four-body T-matrix element to a breakup state with $(\boldsymbol{p}, \boldsymbol{k})$ can be described by

$$T_\varepsilon(\boldsymbol{p}, \boldsymbol{k}, \boldsymbol{P}) = \langle \psi_\varepsilon^{(-)}(\boldsymbol{p}, \boldsymbol{k})\chi_\varepsilon^{(-)}(\boldsymbol{P})|U|\Psi^{(+)}\rangle_{\boldsymbol{\xi}\boldsymbol{R}} \tag{3}$$

with

$$U = U_{nT} + U_{nT} + U_{\alpha T} - V_{PT}^{\mathrm{Coul}} \tag{4}$$

where V_{PT}^{Coul} is the Coulomb interactions between ^{6}He and a target with the arguments are replaced by R. The final-state wave functions, $|\psi_\varepsilon^{(-)}(\boldsymbol{p}, \boldsymbol{k})\rangle$ and $|\chi_\varepsilon^{(-)}(\boldsymbol{P})\rangle$, with the incoming boundary condition are defined by

$$\left[T_R + V_{PT}^{\mathrm{Coul}}(R) - (E_{\mathrm{tot}} - \varepsilon)\right] |\chi_\varepsilon^{(-)}(\boldsymbol{P})\rangle = 0, \tag{5}$$

$$[h_P - \varepsilon]\,|\psi_\varepsilon^{(-)}(\boldsymbol{p}, \boldsymbol{k})\rangle = 0, \tag{6}$$

where $E_{\mathrm{tot}} - \varepsilon = (\hbar P)^2/(2\mu_R)$ and $\varepsilon = (\hbar p)^2/(2\mu_y) + (\hbar k)^2/(2\mu_r)$ for reduced masses μ_R and μ_ξ of coordinates $\boldsymbol{R}$ and $\boldsymbol{\xi}$, respectively. Inserting the approximate complete set defined as

$$1 \approx |\Phi_0\rangle\langle\Phi_0| + \sum_\nu |\Phi_\nu\rangle\langle\Phi_\nu|, \tag{7}$$

we can find that the T-matrix element is well approximated by

$$T_\varepsilon(\boldsymbol{p}, \boldsymbol{k}, \boldsymbol{P}) \approx \sum_{\nu\neq 0} \langle \psi_\varepsilon^{(-)}(\boldsymbol{p}, \boldsymbol{k})|\Phi_\nu\rangle T_\nu, \tag{8}$$

where T_ν is the T-matrix element to a ν-th discrete breakup state Φ_n with an eigenenergy ε_n,

$$T_\nu = \langle \Phi_\nu \chi_{\varepsilon_\nu}^{(-)}(\boldsymbol{P}_\nu)|U|\Psi^{(+)}_{\mathrm{CDCC}}\rangle. \tag{9}$$

Here Eq. (8) is derived by replacing $\boldsymbol{P}$ by $\boldsymbol{P}_n$ in $\chi_\varepsilon^{(-)}(\boldsymbol{P})$. The T_ν are obtainable by CDCC, but it is quite hard to calculate the smoothing factor $\langle\psi_\varepsilon^{(-)}(\boldsymbol{p},\boldsymbol{k})|\Phi_n\rangle$ directly with either numerical integration [7, 14, 15]. Hence, we propose a new way of obtaining the differential cross section with respect to ε without calculating the smoothing factor [16].

Using the exact T matrix, The differential cross section as a function of ε is calculated by

$$\frac{d^2\sigma}{d\varepsilon d\Omega_{\boldsymbol{P}}} = \int d\boldsymbol{p}'d\boldsymbol{k}'\delta(\varepsilon-\varepsilon')|T_{\varepsilon'}(\boldsymbol{p}',\boldsymbol{k}',\boldsymbol{P}')|^2 = \frac{1}{\pi}\mathrm{Im}\mathcal{R}(\varepsilon). \tag{10}$$

Here $\mathcal{R}(\varepsilon)$ is the generalized response function defined as

$$\mathcal{R}(\varepsilon) = \int d\boldsymbol{\xi}d\boldsymbol{\xi}'\langle\Psi^{(+)}(\boldsymbol{\xi})|U|\chi_\varepsilon^{(-)}\rangle_{\boldsymbol{R}}G^{(-)}(\varepsilon,\boldsymbol{\xi},\boldsymbol{\xi}')\langle\chi_\varepsilon^{(-)}|U|\Psi^{(+)}(\boldsymbol{\xi}')\rangle_{\boldsymbol{R}} \tag{11}$$

with the three-body Green's function of ^{6}He

$$G^{(-)}(\varepsilon,\boldsymbol{\xi},\boldsymbol{\xi}') = \lim_{\eta\to+0}\langle\boldsymbol{\xi}|\frac{1}{\varepsilon-h_P-i\eta}|\boldsymbol{\xi}'\rangle. \tag{12}$$

For the sides of $G^{(-)}$ in Eq. (11), we insert an approximate complete set represented by Eq. (7) and then $\mathcal{R}$ is calculated as

$$\begin{aligned}\mathcal{R}(\varepsilon) &= \sum_\nu\sum_{\nu'}\langle\Psi^{(+)}|U|\chi_\varepsilon^{(-)}\Phi_\nu\rangle\langle\Phi_\nu|G^{(-)}(\varepsilon,\boldsymbol{\xi},\boldsymbol{\xi}')|\Phi_{\nu'}\rangle\langle\Phi_{\nu'}\chi_\varepsilon^{(-)}|U|\Psi^{(+)}\rangle,\\ &\approx \sum_\nu\sum_{\nu'}T_\nu^*\langle\Phi_\nu|G^{(-)}(\varepsilon,\boldsymbol{\xi},\boldsymbol{\xi}')|\Phi_{\nu'}\rangle T_\nu\end{aligned} \tag{13}$$

In Eq. (13), there is no the smoothing factor, as expected. Furthermore, the propagator $G^{(-)}$ operates only on spatially damping functions Φ_ν, so that the calculation of $\langle\Phi_\nu|G^{(-)}|\Phi_{\nu'}\rangle$ becomes feasible as show below.

The complex scaling method (CSM) [17, 18, 19, 20] is now applied to evaluating $\langle\Phi_\nu|G^{(-)}|\Phi_{\nu'}\rangle$. The scaling transformation operator $C(\theta)$ and its inverse are defined by

$$\langle\boldsymbol{r},\boldsymbol{y}|C(\theta)|f\rangle = e^{3i\theta}f(\boldsymbol{r}e^{i\theta},\boldsymbol{y}e^{i\theta}), \tag{14}$$

$$\langle f|C^{-1}(\theta)|\boldsymbol{r},\boldsymbol{y}\rangle = \{e^{-3i\theta}f(\boldsymbol{r}e^{-i\theta},\boldsymbol{y}e^{-i\theta})\}^*. \tag{15}$$

Using the operators, one can get

$$\langle\Phi_\nu|G^{(-)}|\Phi_{\nu'}\rangle = \langle\Phi_\nu|C^{-1}(\theta)G_\theta^{(-)}C(\theta)|\Phi_{\nu'}\rangle, \tag{16}$$

where

$$G_\theta^{(-)} = \lim_{\eta\to+0}\frac{1}{\varepsilon-h_P^\theta-i\eta}. \tag{17}$$

with $h_P^\theta = C(\theta)h_PC^{-1}(\theta)$. When $-\pi<\theta<0$, the scaled propagator $\langle\boldsymbol{\xi}|G_\theta^{(-)}|\boldsymbol{\xi}'\rangle$ is a damping function of $\boldsymbol{\xi}$ and $\boldsymbol{\xi}'$. It should be noted that although the scaling angle in general calculations with CSM has been taken as positive, the angle in the present situation becomes negative since $G^{(-)}$ has the incoming boundary condition. Hence, it can be expanded with L^2-type basis functions with high accuracy:

$$G_\theta^{(-)} \approx \sum_i\frac{|\phi_i^\theta\rangle\langle\tilde{\phi}_i^\theta|}{\varepsilon-\varepsilon_i^\theta}, \tag{18}$$

where ϕ_i^θ is an i-th eigenstate obtained by diagonalizing h_P^θ in a modelspace spanned by L^2-type basis functions:

$$\langle\tilde{\phi}_i^\theta|h_P^\theta|\phi_{i'}^\theta\rangle \quad = \varepsilon_i^\theta\delta_{ii'}. \tag{19}$$

In virtue of CSM, thus, we do not need to evaluate exact three-body continuum state $\psi_\varepsilon^{(-)}(\boldsymbol{k},\boldsymbol{p})$ to obtain $\langle\Phi_\nu|G^{(-)}|\Phi_{\nu'}\rangle$.

Inserting Eq. (18) into Eq. (11) through Eq. (16) leads to a useful form of

$$\frac{d^2\sigma}{d\varepsilon d\Omega_{\boldsymbol{P}}} \approx \frac{1}{\pi}\mathrm{Im}\sum_i \frac{T_i^\theta\tilde{T}_i^\theta}{\varepsilon-\varepsilon_i^\theta} \tag{20}$$

with

$$\tilde{T}_i^\theta \equiv \sum_{\nu'}\langle\tilde{\phi}_i^\theta|C(\theta)|\Phi_{\nu'}\rangle T_{\nu'}, \tag{21}$$

$$T_i^\theta \equiv \sum_{\nu} T_\nu^*\langle\Phi_\nu|C^{-1}(\theta)|\phi_i^\theta\rangle. \tag{22}$$

The principal result of this paper is that $C(\theta)$ and $C^{-1}(\theta)$ operate only on the spatially damping function Φ_ν. This makes the calculation of $\langle\tilde{\phi}_i^\theta|C(\theta)|\Phi_{\nu'}\rangle$ and $\langle\Phi_\nu|C^{-1}(\theta)|\phi_i^\theta\rangle$ feasible and makes it possible the convergence of $d^2\sigma/d\varepsilon d\Omega_{\boldsymbol{P}}$ with respect to extending the modelspace, as shown below. In other words, $C(\theta)$ and $C^{-1}(\theta)$ are not allowed to act on a non-damping function such as the plane wave, since the scaled function diverges asymptotically in this case.

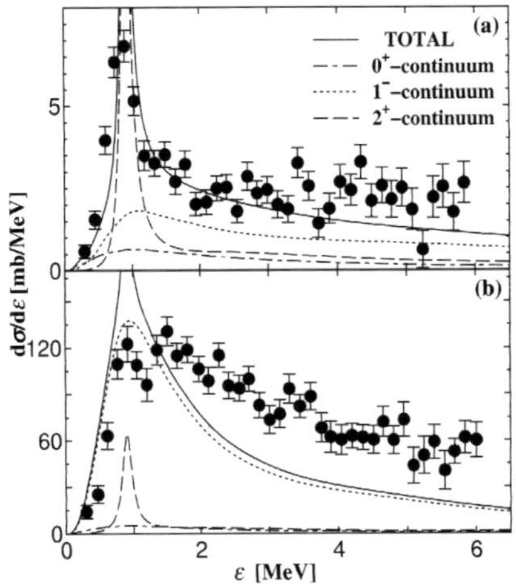

Figure 1. Comparison of the breakup cross section calculated by CDCC (solid line) with experimental data for (a) ^{6}He + ^{12}C scattering at 240 MeV/nucleon and (b) ^{6}He + ^{208}Pb scattering at 240 MeV/nucleon [16]. The dot-dashed, dotted, and dashed lines correspond to contributions of 0^+, 1^-, and 2^+ breakup, respectively. The experimental data are taken from Ref. [21].

We apply the new smoothing approach with CSM to analyses of ^{6}He breakup cross sections on ^{12}C and ^{208}Pb at 240 MeV/nucleon [21]. Figure 1 shows the calculated breakup cross sections compared with the experimental data. These data have already been analyzed by four-body

Journal of Physics: Conference Series **569** (2014) 012036 doi:10.1088/1742-6596/569/1/012036

distorted wave Born approximation (DWBA) [22] and the eikonal approximation [23]. For U_{nT} and $U_{\alpha T}$ in the reaction on ^{208}Pb, we take the same potentials used in Ref. [23]. Meanwhile U_{nT} in the case of ^{12}C is taken from the global nucleon-nucleus potential [24] and the optical potential for α-^{12}C system is evaulated from the ^{12}C + ^{12}C potential at 200 MeV/nucleon [25] by changing the radius parameter from ^{12}C to α. In ^{6}He + ^{12}C scattering at 240 MeV/nucleon, nuclear breakup is dominant, while Coulomb breakup to 1^- continuum is dominant for ^{6}He + ^{208}Pb scattering. For ^{12}C target, we can see the clear peak around $\varepsilon = 1$ MeV corresponding the 2^+-resonance of ^{6}He. The present theoretical result is consistent with the experimental data except for the peak of the 2^+-resonance. Similar overestimations are also seen in the results of four-body DWBA. We expect that the overestimation can be partly solved by taking into account of the resolution of the experimental data. For ^{208}Pb target, the present method underestimates the experimental data at $\varepsilon \gtrsim 2$ MeV. A possible origin of the underestimation are that the inelastic breakup reactions are not included in the present calculation. As mentioned in Ref. [22], the inelastic breakup effect is not negligible, and the elastic breakup cross section calculated with four-body DWBA also underestimates the data. Furthermore, for the present reaction systems, we calculated one- and two-neutron stripping cross sections with ERT. The results will be reported in the forthcoming paper [26].

3. Application of CDCC to nuclear engineering

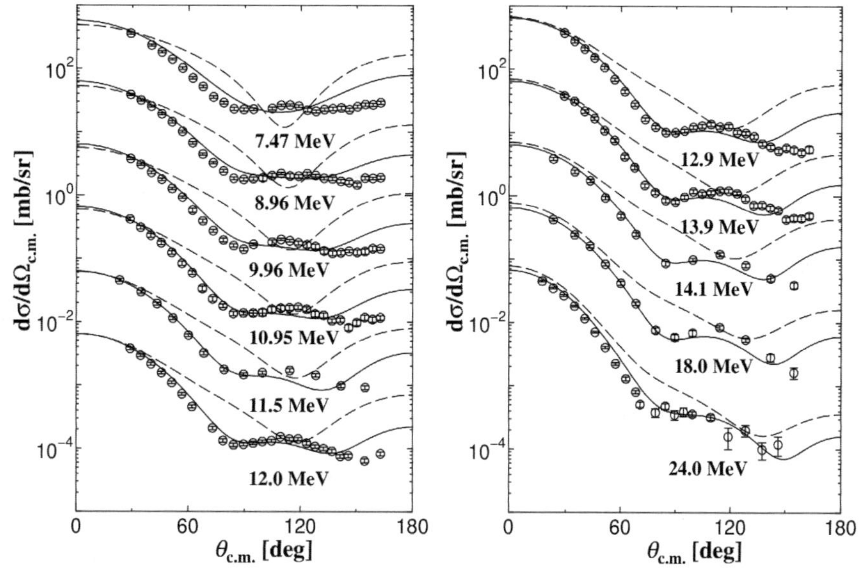

Figure 2. Angular distribution of the elastic differential cross section of n + ^{6}Li scattering for incident energies between 7.47 and 24.0 MeV [27]. Experimental data are taken from Refs. [31, 32, 33].

Since CDCC is a fully quantum-mechanical method, it is applicable to reactions at low incident energies, which are important for nuclear engineering. In this field, Lithium isotopes are important for not only tritium breeding material in DT fusion reactors but also a candidate for target material in the intense neutron source of IFMIF. Recently, accurate nuclear data of nucleon induced reactions on ^{6,7}Li are highly required for incident energies up to 150 MeV.

In Ref. [27], we performed a microscopic analysis of n + ^{6}Li scattering by the three-body CDCC method, in which ^{6}Li is described as a d + α system. As the interaction between n and ^{6}Li, we applied a folding model with the JLM effective interaction [30], where the normalization of the imaginary part, λ_w, is optimized to reproduce the elastic cross sections. Details of the calculation are shown in Ref. [27]. Figure 2 shows the differential elastic cross sections of n + ^{6}Li for incident energies between 7.47 and 24.0 MeV. One sees that the results of the CDCC calculation (the solid lines) are in good agreement with the experimental data. The dashed lines represent the results of a single-channel calculation without couplings to the breakup states. Thus breakup effects are significant to reproduce the angular distributions of the elastic scattering. For all incident energies, we take $\lambda_w = 0.1$ to reproduce the data.

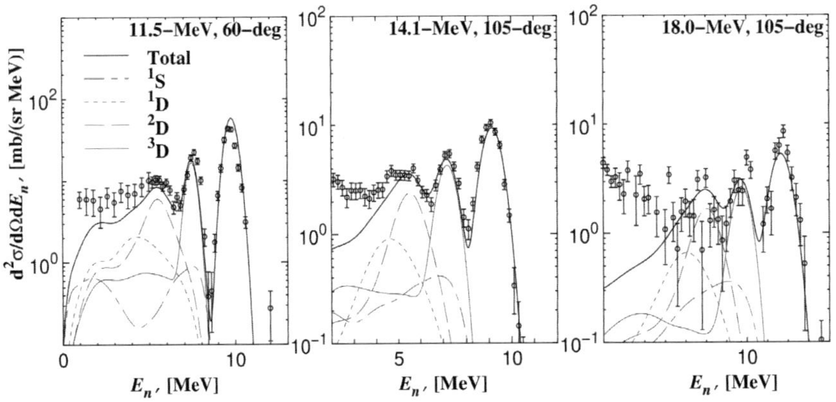

Figure 3. The neutron spectra calculated by CDCC with the JLM interaction comparing with measured data at selected angular points in the laboratory system [27]. The experimental data are taken from Ref. [31].

In Fig. 3, the calculated neutron spectra are compared with the experimental data at selected scattering angles in the laboratory frame and incident energies. Components of the breakup states of ^{1}S, ^{1}D, ^{2}D, and ^{3}D are represented by the dash-dotted, dashed, dotted, and thin solid lines, respectively. One sees that the CDCC calculation with the finite resolution of the experimental apparatus [31] gives a good agreement with experimental data in the high neutron energy region. On the other hand, in the low neutron energy region, which corresponds to highly exited states of ^{6}Li, the calculated cross section considerably undershoot the experimental data. This indicates that experimental data contain contribution from the $(n, 2n)$ process due to a four-body breakup reaction ^{6}Li$(n, nnp)\alpha$, as indicated in Ref. [31]. Estimate of these effects with four-body CDCC will be very important.

4. Summary

CDCC is one of the most reliable methods for studying on unstable nuclei via the breakup reaction. Recently, we propose the four-body CDCC method that can describe three-body breakup processes of the projectile in the scattering. For analyses of ^{6}He reactions, the four-body CDCC well reproduces the breakup cross sections. Thus the four-body CDCC is indispensable for investigating of properties of unstable nuclei.

CDCC with the JLM interaction is expected to be a powerful framework for the data evaluation of the ^{6}Li(n, n') reactions. Once the JLM parameter is determined by an analysis of elastic scattering, evaluation of the inelastic cross sections and neutron breakup spectra can be done with no free adjustable parameters. This is a very important feature of the present framework that enables a quantitative calculation of the cross sections for nuclear engineering studies.

References

[1] Kamimura M, Yahiro M, Iseri Y, Sakuragi Y, Kameyama H and Kawai M 1986 *Prog. Theor. Phys. Suppl.* **89** 1
[2] Austern N, Iseri Y, Kamimura M, Kawai M, Rawitscher G and Yahiro M 1987 *Phys. Rep.* **154** 125
[3] Yahiro M, Ogata K, Matsumoto T and Minomo K 2012 *Prog. Theor. Exp. Phys.* **01A206**
[4] Matsumoto T, Hiyama E, Ogata K, Iseri Y, Kamimura M, Chiba S and Yahiro M 2004 *Phys. Rev.* C **70** 061601(R)
[5] Rodríguez-Gallardo M *et al.* 2005 *Phys. Rev.* C **72** 024007
[6] Matsumoto T, Egami T, Ogata K, Iseri Y, Kamimura M and Yahiro M 2006 *Phys. Rev.* C **73** 051602(R)
[7] Rodríguez-Gallardo M, Arias J M,Gómez-Camacho J, Moro A M, Thompson I J and Tostevin J A, 2009 *Phys. Rev.* C **80** 051601(R)
[8] Yahiro M, Ogata K and Minomo K, 2011 *Prog. Theor. Phys.* **126** 167
[9] Moro A M, Arias J M, Gómez-Camacho J, Martel I, Pérez-Bernal F, Crespo R and Nunes F 2001 *Phys. Rev.* C **65** 011602(R)
[10] Matsumoto T, Kamizato T, Ogata K, Iseri Y, Hiyama E, Kamimura M and Yahiro M 2003 *Phys. Rev.* C **68** 064607
[11] Egami T, Ogata K, Matsumoto T, Iseri Y, Kamimura M, and Yahiro M, 2004 *Phys. Rev.* C **70** 047604
[12] Moro A M, Perez-Bernal F, Arias J M and Gomez-Camacho J, 2006 *Phys. Rev.* C **73** 044612
[13] Hiyama E, Kino Y and Kamimura M 2003 *Prog. Part. Nucl. Phys.* **51** 223
[14] Egami T, Matsumoto T, Ogata K and Yahiro M 2009 *Prog. Theor. Phys.* **121** 789
[15] Matsumoto T, Egami T, Ogata K and Yahiro M 2009 *Prog. Theor. Phys.* **121** 885
[16] Matsumoto T, Katō K and Yahiro M 2010 *Phys. Rev.* C **82** 051602
[17] Aguilar J and Combes J M 1971 *Commun. Math. Phys.* **22** 269, Balslev E and Combes J M 1971 *ibid* **22** 280
[18] Aoyama S, Myo T, Katō K and Ikeda K 2006 *Prog. Theor. Phys.* **116** 1
[19] Myo T, Katō K, Aoyama S and Ikeda K 2001 *Phys. Rev.* C **63** 054313
[20] Myo T, Katō K, Toki H and Ikeda K 2007 *Phys. Rev.* C **76** 024305
[21] Aumann T, *et al.* 1999 *Phys. Rev.* C **59** 1252
[22] Ershov S N, Danilin B V and Vaagen J S 2000 *Phys. Rev.* C **62** 041001(R)
[23] Baye D, Capel P, Descoubemont P and Suzuki Y 2009 *Phys. Rev.* C **79** 024607
[24] Koning A J and Delaroche J P 2003 *Nucl. Phys.* **A713**, 231
[25] Hostachy J Y, *et al.* 1988 *Nucl. Phys.* **A490** 441
[26] Minomo K, Matsumoto T, Egashira K, Ogata K and Yahiro M to be published for *Phys. Rev.* C.
[27] Matsumoto T, Ichinkhorloo D, Hirabayashi Y, Katō K and Chiba S 2011 *Phys. Rev.* C **83** 064611
[28] Chiba S, Fukahori T, Shibata K, Yu B and Kosako K 1997 *Fusion Eng. Des.* **37** 175
[29] Fukahori T, Chiba S, Kishida N, Kawai M, Oyama Y and Hasegawa A 1997 *Proceedings of the International Conference on Nuclear Data for Science and Technology, Trieste, Italy, 1997 (Societa Italiana di fisica)* 899
[30] Jeukenne J P, Lejeune A and Mahaux C 1997 *Phys. Rev.* C**16** 80; 1976 ibid. *Phys. Rep.* **25** 83
[31] Chiba S, *et al.* 1998 *Phys. Rev.* C **58** 2205
[32] Hogue H, *et al.* 1979 *Nucl. Sci. Eng.* **69** 22
[33] Hansen L *et al.* 1988 *Phys. Rev.* C **38** 525

Three-body asymptotic normalization factors for halo nuclei and their determination

R Yarmukhamedov

Institute of Nuclear Physics, Ulughbek, 100214 Tashkent, Uzbekistan

E-mail: rakhim@inp.uz

Abstract. The explicit forms of the true asymptotics of the three-body (312) bound state radial wave functions of halo nuclei a both with two loosely bound external neutrons and with two charged particles, valid in different parts of the asymptotic region of the configuration space, are presented. The results for three-body asymptotic normalization factors for $\alpha+n+n\to{}^6$He and $n+\alpha+\alpha\to{}^9$Be obtained from the comparative analysis of these asymptotic forms with the specific approximate model wave functions for asymptotic regions are also given.

1. Introduction

In this review work, we will briefly present the true asymptotic forms derived in works [1–5] for the three-body (3+1+2) halo nucleus a, which is also denoted by 3+1+2$\to$ a below, and the results for the three-body asymptotic normalization factors(TBANFs) for $\alpha+n+n\to{}^6$He and $n+\alpha+\alpha\to{}^9$Be obtained from the comparative analysis of these asymptotic forms with the specific approximate model wave functions for asymptotic regions. We use $\hbar\ = c\ =1$ throughout the paper.

2. True asymptotic forms of the three-body bound state radial wave functions for halo nuclei

Let us consider the three-body bound system (a = (312)) consisting of two external particles, say, particles 1 and 2, and of a core 3. Let r and ρ be the relative distance between the (1,2)-pair and centers of mass of third particle 3 and of the (1,2)-pair, respectively, q and p be the conjugate to them relative momentums as well as $r_j(q_j)$ be the relative distance (momentum) between centers of mass of the particle j (j=1 and 2) and of the core 3. Besides, we denote $\mu_{ij}=m_im_j/m_{ij}$ and $\mu_{(ij)k}=m_km_{ij}/m$ as the reduced masses of the subsystem (ij) and the $(ij)k$ system, respectively; $R=\sqrt{2\mu_{12}r^2+2\mu_{(12)3}\rho^2}$ as the hyperradius; $\varphi=\mathrm{atan}(\sqrt{2\mu_{(12)3}}\rho/\sqrt{2\mu_{12}}r)$ as the hyperangle; $R^{(j)}(r_2,r_1)=[2\mu_{23}r_2^2+2\mu_{(23)1}(r_2-(-1)^j\lambda_1r_1)^2]^{1/2}=[2\mu_{13}r_1^2+2\mu_{(13)2}(r_1-(-1)^j\lambda_2r_2)^2]^{1/2}$ as a modified hyperradius with $\lambda_j=m_j/(m_j+m_3)$(j=1 and 2); $\eta=Z_1Z_2e^2(\mu_{12}/\varepsilon_a)^{1/2}/\cos\varphi$ as the Coulomb parameter for the subsystem (12) in the three-body (312) bound system and $\varepsilon_a=m_1+m_2+m_3-m$ as the binding energy for the a system in the (3+2+1)-channel, where $m_{ij}=m_i+m_j$, $m=m_i+m_j+m_k$, m_j and Z_je are the mass and the charge of particle j.

The explicit forms for the true three-body asymptotics for the three-body a(=(312)) bound state radial wave functions can be laid down by means of the following theorems:

Theorem 1. For $r \to \infty$ and $\rho \to \infty$ the asymptotics formula

$$\Psi_{a\nu}(r,\rho) \approx C_{a;\,\nu}(\varphi) f_l(\sqrt{\varepsilon_a}R\sin^2\varphi) f_\lambda(\sqrt{\varepsilon_a}R\cos^2\varphi)\frac{e^{-\sqrt{\varepsilon_a}R}}{R^{5/2}} \tag{1}$$

is valid for a neutral case [1] *and that*

$$\Psi_{a\nu}(r,\rho) \approx C_{a;\,\nu}(\varphi) f_l(\sqrt{\varepsilon_a}R\sin^2\varphi) f_\lambda(\sqrt{\varepsilon_a}R\cos^2\varphi)(2\sqrt{\varepsilon_a}R)^{-\eta}\frac{e^{-\sqrt{\varepsilon_a}R}}{R^{5/2}} \tag{2}$$

is valid for the bound system with two (1 and 2) the charged particles [3], where

$$f_{L_o}(x) = \sum_{n=0}^{L_o}\frac{L_o!}{n!(L_o-n)!}\frac{1}{(2x)^n}.$$

Theorem 2. For a neutral case the asymptotic formula

$$\Psi_{a\nu}(r,\rho) \approx C_{a;\,\nu}(\varphi) g_l(\sqrt{\varepsilon_a}R\sin^2\varphi) f_\lambda(\sqrt{\varepsilon_a}R\cos^2\varphi)\frac{e^{-\sqrt{\varepsilon_a}R}}{R^{5/2}} \tag{3}$$

is valid for $r \to \infty$ and $\rho \to 0\,(\rho \neq 0,\ \varphi \to 0)$ and that

$$\Psi_{a\nu}(r,\rho) \approx C_{a;\,\nu}(\varphi) f_l(\sqrt{\varepsilon_a}R\sin^2\varphi) g_\lambda(\sqrt{\varepsilon_a}R\cos^2\varphi)\frac{e^{-\sqrt{\varepsilon_a}R}}{R^{5/2}} \tag{4}$$

is valid for $\rho \to \infty$ and $r \to 0\,(r \neq 0,\ \varphi \to \pi/2)$ [5], where

$$g_L(x) = \frac{1}{2}[f_L(x) + (-1)^{L+1}e^{2x}f_L(-x)] \approx x^{L+1}/(2L+1)!! << 1 \text{ at } x \to 0. \tag{5}$$

In Eqs. (1)–(4), the TBANF $C_{a\nu}(\varphi)$ is determined by [1, 6]

$$C_{a;\,\nu}(\varphi) = (-1)^{l+\lambda+1}\frac{(m\sqrt{\varepsilon_a})^{3/2}}{\sqrt{2}\pi^{5/2}}W_{a;\,\nu}(i\sqrt{\varepsilon_a}\cos\varphi, i\sqrt{\varepsilon_a}\sin\varphi), \tag{6}$$

where $\nu = \{\lambda l L S\}$ and $W_{a\nu}(i\sqrt{\varepsilon_a}\cos\varphi, i\sqrt{\varepsilon_a}\sin\varphi)$ is the on-shell partial vertex function (OSWF) for the virtual decay $a \to 3+1+2$, i.e., values of the partial vertex functions $W_{a\nu}(q,p)$ when all the particles 3, 1 and 2 are on-shell ($\varepsilon = q^2/2\mu_{12} + p^2/2\mu_{(12)3} = -\varepsilon_a$).

Theorem 3. For $r_1 \to \infty$ and $r_2 \to \infty$ the asymptotic formula

$$\Psi_{a\nu_1}(r_1,r_2) \approx \{C^{(1)}_{a;\,\nu_1}(r_1/r_2) f_{l_2}(q_2^{(1)}r_2) f_{l_1}(q_1^{(1)}r_1)\exp[-\sqrt{\varepsilon_a}R^{(1)}(r_2,r_1)]/[R^{(1)}(r_2,r_1)]^{3/2}$$

$$-C^{(2)}_{a;\,\nu_1}(r_1/r_2) f_{l_2}(q_2^{(2)}r_2) f_{l_1}(q_1^{(2)}r_1)\exp[-\sqrt{\epsilon_a}R^{(2)}(r_2,r_1)]/[R^{(2)}(r_2,r_1)]^{3/2}\}/r_2r_1 \tag{7}$$

is valid [2], where $q_1^{(j)} = i\,2\mu_{(23)1}\sqrt{\epsilon_a}|r_1-(-1)^j\lambda_2 r_2|/R^{(j)}$ and $q_2^{(j)} = i\,2\mu_{(13)2}\sqrt{\varepsilon_a}|r_2-(-1)^j\lambda_1 r_1|$ $/R^{(j)}$. In (7), the TBANFs $C^{(j)}_{a;\,\nu_1}(r_1/r_2)$ is determined by

$$C^{(j)}_{a;\,\nu_1}(r_1/r_2) = N(-1)^{l_1+l_2+L}\sum_{l_2'l_1'L'}\xi^{(j)}_{L'}\hat{L}'\sqrt{\hat{l}_1\hat{l}_2\hat{l}_2'\hat{l}_1'}W_{a;\,\nu_1'}(q_1^{(j)},q_2^{(j)})\begin{pmatrix} l_2 & l_2' & L' \\ 0 & 0 & 0\end{pmatrix}\begin{pmatrix} l_1 & l_1' & L' \\ 0 & 0 & 0\end{pmatrix}$$

$$\times\begin{Bmatrix} l_1 & l_1' & L' \\ l_2' & l_2 & L\end{Bmatrix},\quad N = (2\pi)^{-5/2}m_3(m_1m_2m_3/m)^{1/2}\varepsilon_a^{1/4}, \tag{8}$$

where $W_{a;\nu_1}(q_1^{(j)}, q_2^{(j)})$ is the OSVF for the virtual decay $a \to 3+2+1$ but written for the variables q_1 and q_2 at $\varepsilon = -\varepsilon_a$, $\xi_{L'}^{(1)}=1$, $\xi_{L'}^{(2)} = (-1)^{L'}$, the cycle and figured brackets are the $(3j)$-and $(6j)$-symbols, respectively, $\nu_1 = \{l_1 l_2 L s_{12} S\}$, l_j is the relative orbital angular momentum of particle j and a core 3, s_{12} is a total spin of the (1,2)- pair and $\hat{L}_o = 2L_o + 1$. One notes that when $m_3 \to \infty$ (a heavy particle 3 for which $m_j/m_3 <<1$), the asymptotics formula for $\Psi_{a;\nu_1}(r_1, r_2)$ passes to the same form for $\Psi_\nu^{(as)}(r, \rho)$ at $r \to \infty$ and $\rho \to \infty$ given by (1) since for the limit at $m_3 \to \infty$ $R^{(j)} = R$ and $C_{a;\nu_1}^{(j)} = C_{a;\nu}$ [2]. As is seen from Egs.(1)–(4), (7) and (8), the true asymptotic formulas obtained for the three-body bound state radial wave functions $\Psi_{a;\nu}(r, \rho)$ and $\Psi_{a\nu_1}(r_1, r_2)$ in the different asymptotic regions are expressed through the unknown TBANFs $C_{a\,\nu}(\varphi)$ and $C_{a;\nu_1}(r_1/r_2)$ for 3+2+1$\to a$.

One note that the residue of the three-body partial amplitude $M_\nu(q, p)$ for the $1 + 2 + 3 \to 1 + 2 + 3$ scattering at the pole $\varepsilon = \varepsilon_a$ is expressed in terms of the OSVF $W_{a;\nu}(q, p)$ for the virtual processes $a \leftrightarrow 3 + 2 + 1$. Therefore, the OSWF and TBANF are fundamental nuclear characteristics for the three-particle (312) bound system and carry an information on the dynamics of strong two-particle interactions. Consequently, knowledge of the TBANF (or the OSVF) allows to get the information both on the three-body cluster structure of halo nuclei and on types of two-particle (cluster-cluster, cluster-nucleon and nucleon-nucleon) interactions as well as on the mechanism of the two nucleon transfer reactions (for example, the $\alpha(^6\text{He}, \alpha)^6\text{He}$ and $\alpha(^6\text{Li}, \alpha)^6\text{Li}(E^*$=3.532 MeV;$0^+)$ reactions). Besides, a value of OSVF (or TBANF) enables one to calculate reliable values of the three-body bound state radial wave function in the asymptotic regions, which in turn are very important for reliable quantitative verification of behaviour of different approximate model three-body bound state wave functions in the asymptotic region of the configuration space. Therefore, systematic collection of data about TBANFs for different three-body halo nuclei must be extremely encouraged nowadays. For this end, at the present time a comparative analysis of the approximate model wave functions with the obtained asymptotic expressions [1–5] is usually done to extract the TBANF values.

3. Comparative analysis for the approximate model three-body bound state model wave functions for the ^{6}He and ^{9}Be nuclei

In this section, we present the result of the analysis of the asymptotic behaviour of the approximate three-body (αnn) and $(n\alpha\alpha)$ radial wave functions for the ground state of ^{6}He and ^{9}Be nuclei obtained in [2,7–9] and [10], respectively, as well as the results of the TBANF values for $\alpha+n+n \to {}^6$He and $n+\alpha+\alpha \to {}^9$Be obtained in Refs. [2, 5] and in Ref. [3], respectively. The approximated radial wave functions $\Psi_{a;\nu}(r, \rho)$ were derived within the multicluster stochastic variational approach for the $^6\text{He}(\alpha nn)$ nucleus [9] and the $^9\text{Be}(n\alpha\alpha)$ one [10], which is based on two dimensional gaussian basic expansion. At this, in [9] the Sack-Biedeharn-Breit for αN-interaction and the hard-core Reid for the NN potential (M1-model; ε_{He}=0.3048 MeV, $\varepsilon_{He}^{(exp)}$ =0.975 MeV) as well as the Kukulin et al potential for αN-interaction and the hard-core Reid for the NN potential (M2-model; ε_{He}=0.1848 MeV, $\varepsilon_{He}^{(exp)}$=0.975 MeV) were used, whereas, in [10] the Ali-Bodmer $\alpha\alpha$ potential and the Kukulin et al αn potential (M3-model; ε_{Be}=1.76 MeV, $\varepsilon_{Be}^{(exp)}$=1.56 MeV) as well as the Buck-Friedrich-Wheatley $\alpha\alpha$ potential and the Kukulin et al αn potential (M4-Model; ε_{Be}=2.86 MeV, $\varepsilon_{Be}^{(exp)}$=1.56 MeV) were applied. In [2, 7, 8], the radial wave function $\Psi_{a;\nu_1}(r_1, r_2)$ were obtained within the Lagrange-mesh approach by using the Kanada et al for αN-interaction and the Minnesota NN potential.

In [1, 3] and [5], the asymptotic behaviour of the approximate three-body (αnn) and $(n\alpha\alpha)$ bound state radial wave functions of [9, 10] has been compared with the asymptotics formulas given by Eqs. (1)–(4) and information has been obtained about the values of the TBANFs for the $\alpha + n + n \to {}^6$He and $n + \alpha + \alpha \to {}^9$Be as a function of the hyperangle φ. To this end, for

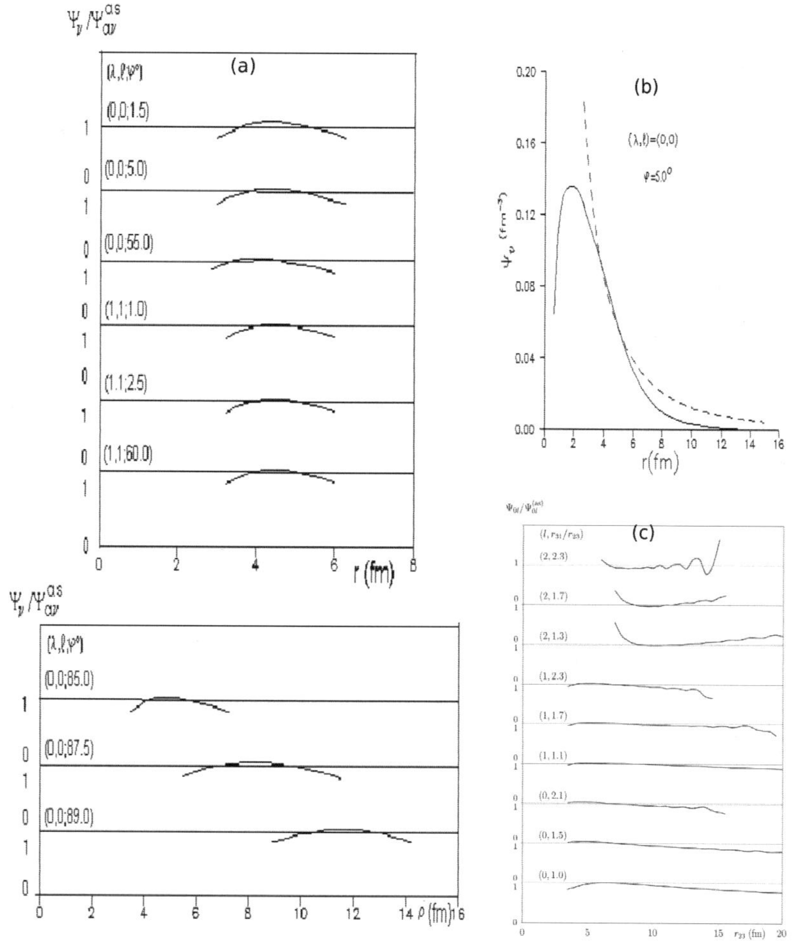

Figure 1. Ratio of the radial wave functions $\Psi_{He;\nu}(r,\rho)$ for M1-model (or $\Psi_{He;\nu_1}(r_1,r_2)$) to the asymptotic expression $\Psi^{(as)}_{He;\,\nu}(r,\rho)$ (or $\Psi^{(as)}_{He;\,\nu_1}(r_1,r_2)$) (a) (or (c), where $r_{31}\equiv r_1$ and $r_{23}\equiv r_2$) obtained in [5]([2]). In (b), the result of a comparison of the numerical radial wave function of [9] for the M1-model (solid line) with the asymptotic wave function (4) (dashed line) calculated in [5] for the ^{6}He(αnn) nucleus at φ=5.0^o.

each fixed value of φ the ratio

$$R_{\lambda lLS}(\varphi)=\frac{\Psi_{\lambda lLS}(r,\rho)}{\Psi^{(as)}_{a\nu}(r,\rho)}C_{a\nu}(\varphi) \qquad (9)$$

was considered. The similar way of testing for the asymptotic behaviour of the approximate three-body (αnn) radial wave functions $\Psi_{He;\,\nu_1}(r_1,r_2)$ was applied in [2] and information has been obtained about the values of the TBANFs $C^{(j)}_{a;\,\nu_1}(r_1/r_2)$ (j=1 and 2).

The analysis performed in Refs. [5] and [3] showed that the approximate wave functions $\Psi_{a;\,\nu}(r,\rho)$ derived in Refs. [9] and [10] for the ^{6}He(αnn) and the ^{9}Be$(n\alpha\alpha)$ nuclei, respectively, have a correct asymptotic behaviour in the fairly narrow range of the asymptotical regions. As an example, this is illustrated in Fig. 1 *a* and *b* only for the ^{6}He(αnn) nucleus. But, as is illustrated in Fig. 1*c*, the approximated radial three-body ^{6}He(αnn) wave functions $\Psi_{a;\nu_1}(r_1,r_2)$ for the ^{6}He [2, 7, 8] have a correct asymptotic behaviour in the fairly wide range of the asymptotical regions.

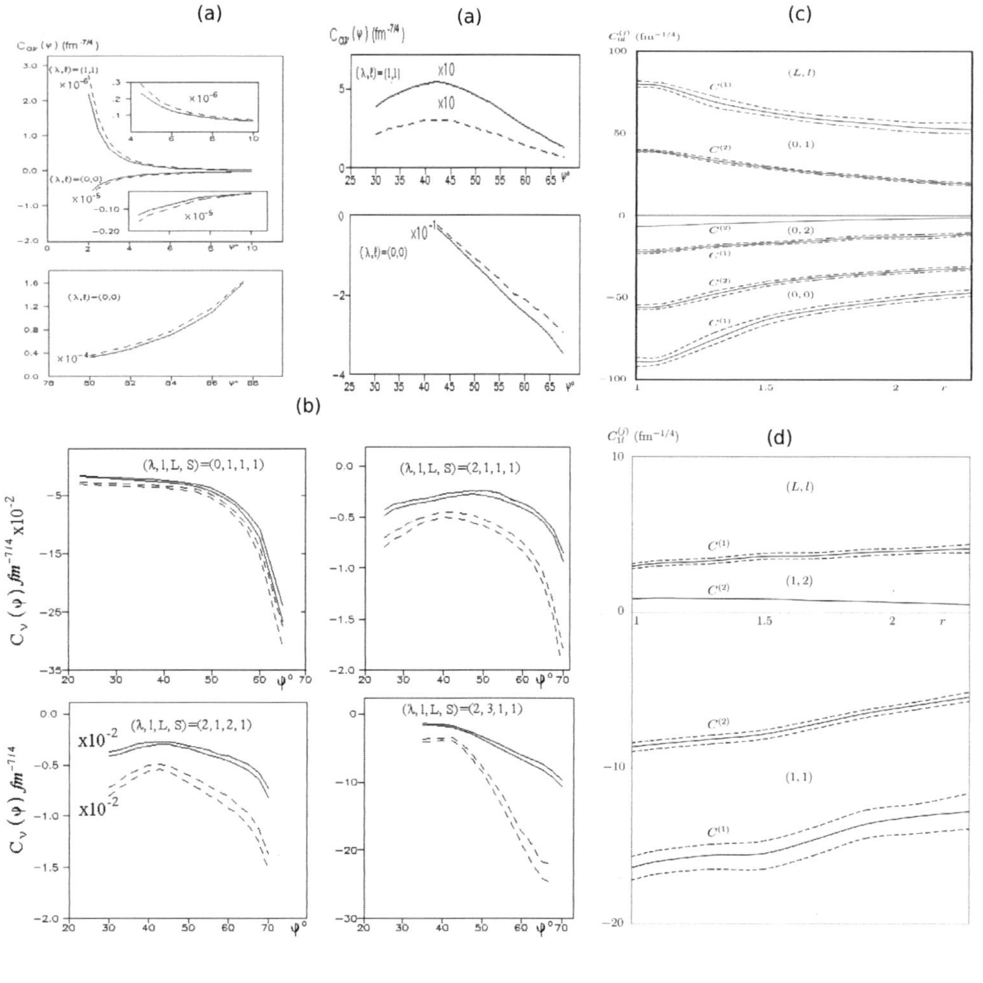

Figure 2. The TBANF values for $\alpha + n + n \to {}^6$He (*a*, *c* and *d*) and $n + \alpha + \alpha \to {}^9$Be (*b*) for different values of the quantum numbers. A detail caption is given in the text

The results for the TBANF values ($C_{^6\text{He};\,\nu}(\varphi)$ and $C_{^9\text{Be};\,\nu}(\varphi)$) recommended in [5] and [3] for the $\alpha + n + n \to {}^6$He and $n + \alpha + \alpha \to {}^9$Be, respectively, are displayed in Figs. 2*a* and 2*b* as a function of the hyperangle φ. In Fig. 2*a*, the solid and dashed lines are related to $\alpha + n + n \to {}^6$He for the potentials of the M1-and M2-models, respectively. In Fig. 2*b*, the solid and dashed curves correspond to $n + \alpha + \alpha \to {}^9$Be obtained for the potentials of the M3-and M4-models, respectively. There the upper and the lower curves present, respectively, the maximum and the minimum values of $C_{^9\text{Be};\,\nu}(\varphi)$. The TBANF values were obtained by matching the approximated radial wave functions with the asymptotical expressions given (1)–(4) for the main components of the quantum numbers (λ, l, L, S). As is seen from this figures the TBANF values are sensitive to the forms of the nuclear αN and $\alpha\alpha$ potentials used. The results for the TBANFs $C^{(j)}_{^6\text{He}\nu_1}(r_1/r_2)$ recommended in [2] are displayed in Figs. 2*c* and *d* as a function of the ratio $r = r_1/r_2$. In Figs. 2*c* and 2*d*, ranges of obtained functions $C^{(j)}_{Ll}(r_1/r_2)$ (delimited by dashed lines) and recommended value (solid lines) as a function of the coordinate ratio r_1/r_2 for $L = S = 0$ and $l_1 = l_2$=0, 1 and 2 (curves in Fig. 2*c*) as well as for $L = S = 1$ and $l_1 = l_2 = 1$ and 2 (curves in Fig. 2*d*).

4. Conclusion

The true asymptotic forms are presented for the three-body (312) bound state wave function of halo nuclei a with two valence neutrons (1=2=n) and with two charged particles (1=2=α) in respect to the Jacobi coordinates and that with two valence neutrons in the relative coordinates of the valence neutrons. These asymptotic expressions contain the unknown TBANF for $3 + 1 + 2 \rightarrow a$ being a new fundamental characteristic three-body (312) nucleus a and relating to the OSVFs for the virtual decay $a \rightarrow 3 + 1 + 2$ given by (6).

The TBANF values for $\alpha + n + n \rightarrow {}^6$He and $\alpha + \alpha + n \rightarrow {}^9$Be recommended in [2, 3, 5] are presented and their sensitivity in the forms of the adopted two-body (αn and $\alpha\alpha$) nuclear potentials is also demonstrated. On the other hand, the TBANFs (OSVFs) for $\alpha+n+n \rightarrow {}^6$He and $\alpha + \alpha + n \rightarrow {}^9$Be can be calculated within the Faddeev's equation in a correct manner.

The TBANFs (OSVFs) are in principle observable quantities. For instance, the TBANFs (or OSVFs) for $\alpha + n + n \rightarrow {}^6$He can be obtained from the exchanged scattering $\alpha({}^6He, \alpha){}^6He$. It would be interesting to compare the present results with experiment. It is possible to get additional information about the form of the two-body potential and to verify an accuracy of the different approximated three-body wave functions as a source of reliable information on the TBANFs. Therefore, systematic collection of information about the TBANF values for different three-body halo nuclei is now encouraged in the future.

Acknowledgments

The work has been supported by the Academy of Sciences of the Republic of Uzbekistan (grant No. FA–F2–F177).

References

[1] Blokhintsev L.D., Ubaydullaeva M.K., Yarmukhamedov R. 1999 *Phys.At.Nucl.* **62** 1289
[2] Yarmukhamedov R.,Baye D., Leclerq-Willain C. 2002 *Nucl.Phys.* A **705** 335.
[3] Blokhintsev L.D., Ubaidullaeva M.K. and Yarmukhamedov R., 2005 *Phys.At. Nucl.* **68** 1372.
[4] Yarmukhamedov R. and Ubaydullaeva M.K. 2008 *Uzbek Math.J.* **4** 15.
[5] Yarmukhamedov R. and Ubaydullaeva M.K. 2009 *Int. J. Mod. Phys.* **18** 1561.
[6] Mukhamedzhanov A.M., Ubaydullaeva M.K., and Yarmukhamedov R. 1993 *Theor. and Math.Phys.* **94** 315.
[7] Baye D., Kruglanski M., Vincke M. 1994 *Nucl.Phys.* A **573** 431.
[8] Baye D. 1997 *Nucl.Phys.* A **627** 305.
[9] Kukulin V.I., Pomerantsev V.N., Razikov Kh.D., Voronchev V.T., Ryzhikh G.G. 1995 *Nucl.Phys.* A **586** 151.
[10] Voronchev V.T., Kukulin V.I., Pomerantsev V.N., and Ryzhikh G.G. 1995 *Few-body System.* **18** 191.

Journal of Physics: Conference Series **569** (2014) 012038 doi:10.1088/1742-6596/569/1/012038

Clustering in atomic nuclei: a mean field perspective

Luis M. Robledo

Departamento de Física Teórica, Universidad Autónoma de Madrid, E-28049 Madrid, Spain

E-mail: luis.robledo@uam.es

Abstract. In this paper the physics of clustering in atomic nucleus as seen from a mean field perspective will be discussed. Special attention is paid to phenomena involving octupole deformation like the α structure of ^{20}Ne or the emission of heavy clusters. The stabilizing role of spin for cluster-like highly deformed states is also discussed in the case of ^{36}Ar.

1. Introduction

Clustering is a fertile field of research in nuclear physics aiming to describe correlation among groups of particles to form aggregates like α particles, the doubly magic nucleus ^{16}O or even ^{14}C [1]. It leads to the understanding of resonances and highly deformed states in light nuclei in terms of cluster aggregates or nuclear molecules as described by the Ikeda diagram [2] of $N = Z$ nuclei, further extended to neutron rich nuclei in Ref [3]. At the heart of clustering is the α particle with it whopping binding energy of 28.3 MeV. However, the binding energy gained by the formation of the α particles inside the nucleus has to compete with an energy loss coming from the increase in the surface of the nucleus as compared to a more compact configuration. It is competition between both effects what determines the cluster nature of a given state. A typical example is the Hoyle state in ^{12}C that is viewed as three α particles located in the vertex of a triangle [4, 5]. Linear chains of α particles lead to hyper-deformed configurations located at a very high excitation energy at low spins. However, at high spins, they get closer to the yrast state and it is easier to characterize them. In heavy systems the existence of clustering is less clear although the phenomenon of cluster emission in the actinides can be regarded as a clustering effect driven by the characteristics of the heavy daughter nucleus.

Among the many theoretical methods used to quantify the physics of clustering we can mention the Anti-symmetrized Molecular Dynamics [6], Fermionic Molecular Dynamics [7] as well as mean field approaches (see [8, 9] for recent applications also dealing with the role of the low density equation of state in clustering). Proposals where the nucleus is depicted as an α particle gas and is described with a simple ansatz for the wave function (THSR wave functions) are also popular [10]. Other alternatives are discussed in the review article in Ref [11]. Last but not least we have to mention the Resonating Group Method of reactions [12]. In this paper, I will use the constrained mean field approach supplemented with some beyond mean field methods (symmetry restoration and configuration mixing) to analyze several physical situations involving clustering: hyper-deformed high spin states in ^{36}Ar with a matter density resembling two ^{16}O plus an α particle, the cluster structure of the first negative parity excited state of ^{20}Ne or the phenomenon of cluster emission characteristic of many actinides.

Journal of Physics: Conference Series **569** (2014) 012038 doi:10.1088/1742-6596/569/1/012038

2. The constrained nuclear mean field

The mean field approach used in the description of the atomic nucleus is based on the Hartree-Fock- Bogoliubov (HFB) approximation [13]. It encompasses in the same framework both the long and short range correlations present in the effective nuclear interaction. In the HFB approximation the elementary excitations are given in terms of the so called Bogoliubov quasiparticles described by creation and annihilation operators which are linear combinations of the creation and annihilation operators of a convenient (typically harmonic oscillator) basis

$$\beta_\mu = \sum_k U_{k\mu} c_k + V_{k\mu} c_k^+, \quad \beta_\mu^+ = \sum_k U_{k\mu}^* c_k^+ + V_{k\mu}^* c_k \tag{1}$$

The Bogoliubov amplitudes U and V satisfy unitary constrains in order to make the quasiparticle creation and annihilation operators preserve fermionic canonical commutation relations. This seemingly too technical comment is relevant because the U and V amplitudes are determined by using the variational principle on the HFB energy and therefore the unitary constraints prevent their immediate use as variational parameters. A more convenient parametrization of the HFB wave function $|\Phi\rangle = \prod_\mu \beta_\mu |-\rangle$is given by the Thouless theorem [13] that establishes that the most general HFB wave function not orthogonal to a reference one $|\Phi_0\rangle$ is given by

$$|\Phi(Z)\rangle = \mathcal{N}(Z, Z^*) \exp(-\sum_{\mu\mu'} Z_{\mu\mu'} \beta_\mu^+ \beta_{\mu'}^+)|\Phi_0\rangle. \tag{2}$$

The factor $\mathcal{N}(Z, Z^*)$ is chosen to normalize the wave function to one. There is a well defined way to compute the U and V amplitudes above as a function of the Z parameters that now are true independent complex parameters. They are determined by the variational principle on the energy $E(Z, Z^*) = \langle \Phi(Z)|\hat{H}|\Phi(Z)\rangle$, where $\hat{H}$ is the phenomenological nuclear hamiltonian already including short range correlations. Popular choices for H are the Gogny, Skyrme and relativistic families of phenomenological interactions. All of them, in a way or another, include density dependent phenomenological terms that have to be handled with care both at the mean field level and beyond. All the calculations in this paper have been performed with the Gogny D1S parametrization [14].

The formation of clusters (or more precisely the existence of cluster-like correlations inside the nucleus) is driven by the nuclear interaction (for instance the huge binding energy of an α particle) and it is difficult to disentangle in a mean field like wave function like the one depicted above. Therefore, the traditional approach to clustering from the mean field side is to look at the spatial density looking for structures resembling cluster aggregates. As those states are usually excited states a simple ground state mean field calculation is not enough and several constraints on quantities characterizing the mass distribution of the nucleus have to be imposed. The energy minimization of the HFB now becomes a constrained minimization that involves the introduction of Lagrange multipliers λ_i and replacing the hamiltonian with

$$\hat{H}' = \hat{H} - \sum_i^{N_c} \lambda_i \hat{C}_i \tag{3}$$

where $\hat{C}_i$ are the operators corresponding to the constrained quantities (plus the constraint on particle numbers). Those are usually taken as the multipoles of the mass distribution $Q_{\lambda\mu} = r^\lambda Y_{\lambda\mu}$ with $\lambda = 2$ (quadrupole) and $\lambda = 3$ (octupole) although it may probe convenient to use other quantities like the neck $n(z_0) = \exp(-(z-z_0)^2/a_0^2)$ or the "slice" $n(z_0, -\infty) = \theta(z_0 - z)$ operators often used in fission calculations. The Lagrange multipliers are fixed by the condition that the gradient of the energy

$$H'^{20}_{\mu\nu} = H^{20}_{\mu\nu} - \sum_i \lambda_i (C_i)^{20}_{\mu\nu} = \frac{\partial E'(Z, Z^*)}{\partial Z_{\mu\nu}} = \langle \hat{H}' \alpha_\mu^+ \alpha_\nu^+ \rangle \tag{4}$$

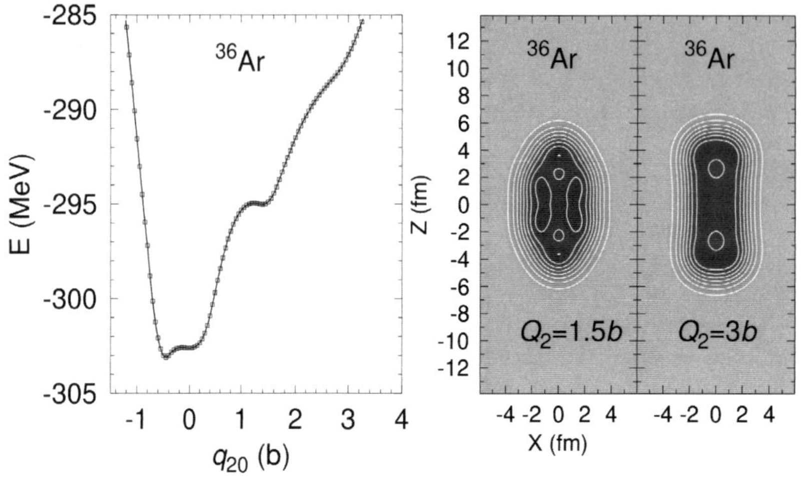

Figure 1. Left panel: the HFB energy of ^{36}Ar is shown as a function of the quadrupole moment. Right panels: contour plots of the matter density at two values of Q_{20}.

has to be orthogonal to the gradient of the constraints

$$(C_i)^{20}_{\mu\nu} = \frac{\partial c_i(Z, Z^*)}{\partial Z_{\mu\nu}} = \langle \hat{C}_i \alpha^+_\mu \alpha^+_\nu \rangle. \tag{5}$$

Here lies the strength of the gradient method to solve the HFB equations: the inclusion of additional constraints do not increase the complexity of the calculation. The present formulation of the problem is a simplified version not including second order curvature terms in the gradient required for a fast and uniform convergence in the solution of the non-linear HFB equations. However the extension of the present formalism to handle that situation is straightforward [15].

3. Clusters at high spin

A common way to search for cluster like structures is to perform a HFB constrained calculation with the quadrupole moment as constraint. If axial symmetry is preserved only linear chains are accessible and the breaking of axial symmetry is required to describe more involved structures like the characteristic triangular shapes of the Hoyle state. A typical example of linear chains is provided by the nucleus ^{36}Ar with $Z = N = 18$ that could show a ^{16}O+^{16}O+^{4}He like structure [16]. In Fig 1 we show the most salient results of our HFB constrained calculation for this nucleus. In the left hand side panel the HFB energy is shown as a function of the quadrupole moment. The nucleus ^{36}Ar shows an oblate ground state but very close in energy to the spherical configuration: this nucleus can be considered as a soft spherical nucleus. At a quadrupole deformation of 1.5 b ($\beta_2 = 0.54$) we have a super deformed minimum which is rather shallow and corresponds to a funny matter distribution that resembles an elongated octagon (see center panel). At even higher quadrupole moments we observe at Q_{20}=3b ($\beta_2 = 0.85$) a shoulder in the energy that could be identify (see the right hand side panel) with a ^{4}He nucleus surrounded by ^{16}O at each side. The only problem with this configuration is that it is only a shoulder in the energy and therefore is unstable at zero spin at the HFB level. However, going to high spin the energy landscape gets modified as a consequence of the rotational energy correction $I(I+1)/(2\mathcal{J})$ and the almost linear increase of the moment of inertia $\mathcal{J}$ with the quadrupole moment. This modification of the energy landscape helps to stabilize the super-deformed shallow minimum as well as the shoulder

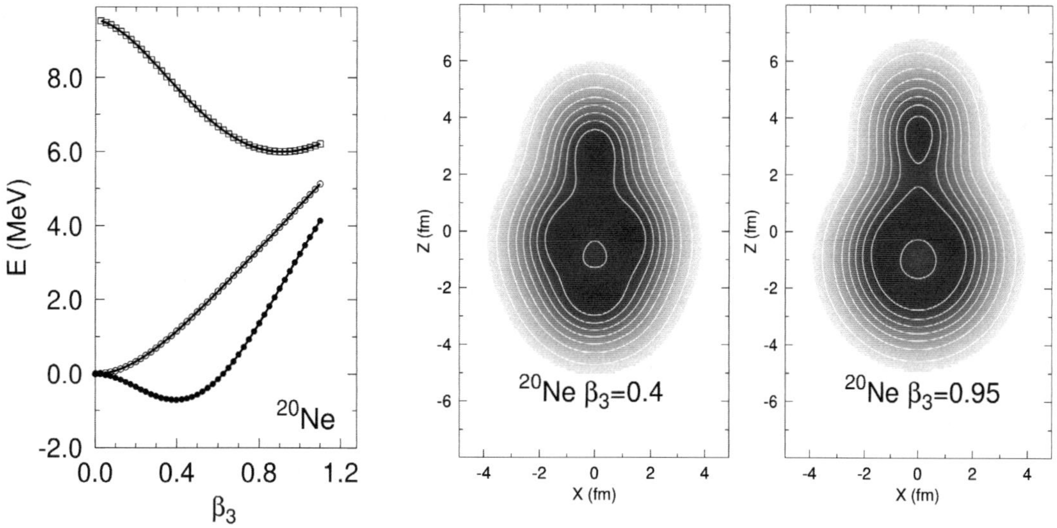

Figure 2. Left panel: the HFB energy (circles), the positive parity projected energy (bullets) and negative parity projected energy (squares) are shown as a function of the octupole moment. Rigth panels: contour plots of the matter density of two relevant configurations are shown. Contour levels values are $0.02fm^{-3}$ apart.

at higher quadrupole moments. In order to analyze the high spin behavior of the nucleus we used in Ref [16] the technique of angular momentum projection to compute the energy as a function of both the quadrupole moment of the intrinsic state and the spin. In addition a Generator coordinate method calculation using the intrinsic quadrupole moment as collective coordinate was carried out on top of the angular momentum projected states. As discussed in Ref [16] the low spin states have an almost spherical intrinsic state. However, at angular momentum $I = 8\hbar$ the lowest state gets shifted to the intrinsic super-deformed configuration. The hyper-deformed intrinsic state corresponding to the exotic linear chain at Q_{20}=3b becomes the lowest energy at $I = 16\hbar$. From these results we conclude that indeed going to high spins helps stabilizing the exotic cluster like structures often observed as shoulders at zero spin in mean field calculations. The same effect is also observed in other nuclei in the region like ^{32}S [17].

4. Cluster structure of ^{20}Ne

The nucleus of ^{20}Ne with its 10 protons and neutrons is a firm candidate to show clustering because it could be portrayed as a ^{16}O plus ^{4}He nuclei. It turns out that this structure is manifest in the matter density of the first negative parity excited state of ^{20}Ne. To describe negative parity states one usually resorts to the octupole degree of freedom [18] due to its breaking of reflection symmetry. The restoration of this symmetry leads to two states, one of positive parity and the other of negative parity that are usually associated to the correlated ground state and the first negative parity excited state, respectively. Extensive studies using this concept lead to the conclusion that just in a handful of nuclei in the actinide and rare earth regions there is a unique intrinsic state for both parities [18]. For the other nuclei each parity has a different optimal intrinsic state that has to be computed using the Variation After Projection

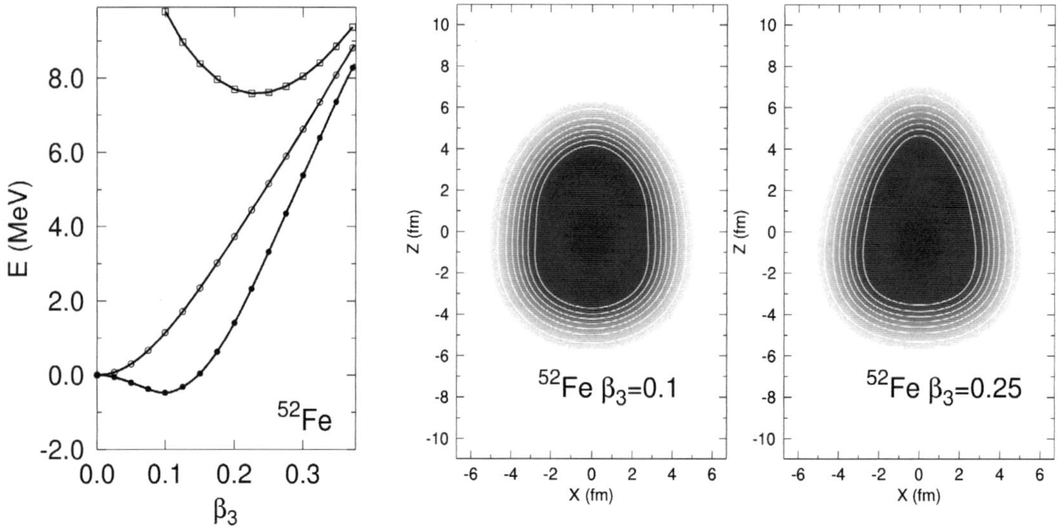

Figure 3. Same as Figure 2 but for ^{52}Fe.

(VAP) method [13] of symmetry restoration. In [18] we implemented the VAP method for parity restoration by first obtaining a set of HFB wave functions with different values of the (axially symmetric) octupole moment $|\Phi(Q_{30})\rangle$ and then computing the energy of the parity projected states [1]

$$E_\pi = \frac{\langle\Phi(Q_{30})|\hat{H}\hat{P}_\pi|\Phi(Q_{30})\rangle}{\langle\Phi(Q_{30})|\hat{P}_\pi|\Phi(Q_{30})\rangle}. \tag{6}$$

The intrinsic configurations (i.e. the intrinsic Q_{30} values) for each parity are determined by looking at the minimum of the projected energies $E_\pi(Q_{30})$. The procedure is exemplified in the left hand side panel of Fig 2 for ^{20}Ne. There we observe three curves that represent the HFB and parity projected energies as a function of the octupole moment computed with the Gogny D1S force. The positive parity curve has a minimum at $\beta_3 = 0.4$ whereas the minimum of the negative parity curve lies at a much higher octupole deformation of $\beta_3 = 0.95$. The excitation excitation energy of the negative parity state is 6.7 MeV and the E3 transition strength to the 0^+ ground state computed with the rotational formula is 12 W.u. The matter distribution of the two intrinsic configurations is depicted in the right hand side panels of Fig 2. For the positive parity intrinsic state we observe the typical density distribution of octupole deformed nuclei with moderate deformation. However, for the negative parity state the intrinsic state has a density that clearly resembles the ^{16}O plus ^{4}He matter distribution. A more quantitative assessment can be obtained by computing the number of particles $N(z, -\infty)$ obtained by integrating the matter distribution from $-\infty$ up to z. This definition takes advantage of the axial symmetry imposed in the calculation.

The clear identification of an α particle in the matter distribution is only possible in light nuclei. For heavier systems it is only possible to look for heavier fragments. To explore this

[1] See Ref [19] for a discussion of the difficulties in computing projected energies with density dependent forces.

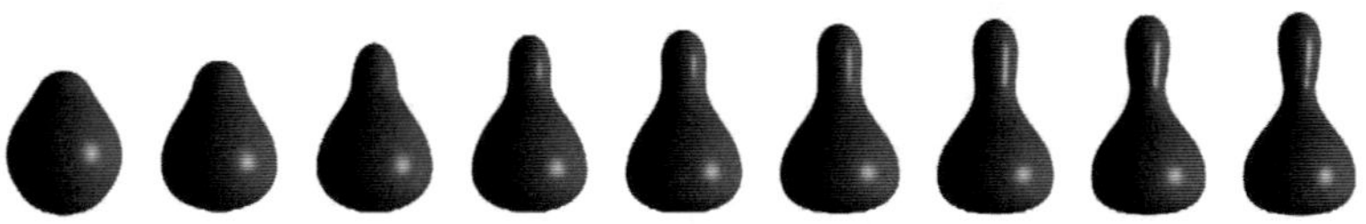

Figure 4. Isosurface plot of the matter density of ^{224}Ra in its evolution from a near ground state configuration with $Q_{30} = 4000\text{fm}^3$ up to a very elongated configuration with $Q_{30} = 52000\text{fm}^3$ in steps of $Q_{30} = 6000\text{fm}^3$. Calculations are performed with the Gogny D1S force.

possibility I have performed calculations in ^{52}Fe which is a ^{12}C away from ^{40}Ca. The results for the parity projected calculations are presented in Fig 3. The main difference is that now the negative parity state lies at a much lower β_3 value than in the ^{20}Ne case and, as a consequence, the matter distribution of the negative parity state has a structure where the ^{12}C is not as prominent as before and therefore it is more difficult to characterize as a cluster structure.

5. Cluster emission

In 1984 Rose and Jones [20], following a prediction by Sandulescu and collaborators [21], reported the results of an experiment where the emission of ^{14}C nuclei by ^{223}Ra was observed for the first time. Since then, the emission of ^{14}C, ^{20}O, ^{23}F, ^{24,26}Ne, ^{28,30}Mg and ^{32}Si light fragments by several isotopes of the actinides Fr, Ra, Ac, Th, U and Pu has been observed -see [22] for a review. This new kind of "radioactivity" lies in between traditional fission and α decay and its energy balance is mostly due to the heavier daughter being close to the doubly magic ^{208}Pb. Cluster emission, a name coined soon after its discovery, was successfully described theoretically by extending models of α decay to handle the emission of heavier clusters - by modifying the pre-formation factor. Before scission the parent nucleus is made of a heavy fragment and a touching lighter one. The mass distribution of such configuration closely resembles the shape of an octupole deformed nucleus with very large deformation parameter β_3. This is the main motivation to attempt the description of "cluster emission" by using the same mean field techniques applied to traditional fission but replacing the quadrupole moment constraint of fission by a constraint on the octupole moment. With this in mind, axially symmetric HFB calculations constraining the shape of the nucleus to have very large values of Q_{30} were carried out. The details are thoroughly described in Ref [23] and section I above. A typical sequence of the evolution of the matter density distributions with increasing octupole moment is depicted in Fig 4 in the ^{224}Ra case.

As a typical example we will consider the nucleus ^{226}Ra. Its HFB energy as a function of the octupole moment is shown in the left panel of Fig 5. The energy sharply increases as a function of Q_{30} until a point at $Q_{30} = 35\text{b}^{3/2}$ where the nucleus breaks apart and the energy starts to decrease as a consequence of Coulomb repulsion (Path 1). The mass split corresponds to the emission of a ^{20}O nucleus. At this point it is also possible to find other configurations with two fragments like the one shown as Path 2. For large octupole moments this is the lowest energy solution and it corresponds to the emission of a ^{14}C cluster. In order to make sure that no additional two fragment solutions are available we carried out a calculation constraining Q_{30} to $50\text{b}^{3/2}$ and the number of particles in the smallest fragment (denoted A_R). The energies obtained are depicted in the right hand side panel of Fig 5 as a function of A_R. We observe the presence of only two minima corresponding to the ^{14}C and ^{20}O fragments, being the ^{14}C the lowest in energy. As discussed in [23] it is possible to compute not only the energy of the nucleus in its path to cluster emission but also the collective inertias (associated with the

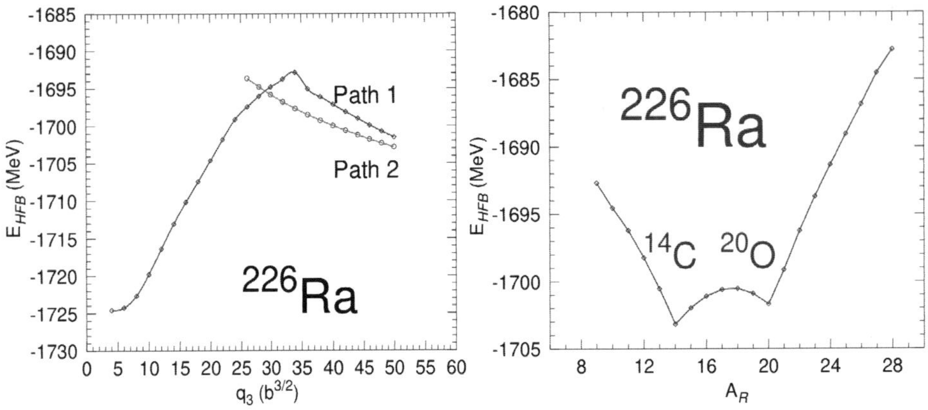

Figure 5. Left panel: HFB energy of ^{226}Ra as a function of the octupole moment. Right panel: once the nucleus splits in two pieces the energy of the configuration can be computed as a function of the mass number of the lighter fragment (A_R).

octupole degree of freedom) what allows the use of the standard techniques in fission to compute lifetimes for spontaneous emission using a WKB like formula. The results for the spontaneous cluster emission half life computed for many actinides are discussed in [23] where it is possible to observe a good agreement with experimental data that extends over several decades.

Acknowledgments
Work supported in part by MICINN grants Nos. FPA2012-34694, FIS2012-34479 and by the Consolider-Ingenio 2010 program MULTIDARK CSD2009-00064.

References
[1] Catford W 2012, *J. Phys: Conference Series* **436** 012095
[2] Ikeda K, Takigawa T, and Horiuchi H 1968 *Prog. Theor. Phys. Suppl.* **465**
[3] von Oerzten W 2001 *Eur. Phys. J. A* **11** 403
[4] Chernykh M, Feldmeier H, Neff T, von Neumann-Cosel P, and Richter A 2007 *Phys. Rev. Lett.* **98** 032501
[5] Furuta T, Hasnaoui K H O, Gulminelli F, Leclerq C, and Ono A 2010 *Phys. Rev. C* **82** 034307
[6] Kanada-En'yo Y, Kimura M, and Ono A 2012 *Prog. Theor. Exp. Phys.* 01A202
[7] Feldmeier H, Schnack J 2000 *Rev. Mod. Phys.* **72** 655; Neff T and Feldmeier 2008 *J. of Phys. Conference Series* **111** 012007
[8] Ebran J P, Khan E, Niksic T and Vretenar D 2012 *Nature* **487** 341
[9] Girod M and Schuck P 2013 *Phys. Rev. Lett.* **111** 132503
[10] Tohsaki A, Horiuchi H, Schuck P and Röpke G 2001 *Phys. Rev. Lett.* **87** 192501
[11] Freer M and Merchant A C 1997 *J. Phys. G: Nucl. Part. Phys.* **23** 261
[12] Baye D and Descouvemont P 1998 *Nucl. Phys. A* **443** 302
[13] Ring P and Schuck P 1980 *The nuclear many body problem* (Springer, Berlin)
[14] Berger J F, Girod M and Gogny D 1984 *Nucl. Phys. A* **428** 23c
[15] Bertsch G F and Robledo L M 2011 *Phys. Rev. C* **84** 014312
[16] Rodriguez-Guzman R, Egido J L and Robledo L M 2004 *Int. J. of Mod. Phys. E* **13** 139
[17] Rodriguez-Guzman R, Egido J L and Robledo L M 2000 *Phys. Rev. C* **62** 054308
[18] Bertsch G F and Robledo L M 2011 *Phys. Rev. C* **84** 054302
[19] Robledo L M 2010 *J. of Phys. G* **37** 064020
[20] Rose H J and Jones G A 1984 *Nature* **307** 245
[21] Sandulescu A, Poenaru D and Greiner W 1980 *Sov. J. Part. Nucl.* **11** 528
[22] Poenaru D N, Nagame Y, Gherghescu R A and Greiner W 2002 *Phys. Rev. C* **65** 054308
[23] Warda M and Robledo L M 2011 *Phys. Rev. C* **84** 044608

Systematics of fine structure in the α decay of deformed odd-mass nuclei

Zhongzhou Ren[1,2,3,4] **and Dongdong Ni**[1,2]

[1]Department of Physics and Key Laboratory of Modern Acoustics, Institute of Acoustics, Nanjing University, Nanjing 210093, China
[2]Joint Center of Nuclear Science and Technology, Nanjing University, Nanjing 210093, China
[3]Kavli Institute for Theoretical Physics China, Beijing 100190, China
[4]Center of Theoretical Nuclear Physics, National Laboratory of Heavy-Ion Accelerator, Lanzhou 730000, China

E-mail: dongdongnick@gmail.com and zren@nju.edu.cn

Abstract.
We present a detailed analysis of the α-decay fine structure in 32 deformed odd-mass nuclei from $Z = 93$ to $Z = 102$. The α-decay half-lives are systematically calculated within the multichannel cluster model (MCCM), which turns out to well reproduce the experimental data and show the neutron deformed shell structure. The branching ratios for various daughter states are investigated in the MCCM and in the WKB barrier penetration approach, respectively. It is found that the MCCM results agree well with the experimental data, while the WKB results have relatively large deviations from the experimental data for the α transitions to the high-lying members of the rotational band.

1. Introduction

Nearly a century ago, Geiger and Nuttall [1] found that the logarithm of the half-life in α decay is inverse proportional to the energy of the outgoing α particle, i.e., the decay energy Q_α. To understand the Geiger-Nuttall rule, the quantum mechanical explanation of α-decay as a quantum tunneling effect was reported independently by Gamow [2] and by Condon and Gurney [3]. Later on, Viola and Seaborg [4] proposed a simple formula for α-decay half-lives of heavy nuclei. The Viola-Seaborg formula is very successful in systematizing the data of α decay and widely used to predict the half-lives of unknown nuclei. Recently, the new Geiger-Nuttall law has been proposed [5] where the effects of the quantum numbers of α-core relative motion are naturally embedded. Since the pioneering work of Gamow [2], α-decay half-lives have been interpreted with improved accuracy for both spherical and deformed α-emitters based on various models and methods [6, 7, 8, 9, 10, 11, 12, 13, 14, 15, 16]. Nowadays the new challenge in α-decay studies is the understanding of the α-decay fine structure observed in heavy deformed nuclei. It has been investigated from both semiclassical and coupled-channel standpoints [17, 18, 19, 20, 21, 22, 23, 24]. The semiclassical methods [17, 18, 19] treat each decay channel as a separate event for one α decay, ignoring the coupling effect of various decay channels. In contrast, the coupled-channel methods [20, 21, 22, 23, 24], based on the three-dimensional Schrödinger equation, employ various channel wave functions with outgoing wave boundary conditions and consider the coupling among the channels.

Recently, we have presented the multichannel cluster model (MCCM) to investigate the α-decay fine structure including half-lives and branching ratios (BRs) [22, 23, 24]. As most α-decay studies, we carried out our investigation following the path from even-even nuclei, to odd-mass nuclei, to odd-odd nuclei. In the coupled-channel study of even-even rotational nuclei, the diagonalization technique [25] and the multipole expansion [26, 27] are separately used to deal with the interaction matrix. Both of them give precise descriptions of the fine structure especially for the BR for excited 4^+ states [22]. In the coupled-channel study of odd-A and odd-odd nuclei, enough decay channels are considered and the calculated results show good agreement with the experimental data [23, 24]. To our knowledge, it was the first coupled-channel description of the fine structure observed in odd-A and odd-odd nuclei. The purpose of this contribution is to present an additional analysis of the α-decay fine structure observed in odd-mass nuclei as a supplement to the previous publication. By comparison with the semiclassical methods, one can discern the reliability of the MCCM based on the coupled-channel approach for the α-decay fine structure.

2. Theoretical models for fine structure observed in α decay

The coupled channel approach has been found to be successful in describing the fine structure observed in α decay. In particular, the MCCM has good applicability for the α decays of well-deformed even-even, odd-mass, and odd-odd nuclei [22, 23, 24]. Within the MCCM, the total wave function of the decaying system is expanded into a sum of partial waves and the cluster radial wave functions satisfy the following coupled equations [20, 21, 22, 23, 24]:

$$\Big[-\frac{\hbar^2}{2\mu}\Big(\frac{d^2}{dr^2}-\frac{\ell(\ell+1)}{r^2}\Big)-(Q_0-E_I)\Big]u_\alpha(r)+\sum_{\alpha'}V_{\alpha,\alpha'}(r)u_{\alpha'}(r)=0, \tag{1}$$

where $\alpha \equiv (n\ell I)$ labels the channel quantum numbers, ℓ is the angular momentum carried by the emitted α particle, Q_0 is the Q_α value for the decay to ground states, E_I is the excitation energy of the daughter states involved in α transitions, and $V_{\alpha,\alpha'}(r)$ is the matrix element of the interaction V taken between channels α and α'. As usual, the deformed potential V consists of the nuclear part of a simple axially deformed Woods-Saxon (WS) form and the Coulomb part in the first-order $\sum_\lambda \beta_\lambda Y_{\lambda 0}(\theta)$ [22]. Also, the multipole expansion of the potential V is used to deal with the interaction matrix elements where the dynamics of rotational core nuclei are considered. The details of the evaluation of the interaction matrix elements can be found in [22, 26, 27].

The coupled equations (1) are solved with outgoing Coulomb-Hankel wave boundary conditions. Moreover, the experimental decay energies $(Q_0 - E_I)$ of each channel are exactly reproduced, and the quantum numbers follow the Wildermuth rule $G = 2n + \ell$ [28] which is an approximation treatment of the Pauli exclusion principle. All these requirements can be simultaneously achieved by adjusting the depth of the nuclear potential. After one obtains the radial wave functions, the partial width of the channel $I\ell$ is given by [22, 23, 24, 26]

$$\Gamma_{I\ell}=\frac{\hbar^2 k_I}{\mu}\frac{|u_{n\ell I}(R)|^2}{G_\ell(k_I R)^2+F_\ell(k_I R)^2}, \tag{2}$$

where R denotes large distances beyond the range of the nuclear potential and beyond the distance where the Coulomb potential can be regarded as spherically symmetric.

In view of the fact that the internal structure of nuclear states has some influences on α transitions as well, the hypothesis of the Boltzmann distribution (BD) for daughter states is proposed $\rho(E_I) = \exp(-cE_I)$ [22, 23, 24], where E_I (in MeV) is the excitation energy of the daughter state I. As early as 1917, Einstein [29] had proposed such a similar hypothesis on the quantum theory of radiation for molecules with a set of discrete states, that is, canonical distribution of states. This hypothesis implies that α-preformation factors have an exponential dependence on the excitation energy of daughter states and show a gradual decline with increasing daughter spin. This is consistent with the other theoretical studies [30, 31]. In order to achieve absolute α-decay half-lives, one also needs to multiply the partial width by an α-preformation factor P_α. Base on the available experimental facts and theoretical analysis [15, 32, 33], the α-preformation factor is kept the same for one certain kind of α emitters (even-even, odd-A, or odd-odd nuclei), making the number of free parameters to a minimum. In this way, one can obtain: $P_\alpha = 0.36$ for even-even nuclei, $P_\alpha = 0.18$ for odd-A nuclei, and $P_\alpha \approx 0.12$ for odd-odd nuclei [22, 23, 24]. Ultimately, the total width is given by $\Gamma = \sum_{I\ell} P_\alpha \rho(E_I)\Gamma_{I\ell}$ and the total α-decay half-life is $T_{1/2} = \hbar \ln 2/\Gamma$. The BR for a daughter state I is expressed as

$$\mathrm{BR}_I^{\mathrm{MCCM}} = P_\alpha \rho(E_I) \sum_\ell \Gamma_{I\ell}/\Gamma \times 100\%. \tag{3}$$

Moreover, one can also describe the α-decay fine structure using the one-dimensional Wentzel-Kramers-Brillouin (WKB) approximation. In this case, each decay channel is treated as a separate event, and the partial widths for various channels are separately calculated with slightly different decay energies and various centrifugal barriers. The WKB barrier penetration probability for the channel $I\ell$ is written as $P_{I\ell} = \exp(-G)$, with

$$G = \frac{2}{\hbar}\int_R^B \sqrt{2\mu\,[V(r) - (Q_0 - E_I)]}\,dr, \tag{4}$$

where $V(r)$ is the sum of the Coulomb and centrifugal potentials, $V(r) = Z_c Z_d e^2/r + \ell(\ell+1)\hbar^2/(2\mu r^2)$, B is the classical turning point satisfying the expression $V(B) = Q_0 - E_I$, and R is the touching radius $R = 1.20A^{1/3}$ fm. It should be particularly noted that the height of the centrifugal barrier at $r = R$ is significantly small relative to the height of the Coulomb barrier at $r = R$,

$$\xi = \frac{\ell(\ell+1)\hbar^2}{2\mu R^2} \Big/ \frac{Z_c Z_d e^2}{R} \simeq 0.002\ell(\ell+1). \tag{5}$$

Using elementary integration techniques and some approximations, one can obtain [8, 24]

$$G = \frac{Z_c Z_d e^2 \pi \sqrt{2\mu}}{\hbar\sqrt{Q_0 - E_I}} - \frac{4\sqrt{2\mu Z_c Z_d e^2 R}}{\hbar}\Big(1 - \frac{\xi}{2}\Big). \tag{6}$$

It is seen that the effect of the centrifugal barrier is small but non-negligible. Combining with the BD hypothesis, the BR for a daughter state I is written as

$$\mathrm{BR}_I^{\mathrm{WKB}} = \frac{\rho(E_I)\sum_\ell P_{I\ell}}{\sum_{I\ell} \rho(E_I) P_{I\ell}} \times 100\%. \tag{7}$$

3. Numerical results and discussion

For odd-mass α emitters, the structure of the decaying state is generally different from that of the ground-state rotational band in the daughter nuclei [34]. As a consequence, there exists competition between α transitions to the ground-state rotational band and to the favored rotational band whose structure is similar to that of the decaying state. In general, the transitions to the favored band play a major role in the α decay of odd-A nuclei [34]. In the following, we will focus on such α transitions. Note that enough channels are considered in numerically integrating the coupled equations (1) for proper convergence. This means that all

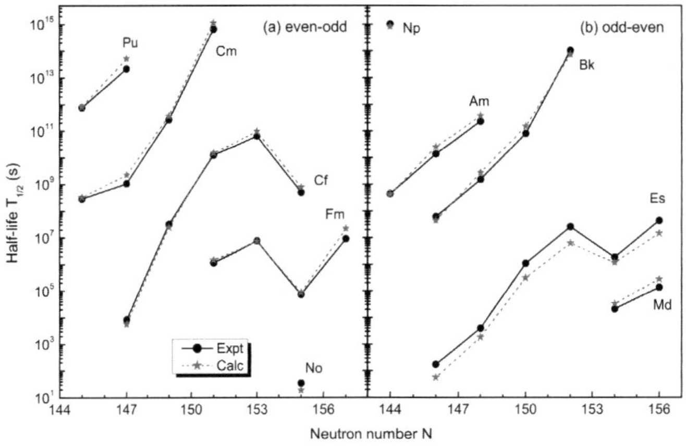

Figure 1. (Color online) Comparison of the calculated α-decay half-lives with the experimental data (a) for deformed even-odd nuclei with $Z = 94 - 102$ and (b) for deformed odd-even nuclei with $Z = 93 - 101$. The deformed neutron shell effect is shown at the neutron number $N = 152$.

partial waves with $\ell \leq 8$ are included in the present calculations [22, 23]. According to this, the number of partial waves required here becomes much larger as compared with the case of even-even nuclei. This is because an α transition $J^{\pi} \rightarrow I^{\pi'}$ in odd-A nuclei allows several decay channels that are characterized by the even ($\pi\pi' = 1$) or odd ($\pi\pi' = -1$) ℓ values in the range of $|J - I| \leq \ell \leq (J + I)$. For example, the α decay of odd-A nuclei from ground $5/2^+$ states to $K = 5/2^+$ rotational bands exhibits 21 decay channels, while the α decay of even-even nuclei from ground 0^+ states to ground-state 0^+ rotational bands exhibits only 5 decay channels.

First, we examine the evaluation of total α-decay half-lives. The comparison of the calculated α-decay half-lives with the experimental data is illustrated in figure 1 for the $Z = 93 - 102$ isotopes. Circles denote the experimental data and stars stand for the theoretical results. One can see that the theoretical results follow the experimental data well, although the half-lives span many orders of magnitude from 10^1 to 10^{15} s. In general, the decay energies of an isotopic chain decrease with increasing neutron number, leading to an increase in the half-life. But there is an abnormal decrease in the half-life across the neutron number $N = 152$, as shown in the Cf, Es, and Fm isotopic chains. This is attributed to the deformed neutron shell closure $N = 152$.

Next, we pay attention to the BRs for various daughter states. Figure 2 displays the comparison of the measured and calculated BRs for the α transitions to seven low-lying members of the favored rotational bands. Here, our calculations are separately performed within the MCCM and in the WKB approximation. In figure 2, the upper panel (a) shows the results calculated within the MCCM and the lower panel (b) shows the results calculated in the WKB approximation. For the MCCM calculations, the calculated results show good agreement with the experimental data including the BRs for the high-lying members of the favored band. Specifically, the values of $\log_{10}$[BR(calc)/BR(expt)] are generally within the range of about ± 0.5, which corresponds to the values of the ratio BR(calc)/BR(expt) within the range of about 0.32−3.2. The largest deviation occurs at the α transition from ^{249}Cf to the fifth member of the favored band. For the WKB calculations, the calculated results agree well with the experimental data for the transitions to the low-lying three members of the favored band. But the deviations from the experimental data become quite large as we proceed to the other four members. For the transition to the seventh member of the favored band, the WKB calculations underestimate

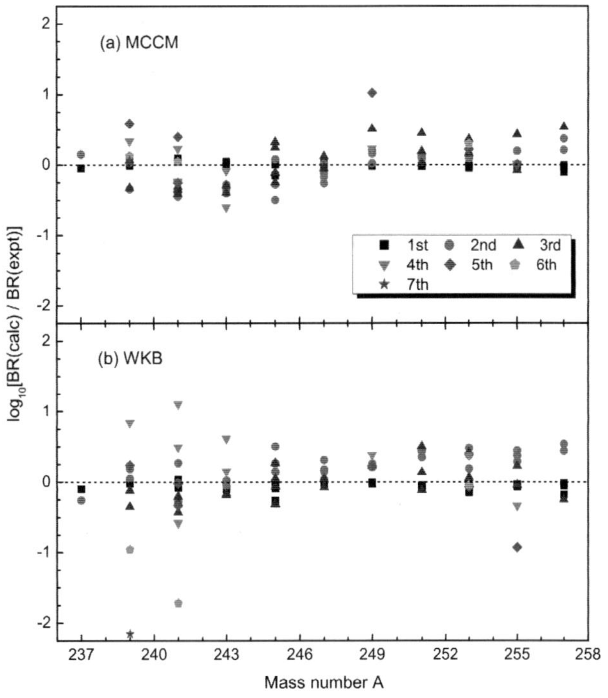

Figure 2. (Color online) Comparison of the calculated and experimental branching ratios as a function of the parent mass number A. Calculations within the MCCM and in the WKB approximation are, respectively, shown in the upper and lower panels. Squares, circles, uptriangles, downtriangles, diamonds, pentagons, and stars represent the transitions to the low-lying seven members of the favored rotational band, respectively.

the BR by about two orders of magnitude. On the whole, the points of all kind gather toward the line $\log_{10}[\mathrm{BR(calc)}/\mathrm{BR(expt)}] = 0$ more evidently within the MCCM than in the WKB approximation. This is particularly considerable for the transitions to the high-lying members of the favored band. In the present study, the standard deviation of the MCCM calculations for 103 BRs is $\sigma = \{\sum_{i=1}^{103}[\log_{10}(\mathrm{BR}^i_{\mathrm{calc}}/\mathrm{BR}^i_{\mathrm{expt}})]^2/102\}^{1/2} = 0.254$ corresponding to a factor of roughly 1.80, while the standard deviation of the WKB calculations is 0.41 corresponding to a factor of about 2.57. This demonstrates that the MCCM is appropriate and successful in describing the α-decay fine structure. Along with the enhancement of experimental sensitivity and the development of experimental techniques, measurements of the BRs would be most welcome to further examine the validity and reliability of the MCCM.

4. Summary

In summary, the α-decay half-lives of 32 odd-mass nuclei with $Z = 93-102$ are calculated within the MCCM and compared with the experimental data. It is found that the theoretical results are in good agreement with the experimental data and the deformed neutron shell effect is reflected at $N = 152$. Moreover, the branching ratios for the transitions to the favored rotational band are evaluated within the MCCM. For comparison, they are calculated in the WKB approximation as well. It turns out that the results obtained from the WKB approximation deviate from the experimental data evidently for the transitions to the high-lying members of the favored band, while the results obtained from the MCCM agree well with the experimental data. This confirms the good reliability of the coupled-channel approach so that coupled-channel predictions are allowed for the α-decay fine structure in the heavier mass region.

Acknowledgments

This work is supported by the National Natural Science Foundation of China (Grant Nos. 10735010, 10975072, 11035001, 11120101005, and 11375086), by 973 National Major State Basic Research and Development of China (Grants Nos. 2013CB834400 and 2010CB327803), by a Project Funded by the Priority Academic Programme Development of JiangSu Higher Education Institutions (PAPD), by the Research Fund of Doctoral Point (RFDP; Grant No. 20100091110028), by the Science and Technology Development Fund of Macau (Grant No. 068/2011/A), and by a China Postdoctoral Science Foundation funded project.

References

[1] Geiger H and Nuttall J M 1911 *Philos. Mag.* **22** 613
[2] Gamow G 1928 *Z. Phys.* **51** 204
[3] Condon E U and Gurney R W 1928 *Nature* **122** 439
[4] Viola V E and Seaborg G T 1966 *J. Inorg. Nucl. Chem.* **28** 741
[5] Ren Y and Ren Z 2012 *Phys. Rev.* C **85** 044608
[6] Buck B, Merchant A C, and Perez S M 1993 *At. Data Nucl. Data Tables* **54** 53
[7] Royer G 2000 *J. Phys. G* **26** 1149
[8] Poenaru D N, Plonski I H, and Greiner W 2006 *Phys. Rev.* C **74** 014312
[9] Xu C and Ren Z 2006 *Phys. Rev.* C **73** 041301(R) Xu C and Ren Z 2006 *Phys. Rev.* C **74** 014304
[10] Denisov V Yu and Ikezoe H 2005 *Phys. Rev.* C **72** 064613
[11] Mohr P 2006 *Phys. Rev.* C **73** 031301(R)
[12] Kelkar N G and Castañeda H M 2007 *Phys. Rev.* C **76** 064605
[13] Chowdhury P R, Samanta C, and Basu D N 2008 *Phys. Rev.* C **77** 044603
[14] Delion D S, Liotta R J, Schuck P, Astier A and Porquet M G 2012 *Phys. Rev.* C **85** 064306
[15] Ni D and Ren Z 2009 *Phys. Rev.* C **80** 051303(R) Ni D and Ren Z 2010 *Phys. Rev.* C **81** 024315
[16] Qian Y, Ren Z and Ni D 2011 *Phys. Rev.* C **83** 044317
[17] Xu C and Ren Z 2006 *Nucl. Phys.* A **778** 1
[18] Denisov V Yu and Khudenko A A 2009 *Phys. Rev.* C **80** 034603
[19] Santhosh K P, Sahadevan S, and Joseph J G 2011 *Nucl. Phys.* A **850** 34
[20] Delion D S, Peltonen S, and Suhonen J 2006 *Phys. Rev.* C **73** 014315
[21] Peltonen S, Delion D S, and Suhonen J 2008 *Phys. Rev.* C **78** 034608
[22] Ni D and Ren Z 2010 *Phys. Rev.* C **81** 064318 Ni D and Ren Z 2011 *Phys. Rev.* C **83** 067302
[23] Ni D and Ren Z 2012 *Phys. Rev.* C **86** 054608
[24] Ni D and Ren Z 2013 *Phys. Rev.* C **87** 027602
[25] Hagino K, Rowley N, and Kruppa A T 1999 *Comput. Phys. Comm.* **123** 143
[26] Kruppa A T, Barmore B, Nazarewicz W, and Vertse T 2000 *Phys. Rev. Lett.* **84** 4549 Barmore B, Kruppa A T, Nazarewicz W, and Vertse T 2000 *Phys. Rev.* C **62** 054315
[27] Suzuki Y and Ohkubo S 2010 *Phys. Rev.* C **82** 041303(R)
[28] Wildermuth K and Tang Y C 1994 *A Unified Theory of the Nucleus* (Academic Press, New York)
[29] Einstein A 1917 *Phys. Z* **18** 121
[30] Stewart T L, Kermode M W, Beachey D J *et al* 1996 *Phys. Rev. Lett.* **77** 36
[31] Delion D S 2009 *Phys. Rev.* C **80** 024310
[32] Hodgson P E and Běták E 2003 *Phys. Rep.* **374** 1
[33] Iriondo M, Jerrestam D and Liotta R J 1986 *Nucl. Phys.* A **454** 252
[34] NNDC of the Brookhaven National Laboratory, http://www.nndc.bnl.gov.

Journal of Physics: Conference Series **569** (2014) 012040 doi:10.1088/1742-6596/569/1/012040

True ternary fission of ^{252}Cf(sf), the collinear decay into fragments of similar size

W von Oertzen[1,2] and A K Nasirov[2]

[1] Helmholtz-Zentrum Berlin, 14109 Berlin, Germany
[2] Joint Institute for Nuclear Research, Dubna, Russia

E-mail: oertzen@helmholtz-berlin.de

Abstract. The ternary decay in ^{252}Cf(sf,fff), with three cluster fragments of different masses (e.g.^{132}Sn,$^{52-48}$Ca,$^{68-72}$Ni), has been observed by the FOBOS group in JINR. This work has established a new decay mode of heavy nuclei, the collinear cluster tripartition, (CCT). This "true ternary fission" of heavy nuclei has been predicted many times in theoretical works during the last decades. In the present report we discuss true ternary fission (FFF) into three nuclei of almost equal size (e.g. Z=98 $\rightarrow Z_i$ = 32, 34, 32) and other fission modes in the same system. The possible fission channels for ^{252}Cf(sf) are predicted from potential-energy (PES) calculations. These PES's show pronounced minima for several ternary fragmentation decays, suggesting a variety of collinear ternary fission modes. The FFF-decays have very similar dynamical features as the previously observed collinear CCT-decays, the central fragment has *very small* kinetic energy. The data of the cited experiment allow the extraction of the yield for some FFF-decays, by using specific gates on the measured parameters.

1. Introduction

Binary fission has been studied intensively over the last decades, for an overview there is the book edited by C. Wagemans: The Nuclear Fission Process [1], covering all important aspects of this process. The term ternary fission, has been used for light particle accompanied binary fission. For such decays, when a third light particle is emitted *perpendicular* to the binary fission axis compilations are available [2]. These must be considered to be emitted from an oblate configuration, the corresponding yields decrease strongly as function of increasing mass(charge) of the third particle [2]. Recent experimental observations and numerous theoretical predictions [3, 4, 5] suggest, however, that in heavy systems and for ternary fragments with larger charge, ternary *collinear* decay from a prolate configuration is preferred. The various ternary decay modes studied here, are in a collinear geometry and are now referred to as "true ternary fission", see also e.g. Zagrebaev *et al.* in ref. [6]. More recent surveys of clustering effects in fission and other binary decays can be found in recent articles by G. Adamian, N. Antonenko and W. Scheid [7], and by D. Poenaru and W. Greiner [8]. Quite important for the discussion of the relative probabilities of oblate to prolate ternary fission is the experiment by Schall *et al.* [9], performed to detect ternary fission from an oblate configuration with three heavier fragments using several detectors covering large (90°) angles each to detect a *triangular shape of the decay-vectors.* This experiment gave a negative result, with a upper limit of the probability for this decay of $1.0 \cdot 10^{-8}$/(binary fission).

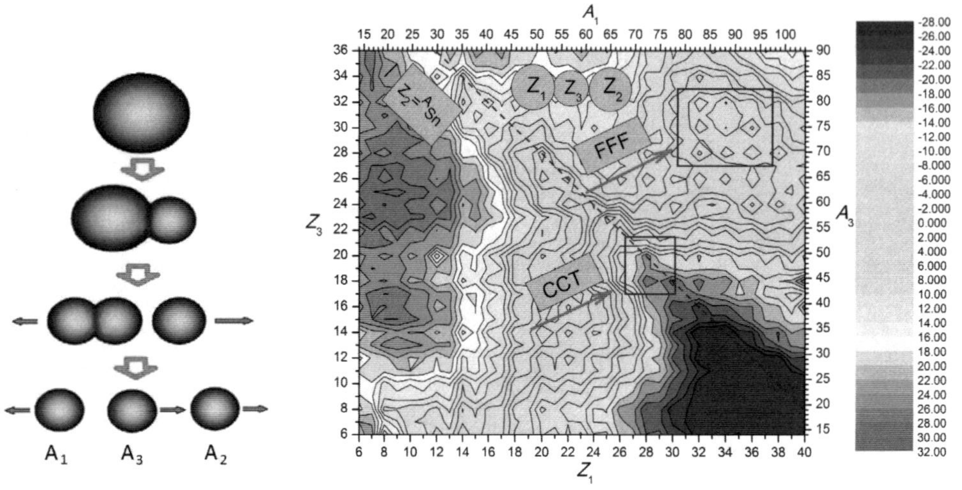

Figure 1. Left side: Schematic illustration of sequential collinear ternary fission into fragments of comparable mass. Right side: Contour plot of the PES as function of two parameters, Z_1, Z_3, showing the preference of the arrangement with $Z = 20$ at the center: (Ni, Ca, Sn).

2. Ternary collinear decays

Ternary fission into fragments with comparable masses is a process, which occurs in heavy nuclei under conditions of large values of the fissility parameter: X, for the ratios $Z^2/A > 31$. The decay into three heavier fragments (true ternary fission) is found to be collinear, as in fact often predicted in the last decades [3, 4, 5]. Recent experiments of two fragment coincidences with two FOBOS-detectors [10] placed at 180°, using the missing mass approach, have established the phenomenon of *collinear cluster tripartition*, the CCT-decay, for the spontaneous decay of ^{252}Cf(sf,fff). A similar ternary fission mode has been observed for neutron induced fission in ^{235}U(n_{th},fff), see refs. [10, 11, 12]. In this fission mode, the CCT, typical fragments are isotopes (clusters, nuclei with closed shells) of Sn, Ni, and Ca. The latter, Ca, as the smallest third particle is positioned along the line connecting Sn and Ni, in this way minimizing the potential energy. This is illustrated in Fig. 2.

In the experiments described in refs. [10, 11], two of the three fragments, moving towards one of the detectors (called arm1) are dispersed in angle in the source backing. It is assumed that due to this small angular dispersion one of the fragments is lost in arm1 on a structure in front of the detectors. Thus binary coincidences are obtained with the missing mass method (A_3, missing Ca isotopes). This new exotic decay can be understood as the breakup of very (prolate deformed) elongated hyper-deformed shapes, as discussed for hyper-deformation in ^{236}U in ref. [13]. The decay is considered with two sequential neck ruptures [12, 14], as illustrated in Fig.1. Actually the central fragment A_3 has extremely low energy (see Fig.5) and is mostly lost.

The main experimental effect in the missing mass method in the binary coincidences is the difference in the counting rates (mass spectra) in the two arms of the coincidence arrangement: Only in one arm1 (see ref. [10]) two fragments of the ternary decay travel through the dispersive media, the source backing and the foils of the start detector. Thus the dispersive effect (angular dispersion of 1-2°) of the two fragments from the ternary decay, is only present in arm1! Thus the missing mass effect appears in the counting rate difference of the mass spectra of arm1 and arm2 (as the difference N(arm1)-N(arm2)).

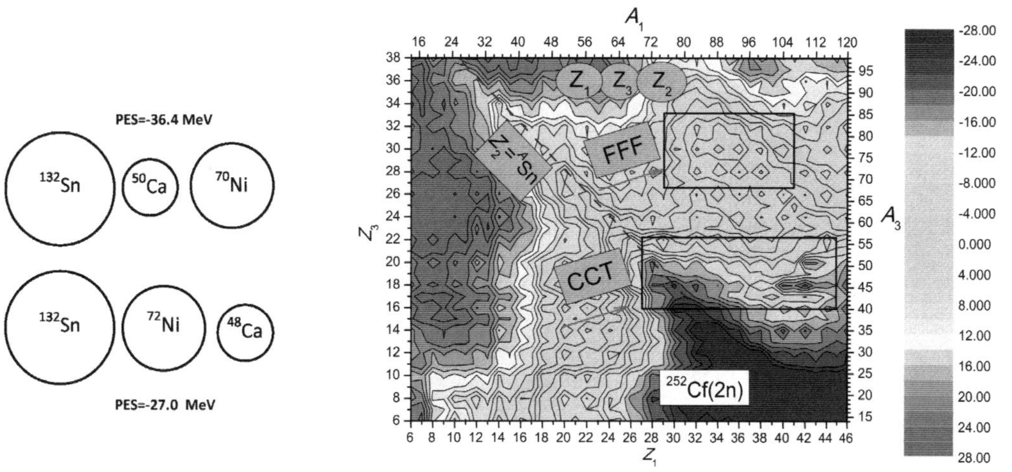

Figure 2. Schematic illustration of the Coulomb energy in ternary fission of ^{252}Cf into three cluster fragments, with isotopes of Sn, Ca and Ni. Right side: The contour map of the potential energies surface (PES) for decays with three nuclei in an arrangement, as indicated in the figure. The PES is shown as function of two parameters Z_1 and Z_3, with the possible variations of the charges and masses of the decay products with values of (A_i/Z_j), for A_3 from 16 - 100 and for A_1 from 18 – 120. This PES has a larger scale, showing the FFF-cases with three fragments of $Z = (32, 34, 32)$ and other cases with $Z_3 = 18$ at the center and $Z_{1,2} = (40,42)$.

3. Potential energy surfaces

The fission processes is considered to be a statistical decay of the total (compound) nucleus, CN, as already discussed in 1939 by N. Bohr and J.A. Wheeler [15]. Actually the ternary decay occurs with two neck ruptures in a short time-sequence, each of the ruptures being governed by the internal barriers and the potential energies surface (PES) in each configuration, as also discussed in ref. [16]. The PES are obtained as described in ref. [17]. The total *phase space* of these statistical decays is determined by:

i) the energy balance and thus the details of the *potential energy surface*, PES, namely, its valleys and hills,

iii)the internal barriers for the two necks,

iv) the Q_{GGG}-values, the latter determining the kinetic energies and the number of possible fragment (isotope) combinations,

v) the excitation energy range in the individual fragments,

vi) the momentum range,

vii) the number of excited states (or the density of states) in each of the fragments, the combinations consisting of 2(3) isotopes, and by

viii) the spin (J) multiplicity in these excited states with spins up to $(6\text{-}8)^+$ (phase space factor $(2J + 1)$).

For the ternary decays of ^{252}Cf, the PES's, the contour-plot in Fig.2 shows distinct minima for various charge combinations with $\Sigma_Z = 98$:

i) for CCT $Z_3 = 20$, and $Z_1 = 28$, this is the main CCT-mode in refs. [10] and

ii) less pronounced for $Z_3 = 28$, and $Z_1 = 20$, as observed in ref. [10, 11]. The complementary fragments with Z_2 are isotopes of Sn ($Z = 50$). The PES shows a pronounced valley with charge $Z_2 = 50$, with the closed shell for the number of protons, because of the dominance of the Coulomb interaction, the proton shells are most important. In addition we observe a

Journal of Physics: Conference Series **569** (2014) 012040 doi:10.1088/1742-6596/569/1/012040

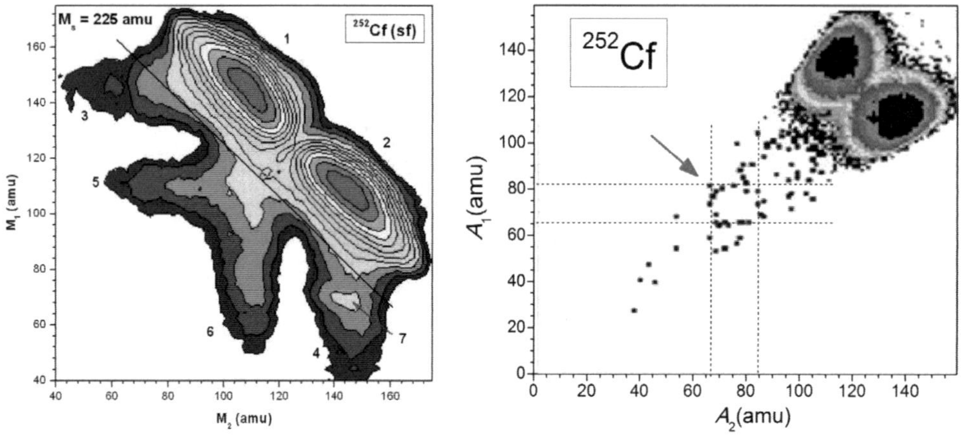

Figure 3. Left side: Binary coincidences of fragments in the spontaneous decay of ^{252}Cf(sf) from the inclusive data measured in ref. [10], the outer fragments (A_1, A_2, M_1, M_2) are registered. Right side: For the selection of FFF decays, a gate with the condition for velocities and momenta, ($V_1 \approx V_2$, and $P_1 \approx P_2$) is chosen. Remnants of the binary fragments in coincidences are seen. The region of the FFF-decays is marked. Missing fragments (A_3), here are isotopes of nuclei with ($Z_3 = 30 - 36$). Points above this region are true coincidences of other symmetric ternary decays, from ref.[18].

pronounced region of minima for the symmetric charge combinations with three comparable fragments, for the FFF-decays (this decay is marked as FFF):
iii) *(Z_1 = 32, 34, 32), Z_3 = (34, 32, 30)*, the fragment Z_2 has an equivalent role as the other two and Z-values Z_2 = 32, 34, occur - we have an almost symmetric ternary decay. Because of this fact, permutations of the labels including Z_2 in the figure of the PES (contour plots in Fig.2) will produce similar results, and a symmetric shape of the coincident events. This is suggested by the rectangular two-dimensional field of the binary coincident events in the favored region (obtained by the reflection of the labels) in Fig. 3.

4. Experimental observations

In the original observation of CCT a wide region of missing fragment masses has been observed, which is manifested as a wide bump in the difference of the counting rates between arm1 and arm2. The yield contains many ternary mass combinations, seen as a large background in this mass region (see in the left side of Fig.3). For the symmetric decay we have to remove the background in the middle mass region. In these experimental data, we have the possibility to select events with the conditions on comparable values of the velocities $V_3 \approx V_2$ and $V_3 \approx V_1$, defining a symmetric decay, see Fig. 3, right side. This choice reduces completely the background due to the scattering tails seen in the inclusive experimental data. This background dominates the inclusive original data shown in the Fig. 3, left side, from ref. [10]. In the experiment the masses and velocities of the fragments are obtained by measurements of time of flight and energy. Because of the symmetry discussed before, the events appear in a rectangular region (for A_1, A_2) in the ranges of masses for all three fragments: (64 – 84) and (64 – 82). This corresponds to the true ternary fission into three equivalent fragments, (FFF) in ref. [18], as predicted by the PES's, Fig. 2.

As a surprise we observe in the PES for ^{252}Cf another region for FFF decays, they are suggested by the minima in the region of charges with $Z_1 = 40 - 42$, combined with values of $Z_3 = 18$, which for Cf (Z = 98) has to be complemented by a fragment with charges $Z_2 =$

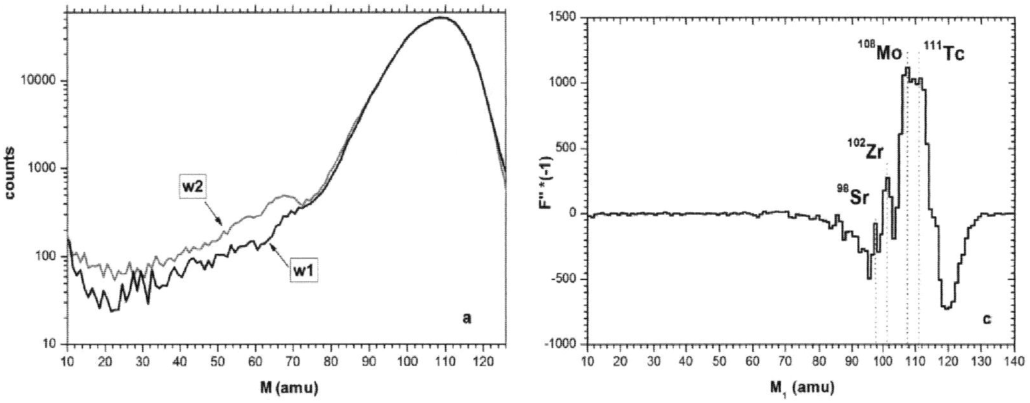

Figure 4. Extraction of the yield of the "Mo-Ar"-mode [22]. The mass spectra, the data are projected to the A_1 axis for arm1 and arm2, left side, measured in ref. [10] for binary coincidences. Right side: the derivative of the difference of the the two spectra, events in the region of ^{106}Mo appear. Missing fragments, (A_3), here are isotopes of Ar with $(Z_3 = 18)$.

40. This is again a rather symmetric decay. The dominating effect comes from the closed shell properties of the central smaller fragment with $Z_3 = 18$ and masses $A_3 = 44$-46, which appears as a closed neutron shell with N = 28 (^{46}Ar). Actually the isotope ^{46}Ar shows strong closed shell effects with a 2^+-state at 1.554 MeV as described in ref.[19]. Also the properties of the *Mo-isotopes* are remarkable, see recent references in ref.[20]. In a binary coincidence study the formation of the pair (^{106}Mo+^{144}Ba, two neutrons missing) has the highest yield. There is a sharp transition from spherical to prolate deformation in the region of A = 102(spherical) to A = 106 (deformed with $Q = 5$ *barn*). Quite important we note, that the favored region with Z_3 = 18, corresponding to (^{46}Ar) appears also as a valley in the PES for ^{236}U, U(Z = 92). Now the complementary charges are $Z_1 = 34$, 36 and $Z_2 = 40$. This case is not as symmetric as the previous one, the persistence of the strong binding in the lighter fragment points to the fact, that the valley of the binding energy as function of N and Z forms a steeper parabola for lighter nuclei, compared to the more shallow valley of the binding energies in heavier nuclei.

Inspecting the PES in Fig. 2 there are other ternary decay modes suggested by minima of the potential energies, with e.g. larger values of the masses/charges of the fragments at the two borders of the total (prolate) nucleus. In addition there are deep valleys, particularly for smaller central fragments, for A_3, such as $A_3 = {}^{22}\mathrm{Ne}(Z_3 = 10)$, ${}^{18--20}\mathrm{O}(Z_3 = 8)$, as well as for C-isotopes, see ref. [18].

In order to extract the yield of the FFF-mode with the (^{46}Ar) as the central fragment, we need a presentation of the data without background. We use as previously the difference between the counting-rates between arm1 and arm2, in the region of interest (Fig.3). As suggested by Kamanin and Pyatkov [22] we can obtain the counting rate for this specific ternary channel covered by the background in the inclusive data. In arm1 the CCT-events are removed from the smooth region, and they appear at a lower mass value of the missing mass, the complement to [A_1 - A_3]. The complement (the missing mass) is stopped at the entrance grid of the ionization chamber. This extra counting rate can be made visible by taking the derivative of the difference in the spectra of N(arm1-arm2) (right side in Fig.4). A rather strong peak appears in the mass region of A_2 = (^{106}Mo - ^{110}Tc).

A final remark on the possibility of triple coincidences. It has often been asked to make experiments, where all three fragments are registered. The kinematics of the process has been studied by Vijayaraghavan *et al.* [14] using the assumption of a sequential decay process. The

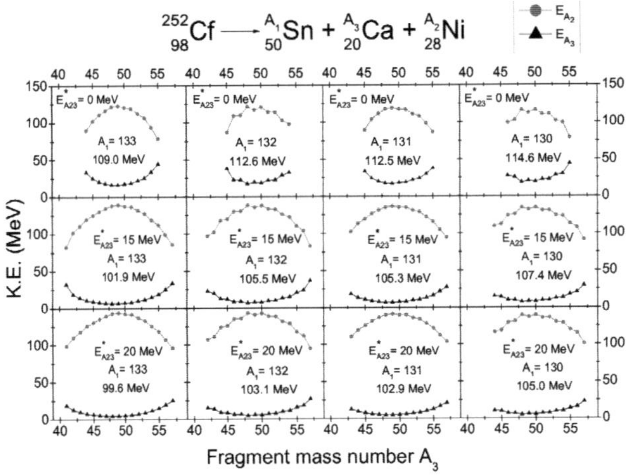

Figure 5. Kinetic energies of fragments in the sequential decay $A \rightarrow (A_1 + A_{23})$, $A_{23}(\mathrm{Ex}) \rightarrow A_3 + A\ _2)$ as obtained in ref. [14]. The kinetic energies of A_2 and A_3 are shown. With different excitation energies in A_{23} of 0.0, 15 MeV and 30 MeV. For only 10 MeV the kinetic energy of A_3 is very small and A_3 can not be detected (figure from ref.[14]).

kinetic energies of all three fragments are obtained, they are shown in Fig.5. as function of increasing excitation energy of the intermediate fragment A_{23}, which is needed for a sizeable probability of fission. The kinetic energy of A_3 becomes very small. Actually it can be expected for most (fff) experiments, that A_3 is absorbed in the target and the target backing.

References

[1] Wagemans C (ed) 1991 *The nuclear fission process*, (CRC Press Inc., Boca Raton, Florida USA)
[2] Gönnenwein F 2004 *Nucl. Phys.* A **734** 213
[3] Diehl D, Greiner W 1974 *Nucl. Phys.* A **229** 29
[4] Poenaru D N, Gherghescu R A and Greiner W 2005 *Nucl. Phys.* A **747** 182
[5] Manimaran K *et al.* 2011 *Phys. Rev.* C **83** 034609
[6] Zagrebaev V I, Karpov A V and Greiner W 2010 *Phys. Rev.* C **81**, 044608 also Zagrebaev V I and Greiner W 2010 in *Lecture Notes in Physics* **818** 267-315 *Clusters in Nuclei*, Vol. 1 Beck C (ed.), Chapter 7, Springer-Verlag (Heidelberg, Berlin)
[7] Adamian G, Antonenko N and Scheid W 2012 in *Lecture Notes in Physics* **848** 165-228 *Clusters in Nuclei*, Vol.2, Beck C (ed), Springer-Verlag Berlin Heidelberg 2012.
[8] Poenaru D and Greiner W 2010 in *Lecture Notes in Physics* **875** 1-56 *Clusters in Nuclei*, Vol.1, Beck C (ed), Springer-Verlag Berlin Heidelberg
[9] Schall P *et al.* 1987 *Phys. Lett.* B **191** 339
[10] Pyatkov Yu V *et al.* 2010 *Eur. Phys. J.* A **45** 29
[11] Pyatkov Yu V *et al.* 2012 *Eur. Phys. J.* A **48** 94
[12] von Oertzen W, Pyatkov Y V and Kamanin D 2013 *Acta Phys. Polonica* **44** 447 Zakopane Conference
[13] Csatlos M, Krasnahorkay A, Thirolf P G, *et al.* 2005 *Phys. Letters* B **615** 213
[14] Vijayaraghavan K R, von Oertzen W and Balasubramaniam M 2012 *Eur. Phys. J.* A **48** 27
[15] Bohr N and Wheeler J A 1939 *Phys. Rev.* **56** 426
[16] Tashkodajev R B, Nasirov A K and Scheid W 2011 *Eur. Phys. J.* A. **47** 136
[17] Nasirov A K *et al.* 2014 *Phys.Scripta* **89**, 054022 (2014).
[18] von Oertzen W and Nasirov A K 2014 *Phys. Letters* B **734**, 234
[19] Fornal B, Broda R, Krolas R W 2000 *et al. Eur. Phys. J.* A **7**, 147
[20] Roddriguez R *et al.* 2010 *Phys. Letters* B **691** 202
[21] Mukhopadhyay S *et al.* 2012 *Phys. Rev.* C **85** 064321
[22] Kamanin D and Pyatkov Y V 2014 in *Lecture Notes in Physics* Vol. **875** 183-246 *Clusters in Nuclei*, Vol.3 Beck C (ed), Chapter 6. Springer(Heidelberg, Berlin)

Investigation of ^{9}Be with THSR wave function

Mengjiao Lyu

E-mail: mengjiao_lyu@163.com.
School of Physics and Key Laboratory of Modern Acoustics, Institute of Acoustics, Nanjing University, Nanjing 210093, China

Zhongzhou Ren

E-mail: zren@nju.edu.cn.
School of Physics and Key Laboratory of Modern Acoustics, Institute of Acoustics, Nanjing University, Nanjing 210093, China
Center of Theoretical Nuclear Physics, National Laboratory of Heavy-Ion Accelerator, Lanzhou 730000, China

Bo Zhou

E-mail: bo@nucl.sci.hokudai.ac.jp.
School of Physics and Key Laboratory of Modern Acoustics, Institute of Acoustics, Nanjing University, Nanjing 210093, China
Research Center for Nuclear Physics (RCNP), Osaka University, Osaka 567-0047, Japan
Nishina Center for Accelerator-Based Science, The institute of Physical and Chemical Research (RIKEN), Wako 351-0198, Japan

Akihiro Tohsaki

Research Center for Nuclear Physics (RCNP), Osaka University, Osaka 567-0047, Japan

Hisashi Horiuchi

Research Center for Nuclear Physics (RCNP), Osaka University, Osaka 567-0047, Japan
International Institute for Advanced Studies, Kizugawa 619-0225, Japan

Abstract. We extend the THSR wave function into N≠Z nucleon ^{9}Be and perform calculations for its ground state. The negative parity of the ground state is correctly described by the introduction of an extra phase factor. The calculation result of the ground state agrees well with other model. We also calculate the density distribution of the extra neutron. The molecular orbit structure of the extra neutron is found to be reproduced correctly with the THSR wave function without prior assumptions.

1. Introduction

Since the discovery of α-cluster in 1930s, the cluster structure has become a fundamental concept in understanding nuclear structures. The light nucleus ^{9}Be, which is a typical three body system consists of two alpha clusters and an extra nucleon, has been used widely in testing different cluster models. This includes nuclear Molecular Orbit (MO) models which considers the extra neutron in ^{9}Be to occupy an orbit constructed with two cluster orbits (LCCO) [1, 2, 3]. These models successfully reproduced physical properties of ^{9}Be and provided good description of its structure. However, these calculations did not illustrate clearly the motion dynamics of clusters inside nucleus.

The Tohsaki-Horiuchi-Schuck-Ropke (THSR) wave function [4], which was first proposed to study 4-n nuclei ^{12}C and ^{16}O, is now considered to provide a new concept of nonlocalized clustering dynamics in nuclei [5, 6]. The nonlocalized motion of clusters inside the nucleus has been tested with both positive and negative bands of nucleus ^{20}Ne [5, 6, 7]. Considering these successful calculation of these 4-N nuclei, the application of the THSR wave function to non-4N nuclei ^{9}Be worth investigation. In this paper, we will apply the THSR wave function into nucleus ^{9}Be, and perform calculations for its ground state.

2. The THSR Wave Function of ^{9}Be with Intrinsic Negative Parity

The THSR wave function of ^{9}Be starts from the integration of Brink wave function Φ^B over generate coordinates,

$$\Phi = \int G_\alpha(\mathbf{R}_1) \int G_\alpha(\mathbf{R}_2) \int G_n(\mathbf{R}_n)\Phi^B(\mathbf{R}_1, \mathbf{R}_2, \mathbf{R}_n) d^3\mathbf{R}_1 d^3\mathbf{R}_2 d^3\mathbf{R}_n \tag{1}$$

where $\Phi^B(\mathbf{R}_1, \mathbf{R}_2, \mathbf{R}_n)$ is the Brink wave function with generate coordinates $\mathbf{R}_1$, $\mathbf{R}_2$ for the two alpha clusters and generate coordinate $\mathbf{R}_n$ for the extra nucleon. The Brink wave function is the total antisymmetrization of nine single nucleon wave functions,

$$\begin{aligned}\Phi^B(\mathbf{R}_1, \mathbf{R}_2, \mathbf{R}_n) = \hat{A}\{&\psi_{\mathbf{R}_1}(r_1)\psi_{\mathbf{R}_1}(r_2)\psi_{\mathbf{R}_1}(r_3)\psi_{\mathbf{R}_1}(r_4) \\ &\times \psi_{\mathbf{R}_2}(r_5)\psi_{\mathbf{R}_2}(r_6)\psi_{\mathbf{R}_2}(r_7)\psi_{\mathbf{R}_2}(r_8)\psi_{\mathbf{R}_n}(r_9)\}.\end{aligned} \tag{2}$$

Here $\hat{A}$ is the total antisymmetrizer, $\phi_{\mathbf{R}}$ is the single nucleon wave function. For the eight nucleons inside two alpha clusters, we use the single nucleon wave function of the Gaussian form,

$$\psi_{\mathbf{R}}(\mathbf{r}) = (\frac{1}{\pi b^2})^{3/4} e^{-(r-\mathbf{R})^2/2b^2} \chi_{\{1,2,3,4\}} \tag{3}$$

where b is the size parameter of the Gaussian function and χ is the corresponding spin and isospin of the nucleon. The function $G_\alpha(\mathbf{R})$ in Eq. 1 is also of Gaussian style,

$$G_\alpha(\mathbf{R}_i) = \exp\left(-\frac{R_{i,x}^2}{\beta_{\alpha,xy}^2} - \frac{R_{i,y}^2}{\beta_{\alpha,xy}^2} - \frac{R_{i,z}^2}{\beta_{\alpha,z}^2}\right), \tag{4}$$

where $\beta_{\alpha,xy}$ and $\beta_{\alpha,z}$ are size parameters of the α-cluster container. The same parameter $\beta_{\alpha,xy}$ is used for the x and y axis to ensure the rotational symmetry about the z-axis. Usually, the THSR wave function is used for systems with intrinsic angular momentum $l = 0$. However this is not suitable for the ground state of ^{9}Be which has intrinsic angular momentum $l = 1$ with negative parity. To describe this negative parity of the ground state of ^{9}Be, an extra phase factor is introduced in the function $G_n(\mathbf{R}_n)$ for the extra neutron as,

$$G_n(\mathbf{R}_n) = \exp\left(-\frac{R_{n,x}^2}{\beta_{n,xy}^2} - \frac{R_{n,y}^2}{\beta_{n,xy}^2} - \frac{R_{n,z}^2}{\beta_{n,z}^2}\right) e^{im\phi_{\mathbf{R}_n}}, \tag{5}$$

where $\beta_{n,xy}$ and $\beta_{n,z}$ are size parameters of the extra neutron container. $\phi_{\mathbf{R}_n}$ in the extra phase factor $e^{im\phi_{\mathbf{R}_n}}$ is the azimuthal angle in spherical coordinates $(R_{\mathbf{R}_n}, \theta_{\mathbf{R}_n}, \phi_{\mathbf{R}_n})$ of vector $\mathbf{R}_n$.

When the spherical coordinates $(R_{n,x}, R_{n,y}, R_{n,z})$ are transformed as,

$$(R_{n,x}, R_{n,y}, R_{n,z}) \to (-R_{n,x}, -R_{n,y}, -R_{n,z}), \tag{6}$$

the sign of phase $\phi_{\mathbf{R}_n}$ and the phase factor $e^{im\phi_{\mathbf{R}_n}}$ will be determined by the choice of parameter m. When choosing $m = 1$, we will obtain

$$\phi_{\mathbf{R}_n} \to -\phi_{\mathbf{R}_n}, \quad e^{im\phi_{\mathbf{R}_n}} \to -e^{im\phi_{\mathbf{R}_n}}. \tag{7}$$

Thus the negative parity of this intrinsic wave function is proved. With similar argument, we can also obtain the traditional THSR wave function with positive parity for the case of $m = 0$. The extra phase factor also lead to a changed z component of orbital angular momentum l_z in the intrinsic wave function. For the ground state of ^{9}Be, we have $l_z = m = 1$ because of the rotational symmetry of the two-α-clusters subsystem.

The wave function Φ is then transformed in coordinates by

$$\mathbf{r}_i \to \mathbf{r}_i - \mathbf{X}_G. \tag{8}$$

This technique ensures that the center-of-mass (c.o.m.) motion is an 0s state of harmonic oscillator so that the spurious c.o.m. motion can be analytically removed.

The angular-momentum projection technique is applied to $|\Phi\rangle$ to obtain correct angular momentum as [8],

$$\begin{aligned} |\Phi^{IM}\rangle &= \hat{P}^I_{MK} |\Phi\rangle \\ &= \frac{2I+1}{8\pi^2} \int d\Omega D^{I*}_{MK}(\Omega)\hat{R}(\Omega) |\Phi\rangle . \end{aligned} \tag{9}$$

where I is the orbital angular momentum of the total system.

The Hamiltonian of ^{9}Be system is calculated as,

$$H = \sum_{i=1}^{9} T_i - T_{c.m.} + \sum_{i<j}^{9} V^N_{ij} + \sum_{i<j}^{9} V^C_{ij} + \sum_{i<j}^{9} V^{ls}_{ij}, \tag{10}$$

where $T_{c.m.}$ is the kinetic energy of the center-of-mass motion. The Volkov No. 2 interaction is taken as the central force of nucleon-nucleon potential [9],

$$V^N_{ij} = \{V_1 e^{-\alpha_1 r^2_{ij}} - V_2 e^{-\alpha_2 r^2_{ij}}\}\{0.6 - 0.4\hat{P}_\sigma \hat{P}_\tau + 0.125\hat{P}_\sigma - 0.125\hat{P}_\tau\}, \tag{11}$$

where V_1, V_2, α_1 and α_2 are parameters. The G3RS term is taken as the spin-orbit part of interaction [10],

$$V^{ls}_{ij} = V^{ls}_0\{e^{-\alpha_1 r^2_{ij}} - e^{-\alpha_2 r^2_{ij}}\}\mathbf{L}\cdot\mathbf{S}\hat{P}_{31}, \tag{12}$$

where V^{ls}_0, α_1, and α_2 are parameters and $\hat{P}_{31}$ is an operator which projects two-body systems into triplet odd states.

Parameter m in the phase factor $e^{im\phi_{\mathbf{R}_n}}$ in Eq. (5) is taken to be $m = 1$ as discussed above to describe the negative parity of the ground state of ^{9}Be. $I = 1$, $M = 1$ and $K = 1$ are used in the angular momentum projection operators. The spin of the extra nucleon is taken to be parallel to the orbital angular momentum.

The parameters $\beta_{\alpha,xy}, \beta_{\alpha,z}, \beta_{n,xy}$, and $\beta_{n,z}$ relates to the size of the containers of clusters. They are determined by variational methods respect to the binding energy. Two different values of parameter b in single nucleon wave functions are used in our calculations. The first one is to treat b as a variational parameter with the optimum value as $b = 1.35$ fm. We also use another value which is the same value $b = 1.46$ fm as used in Ref. [3],

3. Results and Discussions

The binding energy of the ground state of ^{9}Be is calculated with previous wave function and parameters. The Monte Carlo method is adopted to perform numerical calculations of the integration in the wave function. The calculated binding energies of the ground states $3/2^-$ of ^{9}Be with different parameter b are listed in Table 1 and compared with other results. The optimum parameters for the ground state are $\beta_{\alpha,xy} = 0.1$ fm, $\beta_{\alpha,z} = 4.2$ fm, $\beta_{n,xy} = 2.5$ fm and $\beta_{n,z} = 2.8$ fm. When $b = 1.46$ fm as the same value in Ref. [3], the calculated binding energy for the ground state is -54.8 MeV. This value agrees with the result of the Molecular Orbit (MO) Model with model space II in Ref. [3], which shows that our wave function is a good description for the ground state of ^{9}Be. The difference between our result and MO+GCM method is relatively larger but still acceptable. We also choose b as the variational optimum value $b = 1.35$ fm, which is the same as used in previous calculations of ^{8}Be [11]. This agreement of parameter b after variational optimization is natural because the motion of α-clusters in ^{9}Be does not change much with the existence of the relatively small neutron. With this $b = 1.35$, the obtained binding energy of the $3/2^-$ ground state is -56.0 MeV which is 1.2 MeV improved than the previous one and closer to the experimental value -58.2 MeV.

Table 1. The binding energy of the ground state of ^{9}Be calculated with different methods. E denotes the binding energy of the ground state. THSR, MO, and MO+GCM denotes corresponding models.

Model	b (fm)	$E(3/2^-)$ (MeV)
THSR	1.35	-56.0
THSR	1.46	-54.8
MO [3]	1.46	-54.8
MO + GCM [3]	1.46	-56.1
Exp.[12]	—	-58.2

The agreement of the THSR wave function and the Molecular Orbit model implies that the molecular orbit structure is correctly reproduced by the THSR wave function. To clearly show the molecular orbit structure of the extra neutron, the density distribution of the extra neutron is calculated. Here the density distribution is calculated as the opportunity that a nucleon, which occupies the single nucleon state $\psi_{\mathbf{R}_n}$, exist at position r'. The corresponding equation is,

$$\rho(\mathbf{r}'_n) = \langle\Psi| \frac{1}{9}\sum_{i=1}^{9} \hat{P}_{i,n}\delta(\mathbf{r}_i - \mathbf{X}_G - \mathbf{r}'_n)\,|\Psi\rangle \,/\, \langle\Psi|\Psi\rangle \,. \tag{13}$$

In this equation, the states that the ith nucleon occupies the single nucleon state $\psi_{\mathbf{R}_n}$ is picked out by the operator $\hat{P}_{i,n}$. Then the opportunity that the ith nucleon appears at position r'_n is calculated with the δ function. The calculated density distribution of the extra nucleon is shown in Fig. 1. This figure is a cross section at $y = 0$ of the density distribution in total space. There are two parts on this cross section, each covers one side of the two α clusters. Considering the rotational symmetry about the z-axis of the ^{9}Be system, this distribution clearly shows a π-orbit structure and the molecular orbit structure is correctly reproduced. In previous calculations such as in Ref. [2] and Ref. [3], the molecular orbit structure is a presumption of the system. However, this presumption is not used in our calculations where the molecular orbit structure comes naturally from the THSR wave function and the constraint of symmetries.

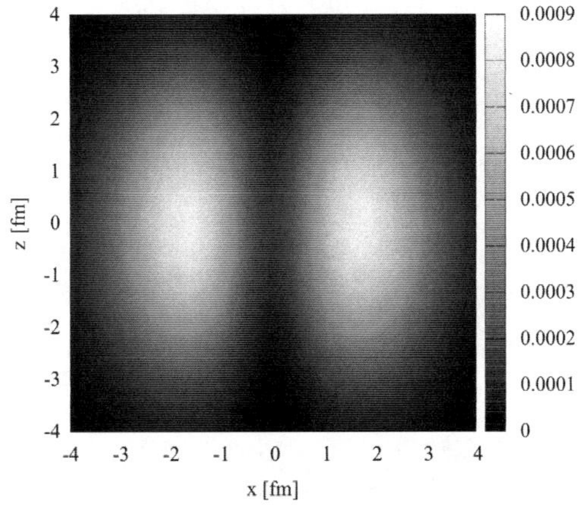

Figure 1. The calculated density distribution of the extra neutron on the $y = 0$ cross section. The density at each point is denoted by gray scale. The unit of the density at each point is fm^{-3}.

4. Conclusion

In this paper, we formulate the THSR wave function for the ground state of ^{9}Be. An extra phase factor is introduced into the THSR wave function to obtain intrinsic negative parity. The calculation result of the ground state agrees well with other calculations. By calculating the density distribution of the extra nucleon, we found clearly that the molecular orbit structure of the extra nucleon is reproduced correctly by the THSR wave function. This shows the power of the THSR wave function because no prior assumptions of molecular orbit are made in our calculations. All this calculations show us the possibility to apply THSR wave function into ^{9}Be and other N$\neq$Z nuclei.

References

[1] Okabe S, Abe Y, and Tanaka H 1977 Prog. Theor. Phys. **57**, 866
[2] von Oertzen W 1996 Z. Phys. A **354**, 37
[3] Itagaki N and Okabe S 2000 Phys. Rev. C **61**, 044306
[4] Tohsaki A, Horiuchi H, Schuck P, and Röpke G 2001 Phys. Rev. Lett. **87**, 192501
[5] Zhou B, Funaki Y, Horiuchi H, Ren Z, Röpke G Schuck P, Tohsaki A, Xu C, and Yamada T 2013 Phys. Rev. Lett. **110**, 262501
[6] Zhou B, Funaki Y, Horiuchi H, Ren Z, Röpke G Schuck P, Tohsaki A, Xu C, and Yamada T 2013 arXiv:1312.1047
[7] Zhou B, Ren Z, Xu C, Funaki Y, Yamada T, Tohsaki A, Horiuchi H, Schuck P, and Röpke G 2012 Phys. Rev. C **86**, 014301
[8] P. Ring and Schuck P 1980 *The Nuclear Many-Body Problem* (New York: Springer-Verlag)
[9] Volkov A B 1965 Nucl. Phys. **74**, 33
[10] Okabe S and Abe Y 1979 Prog. Theor. Phys. **61**, 1049
[11] Funaki Y, Horiuchi H, Tohsaki A, Schuck P, and Röpke R 2002 Prog. Theor. Phys. **108**, 297
[12] Audi G, F. Kondev G, Wang M, Pfeiffer B, Sun X, Blachot J, and MacCormick M 2012 Chin. Phys. C **36**, 1157

Journal of Physics: Conference Series **569** (2014) 012042 doi:10.1088/1742-6596/569/1/012042

Development of MAIKo: the active target with μ-PIC for RI beam experiments

T Furuno[1], T Kawabata[1], S Adachi[1], Y Ayyad[2], T Baba[1], T Hashimoto[2], Y Ishii[1], S Kabuki[1], H Kubo[1], Y Matsuda[2], Y Matsuoka[1], T Mizumoto[1], M Murata[1], H J Ong[2], T Sawano[1], A Takada[1], J Tanaka[2], I Tanihata[2], T Tanimori[1], M Tsumura[1] and H D Watanabe[1]

[1] Department of Physics, Kyoto University, Kyoto 606-8502, Japan

[2] Research Center for Nuclear Physics (RCNP), Osaka University, Osaka 567-0047, Japan

E-mail: furuno@scphys.kyoto-u.ac.jp

Abstract. An active target system MAIKo (μ-PIC based active target for inverse kinematics.) is under development at RCNP. MAIKo is designed to investigate inelastic scattering at forward angles with a radio isotope (RI) beam. The active target is a promising device to study excited states above particle decay thresholds in unstable nuclei. In the present paper, the detailed design of the detector system is described. The first experiment using an accelerated beam was performed to study the detector performance under high counting rate. Preliminary results of the experiment are also discussed.

1. α clustering in neutron-rich nuclei

With the recent developments of new accelerators for RI beam experiments, nuclear physics is extending its interests from stable nuclei toward unstable neutron-rich nuclei. Antisymmetrized molecular dynamics (AMD) calculations predict that excess neutrons play a crucial role in α clustering phenemena [1]. For example, in Be isotopes, it is predicted that ^{12}Be has well-developed α cluster structure in the ground state, whereas ^{10}Be has a shell-model like structure in the ground state. This crossover is caused by valence neutrons occupying the molecular orbitals between two alpha cores which make the system gain biding energy. The study of α cluster structures in neutron-rich nuclei is, thus very impotant to reveal the origin of clustering correlations in nuclear physics.

According to generalized two-center cluster model, which predicts the α-molecular structure in the ground state of ^{12}Be, there exists monopole excitation modes in which 2α cores oscillate each other [2]. These collective cluster-excitation modes should have relatively large monopole strenghts compared to single-particle excitation modes. Therefore, a systematic measurement of monopole strengths is crucial to investigate the α cluster structure in the Be isotopes.

2. Missing mass spectroscopy in RI beam experiments

α-inelastic scattering has been successfully applied to measure monopole strengths of stable nuclei [3, 4]. In order to measure monopole strenghts in unstable nuclei by α-inelastic scattering, such measurements have to be performed under inverse kinematics conditions. Invariant mass spectroscopy is widely used in previous experiments with unstable nuclei. However, this method is not suitable to study cluster states located around particle decay thresholds because of the high multiplicities of the reaction products. On the other hand, missing mass spectroscopy is a powerful tool to explore excited states above decay thresholds, because excitation energies are determined by detecting only one recoil particle.

However, it is not easy to apply this method to inverse kinematics experiments. Figure 1 shows the angular distribution for the cross section for the monopole transition in the ^{12}Be(α,α')^{12}Be scattering, calculated using the distorted-wave-born approximation (DWBA) model. The excitation energy is assumed to be 5 MeV. The energy of the ^{12}Be beam is 62 MeV/u which is the highest energy available at the RCNP cyclotron facility. Since the cross section is larger at forward angles in the center of mass frame, the measurement should be done around the second peak of the cross section ($\theta_{\rm CM} \sim 8^\circ$) to perform multi-pole decomposition analysis. Figure 2 shows recoil α particle energy as a function of recoil angle at different excitation energies. Measurements at $\theta_{\rm CM} \sim 8^\circ$ require the detection of recoil α particles with an energy of about 3 MeV. It is very difficult to detect such a low-energy particles with a normal experimental setup because these particles stop in the target. In addition, the recoil particles have to be detected with high resolution. For example, angular resolution of 6 mrad and energy resolution of 10% in sigma are required to achieve 300-keV resolution of the excitation energy.

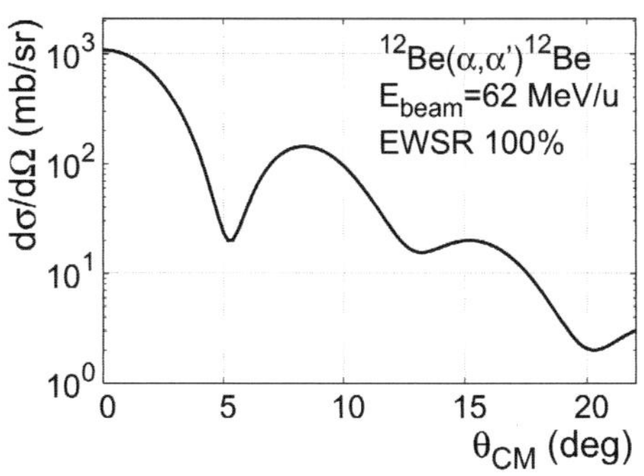

Figure 1: Angular distribution of cross section for the monopole transition calculated by DWBA model.

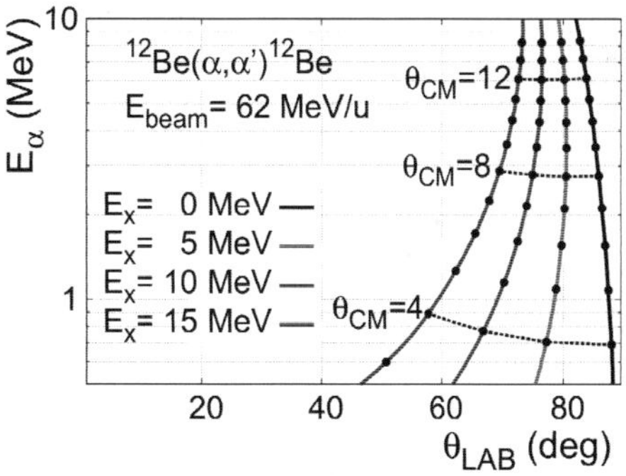

Figure 2: Energies of recoil α particles as a function of recoil angle at different excitation energies.

3. Active target system MAIKo

For the detection of low-energy recoil particles, an active target system MAIKo (μ-PIC based active target for inverse kinematics.) is under development at RCNP.

A sketch of MAIKo is shown in Fig. 3. The detector is based on a time projection chamber (TPC) which provides three-dimensional tracks of charged particles. In the active target mode, the TPC gas act as the target and detector gas at the same time. Since the scattering occurs inside the TPC, the active target can detect even low-energy particles. To investigate α inelastic scattering, the TPC chamber is filled with He gas with a small fraction of quench gas to improve the detector performance. The pressure of the gas can be changed from 400 hPa to atmospheric pressure depending on experiments.

To achieve high position resolution in the TPC, the micro-pixel chamber (μ-PIC) [5] is utilized for the multiplication and detection of the drifted electrons. The μ-PIC is a type of micro-pattern gaseous detectors. The structure of the μ-PIC is drawn in Fig. 4. Electron avalanche simulated by Garfield [6] is also drawn in the figure. Anode pixels surrounded by cathode strips are fabricated on a polyimide substrate with 400 μm pitch. High voltage of about 400 V is applied to the anode. The cathode strips provide one coordinate of the particle trajectory and the anode pixels, which are interconnected on the back plane of the substrate to form strips, provide the

Journal of Physics: Conference Series **569** (2014) 012042 doi:10.1088/1742-6596/569/1/012042

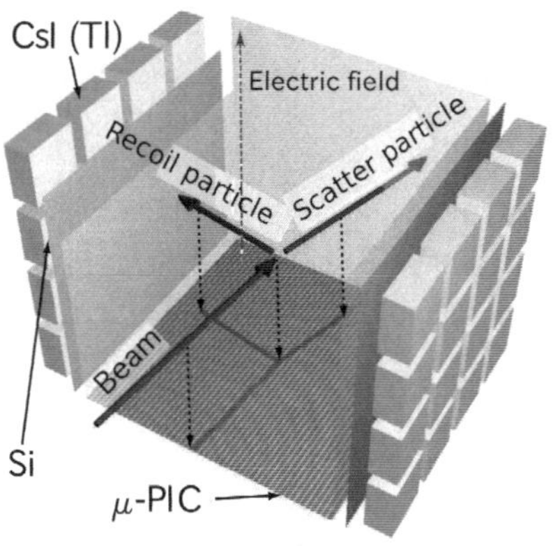

Figure 3: A sketch of MAIKo.

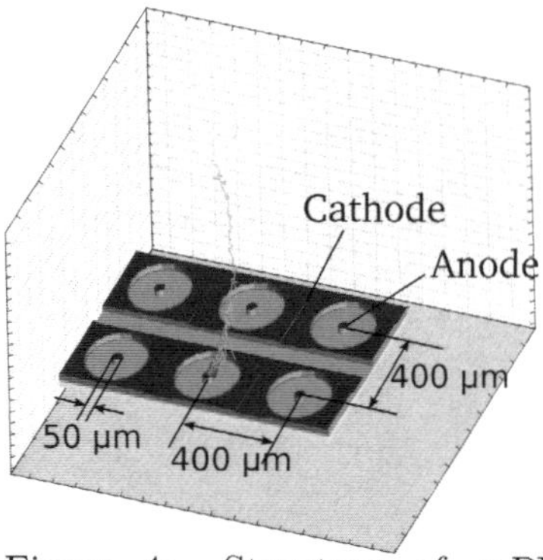

Figure 4: Structure of μ-PIC. Electron avalanche simulated by Garfield [6] is also drawn.

second coordinate. The vertical position of the particle trajectory is determined by the electron drift time to the μ-PIC. If the position resolution of the μ-PIC is assumed to be 300 μm, the tracks of the recoil particles should be longer than 5 cm in order to achieve an angular resolution of 6 mrad. Therefore, the active area of μ-PIC was determined to be 10 $\times$ 10 cm^2.

To make a trigger signal for the scattering event, Si and CsI(Tl) detectors are installed at left, right and downstream sides of the TPC. Each Si detector has 9 $\times$ 6 cm^2 active area with thickness of 325 or 500 μm. The CsI crystals, which are used to stop high-energy particles, have a volume of 3 $\times$ 3 $\times$ 3 cm^3 and are wrapped with ESR films to efficiently collect scintillating photons.

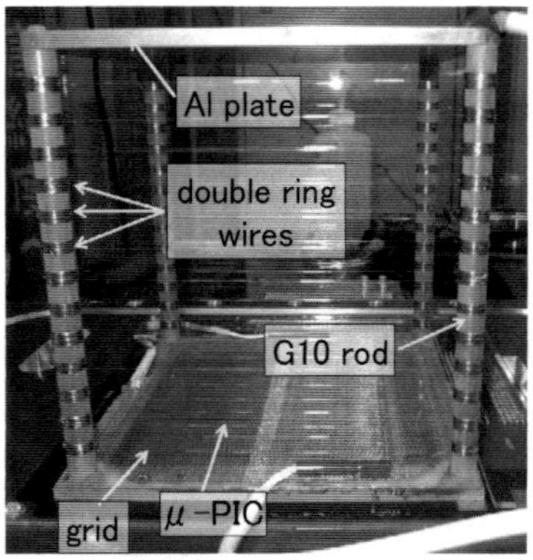

Figure 5: Side view of the TPC.

A picture of the TPC cage to form the electric field is shown in Fig. 5. The volume of the TPC cage is 14 $\times$ 14 $\times$ 14 cm^3. The electric drift field of about 200 V/cm is generated by appling a high voltage to the Al plate and Ni grid (0.15 mm diameter and 0.85 mm pitch). The high voltage supplied to the grid makes it transparent for the electrons drifted from the active volume but opaque for the positive ions produced in the avalanche near the surface of the μ-PIC. Between the Al plate and the grid, the double-ring Be-Cu field wires (diameter of 125 μm) are wounded around four G10 rods with a pitch of 1 cm in height. Each wire is connected with a 10 MΩ resistor. The wire at the upper position and the wire at the lower are connected to the plate and grid, respectively. The uniformity of the field is better than 97% in the active volume of the TPC even with the presence of the Si detectors located 3 cm away from the field wires.

The signals induced on the μ-PIC are read with the 256 anode strips and 256 cathode strips. These signals are amplified and discriminated by the EF2009bal chips, originally developed for the compton camera experiment at Kyoto university [7]. The discriminated signals are synchronized with 100-MHz clock, and the time over threshold information is registered in memory of the VME module. The sum of the analog signals of each 32 consecutive strips are also read by a flash ADC.

4. Performance test experiment

4.1. Purposes of the experiment

In 2013, the first test experiment was performed at RCNP. The purposes of the experiment are as follows: one is to study the beam rate dependency of the TPC capabilities such as gas gain and position resolution. Typical beam intensity in RI beam experiments is around 100 kcps.

The other purpose is to acquire scattering events to develop a tracking algorithm in order to reconstruct the scattering angle and excitation energy from the track and energy information.

4.2. Experimental setup

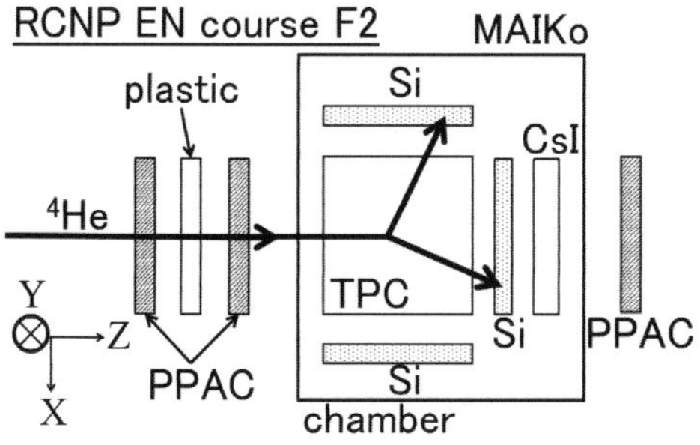

Figure 6: Setup of the experiment.

The experiment was performed at the RCNP EN course. A ^{4}He beam at 12.5 MeV/u was used. The beam energy was adjusted so that the energy loss in the TPC is the same as that of ^{12}Be at 62 MeV/u.

The experimental setup is drawn in Fig. 6. MAIKo was installed at the F2 focal plane of the EN course. Three parallel plate avalanche counters (PPAC) were installed at the upstream of and downstream of the TPC chamber to monitor the position of beam particles. To count the number of beam particles, a plastic scintillator was installed at upstream of the chamber. In the experiment, the Si detectors were used to generate the trigger signal for the scattering events. We tested two gas mixtures, He(93%)+iso-C_4H_{10}(7%) at 430 hPa and He(93%)+CO_2(7%) at 430 hPa, to compare the detector performance.

4.3. Analysis and results

An example of anode track of a beam particle is shown in Fig. 7. Electron drift time of anode and cathode strips provides Y-Z and X-Z projection, respectively. The anode strips were perpendicular to the beam axis. The tracks of the beam particles were fitted to a straight line. Figure. 8 shows the histogram of the slope parameter of the fitted line. The beam intensity was about 1 kcps. The width of the histogram was 2.8 ± 0.5 mrad in sigma. This width is given by the sum of the angular spread of the beam and the angular resolution of the TPC. Therefore, the TPC has angular resolution better than 2.8 mrad under beam intensity of 1 kcps.

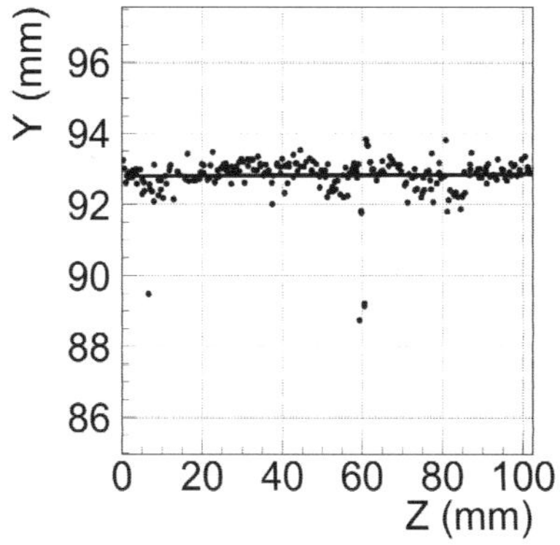

Figure 7: Anode track of a beam particle.

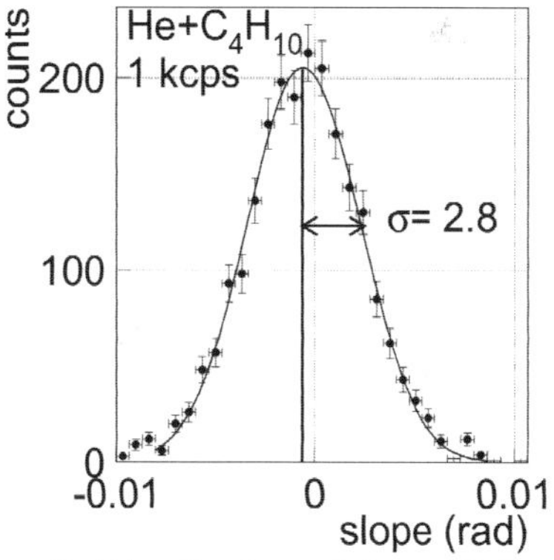

Figure 8: Distribution of slope of the tracks.

Figure 9 shows the rate dependency of the angular resolution for the two gas mixtures. The detector could not be operated with He+CO_2, under beam intesity higher than 100 kcps, due to the discharge of μ-PIC. On the other hand, the TPC with He+iC_4H_{10} was stable even at 1 Mcps. With this gas, the required resolution (6 mrad) was achieved at 300 kcps.

During the experiment, scattering events were successfully aquired with the He(93%)+iso-C_4H_{10}(7%) gas. A typical track of a scattering event is shown in Fig. 10. This event was identified as the ^{4}He+^{4}He elastic scattering from the angle and energy. Detailed analysis of the scattering events is on going.

Journal of Physics: Conference Series **569** (2014) 012042 doi:10.1088/1742-6596/569/1/012042

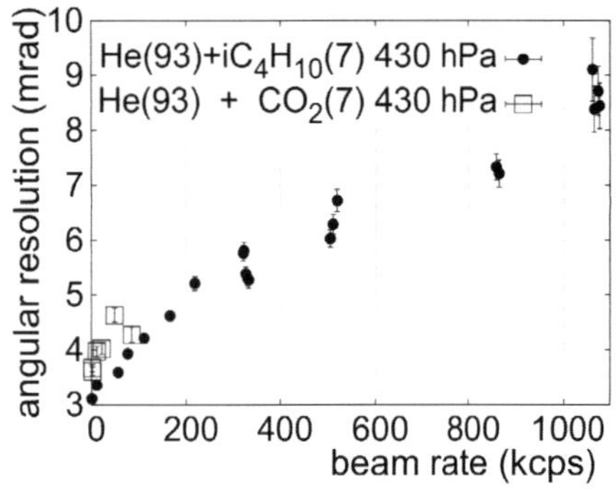

Figure 9: Rate dependence of the TPC angular resolution.

Figure 11 shows an example of $^{12}C(^4He,^4He')^{12}C^*$ inelastic scattering from carbon contained in iso-C_4H_{10}. In this event, ^{12}C was excited to above the three α decay threshold (7.27 MeV) and decayed into three α particles. The width of the track corresponds to the time over threshold and the energy loss in the TPC. The thin tracks in Fig. 11 are the incident and the scattered ^4He whose energy losses in the TPC are relatively small whereas the thick tracks are the decay α particles which lost larger energy. One of the three α particles (α_3 in the figure) stopped in the TPC.

In conclution, MAIKo system can succesfully acquire scattering events under high irradiating rate. This active target will be used to investigate α-clustering phenomena in ^{12}Be at RCNP.

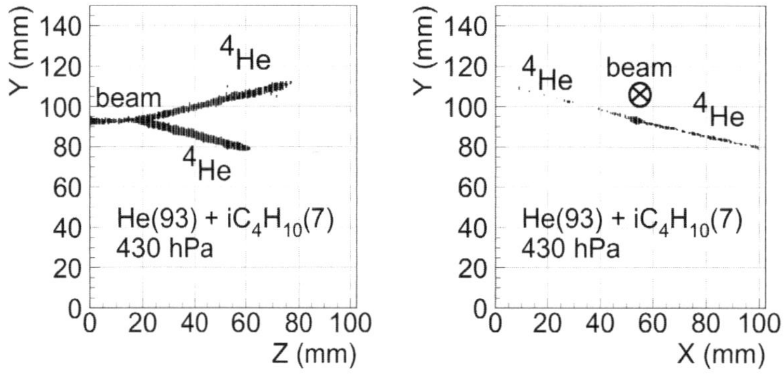

(a) anode (b) cathode

Figure 10: Event display of a scattering event. (a): anode track, (b): cathode track.

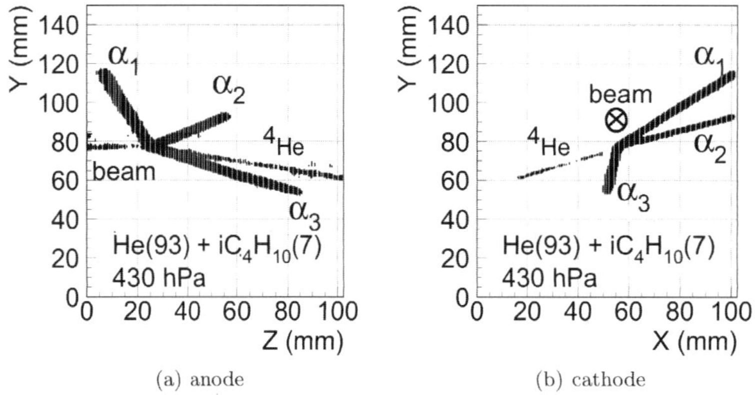

(a) anode (b) cathode

Figure 11: Event display of a three α decay event. (a): anode track, (b): cathode track.

References

[1] En'yo K Y, Kimura M and Ono A 2012 *Prog. Theor. Exp. Phys.* 01A202
[2] Ito M 2011 *Phys. Rev.* C **83** 044319
[3] Ito M *et al* 2002 *Phys. Lett.* B **549** 58
[4] Li T *et al* 2007 *Phys. Rev. Lett.* **99** 162503
[5] Ochi A, Nagayoshi T, Koishi S, Tanimori T, Nagae T and Nakamura M 2001 *Nucl. Inst. and Meth. in Phys. Res.* A **471** 264
[6] Veenhof R 1984 http://garfield.web.cern.ch/garfield/
[7] Mizumura Y *et al* 2014 *JINST* **9** C05045

Coupled-channel picture of the core nucleus for the expansion of the nuclear size in the drip-line region

Hiroshi Masui[1], Kiyoshi Katō[2], Kiyomi Ikeda[3]

[1] Information Processing Center, Kitami Institute of Technology, Kitami, 090-8507, Japan
[2]Nuclear Reaction Data Centre, Hokkaido University, Sapporo, 060-0810, Japan
[3] RIKEN Nishina Center, Wako, 351-0198, Japan

E-mail: hgmasui@mail.kitami-it.ac.jp

Abstract. We study the energy and radius of neutron rich nuclei toward to the drip-line region. We employ the cluster-orbital shell model approach to calculate the radius of the oxygen isotopes. The result shows that the calculated radius underestimates the observed one at ^{23}O, though the binding energy and the drip-line of the oxygen isotopes are reproduced. Therefore, we propose a coupled-channel approach to explain the sudden increase of the radius of the oxygen isotopes at ^{23}O. As a result, the sudden change of the radius can be descried in terms of the competition between the available valence neutron orbits and occupied orbits in the core nucleus. For the ^{23}O case, we show that the excitation to the $1s_{1/2}$-orbit in the core nucleus is a key configuration to reproduce the radius.

1. Introduction

Nuclear size is one of an illustrative quantity for describing the nuclear systems. As the system comes close to the drip-line, and the binding energy becomes small, the observed radius shows a very large value. In general, such the phenomenon is called the "halo" structure of the nuclear system. The halo structure is described by a relatively stiff core nucleus and loosely bound valence nucleon(s) with a very small binding energy. The typical binding energy of the halo nuclei is a few hundred keV, which is small enough compared to the averaged separation energy of the normal nuclei such as ~ 8 MeV.

An exceptional case for the relation between the binding energy and radius is observed on the oxygen isotopes. The drip-line of the oxygen isotopes is determined as ^{24}O and no bound state is observed for $A \geq 25$. For ^{23}O and ^{24}O, an abrupt increase of the r.m.s.radii has been observed [1]. However, recent experiment[2] shows smaller radius for ^{23}O. In the theoretical study, it is not so straightforward to reproduce such the abrupt increase, since the one neutron separation energies of ^{23}O and ^{24}O are not so small compared to the typical "halo" nuclei such as ^{11}Be and ^{11}Li.

We have studied the oxygen isotopes by using the cluster-orbital shell model (COSM) approach [3, 4] in a model space of the ^{16}O core plus valence neutrons. From the calculated results, we show that the sudden increase of the radius at ^{23}O cannot be described by a model space with a closed-core configuration for ^{16}O; $(0s)^4\,(0p)^{12}$ plus valence neutrons, even if we adjust the nucleon-nucleon interaction for valence neutrons to reproduce drip-line at ^{24}O [4]. On the other hand, the radii of $^{16-22}$O are reproduced by using the model space described above. Hence, we consider that the model space with the closed-core configuration plus valence neutrons is not enough to reproduce the general tendency of the radius of the oxygen isotopes from $A = 16$ to 24 within the consistency to the observed binding energy.

In this paper, we propose a coupled-channel picture for the ^{16}O core [5], and the increase of the radius can be explained as the competition between the occupied orbit in the core-excited configuration and the Pauli-blocking effect to the valence neutrons.

2. Formalism and the results

For studying weakly bound systems with the model space of the core and many valence nucleons, we employ the cluster-orbital shell model (COSM) approach [6]. All the coordinates for the valence nucleons are defined so that the origin is the center of mass of the core nucleus and vectors are spanned to each valence nucleon. For the oxygen isotopes, the core nucleus is large enough and the recoil part due to the subtraction of the center of mass motion becomes small. Therefore, we consider the COSM approach is suitable to study the oxygen isotopes and other heavier nuclei in the drip-line region.

2.1. M-scheme cluster-orbital shell model

In the COSM approach, we transform the total A-body Hamiltonian by using the coordinate systems defined from the center of mass of the core as follows:

$$\hat{H} = \sum_{i=1}^{A} \hat{t}_i - \hat{T}_G + \sum_{i<j}^{A} \hat{v}_{ij} = \hat{H}_C + \sum_{i \in V} (\hat{t}'_i + \hat{V}'_i) + \sum_{i<j \in V} (\hat{T}_{ij} + \hat{v}_{ij}) \; . \tag{1}$$

Here, $\hat{t}'_i$ and $\hat{V}'_i$ are one-body kinetic and potential terms for the ith valence nucleon. $\hat{T}_{ij}$ is the recoil part, which comes from the subtraction of the center of mass motion, $\hat{T}_G$. $\hat{H}_C$ is the Hamiltonian for the core nucleus.

To solve the system described by the COSM Hamiltonian (1), we use the Gaussian basis function for the valence nucleons. For the case that the number of valence nucleons is N, the radial part of the basis function is defined as follows:

$$F(r_1, \cdots, r_N) \equiv g_1(r_1) \cdots g_N(r_N) \; , \tag{2}$$

where $g_i(r_i)$ is the Gaussian for the ith valence nucleon with a polynomial, which is proportional to the angular momentum l_i, and its form is $g_i(r) = N_i r^{l_i} \exp(-r^2/(2a_i))$.

In order to simplify the treatment of the angular momentum part and the anti-symmetrization, we employ an m-scheme approach for the angular momentum part of the basis function [4]. We only fix the z-components of the total spin and isospin described as $|MM_T\rangle$ and solve the system using the basis functions with the same fixed z-components as follows:

$$\Phi_V = \sum_m c^{(m)} \Phi^{(m)}_{MM_T} = \sum_m c^{(m)} \mathcal{A}\left\{ F^{(m)}(r_1, \cdots, r_N) \cdot |MM_T{}^{(m)}\rangle \right\} \; . \tag{3}$$

If the basis size is sufficiently large, eigen vectors have a good quantum number for the total spin and isospin, and the coefficients $c^{(m)}$ play the role of the Clebsch-Gordan coefficients.

To discuss the structure of the core and the motion of the valence nucleons on the same footing, we use a semi-microscopic treatment for the interaction between the core and a valence nucleon. The interaction between the core and the ith valence nucleon $\hat{V}'_i$ is constructed by using the folding procedure of the wave function of the core nucleus Φ_C as

$$\hat{V}'_i |\Phi_V\rangle \equiv \sum_{k \in C} \left\langle \Phi_C \middle| \hat{v}_{ik} \middle| \mathcal{A}' \left\{ |\Phi_C\rangle |\Phi_V\rangle \right\} \right\rangle \simeq \hat{V}^d_i + \hat{V}^{ex}_i \; . \tag{4}$$

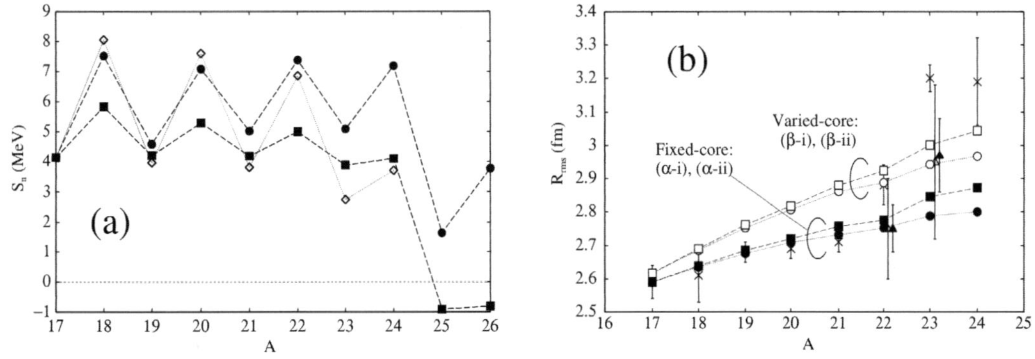

Figure 1. Calculated and experimental (a) neutron separation energies and (b) r.m.s.radii of the oxygen isotopes. Open and solid circles are calculated results by using the parameter set (i) and squares are calculated with parameter set (ii), see text. In (a), open diamonds are experiment [9, 10]. In (b), crosses with bars are experiments [1] and Solid and open triangles with error bars are other experiments [2].

For the nucleon-nucleon interaction, we use the Volkov No.2 [8] interaction and modify the potential strength and parameters to reproduce the drip-line of the oxygen isotopes at ^{24}O.

Using the m-scheme COSM approach, we calculated the binding energy and r.m.s.radius of the oxygen isotopes. We perform calculations using two types of the core-size with respect to the change of the mass number and two different exchange parameters for the nucleon-nucleon interactions of the valence part.

For the core part, we use (α) the fixed core size and (β) the varied one, in which the size parameter has the mass number dependent as the $b_C \sim A^{1/6}$. Here, b_C is the width parameter of the wave function of the core, which is described by using the harmonic oscillator. For the valence part, two exchange parameters are applied to reproduce (i) the energy of ^{18}O ground state, and (ii) the drip-line of the oxygen isotopes at ^{24}O.

The calculated r.m.s.radii using the four different parameter sets for the core and valence nucleons are shown in Figs. 1(a) and (b). As seen from Fig. 1(b), for $^{16-22}$O, the calculated radii using the fixed core size corresponds to the experimental one for both fixed-core cases, (α-i) and (α-ii). On the other hand, for ^{23}O and ^{24}O, the fixed core calculations do not show the sudden increase of the radius at ^{23}O even for the (α-ii) case, and the varied core-size case (β-ii) shows the better agreement compared to the other calculations.

This result suggests that a new mechanism is necessary to be introduced in order to describe the energy and radius of the oxygen isotopes consistently for $^{16-24}$O. Further, it can be considered that the configuration of the core changes drastically from ^{22}O to ^{23}O as the consequence for the valence neutrons added.

2.2. A coupled-channel picture for ^{16}O

In order to reproduce the general tendency of the radius of the oxygen isotopes, and its sudden change at ^{23}O, we introduce a coupled-channel picture to the ^{16}O core [5]. We consider the essential point of this coupled-channel picture is to introduce an excitation of the neutron and proton ($S = 1$ and $T = 0$) in the core nucleus to the $1s_{1/2}$-orbit, see Fig. 2. The tensor induced core-excitation has been developed as the treatment of the tensor-optimized shell model (TOSM) approach [11]. With this condition, high-momentum components are taken into the model space as the size parameter of the orbit in the core, and the small size-parameter of the core is favored

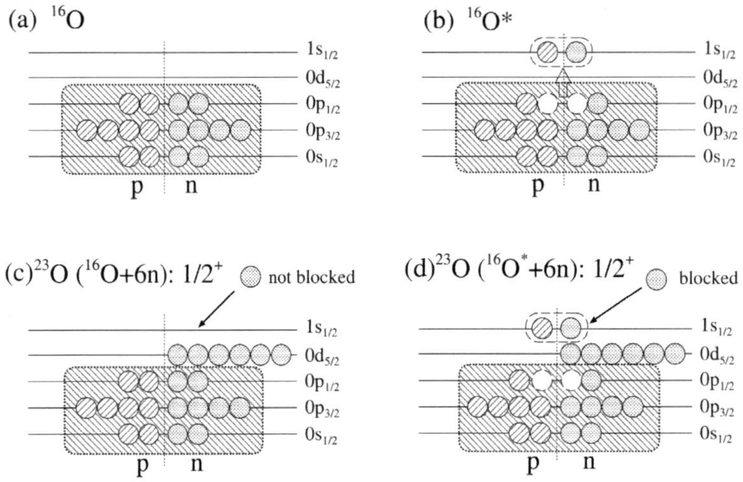

Figure 2. Configurations of the ^{16}O in this model. The core configurations are shown in (a) and (b). The configuration of ^{22}O is shown in (c) and (d).

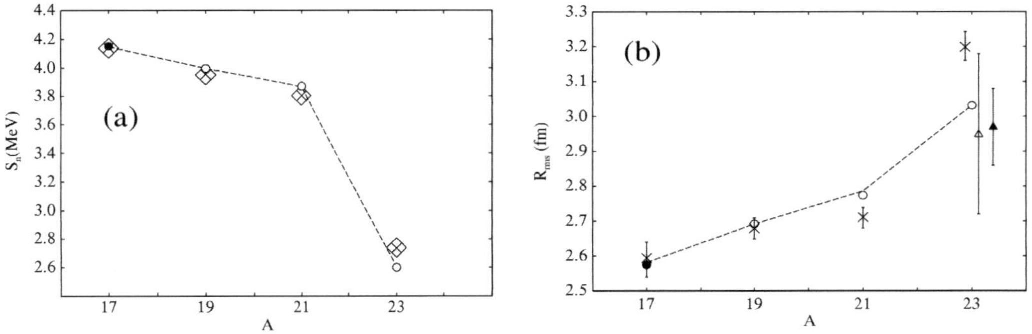

Figure 3. Calculated results of the coupled channel approach for (a) separation energy and (b) r.m.s.radius of the oxygen isotopes. Open and solid circles are our calculation. Crosses with a diamond in (a) are experimental values of the separation energy[9, 10]. Crosses with error bars are experiments[1]. Solid and open triangles with error bars are other experiments[2].

in the $2p2h$-excitation channel.

In this work, we use the same king of mechanism of the TOSM to the ^{16}O-core. To reduce the model space, we restrict ourselves to a two-channel model for describing the core configuration. Here, we consider a neutron and a proton in the $0p_{1/2}$-orbit go up to $1s_{1/2}$-orbit, and other particle-hole excitations, for example the pairing-type one, are assumed to be renormalized into the $0p0h$- or $2p2h$-channel.

We parameterize the coupled-channel Hamiltonian as follows:

$$\hat{H}_{ij} \quad = \quad \{\hat{T}_i + \hat{V}_i\}\delta_{ij} + \Delta_{ij} \; . \qquad (5)$$

The interactions in each channel and coupling parameter Δ_{ij} are adjusted to reproduce the energy and radius of the ground state of ^{17}O. We use the parameter for the diagonal part $\Delta E \equiv \Delta_{22} - \Delta_{11} = 10$ (MeV).

We calculate the energy and r.m.s.radius of the oxygen isotopes for $^{17-23}$O using the coupled-channel approach within the core+n model space. Results are shown in Fig. 3. Here, we adjusted the strength of the core-n interaction (4) so as to reproduce the ^{17}O ground state. For $^{19-23}$O, we use the same potential parameters. Nevertheless, the calculated energies agree with the experimental ones, which is shown in Fig. 3(a).

The mechanism to make different binding energies for heavier oxygen isotopes is as follows. In our model, the core size parameter in the first channel is taken to be proportional to the mass number as $\sim A^{1/6}$. The semi-microscopic potential (4) becomes weak for the large size parameter of the core wave function $|\Phi_C\rangle$. Hence, the binding energy becomes small with the large mass number. For the second channel, the $1s_{1/2}$-orbit of the valence neutron is occupied in the core configuration. The $1s_{1/2}$-component in the second channel of valence neutron is hindered due to the Pauli principle for ^{23}O, while the $1s_{1/2}$-orbit is unoccupied in the first channel. Therefore, the dominant component becomes only the first channel, and binding energy becomes small because of the smallness of the core+n potential with a large core-size parameter.

The calculated radii show a promising result in the view point for reproducing the sudden change of the radius at ^{23}O, see Fig. 3 (b). As mentioned in the calculated energies, due to the Pauli blocking effect for the second channel, the fist channel becomes dominant in ^{23}O. Since the first channel has the mass number dependence for the core size, the calculated radius with the first channel dominance show the large value for ^{23}O, while the second channel has the same amount of the contribution until ^{21}O.

3. Summary and Discussions

We study the radius of the oxygen isotopes and its sudden increase at ^{23}O. The m-scheme COSM approach indicates that the different core configurations are necessary to be introduced to reproduce the sudden increase of the radius at ^{23}O. For this purpose, we proposed a coupled-channel approach with different core configurations.

The coupled-channel approach for the core configuration successfully reproduce the change of the radius. The essential point of the coupled-channel picture is the co-existence of a broad and small core-size in each channels and its competition with the presence of the valence neutrons.

For the future work, we will proceed to perform a microscopic treatment of the core excitation and quantitative estimation of the configurations of the core. The mechanism might be a general for neutron rich nuclei, and the radius and energies of these nuclei can be discussed in this point of view.

References

[1] Ozawa A, Suzuki T and Tanihata I 2001 *Nucl. Phys.* A **693** 32
[2] Kanungo R,et al. 2011 *Phys. Rev.* C **84** 061304(R)
[3] Masui H, Katō K and Ikeda K 2006 *Phys. Rev.* C **73** 034318
[4] Masui H, Katō K and Ikeda K 2009 *Eur. Phys. Jour.* A **42** 535
[5] Masui H, Katō K, and Ikeda K 2012 *Nucl. Phys* A **895** 1
[6] Suzuki Y and Ikeda K: 1988 *Phys. Rev.* C **38** 410
[7] Kaneko T, LeMere M and Tang Y.C 1991 *Phys. Rev.* C **44** 1588
[8] Volkov A B 1965 *Nucl. Phys.* **74** 33
[9] Ajzenberg-Selove F 1986 *Nucl. Phys.* A **460** 1
[10] Ajzenberg-Selove F 1987 *Nucl. Phys.* A**475** 1
[11] Myo T, Kikuchi Y, Katō K, Toki H and Ikeda K 2008 *Prog. Theor. Phys.* **119** 561

Journal of Physics: Conference Series **569** (2014) 012044 doi:10.1088/1742-6596/569/1/012044

β-decay half-lives including first-forbidden contributions for neutron-rich Zn isotopes in the extended QRPA with neutron-proton pairing

Dongdong Ni[1,2] and Zhongzhou Ren[1,2,3,4]

[1]Department of Physics and Key Laboratory of Modern Acoustics, Institute of Acoustics, Nanjing University, Nanjing 210093, China
[2]Joint Center of Nuclear Science and Technology, Nanjing University, Nanjing 210093, China
[3]Kavli Institute for Theoretical Physics China, Beijing 100190, China
[4]Center of Theoretical Nuclear Physics, National Laboratory of Heavy-Ion Accelerator, Lanzhou 730000, China

E-mail: dongdongnick@gmail.com and zren@nju.edu.cn

Abstract. The β decays of neutron-rich Zn isotopes are investigated within the extended quasiparticle random-phase approximation, where neutron-neutron, proton-proton, and neutron-proton pairing correlations are considered in the similar manner. The Brückner G-matrix obtained with the charge-dependent Bonn nucleon-nucleon force is used for the residual particle-particle and particle-hole interactions in addition to the pairing interactions. Contributions from both allowed Gamow-Teller and first-forbidden transitions are considered, and β-decay half-lives together with β-delayed neutron emission probabilities are calculated. The calculated results are found to agree well with the available experimental data.

1. Introduction

Beta decay plays an important role in understanding nuclear structure and astrophysics [1, 2, 3]. Continual attention has been paid from both experimental and theoretical sides. One of the popular theoretical models is the quasiparticle random-phase approximation (QRPA). Within the QRPA framework, pairs of like nucleons, proton with proton (pp) or neutron with neutron (nn), have been widely considered, but treatments of neutron-proton (np) pairing correlations are less documented. It is usually expected that spin-orbit and Coulomb forces become stronger and restrain np pairing in heavy nuclei with $N > Z$. But the recent theoretical study [4] suggests that enhanced np interactions occur in heavy nuclei along a trajectory of approximately equal numbers of valence protons and neutrons. Such a phenomenon is understood in terms of parameter-free spatial overlaps of proton and neutron Nilsson orbitals with $\Delta K[\Delta N, \Delta n_z, \Delta\Lambda]$=0[110].

There are two types of np pairing, namely, the isoscalar $T = 0$ and isovector $T = 1$ pairing interactions. Some QRPA calculations deal with the isoscalar np pairing interaction in the QRPA particle-particle channel by finite-range or zero-range forces [5, 6, 7, 8, 9]. Some deal with the isovector np pairing interaction in the Hartree-Fock-Bogoliubov (HFB) equations besides pp and nn pairing interactions and the isoscalar np pairing contributions are added by renormalizing the pairing potential gaps [10, 11, 12]. Based on the second approach, we have recently investigated allowed Gamow-Teller (GT) transitions of r-process waiting-point nuclei around the neutron magic numbers within the extended QRPA with np pairing [2]. As a further improvement, the present study reports on the investigation of both GT and first-forbidden (FF) transitions for neutron-rich Zn isotopes within the extended QRPA. In particular, very neutron-rich Zn isotopes are in the vicinity of the $N = 50$ shell closure and of great interest to describe the rapid neutron-capture (r-process). For example, ^{82}Zn has been produced and separated at the ISOLDE-CERN facility and its β-decay half-life and β-delayed neutron emission probability have been remeasured with high precision [13].

2. Theoretical framework

The partial half-life $t_{1/2}$ for β transitions is obtained by the expression $ft_{1/2} = 6170$ s with [14, 15, 16, 17]

$$f = \int_1^{W_0} dW C(W) F(Z, R, W)(W_0 - W)^2 W \sqrt{W^2 - 1}, \tag{1}$$

where W_0 and W are the maximum energy and the total energy (including the rest energy) of the β particle, Z and R are separately the atomic number and the nuclear radius of the daughter nucleus, $C(W)$ is the so-called shape factor depending on nuclear transition matrix elements, and $F(Z, R, W)$ is the Fermi function which accounts for the Coulomb interaction between the charged β particle and the residual daughter nucleus. For GT transitions, the shape factor has the simple form without dependence on the β energy, $C(W) = B(GT) = (g_A/g_V)^2 |\langle f \| \vec{\sigma}\vec{\tau}_+ \| i \rangle|^2/(2J_i + 1)$. In contrast, the shape factor of FF transitions depends on the β energy W. If only dominant terms are considered, it can be written as $C(W) = k(1 + aW + b/W + cW^2)$. The details of k, ka, kb and kc can be found in [14, 15, 16, 17] according to the treatment by Behrens and Bühring [18].

To manipulate the dynamics of the decaying system, the main focus lies on the calculation of the shape factor $C(W)$. Our approach is based on the extended QRPA with np pairing [2, 10, 11, 12]. The first step of our calculations is to deal with the $T = 1$, $J = 0$ pp, nn, and np pairing correlations using the HFB equations, which turn the simple single-particle mean-filed description to the quasiparticle picture for the decaying ground state. The quasiparticle states are formed by the admixture of proton and neutron single-particle states. Then the particle-hole and particle-particle excitations are considered using the extended QRPA equations, which mix the pure quasiparticle states for the excitation states of residual daughter nuclei. It should be noted that the two-body interaction matrix elements are evaluated based on the Brückner G-matrix with charge-dependent Bonn nucleon-nucleon forces. The details of the calculations are described in [19, 20].

After solving the extended QRPA equation, one can obtain the forward and backward-going amplitudes X, Y, as well as the QRPA energies ω. The amplitude for β^- transitions from the ground state $|\tilde{0}\rangle$ of an even-even nucleus to the mth phonon state $|mJ^\pi\rangle$ of the neighboring odd-odd nucleus is expressed by [2, 10, 11, 12]

$$\beta^- \equiv \langle mJ^\pi | \beta_J^- | \tilde{0} \rangle = \sum_{a\alpha' < b\beta'} \langle p | T_J | n \rangle \Big[X^m_{(a\alpha' b\beta')} (u_{a\alpha' p} v_{b\beta' n} - u_{b\beta' p} v_{a\alpha' n}) + Y^m_{(a\alpha' b\beta')} (v_{a\alpha' p} u_{b\beta' n} - v_{b\beta' p} u_{a\alpha' n}) \Big], \tag{2}$$

where the one-body operators T_J are associated with the GT transition operator ($\vec{\sigma}$) and the FF transition operators ($[\vec{r} \times \vec{\sigma}]^\lambda$ with λ =0, 1, 2 and $\vec{r}$, as well as relativistic vector operator $\vec{\alpha}$ and axial charge operator γ_5). The detailed expressions of β^- can be found in [17].

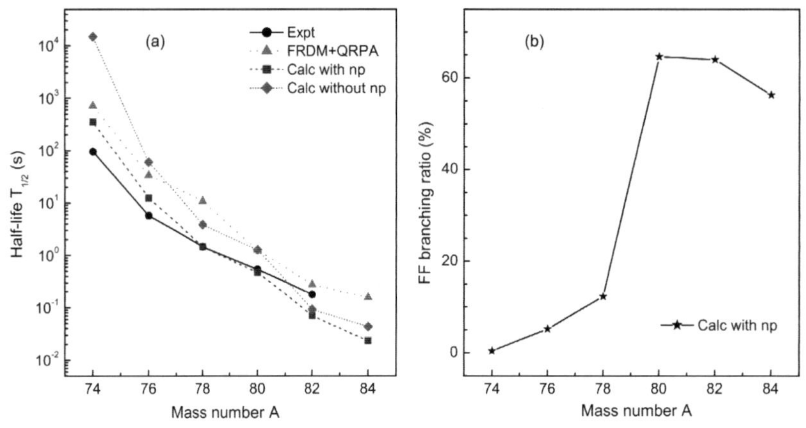

Figure 1. (a) Comparison of the calculated half-lives with the available experimental data (labeled by Expt) for the β decay of neutron-rich Zn isotopes. The theoretical results of Möller *et al* (denoted by FRDM + QRPA) are also shown for comparison and our calculations are separately performed with and without np pairing (denoted by Calc with np and Calc without np). (b) Contributions of FF transitions to the decay width where np pairing correlations are considered in the calculations.

3. Numerical results and discussion

Using the formalism described above, we have performed a detailed investigation on β-decay half-lives of neutron-rich Zn isotopes. The parameters used in the calculations are clearly explained in [2] and here we do not repeat it. Attention should be paid to the quenching effect which is usually taken into account by the quenching of the axial vector coupling constant. In view of the fact that the inclusion of FF transitions reduces calculated β-decay half-lives, there is a slightly difference between the quenching factors in the previous study of GT transitions [2] and in this contribution of both GT and FF transitions. Here, the quenching factor $g_A/g_A^{\mathrm{free}} = 0.75$ is used for both GT and FF transitions except for the 0^- FF case. For the case of 0^- FF transitions, the tensor part of the transition operator is enhanced due to meson exchange current effects [15]. Therefore, following the analysis of [15], the enhancement factor $\epsilon = g_A/g_A^{\mathrm{free}} = 2.0$ is employed for the tensor part while the factor ϵ is fixed at 1.0 for the scalar part. In our calculations, the experimental data of pairing gaps, β-decay energies, and neutron separation energies are taken either from [21, 22] when available, or from the KUTY mass formula [23]. To estimate our theoretical results, the experimental β-decay half-lives are taken from [21] and some new data with high precision are taken from [13].

First, let us focus on the calculation of β-decay half-lives. Figure 1(a) shows the numerical results of the calculated half-lives, compared with the experimental data (labeled by Expt). In 2003, Möller *et al* investigated GT transitions within the FRDM plus QRPA and included FF transitions by the gross theory instead of microscopic calculations [1]. For comparison, the theoretical results of Möller *et al* (labeled by FRDM+QRPA) are also shown. Our calculations are separately performed with and without np pairing (denoted by Calc with np and by Calc without np). One can see that the calculations without np pairing yield longer half-lives as compared with the experimental data except for ^{82}Zn. And this behavior is particularly evident for the long-lived isotopes near the β-stability line. After np pairing correlations are taken into account, the calculations give shorter half-lives and reproduce the experimental half-lives well. The slightly large deviation from the experimental data occurs at ^{82}Zn, for which the recent half-life measurement is known as 178(12) ms [13] while the calculated half-life is obtained as 70 ms. This may be due to large uncertainties of the theoretical β-decay energy used in the calculations [22, 23] and also because it is not good enough for the Woods-Saxon mean-field model to accurately describe the single-particle levels in exotic nuclei far from the β-stability line. In addition, the FRDM+QRPA results have a clear tendency to overestimate the experimental half-lives and yield longer half-lives as compared with the calculations with np pairing. But the FRDM+QRPA result for ^{82}Zn is close to the experimental data [1].

In order to discern the competition between GT and FF transitions, we also illustrate the contributions of FF transition to the decay width in figure 1(b). For the isotopes with $A \leq 78$, the contributions of FF transitions are small. However, there is a strong increase in the branching ratios across ^{78}Zn. As a result, for ^{80,82}Zn the calculated half-lives presented here are shorter by a factor of roughly two as compared with the previous study for GT transitions [2]. Moreover, β-delayed neutron emission probabilities P_n are calculated for the neutron-rich Zn isotopes as well. For the β decays of $^{74-78}$Zn, the β-decay energies are generally smaller or comparable with respect to the neutron separation energies of the daughter nuclei, which leads to a low degree of neutron emission [21]. Both the theoretical calculations and the experimental measurements suggest $P_n = 0$ for them. For the β decays of ^{80,82}Zn, the experimental P_n values are known as 1.0(0.5)% and 57.57(12.34)%, respectively [13, 21]. The present calculations with np pairing yield the values of 4.8% and 37.0%, which are comparable with the FRDM+QRPA results 6.2% and 44.8% [1].

Acknowledgments
This work is supported by the National Natural Science Foundation of China (Grant Nos. 10735010, 10975072, 11035001, 11120101005, and 11375086), by 973 National Major State Basic Research and Development of China (Grants Nos. 2013CB834400 and 2010CB327803), by a Project Funded by the Priority Academic Programme Development of JiangSu Higher Education Institutions (PAPD), by the Research Fund of Doctoral Point (RFDP; Grant No. 20100091110028), and by a China Postdoctoral Science Foundation funded project.

References

[1] Möller P, Pfeiffer B and Kratz K L 2003 *Phys. Rev.* C **67** 055802
[2] Ni D and Ren Z 2014 *J. Phys. G: Nucl. Part. Phys.* **41** 025107
[3] Ren Y and Ren Z 2014 *Phys. Rev.* C **89** 064603
[4] Bonatsos D, Karampagia S, Cakirli R B, Casten R F, Blaum K and Susam L A 2013 *Phys. Rev.* C **88** 054309
[5] Vogel P and Zirnbauer M R 1986 *Phys. Rev. Lett.* **57** 3148
[6] Engel J, Bender M, Dobaczewski J, Nazarewicz W and Surman R 1999 *Phys. Rev.* C **60** 014302
[7] Nikšió T, Marketin T, Vretenar D, Paar N and Ring P 2005 *Phys. Rev.* C **71** 014308
[8] Ni D and Ren Z 2012 *J. Phys. G: Nucl. Part. Phys.* **39** 125105
[9] Ni D, Ren Z and Zhi Q 2012 *Sci. China-Phys. Mech. Astron.* **55** 2397
[10] Cheoun M K, Bobyk A, Faessler A, Šimkovic F and Teneva G 1993 *Nucl. Phys.* A **561** 74
[11] Cheoun M K, Bobyk A, Faessler A, Šimkovic F and Teneva G 1993 *Nucl. Phys.* A **564** 329
[12] Cheoun M K, Faessler A, Šimkovic F, Teneva G and Bobyk A 1995 *Nucl. Phys.* A **587** 301
[13] Gheorghe I (ISS441 Collaboration) 2014 *11th Russbach School on Nuclear Astrophysics* 9-15 March (Salzburg)
[14] Millener D J, Alburger D E, Warburton E K and Wilkinson D H 1982 *Phys. Rev.* C **26** 1167
[15] Suzuki T, Yoshida T, Kajino T and Otsuka T 2012 *Phys. Rev.* C **85** 015802
[16] Zhi Q, Caurier E, Cuenca-García J J, Langanke K *et al* 2013 *Phys. Rev.* C **88** 025803
[17] Ni D and Ren Z 2014 *Phys. Rev.* C **89** 064320
[18] Behrens H and Bühring W 1971 *NucL. Phys.* A **162** 111
[19] Sommermann H M, Müther H, Tam K C, Kuo T T S and Faessler A 1981 *Phys. Rev.* C **23** 1765
[20] Hjorth-Jensen M, Kuo T T S and Osnes E 1995 *Phys. Rep.* **261** 125
[21] NNDC of the Brookhaven National Laboratory, http://www.nndc.bnl.gov/.
[22] Wang M, Audi G, Wapstra A H, Kondev F G *et al* 2012 *Chinese Phys.* C **36** 1603
[23] Koura H, Tachibana T, Uno M and Yamada M 2005 *Prog. Theor. Phys.* **113** 305

Journal of Physics: Conference Series **569** (2014) 012045 doi:10.1088/1742-6596/569/1/012045

α + ^{15}O cluster structures in ^{19}Ne and the α resonant scattering

R. Otani, M. Iwasaki, M. Tomita, and M. Ito

Department of Pure and Applied Physics, Kansai University, 3-3-35 Yamatecho, uita, Osaka 564-8680, Japan.

E-mail: k592822@kansai-u.ac.jp

Abstract. The rotational bands of α + ^{15}O(1/2$^-$) in ^{19}Ne are calculated by employing a simple potential model. The $\alpha-^{15}$O interaction potential is constructed from the scattering calculation of its mirror system, α + ^{15}N. The resonance states are identified by imposing the absorbing boundary condition. The present calculations predict the sequence of the spin-orbit splitting states in the unbound region of ^{19}Ne, which are generated from the higher spin states in the ^{20}Ne nucleus. The excitation function of the α + ^{15}O elastic scattering is also predicted, and the appropriate condition to observe the resonances is also discussed.

1. Introduction

Cluster structures are well known to appear in the excited states of light nuclear systems. A typical example can be seen in ^{20}Ne. The α + ^{16}O cluster model is successful in reproducing the ground and excited rotational band structure in ^{20}Ne [1]. In ^{19}F, which is one proton deficient system of ^{20}Ne, the formation of the α + ^{15}N cluster structures are deeply analyzed [2, 3]. On the contrary, the α + ^{15}O cluster structure, corresponding to the neutron deficient system of ^{20}Ne, still remains unclear although the pioneering works can be seen in Refs. [2, 3].

In the present study, we investigate the α + ^{15}O structure in the ^{19}Ne systems by employing a simple potential model. A potential parameter set is determined from the calculation of the elastic scattering of the respective mirror system, α + ^{15}N. From the nuclear potential of α + ^{15}N, the energy levels of α + ^{15}O are calculated by adding Coulomb force. In addition to the bound energy levels, the unbound resonant states are identified by the absorbing boundary condition [4, 5], and the rotational band structures are discussed in connection to the ^{20}Ne = α + ^{16}O system, which is a typical example of the α cluster system.

Moreover, the excitation function of the ^{15}O resonant scattering by an α target are also investigated. This resonant scattering will be available in future experiment, and we predict the excitation function under the possible conditions reported in the recent experiment [6].

2. Theoretical Framework

The Woods Saxon (WS) potential is assumed for the nuclear potential in α + ^{15}O, and its parameter set is fixed in the following two steps. First, the scattering problem of α + ^{15}N is solved in the range of $E_\alpha = 6 \sim 55$ MeV. In the scattering calculation, the complex WS potential is employed, and we consider only the central part of the complex potential. The WS parameters of the real part is fixed over a whole energy range of the scattering calculation, while the parameters in the imaginary part are set to be energy dependent. All the parameters are optimized so as to reproduce the α + ^{15}N elastic scattering as much as possible. The obtained parameters of the real part of WS are $V = -169.21$ MeV, $R = 2.52$ fm and a= 0.796 fm. The radial form of the potential calculated with this parameter set is similar to the folding type interaction used in the α + ^{16}O OCM calculation [1]. The Coulomb potential is calculated by assuming the uniform charge distribution with the same radius R as the nuclear potential.

Secondly, the spin-orbit interaction is introduced in the calculation of the energy levels of the α + ^{15}O system. The strength of the spin-orbit potential is chosen to reproduce the energies of the bound levels of the $5/2^-$ and $3/2^-$ states, which are considered to be the pair levels in the spin-orbit splitting. The parameters of the radius R_{LS} and the diffuseness a_{LS} are taken to be the same values as those of the central potential, while the strength parameter, V_{LS} is tuned. The resultant strength of the spin-orbit interaction is quite small, $V_{LS} = -0.3$ MeV, because the spin-orbit splitting of these two levels is fairly small (~300 keV),

In the calculation of the energy spectra, we apply the basis expansion method. The shifted aussian, having a functional form of $re^{-\nu(r-S)^2}$, is employed as the basis function in the calculation of the reduced radial wave function of $rR(r)$. In the basis function, the width parameter ν is 3.8 fm^{-2}, while the distance parameter S ranges from $S_{min} = 0.1$ fm to $S_{max} = 40.1$ fm with a constant mesh of $\Delta S = 0.5$ fm.

Above the α decay threshold, the absorbing boundary condition (ABC) [4, 5] is applied to identify the resonance parameters, such as the resonance energy E_R and the decay width Γ_R. As for the absorbing potential, we employ the shifted polynomial function [4, 5], which starts at r_a = 12 fm. The power of the polynomial is taken to be $\beta = 2$ or 4, depending on the energy position of the resonances [4]. The strength of η is determined to reduce the errors arising from the refrection by the absorber as much as possible [4].

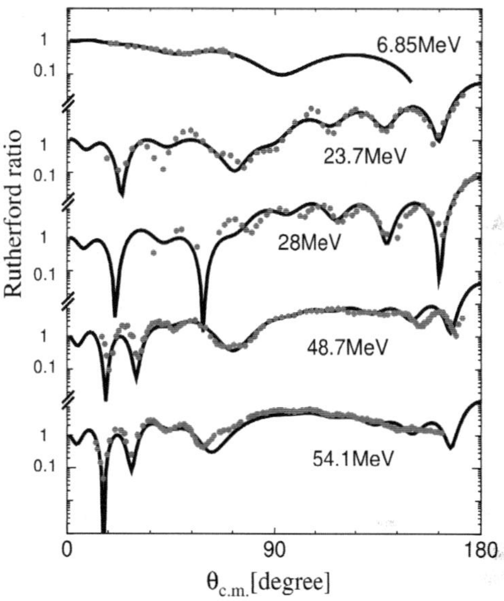

Figure 1. Angular distributions of the α + ^{15}N elastic scattering. See text for details.

The excitation function of the α + ^{15}O resonant scattering is also calculated. In solving the scattering problem, we apply the Kohn-Hulthén-Kato (KHK) variational method [4, 7]. In the present calculation, the computational procedure is the same as the method with the trial function (II) in Ref. [7]. The width parameter is ν = 4.0 fm^{-2}, while the matching radius R_C is taken to be 12.0 fm. The distance parameters are $S_{min} = 0.1$ fm, $S_{max} = 14.1$ fm and $\Delta S = 0.35$ fm.

3. Results

3.1. Elastic scattering of the α + ^{15}N system

Figure 1 shows the results of the angular distributions of the α + ^{15}N system. The theoretical calculations, shown by the solid curve, nicely reproduce the global oscillating pattern of the experimental distribution, plotted by the solid circles. The volume integral of the imaginary potential per nucleon pair reveals the increase tendency above higher energy than $E_\alpha = 23.7$ MeV, while the value at $E_\alpha = 6.85$ MeV deviates from the increase tendency. The determination of the parameters at $E_\alpha = 6.85$ MeV a little di cult because the observed angular range is limited to the forward region of $\theta_{c.m.} = 50° \sim 80°$ at this energy.

Journal of Physics: Conference Series **569** (2014) 012045 doi:10.1088/1742-6596/569/1/012045

3.2. Energy spectra of ^{19}Ne

The energy spectra of α + ^{15}O is calculated from the nuclear potential of α + ^{15}N, which is determined from the elastic scattering. The imaginary potentials are switched off in the calculation of the energy spectra. In the following calculations, we focus on the energy levels, which have the total quanta of $N \geq 8$ in the wave function of the α − ^{15}O relative motion. Here N corresponds to the total oscillator quanta in the Harmonic Oscillator (HO) potential, and N is given by $N = 2n + L$ with the radial node number n and the relative spin L in the α − ^{15}O relative wave function. This restriction of N is based on the auli's exclusion principle in the ^{20}Ne = α + ^{16}O system [1].

In Fig. 2, the calculated spectra of the negative parity states in ^{19}Ne with N = 8 (middle levels) are compared with the spectra of ^{20}Ne = α + ^{20}Ne (right levels). Here the spectra of ^{20}Ne are calculated from the WS potential, which simulates the folding type potential in Ref. [1]. The resonant levels above the α threshold are identified by the absorbing boundary condition [4].

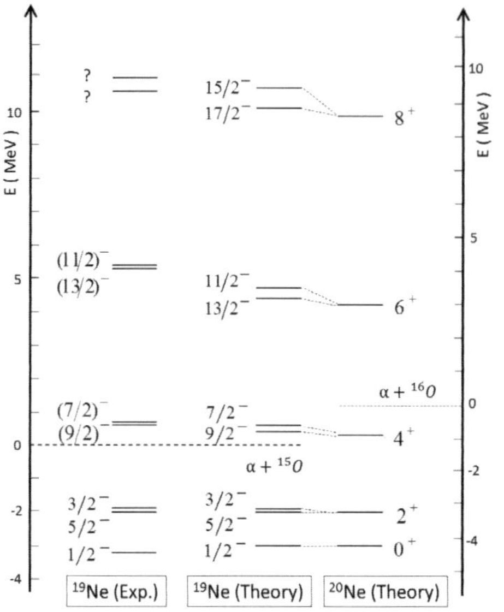

Figure 2. Energy spectra in ^{19,20}Ne. The left and middle levels show the experimental and theoretical spectra of ^{19}Ne, respectively, while the right spectra shows the theoretical calculation of ^{20}Ne.

Since the ^{15}O core contains one neutron $0p_{1/2}$ hole, the spin-orbit splitting in ^{19}Ne are generated from the coupling of the α−^{16}O relative motion in ^{20}Ne and the neutron hole in ^{16}O, as indicated by the dotted line attached to the energy levels. The present calculation seems to nicely reproduce the observed spectra (left levels) although the experimental information on the highly excited states is still insu cient. We have also performed the calculation of the positive parity states with N = 9 although the results are not shown here. The spin-orbit splittings form the N = 9 bands in ^{20}Ne = α + ^{16}O are clearly confirmed.

3.3. Excitation function of the resonant α + ^{15}O scattering

In recent experiments, an α resonant scattering by the β-unstable nucleus has been developed by employing the technique of the inverse kinematics [6]. The α resonant scattering is one of useful tools in identifying the resonance energy and the α decay width, which are essential in astrophysical calculations. The scattering of the ^{15}O beam will be available in future experiments. Thus, the evaluation of its excitation function is very important before the realistic experiments. We calculate the excitation function under the same condition as in the recent experiment of the α + ^{7}Li scattering [6] and confirm whether the experimental setup, such as the energy resolution and the detection angle, is valid or not for the identification of the resonant structures.

The calculated excitation function is shown by the solid curve in Fig. 3. In this calculation, the scattering angle is fixed to $\theta_{c.m.} = 170°$, which is almost the same condition to the experimental setting [6]. The energy step in the calculation is taken to be 0.01 MeV, which is much smaller than the experimental resolution, about 0.1 MeV [6]. In this figure, the resonance energies are plotted by the solid circles, and the decay width of the individual resonances are shown by the error bars. In the higher energy region of $E_{c.m.} \geq 2$ MeV, the three peaks of the positive parity resonances can be seen: $E_{c.m.} \sim 2.8$ MeV ($3/2^+$ and $1/2^+$), $E_{c.m.} \sim 4.0$ MeV ($7/2^+$ and $5/2^+$), and $E_{c.m.} \sim 8.0$ MeV ($11/2^+$ and $9/2^+$). The width of these resonances ranges from 0.2 MeV to 1 MeV, and the resonance enhancements appear with the smooth energy variation. On the contrary, the resonance width of the negative parity states, $13/2^-$ and $11/2^-$, are extremely sharp, and two sharp peaks are superposed on the broad bump appearing around $E_{c.m.} \sim 4$ MeV. There are two resonances of $9/2^-$ and $7/2^-$ in the lower energy region of $E_{c.m.} \leq 2$ MeV but these resonances are completely masked by the enhanced back ground, arising from the kinematic factor of $E_{c.m.}^{-1}$ in the cross section.

We investigate the sensitivity of the excitation function to the energy resolution and the detection angle. If we smear the excitation function by the finite width of the experimental energy resolution, typically $\Delta E = 0.1$ MeV [6], the broad peaks for the positive parity resonances are almost unchanged, but the sharp peak structures of the negative parities are completely disappeared. We also vary the detection angle in the range of $\theta_{c.m.} = 40°$ to $\theta_{c.m.} = 180°$. We have confirmed that the peak to valley structure becomes the most prominent in the angular region of $\theta_{c.m.} = 170° \sim 180°$. Therefore, the experimental setup [6], such as the energy resolution ($\Delta E = 0.1$ MeV) and the detection angle ($\theta_{c.m.} \sim 170°$), is considered to be optimal for the observations of the positive parity resonances.

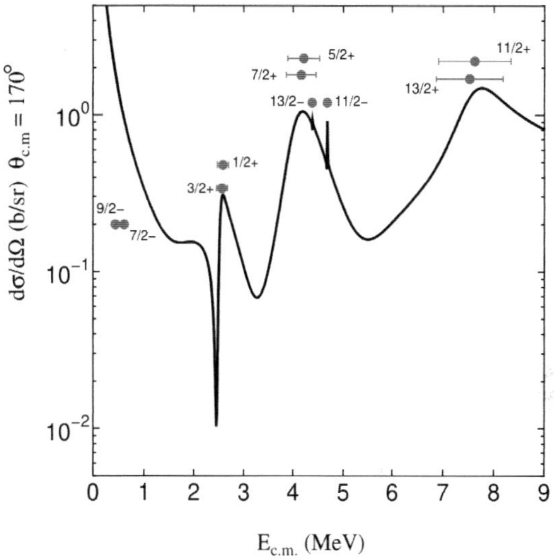

Figure 3. Excitation function of the α + ^{15}O elastic scattering at $\theta_{c.m.} = 170°$. The solid curve shows the strength of the elastic scattering, while the solid circles show the energies of the resonant states identified by the ABC method. The error bars attached to the resonant levels represent the decay width.

4. Summary

In summary, we have calculated the energy levels of α + ^{15}O in ^{19}Ne, and the excitation function in the resonant scattering of ^{15}O by an α target. The interaction potential of α − ^{15}O is determined from the elastic scattering of the respective mirror system, α + ^{15}N. The present calculation predicts the rotational band structure, generated from the spin-orbit splitting of the one neutron-hole state in the α + ^{16}O cluster structure. In the α + ^{15}O resonant scattering, the present calculation strongly support the possible observation of the positive parity states because their decay width matches the recent experimental setting. The observation of the excitation function and the identification of the higher resonances, predicted in the present calculations, are strongly desired in future experiments.

References

[1] Kruppa A T and Kato K 1990 *Prog. Theor. Phys.* **84** 1145
[2] Buck B and Pilt A A 1977 *Nucl Phys.* A **280**
[3] akuda T and Nemoto F 1979 *Prog. Theor. Phys.* **62** 1274
[4] Ito M and Yabana K 2005 *Prog. Theor. Phys.* **113** 1047
[5] Riss U V and Meyer H-D 1993 *Jour. Phys.* B**26** 4503
[6] Yamaguchi H, Hashimoto T, Hayakawa , Binh D N, Kahl D, Kubono , Wakabayashi Y, Kawabata T, and Teranishi T 2011 *Phys. Rev.* C**83** 034306
[7] Kamimura M, 1977 *Prog. Theor. Phys. Suppl.* **62**, 236

3rd International Workshop on "State of the Art in Nuclear Cluster Physics" IOP Publishing
Journal of Physics: Conference Series **569** (2014) 012046 doi:10.1088/1742-6596/569/1/012046

The structure change of the ^{28}Si core in ^{32}S+α cluster structure

Y. Yoshida[1], Y. Kanada-En'yo[2], and F. Kobayashi[3]

[1]Yukawa Institute for Theoretical Physics, Kyoto University, Kitashirakawa Oiwake-Cho, 606-8502 Kyoto, Japan
[2]Department of Physics, Graduate chool of cience, Kyoto University, Kitashirakawa Oiwake-Cho, 606-8502 Kyoto, Japan
[3]Graduate chool of cience and Technology, Niigata University, Niigata 950-2181, Japan

E-mail: `yyuta@yukawa.kyoto-u.ac.jp`[1]

Abstract. The ^{28} i+α cluster structure was studied. We construct the ^{28} i+α cluster model, where structure change of the ^{28} i core and the α-cluster breaking are taken into account. It was found that, in the ^{28} i+α cluster system having α-cluster at the surface of the ^{28} i core, the structure change of the ^{28} i core is energetically rather important while the α-cluster breaking is not significant.

1. Introduction

The nuclear clustering plays important roles in ground and excited states of light nuclei. Excited states having core+α cluster structure called the α-cluster excited state are known to exist in such nuclei as ^{20}Ne [1] and ^{16}O [2], and they are suggested also in ^{40}Ca and ^{44}Ti [3, 4]. Meanwhile, researches of the α-cluster excited state in the middle of *sd*-shell are still limited, and the existence of α-cluster excited states in this region is an open problem. Recently, ^{28}Si+α excited states of ^{32}S have been suggested in the experiments of α scattering on ^{28}Si and inelastic scattering of ^{32}S [5, 6]. In order to understand structure of the ^{28}Si+α excited states by solving the core-α relative motion, it is important to take into account the structure change of the core nuclei and the breaking of the α-cluster at the nuclear surface. However, in conventional cluster models, inner clusters are usually assumed and the cluster excitation effects are not considered. Our final goal is to study the ^{28}Si+α excited states of ^{32}S. To this end, for the first step, we investigate the structure change of the ^{28}Si core and the α-cluster breaking in the ^{28}Si+α cluster structure.

2. Framework

We apply extended cluster model to take into account the structure change of the ^{28}Si core and the α-cluster breaking. We first explain Brink-Bloch α-cluster model (conventional cluster model) [7], and later, we describe the extended model for treating the α-cluster breaking and the structure change of the ^{28}Si core.

2.1. *Brink-Bloch α-cluster model*

In the Brink-Bloch α-cluster model [7], each single-particle wave function φ_i is expressed by a aussian wave packet:

$$\varphi_i = \left(\frac{2\nu}{\pi}\right)^{\frac{3}{4}} \exp\left[-\nu\left(\mathbf{r}-\mathbf{R}_i\right)^2\right]\chi_i\tau_i, \tag{1}$$

where χ_i is the spin part, τ_i is the isospin part, $\mathbf{R}_i$ is the center of the aussian wave packet, and ν is the width parameter. An α-cluster is described four nucleons (spin up neutron and proton and spin down neutron and proton) expressed by the aussian wave packets at the same position. The total wave function Φ of A-particle system is described by the antisymmetrized single-particle wave functions of α-clusters.

$$\Phi = \mathcal{A}\left[\varphi_1\cdots\varphi_A\right], \tag{2}$$

where $\mathcal{A}$ is the antisymmetrizing operator.

2.2. *Extended cluster model with α-cluster breaking*

For description of the α-cluster breaking, we apply the method proposed by Itagaki et al [8] to introduce the effect of the spin-orbit potential. In this method, a parameter is introduced to break the α-cluster so as to gain the spin-orbit interaction. We add a momentum to nucleons of the α-cluster that fixes vertical direction with spin and $\mathbf{R}_i$.

$$\mathbf{R}_i \to \tilde{\mathbf{R}}_i = \mathbf{R}_i + i\lambda_\alpha\frac{(\mathbf{e}_{spin,i})\times(\mathbf{e}_{R_i})}{\sqrt{\nu}}, \tag{3}$$

where λ_α represents the degree of α-cluster breaking. If λ_α is zero, this model becomes the conventional cluster model and describes the intrinsic spin saturated state, where the expectation value of the spin-orbit interaction vanishes. When λ_α is positive, spin-up and spin-down nucleons in the α-cluster have finite momenta with the opposite directions so as to gain the spin-orbit interaction, and the α-cluster has breaking. Indeed for a finite positive value of λ_α, the single nucleons in the α-clusters are boosted to have angular momentum parallel to their intrinsic spin, and therefore they feel the spin-orbit interaction from the core attractively.

2.3. *Extended cluster model with structure change of ^{28}Si core*

To describe ^{28}Si core and its structure change we adopt the extended 7α-cluster model where the parameter Λ_c for the cluster breaking is incorporated to take into account the spin-orbit interaction effect. The present extended 7α-cluster Brink model is based on the studies of ^{28}Si with the Brink-Bloch 7α-cluster model [9] and also with the AMD (antisymmetrized molecular dynamics) method [10]. This extended 7α-cluster Brink model can describe both the jj-coupling shell model limit at $\Lambda_c = 1$ that is described the spherical $(0d_{5/2})^{12}$ configuration, and the oblate deformation limit of the shell model configuration at $\Lambda_c = 0$.

3. Results

Using the extended cluster model, the energy expectation value of a ^{28}Si+α system is calculated as a function of the inter-cluster distance (R). By analyzing the Λ_c and λ_α dependence of the energy, we investigate how the core structure change and the α-cluster breaking affect to the energy of the ^{28}Si+α system. The total wave function of the extended cluster model is projected to the $J^\pi = 0^+$ state. As for the effective nuclear interaction, the Volkov No. 2 (Majorana parameter $M = 0.67$) [11] for the central force and 3SR ($V = 2000$ MeV) [12] for the spin-orbit force are adopted. The Coulomb force is approximated seven range aussians.

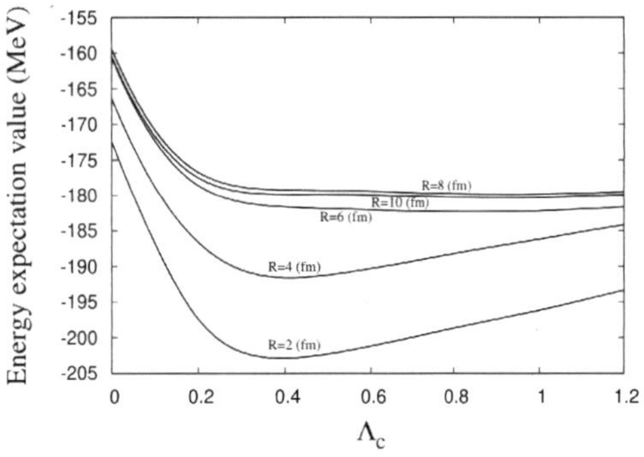

Figure 1. Energy expectation value of ^{28}Si+α as a function of Λ_c for the structure change of the ^{28}Si core. Lines shows the energies for the inter-cluster distance R=2, 4, 6, 8, 10 fm.

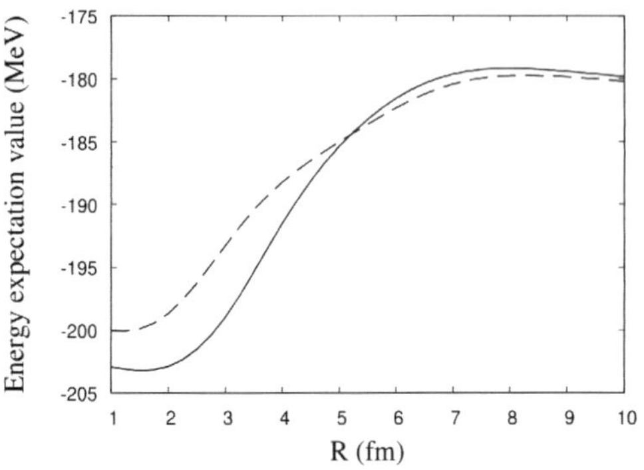

Figure 2. Energy expectation value of ^{28}Si+α as a function of the inter-cluster distance R. The parameter Λ_c for the ^{28}Si-cluster is fixed to Λ_c=0.38 (oblate-type:solid) and Λ_c=0.80 (spherical-type:dashed).

3.1. Structure change of ^{28}Si core
We here discuss the calculation without the α-cluster breaking ($\lambda_\alpha = 0$) to focus only on the structure change of the ^{28}Si core. We show the energy expectation value of ^{28}Si+α as a function of the parameter Λ_c for the structure change of the ^{28}Si core in Fig.1. It is found that, in the case of the large inter-cluster distance $R = 8, 10$ fm, the energy almost degenerates in a wide Λ_c region of $\Lambda_c \geq 0.3$ because the oblate and spherical states degenerates in the isolate ^{28}Si without an α-cluster. On the other hand, in the small R region, the oblately deformed ^{28}Si core (around $\Lambda_c = 0.38$) tends to be energetically favored than the spherical ^{28}Si core. This suggests that the core structure change occurs when the α-cluster exists at the surface of the ^{28}Si core. The origin structure change of the ^{28}Si core can be understood by the auli blocking effect. Namely, the ^{28}Si core favors to deform oblately to avoid overlapping with the α-cluster at the surface.

We also show the energy expectation value of ^{28}Si+α as a function of the inter-cluster distance R in Fig.2. Here the energies for the fixed ^{28}Si core cases of the oblale ^{28}Si with Λ_c=0.38 and the spherical ^{28}Si with Λ_c=0.80 are shown. As mentioned above, the oblately deformed ^{28}Si core is favored than the spherical state in the small R region, while they almost degenerate to each other at the large distance region.

If the α-cluster excited states exist in ^{32}S, they are expected to have the α-cluster amplitude at the nuclear surface. The present result suggests that the structure change of the ^{28}Si core may occur because of the α-cluster at the surface and it may give some effect on the α-cluster excited state of ^{32}S.

3.2. α-cluster breaking
We investigate the α-cluster breaking effect on the energy expectation value of ^{28}Si+α by analyzing the λ_α dependence of the energy. For ^{28}Si+α with a fixed inter-cluster distance R, the optimized λ_α for the minimum energy is small as $\lambda_\alpha \sim 0$ in $R > 3$ fm region indicating that the α-cluster breaking may be minor effect in ^{28}Si+α system (hardly energy difference at $R \geq 4$ region). The reason for the unflavored α-cluster breaking is as follows. In general, in the mechanism of the α-cluster breaking caused by the spin-orbit potential at the nuclear surface, nucleons in the α-cluster occupy ls-favored orbits rather than to form the α-cluster. However, in the ^{28}Si+α system, nucleons in the ^{28}Si core already fills the ls-favored orbits, $0d_{5/2}$ orbits, and

they block the α-cluster breaking. In other words, the α-cluster breaking tends to be suppressed at the surface of the ^{28}Si core because of the auli blocking effect from the core nucleons.

4. Summary and outlook

We calculate the energy expectation value of a ^{28}Si+α system using the extended cluster model. By analyzing the core structure change and the α-cluster breaking in the ^{28}Si+α system, we found that the structure change of the ^{28}Si core is energetically rather important while the α-cluster breaking is not significant when the α-cluster exists at the surface of the ^{28}Si core. In the ^{28}Si core, the oblately deformed state tends to be favored than the spherical sub-shell closed state because of the α-cluster at the surface though they almost degenerate in the isolate ^{28}Si without an α-cluster. We also found that the degree of α-cluster breaking depends on the core nuclei and it is minor in ^{28}Si+α system compared with ^{16}O+α system. The present result suggest that the structure change of the ^{28}Si cluster may give important effect on the α-cluster excited state of ^{32}S and that the α-cluster breaking effect may be relatively minor. We are now proceeding to perform further calculations of the generator coordinate method by superposing the ^{28}Si+α wave functions to study the α-cluster excited states of ^{32}S.

References

[1] Nemoto F and Bandō H 1972 *Prog. Theor. Phys.* **47** 1210
[2] uzuki Y 1976 *Prog. Theor. Phys.* **55** 1751
[3] Michel F, Reidemeister G and Ohkubo 1987 *Phys. Rev.* C **35** 1961
[4] Yamaya T, Oh-ami , Fujiwara M, Itahashi T, Katori K, Tosaki M, Kato , Hatori and Ohkubo 1990 *Phys. Rev.* C **42** 1935
[5] Lönnroth T, et. al. 2010 *Eur. Phys. J A* **46** 5
[6] Itoh M, et al., 2013 *Phys. Rev.* C **88** 064313
[7] Brink D M 1966 *International School of Physics "Enrico Fermi" Course XXXVI* (Academic, New York) p 247
[8] Itagaki N, Cseh J and Poszajczak M 2011 *Phys. Rev.* C **83** 014302
[9] Bauhoff W, chultheis H and chultheis R 1982 *Phys. Rev.* C **26** 1725
[10] Kanada-En yo Y 2005 *Phys, Rev.* C **71**, 014303
[11] Volkov A B 1965 *Nucl. Phys.* **74** 33
[12] Yamaguchi N, Kasahara T, Nagata and Akaishi Y 1979 *Prog. Theor. Phys.* **62** 1018

Journal of Physics: Conference Series **569** (2014) 012047 doi:10.1088/1742-6596/569/1/012047

Study of cluster structure in ^{13}C with AMD+HON-constraint method

Yohei Chiba and Masaaki Kimura

Department of Physics, Hokkaido University, 060-0810 apporo, Japan

E-mail: chiba@nucl.sci.hokudai.ac.jp

Abstract. The $3\alpha + n$ cluster states of ^{13}C are discussed on the basis of antisymmetrized molecular dynamics with the constraint on the harmonic oscillator quanta. We predict two different kinds of the cluster states, the hoyle analogue state and the linear-chain state. The former is understood as the 0_2^+ state (Hoyle state) of ^{12}C accompanied by a valence neutron occupying the s-wave. The latter constitute the parity doublet bands of $K^\pi = 1/2^\pm$ owing to its parity asymmetric intrinsic structure.

1. Introduction

It is known and expected that two different types of 3α cluster states, gas-like [1, 2] and linear-chain states [3, 4, 5], appear in ^{12}C. Naturally, their analog states accompanied by a valence neutron are expected in ^{13}C, and how the valence neutron will modify those 3α cluster structure is very interesting issue to understand the interplay between the clustering of the core nucleus and the single particle motion of the valence neutron. For example, it was pointed out that a p-wave valence neutron attached to the 0_2^+ state will feel very weak spin-orbit interaction due to the very low matter density. As a result, the neutron p-waves coupled to the ^{12}C(0_2^+) will show very reduced spin-orbit splitting [6]. However, if such cluster states really exist or not is unknown. Another example is a stabilization of the linear-chain configuration. The valence neutrons may stabilize the linear-chain configuration [7, 8, 9, 10] as they were in the case of Be isotopes. ^{13}C is a very suitable system to investigate the stabilization mechanism of the linear chain.

For those purpose, we investigated the excited states of ^{13}C on the basis of antisymmetrized molecular dynamics (AMD) [11, 12]. Our aims in this study are (1) the search for the hoyle analogue states with s or p-wave valence neutron, and (2) the search for the linear-chain states.

2. Theoretical framework

We start from the microscopic Hamiltonian with an effective NN interaction of ogny D1S [13], and employ the parity-projected wave function as the variational wave function,

$$\Phi^\pi = \frac{1+\pi\hat{P}_x}{2}\Phi_{int}, \quad \Phi_{int} = \mathcal{A}\{\varphi_1, \varphi_2, ..., \varphi_A\}, \tag{1}$$

$$\varphi_i = \exp\Big\{-\sum_{\sigma=x,y,z}\nu_\sigma\Big(r_\sigma - \frac{Z_{i\sigma}}{\sqrt{\nu_\sigma}}\Big)^2\Big\} \otimes (a_i\chi_\uparrow + b_i\chi_\downarrow) \otimes (\text{proton or neutron}). \tag{2}$$

Here φ_i is the single particle wave packet whose spatial part is represented by the deformed aussian [14], and the variational parameters $\boldsymbol{Z}_i$, a_i, b_i and ν_σ are optimized to minimize the total energy under the constraint. In many AMD studies, the constraint on the quadrupole deformation parameters which is very effective to describe the low-lying yrast states are often used. However, it is not su cient to describe the non-yrast states, in particular, cluster states. Therefore, we have extended the constraint on the harmonic oscillator quanta used in Ref. [15]. We introduce the following quantities,

$$N = N_x + N_y + N_z, \quad \lambda = N_z - N_y, \quad \mu = N_y - N_x, \quad (N_z \geq N_y \geq N_x), \tag{3}$$

where N_x, N_y, N_z represents the expectation values of the harmonic oscillator quanta in each direction. The constraint on the values of N, λ and μ which call HON constraint generates the wave functions with various excited nucleon configurations. As we will show below, HON constraint is found very effective to generate various cluster structures without the assumption on nuclear structure.

After the variational calculation, we perform the angular momentum projection and generator coordinate method (CM) by superposing the wave functions with different values of N, λ, μ,

$$\Psi^{J\pi}_{M\alpha} = \frac{2J+1}{8\pi^2}\sum_{Ki} g^{J\pi}_{Ki\alpha}\int d \ \ D^{J*}_{MK}(\ )\hat{R}(\ )\Phi^{\pi}(N_i, \lambda_i, \mu_i), \tag{4}$$

and the coe cients $g^{J\pi}_{Ki\alpha}$ and eigenenergies $E^{J\pi}_{\alpha}$ are determined by the Hill-Wheeler equation. To investigate the internal structure of the excited states, we calculate the spectroscopic factor (S-factor) that is defined as the volume integral of the multipole component $\varphi^{\pi_1\pi_2}_{jl}(r)$ of the overlap function $\varphi(\boldsymbol{r})$ between the ^{12}C and ^{13}C wave functions,

$$\varphi(\boldsymbol{r}) = \sqrt{A+1}\langle\Phi^{J_2\pi_2}_{M_2}(^{12}\mathrm{C})|\Psi^{J_1\pi_1}_{M_1}(^{13}\mathrm{C})\rangle = \sum_{jl}\langle J_1M_1|J_2M_2jM_1-M_2\rangle\varphi^{\pi_1\pi_2}_{jl}(r)[Y_l\otimes\chi]_{jM_1-M_2},$$

$$S^{\pi_1\pi_2}_{jl} = \int_0^\infty dr\ r^2|\varphi^{\pi_1\pi_2}_{jl}(r)|^2. \tag{5}$$

3. Results

By applying the HON constraint, various $3\alpha + n$ cluster structures which constitute the non-yrast states are generated as shown in Fig. 1. The fact that these cluster structures are obtained without any assumption clearly demonstrates that the clustering is an essential degree-of-freedom of nuclear excitation together with the single-particle and collective motions. The panels (a)-(d) show the triangular configurations of $3\alpha + n$ which constitute the $1/2^+_2$ state analogous to the 0^+_2 state of ^{12}C, and the panel (e) shows the linear-chain configuration of 3α particles with a valence neutrons which constitutes the doublet band of $K^\pi = 1/2^\pm$ as discussed below.

By performing the CM calculation, we obtained many yrast and non-yrast states shown in Fig. 2 (a). It should be noted that many non-yrast states corresponding to the observations are described by the HON constraint, although their excitation energies are overestimated by 5 MeV in average. Among those states we focus on the $1/2^+_2$ state at 15.7 MeV and the $K^\pi = 1/2^\pm$ bands built on the $1/2^\pm$ states at 18.0 and 14.9 MeV, respectively.

3.1. The hoyle analogue $1/2^+_2$ state

To search for the hoyle analogue states with a s or p-wave valence neutron, we have investigated the $^{12}\mathrm{C}(0^+_2)\otimes n(s_{1/2}, p_{1/2}, p_{3/2})$ S-factors for all of the obtained $1/2^+$, $1/2^-$ and $3/2^-$ states. As a result, it was found that the $^{12}\mathrm{C}(0^+_2)\otimes n(p_{1/2}, p_{3/2})$ S-factors are strongly fragmented into many states (it is 0.2 at most for a single state) and does not exist as a single-particle states.

Journal of Physics: Conference Series **569** (2014) 012047 doi:10.1088/1742-6596/569/1/012047

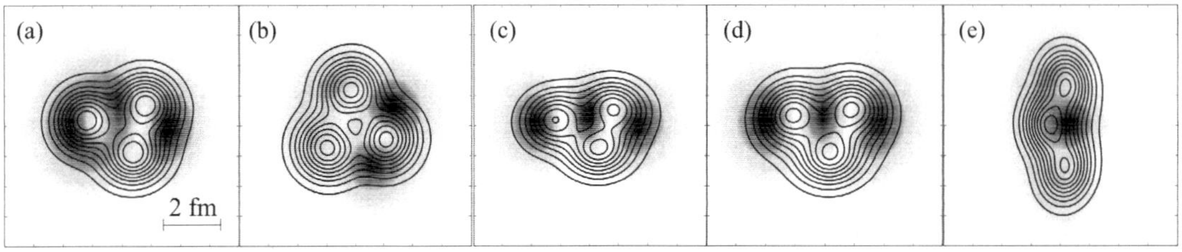

Figure 1. The density distributions of the $3\alpha + n$ cluster states obtained by applying the HON constraint. The contour lines show the proton density distributions that are almost same as those of the 3α cluster core, and the color plots show the valence neutron orbits. (a)-(d) are the dominant component of the $1/2_2^+$ state, while (e) constitutes the doublet bands of the $K^\pi = 1/2^\pm$.

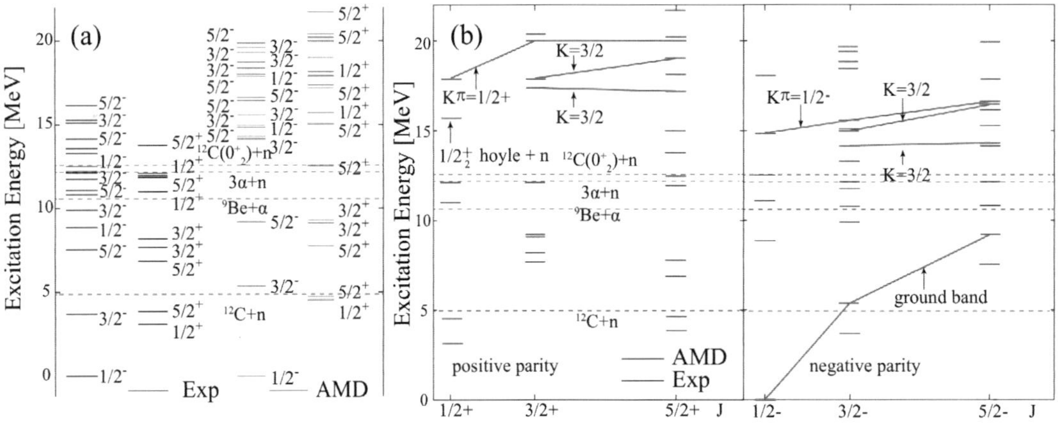

Figure 2. The panel (a) compares the calculated and observed spectra up to $5/2^\pm$ states. The panel (b) shows the spectra as function of the angular momentum for each parity. The lines in panel (b) are the guide of eyes to show the ground band and linear-chain parity doublet.

Therefore, we conclude that the p-wave states coupled to the hoyle state discussed in Ref. [6] may not exist. On the other hand, the $1/2_2^+$ state carries 0.5 of the $^{12}\mathrm{C}(0_2^+) \otimes n(s_{1/2})$ S-factor, and hence, it is understood as a single-particle s-wave state coupled to the Hoyle state. Indeed, similar to the Hoyle state, this state cannot described by a single Slater determinant but a linear combination of many Slater determinants with different cluster configurations as illustrated in Fig. 1 (a)-(d). Furthermore, this state has a larger radius (2.78 fm) compared to the ground state (2.52 fm) indicating analogous nature to the Hoyle state. However, it also should be noted that the gas-like nature is distorted considerably due to the interaction between the core and valence neutrons. For example, the radius of the $1/2_2^+$ is much smaller than the Hoyle state (2.90 fm), and it also couples to the other channels such as $^{12}\mathrm{C}(0_1^+) \otimes n(s_{1/2})$.

3.2. $3\alpha + n$ linear-chain doublet bands

Different from the $1/2_2^+$ state, the $1/2_3^+$ state is well described by a single Slater determinant shown in Fig. 1 (e) indicating the strong-coupling between 3α and valence neutron, and rigid rotor nature. Indeed a rotational band with $K^\pi = 1/2^+$ is built on this state that has very enhanced intra-band $E2$ transition probabilities. Therefore, we conclude that this band is a

$3\alpha + n$ linear-chain band with very strong deformation. Furthermore, this linear-chain wave function has parity asymmetry, and hence, constitute the parity-doublet band. As shown in Fig. 2 (b), $K = 1/2$ rotational bands with very large moment-of-inertia appear in both parity, and both of them are well described by a single Slater determinant shown in Fig. 1 (e).

4. Summary

In summary, we have investigated the $3\alpha + n$ cluster states on the basis of AMD calculation. By introducing HON constraint, $3\alpha + n$ cluster states are described well. We predict two different kinds of the cluster states, the hoyle analogue state and the linear-chain state. By the analysis of the S-factor, we found that the $1/2_2^+$ state is understood as the 0_2^+ state (Hoyle state) of ^{12}C accompanied by a valence neutron occupying the s-wave. We also conclude that the p-wave states coupled to the hoyle state may not exist as a single state. The linear-chain state appears as a $K = 1/2$ rotational bands in both parity owing to its parity asymmetric structure.

Acknowledgments

The numerical calculations were performed on the HITACHI SR16000 at KEK and IT . One of the authors (M.K.) acknowledges that this work is supported by the rants-in-Aid for Scientific Research on Innovative Areas from ME T (rant No. 2404:24105008) and JS S KAKENHI rant No. 25400240.

References

[1] Tohsaki A, Horiuchi H, chuck P and Röpke G 2001 *Phys. Rev. Lett.* **87** 192501
[2] Funaki Y, Tohsaki A, Horiuchi H, chuck P and Röpke 2003 *Phys. Rev.* C **67** 051306
[3] Morinaga H 1956 *Phys. Rev.* **101** 254
[4] Kanada-En yo Y 1998 *Phys. Rev. Lett.* **81** 5291
[5] Chernykh M, Feldmeier H, Neff T, von Neumann-Cosel P and Richter A 2007 *Phys. Rev. Lett.* **98** 032501
[6] Yamada T, Horiuchi H and chuck P 2006 *Mod. Phys. Lett.* A **21** 2373
[7] Itagaki N, Okabe , Ikeda K and Tanihata I 2001 *Phys. Rev.* C **64** 014301
[8] von Oertzen W, *et al. Eur. Phys. J.* A **21** 193
[9] uhara T and Kanada-En yo K 2010 *Phys. Rev.* C **82** 044301
[10] Furutachi N and Kimura M 2011 *Phys. Rev.* C **83** 021303(R)
[11] Kanada-En yo Y, Kimura M and Horiuchi H 2003 *C. R. Physique* **4**, 497
[12] Kanada-En yo Y, Kimura M and Ono A 2012 *PTEP* **2012** 01A202
[13] Berger J F, Girod M and Gogny D 1991 *Comput. Phys. Comm.* **63** 365
[14] KimuraM 2004 *Phys. Rev.* C **69** 044319.
[15] Kanada-En yo Y and Kimura M 2005 *Phys.Rev.* C **72** 064322

Journal of Physics: Conference Series **569** (2014) 012048 doi:10.1088/1742-6596/569/1/012048

Breakup dynamics in ^{6}Li elastic scattering with four-body and three-body CDCC

S Watanabe[1], T Matsumoto[1], K Minomo[2], K Ogata[2] and M Yahiro[1]

[1] Department of Physics, Kyushu University, Fukuoka 812-8581, Japan
[2] Research Center for Nuclear Physics (RCNP), Osaka University, Ibaraki 567-0047, Japan

E-mail: s-watanabe@phys.kyushu-u.ac.jp

Abstract. We have investigated projectile breakup effects on ^{6}Li + ^{209}Bi elastic scattering near the Coulomb barrier with the four-body and three-body continuum-discretized coupled-channels methods. In the analysis, the elastic scattering is well described by the p + n + ^{4}He + ^{209}Bi four-body model. Four-body dynamics of the elastic scattering is precisely investigated and we then found that d (p + n) in ^{6}Li may hardly break up during the scattering.

1. Introduction

In reactions of weakly-bound nuclei, projectile breakup processes are essential and we need to treat the effect properly in the calculation. The continuum-discretized coupled-channels method (CDCC) has been proposed as an accurate method of treating breakup effects [1, 2]. Nowadays, CDCC has become a standard model known as three-body CDCC in which the total system is assumed to be a three-body system (two-body projectile + target). Furthermore, three-body CDCC has been extended to four-body CDCC in which the total system is assumed to be four-body system (three-body projectile + target) [3]. Thus, CDCC is a powerful method to describe not only three-body scattering [4] but also four-body scattering [5]. The recent developments of CDCC are shown in Ref. [6, 7, 8].

^{6}He + ^{209}Bi scattering near the Coulomb barrier was analyzed with three-body CDCC based on a ^2n + ^{4}He + ^{209}Bi three-body model [9]; that is to say a pair of extra neutrons in ^{6}He was treated as a single particle, dineutron (^2n). The three-body CDCC calculation, however, did not reproduce the angular distribution of the measured elastic cross section and overestimated the measured total reaction cross section by a factor of 2.5. This problem has been solved by four-body CDCC in which the total system is described as a n + n + ^{4}He + ^{209}Bi four-body system [3]. Also for ^{6}Li + ^{209}Bi scattering near the Coulomb barrier, three-body CDCC was applied [9]. In the three-body CDCC, a d + ^{4}He + ^{209}Bi three-body model was assumed. However, the calculation could not reproduce the data without normalization factors for the potential between ^{6}Li and ^{209}Bi. These studies [3, 9] strongly suggest that ^{6}Li + ^{209}Bi scattering should also be treated with four-body CDCC.

In this work, we analyze ^{6}Li + ^{209}Bi elastic scattering at 29.9 and 32.8 MeV with four-body CDCC by assuming the p + n + ^{4}He + ^{209}Bi four-body model. This is the first application of four-body CDCC to ^{6}Li scattering. We deal with four-body dynamics explicitly and compare with the results of three-body CDCC. This poster presentation is mainly based on our recent work of Ref. [10].

2. Theoretical framework and Model Hamiltonian

Breakup dynamics of the ^{6}Li + ^{209}Bi scattering is governed by the Schrödinger equation

$$(H - E)\Psi = 0 \tag{1}$$

for the total wave function Ψ, where E is a total energy of the system. In order to clarify the difference between four-body (p + n + ^{4}He + ^{209}Bi) and three-body (d + ^{4}He + ^{209}Bi) dynamics, we set two types of the model Hamiltonian. One is for four-body CDCC defined by

$$H^{(4b)} = K_R + U_n(R_n) + U_p(R_p) + U_\alpha(R_\alpha) + \frac{e^2 Z_{\mathrm{Li}} Z_{\mathrm{Bi}}}{R} + h, \tag{2}$$

where h denotes the internal Hamiltonian of ^{6}Li described by three-cluster model, R is the center-of-mass coordinate of ^{6}Li relative to ^{209}Bi, K_R stands for the kinetic energy and U_x describes the nuclear part of the optical potential between x and ^{209}Bi as a function of the relative coordinate R_x. Since the Coulomb breakup effect is negligible for the ^{6}Li elastic scattering [9], we approximate the Coulomb part of p-^{209}Bi and ^{4}He-^{209}Bi interactions by $e^2 Z_{\mathrm{Li}} Z_{\mathrm{Bi}}/R$, where Z_{A} is the atomic number of the nucleus A. The other is for three-body CDCC defined by

$$H^{(3b)} = K_R + U_d(R_d) + U_\alpha(R_\alpha) + \frac{e^2 Z_{\mathrm{Li}} Z_{\mathrm{Bi}}}{R} + h', \tag{3}$$

where h' denotes the internal Hamiltonian of ^{6}Li described by two-cluster model.

In general, the total wave function Ψ is expanded in terms of the orthonormal set of eigenstates ϕ of h (h')

$$\Psi(\boldsymbol{R}, \boldsymbol{\xi}) = \chi_0(\boldsymbol{R})\phi_0(\boldsymbol{\xi}) + \int_0^\infty d\varepsilon \chi_\varepsilon(\boldsymbol{R})\phi_\varepsilon(\boldsymbol{\xi}), \tag{4}$$

where $\boldsymbol{\xi}$ is the Jacobi coordinate in ^{6}Li, subscripts 0 and ε denote the ground state and the continuum state with the internal energy ε of ^{6}Li respectively. The expansion coefficient χ_0 (χ_ε) describes the relative motion between ^{209}Bi and ^{6}Li in its ground state (continuum state with ε). In CDCC, the continuum state is truncated at an upper limit $\varepsilon_{\mathrm{max}}$, and the continuum up to $\varepsilon_{\mathrm{max}}$ is discretized into a finite number of discrete states

$$\Psi(\boldsymbol{R}, \boldsymbol{\xi}) \approx \chi_0(\boldsymbol{R})\phi_0(\boldsymbol{\xi}) + \int_0^{\varepsilon_{\mathrm{max}}} d\varepsilon \chi_\varepsilon(\boldsymbol{R})\phi_\varepsilon(\boldsymbol{\xi}) \approx \sum_{i=0}^{i_{\mathrm{max}}} \hat{\chi}_i(\boldsymbol{R})\hat{\phi}_i(\boldsymbol{\xi}). \tag{5}$$

Readers are directed to Ref. [10] for more details.

In Eqs. (2) and (3), the optical potentials (U_n, U_p, U_α, and U_d) are taken from Refs. [11, 12, 13]. The proton optical potential U_p is assumed to be the same as U_n, and parameters of U_n are refitted to reproduce experimental data on n + ^{209}Bi elastic scattering [14], where only the central interaction is taken for simplicity. We confirmed that these optical potentials well reproduce the experimental data for each subsystem of ^{6}Li + ^{209}Bi as shown in Fig. 1 [(a) n + ^{209}Bi scattering at 5 MeV, (b) ^{4}He + ^{209}Bi scattering at 19–22 MeV, (c) d + ^{209}Bi scattering at 12.8 MeV].

3. Results

Figure 2 shows the angular distribution of elastic cross section for ^{6}Li + ^{209}Bi scattering at 29.9 MeV and at 32.8 MeV. The dotted line shows the result of three-body CDCC calculation with the d-optical potential U_d^{OP}. This calculation is similar to the one in the previous study [9] and underestimates the measured cross section. The solid (dashed) line, meanwhile, stands for

Journal of Physics: Conference Series **569** (2014) 012048 doi:10.1088/1742-6596/569/1/012048

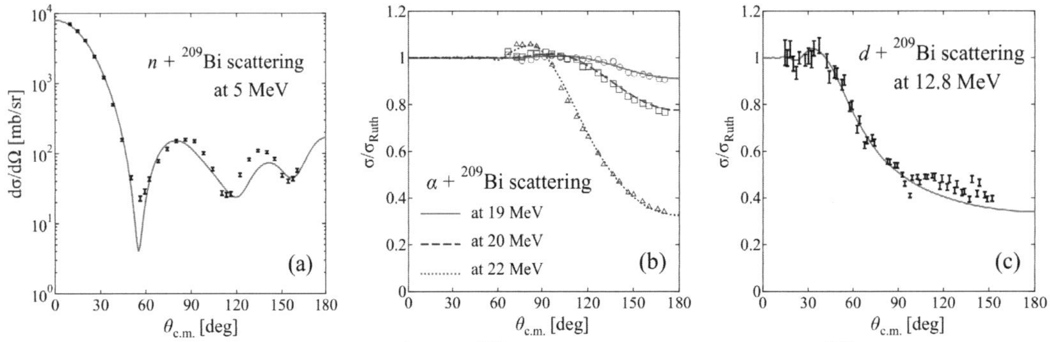

(a) n + ^{209}Bi at 5 MeV (b) ^{4}He + ^{209}Bi at 19–22 MeV (c) d + ^{209}Bi at 12.8 MeV

Figure 1. Angular distribution of the elastic cross section for (a) n + ^{209}Bi scattering at 5 MeV, (b) ^{4}He + ^{209}Bi scattering at around 20 MeV, and (c) d + ^{209}Bi scattering at 12.8 MeV. The experimental data are taken from Refs. [14, 12, 13].

the result of four-body CDCC calculation with (without) projectile breakup effects. In CDCC calculations without ^{6}Li-breakup, the model space is composed only of the ^{6}Li ground state ($\hat{\phi}_0$). The solid line reproduces the measured cross section but the dashed line does not, indicating the projectile breakup effects are thus significant. As just described, the present ^{6}Li scattering is well described by the p + n + ^{4}He + ^{209}Bi four-body model.

In order to understand breakup dynamics in the ^{6}Li scattering, we investigate how d breakup affects the cross section in the framework of three-body CDCC. As a calculation in the limit of no d-breakup effect, the optical potential between d and ^{209}Bi ($U_d^{\rm OP}$) should be replaced by the single-folding potential ($U_d^{\rm SF}$) which is obtained by folding U_n and U_p with the ground-state deuteron density. Note that we use the same U_n and U_p as for four-body CDCC (see Eq. 2). In Fig. 2, the dot-dashed line shows the result of the three-body CDCC calculation with $U_d^{\rm SF}$. The result well simulates that of four-body CDCC calculation (the solid line). This result suggests d breakup is suppressed in the ^{6}Li scattering.

Finally, we would like to mention about d + ^{209}Bi scattering at 12.8 MeV which is the same reaction as in Fig. 1(c). For this scattering, we can apply three-body CDCC in which the p + n + ^{209}Bi model is assumed and both Coulomb and nuclear breakup effects are taken into account. The results are as follows. The CDCC calculation with breakup effects reproduces the experimental data but the CDCC calculation without breakup (one-channel calculation with $U_d^{\rm SF}$) does not. d-breakup effect is thus quite important for "normal" d scattering. As already mentioned, the one-channel calculation with $U_d^{\rm OP}$ well reproduces the data [see Fig. 1(c)] and this is because the optical potential ($U_d^{\rm OP}$) implicitly includes the d-breakup effect. For ^{6}Li + ^{209}Bi scattering, the reason why three-body CDCC with $U_d^{\rm OP}$ does not work may be because we have overcounted the d-breakup effect which is almost absent in d in ^{6}Li scattering.

4. Summary

The ^{6}Li + ^{209}Bi scattering at 29.9 MeV and 32.8 MeV near the Coulomb barrier is well described by four-body CDCC based on the p + n + ^{4}He + ^{209}Bi model. This is the first application of four-body CDCC to ^{6}Li scattering. Through the three-body CDCC analysis based on d + ^{4}He + ^{209}Bi model, we found that three-body CDCC can reproduce the measured cross section if the single-folding potential $U_d^{\rm SF}$ is taken, although three-body CDCC with the optical potential $U_d^{\rm OP}$ cannot. This result suggests that d in ^{6}Li hardly breaks up during the scattering. In other words, the failure of three-body CDCC with $U_d^{\rm OP}$ might come from overcounting the d-breakup

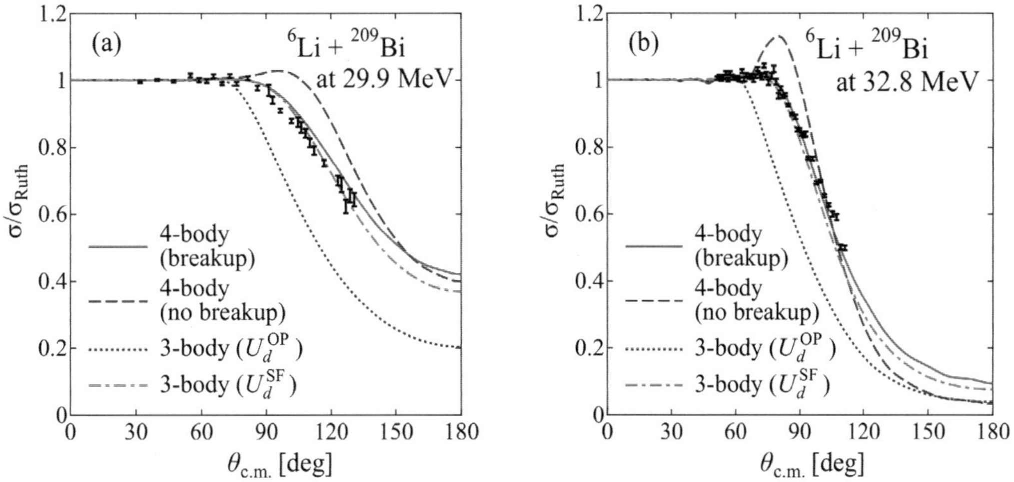

(a) ^{6}Li + ^{209}Bi scattering at 29.9 MeV (b) ^{6}Li + ^{209}Bi scattering at 32.8 MeV

Figure 2. Angular distribution of the elastic cross section for ^{6}Li + ^{209}Bi scattering at 29,9 MeV (a) and at 32.8 MeV (b) [10]. The cross section is normalized by the Rutherford cross section. The dotted (dot-dashed) line stands for the result of three-body CDCC calculation in which $U_d^{\rm OP}$ ($U_d^{\rm SF}$) is taken as U_d. The solid (dashed) line represents the result of four-body CDCC calculations with (without) breakup effects. The experimental data are taken from Refs. [15, 16].

effect which is almost absent in d in ^{6}Li scattering. However, we need to discuss carefully whether the d breakup is always suppressed in the ^{6}Li scattering. Further analysis such as energy and target dependences of d breakup should be done.

References

[1] Kamimura M, Yahiro M, Iseri Y, Sakuragi Y, Kameyama H and Kawai M 1986 *Prog. Theor. Phys. Suppl.* **89** 1
[2] Austern N, Iseri Y, Kamimura M, Kawai M, Rawitscher G and Yahiro M 1987 *Phys. Rep.* **154** 125
[3] Matsumoto T, Egami T, Ogata K, Iseri Y, Kamimura M and Yahiro M 2006 *Phys. Rev.* C **73** 051602(R)
[4] Matsumoto T, Kamizato T, Ogata K, Iseri Y, Hiyama E, Kamimura M and Yahiro M 2003 *Phys. Rev.* C **68** 064607
[5] Cubero M *et al.* 2012 *Phys. Rev. Lett.* **109** 262701
[6] Yahiro M, Ogata K, Matsumoto T and Minomo K 2012 *Prog. Theor. Exp. Phys.* 01A206
[7] Yahiro M, Matsumoto T Minomo K, Sumi T and Watanabe S 2012 *Prog. Theor. Phys. Suppl.* **196** 87
[8] Ogata K 2012 *Prog. Theor. Phys. Suppl.* **196** 203
[9] Keeley N, Cook J M, Kemper K W, Roeder B T, Weintraub W D, Maréchal F and Rusek K 2003 *Phys. Rev.* C **68** 054601
[10] Watanabe S *et al.* 2012 *Phys. Rev.* C **86** 031601
[11] Koning A J and Delaroche J P 2003 *Nucl. Phys.* A **713** 231
[12] Barnett A R and Lilley J S 1974 *Phys. Rev.* C **9** 2010
[13] Budzanowski A, Freindl L, Grotowski K, Rzeszutko M, Słapa M, Szmider J and Hodgson P 1963 *Nucl. Phys.* **49** 144
[14] Annand J, Finlay R and Dietrich P 1985 *Nucl. Phys.* A **443** 249
[15] Aguilera E F *et al.* 2000 *Phys. Rev. Lett.* **84** 5058
[16] Aguilera E F *et al.* 2001 *Phys. Rev.* C **63** 061603

Journal of Physics: Conference Series **569** (2014) 012049 doi:10.1088/1742-6596/569/1/012049

Strength function under the absorbing boundary condition

M. Iwasaki, R. Otani, and M. Ito

Department of Pure and Applied Physics, Kansai University, 3-3-35 Yamatecho, uita, Osaka 564-8680, Japan

E-mail: k036052@kansai-u.ac.jp

Abstract. The strength function of the linear response by the external field is calculated in the formalism of the absorbing boundary condition (ABC). The dipole excitation of a schematic two-body system is treated in the present study. The extended completeness relation, which is assumed on the analogy of the formulation in the complex scaling method (C M), is applied to the calculation of the strength function. The calculation of the strength function is successful in the present formalism and hence, the extended completeness relation seems to work well in the ABC formalism. The contributions from the resonance and the non-resonant continuum is also analyzed according to the decomposition of the energy levels in the extended completeness relation.

1. Introduction

In unstable nuclei, one of interesting subjects is the dynamics of continuum states above the particle decay threshold. Neutron-drip line nuclei are weak binding systems and hence, they are easily excited to unbound continuum states. Therefore, the breakup reaction is an important tool to investigate the dynamical property of neutron-excess systems.

Theoretically, the so-called strength function is important to handle the breakup of weak binding systems. The complex scaling method (CSM) is one of powerful methods to calculate the strength function [1] because, in CSM, the scattering problem of the unbound states can be transformed into the bound-state-like problem. In CSM, the so-called extended completeness relation is basic and important principle in the calculation of the strength function. In fact, we can see the successful application of CSM in the three body problem [2].

The method of the absorbing boundary condition (ABC) is another powerful tool to treat the resonant and continuum states [4, 5, 6, 7]. In the present study, we assume the extended completeness relation on the wave functions under the absorbing boundary and demonstrate that the ABC method can describe the strength function in a similar manner to CSM.

2. Framework

In the present calculation, we consider a schematic two-body problem, which interacts by a simple aussian potential. The detailed settings of the Hamiltonian H is same as those shown in Ref. [1]. First, we calculate the energy eigenvalues and their wave functions by applying the basis expansion method. The shifted aussian is used as the basis function [3]. The distance parameter S in the shifted aussian is set in the range from $S_{min} = 0.1$ fm to $S_{max} = 32.1$ fm with a constant mesh of $\Delta S = 0.4$ fm. In the diagonalization of the Hamiltonian, we introduce the absorbing potential $-i\eta W$ with the strength η. Here the functional form of the absorbing potential is set to be the shifted polynomial in the radial coordinate r, $W(r) = \theta(r-r_a)(r-r_a)^\beta$, with $r_a = 7$ fm and $\beta = 4$ [4].

Secondly, we calculate the strength function for the two body breakup (or capture) by the dipole field, according to the following prescription. The strength function $\mathcal{S}_\lambda(E)$ is generally defined by the imaginary part of the response function, $\mathcal{R}_\lambda(E)$, which are given by

$$\mathcal{S}_\lambda(E) = -\frac{1}{\pi}\Im\left[\mathcal{R}_\lambda(E)\right] \tag{1}$$

$$\mathcal{R}_\lambda(E) = \int\int d\mathbf{r}d\mathbf{r}'\psi_i^*(\mathbf{r})\hat{O}_\lambda^\dagger\mathcal{G}(E,\mathbf{r},\mathbf{r}')\hat{O}_\lambda\psi_i(\mathbf{r}') \ . \tag{2}$$

Here ψ_i and $\hat{O}_\lambda$ are an initial state and the external field with a multipole λ. $\mathcal{G}(E,\mathbf{r},\mathbf{r}')$ denotes the coordinate representation of the reen function, which is given by

$$\mathcal{G}(E,\mathbf{r},\mathbf{r}') = \left\langle \mathbf{r}\left|\frac{1}{E-H}\right|\mathbf{r}'\right\rangle \ . \tag{3}$$

In the calculation employing the ABC method, the absorbing potential, $-i\eta W$ is added to the total Hamiltonian H in the reen function, Eq. (3) and hence, $H \to H^\eta = H - i\eta W$.

We assume the extended completeness relation for the energy levels calculated by diagonalizing the Hamiltonian with an absorber, $H - i\eta W$,

$$\hat{1} = \sum_B^{n_B} |\psi_B><\tilde{\psi}_B| + \sum_R^{n_R(\eta)} |\psi_R^\eta><\tilde{\psi}_R^\eta| + \sum_C^{n_C(\eta)} |\psi_C^\eta><\tilde{\psi}_C^\eta|. \tag{4}$$

Here ψ_i^η shows the wave function calculated under the absorbing boundary with the strength η. The subscript of i denotes a kind of energy levels: $i =$ B (bound states), R (resonant states), C (continuum states). In Eq. (4), tilde in ψ_i means that the complex conjugate is not performed. By inserting the extended completeness relation in Eq. (4) into the response function in Eq. (2), the strength function can be easily calculated from the eigenstates with the absorbing boundary. This formulation is similar technique proposed in CSM [1].

3. Results

In Fig. 1, the strength functions calculated from the extended completeness relation in Eq. (4) are shown by the solid curves. The strength of the absorber is set to be $\eta = 10^{-5}$. To check the ABC calculation, we also calculate the strength function from the scattering wave function, $\psi_E^{(+)}$, at a scattering energy E. The explicit form of the strength function with the scattering wave function of $\psi_E^{(+)}$ becomes

$$\mathcal{S}_\lambda^{(+)}(E) = \left|<\psi_E^{(+)}|\hat{O}_\lambda|\psi_i>\right|^2 \ . \tag{5}$$

Here the scattering wave function $\psi_E^{(+)}$ is solved under the normal scattering boundary condition, the incoming wave plus outgoing wave with the amplitude of the scattering matrix. The results of Eq. (5) are shown by the dotted curves. In the scattering calculation, we have applied the Fox- oodwin method with a mesh of $\delta = 0.01$ fm and the matching distance of $r_m = 12$ fm.

Because the normalization condition in the ABC calculation is different from the scattering calculation, the peak magnitude of the latter calculation is re-normalized to that of the former calculation. In both panels, we can confirm the strength functions calculated by ABC completely reproduce the results of the scattering solution. Thus, the assumption of extended completeness relation to the ABC method is considered to nicely work in the pragmatic calculation of the strength function. This result is consistent to the previous study of CSM [1].

According to Eq. (4), the strength function can be decomposed into three parts, the bound (ψ_B^η), resonance (ψ_R^η) and continuum (ψ_C^η) states. In the present ABC calculation, two resonances are separated from the non-resonant continuum in the complex energy plane. The energy and width of these two resonances are the same as those calculated by using CSM [1].

The individual contributions from the resonances and non-resonances are shown in Fig. 2. In this figure, the contributions from the first resonance (dashed curve) and the second resonance (dotted curve) are dominant, but the continuum states (dotted-dashed curve) gives considerable contribution in the total strength function (solid curve), especially at the higher energy region of $E \geq 2$ MeV.

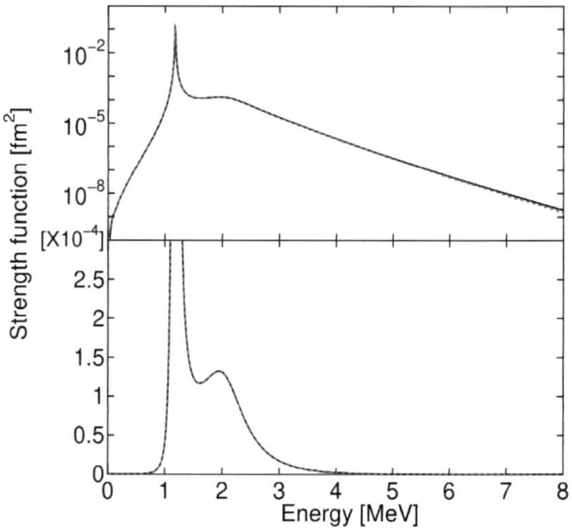

Figure 1. Strength function calculated by using the ABC method. The top and bottom panels are the plot in the linear and logarithmic scales, respectively. See text for details.

It should be noticed that the decomposed strengths are not necessarily positive values although the total strength must be defined by the positive value because it is an observable quantity. In the low energy region of $E \leq 1$ MeV, the positive contribution from the second resonance (dotted curve) cancels out the negative contribution from the first resonance (dashed curve). As a result of this cancellation, the total strength becomes almost zero value in the lower energy region. Moreover, the contribution from the continuum states are negative in a whole energy region. Thus, the continuum contributions considerably reduce the resonant contribution at the higher energy region.

These features of the cancellation are consistent to the previous study by using CSM [1]. In the CSM calculation, the two resonances are separated from the continuum line when the rotation angle passes $\theta = 10°$. At this rotation angle, the total strength function is decomposed into the contributions from two resonances and the continuum states. The individual contributions in ABC, presented in Fig. 2, are almost same as those obtained in the CSM calculation [1].

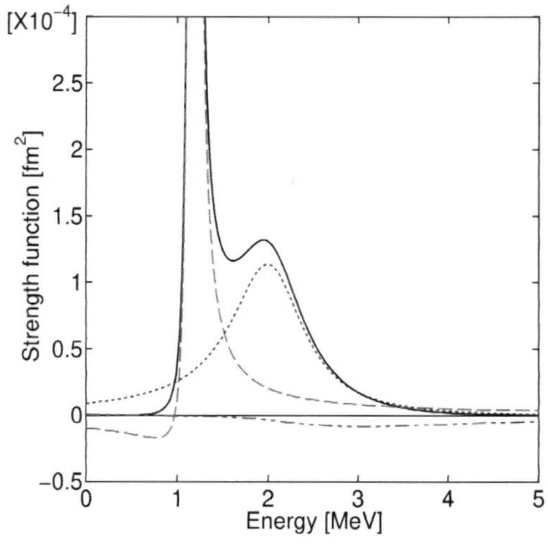

Figure 2. Decomposed strength function. See text for details.

In the ABC calculation, the strength of the absorber, η, is an external parameter in calculating the resonant and continuum states. The imaginary part of the resonant states is almost invariant with respect to the variation of η if once the resonant levels are separated from the continuum. However, the maximum value of the imaginary part of the continuum states depends on η and hence, they are controlled by this parameter [4, 5, 6].

The sensitivity of the strength function to η is investigated in Fig. 3. In this figure, the various strength functions obtained by changing η in the absorbing potential. The solid curve shows the strength function obtained from the ABC method, while the dashed curve shows the strength function calculated from the scattering wave function in the final state, according to Eq. (5).

In the individual panels of Fig. 3, η denotes the strength of the absorber used in the ABC calculation. Many spikes appear in the case of the smaller strength, $\eta = 10^{-8}$ (top panel), but the appearance of the spikes is suppressed more and more as η becomes stronger: $\eta = 10^{-7}$ (second panel from the top) and $\eta = 10^{-6}$ (third panel). Finally, the smooth strength function is obtained at $\eta = 10^{-5}$ (bottom panel). The smoothed strength function reproduces the strength function calculated from the scattering wave function in Eq. (5).

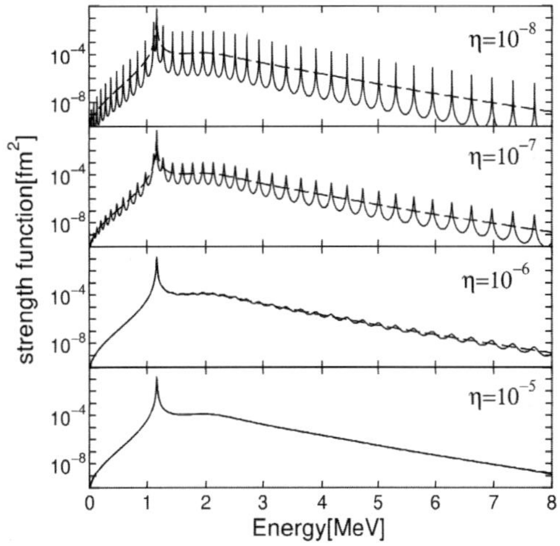

Figure 3. η dependence of the strength functions. See text for details.

We have checked the η dependence of the smoothed strength function. In the range of $\eta = 4.0\times10^{-6}$ to $\eta = 4.0\times10^{-4}$, the distribution of the strength function keeps its original form shown in the bottom panel of Fig. 3. Thus, the strength function is insensitive and stable to the variation of η over about two order of the magnitudes after the spikes are completely smeared out. If η becomes further stronger, the distribution of the ABC strength function begins to deviate from the result of the scattering calculation. This is because the reflection effect from the absorber becomes prominent in the case that the strength of the absorber is much stronger than the optimal value.

4. Summary

In the present study, we have assumed the extended completeness relation, proposed in the study of CSM, to the energy eigenstates under the absorbing boundary condition. The schematic two-body problem has been handled, and the extended completeness relation is applied to the calculation of the strength function. We have found that the pragmatic calculation of the strength function is successful, and the obtained results are completely consistent to the previous study of CSM [1].

In the present report, we have shown the decomposition of the strength function into the resonances and the non-resonant continuum according to the extended completeness relation. In the ABC method, however, the non-resonant states are classified into three type of continuum [4, 5]. The further analysis on the continuum states is very interesting because the feature of the "triple-continuum states" in ABC is in marked contrast to the "single continuum state" in CSM. The analysis of the continuum state is now under progress.

References

[1] Myo T, Ohnishi A and Kato K 1998 *Prog. Theor. Phys.* **99** 801
[2] Myo T, Kato K, Aoyama and Ikeda K 2001 *Phys. Rev.* C **63** 054313
[3] Horiuchi H 1977 *Prog. Theor. Phys. Suppl.* **62** 90 (and references therein.)
[4] Ito M and Yabana K 2005 *Prog. Theor. Phys.* **113** 1047
[5] Riss U V and Meyer H-D 1993 *Jour. Phys.* B **26** 4503
[6] Riss U V and Meyer H-D 1995 *Jour. Phys.* B **28** 1475, 1998 *Jour. Phys.* B **31** 2279
[7] Masui H and Ho Y K 2002 *Phys. Rev.* C **65** 054305

Multiple-scattering effects in proton- and alpha-nucleus reactions with Glauber theory

S. Hatakeyama[1], S. Ebata[2], W. Horiuchi[1], M. Kimura[1]

[1] Department of Physics, Hokkaido University, apporo 060-0810, Japan.
[2] Meme Media Laboratory, Hokkaido University, apporo 060-0813, Japan.

E-mail: `hatakeyama@nucl.sci.hokudai.ac.jp`

Abstract. We study the total reaction and elastic differential cross sections for proton-nucleus and ^{4}He-nucleus reactions in the framework of the Glauber theory which describes multiple-scattering processes. The input wave functions are obtained using the kyrme-Hartree-Fock method and prepared for a wide range of mass numbers, O, Ca, Ni, n, and Pb isotopes. The theory reproduces experimental data very well. An effect of the multiple scattering is discussed by comparing with a standard optical-limit approximation. We see that the multiple-scattering effects play a crucial role, especially in enhancing the elastic differential cross sections at large scattering angles.

1. Introduction

A study of new unstable nuclei has become possible in new radioactive beam facilities. In order to relate nuclear structure with observables, we need reaction theory which reflects the nuclear structure. The lauber theory is widely used to evaluate cross sections in high-energy nuclear reactions [1]. In the lauber theory, the scattering amplitude of two colliding nuclei with mass number A_P and A_T is obtained by evaluating a $[3 \times (A_P + A_T - 2)]-$fold multiple integral induced by an $(A_P + A_T - 2)-$body multiple-scattering operator. The optical-limit approximation (OLA) is a standard way to approximate the scattering amplitude. The complicated multiple-scattering operator is approximated to a one-body (two-body) operator in proton-nucleus (nucleus-nucleus) reactions, and thus the wave function reduced to a one-body density. However, the use of one-body densities may wash out some pieces of structural information. In fact, the multiple-scattering effects play an important role in the nuclear reaction involving unstable nuclei [2, 3, 4, 5, 6]. In this paper, we extend the application range of the lauber theory to a wide range of mass numbers and discuss the effect of the multiple-scattering processes to the cross sections.

The paper is organized as follows: In Section 2, we explain how to calculate the scattering amplitude in the lauber theory and its approximations. Section 3 presents our results of the total reaction and elastic differential cross sections. A role of multiple-scattering effects is discussed in this section. A summary is given in Section 4.

2. Formalism

The lauber theory is formulated based on eikonal and adiabatic approximations [1]. The final state wave function ψ_f is expressed by a product of the phase-shift function and the initial wave function, ψ_i:

$$\psi_f = e^{i\chi}\psi_i. \tag{1}$$

The total reaction cross sections and elastic differential cross sections with phase-shift function are calculated by

$$\sigma_R = \int \left(1 - |e^{i\chi(\boldsymbol{b})}|^2\right) d\boldsymbol{b}, \tag{2}$$

$$\frac{d\sigma}{d} = \left|\frac{k}{2\pi}\int e^{i\boldsymbol{q}\cdot\boldsymbol{b}}\left(e^{i\chi(\boldsymbol{b})} - 1\right) d\boldsymbol{b}\right|^2, \tag{3}$$

where k is the wave number of the relative motion between the two nuclei, $\boldsymbol{q}$ is a momentum transfer vector, and $\boldsymbol{b}$ is an impact parameter vector. In a proton-nucleus (pT, target: T) reaction, the phase-shift function of the pT reaction is

$$e^{i\chi_{pT}(\boldsymbol{b})} = \left\langle \psi_T(\boldsymbol{r}_1, \cdots, \boldsymbol{r}_{A_T}) \middle| \prod_{j=1}^{A_T} e^{i\chi_{pN}(\boldsymbol{b}-\boldsymbol{s}_j)} \middle| \psi_T(\boldsymbol{r}_1, \cdots, \boldsymbol{r}_{A_T}) \right\rangle, \tag{4}$$

where $e^{i\chi_{pN}}$ is a proton-nucleus (pN) phase-shift function, $\boldsymbol{s}$ is a two-dimensional vector perpendicular to the beam direction, and ψ_T is the wave function of the target nucleus. The NN phase-shift function employed here is determined so as to reproduce the NN scattering properties [7]. It should be noted that the above equation includes the multiple integral.

The optical-limit approximation (OLA) offers the most simple expression, which is obtained by taking only the first-order term of the cumulant expansion of Eq. (4) [1], that is

$$e^{i\chi_{pT}^{\rm OLA}(\boldsymbol{b})} = \exp\left\{-\int d\boldsymbol{r}\rho_T(\boldsymbol{r})(1 - e^{i\chi_{NN}(\boldsymbol{b}-\boldsymbol{s})})\right\}, \tag{5}$$

where the ρ_T is a one-body density of the target nucleus.

In a nucleus-nucleus (PT, projectile: P) reaction, we employ the nucleon target formalism in the lauber theory (NT) proposed in Ref. [8]. In the NT , the pN phase-shift function in Eq. (4) is simply replaced with a PN one,

$$e^{i\chi_{PT}(\boldsymbol{b})} = \left\langle \psi_T(\boldsymbol{r}_1, \cdots, \boldsymbol{r}_{A_T}) \middle| \prod_{j=1}^{A_T} e^{i\chi_{PN}(\boldsymbol{b}-\boldsymbol{s}_j)} \middle| \psi_T(\boldsymbol{r}_1, \cdots, \boldsymbol{r}_{A_T}) \right\rangle. \tag{6}$$

The phase-shift function of PN in the NT is obtained in the spirit of the OLA by

$$e^{i\chi_{PN}(\boldsymbol{b})} = 1 - \exp\left\{-\int d\boldsymbol{r}'\rho_P(\boldsymbol{r}')(1 - e^{i\chi_{NN}(\boldsymbol{b}+\boldsymbol{s}')})\right\}, \tag{7}$$

where ρ_P is the one-body density in the projectile nucleus. The phase-shift function in the OLA is given by

$$e^{i\chi_{PT}^{\rm OLA}(\boldsymbol{b})} = \exp\left\{-\iint d\boldsymbol{r}'d\boldsymbol{r}\rho_P(\boldsymbol{r}')\rho_T(\boldsymbol{r})(1 - e^{i\chi_{NN}(\boldsymbol{b}+\boldsymbol{s}'-\boldsymbol{s})})\right\}. \tag{8}$$

The wave function of those nuclei are generated by the Skyrme-Hartree-Fock (HF) method on three-dimensional coordinate space [9]. The SkM* interaction [10] is employed. The wave function is expressed in a Slater determinant that greatly reduces the computational cost of evaluating the phase-shift functions defined in Eqs. (4) and (6) [11, 12]. Since the multiple-scattering operator is written by a product of one-body operators, the multiple integration is reduced to three-dimensional. We now systematically investigate the cross sections for a wide range of mass number without *ad hoc* parameters. In this study, the wave functions for $^{16-24}$O, $^{40-70}$Ca, ^{56,58}Ni, $^{100-140}$Sn, and $^{190-214}$ b are analyzed.

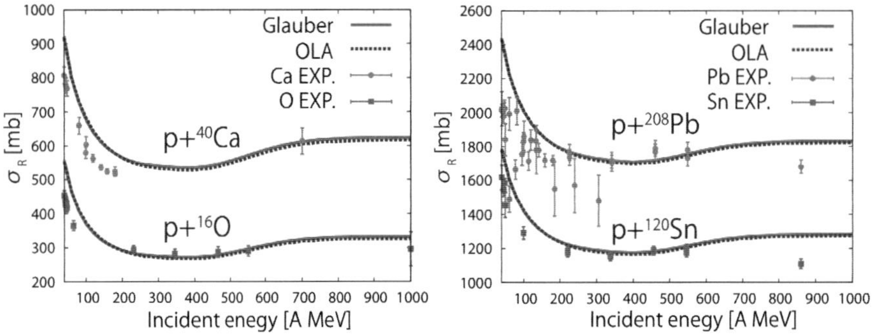

Figure 1. Total reaction cross sections of proton-nucleus reactions incident at 40-1000A MeV calculated with the lauber theory and the OLA on (left) ^{16}O and ^{40}Ca, and (right) ^{120}Sn and ^{208} b targets. The experimental data are taken from [13, 14, 15, 16, 17, 18, 19, 20, 21, 22, 23, 24, 25, 26, 27].

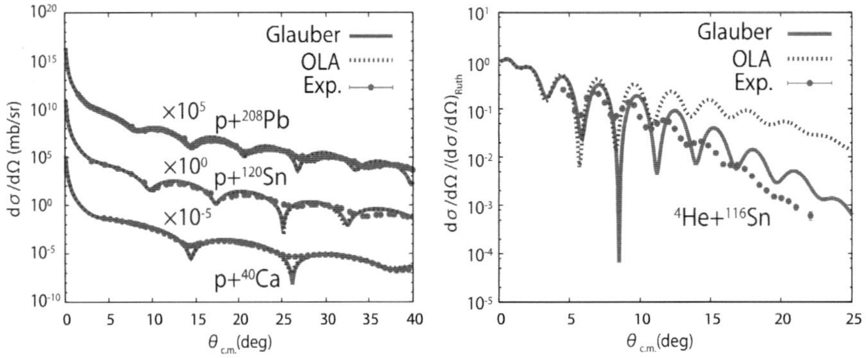

Figure 2. (Left) Elastic differential cross sections of proton-nucleus reactions on ^{40}Ca, ^{120}Sn, and ^{208} b targets at 300A MeV calculated with the lauber theory and the OLA. (right) Same as the left panel but for a ^{4}He-^{116}Sn reaction at 120A MeV. The experimental data are taken from [30, 31, 32, 33, 34, 35].

3. Results

We show our results of the total reaction cross sections, σ_R, and elastic differential cross sections for proton-nucleus and ^{4}He-nucleus reactions. Fig. 1 displays σ_R of proton-nucleus reactions on ^{16}O, ^{40}Ca, ^{120}Sn, and ^{208} b targets. The calculated cross sections at the energy higher than 200A MeV are in reasonable agreement with the experimental data. At the energy lower than 200A MeV, the theory overestimate the data. At such a low energy, the validity of the lauber model is questioned and the Fermi motion and the auli blocking effects may not be negligible [28, 29]. In most energies the lauber theory gives larger cross sections than those of the OLA due to the multiple-scattering effects.

Fig. 2 shows the elastic differential cross sections of proton-nucleus reactions on ^{40}Ca, ^{120}Sn, and ^{208} b targets, and for a ^{4}He-^{116}Sn reaction. The calculations include Coulomb effect. For the proton-nucleus reactions, as displayed in the left panel of Fig. 2 both the lauber theory and the OLA reproduce the experimental data very well and give virtually the same results. For the ^{4}He-^{116}Sn reaction, the two calculated cross sections give the same results at small scattering angles up to 5° but deviate significantly at larger angles as shown in the right panel of Fig. 2. The lauber theory gives a much better description at larger scattering angles than the OLA.

4. Summary

We have applied the lauber theory to high-energy proton-nucleus and nucleus-nucleus reactions and calculate the total reaction and elastic differential cross sections. The wave functions are generated by the Skyrme-Hartree-Fock method for a wide range of mass number. The theory agrees with the experimental data very well in the energy higher than 200A MeV. Our systematic analysis shows the effects of multiple scattering to the cross sections by comparing with the standard optical-limit approximation. The effects are small when the proton-nucleus reactions are used. In the ^{4}He-^{116}Sn reaction, the effects are large that the multiple scattering processes significantly enhance the elastic differential cross sections at large scattering angles. A detailed analysis is in progress and will be reported elsewhere soon.

Acknowledgments

This work was supported in part by JS S KAKENHI rants No. 25800121.

References

[1] Glauber R J, 1959 *in Lectures on Theoretical Physics* vol 1, ed W E Brittin and L C Dunham (New York: Interscience) p 315
[2] Bertsch G F, Esbensen H and ustich A 1990 *Phys. Rev.* C **42** 758
[3] Ogawa Y, Yabana K and uzuki Y 1992 *Nucl. Phys.* A **543** 722
[4] Al-Khalili J *et al* 1996 *Phys. Rev. Lett.* **76** 3903 ; Al-Khalili J *et al* 1996 *Phys. Rev.* C **54** 1843
[5] Yabana K, Ogawa Y and uzuki Y 1992 *Phys. Rev.* C **45** 2909
[6] Al-Khalili J , Tostevin J A and Brooke J M 1997 *Phys. Rev.* C **55** R1018
[7] Abu-Ibrahim B, Horiuchi W, Kohama A and uzuki Y 2008 *Phys. Rev.* C **77** 034607
[8] Abu-Ibrahim B and uzuki Y 2000 *Phys. Rev.* C **62** 034608
[9] Vautherin D and Brink D M 1972 *Phys. ReV.* C **5** 626
[10] Bartel J, Quentin P, Brack M, Guet C and Håkansson H B 1982 *Nucl. Phys.* A **386** 79
[11] Abu-Ibrahim B, Iwasaki and Horiuchi W 2009 *J. Phys. Soc. Jpn.* **78** 044201
[12] Bassel R H and Wilkin C 1968 *Phys. Rev.* **174** 1179
[13] Carlson R F, Cox A J, Nimmo J R and Davison N E *et al* 1975 *Phys. Rev.* C **12** 1167
[14] Renberg P U *et al* 1972 *Nucl. Phys.* A **183** 81
[15] Ingemarsson A, Nyberg J and Renberg P U *et al* 1999 *Nucl. Phys.* A **653** 341
[16] laus I, Margaziotis D J, Carlson R F, Van Oers W T H and Richardson J R 1975 *Phys. Rev.* C **12** 1093
[17] Auce A, Igemarsson A and Johansson R *et al* 2005 *Phys. Rev.* C **71** 064606
[18] Johansson A, vanberg U and undberg O 1961 *Arkiv Fysik* **19** 527
[19] Kirkby P and Link W T 1966 *Can. J. Phys.* **44** 1847
[20] Anderson B D, Bevington P R and Cverna F H *et al* 1979 *Phys. Rev.* C **19** 905
[21] Menet J J H, Gross E E, Malanify J J and Zucker A 1971 *Phys. Rev.* C **4** 1114
[22] Carlson R F, Cox A J, Eliyakut-Roshko T and van Oers W T H 1995 *Can. J. Phys.* **73** 512
[23] Chen F F, Leavitt C P and hapiro A M 1955 *Phys. Rev.* **99** 857
[24] Meyer V, Eisberg R M and Carlson R F 1960 *Phys. Rev.* **117** 1334
[25] Goloskie R and trauch K 1962 Nucl. Phys. **29** 474
[26] Cassels J M and Lawson J D 1954 *Proc. Phys. Soc.* A **67** 125
[27] Millburn G P, Birnbaum W, Crandall W E and checter L 1954 *Phys. Rev.* **95** 1268
[28] DiGiacomo N J, DeVries R M and Peng J C 1980 *Phys. Rev. Lett.* **45** 527 ; 1981 *Phys. Lett.* B **101** 383
[29] Takechi M, Fukuda M and Mihara M *et al* 2009 *Phys. Rev.* C **79** 061601(R)
[30] Kelly J J, Boberg P and Feldman A E *et al* 1991 *Phys. Rev.* C **44** 2602
[31] Terashima , akaguchi H and Takeda H *et al* 2008 *Phys. Rev.* C **77** 024317
[32] Takeda H 2003 *Memoirs of the Faculty of Science, Kyoto University - Series of physics, astrophysics, geophysics and chemistry* **44-1** 1
[33] Hutcheon D A, Olsen W C and herif H *et al* 1988 *Nucl. Phys.* A **483** 429
[34] Zenihiro J, akaguchi H and Murakami T *et al* 2010 *Phys. Rev.* C **82** 044611
[35] Bonin B, Alamanos N and Berthier B *et al* 1985 *Nucl. Phys.* A **445** 381

Feasibility study to search for the rare γ-decay mode in ^{12}C

M Tsumura[1], T Kawabata[1], S Adachi[1], T Baba[1], T Furuno[1], Y Ishii[1], M Murata[1], F Kobayashi[1], Y Kanada-En'yo[1], A Tamii[2], T Hashimoto[2], K Hatanaka[2], Y Matsuda[2], K Miki[2], C Iwamoto[2], T Ito[2], M Miura[2], J Zenihiro[3], S Kubono[3], M Itoh[4], S Ando[4], Y Maeda[5], S Sakaguchi[6], H Akimune[7], H Fujimura[8] and I Ou[9]

[1] Department of Physics, Kyoto University, akyo, Kyoto, 506-8502, Japan.
[2] Reserch Center for Nuclear Physics, Osaka University, Ibaraki, Osaka, 567-0047, Japan.
[3] RIKEN, Wako, aitama, 351-0198, Japan.
[4] Cyclotron and Radioisotope Center, Tohoku University, endai, Miyagi, 980-8578, Japan.
[5] Faculty of Engineering, University of Miyazaki, Miyazaki, 889-2155, Japan.
[6] Department of Physics, Kyusyu University, Fukuoka, 812-8581, Japan.
[7] Department of Physics, Konan University, Kobe, Hyogo, 658-8501, Japan.
[8] Wakayama Medical University, Wakayama, 640-8441, Japan.
[9] Department of Physics, Okayama University, Okayama, 700-0082, Japan.

E-mail: tsumura@scphys.kyoto-u.ac.jp

Abstract. Excited states in ^{12}C nuclei play a very important role in the nucleosynthesis in the universe. ^{12}C nuclei are synthesized by the triple α reaction. Normally this process proceeds via the 0_2^+ state (Hoyle state) at $E_x = 7.65$ MeV in ^{12}C, however, at high temperature $T > 10^9$, the highly excited 3α resonant states such as the 3_1^- state at $E_x = 9.64$ MeV and the 2_2^+ state at $E_x = 9.84$ MeV play a part of the triple α reaction. Unfortunately, the γ-decay probability of the 3_1^- state has not been determined, therefore we planned to determine the γ-decay probability using the inverse kinematic reaction ^{1}H(^{12}C, ^{12}Cp) without measuring γ-rays. The test experiment was carried out at the cyclotron facility in RCNP. The results of the test experiment are reported in the present paper.

1. Introduction

1.1. ^{12}C *synthesis in the universe*

^{12}C nuclei are synthesized by the triple α reaction in the universe. Since heavier elements are synthesized via ^{12}C, the triple α reaction plays a crucial role in the nucleosynthesis in the universe.

In the triple α reaction, an α particle collides with a 2α resonant state, and consequently, excited states of ^{12}C are produced as 3α resonant states as shown in Fig. 1. After that, most of these excited ^{12}C nuclei decay back into 3α, but a tiny fraction of these excited ^{12}C nuclei decays into the ground state of ^{12}C by emitting γ-rays. Therefore, γ-decay probabilities are the key parameters in the ^{12}C synthesis. For example, the γ-decay probability of the Hoyle state (the 0_2^+ state at $E_x = 7.65$ MeV) is known as $4.4(1) \times 10^{-4}$ and the triple α process proceeds via this state at the normal stellar temperature. However, at high temperature $T > 10^9$ K, a contribution of the highly excited 3α resonant states such as the 3_1^- state at $E_x = 9.64$ MeV and the 2_2^+ state at $E_x = 9.84$ MeV [1] becomes important [2].

Unfortunately, the γ-decay probabilities of these states are still unknown, and it causes uncertainty to calculate the triple α reaction rate. Thus we have planned to determine the

γ-decay probability of the 3_1^- state.

1.2. γ-decay probability of the 3_1^- state

It is di cult to measure the γ-decay probability of the 3_1^- state because it is very small. The present knowledge about the total decay width Γ, γ-decay width Γ_γ and γ-decay probability Γ_γ/Γ of the 3_1^- state is summarized in Table 1 [3].

The γ-decay width to the ground state ($\Gamma_{\gamma0}$) was obtained from the inelastic electron scattering. Since Γ_γ must be larger than $\Gamma_{\gamma0}$, the lower limit of Γ_γ of the 3_1^- state is $\Gamma_{\gamma0}$ = 0.31(4) meV. The upper limit was reported in the measurement of the ^{12}C(α, α' ^{12}C) reaction [4]. In this previous measurement, only the upper limit of Γ_γ was determined because the ^{13}C contaminants in the isotopically enriched ^{12}C target caused serious background. Therefore, we decided to use the inverse kinematic reaction ^{1}H(^{12}C, ^{12}Cp), because no ^{13}C contaminants are included in ^{12}C beam.

Table 1. Total decay width Γ, γ-decay width Γ_γ and γ-decay probability Γ_γ/Γ of the 3_1^- state [3].

Γ	Γ_γ	Γ_γ/Γ
34(5) keV	$>$ 0.31(4) meV $<$ 14 meV(1σ C.L.)	$> 9.1\times10^{-9}$ $< 4.1\times10^{-7}$

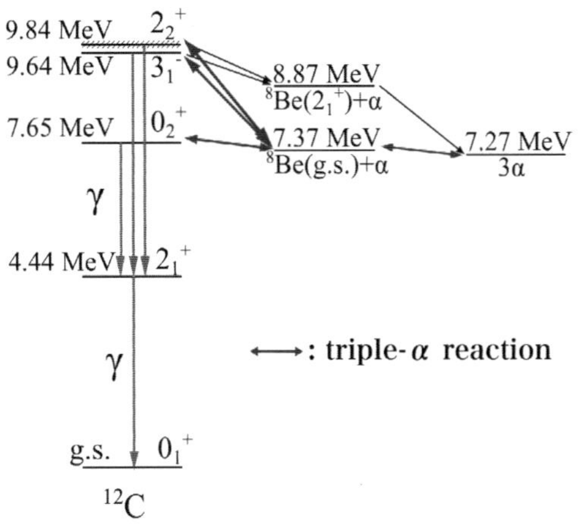

Figure 1. Level diagram related to the triple α reaction.

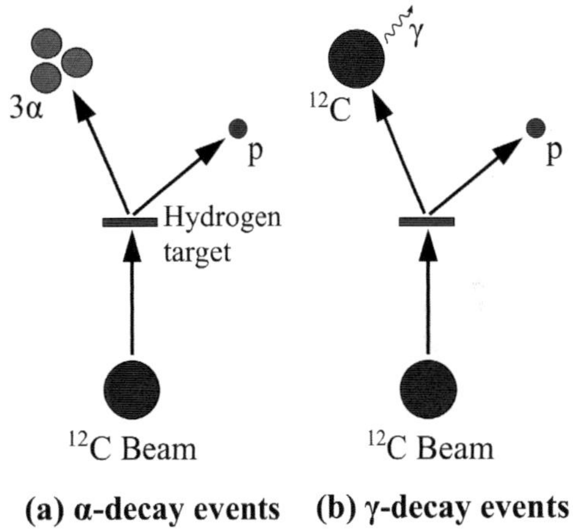

Figure 2. (a) α-decay events and (b) γ-decay events in the ^{1}H(^{12}C, ^{12}Cp) reaction.

2. Experimental consideration

2.1. Experimental procedure

A hydrogen target is bombarded with a ^{12}C beam as shown in Fig. 2. The excitation energy of the inelastically scattered ^{12}C is determined from the energy and angle of the recoiled proton. Most of the ^{12}C nuclei excited to the 3_1^- state decay into 3α as shown in Fig. 2(a), but a tiny fraction of them decays into the ground states by emitting γ-rays as shown in Fig. 2(b). Therefore, if the scattered ^{12}C is detected in coincidence with the recoiled proton, the γ-decay event is identified without the γ-ray detection and the γ-decay probability Γ_γ/Γ is obtained from the ratio of the number of all the inelastic events to the number of the γ-decay events.

Before the physics run, we carried out a test experiment for eight hours using the CH_2 target in order to study the feasibility of this plan.

Journal of Physics: Conference Series **569** (2014) 012051 doi:10.1088/1742-6596/569/1/012051

2.2. *Experimental setup*

The experiment will be performed at the cyclotron facility in Research Center for Nuclear hysics, Osaka University. A solid hydrogen target [5] with a thickness of 0.5 mm is bombarded with a 250-MeV ^{12}C beam as shown in Fig. 3.

The recoiled proton is detected and particle-identified by the Si+CsI telescope in the scattering chamber. The front and rear sides of the Si detector are segmented into the 32 vertical and 16 horizontal strips, respectively. The recoiled angle of the proton is obtained from the double-sided Si strip detector. The excitation energy of the scattered ^{12}C is determined by the energy and angle of the recoiled proton.

The scattered ^{12}C is analyzed by the rand Raiden spectrometer and detected by the two multi-wires drift chambers (VDC1 and VDC2) and two plastic scintillators at the focal plane of the spectrometer. The thicknesses of the plastic scintillators are 1 mm at the upstream and 10 mm at the downstream. The scattered ^{12}C cannot penetrate the 1-mm scintillator, while α particles can penetrate and stop at the 10-mm scintillator. Therefore, the trigger signal for the ^{12}C detection was generated using the anti-coincidence technique between the two scintillators.

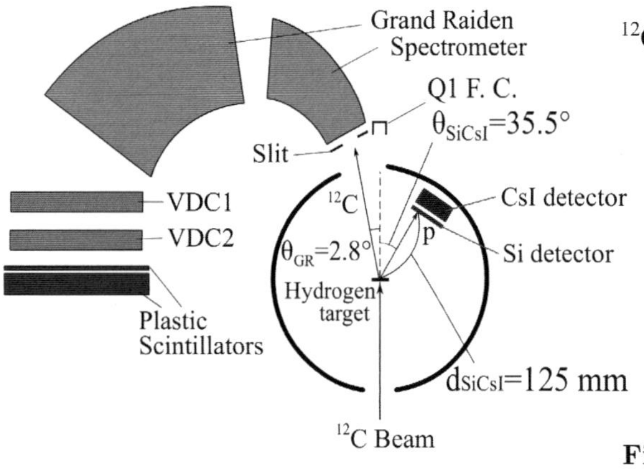

Figure 3. Schematic view of the experimental setup.

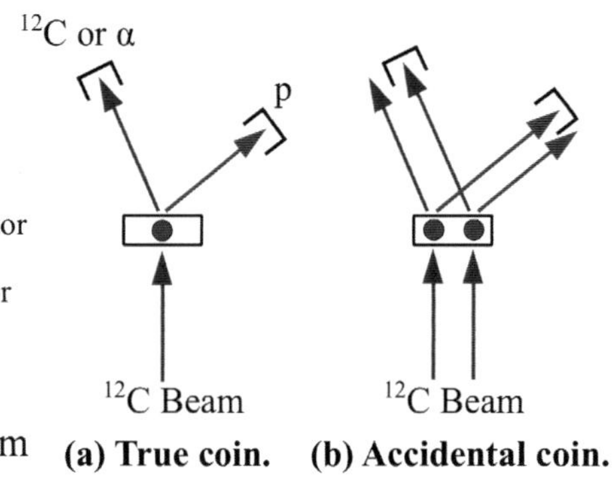

Figure 4. Schematic diagram of (a) true coincidence event and (b) accidetal coincidence event.

2.3. *Improvement of the signal to noise ratio*

Accidental coincidence events are expected to cause serious backgrounds. A true-coincidence event [Fig. 4(a)] is an event in which a recoiled proton and a scattered ^{12}C derived from the same reaction are detected in coincidence, while an accidental coincidence event [Fig. 4(b)] is an event in which a recoiled proton and a scattered ^{12}C derived from the different reactions are accidentally detected at the same time. The number of the accidental coincidence events can be estimated by making random coincidence between different beam bunches, but a special treatment to remove the accidental coincidence events is needed because the true coincidence events are extremely rare. In addition, contaminations the target also cause serious background.

A detailed Monte Carlo simulation was performed to estimate the signal to noise (S N) ratio and it was found that the S N ratio was less than 1 250 if no special treatment to remove backgrounds was done. Therefore we planned to introduce some tricks to improve the S N ratio, for example, a thin solid hydrogen target for impurities reduction, an event-tagging Si+CsI telescope for removal of accidental coincidence events, and so on.

3. Results of the test experiment

The test experiment was carried out at RCN using the CH_2 target instead of the solid hydrogen target. Figure 5(a) shows the excitation energy spectrum including all the inelastic events. The solid line indicates the spectrum with the CH_2 target, and the hatched histogram presents the spectrum with the natural carbon target. The prominent peaks of the ground state, the 2_1^+ state at $E_x = 4.44$ MeV, the 0_2^+ state at $E_x = 7.65$ MeV, and the 3_1^- state at $E_x = 9.64$ MeV are clearly observed in Fig. 5(a). The excitation energy resolution about the 3_1^- state is 220 keV in the standard deviation σ, and the observed number of the inelastic events exciting the 3_1^- state is $5.90(4) \times 10^6$ within the $\pm 2\sigma$ width shown by the vertical dashed lines in Fig. 5.

Figure 5(b) shows the excitation energy spectrum of the γ-decay events. The peaks of the 2_1^+ state at $E_x = 4.44$ MeV and the 1_1^+ state at $E_x = 12.71$ MeV are clearly observed. Figure 6 is the same spectrum as Fig. 5(b) but expanded around $E_x = 9.64$ MeV. In Fig. 6, no peaks of the 3_1^- state is observed but the continuous background is observed. At present, this continuous background limits the sensitivity to the γ-decay probability Γ_γ/Γ in the test experiment. The number of the observed γ-decay events within $\pm 2\sigma$ around the 3_1^- state is 48(12), therefore the sensitivity of the γ-decay probability Γ_γ/Γ is $2.0(5) \times 10^{-5}$ in the test experiment. The sensitivity was estimated by considering the statistical error and the geometrical acceptance of the detectors. On the other hand, the γ-decay probability Γ_γ/Γ of the 1_1^+ state at $E_x = 12.71$ MeV is found to be $2.07(14) \times 10^{-2}$. This result is quite close to the known value $2.21(7) \times 10^{-2}$ [3] and it endorses that the present method using inverse kinematic reactions is useful to determine γ-decay probabilities above particle-decay thresholds.

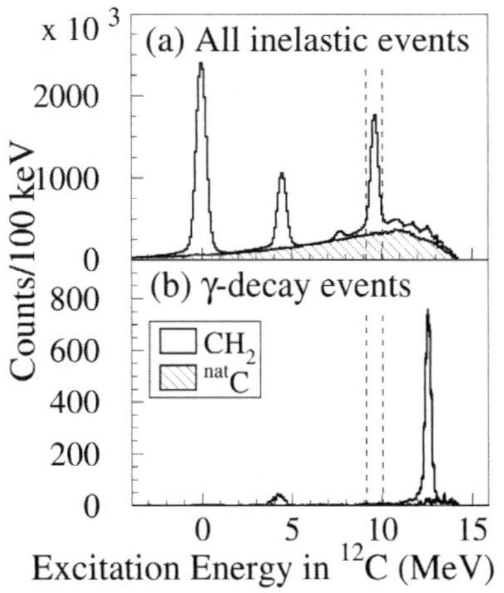

Figure 5. Excitation energy spectra including (a) all the inelastic events and (b) the γ-decay events only.

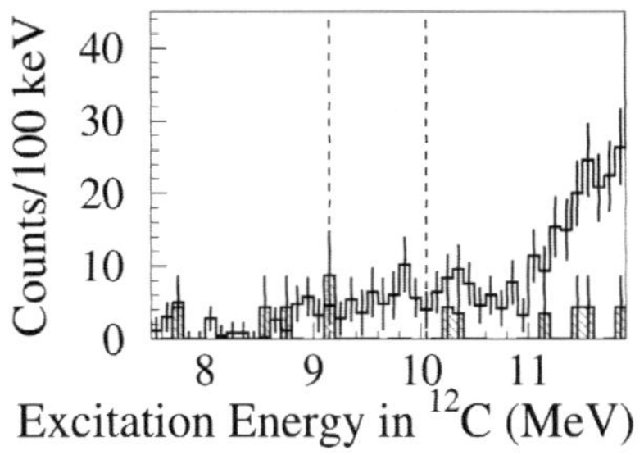

Figure 6. Excitation energy spectrum expanded around $E_x = 9.64$ MeV in Fig. 5(b).

4. Summary

We planned to determine the γ-decay probability Γ_γ/Γ of the 3_1^- state in ^{12}C using the inverse kinematic reaction ^{1}H(^{12}C, ^{12}Cp) for the precise determination of the triple α reaction rate, and performed the test experiment. As a result, the γ-decay probability Γ_γ/Γ for the 1_1^+ state in ^{12}C was successfully determined without measuring γ-rays. However, the sensitivity to the γ-decay probability Γ_γ/Γ in the present measurement is $2.0(5) \times 10^{-5}$, and it is not good enough to determine the γ-decay probability Γ_γ/Γ of the 3_1^- state. The origin of the continuous background must be examined.

References

[1] Itoh M, Akimune H, Fujiwara M, Garg U, Hashimoto N, Kawabata T, Kawase K, Kishi , Murakami T, Nakanishi K, Nakatsugawa Y, Nayak B, Okumura , akaguchi H, Takeda H, Terashima , Uchida M, Yasuda Y, Yosoi M, and Zenihiro J 2011 *Phys. Rev.* C **84** 1–6

[2] Hoyle F and Fowler W A 1960 *Astrophys. J* **132** 565

[3] Anzenberg- elove F 1990 *Nucl. Phys.* A **506** 158

[4] Chamberlin D, Bodansky D, Jacobs W W and Oberg D L 1974 *Phys. Rev.* C **10** 909–911

[5] Matsuda Y, akaguchi H, Zenihiro J, Ishimoto , uzuki , Otsu H, Ohnishi T, Takeda H, Ozeki K, Tanaka K, Terashima , Maeda Y, Kobayashi T, Koreeda A and Kamei K 2011 *Nucl. Instrum. Methods. Phys. Res.* A **643** 6–10

Journal of Physics: Conference Series **569** (2014) 012052 doi:10.1088/1742-6596/569/1/012052

The thick target inverse kinematics technique with a large acceptance silicon detector array

J Walshe[1], M Freer[1], C Wheldon[1], L N Achouri[2], N I Ashwood[1], W N Catford[3], I C Celik[3], N Curtis[1], F Delaunay[2], B Fernández-Domínguez[4], L Grassi[5], Tz Kokalova[1], M Marquês[2], N A Orr[2], L Prepolec[5], V Scuderi[6], N Soić[5] and V Tokić[5]

[1]School of Physics and Astronomy, University of Birmingham, Birmingham, B15 2TT, UK.
[2]Laboratoire de Physique Corpusculaire de Caen, 6 bd Maréchal Juin, 14050 Caen, France
[3]Department of Physics, University of Surrey, Guildford, Surrey, GU2 7XH, UK.
[4]Universidade de Santiago de Compostela, 15782 Santiago de Compostela, Spain
[5]Institut Ruđer Bošković, Bijenička cesta 54, 10000 Zagreb, Croatia
[6]INFN - Laboratori Nazionali del Sud, via S.Sofia 62, 95125 Catania, Italy

E-mail: `j.walshe@pgr.bham.ac.uk`

Abstract. An experimental technique for studying elastic scattering using a thick gas target is described, with a measurement of the $\alpha(^{24}\mathrm{Ne},\alpha)$ reaction used as an example. Advantages such as ease, detector efficiency, and the possibility of measuring the cross section at 180° in the centre-of-mass are discussed. It is shown that a resolution of tens of keV is practical at zero degrees, and that the dominant contribution to the resolution for large angles is angular straggling of the beam in the entrance window. The use of helium gas as the target allows direct measurement of α-cluster states.

1. Introduction

Traditional methods of studying elastic nuclear scattering use thin foil targets, and the elastically scattered beam-like particles are identified by energy discrimination. Counting the events in this elastic peak gives the cross section for a single centre-of-mass (c.m.) energy; in order to construct an excitation function the beam energy must be changed and the measurement repeated, often hundreds of times. While effective, this method is only practical with relatively simple accelerators – most commonly tandem Van De Graaffs – and so experiments of this type are limited by the beams available at such facilities. In particular, modern radioactive beam facilities provide access to much new physics, but as the creation and acceleration of radioactive nuclei is typically a complex process the retuning of these beams can take several hours. Studies using these beams must then use a different technique.

2. Method

The thick target inverse kinematics technique, first described in [1], allows continuous measurement of an excitation function using a single beam energy. The set-up of the target chamber is shown in figure 1. In this work the target was helium gas, which filled the chamber and is described as thick because the pressure is tuned such that the beam is stopped within the

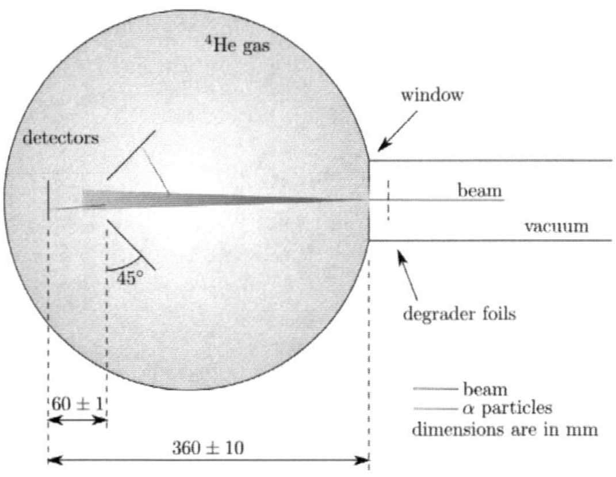

Figure 1. Schematic of the experimental chamber.

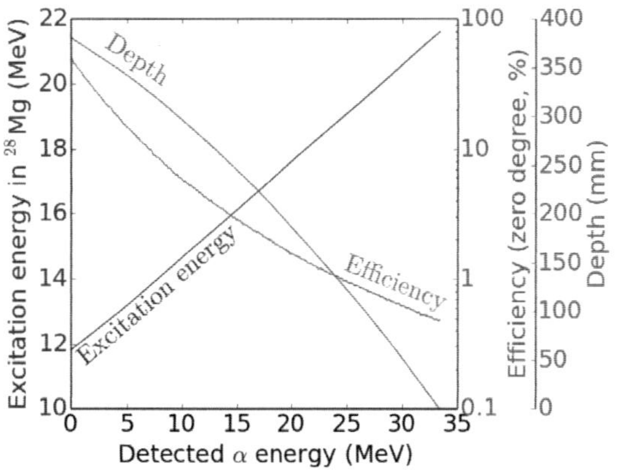

Figure 2. Calculation of the kinematics and energy loss allows reconstruction of the reaction from the detected particle energy.

gas volume. A thin window, in this case a 5 μm Havar foil, separated the gas from the beam line. A beam of ^{24}Ne with an energy of 3.8 MeV/A and intensity of 3×10^5 pps was provided by the SPIRAL facility at GANIL. As the beam is stopped in the target, nuclear reactions take place over a continuous range of energies with the maximum defined by the beam energy immediately after the window and the minimum limited only by Rutherford scattering. While reactions are produced simultaneously over this energy range, it is useful to note that each c.m. energy is localised to a point in the chamber.

In a thick target experiment, the detected particles must be lighter than the beam nuclei to escape the target gas and reach the detectors. Normally, experiments are performed in inverse kinematics where the beam nuclei are heavier than the target nuclei, and the target-like particles can be detected. An important feature of this method is that the scattered light particles are confined to forward angles in the laboratory (for elastic scattering), which means that the efficiency of appropriately positioned detectors is doubled. In addition, as the beam is absorbed, a detector can be placed at zero degrees in the laboratory, corresponding to 180° in the c.m. frame; at this point the Rutherford scattering cross-section is at a minimum and the resonance scattering cross-section is maximal.

The detector array is placed within the gas volume such that the beam does not reach the detectors but the scattered target-like particles do. In the current work the array consisted firstly of a square double-sided silicon strip detector at zero degrees with 256 pixels, each 3 mm square. In addition, a 'lampshade' array of six wedge shaped silicon detectors (Micron YY1 design[2]) captured events at large angles. These detectors were single sided, with 16 curved strips each covering an approximately constant polar angle range. The lampshade provided full azimuthal coverage, and the maximum θ measured was at least 44° in the c.m. frame, increasing to full coverage for lower energy events that take place closer to the array. Finally three additional Havar foils, each 2 μm thick, were available upstream of the window to allow the beam energy to be easily reduced in approximately 10 MeV steps.

3. Data reconstruction

In a thin target experiment the c.m. interaction energy can be straight forwardly reconstructed kinematically from the energy of the beam. For a thick target experiment, the energy of the detected particle is used but the energy loss of the beam and target through the gas must be

Journal of Physics: Conference Series **569** (2014) 012052 doi:10.1088/1742-6596/569/1/012052

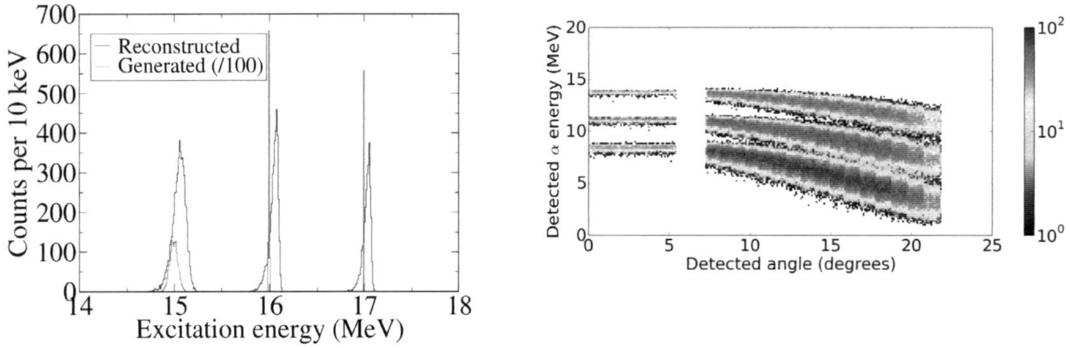

Figure 3. Simulated excitation function for ^{28}Mg using REX. (left) The zero degree detector showing (red) input data reduced by a factor of 100 and (black) reconstructed data. (right) The entire array.

taken into account. Figure 2 shows a calculation for the ^{24}Ne+α reaction, which can be used to find the properties of the interaction from only the energy of the detected α particle. However, in order to do this reconstruction, elastic scattering must be assumed – there is no way to distinguish elastic and inelastic events using only one beam energy. It is known though that the very highest energy events must be elastic; by leaving one of the reaction nuclei in an excited state a known minimum amount of energy must be removed from the available kinetic energy. In the case of ^{24}Ne+α, the first excited state of ^{24}Ne is at 1.98 MeV, so events within this energy of the maximum of the spectrum must be elastic. Then, by stepping the beam energy down (using additional Havar foils as described above) a larger 'clean' region can be produced.

4. Experimental characteristics

A new in-house Monte-Carlo code, the Resonant Excitation Function Simulation (REX), was used to simulate the experiment and identify the effects that contribute most to the resolution and efficiency of the data. This code generates events using a given excitation function and includes all of the experimental effects such as energy loss, energy and angular straggling, and the geometry of the detector array. The output is a list of detected events that can be analysed in the same way as experimental data in order to test the reconstruction method.

Figure 3 shows how well the true excitation function simulated using REX can be reconstructed using the method outlined above. The input parameters for the simulation were three states in ^{28}Mg at 15, 16 and 17 MeV, with widths of 100, 10 and 1 keV respectively; the detector set-up was identical to the experiment as described above with a ^{24}Ne beam. The reconstructed widths were 147, 70 and 68 keV. As the measured peak width does not change for the 1 and 10 keV peaks this minimum width represents the overall resolution of the method, approximately 70 keV FWHM. Importantly, this is better than the intrinsic energy resolution of the silicon detectors. The transformation of the data into the c.m. frame improves the resolution by a factor of approximately 3.5, and this effect scales with the c.m. velocity so that for heavier beams the resolution will be better.

Figure 3 also shows the data from the same simulation including the data from the lampshade array. The detected α-particle energy is plotted against the angle of the strip with respect to the entrance window (*i.e.*, these data are shown in the laboratory frame). The kinematic fall-off of the angular distributions can be seen, as well as the change in resolution with angle. Examination of the simulations reveals the reason for the latter effect. A number of factors are involved, but

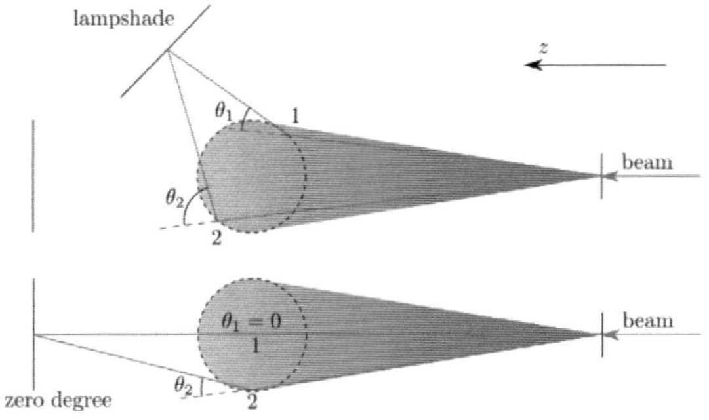

Figure 4. The effects of angular and energy straggling. See text for details.

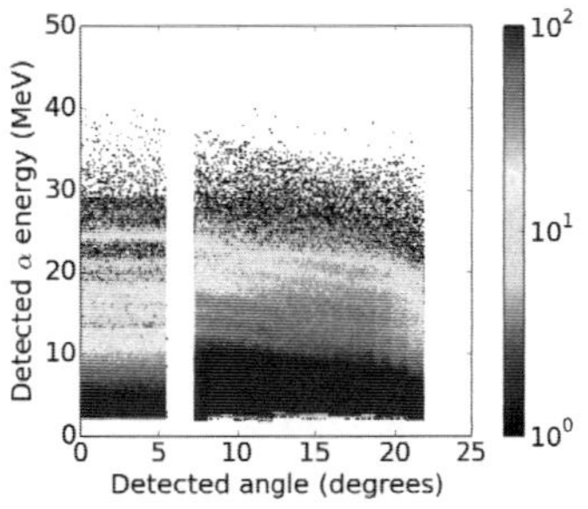

Figure 5. Experimental data for ^{28}Mg from the entire detector array.

the most important are energy and angular straggling of the beam caused by the entrance window and, to a lesser extent, the gas, as shown in figure 4. Events for a given c.m. energy, which can nominally be associated with a given depth in the gas, can in fact take place within a finite area, denoted roughly by the dashed circle. Points 1 and 2 represent two extremes. For events in the zero degree detector the scattering angle, θ, varies only a small amount, and thus the resolution is best in this detector. For the lampshade, however, the range of scattering angles available for a given c.m. energy and a given detector strip is large, and so there is significant kinematic broadening of the data.

For a ^{24}Ne beam at 90 MeV passing through 5 μm of Havar, the energy loss is approximately 20 MeV with 100 keV of straggling. This corresponds to a position straggling in the beam direction of the order of millimetres. The angular straggling after the window is approximately one degree, which produces a position straggling perpendicular to the beam direction approaching 5 mm for low energy events. Thus the angular straggling is the dominant contributor to the resolution of the array at large angles.

5. Data

Figure 5 shows experimental data taken using this technique for the $\alpha(^{24}\text{Ne},\alpha)$ reaction. As in the simulated data, states are seen as lines in the zero degree detector, with angular distributions present in the lampshade array. The resolution at larger angles means that the angular distributions of states cannot always be separated, even if they can be at zero degrees. For some isolated states, however, the angular distributions are clear.

6. Conclusions

The Thick Target Inverse Kinematics technique provides an effective method of studying elastic scattering which can be used with stable or radioactive beams. It allows the cross section to be measured at 180° in the c.m. frame with a resolution of the order of 50 keV FWHM. Away from this angle the resolution is worse, primarily due to the geometrical effect of angular straggling, most of which is caused by the entrance window. The use of helium gas as a target is a powerful method of studying α-cluster states.

References

[1] Artemov K *et al.* 1990 *Sov. J. Nucl. Phys.* **52** 408–411 ISSN 0038-5506
[2] Davinson T *et al.* 2000 *Nucl. Instrum. Meth. A* **454** 350–358

Journal of Physics: Conference Series **569** (2014) 012053 doi:10.1088/1742-6596/569/1/012053

Alpha clustering in ^{18}F

S Bailey[1], M Freer[1], Tz Kokalova[1], S Cruz[1,2], H Floyd[1] and D J Parker[1]

1 University of Birmingham, Birmingham, B15 2TT, UK

E-mail: `s.c.bailey@pgr.bham.ac.uk`

Abstract. We review some of the key experimental and theoretical studies of α-clustering in ^{18}F. Particular attention is given to the $4p$-$2h$ nature of such α-clustered states, and the interaction between the holes and clusters is examined in terms of both weak and strong-coupling regimes. The experimental work focuses on α-transfer spectroscopy and α resonant scattering as tools for investigating α-clustering.

1. Introduction

Alpha clustering is a well established concept in the structure of light α-conjugate nuclei [1]. Further to this, it has been shown that neutron-rich nuclei exhibit similar cluster structures with the additional neutrons playing the role of valence particles, contributing to the binding of the clusters in a way that resembles electrons in atomic molecules [2]. This has been investigated in detail recently both from theoretical and experimental perspectives, finding a lot of success in the $\alpha + x\mathrm{n} + \alpha$ structures in Be isotopes [3, 4] and the $\alpha + x\mathrm{n} + {}^{16}\mathrm{O}$ structures in Ne isotopes [5, 6, 7]. There has been comparatively little work done, however, to investigate the persistence of cluster structure in isotopes which deviate from α-conjugate nuclei in ways other than the addition of valence neutrons.

^{18}F is of interest from a clustering perspective because of the predicted $\alpha + {}^{14}\mathrm{N}$ structure, and how this compares with the $\alpha + {}^{16}\mathrm{O}$ structure observed in ^{20}Ne [7]. It is thought that the cluster structure is amplified in ^{20}Ne by the large shell gap above the p shell, which provides a clear separation between the ^{16}O core and the α-cluster. The $\alpha + {}^{14}\mathrm{N}$ structure is therefore expected to manifest itself in $4p$-$2h$ excitations of ^{18}F [8], displayed in Fig. 1. This produces a shell structure very similar to that of ^{20}Ne, again with a clear shell gap separating the α-cluster from the core, and deviating from ^{20}Ne only by the addition of one proton hole and one neutron hole in the p shell. It is the interplay between these proton and neutron holes and the cluster-core structure that motivates the present work on ^{18}F.

2. Hole-particle coupling

The regime which governs the coupling between the p shell holes and the s-d shell particles in $4p$-$2h$ states is extremely important for understanding the structure of ^{18}F. This coupling can either be described as weak-coupling or strong-coupling. In the weak-coupling picture, the particle wavefunctions are assumed to be similar to the eigenfunctions of ^{20}Ne, and the

[2] Present Address: Department of Physics and Astronomy, The University of British Columbia, 6224 Agricultural Road, Vancouver, BC V6T 1Z1, Canada

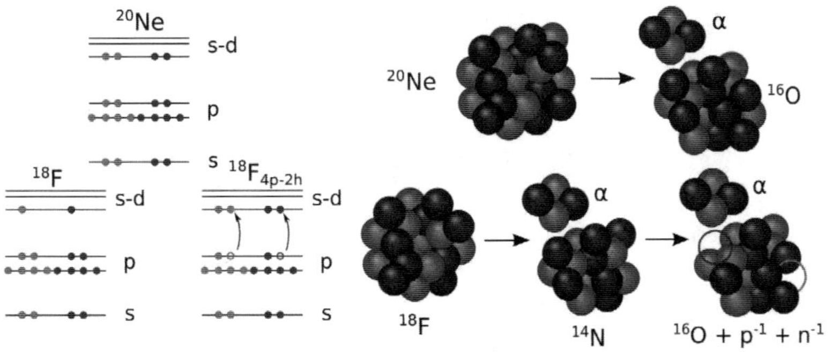

Figure 1. (Colour online) ^{18}F and ^{20}Ne from shell model (left) and cluster model (right) perspectives.

hole wavefunctions like that of the ^{14}N ground state [8, 9, 10]. This is interpreted as weak coupling since the interaction is expected to be well described as a small perturbation on these wavefunctions, leading to a structure very similar in nature to ^{20}Ne with some additional core excitations.

The strong-coupling model takes the opposite approach, asserting that the interaction between the holes and particles is too strong for the wavefunctions to be approximated in such a way. Instead more robust techniques are required, leading to more varied and exotic deformations and structures.

One of the earliest experimental investigations of hole-particle states in ^{18}F was performed in 1968 by Middleton *et al.* [8] using the ^{14}N(^{7}Li,t)^{18}F α-transfer reaction. It is expected that α-transfer reactions preferentially excite states formed by transferring all four nucleons into the same shell, leading to an amplified cross-section for $4p$-$2h$ states. In this work it was argued that this allowed the $4p$-$2h$ excitations to be identified by selecting states which were strongly populated by this reaction but weakly populated by one or two nucleon transfer reactions. However this may be an overly simplistic interpretation since it ignores the effects that the matching conditions have on the cross-section [11]. Additionally the nonzero spins of the nuclei involved in this reaction often lead to ambiguous angular distributions, making experimental J^{π} assignments very difficult [12].

Middleton *et al.* [8] reported six states to be likely $4p$-$2h$ candidates at 1.70, 2.52, 3.36, 4.23, 5.30 and 6.55 MeV using this technique. They go on to argue that these states can be well described by the weak-coupling model, coupling the ^{20}Ne ground state band to the $(1p_{1/2})_{1^+}{}^{-2}$ hole configuration. The application of this model led to the following spin-parity assignments for the observed $4p$-$2h$ states, respectively: $J^{\pi} = 1^+$, 2^+, 3^+, 1^+, 3^+ and $(4^+, 5^+)$.

This weak-coupling description of the $4p$-$2h$ states was challenged following an extensive study of the structure of ^{18}F in 1973 by Rolfs *et al.* [13]. In this work the hole-particle states were investigated in detail using the ^{14}N(α,γ)^{18}F, ^{17}O(p,γ)^{18}F and ^{16}O(^{3}He,pγ)^{18}F reactions [13, Part III]. Following these measurements 5 states were identified as being members of a K^{π} = 1^+ rotational band of predominantly $4p$-$2h$ nature, at 1.701, 2.523, 3.358, 5.298 and 6.567 MeV. Spin-parity assignments were made based on measured branching ratios and angular distributions to be respectively $J^{\pi} = 1^+$, 2^+, 3^+, 4^+ and 5^+. The reduced α-widths were extracted for the 4^+ and 5^+ members, and in both cases were found to be exceptionally large.

It is clear that these states are the same states as those measured by Middleton *et al.* [8], with the exception that the 4.23 MeV state was found by Rolfs *et al.* to be unlikely to be $4p$-$2h$ in nature, and was instead assigned to be a $J^{\pi} = 2^{(-)}$ state of 3p-1h structure [13, Part IV]. Additionally the spin-parity assignment for the 5.298 MeV state disagrees with the assignment

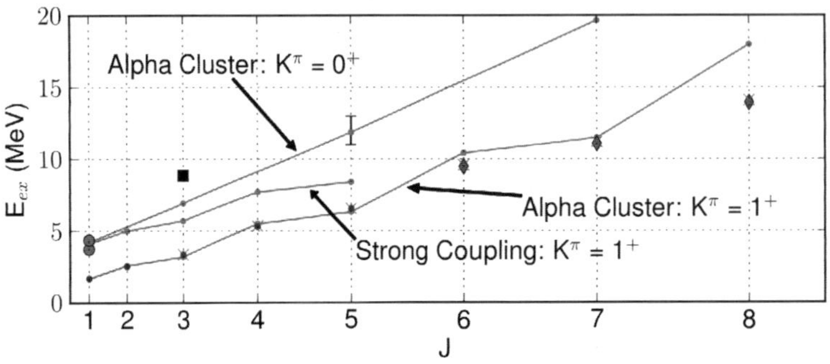

Figure 2. (Colour online) ^{18}F rotational bands, x-axis scaled as J(J + 1). Theoretical calculations displayed as lines (annotated), data displayed as points. Data from Rolfs *et al.* [13, Part III] (blue dots), Cobern *et al.* [12] (green diamonds), Etchegoyen *et al.* [15] (red crosses), Bailey *et al.* [16] (black squares/bars) and the Buck *et al.* mixed states assignment, see text [17] (red circles)

made based on the weak-coupling model by Middleton *et al.* [8]. These considerations led Rolfs *et al.* to reject the weak-coupling model in favour of the strong-coupling model.

A comparison was made between this band and a strong-coupling calculation performed in 1965 by Bassichis *et al.* [14], which predicted a $K^\pi = 1^+$, $4p$-$2h$ rotational band in ^{18}F. The gradient of the predicted band was found to be in excellent agreement with the measurements and it reproduced the observed zig-zagging in excitation energy, however the predicted band was shifted ≈ 2.5 MeV higher in energy. Rolfs *et al.* [13, Part III] speculated that based on the exceptionally large reduced α-widths observed for the 4^+ and 5^+ members, α-particle clustering may in fact play a prominent role in the structure of this band, and perhaps the explicit inclusion of α-clustering in the microscopic description of the band may reconcile this energy shift.

Based on these results it seems likely that the strong-coupling regime better describes the hole-particle coupling in ^{18}F, since the work by Rolfs *et al.* [13, Part III] is more extensive, allowing the spin-parity assignments to be made based on experimental measurements. Further to this, more α-transfer measurements have been made by Cobern *et al.* [12] and Etchegoyen *et al.* [15], extending the $K^\pi = 1^+$ rotational band observed by Rolfs *et al.* [13, Part III] from $J^\pi = 5^+$ up to 8^+. This rotational band is displayed in Fig. 2, with the band calculated using the strong-coupling model by Bassichis *et al.* [14] for comparison.

3. Alpha clustering

Based on the work by Rolfs *et al.* [13, Part III] it seems likely that there is a large α-cluster component to the structure of the $K^\pi = 1^+$ rotational band. A semi-microscopic calculation of α-cluster states in ^{18}F was performed in 1979 by Buck *et al.* [17]. In this work all four nucleons from which the α-particle is built up are placed in the *s*-*d* shell, enforcing the experimentally determined $4p$-$2h$ structure discussed previously. This leads to the ^{14}N core wavefunction being given by two p shell holes in the ^{16}O closed shell. Further details regarding this model can be found in Ref. [17].

This model produced two distinct rotational bands, one with $K^\pi = 1^+$ and one with $K^\pi = 0^+$. The states in the $K^\pi = 1^+$ band agreed exceptionally well with the experimentally determined states up to $J^\pi = 5^+$, however above this they began to diverge, with the calculated levels shifted higher in energy compared with the experimentally observed levels. This can be seen in Fig. 2. It was however explained by Buck *et al.* that this may be due to the way the nuclear potential was modelled, since similar discrepancies have arisen in other calculations involving the same potential [17].

The extremely good agreement between observed and calculated levels in the $K^\pi = 1^+$ band leads us to believe that this is indeed an α-clustered $4p$-$2h$ rotational band. However if this model is correct, it should also be possible to identify the $K^\pi = 0^+$ rotational band. Buck *et al.* [17] assigned the 1^+ state in this band to two experimentally observed 1^+ states at 3.724 and 4.361 MeV. It is argued that this is due to mixing between this state and a $2p$ shell model state predicted to exist at a similar energy, causing the $4p$-$2h$ strength to be shared between both states. This hypothesis is confirmed to an extent experimentally by the slightly reduced cross-section for these states in α-transfer measurements [8]. However Buck *et al.* were unable to confidently assign any of the other members of this band.

4. ^{14}N + α resonant reaction

More recently a study of ^{18}F was performed in 2014 by Bailey *et al.* [16] using the α + ^{14}N resonant reaction to populate α-clustered states in the compound nucleus ^{18}F. The reduced widths of these states were then extracted by performing an R-Matrix fit simultaneously of these data and data from other sources covering the d + ^{16}O and p + ^{17}O channels [18, 19]. Bailey *et al.* [16] proceeded to assign the 3^+ state in the $K^\pi = 0^+$ band to a state at E_{ex} = 8.858 MeV, and also provided tentative evidence for the existence of the broad 5^+ state in the region between 11 and 13 MeV. The 3^+ assignment was made due its unusually large α-reduced width, however the state was predicted at 6.92 MeV, almost 2 MeV lower than observed. These assignments are compared with the calculated $K^\pi = 0^+$ band in Fig. 2.

5. Conclusion

It is clear that α-clustering plays a significant role in the structure of ^{18}F, with perhaps the strongest evidence coming from the excellent agreement between the $K^\pi = 1^+$ rotational band calculated using a semi-microscopic α-cluster model [17] and experimental observations made using α-transfer reactions [13, 12, 15]. However the $K^\pi = 0^+$ rotational band is still not fully understood, and there has yet to be any work done to investigate negative-parity α-clustered rotational bands.

References

[1] Freer M and Merchant A C 1997 *J. Phys. G: Nucl. Part. Phys.* **23** 261–322
[2] von Oertzen W, Freer M and Kanada-En'yo Y 2006 *Physics Reports* **432** 43 – 113
[3] Hiura J and Tamagaki R 1972 *Progress of Theoretical Physics Supplement* **52** 25–88
[4] Itagaki N and Okabe S 2000 *Phys. Rev. C* **61**(4) 044306
[5] von Oertzen W 2001 *Eur. Phys. J. A* **11** 403–411 ISSN 1434-6001
[6] Rogachev G V, Goldberg V Z, Lönnroth T, Trzaska W H, Fayans S A, Källman K M, Kolata J J, Mutterer M, Rozhkov M V and Skorodumov B B 2001 *Phys. Rev. C* **64**(5) 051302
[7] Richards H T 1984 *Phys. Rev. C* **29**(1) 276–283
[8] Middleton R, Polsky L M, Holbrow C H and Bethge K 1968 *Phys. Rev. Lett.* **21**(19) 1398–1401
[9] Arima A, Horiuchi H and Sebe T 1967 *Physics Letters B* **24** 129 – 131 ISSN 0370-2693
[10] Benson H and Flowers B 1969 *Nuclear Physics A* **126** 332 – 354 ISSN 0375-9474
[11] Brink D 1972 *Physics Letters B* **40** 37 – 40 ISSN 0370-2693
[12] Cobern M E and Parker P D 1977 *Phys. Rev. C* **15**(5) 1929–1932
[13] Rolfs C, Berka I, Trautvetter H, Azuma R, Kieser W, Litherland A and Charlesworth A 1973 *Nuclear Physics A* **199** 257 – 350 ISSN 0375-9474
[14] Bassichis W H, Giraud B and Ripka G 1965 *Phys. Rev. Lett.* **15**(25) 980–982
[15] Etchegoyen M, Sinclair D, Etchegoyen A and Moreno E 1983 *Nuclear Physics A* **402** 87 – 113 ISSN 0375-9474
[16] Bailey S, Freer M, Kokalova T, Cruz S, Floyd H and Parker D J 2014 *Phys. Rev. C* **90**(2) 024302
[17] Buck B, Merchant A and Rowley N 1979 *Nuclear Physics A* **327** 29 – 52 ISSN 0375-9474
[18] Terwagne G, Genard G, Yedji M and Ross G G 2008 *J Appl. Phys.* **104** 084909
[19] Seiler R F, Jones C H, Anzick W J, Herring D F and Jones K W 1963 *Nucl. Phys.* **45** 647–656

Journal of Physics: Conference Series **569** (2014) 012054 doi:10.1088/1742-6596/569/1/012054

Cluster properties of nuclear states in the modern shell model approach

Alexander Volya[1] **and Yury M. Tchuvil'sky**[2]

[1] Department of Physics, Florida State University, Tallahassee, FL 32306, USA
[2] Skobeltsyn Institute of Nuclear Physics, Lomonosov Moscow State University, 119991 Moscow, Russia

Abstract. In this presentation we summarize our progress in the study of α-clustering in the shell model configuration interaction approach. We put forward Cluster-Nucleon Configuration Interaction Model where the study of clustering is facilitated by the SU(3) symmetry of the cluster channels and by Orthogonality Condition Model. Pioneering methods and results concerning α spectroscopic factors in *sd*-shell nuclei and in ^{16}O treated in *p-sd* shell are presented. Comparison with experimental data is in favor of the approach.

1. Introduction

Numerous theoretical techniques have been developed for addressing clustering phenomena in nuclei, and the present proceedings volume offers an exhaustive overview. However, many of these techniques focus on the structure of highly clustered nuclear states, often connecting them to experimental results using observables that are not directly related to clustering such as nuclear moments of inertia, quadrupole moments, gamma-transitions, etc. At the same time, a large body of experimental information concerning clustering, primarily α, is accumulated from measurements of spontaneous radioactivity, elastic scattering of composite particles, transfer and knock-out nuclear reactions. Recent results and rich prospects offered by the new thick ^{4}He target inverse kinematics elastic scattering techniques provide comprehensive experimental data on distribution of clustering strength in light nuclei; see Refs. [1–5] for example. This data shows both strongly and weakly clustered states, therefore a quantitative theoretical description of physics involving both nucleonic and cluster degrees of freedom is required.

Magic- and light-cluster channels are distinguished by a specific – "elongated" – Elliott's SU(3) irreducible representations of the type $(n, 0)$. A symmetry-based approach to clustering is discussed in Ref. [6]. Weakly clustered states and single-nucleon dynamics, however, are better accessible in the microscopic shell model approach, where clustering questions have been pursued, Ref. [7–14]. In our contribution we put forward a technique that combines approaches in both directions: it is based on the shell model technique, and at the same time, takes into account the symmetry properties of cluster configurations. We implement large scale modern configuration interaction shell model with realistic effective Hamiltonians. It is a well established microscopic approach, where within the same formalism, a high quality description and good predictive power is achieved for numerous single-particle as well as collective nuclear properties [15-18]. In the nuclear shell model cluster degrees of freedom are not introduced explicitly. Thus, the emergence of clustering and interplay of cluster and single nucleon degrees of freedom can be studied. Recent advances in computational techniques and exponential growth of computational power [15, 19, 20] facilitate work in this direction.

2. Formalism

Shell model configuration interaction approach and SU(3)-symmetric structures

In the shell model approach each many-nucleon state is built from the single-particle basis states as a linear combination of Slater determinants

$$|\Psi\rangle \equiv \Psi^\dagger|0\rangle = \sum_{\{1,2,3,\dots A\}} \langle 1,2\dots A|\Psi\rangle\, a_1^\dagger a_2^\dagger \dots a_A^\dagger|0\rangle. \tag{1}$$

Here index $1 \equiv \{n,l,j,m\}$ denotes the single-particle j-scheme basis state built from the radial harmonic oscillator wave function (WF) $\varphi_{n_1,l_1}(r) \equiv \langle r|1\rangle \equiv \langle r|a_1^\dagger|0\rangle$; the operator $a_1^\dagger$ is the corresponding fermion creation operator in the second quantization; $\langle 1,2\dots A|\Psi\rangle$ is the numeric coefficient determining the weight of each Slater determinant in the linear superposition (1).

In order to address clustering, multi-nucleon substructures related to a certain irreducible representation of the SU(3) group are constructed. In the present paper we discuss α-clustering. In the following, $\{\mathfrak{n}_i^{\alpha_i}\}$ denotes configuration where α_i is the number of particles in the major oscillator shell $\mathfrak{n}_i$; L, S, and T are orbital, spin, and isospin quantum numbers; (λ,μ) is the SU(3) Elliott's symbol; and the Young frame $[f]$ classifies the permutation symmetry. The four-nucleon states

$$|\Psi_{(n,0):L}\rangle \equiv \Psi^\dagger_{(n,0):L}|0\rangle \equiv |\{\mathfrak{n}_i^{\alpha_i}\}[f]=[4](n,0):L, S=0, T=0\rangle \tag{2}$$

are constructed by diagonalization of some linear combination of the SU(3) Casimir operator of the second rank, L^2, T^2, S^2, and other operators as needed in the basis of four-nucleon shell-model states. The states (2) are normalized as $\langle\Psi_{(n,0):L}|\Psi_{(n,0):L}\rangle = 1$.

The advantage of the second quantization formalism implemented in the modern polymorphic configuration interaction techniques is the full correspondence between states $|\Psi\rangle$ and the creation and annihilation operators $\Psi^\dagger$ and Ψ. Thus, many-body overlap integrals needed to evaluate fractional parentage coefficients (FPCs)

$$\mathcal{F}_{nl} \equiv \langle\Psi_P|\hat{\mathcal{A}}\left\{\Psi_{(n,0):l}\,\Psi_D\right\}\rangle \equiv \langle 0|\Psi_P\{\Psi_{(n,0):l}\Psi_D\}^\dagger|0\rangle \tag{3}$$

can be readily computed between arbitrary states $|\Psi_P\rangle$ and $|\Psi_D\rangle$ of the type described in Eq. (1). Here $\hat{\mathcal{A}}$ is the antisymmetrization operator. Some examples of FPCs and channel norms for selected states with SU(3) symmetry are shown in Tab. 1. In the last column the channel norms are shown, which can be seen as a measure of bosonic enhancement. Indeed, if four-nucleon $L=0$ operators $\Psi^\dagger$ and Ψ are thought of as boson creation and annihilation operators, then $\Psi\Psi^\dagger = 1 + N_b$, where N_b is the boson number operator. The numbers in the last column in Tab. 1 are less than 2, showing that four-nucleon configurations (2) are not true bosons. However, since larger shells can accommodate more "bosons" the non-bosonic effects are reduced and the norm gets closer to 2.

Cluster form factors and spectroscopic factors

The basic measure of clustering – cluster form factor (CFF), also commonly referred to as the spectroscopic amplitude – is defined as:

$$\phi_l(\rho') = \langle\Psi'_P|\hat{\mathcal{A}}\{\Psi'_D\,\frac{\delta(\rho-\rho')}{\rho^2}\,Y_{lm}(\Omega_\rho)\Psi'_\alpha\}\rangle, \tag{4}$$

where Ψ'_P, Ψ'_D, and Ψ'_α are internal, translationally invariant, free of the center of mass (c.m.) coordinate WFs of the parent (P) nucleus, the daughter (D) nucleus, and the α-cluster,

n	Ψ_P	Ψ	$\left\vert\langle\Psi_P\vert\Psi^\dagger\vert\Psi\rangle\right\vert^2$	$\langle\Psi\vert\Psi\Psi^\dagger\vert\Psi\rangle$
1	$(p)^8\,(0,4)$	$(p)^4\,(4,0)$	1.42222*	1.42222
2	$(sd)^8\,(8,4)$	$(sd)^4\,(8,0)$	0.487903	1.20213
3	$(fp)^8\,(16,4)$	$(fp)^4\,(12,0)$	0.292411	1.41503
4	$(sdg)^8\,(24,4)$	$(sdg)^4\,(16,0)$	0.209525	1.5278

Table 1. Fractional parentage coefficients and channel norms for selected SU(3) states. All WF and operators are $L = 0$. *For p-shell this result agrees with the value of $64/45 = 1.42222$ found in the literature [21].

respectively. Here and below we use primed notation to distinguish these WFs from those of the shell model type (1) that implicitly depend on the c.m. motion. The coordinate ρ is the Jacobi radial coordinate of the relative cluster-daughter motion; a proper coupling to a relative angular momentum l is established.

The CFFs are computed using the shell model approach. In our calculations the parent and daughter states are computed implementing Glockner-Lawson procedure [22] leading the c.m. motion being in the lowest oscillator state $\varphi_{00}(R)$. The oscillator frequency parameter in the WF depends, in the usual way, on the mass number. Of the WF $|\Psi_{(n,0):l}\rangle$, we are interested in the component where the c.m. of the α-particle is in the oscillator state $\varphi_{nl}(R_\alpha)$. We assume that the α-particle's WF is represented by the lowest four-nucleon oscillator function written through the Jacobi coordinates

$$|\Psi'_\alpha\rangle \equiv |(0s)^4[f] = [4](\lambda,\mu) = (0,0) : L = 0, S = 0, T = 0\rangle. \tag{5}$$

Then the component of interest, referred to as the cluster coefficient, is known analytically [8, 9, 23],

$$X_{nl}^{(\lambda 0)} \equiv \langle\Psi_{(\lambda,0):l}|\varphi_{nl}(R_\alpha)\,\Psi'_\alpha\rangle = \sqrt{\frac{1}{4^n}\frac{n!}{\prod_i(\mathfrak{n}_i!)^{\alpha_i}}\frac{4!}{\prod_i \alpha_i!}}. \tag{6}$$

By expanding the parent state using FPC (3) and by decoupling intrinsic and c.m. variables, the CFF in Eq. (4) is expanded in oscillator states as [7, 8, 9, 11]

$$\phi_l(\rho) = \sum_n \mathcal{C}_{nl}\,\varphi_{nl}(\rho), \quad \mathcal{C}_{nl} = X_{nl}\,\mathcal{F}_{nl}\,\mathcal{R}_n \quad \text{where} \quad \mathcal{R}_n = (-1)^n[(m_d+m_\alpha)/m_d]^{n/2}, \tag{7}$$

is a recoil factor that comes from recoupling the c.m. variables R_α and R_D into their relative coordinate ρ and the parent c.m. coordinate R_P.

Identifying the CFF in Eq. (7) with the observable spectroscopic factors (SFs) $\mathcal{S}_l = \sum_n |\mathcal{C}_{nl}|^2$ has been common in the literature. However, it was argued in Refs. [24, 25] that the matching of $\phi_l(\rho)$ with the two-body cluster-nucleus solution is not appropriate. Instead, one should use the channel WF in the form of Resonating Group Model or, for easier reduction to a two-body problem, in the form of Orthogonality Condition Model (OCM) [26]. Therefore, the CFF should be redefined as

$$f_l(\rho) \equiv \hat{\mathcal{N}}_l^{-1/2}\phi_l(\rho), \quad \text{where the norm operator} \quad \hat{\mathcal{N}}_l\phi_l(\rho) \equiv \int \mathcal{N}_l(\rho',\rho)\phi_l(\rho)\rho^2 d\rho \tag{8}$$

contains an overlap norm kernel

$$\mathcal{N}_l(\rho',\rho'') = \langle\hat{\mathcal{A}}\{\Psi'_D\,\frac{\delta(\rho-\rho')}{\rho^2}Y_{lm}(\Omega_\rho)\Psi'_\alpha\}|\hat{\mathcal{A}}\{\Psi'_D\,\frac{\delta(\rho-\rho'')}{\rho^2}Y_{lm}(\Omega_\rho)\Psi'_\alpha\}\rangle \tag{9}$$

The validity and importance of this new definition are discussed in details in Refs. [27, 28]. We construct and diagonalize the norm kernel operator as a matrix in the oscillator basis,

$$\langle \varphi_{n'l}|\hat{\mathcal{N}}_l|\varphi_{nl}\rangle = \mathcal{R}_{n'}\mathcal{R}_n X_{n'l} X_{nl} \,\langle \hat{\mathcal{A}}\left\{\Psi_{(n',0):l}\,\Psi_D\right\}|\hat{\mathcal{A}}\left\{\Psi_{(n,0):l}\,\Psi_D\right\}\rangle. \tag{10}$$

This leads to a new definition of the SF:

$$S_l \equiv \int \rho^2 d\rho\, |f_l(\rho)|^2 = \sum_k \frac{1}{N_{kl}} \left|\sum_n \langle kl|\varphi_{nl}\rangle\, \mathcal{C}_{nl}\right|^2, \tag{11}$$

where $|kl\rangle$ is an eigenvector and N_{kl} is the associated eigenvalue of the norm kernel $\hat{\mathcal{N}}_l|kl\rangle = N_{kl}|kl\rangle$, both corresponding to angular momentum l. In this form the SFs are normalized; for any given parent nucleus the sum of all SF for a given partial wave l and to a particular daughter state equals to the number of channels (characterized by different values of n in four-nucleon functions $\Psi_{(n,0):l}$) involved. In the one-channel case (such examples are considered in the next section) using the completeness of the parent states, $\sum_i |\Psi_{P_i}\rangle\langle\Psi_{P_i}| \equiv 1$, the single diagonal matrix element for the norm (10) can be expressed as

$$N_{nl} = \mathcal{R}_n^2 X_{nl}^2 \sum_i (\mathcal{F}_{nl}^i)^2 = \sum_i \mathcal{S}_l^i, \quad \text{thus} \quad S_l^i = \mathcal{S}_l^i / \sum_{i'} \mathcal{S}_l^{i'} = (\mathcal{F}_{nl}^i)^2 / \sum_{i'} (\mathcal{F}_{nl}^{i'})^2. \tag{12}$$

In what follows we demonstrate this approach using some realistic examples. We refer to our approach as Cluster-Nucleon Configuration Interaction Model (CNCIM). Additional details and applications can be found in Refs. [5, 29].

3. Applications

Study of the ground state α-clustering in sd-shell nuclei

There is a large body of both theoretical [10, 12] and experimental [30–33] work exploring the α strength in low-lying states of *sd*-shell nuclei. We performed calculations of α-particle SFs for ground state to ground state transitions of even-even *sd*-shell nuclei. The USDB [34] Hamiltonian was used. Within this model only one four-nucleon operator with SU(3) quantum numbers (8,0) contributes. Thus, the relationship (12) holds.

Our results, when compared to the experimental data in Tab. 2, highlight the merits of the method. The long-standing problem that the values of the old SFs (third column in Tab. 2) are not only several times smaller than the ones extracted from various experiments (columns 4–6 in Tab. 2) but also rapidly decrease with the increasing nuclear mass appears to be resolved in our approach, as indicated by the new SFs in the second column. Furthermore, the measured SFs and their tendencies to drop down towards the middle of the *sd*-shell and to increase at the edges are well-reproduced. It is noteworthy that the values of the "old" spectroscopic factors obtained by us and the ones presented in Ref. [10] are close – the difference is not larger than 12 percent.

In Fig. 1 we explore the dependence of the new SFs on the pairing interaction. Here, all the isovector pairing ($T = 1$, $J = 0$) matrix elements are scaled, and the SFs are shown as a function of this scaling. The role of pairing and its relation to α-clustering is not trivial in many-body systems. However, in this example, it appears that increased pairing also favors clustering. Moreover, for strong pairing almost all the α-clustering strength is concentrated in the ground state.

Study of α-clustering in ^{16}O

A more complicated investigation in the framework of the CNCIM is summarized in Tab. 3. Here we examine α-clustering of the ground and multiple excited states in ^{16}O relative to channels

$A_P - A_D$	S_0	$\mathcal{S}_0$	S_0 [35]	S_0 [36]	S_0 [37][a]
^{20}Ne-^{16}O	0.755	0.173	1.0	0.54	1
^{22}Ne-^{18}O	0.481	0.085			0.37
^{24}Mg-^{20}Ne	0.411	0.091	0.76	0.42	0.66
^{26}Mg-^{22}Ne	0.439	0.068			0.20
^{28}Si-^{24}Mg	0.526	0.080	0.37	0.20	0.33
^{30}Si-^{26}Mg	0.555	0.061			0.55
^{32}S-^{28}Si	0.911	0.082	1.05	0.55	0.45
^{34}S-^{30}Si	0.974	0.062			
^{36}Ar-^{32}S	0.986	0.061			
^{38}Ar-^{34}S	0.997	0.030			1.30
^{40}Ca-^{36}Ar	1	0.037	1.56	0.86	1.18

Table 2. Ground state to ground state α-particle SFs, "new" S_0 and "old" $\mathcal{S}_0$ and the experimental SFs extracted from the cross sections of (p,pα) [35, 36] and (^{6}Li,d) [37] reactions. [a] The values of the SFs are normalized by the SF of ^{20}Ne. Taking into account that the experimental absolute value of SF in ^{20}Ne according to [35] is very close to 1.0 (see column 4), they may be considered as absolute ones.

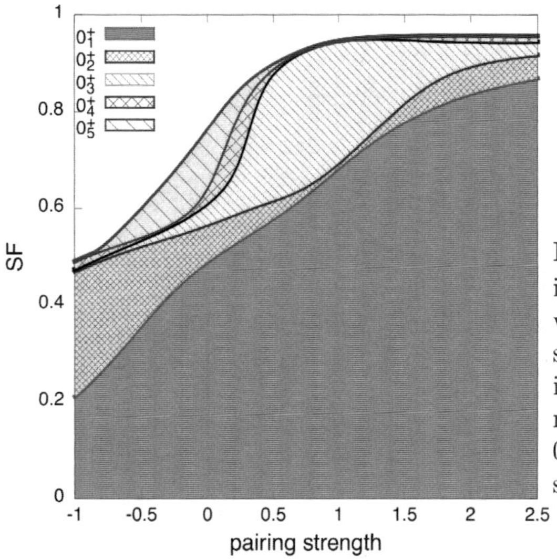

Figure 1. SF (new) for the lowest 0^+ states in ^{28}Si for the $l = 0$ channel of the α-particle with ^{24}Mg in the ground state. The SFs are stacked because their total sum equals to unity in this model space. The USDB [34] shell model Hamiltonian is used with all $T = 1$; $J = 0$ pairing matrix elements being scaled. The scaling parameter is shown on the x-axis.

involving ^{12}C nucleus in the ground state. Both parent and daughter systems are treated in the unrestricted *p-sd* configuration space with the effective interaction Hamiltonian from [19]. The study in Ref. [19] suggests that this effective Hamiltonian describes well the multi-particle correlations in ^{16}O which makes it a good choice for exploring clustering. The *p-sd* valence space permits the following SU(3)-classified four-nucleon configurations:

$$\Phi_{(n,0):l}(R) = |R; (0p)^q (2s-1d)^{4-q}\,[4](n,0) : l, S=0, T=0\rangle, \quad (13)$$

where $q = 0, 1...4$; $n = 8 - q$; $l = n, n-2, ..., 1$ or 0; and $\pi = (-1)^l$.

In our study the spectrum of ^{16}O has been calculated in a broad energy region. Experimentally known characteristics – spin, parity and α decay reduced widths θ^2_α – of more than sixty states are reasonably described. For the sake of brevity in Tab. 3 we exemplify our results only for the states with SFs $S_\alpha > 0.1$.

J_i^π	E(sm)	S_l	E(exp)	θ_α^2
0_1^+	0.000	0.794	0.000	0.86^a
3_1^-	5.912	0.663	6.13	0.41^a
0_2^+	6.916	0.535	6.049	0.40^a
1_1^-	7.632	0.150	7.117	0.14
2_1^+	8.194	0.500	6.917	0.47^a
2_2^+	9.988	0.349	9.844^b	0.0015
4_1^+	10.320	0.313	10.356	0.44
0_3^+	10.657	0.216	11.26	0.77
2_3^+	11.307	0.158	11.52^b	0.033
4_2^+	11.334	0.203	11.097^b	0.0014

continued				
J_i^π	E (sm)	S_l	E(exp)	θ_α^2
2_4^+	12.530	0.123	$—^c$	
6_1^+	13.286	0.465	14.815	0.17
4_3^+	13.308	0.160	14.62	0.19
3_3^-	13.733	0.144	14.1	0.21
2_6^+	14.646	0.102	14.926^b	0.0098
1_4^-	15.298	0.174	9.585	0.67
4_5^+	15.474	0.152	16.844	0.13
5_1^-	15.945	0.289	14.66	0.55
6_2^+	16.304	0.415	16.275	0.43

Table 3. The α-particle SF for states in ^{16}O. See text for additional details.
[a] Recalculated value of the SF from [38] (see the text).
[b] Identified states are, probably, of different nature.
[c] No experimental analog has been found.

Tab. 3 is based on the theoretically calculated spectrum of ^{16}O. We attempted to identify each theoretically predicted state with an experimentally known state. Experimentally observed excitation energy E(exp) and α spectroscopic strength θ_α^2 are listed in the last two columns. The most part of the experimental information is taken from the spectroscopic tables [39, 40]; α decay reduced widths were calculated using ordinary formulas typical for resonance reaction theory. For evaluation of the SFs of sub-threshold states the experimental data from (^{6}Li,d) reaction [38] are used, where SFs relative to 4_1^+ 10.356 MeV are presented. Taking into account some inconsistencies in determination of absolute values of the sub-threshold SFs, we rescale this data using an over-threshold reference state with a known α decay width. It should be noted that in this compilation of data the accuracy of the θ_α^2 values is very limited. Thus, in establishing theory-experiment correspondence in Tab. 3 an agreement within a factor of 3 to 4 in SF is the primary criterion, a theory-experiment agreement in excitation energy within about 1 MeV is considered secondary. Overall, we find the accord between theory and experiment displayed in Tab. 3 encouraging. Indeed, about $^2/_3$ of theoretical results obtained without introducing any parameters or fitting procedures turn out to be supported by the experimental data. Many states with lower α SFs (not listed in Tab. 3) are also reproduced by this theory. For most levels observed in experiments theoretical partners may be found. Other properties of the ^{16}O states that include electric quadrupole transitions and possible identification of rotational bands are also well-described. The details can be found in Ref. [29]. Discrepancies between theory and observations in Tab. 3 shed new light on the nature of states and provide guidance for further development of microscopic approaches to nuclear clustering. The expansion of the basis is the most obvious direction towards an improvement of the method. The lack of configurations from the pf-shell appears to be an origin of the serious discrepancy in the energy and the width of the 1_4^- E(exp)=9.585 MeV.

4. Summary

In this contribution we summarize our recent progress with the Cluster-Nucleon Configuration Interaction Model. The approach expands the modern configuration interaction versions of the nuclear shell model towards clustering. Large configuration-space studies, interplay between cluster and nucleon degrees of freedom, cluster strength distribution at high excitation energies, and numerous opportunities for extension and development are the particularly appealing sides of this technique. Here we provide a brief description of CNCIM and discuss the SU(3) properties of α-cluster channels. We demonstrate the effects that follow from the Orthogonality Condition Model which appear to resolve a long-standing problem of underestimation of α spectroscopic factors in *sd*-shell nuclei. We briefly touch on an issue of interplay between nuclear pairing and clustering. The ^{16}O nucleus is selected as an example of a large scale study. Comparison of theoretical results and experimental data represents an affirmative assessment to the approach.

Acknowledgments
We thank T. Dytrych and G. Rogachev for motivating discussions. This material is based upon work supported by the U.S. Department of Energy Office of Science, Office of Nuclear Physics under Award Number DE-SC-0009883.

[1] Lönnroth T *et al.* 2010 *Eur. Phys. J. A* **46** 5
[2] Norrby M *et al.* 2011 *Eur. Phys. J. A* **47** 73; 96
[3] Artemov K *et al.* 1990 *Sov. J. Nucl. Phys.* **52** 406
[4] Goldberg V and Pakhomov A 1993 *Phys. At. Nucl.* **56** 1167
[5] Avila M, Rogachev G, Goldberg V, Johnson E, Kemper K *et al.* 2014 (*Preprint* 1406.6734)
[6] Gnilozub I A, Kurgalin S D and Tchuvl'sky Yu M 2013 *J. Phys.: Conf. Ser.* **436** 012034
[7] Smirnov Y F and Chlebowska D 1961 *Nucl. Phys.* **26** 306
[8] Ichimura M, Arima A, Halbert E C and Terasawa T 1973 *Nucl. Phys.* **A204** 225
[9] Smirnov Y F and Tchuvil'sky Yu M 1977 *Phys. Rev. C* **15** 84
[10] Chung W, van Hienen J, Wildenthal B H and Bennett C L 1978 *Phys. Lett. B* **79** 381
[11] Nemetz O F, Neudatchin V G, Rudchik A T, Smirnov Y F and Tchuvil'sky Yu M 1988 *Nucleon Clusters in Atomic Nuclei and Multi-Nucleon Transfer Reactions* (Kiev: Naukova Dumka) p 295
[12] Grigorescu M, Brown B A and Dumitrescu O 1993 *Phys. Rev. C* **47** 2666
[13] Tchuvil'sky Yu M, Kurowsky W W, Sakharuk A A and Neudatchin V G 1995 *Phys. Rev. C* **51** 784
[14] Navratil P 2004 *Phys. Rev. C* **70** 054324
[15] Caurier E, Martinez-Pinedo G, Nowacki F, Poves A and Zuker A P 2005 *Rev. Mod. Phys.* **77** 427
[16] Talmi I 2003 *Fifty years of the nuclear shell model-the quest for the effective interaction* (*Advances in Nuclear Physics* vol 27) (Kluwer Academic Publishers)
[17] Brown B A 2001 *Prog. Part. Nucl. Phys.* **47** 517
[18] Volya A 2009 *Phys. Rev. C* **79** 044308
[19] Utsuno Y and Chiba S 2011 *Phys. Rev. C* **83** 021301
[20] Shirokov A M and Mazur A I 2014 *Nuclear Theory in the Supercomputing Era T 2013(NTSE-2013)* URL http://ntse-2013.khb.ru/proceedings/
[21] Rotter I 1965 *Ann. der Phys.* **16** 242
[22] Lawson R D 1980 *Theory of the shell model* (Oxford: Clarendon Press) ISBN 0198515162
[23] Smirnov Y F and Tchuvil'sky Yu M 1983 *Czech. J. Phys.* **33** 1215
[24] Fliessbach T and Mang H 1976 *Nucl. Phys. A* **263** 75
[25] Fliessbach T and Manakos P 1977 *J. Phys. G* **3** 643
[26] Saito S 1969 *Prog. Theor. Phys.* **41** 705
[27] Lovas R G, Liotta R J, Insolia A, Varga K and Delion D S 1998 *Phys. Rep.* **294**
[28] Kadmensky S G, Kurgalin S D and Tchuvil-sky Yu M 2007 *Phys. Part. Nucl.* **38** 699
[29] Volya A and Tchuvil'sky Yu M 2014 *World Scientific* IASEN2013 Conference Proceedings
[30] Oelert W, Chung W, Djaloeis A, Mayer-Böricke C and Turek P 1980 *Phys. Rev. C* **22** 408
[31] Oelert W, Chung W, Betigeri M, Djaloeis A, Mayer-Böricke C and Turek P 1979 *Phys. Rev. C* **20** 459
[32] Fulbright H W 1979 *Ann. Rev. Nucl. Part. Sci.* **29** 161
[33] Anantaraman N, Fulbright H W and Stwertka P M 1980 *Phys. Rev. C* **22** 501
[34] Brown B A and Richter W A 2006 *Phys. Rev. C* **74** 034315
[35] Carey T, Roos P, Chant N, Nadasen A and Chen H 1981 *Phys. Rev. C* **23** 576(R)
[36] Carey T, Roos P, Chant N, Nadasen A and Chen H 1981 *Phys. Rev. C* **29** 1273
[37] Anantaraman N *et al.* 1975 *Phys. Rev. Lett.* **35** 1131
[38] Becchetti F D, Overway D, Jänecke J and Jacobs W W 1980 *Nucl. Phys. A* **344** 336
[39] Tilley D R, Weller H R and Cheves C M 1993 *Nucl. Phys. A* **564** 1
[40] Evaluated nuclear structure data file URL http://www.nndc.bnl.gov/ensdf/

Isovector pairing in self-conjugate nuclei in a formalism of quartets

M Sambataro[1] **and N Sandulescu**[2]

[1] Istituto Nazionale di Fisica Nucleare - Sezione di Catania, Via S. Sofia 64, I-95123 Catania, Italy

[2] National Institute of Physics and Nuclear Engineering, P.O. Box MG-6, Magurele, Bucharest, Romania

E-mail: michelangelo.sambataro@ct.infn.it, sandulescu@theory.nipne.ro

Abstract. The isovector proton-neutron pairing in self-conjugate nuclei is treated in a formalism of quartets. Quartets are four-body correlated structures built from two neutrons and two protons coupled to total isospin $T = 0$. The ground state of the isovector pairing Hamiltonian is described as a product of quartets. We review both the case in which the quartets are constrained to be all identical and the case in which they are allowed to be distinct from one another. The quality of the two approaches is tested by making comparisons with exact shell model calculations for $N = Z$ nuclei with valence nucleons outside the ^{16}O, ^{40}Ca, and ^{100}Sn cores. We consider both spherical and axially deformed mean fields. Both approaches are found to be very accurate. In the applications to a deformed mean field, in particular, the formalism with distinct quartets gives rise to results which are basically exact.

1. Introduction

In self-conjugate nuclei protons (p) and neutrons (n) occupy the same orbitals. Owing to this fact and to the charge independence of the nuclear force, one expects that in these nuclei the isovector pn pairing should exist on an equal footing with the isovector pp and nn pairing. Extending traditional approaches to pairing like BCS or projected-BCS to pn systems has not led to a satisfactory description of the isovector pairing [1, 2, 3, 4]. An important limitation of these approaches is that they violate isospin symmetry. However, restoring this symmetry has not been sufficient to obtain a proper description of the isovector pairing correlations even in the presence of particle number conservation [5]. This fact indicates the need for a more general isospin conserving formalism which goes beyond the BCS-based approximations.

The isovector pairing ground state in BCS-like approaches is decribed in terms of collective pp, nn and pn pairs. Recently, an alternative approach has been proposed [6] which is instead based on four-body correlated structures formed by two protons and two neutrons coupled to a total isospin $T = 0$ (quartets). In this approach, named quartet condensation model (QCM), the ground state of a self-conjugate nucleus is simply described as a product of identical quartets. The structure of these quartets is determined by minimizing the ground state energy. This approach has been afterwards generalized by allowing the quartets to be all distinct from one another [7]. In this case, in order to construct the quartets, use is done of an iterative procedure based on the diagonalization of the Hamiltonian in restricted spaces. In the following, this approach will be referred to as quartet model (QM). Both QCM and QM have been tested by

Journal of Physics: Conference Series **569** (2014) 012055 doi:10.1088/1742-6596/569/1/012055

making comparisons with exact shell model calculations for $N = Z$ nuclei with valence nucleons outside the ^{16}O, ^{40}Ca, and ^{100}Sn cores for both spherical and axially deformed mean fields. These comparisons will be illustrated in section 3. Before that, in section2, we will shortly present the quartet formalism. Finally, in section 4, we will give some conclusions.

2. The quartet formalism

The isovector pairing Hamiltonian has the form

$$H = \sum_i \epsilon_i (N_i^\nu + N_i^\pi) + \sum_{i,j} V_{ij} \sum_\tau P_{i,\tau}^+ P_{j,\tau} \tag{1}$$

with the operators N_i^ν and N_i^π being, respectively, the neutron and proton number operators, ϵ_i the single-particle energies and

$$P_{i,\tau}^+ = [a_i^+ a_{\bar{i}}^+]_\tau^{T=1}. \tag{2}$$

In the pair creation operator (2), $\bar{i}$ stands for the time conjugate of the state i and τ denotes the three projections of the isospin $T = 1$ corresponding to nn ($\tau = 1$), pp ($\tau = -1$) and pn ($\tau = 0$) pairs.

To describe the ground state of the Hamiltonian (1) for systems with $N = Z$ we shall use as building blocks not collective pairs, as done in BCS-type models, but collective quartets formed by two neutrons and two protons. To such a purpose, let's first introduce a set of non-collective quartets composed by two isovector pairs coupled to $T = 0$

$$A_{ij}^+ = [P_i^+ P_j^+]^{T=0} = \frac{1}{\sqrt{3}} (P_{i,1}^+ P_{j,-1}^+ + P_{i,-1}^+ P_{j,1}^+ - P_{i,0}^+ P_{j,0}^+). \tag{3}$$

The collective quartets are defined as linear combinations of the quartet operators (3)

$$Q_\nu^+ = \sum_{i,j} q_{ij}^{(\nu)} A_{ij}^+. \tag{4}$$

By construction, each quartet has a total isospin $T = 0$. In addition, if the Hamiltonian (1) has spherical (axial) symmetry, Q_ν^+ has also $J = 0$ ($J_z = 0$), where J and J_z are the total angular momentum and its projection on the symmetry axis.

For a self-conjugate nucleus with $N = Z \equiv 2n$, the QCM assumes that the ground state is formulated in terms of just one collective quartet, say Q^+, as [6]

$$|\Psi^{(QCM)}\rangle = (Q^+)^n |0\rangle. \tag{5}$$

The mixing amplitudes defining the quartet Q^+ are determined by minimizing the expectation value of H in the state $|\Psi^{(QCM)}\rangle$ properly normalized. Assuming that the coefficients q_{ij} of Q^+ have a separable form, i.e. $q_{ij} \equiv q_i q_j$, allows the use of recurrence relations which greatly facilitate the evaluation of this expectation value [6]. With this approximation the collective quartet can be written as

$$Q^+ = 2\Gamma_1^+ \Gamma_{-1}^+ - (\Gamma_0^+)^2, \tag{6}$$

where $\Gamma_\tau^+ = \sum_i q_i P_{i,\tau}^+$. Due to the isospin invariance, all the collective pairs have the same mixing amplitudes q_i.

According to QM, instead, the isovector pairing ground state has the form [7]

$$|\Psi^{(QM)}\rangle = \prod_{\nu=1}^{n} Q_\nu^+ |0\rangle. \tag{7}$$

In order to search for the most appropriate $q_{ij}^{(\nu)}$'s we make use of an iterative variational procedure which is derived from an analogous treatment of like-particle pairing in terms of independent pairs [8]. The procedure consists of a sequence of diagonalizations of the Hamiltonian in spaces of a rather limited size. Each diagonalization allows to update one quartet while leaving the others unchanged and it drives the ground state toward its minimum. No approximations on the mixing amplitudes $q_{ij}^{(\nu)}$ are introduced in this case.

3. Results

To test the accuracy of the quartet approaches we have performed calculations for three sets of $N = Z$ nuclei with valence nucleons outside the cores ^{16}O, ^{40}Ca, and ^{100}Sn. We have first considered the case of spherically symmetric single-particle states and isovector pairing forces extracted from the $(T = 1, J = 0)$ part of standard shell model interactions. Details about the single-particle energies employed in the calculations are given in Refs. [6, 7]. The results for the pairing correlation energies, defined as the difference between the ground state energies obtained without and with the pairing force, are given in Table 1.

Table 1. Correlation energies for spherical single-particle states and pairing forces extracted from standard shell model interactions. The results are shown for the exact diagonalizations, the QM and the QCM. In brackets we give the errors relative to the exact results.

	Exact	QM	QCM
^{20}Ne	9.174	9.174 (-)	9.170 (0.04%)
^{24}Mg	14.461	14.458 (0.02%)	14.436 (0.17%)
^{28}Si	15.787	15.780 (0.04%)	15.728 (0.37%)
^{32}S	15.844	15.844 (-)	15.795 (0.31%)
^{44}Ti	5.965	5.965 (-)	5.964 (0.02%)
^{48}Cr	9.579	9.573 (0.06%)	9.569 (0.10%)
^{52}Fe	10.750	10.725 (0.23%)	10.710 (0.37%)
^{104}Te	3.832	3.832 (-)	3.829 (0.08%)
^{108}Xe	6.752	6.752 (-)	6.696 (0.83%)
^{112}Ba	8.680	8.678 (0.02%)	8.593 (1.00%)

In this table the correlation energies predicted by QM and QCM are compared to the exact results. We notice that for the systems with one quartet outside the closed core the state (7) is by construction exact. This is not the case for the quartet condensate (5) because of the factorization approximation $q_{ij} = q_i q_j$. Both QM and QCM results are characterized by very small errors, with those of QM being always smaller. This reflects the gain in correlation energy obtained in QM by allowing the quartets to be different.

We have also tested the accuracy of the quartet approaches for an isovector pairing interaction acting on the single-particle spectrum corresponding to an axially deformed mean field. The single-particle energies have been extracted from axially deformed Skyrme-HF calculations performed with the force SLy4 and neglecting the Coulomb interaction. For the isovector pairing force we have taken a state-independent interaction with strength $g = 24/A$. As in the case of spherical symmetry, the pairing has been applied to the nucleons outside the cores ^{16}O, ^{40}Ca, and ^{100}Sn. In the calculations we have considered, respectively, the lowest seven, nine, and ten HF single-particle states above the cores just mentioned. The number of these states has been chosen such as to keep the total degeneracy of the model space approximately the same as in the case of spherical calculations. The results are presented in Table 2. One still observes a

Table 2. Correlation energies calculated for axially deformed single-particle states and a state-independent isovector pairing force. The results are shown for the exact diagonalizations, the QM and the QCM. In brackets we give the errors relative to the exact results.

	Exact	QM	QCM
^{20}Ne	6.5505	6.5505	6.539 (0.18%)
^{24}Mg	8.4227	8.4227	8.388 (0.41%)
^{28}Si	9.6610	9.6610	9.634 (0.28%)
^{32}S	10.2629	10.2629	10.251 (0.12%)
^{44}Ti	3.1466	3.1466	3.142 (0.15%)
^{48}Cr	4.2484	4.2484	4.227 (0.50%)
^{52}Fe	5.4532	5.4531	5.426 (0.50%)
^{104}Te	1.0837	1.0837	1.082 (0.16%)
^{108}Xe	1.8696	1.8696	1.863 (0.35%)
^{112}Ba	2.7035	2.7034	2.688 (0.57%)

very good performance of QCM and an even improved one by QM whose predictions are, in this case, basically exact. The same high precision is observed for the occupation probabilities.

4. Conclusions

We have provided a description of the ground state of an isovector pairing Hamiltonian in a formalism of quartets. We have discussed two approaches, one describing the ground state as a product of identical quartets and the other allowing the quartets to be all distinct from one another. We have tested these approaches by carrying out a number of comparisons with exact shell model calculations. Both approaches have been found to be very accurate. The one corresponding to distinct quartets, in particular, has provided results basically exact in the applications to an axially deformed mean field. These results strongly support the conclusion that $T = 0$ quartets are appropriate tools for the treatment of the isovector pairing in self-conjugate nuclei.

Acknowledgments

N.S. acknowledges the support from UEFISCDI of Romanian Ministry of Education and Research through Grant IDEI No 57.

References

[1] Engel J, Pittel S, Stoitsov M, Vogel P and Dukelsky J 1997 *Phys. Rev.* C **55** 1781
[2] Satula W and Wyss R 1997 *Phys. Lett.* B **393** 1
[3] Civitarese O, Reboiro M and Vogel P 1997 *Phys. Rev.* C **56** 1840
[4] Sandulescu N, Errea B and Dukelsky J 2009 *Phys. Rev.* C **80** 044335
[5] Chen H -T, Muther H and Faessler A 1978 *Nucl. Phys.* A **297** 445
[6] Sandulescu N, Negrea D, Dukelsky J and Johnson C W 2012 *Phys. Rev.* C **85** 061303(R)
[7] Sambataro M and Sandulescu N 2013 *Phys. Rev.* C **88** 061303(R)
[8] Sambataro M 2012 *Phys. Rev.* C **85** 064326

Journal of Physics: Conference Series **569** (2014) 012056 doi:10.1088/1742-6596/569/1/012056

Cluster approach to the structure of heavy nuclei

T.M. Shneidman[1,2,3],**G.G. Adamian**[2], **N.V. Antonenko**[2], **R.V. Jolos**[2], **Shan-Gui Zhou**[1]

[1]State Key Laboratory of Theoretical Physics, Institute of Theoretical Physics, Chinese Academy of Sciences, Beijing 100190, China
[2]Bogoliubov Laboratory of Theoretical Physics, Joint Institute for Nuclear Research, Dubna, 141980, Russia
[3]Kazan Federal University, Kazan 420008, Russia

E-mail: shneyd@theor.jinr.ru

Abstract. We developed a cluster model which allows to take into account both shape deformation parameters and cluster degrees of freedom. The important ingredient of the model is the dinuclear system concept in which the wave function of the nucleus is treated as a superposition of a mononucleus and two-cluster configurations. The model is applied to describe the multiple negative-parity bands in the deformed actinides.

1. Introduction

In the even–even isotopes of actinides and also in the heavy Ba and Ce isotopes the low-lying negative parity states are observed in the rotational or quasirotational bands. While the formation of the positive–parity bands is connected in general to the quadrupole collective motion, the lowering of the states with negative-parity is a signature of the reflection asymmetric collective mode.

There are several approaches to treat the collective motion related to the reflection asymmetric degrees of freedom (see [1] and references therein). In present work, we use the ideas of the model developed in [2] that the collective oscillations of the nucleus in the mass asymmetry degrees of freedom lead to the formation with some probability of the cluster-type configurations. The contribution of the asymmetric cluster configurations to the intrinsic nuclear wave function simultaneously creates deformations with even and odd–multipolarities. The idea that the motion in mass asymmetry is responsible for the reflection–asymmetric deformations of the medium mass and heavy nuclei is based on the fact that these nuclei are good alpha-emitters.

In [2], only the lowest negative parity bands (K^{π}=0^-) have been considered. However, there are experimental evidences of the negative-parity collective states characterized by non zero values of K (see, for example, [3]). Moreover, the recent investigations [4] have indicated that the reflection-asymmetric degrees of freedom can also contribute significantly to the structure of positive parity rotational band built upon excited 0_2^+ state which are observed in the excitation spectra of many actinides.

To treat these collective excitations in the framework of our cluster approach, we modify the cluster model [2] to take into account the intrinsic excitations of clusters produced by the motion in mass-asymmetry degrees of freedom. In [5], we applied the modified model to treat the properties of ^{220}Th. This is a peculiar case of the nucleus characterized by the strong reflection-asymmetric deformation and relatively small quadrupole deformation. The aim of this work is to treat the situation when the nucleus has both strong quadrupole and strong reflection-asymmetric deformations. The model is applied to the description of the multiple reflection-asymmetric bands in heavy isotopes of actinides.

2. Model

Instead of parametrizing nuclear shapes in terms of multipole deformation parameters, we use the degrees of freedom related to the dinuclear system (DNS). Under the expression of dinuclear system we understand the system of two touching fragments (A_1,Z_1) and (A_2,Z_2) with $A_1 + A_2 = A$ and $Z_1 + Z_2 = Z$ kept together by the molecular-type nucleus-nucleus potential. The special case of the DNS in which one fragment has zero mass is denoted as a mononucleus. Each dinuclear system configuration can be characterized by the mass-asymmetry $\xi = 2A_2/A$, charge-asymmetry $\xi_Z = 2Z_2/Z$ and the vector of the relative distance $\mathbf{R} = (R, \theta_R, \phi_R)$.

The main idea of the DNS model is that the intrinsic nuclear wave function can be presented as a superposition of the mononucleus and different dinuclear configurations. The mononucleus is taken to be quadrupole-deformed; thus, in the DNS model, the reflection-asymmetric deformation of the nucleus related solely to the non zero weight of the asymmetric dinuclear systems.

Applying the model to the description of the low-lying excitations in actinides, we presume that the main source of the reflection asymmetric deformation is the contribution of the DNS with an alpha-particle as a light cluster. This idea is related to the experimental observation that the actinides are good α-emitters. Thus, there is a significant probability to form an α–cluster in the surface region of the nucleus. Due to the stable closed-shell structure of the α–cluster, it is reasonable to presume that the ground-state and low-lying excited states of actinides can have a significant contribution of $\alpha + (A-4, Z-2)$ cluster structures (α-cluster DNS) without appreciable internal excitations of the heavy fragment.

The degrees of freedom chosen to characterize a system with quadrupole deformed heavy cluster and spherical light cluster are related to the transfer of nucleons between the fragments, to the rotation of the heavy fragment, and to the relative motion of fragments. The charge asymmetry ξ_Z is not considered as an independent collective variable, and fixed to minimize the symmetry energy of the DNS for each value of ξ.

The Hamiltonian of the model has the form

$$\hat{H}_0 = -\frac{\hbar^2}{2B_\xi}\frac{1}{\xi^{3/2}}\frac{\partial}{\partial\xi}\xi^{3/2}\frac{\partial}{\partial\xi} + \frac{\hbar^2}{2\Im_h(\xi)}\hat{l}_h^2 + \frac{\hbar^2}{2\mu(\xi)R_m^2}\hat{l}_0^2 + \frac{\hbar\omega_R(\xi)}{2} + U(\xi, R_m, \beta_0, \epsilon). \quad (1)$$

Here, we assumed that the value of the internuclear distance $R = R_m(\xi, \beta_0, \epsilon)$ is set to be equal to the position of the minimum of the potential in R for given values of ξ, relative orientation of the fragments defined by angle ϵ and the deformation β_0 of the heavy cluster. The calculations of the potential energy [6] show that the minimum corresponds to the touching configuration of the clusters. $\hbar\omega_R(\xi)/2$ is the energy of the zero-order vibrations in R around the equilibrium value R_m. We treat mass-asymmetry ξ as a continuous variable.

In eq. (1), $\mu \approx m_0 A\xi/2$ is the reduced mass of the DNS ($\xi \ll 1$), m_0 is the nucleon mass, and $\Im_h(\xi)$ is the moment of inertia of the heavy fragment. The $\hat{l}_i^2$ are the operators of the square of the angular momentum of the rotation of the heavy fragment ($i = h$) and of the relative rotation of the two clusters ($i = 0$): $\hat{l}_i^2 = -\frac{1}{\sin\theta_i}\frac{\partial}{\partial\theta_i}\sin\theta_i\frac{\partial}{\partial\theta_i} - \frac{1}{\sin^2\theta_i^2}\frac{\partial^2}{\partial\phi_i^2}$. The angles $\Omega_i = (\theta_i, \phi_i)$ describe the orientation of the symmetry axis of the heavy fragment ($i = h$) and of the vector $\mathbf{R}$ ($i = 0$) with respect to the laboratory frame. The angle of relative orientation ϵ can be related to Ω_0 and Ω_h by the expression $\sin^2\epsilon = (2/3)(1 - \sqrt{5}\,[Y_2(\Omega_h) \times Y_2(\Omega_0)]_{(00)})$.

The calculation of the mass parameter B_ξ is given in [7]. In this work, we take $B_\xi = (10-20)\times 10^5 m_0$ fm^2. One should note that this value is larger than those used in our previous calculations [2] because of the fact that additional degrees of freedom are considered and renormalize of the mass tensor.

The potential energy $U(\xi, R_m, \beta_0, \epsilon)$ is determined as

$$U(\xi, R_m, \beta_0, \epsilon) = B_1(\xi) + B_2(\xi) + V(\xi, R_m, \beta_0, \epsilon), \tag{2}$$

where the quantities B_1, B_2 are the binding energies of the clusters forming the DNS. The experimental ground–state masses [8], if available, are used in the calculations. If not, the predictions of [9] are used. Shell effects and pairing correlations are included in the binding energies. The nucleus-nucleus potential $V(R = R_m, \xi, \beta_0, \epsilon)$ is calculated using the procedure described in [6].

The potential energy as a function of relative orientation of the fragments has a minimum which corresponds to the pole-to-pole orientation (ϵ=0). We approximate the ϵ–dependence of the potential energy by the second order expansion in the Legendre polynomials as

$$\begin{aligned} U(\xi, R_m, \beta_0, \epsilon) &= U_0(\xi, R_m, \beta_0) + C_0(R_m, \beta_0)\xi \sin^2\epsilon \\ &= (U_0 + (2/3)C_0\xi) - (2/3)\sqrt{5}C_0\xi\,[Y_2(\Omega_h)\times Y_2(\Omega_0)]_{(00)}, \end{aligned} \tag{3}$$

where $U_0(\xi, R_m, \beta_0) = U(\xi, R_m, \beta_0, \epsilon = 0)$. Since $\xi \ll 1$, we expand the interaction term in (3) leaving only term linear in ξ, that is $C_0(R_m, \xi, \beta_0) \approx C_0(R_m, \beta_0)\xi$. The value of $C_0(R_m, \beta_0)$ is fixed by fitting ϵ–dependence of the potential energy U calculated with eq. (2). The potential energy of the mononucleus configuration (ξ=0) which can not be calculated with eq. (2) is fitted to reproduce the correct value of the nuclear binding energy as the lowest eigenstate of the Hamiltonian (1).

The Hamiltonian (1) is diagonalized on the set of basis functions

$$\Phi^{l_1,l_2,n}_{LM,\pi} = F_n(\xi)\,[Y_{l_1}(\Omega_h)\times Y_{l_2}(\Omega_{\mathbf{R}})]_{(LM)}, \tag{4}$$

where n=0,1,2,..., l_1=0,2,4,..., l_2=0,1,2... The angular part of the wave function (4) is given by the bipolar spherical harmonics, which provide the proper transformation with respect to the rotation and space inversion. Since we have assumed that the heavy fragment is the axially–symmetric quadrupole rotator the quantum number l_1 can take only even values. The parity of the state is then determined as $\pi = (-1)^{l_2}$.

3. Calculated results

The calculations of the energy spectra are presented in fig. 1 together with available experimental data for ^{230}Th,^{232}U, ^{238}U, and ^{240}Pu nuclei. The experimental energies, spins, and parities are taken from [10]. One can see that the structures of the spectra are very similar for all considered nuclei. To elucidate the peculiarities of the calculated spectra, one can examine which basis function from the set (4) mostly contributes to the wave functions of the states of different rotational bands.

For the ground-state rotational band, the angular part of the wave-function of the state with angular momentum L is close to the function $[Y_{l_1=L}(\Omega_h)\times Y_{l_2=0}(\Omega_0)]_{(L,M)}$. Since $\pi = (-1)^{l_2}$, the ground-state band contains only the states of positive parity with even angular momentum L. The wave functions of the lowest negative parity states have the dominant component $[Y_{l_1=L}(\Omega_h)\times Y_{l_2=1}(\Omega_0)]_{(L+1,M)}$. They form a rotational band which contain the states of odd angular momentum and negative parity. This band can be interpreted as $K^\pi = 0^-$ band. One should note, however, that each eigenfunction of Hamiltonian (1) is a superposition of

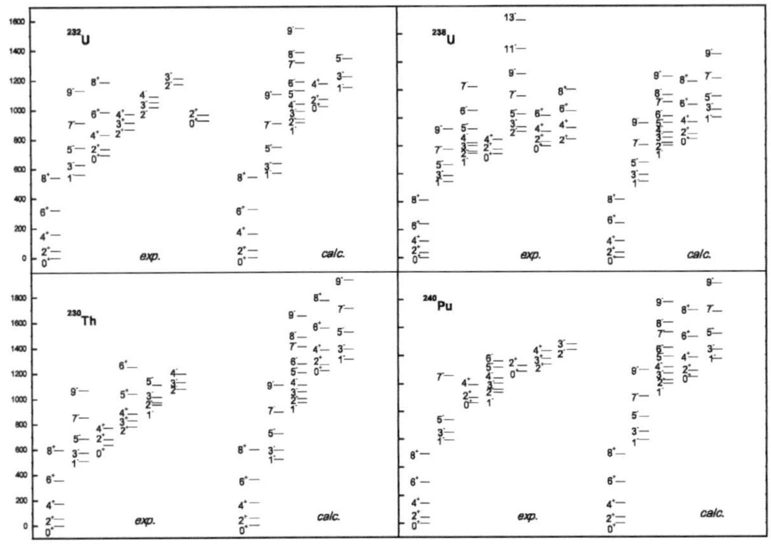

Figure 1. Calculated and experimental low–energy level schemes of ^{230}Th, ^{232}U, ^{238}U, ^{240}Pu. Experimental energies, spin and parity assignments are taken from [10].

the states with different K values, thus K can be considered as a quantum number only approximately. The next lowest band of negative parity is formed from the states with main contribution of $[Y_{l_1=L}(\Omega_h) \times Y_{l_2=1}(\Omega_0)]_{(L-1,M)}$ component for odd angular momentum and $[Y_{l_1=L}(\Omega_h) \times Y_{l_2=1}(\Omega_0)]_{(L,M)}$ component for even angular momentum. This band can be approximately interpreted as $K^\pi = 1^-$. In this case the angular momentum carried by the relative rotation is directed oppositely with the total angular momentum, thus, shifting up the energy of this rotational band with respect to the 0^- band. The first excited 0^+ state is obtained as a lowest excitation in the mass-asymmetry coordinate.

As seen in fig. 1, the agreement between the calculated excitation spectra and corresponding experimental rotational bands are rather good for all presented nuclei. One can notice, however, that some of the bands, measured experimentally, namely, K^π=0^+,2^+, and 2^- rotational bands, are absent in the calculated spectra. The reason for this is that in the present version of the model we have considered the heavy fragment with stable axially-symmetric quadrupole deformation. Thus, we may expect, that if we add into consideration the $\beta-$ and $\gamma-$ vibrations of the heavy fragment, the calculated spectra of actinides will contain additional 0^+ and 2^+ rotational bands, which are build on the β and γ-vibrational excitations, respectively. We can also expect, the appearance of additional low–lying 0^- and 2^- bands, as a negative parity doublets to the β and γ-vibrational bands. Note that the alternative way to explain the low-lying 2^- bands in actinides is based on the concept of tetrahedral deformation [11, 12].

The interesting quantity is the parity splitting defined as the energy shift between the states of 0^- and 0^+ rotational bands. The dependences of the experimental and calculated values of a parity splitting in the ground state and the first negative parity bands (they are treated as a unified alternating parity band) on angular momentum are shown in fig. 2. The parity splitting is defined by the expression [3]:

$$S(I^-) = E(I^-) - \frac{(I+1)E^+_{(I-1)} + IE^+_{(I+1)}}{2I+1}, \tag{5}$$

which provides zero value of parity splitting for the rotational band of nucleus with rigid octupole deformation. Having maximum value in the beginning of the band, the parity splitting decreases with increasing angular momentum. In our model the value of parity splitting is determined by

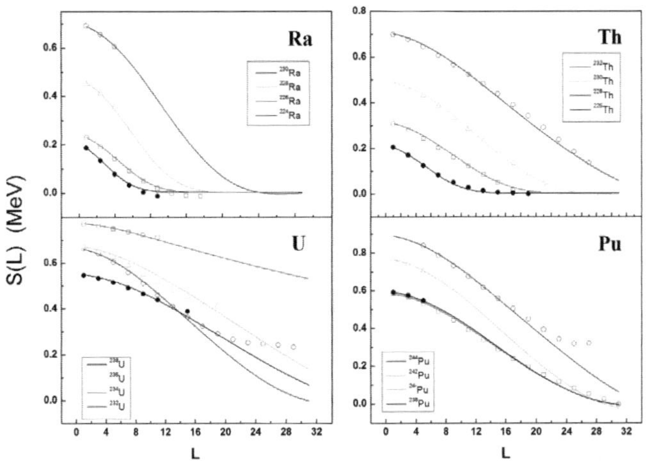

Figure 2. Calculated (lines) and experimental (open circle of the same color) values of parity splitting (see Eq.(5)). Experimental values are taken from [10].

the relative weights of the cluster components and the mononucleus in the wave function. Due to the larger moment of inertia the weight of α-particle DNS increases with angular momentum that decreases the parity splitting.

4. Conclusion

We have suggested a cluster interpretation of the properties of the multiple negative parity bands in deformed even–even actinides. The collective motion related to the cluster degree of freedom leads to the admixture of the very asymmetric cluster configurations to the intrinsic nucleus wave function. To take into account the reflection asymmetric modes with nonzero values of K, the rotational excitations of the heavy cluster are taken into account. The resulting energy spectrum consists of the ground state band, the excited 0^+ band and several negative parity rotational bands which can be approximately interpreted as $K^\pi = 0^-$ and $K^\pi = 1^-$ bands. The angular momentum dependence of the parity splitting is described. The results of calculations are in good agreement with the experimental data.

Acknowledgments
This work was supported by the RFBR (Moscow) and NSFC (Beijing). T.M.S. acknowledges the support from the CAS Fellowship for the the Young International Scientists (2013Y1JA0003) and Russian Government Subsidy Program of the Competitive Growth of Kazan Federal University.

References

[1] Butler P and Nazarewicz W 1996 *Rev. Mod. Phys.* **68** 349
[2] Shneidman T M, Adamian G G, Antonenko N V, Jolos R V and Scheid W 2003 *Phys.Rev.* C **67** 014313
[3] Reviol W *et al.* 2006 *Phys. Rev.* C **74** 044305
[4] Spieker M *et al.* 2013 *Phys. Rev.* C **88**, 041303(R)
[5] Shneidman T M, Adamian G G, Antonenko N V, Jolos R V and Scheid W 2011 *Eur. Phys. J.* A **47** 34
[6] Adamian G G, Antonenko N V, Jolos R V, Ivanova S P and Melnikova O I 1996 *Int. J. Mod. Phys.* E **5** 191
[7] Adamian G G, Antonenko N V and Jolos R V 1995 *Nucl. Phys.* A **584** 205
[8] Audi G, Bersillon O, Blachot J, and Wapstra A H 2003 *Nucl. Phys.* A **729** 3
[9] Moller P, Nix J R, Myers W D, Swiatecki W J 1995 *At. Data Nucl. Data Tables* **59** 185
[10] http://www.nndc.bnl.gov/nndc/ensdf
[11] Chen Y S, Sun Y and Gao Z C 2008 *Phys. Rev.* C 77 061305(R)
[12] Zhao J, Lu B N, Zhao E G and Zhou S G 2012 *Phys. Rev.* C 86 057304

Journal of Physics: Conference Series **569** (2014) 012057 doi:10.1088/1742-6596/569/1/012057

Neutrino Emissivities from Deuteron-Breakup and Formation in Supernovae

Satoshi X. Nakamura

Department of Physics, Osaka University, Toyonaka, Osaka 560-0043, Japan

E-mail: nakamura@kern.phys.sci.osaka-u.ac.jp

Abstract. Recently it was pointed out that there are abundant light clusters, such as deuteron, triton and helium, in supernova environment. An interesting question is how much neutrino emissions from these light clusters affect supernova explosion mechanism. To address this question through a supernova simulation, neutrino emissivities from these light clusters are necessary input. The deuteron is the simplest cluster, and occupy a substantial portion of the light cluster abundance. Thus in this work, we study neutrino emissions from electron/positron capture on the deuteron and the nucleon-nucleon fusion processes in the surface region of a supernova core. We evaluate these weak processes using one-nucleon impulse current supplemented by two-nucleon exchange-currents, and nuclear wave functions generated by a high precision nucleon-nucleon potential. We present the neutrino emissivities from the deuteron calculated for typical profiles of core-collapsed supernovae. These novel neutrino emissivities are compared with the standard neutrino emission mechanisms.

1. Introduction

Neutrino reactions in the dense matter of a supernova core are crucial for understanding the supernova explosion mechanism. The emission of neutrinos acts as a cooling mechanism of the central core, and a portion of the emitted neutrinos are subsequently absorbed by the material behind the shock wave and thereby act as a heating agent.

Recent supernova simulations have shown that all of light clusters can appear copiously in the regions between the supernova core and the shockwave [1, 2]. The light clusters with mass number $A=2$ and 3 can be abundant in hot and moderately dense matter ($< 10^{13}$ g/cm^3) under nuclear statistical equilibrium [1, 3] and should be considered in studying the supernova explosion mechanism. Their appearance gives a new contribution to the neutrino emission and absorption. For example, the neutrino absorption on deuterons as an additional heating mechanism on top of the neutrino reactions on nucleons and ^{4}He was studied in Ref. [4].

In this work [5], we study neutrino emissions from deuteron breakup/formation processes (DBF for short) in the surface region of a proto-neutron star, where neutrino emissions act as a cooling mechanism. We present the first evaluation of the neutrino emissivities from electron/position-capture on the deuteron (deuteron breakup) and from the nucleon-nucleon weak fusion processes (deuteron formation); see (1)-(5) below. The neutrino emissivities arising from these deuteron reactions will be compared with those coming from the "conventional" processes; by "conventional" processes we mean the neutrino emission processes which have been previously considered in the literature. The neutrino emissivities reported here are expected to be useful for numerical simulations of supernova explosion and proto-neutron star cooling.

Theoretical treatments of electroweak processes in two-nucleon systems have been well developed. One approach is the standard nuclear physics approach (SNPA) that involves nuclear wave functions derived from high-precision phenomenological nuclear potentials, and one-nucleon and two-nucleon electroweak currents. SNPA has been well tested by analyses of photo-reactions, electron scattering, and muon capture on the two-nucleon systems. We adopt SNPA in this work.

This article is arranged as follows. The theoretical framework for calculating the neutrino emissivities due to DBF is outlined in section 2, and the numerical results are presented in section 3. A summary is given in section 4.

2. Calculation of neutrino emissivities

We consider neutrino emissions via deuteron breakup/formation (DBF):

$$
\begin{aligned}
d+e^- &\rightarrow n+n+\nu_e\ , & (1)\\
d+e^+ &\rightarrow p+p+\bar{\nu}_e\ , & (2)\\
n+n &\rightarrow d+e^-+\bar{\nu}_e\ , & (3)\\
p+p &\rightarrow d+e^++\nu_e\ , & (4)\\
p+n &\rightarrow d+\nu+\bar{\nu}\ . & (5)
\end{aligned}
$$

The reactions (1) and (2) are deuteron breakup via e^-/e^+-capture, whereas the reactions (3), (4) and (5) are deuteron formation through nucleon-nucleon scattering. The first four reactions that are caused by the charged-current (CC), can only emit ν_e or $\bar{\nu}_e$, whereas the last reaction occurring via the neutral-current (NC) gives rise to $\nu\bar{\nu}$ pair-emission of all three flavors. The Hamiltonian for low-energy semi-leptonic weak processes is, to good accuracy, given by the product of the hadron current and the lepton current with the Fermi weak coupling constant. The hadronic weak currents are combinations of the vector current and the axial-vector current. The nuclear weak currents consist of one-nucleon [impulse-approximation (IA)] terms and two-nucleon meson-exchange current (MEC) terms. In this work we consider the pion and rho meson-exchange currents. The validity of this approach has been well tested for the vector current by comparing the model predictions with, e.g., the measured $n+p\rightarrow d+\gamma$ reaction data. As for the strength of the axial-vector exchange current, we follow the standard practice to adjust its strength to reproduce the experimental triton beta decay rate [6]. Detailed descriptions of the model for the nuclear currents used in the present work are given in Refs. [7, 8], where the basic formulation and relevant input parameters are explained.

In this article, we evaluate the emissivities for the DBF processes, (1) - (5). The emissivities for neutrinos and anti-neutrino are denoted by Q_ν and $Q_{\bar{\nu}}$, respectively. They are given by integrating the transition probability over the momenta $\vec{p}_{i,k}$ and $\vec{p}_{f,l}$ with a weighting factor representing the momentum distributions:

$$Q^\alpha_{\nu(\bar{\nu})} = \frac{(2\pi)^4}{s_i s_f}\int\prod_k[\frac{d\vec{p}_{i,k}}{(2\pi)^3}]\prod_l[\frac{d\vec{p}_{f,l}}{(2\pi)^3}]\delta^{(4)}(\sum_{l'}p_{f,l'}-\sum_{k'}p_{i,k'})\omega_{\nu(\bar{\nu})}\sum_{i,f}|<f|H^\alpha_W|i>|^2\ \Xi \quad (6)$$

where $\omega_{\nu(\bar{\nu})}$ is the energy of the emitted neutrino (anti-neutrino) and s_i and s_f are symmetry factors when two-nucleons in the initial or final state are identical particles. The summation $\sum_{i,f}$ is taken over the spin states of the final particles and the average over the initial spin states. The momenta of the initial (final) particles, labeled by k (l), are $\vec{p}_{i,k}$ ($\vec{p}_{f,l}$). Ξ represents the occupation probability of incoming particles and the Pauli blocking of outgoing particles:

$$\Xi = \prod_{k=\text{initial particle}} f_k(p_k) \prod_{l=\text{final fermion}\neq\nu(\bar{\nu})}(1-f_l(p_l)) \quad (7)$$

with $f_k(p_k) = 1/(\exp((e_k(p_k) - \mu_k)/k_BT) \pm 1)$ where one should use "+" ("−") for a fermion (boson); e_k (μ_k) is the energy (chemical potential) of the particle of the k-th kind, and k_BT is the temperature multiplied by the Boltzmann constant. Note that we have not included the Pauli blocking factor for the final-state neutrino (anti-neutrino) in Ξ, and that we have ignored the bose enhancement factor $(1+f_d)$ for deuterons, because f_d is sufficiently small. More explicit expressions for the emissivities for the various processes under consideration are given in Ref. [5].

3. Neutrino emissivities

3.1. Supernova profiles

In order to study the consequences of neutrino emissions due to DBF for the supernova-explosion mechanism, we calculate neutrino emissivities for a given profile of a core-collapse supernovae, and compare the emissivities due to DBF with those arising from the conventional processes. To this end, we consider two representative profiles of a supernova core, Compositions I and II.

Composition I is the one obtained in Ref. [9] in simulating gravitational collapse and core bounce for a 15 $M_\odot$ star ($M_\odot$: solar mass). This composition, which represents a typical situation of the post-bounce phase with a stalled shock wave, has been obtained from a numerical simulation adopting the Shen equation of state (EOS) [10]. Composition I includes only nucleons, ^{4}He and a single heavy nucleus in the Shen EOS. Figure 1 shows the temperature (T), the density (ρ) and the electron fraction (Y_e) as functions of the distance r from the supernova center, pertaining to a snapshot at 150 ms after the core bounce.

To assess the significance of the new additional emissivities due to DBF, we consider Composition II, which includes the mass fractions of the light clusters obtained from the nuclear statistical equilibrium model [1], i.e., nucleons, deuterons, tritons, ^{3}He, ^{4}He and other nuclei are taken into account. We remark that Compositions I and II share the same data for the profiles of ρ, T and Y_e shown in Fig. 1. The nucleon chemical potentials needed to calculate the emissivities are also taken from the Shen EOS. We consider the surface region of a proto-neutron

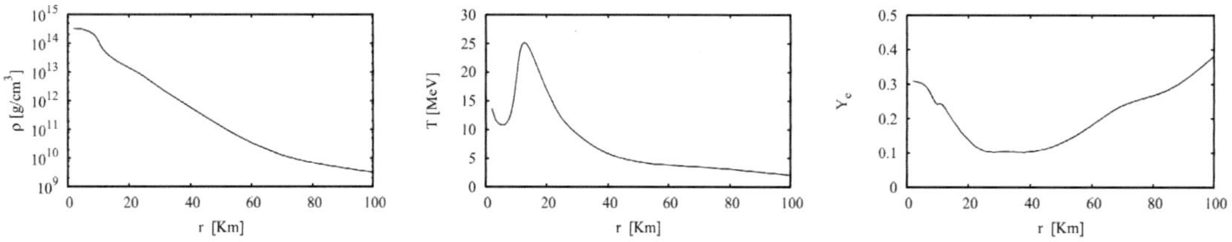

Figure 1. The density(left panel), temperature (middle panel) and electron fraction (right panel) distributions pertaining to a snapshot at 150 ms after the core bounce taken from Ref. [1]. The horizontal axis, r, is the distance from the supernova center.

star ($r > 20$ km, $\rho < 10^{13}$ g/cm^3) where the neutrino-sphere exists between the surface of the nascent proto-neutron star and the shock wave, and neutrino cooling and heating are important.

3.2. Emissivity from the surface region of a proto-neutron star

To set the stage for examining the possible influences of ν_e-emissivities due to DBF, we first present ν_e-emissivities due to the conventional reactions calculated for Composition II. Nucleon-nucleon bremsstrahlung is calculated for Composition I. The left panel in Fig. 2 shows the emissivities arising from e^-p capture and e^-e^+ annihilation, while the right panel gives the emissivities due to nucleon-nucleon bremsstrahlung. The figure indicates that e^-p capture gives

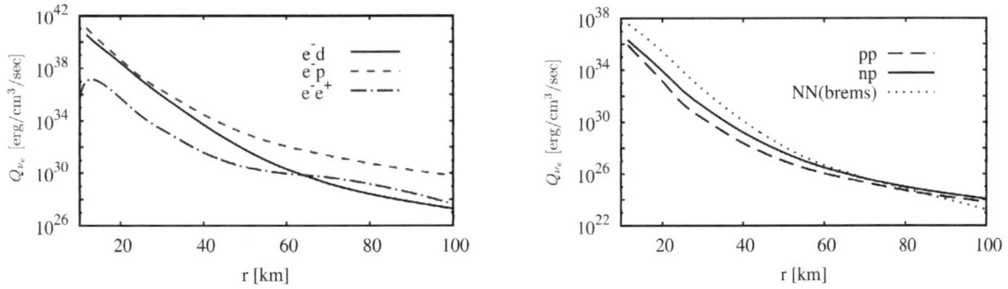

Figure 2. The ν_e-emissivities are shown as functions of the distance r from the center of the supernova evaluated with composition II except NN bremsstrahlung. In the left panel, the neutrino emissivities due to e^- captures on deuteron (1) and proton, and e^+e^- annihilation are shown in solid, dashed and dash-dotted curves, respectively. In the right panel, the emissivities due to the pp and np fusion processes, (4) (5), and NN bremsstrahlung are shown in long-dash, solid and dotted curves, respectively. The emissivities due to the NN bremsstrahlung and e^+e^--annihilation are taken from Ref. [9].

a dominant contribution, and nucleon-nucleon bremsstrahlung and the e^-e^+ annihilation process give only minor contributions to the emissivity.

The left panel in Fig. 2 also gives the neutrino emissivity due to e^--capture on the deuteron (1). The neutrino emissivity due to e^--capture on the deuteron is almost the same magnitude as that on the proton around $r \approx 30$km. In this region, the mass fraction of deuteron is comparable or even larger than that of proton. We will discuss below the role of including the composition of light clusters in the neutrino emissivity.

We see in Fig. 2 (right) that the neutrino emissivities from deuteron formation (4) and (5) are orders of magnitude smaller than those from e^--captures and the e^-e^+ annihilation. However, the neutrino emissivities from deuteron formations become comparable to the $\nu\bar{\nu}$ emissivity from nucleon-nucleon bremsstrahlung for distances closer to 100 km in the cooling region.

As shown in Ref. [5], the electron capture cross section of deuteron is smaller than that of proton. This indicates that the effective neutrino emissivity *per proton* via e^--captures on the proton is reduced if a substantial amount of protons in a supernova are bound in deuterons (and tritons). In fact, this is seen in Fig. 3. The left panel in Fig. 3 shows that the free proton abundance for Composition II(solid curve) is smaller than that for Composition I(short dashed curve) by a factor of ~2, where deuteron(long dashed curve) is abundant. The total emissivity due to the e^--capture on protons in this region is effectively reduced by up to 40% as shown in the right panel of Fig. 3, where the solid curve shows ratio of the neutrino emissivity from e^--captures on the proton and deuteron for Composition II to that from e^--captures on the proton for Composition I. Meanwhile, as can be seen from the short dashed curve in the right panel of Fig. 3, the total anti-neutrino emissivity is only slightly affected by the deuteron abundance, mainly due to the dominant abundance of the neutron. The reduction of the total neutrino emissivity indicates that, in considering the neutrino emissivities due to electron-captures, it is important to take due account of the abundances of deuterons and other light clusters.

4. Summary

Neutrino emissions from $e^{\mp}$-capture on the deuteron and from deuteron formation in nucleon-nucleon weak-fusion processes have been studied as new neutrino emission mechanisms in supernovae. These weak processes are evaluated with the standard nuclear physics approach,

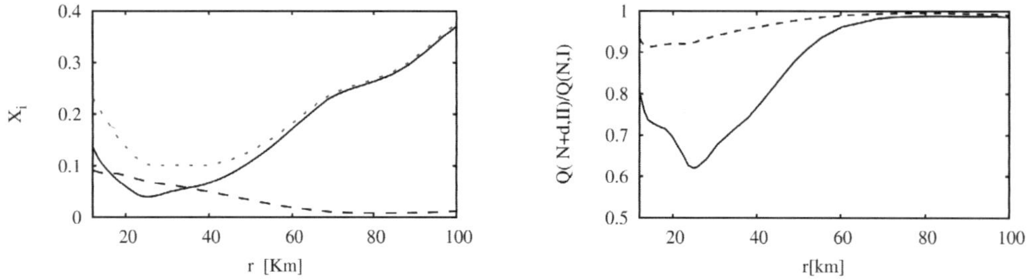

Figure 3. Left panel: Mass fraction of proton (solid curve) and deuteron (long dashed curve) for Composition II and that of proton(short dashed curve) for Composition I. Right panel:The ratio of the neutrino emissivity due to e^--capture(solid curve) and e^+-capture(short dashed curve) calculated for Composition II (emissivity from deuterons and nucleons) to that calculated for Composition I (emissivity from nucleons only).

which consists of the one-nucleon impulse current and two-nucleon exchange current and nuclear wave functions derived from high-precision phenomenological NN potentials. It is found that the contribution of the two-nucleon meson-exchange current is only a few % for the $e^{\mp}$-capture reactions, while it can be as large as the one-nucleon current contribution for the NN fusion reaction at higher energies. The consequences of these new neutrino-emission channels have been examined for representative profiles of core-collapse supernovae at 150 ms after core bounce. The cross section due to the $e^{\mp}$ capture reaction on the deuteron is smaller than that on the free nucleon. Therefore, as Fig. 3 indicates, the total neutrino emissivity due to electron capture on protons and deuterons is suppressed when an appreciable amount of protons in a supernova are bound inside deuterons. This results in a smaller neutrino luminosity and the lower efficiency of neutrino heating behind a stalled shock wave. Therefore, this new process may contribute unfavorably towards a successful supernova explosions. It might lead to a slower speed of the deleptonization and, hence, a slower evolution of nascent proto-neutron stars. On the other hand, as seen in Fig. 2, neutrino emission via deuteron formation can be comparable to nucleon-nucleon bremsstrahlung in the outer region. This implies that there might exist situations in which the deuteron-formation weak processes are the main channels for neutrino emission.

Acknowledgments

The author thanks to Shota Nasu, Toru Sato, Kohsuke Sumiyoshi, Fred Myhrer and Kuniharu Kubodera for their collaboration. This work was supported by MEXT KAKENHI Grant Number 25105010.

References

[1] Sumiyoshi K and Röpke G 2008 Phys. Rev. C **77** 055804
[2] Furusawa S, Nagakura H, Sumiyoshi K and Yamada S 2013 Astrophys. J. **774** 78
[3] Furusawa S, Sumiyoshi K, Yamada S and Suzuki H 2013 Astrophys. J. **772** 95
[4] Nakamura S X, Sumiyoshi K and Sato T 2009 Phys. Rev. C **80** 035802
[5] Nasu S, Nakamura S X, Sato T, Sumiyoshi K, Myhrer F and Kubodera K arXiv:1402.0959
[6] Schiavilla R et al. 1998 Phys. Rev. C **58** 1263
[7] Nakamura S, Sato T, Gudkov V and Kubodera K 2001 Phys. Rev. C **63** 034617
[8] Nakamura S, Sato T, Ando S, Park T-S, Myhrer F, Gudkov V and Kubodera K 2002 Nucl. Phys. **A707** 561
[9] Sumiyoshi K, Yamada S, Suzuki H, Shen H, Chiba S and Toki H 2005 Astrophys. J. **629** 922
[10] Shen H, Toki H, Oyamatsu K and Sumiyoshi K 1998 Prog. Thoer. Phys. **100** 1013; 1998 Nucl. Phys. **A637** 435

Journal of Physics: Conference Series **569** (2014) 012058 doi:10.1088/1742-6596/569/1/012058

Equation of state for nuclear matter in core-collapse supernovae by the variational method

H. Togashi[1,2], Y. Takehara[3], S. Yamamuro[3], K. Nakazato[3], H. Suzuki[3], K. Sumiyoshi[4] and M. Takano[2,5]

[1] Institute for Physical and Chemical Research (RIKEN), 2-1 Hirosawa, Wako, aitama 351-0198, Japan

[2] Research Institute for cience and Engineering, Waseda University, 3-4-1 Okubo hinjuku-ku, Tokyo 169-8555, Japan

[3] Department of Physics, Faculty of cience and Technology, Tokyo University of cience, Yamazaki 2641, Noda, Chiba 278-8510, Japan

[4] Numazu College of Technology, Ooka 3600, Numazu, hizuoka 410-8501, Japan

[5] Department of Pure and Applied Physics, Waseda University, 3-4-1 Okubo hinjuku-ku, Tokyo 169-8555, Japan

E-mail: hajime.togashi@riken.jp

Abstract. We construct a new nuclear equation of state (EO) for core-collapse supernova (N) simulations using the variational many-body theory. For uniform nuclear matter, the EO is constructed with the cluster variational method starting from the realistic nuclear Hamiltonian composed of the Argonne v18 two-body potential and the Urbana I three-body potential. The masses and radii of neutron stars calculated with the obtained EO at zero temperature are consistent with recent observational data. For non-uniform nuclear matter, we construct the EO in the Thomas-Fermi approximation. In this approximation, we assume a functional form of the density distributions of protons, neutrons, and alpha-particles, and minimize the free energy density in a Wigner- eitz cell with respect to the parameters included in the assumed density distribution functions. The phase diagram of hot nuclear matter at a typical temperature is reasonable as compared with that of the hen EO .

1. Introduction

The equation of state (EOS) for hot nuclear matter is one of the crucial ingredients in the studies of core-collapse supernovae (SNe). The EOS for dense uniform matter plays important roles in the core bounce and the subsequent formation of the proto-neutron star, whereas that for non-uniform matter is indispensable for calculating various weak reactions. The EOS for nuclear matter applicable to numerical simulations of core-collapse SNe must cover an extremely wide range of baryon number densities n_{B}, proton fractions Y_{p}, and temperatures T. Because the construction of such EOS is laborious, the nuclear EOSs available for SN simulations are limited. A typical SN-EOS is the Lattimer-Swesty EOS [1], in which a Skyrme-type effective interaction is employed for uniform matter whereas a compressible liquid drop model is employed for non-uniform matter. The Shen EOS [2, 3], which is another typical SN-EOS, is based on the relativistic mean field theory for uniform matter and the Thomas-Fermi (TF) calculation for non-uniform matter. Recently, some new nuclear EOSs applicable to SN simulations have been proposed. However, all the SN-EOSs presently available are based on phenomenological models for uniform matter. It would be desirable to apply nuclear EOSs based on quantum many-body theories using bare nuclear forces to SN simulations.

Under these circumstances, we are constructing a new nuclear EOS applicable to numerical simulations of core-collapse SNe using the variational method with bare nuclear forces. We have constructed the EOS for uniform asymmetric nuclear matter for arbitrary proton fractions using the cluster variational method [4, 5, 6]. In the present study, we calculate the EOS for non-uniform nuclear matter in the TF approximation, and will prepare a complete EOS table available for SN numerical simulations. In this paper, we report the current status of our project.

2. Equation of state for uniform nuclear matter

In this section, we review the variational calculations for uniform nuclear matter reported in Refs. [4, 5, 6]. As in the Fermi Hypernetted Chain (FHNC) variational calculations by Akmal et al. [7], we start from the nuclear Hamiltonian composed of the Argonne v18 (AV18) two-body potential and the Urbana I (UI) three-body potential.

At zero temperature, we express the energy per nucleon of uniform asymmetric nuclear matter E/N as the sum of the two-body energy per nucleon E_2/N and the three-body energy per nucleon E_3/N. The two-body energy per nucleon E_2/N is expressed as the expectation value of the two-body Hamiltonian with the Jastrow wave function in the two-body cluster approximation. Then, E_2/N is minimized with respect to the spin-isospin-dependent central, tensor, and spin-orbit correlation functions included in the Jastrow wave function. In this minimization, we impose the extended Mayer's condition and the healing distance condition to ensure that E_2/N of symmetric nuclear matter and neutron matter reproduce the results obtained by Akmal et al. [7] using the more sophisticated FHNC variational calculations. The three-body energy per nucleon E_3/N is expressed with use of the expectation value of the three-body Hamiltonian with the Fermi-gas wave function. Four parameters included in E_3/N are determined so that the total energy per nucleon E/N reproduces the empirical saturation properties. Furthermore, they are fine-tuned so that the TF calculations of isolated atomic nuclei with the obtained E/N reproduce the gross features of experimental data [5]. As a result, we obtain the following values: the saturation density $n_0 = 0.16$ fm^{-3}, saturation energy $E_0/N = -16.09$ MeV, incompressibility $K = 245$ MeV, and symmetry energy $E_{\rm sym} = 30.0$ MeV. These values are reasonable as compared with those suggested in the experimental and observational studies [8, 9, 10].

The obtained E/N for neutron matter and symmetric nuclear matter are in good agreement with the results of the FHNC calculations by Akmal et al. [7]. Furthermore, we have applied the obtained E/N to neutron stars. The masses and radii of neutron stars with the present EOS are consistent with recent observational data [11, 12, 13].

For uniform nuclear matter at finite temperatures, the free energies per nucleon F/N are calculated with an extension of the variational method by Schmidt and andharipande [14, 15]. In this method, F/N is expressed with use of the averaged occupation probabilities of single particle states for protons $f_{\rm p}(k, m_{\rm p}^*)$ and for neutrons $f_{\rm n}(k, m_{\rm n}^*)$, respectively. Here, m_i^* $(i = {\rm p, n})$ are the effective masses. Then, F/N is minimized with respect to $m_{\rm p}^*$ and $m_{\rm n}^*$.

The obtained F/N for neutron matter and symmetric nuclear matter are in good agreement with the results of the FHNC calculations by Mukherjee [16], which are extensions of the results by Akmal et al. Other thermodynamic quantities derived from F/N, such as the internal energies, entropies, and pressures are also reasonable.

Using this nuclear EOS for uniform matter, we have performed general-relativistic spherically-symmetric numerical simulations of core-collapse SNe [17]. For the EOS of non-uniform matter, the low-density part of the Shen EOS is adopted. In the simulation with neutrino transfers, the stellar explosion is not successful because of the energy loss by neutrino emissions. This result is consistent with other modern spherical SN simulations. Furthermore, the stellar core with the variational EOS at the bounce is more compact than that with the Shen EOS. This implies that the variational EOS is softer than the Shen EOS. This relative softness of the variational EOS is consistent with the fact that the incompressibility of the variational EOS is smaller than that of the Shen EOS, $K = 281$ MeV.

The EOS for uniform nuclear matter obtained in this way with our variational method is reasonable over a wide range of $n_{\rm B}$, $Y_{\rm p}$, and T. At extremely low densities, however, F/N tends to be slightly lower than expected because of the formation of deuteron clusters. This deuteron clustering is an interesting phenomenon because it is caused by many-body calculations with realistic nuclear forces. However, this clustering is inappropriate for a systematic description of the non-uniform phase of SN matter. In fact, we will treat non-uniform matter by the TF method, in which F/N for uniform matter is necessary. Therefore, in this study, we slightly modify the healing distance condition employed in the calculation of E_2/N to remove the deuteron clustering. Details of this treatment will be reported elsewhere [18].

3. Equation of state for non-uniform nuclear matter

In this section, we construct an EOS for non-uniform nuclear matter that is consistent with the EOS for uniform nuclear matter reported above. For this purpose, we adopt the TF approximation as is conducted by Shen et al. [2, 3], or originally by Oyamatsu in a study of neutron star crusts [19]. In this approximation, non-uniform matter is regarded as a mixture of free neutrons, free protons, alpha-particles, and a single heavy nucleus that is located at the center of a Wigner-Seitz (WS) cell in a body-centered cubic (BCC) lattice. The nucleus is assumed to be spherical, and the number densities of nucleons $n_i(r)$ ($i = {\rm n}$ for neutron and $i = {\rm p}$ for proton) in the WS cell are parameterized as follows:

$$n_i(r) = \begin{cases} (n_i^{\rm in} - n_i^{\rm out})[1 - (r/R_i)^{t_i}]^3 + n_i^{\rm out} & (0 \le r \le R_i), \\ n_i^{\rm out} & (R_i \le r \le R_{\rm cell}). \end{cases} \tag{1}$$

Here, $R_{\rm cell}$ is the radius of the WS cell. The number density of alpha-particles $n_\alpha(r)$ is given by

$$n_\alpha(r) = \begin{cases} -n_\alpha^{\rm out}[1 - (r/R_{\rm p})^{t_{\rm p}}]^3 + n_\alpha^{\rm out} & (0 \le r \le R_{\rm p}), \\ n_\alpha^{\rm out} & (R_{\rm p} \le r \le R_{\rm cell}). \end{cases} \tag{2}$$

Using the above density distribution functions, the free energy of a WS cell is expressed as

$$F = \int d\boldsymbol{r} f(n_{\rm p}(r), n_{\rm n}(r), n_\alpha(r)) + F_0 \int d\boldsymbol{r} |\nabla(n_{\rm p}(r) + n_{\rm n}(r))|^2$$
$$+ \frac{e^2}{2} \int d\boldsymbol{r} \int d\boldsymbol{r}' \frac{[n_{\rm p}(r) + 2n_\alpha(r) - n_{\rm e}][n_{\rm p}(r') + 2n_\alpha(r') - n_{\rm e}]}{|\boldsymbol{r} - \boldsymbol{r}'|} + c_{\rm bcc}\frac{(Z_{\rm non}e)^2}{a}. \tag{3}$$

The first term on the right-hand side of Eq. (3) is the bulk term, where $f(n_{\rm p}(r), n_{\rm n}(r), n_\alpha(r))$ is the local free energy density expressed as the sum of the contribution from nucleons and that from alpha-particles. The former contribution is obtained in the previous section, whereas, in the calculations of the latter, the alpha-particle is treated as a classical particle with a fixed volume. The second term on the right-hand side of Eq. (3) is the gradient term. F_0 is chosen to be 68.00 MeV fm^{-5} so that the TF calculation for isolated atomic nuclei reproduces the gross features of their empirical masses and radii [5]. The third term represents the Coulomb energy. Here, the uniform density distribution of electrons is assumed and their number density is given by $n_{\rm e} = n_{\rm B}Y_{\rm p}$. The last term on the right-hand side of Eq. (3) is the correction term for the BCC lattice. Here a is the lattice constant defined as $a^3 = V_{\rm cell} = 4\pi R_{\rm cell}^3/3$, $Z_{\rm non}$ is the non-uniform part of the charge number per cell, and $c_{\rm bcc} = 0.006562$ [19]. Then, the average free energy density of the WS cell $F/V_{\rm cell}$ is minimized with respect to the parameters a, $n_{\rm n}^{\rm in}$, $n_{\rm n}^{\rm out}$, $R_{\rm n}$, $t_{\rm n}$, $n_{\rm p}^{\rm in}$, $n_{\rm p}^{\rm out}$, $R_{\rm p}$, $t_{\rm p}$, and $n_\alpha^{\rm out}$ with $n_{\rm B}$, $Y_{\rm p}$, and T being fixed.

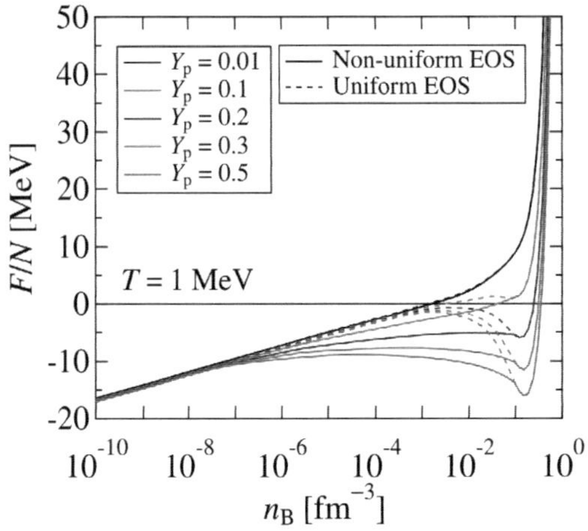

Figure 1. Free energies per nucleon F/N at $T = 1$ MeV for various proton fractions Y_p as functions of the nucleon number density n_B. F/N for uniform matter are also shown.

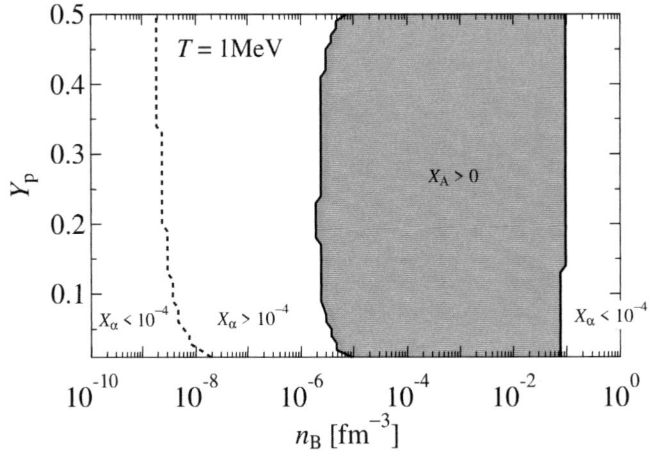

Figure 2. hase diagram of hot nuclear matter at $T = 1$ MeV. The shaded region represents the non-uniform phase where the mass fraction of heavy nuclei is given as X_A. The dashed line is the boundary where the mass fraction of alpha-particles $X_\alpha = 10^{-4}$.

The obtained F/N is shown in Figure 1, together with the results for uniform matter. It is seen that F/N of non-uniform matter becomes lower than F/N for uniform matter in the region $10^{-7}\text{fm}^{-3} \lesssim n_B \lesssim 10^{-1}\text{fm}^{-3}$, and the critical density at which the non-uniform phase disappears is about 10^{-1} fm^{-3} regardless of Y_p.

Figure 2 shows the obtained phase diagram of hot nuclear matter at $T = 1$ MeV. The shaded region represents the non-uniform phase and the dashed line shows the boundary where the mass fraction of alpha-particles $X_\alpha = 10^{-4}$. It is seen that, at $n_B \lesssim 10^{-9}$ fm^{-3}, X_α is negligibly small. As n_B increases, alpha-particles and heavy nuclei appear in nuclear matter. The non-uniform phase appears at a density of $10^{-6} - 10^{-5}\text{fm}^{-3}$, which is higher than the value of $n_B \sim 10^{-7}$ fm^{-3} found in Fig. 1. This discrepancy is due to the mixing of alpha-particles in the uniform phase. In Fig. 1, when we consider the alpha-mixing in uniform matter, the corresponding F/N becomes lower, and the density range for non-uniform matter becomes smaller. The critical density between uniform and non-uniform phases is about 10^{-1} fm^{-3}, which is largely independent of $Y_{\rm p}$, as indicated above. It is also confirmed that the present phase diagram based on the variational EOS is close to that based on the Shen EOS [3]. Here we note that the mass numbers and the proton numbers of heavy nuclei appearing in the non-uniform phase are larger than those in the Shen EOS, though the corresponding data are not explicitly shown in this paper. This difference may be related to the fact that the symmetry energy of the present EOS is smaller than that of the Shen EOS.

4. Summary and concluding remarks

In this paper, we reported the current status of our project to construct a new SN-EOS based on realistic nuclear forces. We first constructed the EOS for uniform asymmetric nuclear matter based on the AV18 and UI potentials using the cluster variational method. Furthermore, the present variational EOS was found to be softer than the Shen EOS in core-collapse SN simulations. Using the free energies for uniform matter, we constructed the EOS for non-uniform matter in the TF approximation. The obtained phase diagram at $T = 1$ MeV is reasonable as compared with that of the Shen EOS. A more systematic study of the EOS for non-uniform matter over a wide range of $n_{\rm B}$, $Y_{\rm p}$, and T, including the chemical composition in nuclear matter, is now in progress, and will be reported elsewhere in the near future [18].

Acknowledgments

The numerical computations in this work were carried out on SR16000 at IT in Kyoto University, on the supercomputer (NEC S 8R) at the Research Center for Nuclear hysics, Osaka University, on the U and SR16000 at the High Energy Accelerator Research Organization (KEK), and on SR16000 at the Information Technology Center of the University of Tokyo. This work is supported by JS S (Nos. 22540296, 23224006, 24244036, 25400275, 26870615), the rant-in-Aid for Scientific Research on Innovative Areas of ME T (Nos. 20105003, 20105004, 20105005, 24105008, 26104006, 26105515), and RIKEN iTHES roject.

References

[1] Lattimer J M and westy F D 1991 *Nucl. Phys. A* **535** 331.
[2] hen H, Toki H, Oyamatsu K and umiyoshi K 1998 *Prog. Theor. Phys.* **100** 1013.
[3] hen H, Toki H, Oyamatsu K and umiyoshi K 2011 *Astrophys. J. Suppl.* **197** 20.
[4] Kanzawa H, Oyamatsu K, umiyoshi K and Takano M 2007 *Nucl. Phys. A* **791** 232.
[5] Kanzawa H, Takano M, Oyamatsu K and umiyoshi K 2009 *Prog. Theor. Phys.* **122** 673.
[6] Togashi H and Takano M 2013 *Nucl. Phys. A* **902** 53.
[7] Akmal A, Pandharipande V R and Ravenhall D G 1998 *Phys. Rev. C* **58** 1804.
[8] Colo G, Giai N V, Meyer J, Bennaceur K and Bonche P 2004 Phys. Rev. C **70** 024307.
[9] Piekarewicz J 2010 J. Phys. G **37** 064038.
[10] teiner A W, Hempel M and Fischer T 2013 Astrophys. J. **774** 17.
[11] Demorest P B, Pennucci T, Ransom M, Roberts M E and Hessels J W T 2010 *Nature* **467** 1081.
[12] Antoniadis J et al. 2013 *Science* **340** 6131.
[13] teiner A W, Lattimer J M and Brown E F 2010 *Astrophys. J.* **722** 33.
[14] chmidt K E and Pandharipande V R 1979 *Phys. Lett. B* **87** 11.
[15] Mukherjee A and Pandharipande V R 2007 *Phys. Rev. C* **75** 035802.
[16] Mukherjee A 2009 *Phys. Rev. C* **79** 045811.
[17] Togashi H, Takano M, umiyoshi K and Nakazato K 2014 *Prog. Theor. Exp. Phys.* **2014** 023D05.
[18] Togashi H et al., in preparation.
[19] Oyamatsu K 1993 *Nucl. Phys. A* **561** 431.

Journal of Physics: Conference Series **569** (2014) 012059 doi:10.1088/1742-6596/569/1/012059

The influence of inelastic neutrino interactions with light clusters on core-collapse supernova simulations

Shun Furusawa, Hiroki Nagakura, Kohsuke Sumiyoshi, Shoichi Yamada

Center for Computational Astrophysics, National Astronomical Observatory of Japan, Mitaka, Tokyo, 181-8588, Japan

E-mail: furusawa@cfca.jp

Abstract. We perform numerical experiments to investigate the influence of inelastic neutrino reactions with light clusters in hot nuclear matter on core-collapse supernova simulations. These interactions have been neglected in most hydrodynamical supernova simulations. The neutrino absorptions and inelastic interactions with deuterons, tritons, helions and alpha particles are taken into account in the hydrodynamical simulations in addition to the ordinary charged-current interactions with nucleons. Axial symmetry is assumed but no equatorial symmetry is imposed. The time evolutions of shock waves are calculated with a simple light-bulb approximation for the neutrino transport and a multi-nuclei equation of state. We show that the heating rates of deuterons reach as high as $\sim$ 10% of those of nucleons around the bottom of the gain region. On the other hand, alpha particles heat the matter near the shock wave, which is important when the shock wave expands and density and temperature of matter become low. It is also found that the models with heating by light clusters have different evolutions from those without it in non-linear evolution phase. The matter in the gain region has various densities and temperatures and there appear regions that are locally rich in deuterons and alpha particles. These results indicate that the inelastic reactions of light clusters, especially deuterons, should be incorporated in the simulations of core-collapse supernovae.

1. Introduction

The mechanism of core-collapse supernovae is not clearly understood at present because of its intricacy. The neutrino driven mechanism is considered to be the most promising scenario for the core-collapse supernova and it is considered the so-called standing accretion shock instability (SASI) and the convection in multi-dimension are helpful to increase the e ciency of neutrino heating. In addition to these hydrodynamical effects, the inelastic neutrino interactions and equation of states (EOS) of nuclear matter are also important for the core-collapse supernovae.

The inelastic interactions between neutrinos and nuclei have been neglected in most hydrodynamical simulations of the neutrino heating phase after the bounce and shock stall. Ohnishi et al. showed that the inelastic neutrino interactions with alpha particles are helpful to revive the shock in 2D simulations if the neutrino luminosity is close to the critical value, which is the threshold for a shock revival [1]. However, the shocked matter is certainly composed not only of nucleons and alpha particles but also of deuterons, tritons and helions [2]. The energy transfer cross sections of deuterons are comparable to those of nucleons and ten times greater than those alpha particles [3].

EOS is another important input physics and its influences on the dynamics of core-collapse supernovae were investigated. There are currently the two EOS's widely used for the simulations of core-collapse supernovae, in which only a single representative nucleus and alpha particles are incorporated to approximate the ensemble of heavy and light nuclei, respectively. We have developed an EOS including a large number of nuclei based on the liquid drop model for heavy nuclei with shell effects and nuclear pasta phases, the quantum approach for light nuclei and relativistic mean field theory for free nucleons [4, 5], This EOS can provide the mass fractions of each light nuclei more reliably.

The aim of this article is to investigate effects of these light nuclei in the new EOS on an aspect of supernova dynamics. We explore the impacts of the inelastic neutrino reactions with light nuclei on the SASI. We perform several hydrodynamical simulations of the post-bounce phase in 1D and 2D. In addition to cooling and heating by nucleons, we incorporate the neutrino heating reactions with light nuclei.

2. Models

The basic set-up of dynamical simulations is the same as that of Ohnishi et al. and Nagakura et al. except for the heating reactions of light nuclei [1, 6, 7]. We perform 2D simulations assuming axial symmetry. Spherical coordinates are used and no equatorial symmetry is assumed. We utilize 300 radial mesh points to cover $r_{in} \leq r \leq r_{out} = 500$ km, where r_{in} is the inner boundary and is set to be the radius of the neutrino sphere of ν_e in the initial state. The 60 angular mesh points are adopted to cover the whole meridian section. The mass of a central object is assumed to be constant and set to be $1.4M_\odot$. Interactions between neutrinos and nucleons are taken into account in the energy equation and the evolution of electron fraction. The calculations of the contributions of nucleons to the energy and electron fraction are done just in the same way as in Ohnishi et al. [6]. The heating rates of deuterons, tritons, helions and alpha particles are also taken into account in the energy equation and they are the new elements in this work.

The heating rates of light nuclei are calculated from the analytic formula [8]. The energy transfer cross sections of deuterons, tritons, tritons alpha particles are available from some references [3, 8, 9, 10, 11]. The cooling reactions involving light nuclei are ignored since the reaction rates are not available at the moment and we focus only the influences of the heating reactions of light nuclei in this paper. The neutrino transport is solved by the simple light bulb approximation. We assume that the temperatures of ν_e, $\bar{\nu_\mathrm{e}}$ and ν_μ emitted from the neutrino spheres are constant and set to be $(T_{\nu_\mathrm{e}}, T_{\bar{\nu}_\mathrm{e}}, T_{\nu_\mu}) = (4, 5, 10)$ in MeV. The luminosities of ν_e and $\bar{\nu_\mathrm{e}}$ are assumed to have the same value: $L_{\nu_\mathrm{e}} = L_{\bar{\nu}_\mathrm{e}} = L$. The luminosity of ν_μ is set to be $L_{\nu_\mu} = 0.5 \times L$.We employ the multi-nuclei EOS, which gives not only thermodynamical quantities but also the abundance of all nuclei [4, 5].

For the first step of the calculations, we prepare the initial conditions of the stars, which are spherical symmetric steady accretion flows [6, 7]. The neutrino inelastic interactions with light nuclei are included in these calculations for the steady states. We start the dynamical simulations, adding the perturbations of 1 , which are proportional to cosθ, to the initial radial velocities. We investigate the influences of light nuclei on the dynamics under different circumstances, varying the luminosity L and mass accretions rate $\dot{M}$. We define the normalized neutrino luminosity $L_{52} \equiv L/(10^{52}\mathrm{erg/s}^{-1})$ and mass accretion rate $\dot{M}_{sun} \equiv -\dot{M}/(M_\odot \mathrm{s}^{-1})$. The detail of the model is described in Furusawa et al. [12].

3. Results

Figure 1 displays the time evolutions of average shock radii for four models, in which all light nuclei, only deuterons, only alpha particles and no light nuclei are taken into account in the heating sources, respectively. The models without deuteron heating do not succeed in the shock revival for $L_{52} = 5.1$, whereas the other two models do though it takes long time. For $L_{52} = 5.2$,

Journal of Physics: Conference Series **569** (2014) 012059 doi:10.1088/1742-6596/569/1/012059

on the other hand, all models produce shock revival. We can see that the heating by deuterons and alpha particles both reduce the time to shock revival. Note that this may be too naive, since the time to shock revival is known to be sensitive to various ingredients such as the initial perturbations when the neutrino luminosity is set to be close to the critical luminosity. We, however, stress that the heating by light nuclei, especially deuterons, brings the clear changes to the evolutions of shock waves in most cases.

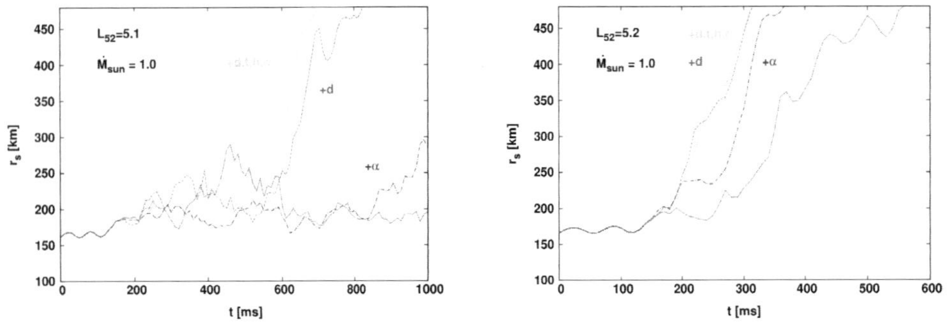

Figure 1. The time evolutions of shock radius of the models with heating by all light nuclei (cyan solid lines), only deuterons (red dashed lines), only alpha particles (blue dashed-dotted lines) and no light nuclei (black dotted lines) for L_{52} = 5.1 and 5.2 with $\dot{M}_{sun}$=1.0.

We now focus on the model including the heating by all light nuclei with $L_{52} = 5.2$ and $\dot{M}_{sun} = 1.0$ to explore the role of light nuclei in the evolution of the shock wave. The shock oscillation grows linearly by t ∼150 ms as shown in Fig. 1. The distributions of nucleons and light nuclei are almost spherically symmetric at t=100 ms. The heating rates of light nuclei are large in the narrow region near the quasi-steady shock wave until t=100 ms. At t=200 and 300 ms, however, the shock waves are deformed and have reached the non-linear regime of SASI. In some regions, the light nuclei are abundant indeed and their heating is e cient accordingly. Figure 2 plots the pairs of (ρ, T) obtained along 5 different radial rays. Although the initial distributions in (ρ, T) plane (the black symbols) are not located in the regions that are rich in light nuclei, the turbulence in the non-linear SASI broadens the distributions. Figure 3 shows the mass fractions and the heating rates of nuclear species along the radial ray with $\theta = 180°$ at t = 200 ms and the one with $\theta = 0°$ at t = 300 ms. The heating rate of deuterons becomes as high as ~ 10 of that of nucleons at t=200 ms around the bottom of the gain region. The shock wave at t = 200 ms moves northwards ($\theta = 0°$) and the matter in the southern ($\theta = 180°$) with low entropies goes down deep into the central regions. The orange in the top panel of Fig. 2 also indicates that the matter in the southern side ($\theta = 180°$) has low entropies at the inner part of the gain regions, resulting in more deuterons than the matter in other parts. At t=300 ms, the shock wave reach at $\sim$ 400 km and the heating by alpha particle is dominant. We can see in the bottom panel of Fig.8 that the matter along the radial ray with $\theta = 0°$ has also lower entropies and as a consequence deuterons and alpha particles are abundant at the regions of high and low densities, respectively. Both at t = 200 and 300 ms, the deuterons have the heating rates comparable to those of nucleons near the bottom of the gain regions. The heating rates of alpha particles are ~ 10 of those of nucleons around the shock wave.

Journal of Physics: Conference Series **569** (2014) 012059 doi:10.1088/1742-6596/569/1/012059

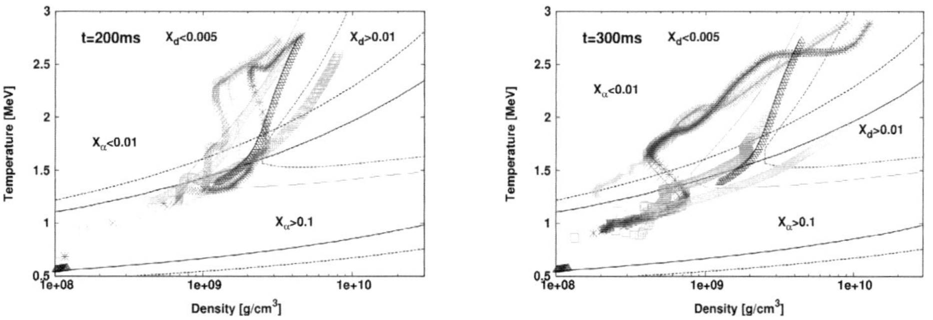

Figure 2. The lines shows contours of mass fractions of deuteron X and alpha particle X_α (red thin lines for $X = 0.005$, red dashed lines for $X = 0.01$, blue dashed lines for $X_\alpha = 0.01$ and blue solid lines for $X_\alpha = 0.1$). The symbols show the densities and temperatures in the gain regions at $t = 200$ and 300 ms for the radial rays $\theta = 0°$ (cyan), 45° (magenta), 90° (green), 135° (brown) and 180° (orange). The black symbols are those of the initial spherical symmetric state ($t = 0$ ms).

4. Discussion

We have investigated the influences of the inelastic interactions of neutrinos with light nuclei on the dynamics in the post-bounce phase of core-collapse supernovae. We have done numerical simulations of SASI with the assumption of axial symmetry for some representative combinations of the luminosity and mass accretion rate. We have not solved the dynamics of the central part of the core and replaced it with the suitable boundary conditions and have started the simulations from spherically symmetric steady state, adding some perturbations to the radial velocity. The neutrino transport has been handled by the simple light-bulb approximation. In addition to the ordinary heating and cooling reactions with nucleons, we have taken into account the heating reactions with four light nuclei for the first time. The abundance of light nuclei is provided by the multi-nuclei EOS together with other thermodynamical quantities.

We have found that the evolutions of shock waves are influenced by the light nuclei heating and the deuterons and alpha particles have different roles in dynamical simulations. We have found that the integrated heating rates of deuterons and alpha particles become high at different phases : the heating rates of deuterons are the highest when the shock wave shrinks and the matter compression is the greatest. The heating by alpha particles has an impact on the shock revival particularly when the shock wave has large radii and the matter becomes of low entropies. The dynamics are more sensitive to the inclusion of the heating by light clusters. because SASI in the non-linear regime make the gain regions more inhomogeneous and there appear the regions that have densities and temperatures favorable for the existence of light nuclei. The heating rates of light nuclei reaches about 10 of that of nucleons locally. As a consequence, the dynamics of the shock revival is influenced by the heating via light nuclei, especially deuterons. Although they are never dominant heating sources, these heating reactions of light nuclei should be included if one were to estimate the critical luminosity and or explosion energy quantitatively accurately.

The numerical simulations in this paper are admittedly of experimental nature and the numbers we have obtained may be subject to change in more realistic simulations. There is also a room for improvement in our experimental computations. We need more systematic investigations, varying not only the neutrino luminosity and mass accretion rate but also the neutrino temperature, mass of a central object and initial perturbation. The cooling reactions

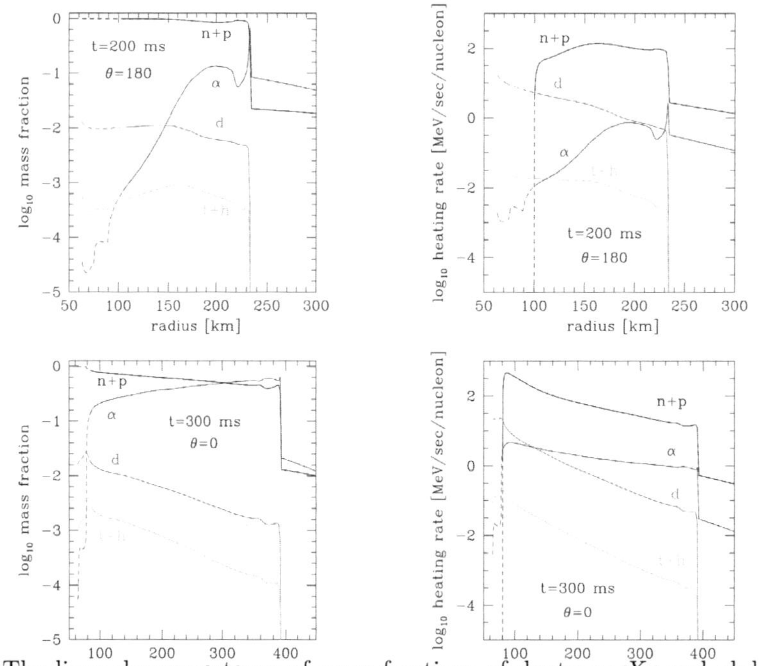

Figure 3. The lines shows contours of mass fractions of deuteron X and alpha particle X_α (red thin lines for X = 0.005, red dashed lines for X = 0.01, blue dashed lines for $X_\alpha = 0.01$ and blue solid lines for $X_\alpha = 0.1$). The symbols show the densities and temperatures in the gain regions at $t = 200$ and 300 ms for the radial rays $\theta = 0°$ (cyan), 45° (magenta), 90° (green), 135° (brown) and 180° (orange). The black symbols are those of the initial spherical symmetric state ($t = 0$ ms).

of light nuclei should be incorporated in the calculations. These issues are currently undertaken and will be reported elsewhere.

A part of the numerical calculations were carried out on Cray C30 and C cluster at Center for Computational Astrophysics, National Astronomical Observatory of Japan.

References

[1] Ohnishi N, Kotake K and Yamada 2007, *Astrophys. J.*, **667**, 375
[2] umiyoshi K and Röpke G 2008, *Phys. Rev.* C, **77**, 055804
[3] Nakamura , umiyoshi K and ato T 2009, *Phys. Rev.* C, **80**, 035802
[4] Furusawa , Yamada , umiyoshi K and uzuki H 2011, *Astrophys. J.*, **738**, 178
[5] Furusawa , umiyoshi K, Yamada and uzuki H 2013, *Astrophys. J.*, **772**, 95
[6] Ohnishi N, Kotake K and Yamada 2006, *Astrophys. J.*, **641**, 1018
[7] Nagakura H, Yamamoto Y and Yamada 2013, *Astrophys. J.*, **765**, 123
[8] Haxton W C 1988, *Phys. Rev. lett.* , **60**, 1999
[9] O Connor E, Gazit D, Horowitz C J, chwenk A and Barnea N 2007, *Phys. Rev.* C, **75**, 055803
[10] Arcones A, Martínez-Pinedo G, O Connor E, chwenk A, Janka H-T, Horowitz C J and Langanke K 2008, *Phys. Rev.* C, **78**, 015806
[11] Gazit D and Barnea N 2004, *Phys. Rev.* C, **70**, 8801
[12] Furusawa , Nagakura H, umiyoshi K and Yamada 2013, *Astrophys. J.*, **774**, 78

Dynamics of nuclear four- and five-body systems with correlated Gaussian method

W. Horiuchi[1] **and Y. Suzuki**[2,3]

[1]Department of Physics, Hokkaido University, apporo 060-0810, Japan
[2]Department of Physics, Niigata University, Niigata 950-2181, Japan
[3]RIKEN Nishina Center, 2-1 Hirosawa, Wako, aitama 351-0198, Japan

E-mail: whoriuchi@nucl.sci.hokudai.ac.jp

Abstract. We report our recent applications of the correlated Gaussian (CG) method to nuclear four- and five-body systems: (I) pin-dipole response functions of ^{4}He and (II) ^{16}O as a ^{12}C$+n+n+p+p$ five-body model. The CG is flexible to describe complex few-nucleon dynamics. The above examples actually demonstrate the power of the CG, giving a simultaneous description of both four-nucleon bound and unbound states using a realistic nuclear force, and both shell- and cluster-configurations in the ground and first excited 0^+ states of ^{16}O.

1. Introduction

Since the wave function has all information of a non-relativistic quantum system, we try to solve the many-body Schrödinger equation as precise as possible within present computational resources. A correlated aussian (C) method is powerful and flexible to describe complex few-body dynamics, e.g., describing different types of structure and correlated motion of the particles, and accurately describing the tail of the wave function in the asymptotic region. The method has been applied to not only in nuclear physics but also in other quantum mechanical fields (See recent review [1]). Here we present our progress and recent applications of the C method to nuclear four- and five-body systems.

In the first topic, we calculate the spin-dipole (SD) response functions of ^{4}He [2] with a full four-body calculation using realistic nuclear forces. The excited states of ^{4}He are all in the continuum and are treated properly with the complex scaling method. The SD operator belongs to the first-forbidden transition induced by the weak interaction and is expected to play a significant role for neutrino-^{4}He reactions. The SD operator can change the spin wave function of the ground state and has three possible multipoles. We show that these properties of the SD operator can be used to probe the role of noncentral forces, especially the tensor force.

The second topic is a ^{12}C$+n+n+p+p$ five-body calculation for the spectrum of ^{16}O [3]. A simultaneous description of the ground and first excited 0^+ states of ^{16}O is one of the outstanding and challenging problems in nuclear theory. The ground state of ^{16}O has been recognized as having a doubly closed shell model configuration, and thus the first excited state is expected to have the negative parity state in accordance with the particle-hole excitation but actually observed as 0^+ state. Many theoretical works are devoted to understand this 'mysterious' state, however, the state has not been reproduced microscopically even in modern large scale calculations [4]. Since a fully microscopic calculation is not practical at present, it is interesting to perform the five-body calculation without assuming an alpha cluster.

The paper is organized as follows: Section 2 briefly explain the C method used in this paper. Two recent applications of the nuclear four- and five-body systems are presented in Section 3. In Section 3.1 the calculation of the SD strength functions are presented. We show the possibility to probe the tensor correlations with the SD excitations. Section 3.2 presents converged results of a ^{12}C+four-nucleon five-body calculation and discusses the coexistence of shell- and cluster-configurations in the spectrum of ^{16}O. Summary and conclusions are given in Section 4.

2. Correlated Gaussian method

A trial wave function in a variational method for nuclear few-body problems has to be flexible enough to cope with the short-range repulsion and strong tensor force as well as different types of structure and correlated motion of the particles. A total wave function with the angular momentum J, its z component M_J, parity π, and isospin quantum numbers T, M_T is expressed as a combination of various components of orbital and spin angular momenta. Here the basis function we use takes a general form in LS coupling scheme

$$\Phi^{\pi}_{(LS)JM_J,TM_T} = \mathcal{A}\,[\phi^{\pi}_{L} \times \chi_S]_{JM_J}\,\eta_{TM_T}, \tag{1}$$

where $\mathcal{A}$ is the antisymmetrizer, and the symbol $[\phi^{\pi}_{L} \times \chi_S]_{JM_J}$ stands for angular momentum coupling. The total spin (isospin) function χ_S (η_T) is constructed by a successive coupling of all spin (isospin) functions of a nucleon.

For the spatial part of the basis function, ϕ^{π}_{L}, we use the C . Let $\boldsymbol{x}$=($\boldsymbol{x}_i$) denote a set of the Jacobi coordinates excluding the center-of-mass coordinate. We express ϕ^{π}_{L} as a combination of C [5, 6] and the two global vectors [7]

$$\phi^{\pi}_{(L_1L_2)LM_L}(A, u_1, u_2) = \exp(-\tilde{\boldsymbol{x}}A\boldsymbol{x})[\mathcal{Y}_{L_1}(\tilde{u}_1\boldsymbol{x}) \times \mathcal{Y}_{L_2}(\tilde{u}_2\boldsymbol{x})]_{LM_L} \tag{2}$$

with a solid harmonic $\mathcal{Y}_{\ell}(\boldsymbol{r}) = r^{\ell}Y_{\ell}(\hat{\boldsymbol{r}})$. Here $\tilde{\boldsymbol{x}}A\boldsymbol{x} = \sum_{i,j} A_{ij}\boldsymbol{x}_i\cdot\boldsymbol{x}_j$ and $\tilde{u}_i\boldsymbol{x} = \sum_k (u_i)_k\boldsymbol{x}_k$, where the positive-definite symmetric matrix A, the vectors of u_1 and u_2 are variational parameters. It is noted that all coordinates are explicitly correlated through them. Both u_1 and u_2 define the global vectors, $\tilde{u}_1\boldsymbol{x}$ and $\tilde{u}_2\boldsymbol{x}$, that are responsible for the rotational motion, as shown in Eq. (2). When a realistic nucleon-nucleon (NN) potential is used, the tensor force mixes different angular momentum components in the wave function. With the two global vectors, any L^{π} states but 0^- can be constructed with a suitable choice of L_1 and L_2. The power of the C basis of type (2) has been demonstrated by many examples [7, 8, 1]. An advantage of the C is that it keeps its functional form under a linear transformation of the coordinates [5, 6], which is a key for describing both cluster and delocalized structure in a unified manner.

3. Application to nuclear four- and five-body systems

3.1. Electroweak response functions of ^{4}He

The Hamiltonian of four-nucleon system consists of two- and three-nucleon forces. As an NN potential, we employ the Argonne $v8'$ [9] (AV8$'$) potential which contain central, tensor and spin-orbit components. The central component includes a strong short-ranged repulsion near the origin. We add a phenomenological three-body interaction (3NF) [10] that is set to reproduce the binding energies of the three- and four-nucleon bound states.

A spin-dipole (SD) response function of interest is given by

$$S(p,\lambda,E) = \mathcal{S}_{f\mu}\,|\langle\Psi_f|\sum_{i=1}^{4}[\boldsymbol{\rho}_i \times \boldsymbol{\sigma}_i]_{\lambda\mu}T_i^p|\Psi_0\rangle|^2\,\delta(E_f - E_0 - E) \quad \text{with} \quad \boldsymbol{\rho}_i = \boldsymbol{r}_i - \boldsymbol{x}_4, \tag{3}$$

where $\boldsymbol{r}_i$ is ith nucleon coordinate, $\boldsymbol{x}_4$ is the center-of-mass coordinate of the four-nucleon system, and $\boldsymbol{\sigma}_i$ is ith nucleon spin. The SD operator is specified by $\lambda\mu$ with the angular momentum coupling of $\boldsymbol{\rho}_i$ and $\boldsymbol{\sigma}_i$. The superscript p of T_i^p distinguishes different types of isospin operators, isoscalar (p =IS) and isovector (p =IV0). The Ψ_0 (Ψ_f) is the ground (final) state of ^{4}He with the energy E_0 (E_f), and $\mathcal{S}_{f\mu}$ denotes a summation over all final states and μ. For the continuum wave function, we use a square-integrable ($\mathcal{L}^2$) basis function which does not satisfy a proper boundary condition of continuum state. To treat it properly, we use the complex scaling method (CSM) which is a widely used method for calculating the strength function in a nuclear system [11]. The accuracy of the CSM calculation crucially depends on how completely the $\mathcal{L}^2$ basis functions are prepared. We attempt at constructing the basis paying attention to two points: the sum rule of the electroweak strength functions and the decay channels [12, 2].

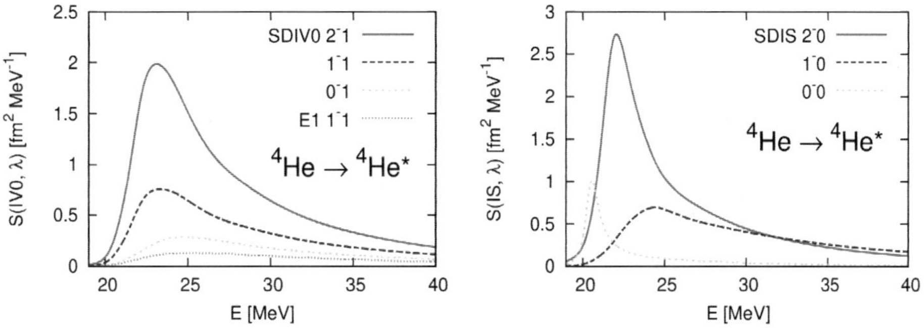

Figure 1. Left: Isovector SD strength functions of IV0 type and $E1$ strength function for ^{4}He as a function of excitation energy. The AV8$'$ + 3NF potential is used. Right: The same plot as the left but isoscalar SD strength functions of IS type. The figures are plotted based on Ref. [2].

The left panel of Fig. 1 plots the SD strength functions of IV0 type with the AV8$'$ + 3NF potential. For the sake of comparison, the electric dipole ($E1$) strength function is also plotted by choosing the $E1$ operator as $\sum_{i=1}^{N} \boldsymbol{\rho}_{i\mu}\frac{1}{2}(1-\tau_z(i))$. As seen in the figure, the three SD strength functions show narrower widths at their peaks than the $E1$ strength function. Moreover their peak positions including the $E1$ case well correspond to the observed excitation energies of the four $T = 1$ negative-parity states of ^{4}He [13].

We display in the right panel of Fig. 1 the IS SD strength functions that reflect the $J^\pi T = \lambda^- 0$ continuum states of ^{4}He. These IS SD strength functions, especially for the 0^- and 2^- cases, show much narrower distribution than the IV0 strength functions. These peak energies again appear to correspond to the observed $T = 0$ negative-parity levels in ^{4}He. A close comparison between the IS and IV0 strength functions indicates that the 0^- case is noteworthy compared to the 1^- and 2^- cases in that the difference in the peak positions of the same J^- becomes much larger. As discussed in Refs. [8, 14], the reason for this is understood by analyzing the role played by the tensor force.

The resonance energy may be identified as the energy where the peak is located. We also estimate the decay width of the resonance by the difference of two excitation energies at which the strength becomes half of the maximum strength at the peak. The agreement between theory and experiment is very satisfactory [2]. The average deviation of the calculated resonance energies from experiment is less than 0.4 MeV for ^{4}He despite the fact that most of their widths are larger than 5 MeV. The estimated width is also in reasonable agreement with experiment.

Here we discuss the non energy-weighted sum rule (NEWSR) for the SD operator. The use of the closure relation enables us to express the NEWSR to the expectation value of the operator

Journal of Physics: Conference Series **569** (2014) 012060 doi:10.1088/1742-6596/569/1/012060

$\sum_\mu \mathcal{O}^{p\dagger}_{\lambda\mu}\mathcal{O}^{p}_{\lambda\mu}$ with respect to the ground state Ψ_0. It is convenient to express that operator as a scalar product of the space-space and spin-spin tensors, $\mathcal{Q}^p_{(\kappa)0} = \sum_{i,j=1}^{A} \left([\boldsymbol{\rho}_i \times \boldsymbol{\rho}_j]_\kappa \cdot [\boldsymbol{\sigma}_i \times \boldsymbol{\sigma}_j]_\kappa\right) T^{p\dagger}_i T^p_j$, where the rank κ can be 0, 1, and 2. With use of the above equation, the expression of the NEWSR is

$$m_0(p,\lambda) \equiv \int_0^\infty S(p,\lambda,E)dE = \sum_{\kappa=0}^{2} U_{\lambda\kappa}\langle \mathcal{Q}^p_{(\kappa)0}\rangle. \tag{4}$$

The NEWSR is fully satisfied: the deviation of the NEWSR from the right hand side of Eq. (4) is less than 1 . This means that the basis functions prepared for the description of the SD excitation are su cient enough to account for all the strength in the continuum. Among the three expectation values of $\langle \mathcal{Q}^p_{(\kappa)0}\rangle$ in Eq. (4), the $\kappa = 0$ term gives a dominant contribution to the NEWSR because the major component of the ground state of ^{4}He is $S = 0$ and it has a non-vanishing expectation value only for $\mathcal{Q}^p_{(0)0}$. In this limiting case $m_0(p,\lambda)$ is proportional to $U_{\lambda 0}$. Therefore the λ-dependence of the NEWSR turns out to be $1:3:5$ for $\lambda = 0, 1, 2$, independently of p. The deviation from this ratio is due to the contributions of other $\mathcal{Q}^p_{(\kappa)0}$ terms, especially the $\kappa = 2$ term. Since the admixture of the $S = 2$ component is primarily determined by the tensor force, the deviation reflects the tensor correlations in the ground state. The term with $\kappa = 2$ is particularly interesting because it contains the tensor operator characteristic of the one-pion-exchange potential. Since $\langle \mathcal{Q}^p_{(2)0}\rangle$ is negative for $p =$ IS, the ratio further increases from $1:3:5$, which is approximately $1:4:7$, whereas it is positive for $p =$ IV0 and the ratio approximately reduces to $1:2:4$.

As discussed above, $\langle \mathcal{Q}^p_{(\kappa)0}\rangle$ plays a central role to determine the NEWSR for the SD strength functions. Inverting the right hand side of Eq. (4) makes it possible to express $\langle \mathcal{Q}^p_{(\kappa)0}\rangle$ as a sum, over the multipole λ, of the NEWSR, $\langle \mathcal{Q}^p_{(\kappa)0}\rangle = \sum_{\lambda=0}^{2} U^{-1}_{\kappa\lambda} m_0(p,\lambda)$, where U^{-1} is the inverse matrix of U. If the NEWSR for all λ are experimentally measured, the above equation indicates that $\langle \mathcal{Q}^p_{(\kappa)0}\rangle$ for all κ can be determined from experiment.

3.2. A ^{12}C+$n+n+p+p$ five-body model for ^{16}O

Let us move to the ^{16}O problem. A five-body system we consider here is characterized by the Hamiltonian which consists of an NN potential V_v for valence nucleons and a ^{12}C-nucleon (CN) potential V_{cv}. A central Minnesota (MN) potential [15] is employed as the NN potential. To fine tune the binding energy of the α particle, the potential strengths are multiplied by 0.9814. Symmetrized Woods-Saxon and its derivative forms are assumed for the CN potential, and their parameters are determined to reproduce the low-lying spectrum of ^{13}C with $1/2^-$, $1/2^+$, and $5/2^+$. Our CN potential is deep enough to accommodate some redundant or auli-forbidden states. To eliminate such states we impose the orthogonality constraint for the relative motion of the valence nucleons, which is practically achieved by adding a pseudo potential to the Hamiltonian [16]. Here we assume the harmonic oscillator (HO) wave functions of $0s_{1/2}$ and $0p_{3/2}$ as the occupied orbits of ^{12}C. The HO frequency is set to be 16.0 MeV, which reproduces the size of the ^{12}C ground state. The core excitation is ignored in this calculation.

We express the wave function in terms of a linear combination of many C states of Eq. (2). Each basis element contains so many variational parameters that discretizing them on grids leads to an enormous dimension of at least 10^{10}. Thus we test a number of candidate bases with the stochastic variational method [5, 6], choose the best one among them and increase the basis dimension one by one until a convergence is reached. This procedure costs expensively for computer time but no other viable methods are at hand to get converged solutions for the present problem.

The left panel of Fig. 2 displays the calculated energy curves of the ground (0_1^+) and first excited (0_2^+) states of ^{16}O as a function of the number of basis functions. Convergence is very slow and more than 9000 basis states are needed. This number is very large compared to a four-nucleon system with the same MN interaction that requires only few tens of basis states [5]. The most basis states are used for eliminating the forbidden states. This is because, when the valence nucleons are strongly correlated, the basis states may have a large overlap with the forbidden states, and thus more basis states are needed to get a converged solution. After the convergence two 0^+ states appear below ^{12}C+α threshold and their energies are both remarkably close to experiment.

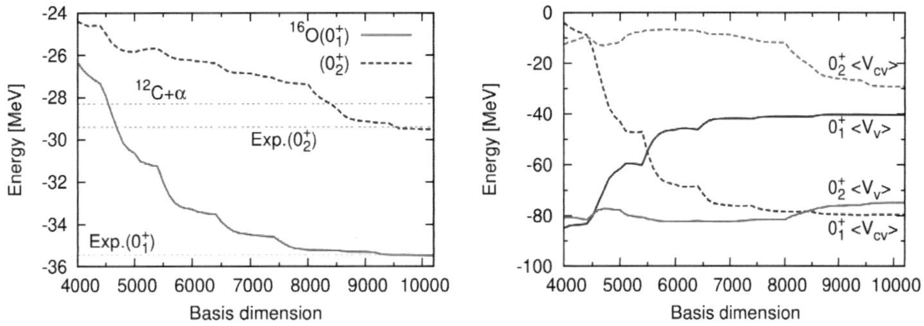

Figure 2. Left: Energy curves for the ground and excited 0^+ states of ^{16}O. The calculated ^{12}C+α thresholds and experimental energies are plotted with thin lines. The figure is plotted based on the data of Ref. [3]. Right: Expectation values of V_{cv} and V_v of the two 0^+ states of ^{16}O as a function of the number of basis dimensions.

We found that the total energies for the ground and first excited 0^+ states of ^{16}O are in good agreement with experiment. The decomposition tells us information on the correlated structure of the valence nucleons. The right panel of Fig. 2 plots the expectation values of V_{cv} and V_v as a function of the number of basis dimensions. The energy curve for the 0_1^+ state changes drastically at 4000 to 6000 basis states where the energy crosses over the ^{12}C+α threshold. We expect a phase transition from the cluster structure to shell-model like structure at this basis dimension. Beyond 8000 basis states where the energies are close to the convergence, the $\langle V_{cv}\rangle$ value turns out to be dominant. In the 0_1^+ state, the CN potential gives the major contribution to the energy $\langle V_{cv}\rangle = -79.55$ MeV, whereas $\langle V_v\rangle$ is approximately a half of it (-40.41 MeV). It is found that the $\langle V_v\rangle$ value is approximately a half of the NN contribution of the free alpha particle obtained with the MN potential (-85.22 MeV). The alpha particle is influenced and strongly distorted by the attraction, kinetic energy, and auli constraint from the core nucleus.

In the case of the 0_2^+ state, in contrast to the ground state, $\langle V_v\rangle$ is large, keeping a constant value shown in the right panel of Fig. 2. The converged $\langle V_{cv}\rangle$ is small -29.22 MeV compared to the contribution of the valence nucleons $\langle V_v\rangle = -74.92$ MeV, which is very close to that of the free alpha particle. This suggests that the 0_2^+ state has a well developed ^{12}C+α structure as predicted by the cluster model [17]. It is interesting to note that $\langle V_{cv}\rangle$ of the 0_1^+ state is similar to $\langle V_v\rangle$ of the 0_2^+ state. Due to the balance of the core-nucleon and the valence nucleons contributions, the two 0^+ states appear at closely to each other, and thus the two different aspects coexist in the spectrum of ^{16}O.

For more quantitative discussion of the clustering degree, we calculate the ^{12}C+α

spectroscopic amplitude of ^{16}O defined by

$$y(r) = \frac{1}{r^2} \left\langle \phi_\alpha \phi_{I=0}(^{12}\mathrm{C}) \delta(R - r) Y_{00}(\hat{\boldsymbol{R}}) \middle| \Psi_{00}(^{16}\mathrm{O}) \right\rangle, \tag{5}$$

where $\Psi_{00}(^{16}\mathrm{O})$, $\phi_{I=0}(^{12}\mathrm{C})$, and ϕ_α are the wave functions of ^{16}O, ^{12}C core, and the free alpha particle obtained by a four-body calculation with the MN potential, respectively; and $\boldsymbol{R}$ is a distance operator between the center-of-mass of the alpha and ^{12}C. The left panel of Fig. 3 displays the calculated spectroscopic amplitudes for the two 0^+ states of ^{16}O. At the short distances, we see some oscillations due to auli principle of ^{12}C+α. For the 0_1^+ state, the highest peak is located at around 4 fm. That is smaller than the touching distance of ^{12}C+α ($\sim$ 5.9 fm). In such a case, the alpha particle may be distorted by the core nucleus. In fact, a spectroscopic factor $S_\alpha^2 = \int_0^\infty (ry(r))^2 dr$ is small (0.105), whereas the ^{12}C+α cluster model calculation gives 0.300 [17]. In this study, we do not assume the alpha cluster, and thus the distortion of the alpha cluster shows up naturally. In the 0_2^+ state, the amplitude is much larger and peaked at the touching distance of ^{12}C+α, showing very long tail which suggests the well developed cluster structure. The spectroscopic factor is 0.680 which is in agreement with the value of the ^{12}C+α calculation, 0.679. The phase space of the first excited state of ^{16}O is exhausted by the ^{12}C+α cluster component.

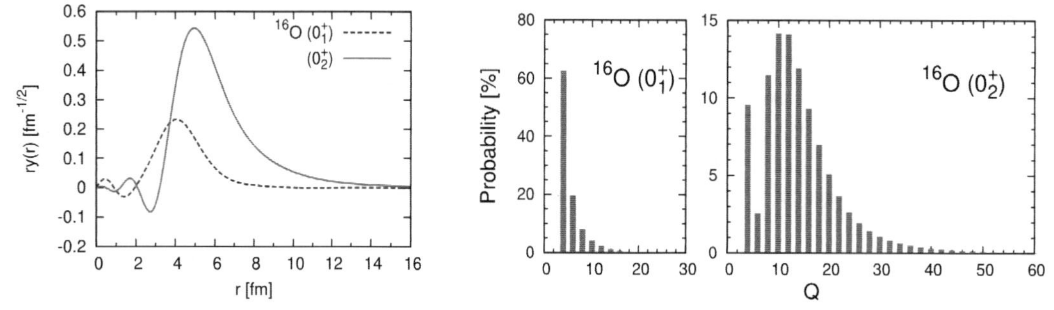

Figure 3. Left: Spectroscopic amplitude of the ground and excited 0^+ states of ^{16}O. Right: Decomposition of the ground and first excited states of ^{16}O wave functions into components having the harmonic oscillator quanta Q. The figures are plotted based on the data of Ref. [3].

Why is it so di cult to describe the 0_2^+ state with the ordinary shell model truncation? To answer this, it is useful to calculate the probability distribution of the harmonic oscillator (HO) quanta Q in our wave functions. The right panel of Fig. 3 plots the histogram of the HO quanta for the ground and first excited states of ^{16}O. Oscillator frequency is set to be the same as the forbidden states in the ^{12}C core. For the ground state, the distribution starts with $Q = 4$, which may correspond to the configuration of $(0p_{1/2})_\nu^2(0p_{1/2})_\pi^2$. The distribution diminishes rapidly with increasing Q. Since a few major shell configurations are required to cover more than 90 of the model spaces, the state may be described with the shell model picture.

As predicted by the cluster model [17], the distribution of the first excited state of ^{16}O is quite different from these states. Since the ground state exhausts the $Q = 4$ component, the component for the excited state becomes small approximately 10 . The HO quanta are again peaked at $Q = 12$ and widely distributed showing still undiminished probability at $Q = 30$. If we can cover the HO model space up to the second peak position $Q = 12$, accumulated probability is still 47 . Therefore, it is practically di cult to describe such a state by the HO basis with the usual major shell truncation. One needs at least $Q = 26$ to cover more than 90 of our model space which is not feasible at present.

4. Summary and conclusions

Recent applications to nuclear four- and five-body problems have demonstrated the power of the explicitly correlated aussian (C) method. Once the Hamiltonian is determined, the wave function of the system is accurately obtained and the model has no adjustable parameter.

First, the spin-dipole (SD) responses of ^{4}He are investigated based on the four-body calculation using a realistic nuclear force. Both the wave function of the ground state and the SD excitation are described accurately using the C . The continuum states including two- and three-body decay channels are described with aid of the complex scaling method.

Experimental data that can directly be compared to the calculation are presently only the resonance parameters of the negative-parity levels of $A = 4$ nuclei. We find that peak position and width of the SD strength function corresponds well with experimental resonant energy and decay width, respectively. The non energy-weighted sum rule (NEWSR) of the SD strength function is investigated by relating it to the expectation values of three scalar products of the space-space and spin-spin tensors with respect to the ground state of ^{4}He. It turns out that our model space satisfies the NEWSR for each SD operator perfectly. The tensor operator of rank 2, $\mathcal{Q}^p_{(2)0}$, is sensitive to the D-state correlation in the ground state induced by the tensor force, and it is mainly responsible for distorting the ratio of the NEWSRs for the multipolarity $\lambda = 0, 1, 2$ from the uncorrelated ratio of $1 : 3 : 5$. An experimental observation of this ratio is desirable since it may lead us to reveal the degree of tensor correlations in the ground state.

In the second topic, we successfully describe the ground and first excited 0^+ states of ^{16}O in a single scheme, a ^{12}C+four-nucleon model. The ^{12}C-nucleon and NN potentials are chosen to reproduce low-energy properties of each subsystem. We solve the full five-body Schödinger equation by superposing many C states prescribed with the stochastic variational method.

The calculated total binding energies for the ground state and first excited states of ^{16}O are obtained consistently with the experimental values. The ground state exhibits a shell model like structure. However, our analyzes suggest the first excited states has a well developed ^{12}C+α cluster structure. The probability distributions of the harmonic oscillator quanta also support our pictures and show the di culty to describe the 0^+_2 state with the ordinary major shell truncation of the harmonic oscillator basis. ossible and promising application is to extend our five-body model with heavier doubly closed shell core, that is, for example, ^{20}Ne ^{44,52}Ti, and ^{212} o as is an interesting analogy to the ^{16}O problem.

Acknowledgments

This work was in part supported by JS S KAKENHI rants Nos. 21540261, 24550261, and 25800121.

[1] Mitroy J *et al.* 2013 *Rev. Mod. Phys.* **85** 693-749
[2] Horiuchi W and uzuki Y 2013 *Phys. Rev.* C **87** 034001
[3] Horiuchi W and uzuki Y 2014 *Phys. Rev.* C **89** 011304(R)
[4] Maris P, Vary J P and hirokov A M 2009 *Phys. Rev.* C **79** 014308
[5] Varga K and uzuki Y 1995 *Phys. Rev.* C **52** 2885
[6] uzuki Y and Varga K 1998 *Stochastic Variational Approach to Quantum-Mechanical Few-Body Problems, Lecture Notes in Physics Monographs M54 (Springer, Berlin)*
[7] uzuki Y, Horiuchi W, Orabi M and Arai K 2008 *Few-Body Syst.* **42** 33
[8] Horiuchi W and uzuki Y 2008 *Phys. Rev.* C **78** 034305
[9] Pudliner B , Pandharipande V R, Carlson J, Pieper C and Wiringa R B 1997 *Phys. Rev.* C **56** 1720
[10] Hiyama E, Gibson B F and Kamimura M 2004 *Phys. Rev.* C **70** 031001(R)
[11] Aoyama , Myo T, Katō K and Ikeda K 2006 *Prog. Theor. Phys.* **116** 1
[12] Horiuchi W, uzuki Y and Arai K 2012 *Phys. Rev.* C **85** 044606
[13] Tilley D R, Weller H R and Hale G M 1992 *Nucl. Phys.* **A 541** 1
[14] Horiuchi W and uzuki Y 2013 *Few-Body Syst.* **54** 2407
[15] Thompson D R, LeMere M and Tang Y C 1977 *Nucl. Phys.* **A 286** 53
[16] Kukulin V I and Pomerantsev V N 1978 *Ann. Phys. (N.Y.)* **111** 30
[17] uzuki Y 1976 *Prog. Theor. Phys.* **55** 1751; **56** 111

Emergence of cluster structures and collectivity within a no-core shell-model framework

K. D. Launey[1], A. C. Dreyfuss[1], J. P. Draayer[1], T. Dytrych[1], R. Baker[1]

[1]Department of Physics and Astronomy, Louisiana State University, Baton Rouge, LA 70803, USA

Abstract. An innovative symmetry-guided concept, which capitalizes on partial as well as exact symmetries that underpin the structure of nuclei, is discussed. Within this framework, *ab initio* applications of the theory to light nuclei reveal the origin of collective modes and the emergence a simple orderly pattern from first principles. This provides a strategy for determining the nature of bound states of nuclei in terms of a relatively small fraction of the complete shell-model space, which, in turn, can be used to explore ultra-large model spaces for a description of alpha-cluster and highly deformed structures together with the associated rotations. We find that by using only a fraction of the model space extended far beyond current no-core shell-model limits and a long-range interaction that respects the symmetries in play, the outcome reproduces characteristic features of the low-lying 0^+ states in ^{12}C (including the elusive Hoyle state and its 2^+ excitation) and agrees with *ab initio* results in smaller spaces. This is achieved by selecting those particle configurations and components of the interaction found to be foremost responsible for the primary physics governing clustering phenomena and large spatial deformation in the ground-state and Hoyle-state rotational bands of ^{12}C. For these states, we offer a novel perspective emerging out of no-core shell-model considerations, including a discussion of associated nuclear deformation, matter radii, and density distribution. The framework we find is also extensible to negative-parity states (e.g., the 3_1^- state in ^{12}C) and beyond, namely, to the low-lying 0^+ states of ^{8}Be as well as the ground-state rotational band of Ne, Mg, and Si isotopes. The findings inform key features of the nuclear interaction and point to a new insight into the formation of highly-organized simple patterns in nuclear dynamics.

1. Introduction

Approximate symmetries in atomic nuclei that favor large deformation in nuclear dynamics have been long recognized [1, 2, 3, 4, 5, 6, 7, 8, 9, 10], and only recently confirmed in the *ab initio* symmetry-adapted no-core shell model (SA-NCSM) [11]. This study unveiled the orderly patterns associated with such symmetries in nuclear wavefunctions from first principles, without *a priori* symmetry constraints. These symmetries are utilized and further understood in the framework of the microscopic no-core symplectic model (NCSpM) [12], which combines a long-range many-nucleon interaction that respects the symmetries in play together with a symmetry-mixing realistic nucleon-nucleon interaction derived from meson exchange theory (e.g., CD-Bonn [13]), chiral effective field theory (e.g., N^{3}LO [14]), or J-matrix inverse scattering theory (JISP16 [15]). The findings of the present study point to a new insight, namely, understanding the mechanism and the primary physics responsible for the emergence of such simple structures from a no-core shell-model perspective.

One of the most successful particle-driven models is the *ab initio* no-core shell model (NCSM) [16], which can accommodate any type of inter-nucleon interaction, including modern two- and three-nucleon realistic interactions. Specifically, for a general problem, the NCSM adopts the intrinsic non-relativistic nuclear plus Coulomb interaction Hamiltonian defined as follows:

$$H = T_{\rm rel} + V_{NN} + V_{NNN} + \ldots + V_{\rm Coulomb}, \tag{1}$$

where the V_{NN} nucleon-nucleon and V_{NNN} 3-nucleon interactions are included along with the $V_{\rm Coulomb}$ Coulomb interaction between the protons. The Hamiltonian may include additional terms such as multi-nucleon interactions among more than three nucleons simultaneously and higher-order electromagnetic interactions such as magnetic dipole-dipole terms. It adopts the harmonic oscillator (HO) single-particle basis characterized by the $\hbar\Omega$ oscillator strength and retains many-body basis states of a fixed parity, consistent with the Pauli principle, and limited by a cutoff $N_{\rm max}$. The $N_{\rm max}$ cutoff is defined as the maximum number of HO quanta allowed in a many-body basis state above the minimum for a given nucleus. It divides the space in "horizontal" HO shells and is dictated by particle-hole excitations (this is complementary to the NCSpM, which divides the space in vertical slices selected by collectivity-driven rules). It seeks to obtain the lowest few eigenvalues and eigenfunctions of the Hamiltonian (1). The NCSM has achieved remarkable descriptions of low-lying states from the lightest nuclei up through ^{12}C, ^{16}O, and ^{14}F [16, 17], and is further augmented by several techniques, such as NCSM/RGM [18], Importance Truncation NCSM [19] and Monte Carlo NCSM [20]. This supports and complements results of other first-principle approaches, such as Green's function Monte Carlo (GFMC) [21], Coupled-cluster (CC) method [22], In-Medium SRG [23], and Lattice Effective Field Theory (EFT) [24].

However, using the established NCSM methods is not enough to reach the physics regime necessary for a description of largely deformed nuclear states, such as the ^{12}C Hoyle state that was predicted based on observed abundances of heavy elements in the universe [25], and which has attracted much recent attention both in theory (e.g., see [26, 27, 24]) and experiment ([28, 29, 30, 31, 32, 33, 34, 35]). In this study, we address this problem, within a no-core shell-model framework, by utilizing a small subset of symplectic Sp(3, $\mathbb{R}$) basis states [6, 7] (with the complete set yielding results equivalent to those of the NCSM), an Sp(3, $\mathbb{R}$)-preserving part of the long-range inter-nucleon interaction [12], together with the bare JISP16 interaction.

2. Collectivity and clustering in light nuclei

2.1. Orderly patterns unveiled in the ab initio SA-NCSM

The next-generation *ab initio* symmetry-adapted no-core shell model (SA-NCSM) [11] combines the first-principle concept of the NCSM with symmetry-guided considerations of the collectivity-driven models. For the first time, we show the emergence – in the framework of the SA-NCSM from first principles – of orderly patterns that favor large deformation/low spin in nuclear wavefunctions [11]. These patterns are linked to the SU(3) and its embedding symplectic Sp(3, $\mathbb{R}$) symmetry[1]. For example, the *ab initio* $N_{\rm max} = 8$ SA-NCSM results with the bare N^{3}LO (similarly, for JISP16) realistic interaction for the ^{8}Be 0^+ ground state (*g.st.*) and its rotational band reveal the dominance of the $0\hbar\Omega$ component with the foremost contribution coming from the leading (4 0) $S = 0$ irrep (Fig. 1a). Furthermore, we find that important SU(3) configurations are then organized into structures with Sp(3, $\mathbb{R}$) symplectic symmetry, that is, the (4 0) symplectic irrep gives rise to (6 0), (4 1), and (2 2) configurations in the $2\hbar\Omega$ subspace and so on, and those configurations indeed realize the major components of the wavefunction in this subspace. Similar results are observed for other p-shell nuclei, such as ^{6}Li, ^{6}He, and ^{12}C. This further confirms the significance of the symplectic symmetry to nuclear dynamics.

[1] Deformation-related $(\lambda\,\mu)$ quantum numbers of SU(3) can be linked to nuclear shapes, e.g., (0 0), (λ 0), and (0 μ) describe spherical, prolate, and oblate shapes, respectively

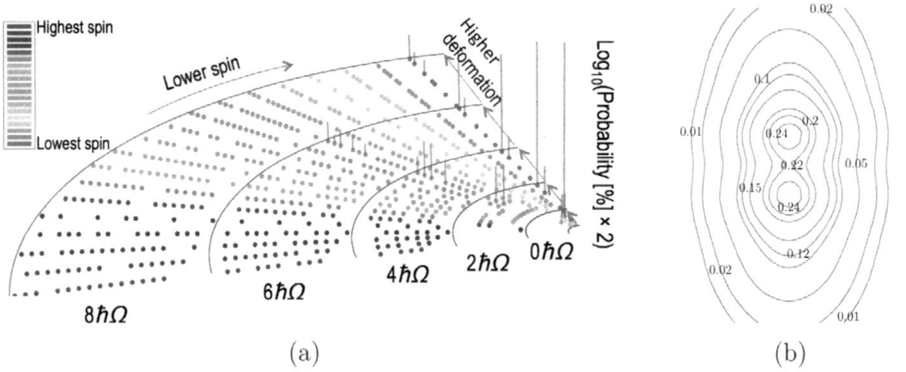

Figure 1. (a) Probability distribution across Pauli-allowed intrinsic spin and deformation configurations for the *ab initio* SA-NCSM 0^+ *g.st.* of ^{8}Be for $N_{\max} = 8$ and $\hbar\Omega = 25$ MeV with the N^{3}LO interaction. The concentration of strengths to the far right demonstrates the dominance of collectivity in the calculated eigenstates. This novel feature enables the SA-NCSM – by using symmetry-dictated subspaces – to reach new domains currently inaccessible by *ab initio* calculations, such as isotopes of Ne, Mg, and Si. (b) NCSpM matter-density profile of ^{8}Be for $N_{\max} = 20$ and $\hbar\Omega = 20$ MeV [with H of Eq. (2) and JISP16 turned off].

The significance of the symplectic Sp(3, $\mathbb{R}$) symmetry for a microscopic description of a quantum many-body system of interacting particles has been recognized by Rowe and Rosensteel [6, 7]. Indeed, the 21 generators of Sp(3, $\mathbb{R}$) are directly related to the particle momentum ($p_{s\alpha}$) and coordinate ($r_{s\alpha}$) operators, with $\alpha = x$, y, and z for the 3 spatial directions and s labeling an individual nucleon, and realize important observables. Namely, Sp(3, $\mathbb{R}$)-preserving operators include: the many-particle kinetic energy $\sum_{s,\alpha} p_{s\alpha}^2/2m$, the HO potential, $\sum_{s,\alpha} m\Omega^2 r_{s\alpha}^2/2$, the mass quadrupole moment $Q_{(2\mathcal{M})} = \sum_s q_{(2\mathcal{M})s} = \sum_s \sqrt{16\pi/5} r_s^2 Y_{(2\mathcal{M})}(\hat{\mathbf{r}}_s)$, and angular momentum L operators, together with multi-shell collective vibrations and vorticity degrees of freedom for a description from irrotational to rigid rotor flows. The symplectic Sp(3, $\mathbb{R}$) symmetry underpins the symplectic shell model, which provides a microscopic formulation of the Bohr-Mottelson collective model [1] and is a multiple oscillator shell generalization of the successful Elliott SU(3) model [2]. The symplectic model with Sp(3, $\mathbb{R}$)-preserving interactions[2] has achieved a remarkable reproduction of rotational bands and transition rates without the need for introducing effective charges, while only a single Sp(3, $\mathbb{R}$) configuration is used [7, 9]. A shell-model study in a symplectic basis that allows for mixing of Sp(3, $\mathbb{R}$) configurations due to pairing and non-degenerate single-particle energies above a ^{16}O core [8] has found that using only seven Sp(3, $\mathbb{R}$) configurations is sufficient to achieve a remarkable reproduction of the ^{20}Ne energy spectrum as well as of $E2$ transition rates without effective charges.

2.2. Formation of clusters in ^{12}C in the hybrid NCSpM model

In this study, a fully microscopic no-core symplectic shell model, NCSpM, is explored to study low-lying 0^+ states and unnatural-parity states in ^{12}C. As a new feature, the model utilizes a hybrid interaction, namely, a schematic many-nucleon $Q \cdot Q$-type interaction [12] – which is Sp(3, $\mathbb{R}$)-preserving – plus the bare JISP16 nucleon-nucleon interaction – which is symmetry-breaking. The JISP16 has been successfully employed in *ab initio* NCSM calculations for the ^{12}C *g.st.* rotational band [36]. And while the coupling constants of the two-body part of the

[2] An important Sp(3, $\mathbb{R}$)-preserving interaction is $\frac{1}{2}Q \cdot Q = \frac{1}{2}\sum_s q_s \cdot (\sum_t q_t)$, as this realizes the physically relevant interaction of each particle with the total quadrupole moment of the nuclear system.

schematic interaction can be tied to realistic nucleon-nucleon interactions, we include many-nucleon interactions of a simple form ($e^{-\gamma Q\cdot Q}$) controlled by a parameter γ, the only adjustable parameter in the model. The study has revealed the importance of ultra-large shell-model spaces that are imperative to provide a successful description of the ^{12}C Hoyle state rotational band and the 3^- state, as well as low-lying states in nuclei from Be (Fig. 1b) to Si [12, 37].

The NCSpM employed within a full model space up through a given $N_{\max}$ coincides with the NCSM for the same $N_{\max}$ cutoff. However, in the case of the NCSpM, the symplectic irreps divide the space into 'vertical slices' that are comprised of basis states of a definite deformation $(\lambda\,\mu)$. Hence, the model space can be reduced to only a few important configurations that are chosen among all possible Sp(3, $\mathbb{R}$) irreps within the $N_{\max}$ model space. The NCSpM, while selecting the most relevant symplectic configurations, is employed to provide shell model calculations beyond current NCSM limits, namely, up through $N_{\max} = 20$ for ^{12}C, the model spaces we found sufficient for the convergence of results [12].

We employ a microscopic Hamiltonian given as,

$$H = T_{\rm rel} + V_{NN} + V_{mN}^{\rm eff} + \ldots + V_{\rm Coulomb}, \tag{2}$$

where $T_{\rm rel}$ is the relative kinetic energy. The V_{NN} is taken to be the bare JISP16 nucleon-nucleon interaction, which is turned on only among bandheads of symplectic irreps, introducing "horizontal" mixing of all the states (up through the $N_{\max}$ cutoff) within the symplectic vertical slices. The $V_{mN}^{\rm eff}$ effective many-nucleon interaction is taken to be,

$$V_{mN}^{\rm eff} = V_0 + \frac{\chi}{2}\frac{\left(e^{-\gamma(Q.Q-\langle Q.Q\rangle_n)} - 1\right)}{\gamma}, \tag{3}$$

which simplifies to the two-body interaction, $V_0 - \frac{\chi}{2}(Q.Q - \langle Q.Q\rangle_n)$, in the $\gamma \to 0$ limit. The spherical HO potential, $V_0 = \sum_{i=1}^{A} \frac{m\Omega^2 \mathbf{r}_i^2}{2}$, and the $Q\cdot Q$ quadrupole-quadrupole interaction directly follow from the second and third term, respectively, in the long-range expansion of any two-body central force, e.g., like the Yukawa radial dependence, $V^{(2)} = \sum_{i<j} V(r_{ij}/a) = \sum_{i<j}(\xi_0 + \xi_2 r_{ij}^2/a^2 + \xi_2 r_{ij}^4/a^4 + \ldots)$ [38], for a range parameter a. The $Q\cdot Q$ interaction is not restricted to a single shell. In addition, its average contribution, $\langle Q\cdot Q\rangle_n$, within an $n\hbar\Omega$ subspace is removed [39], where $\langle Q\cdot Q\rangle_n$ is the trace of $Q\cdot Q$ divided by the space dimension for a fixed number n of HO excitation quanta. This helps eliminate a considerable effect of $Q\cdot Q$ on the HO shell structure, while retaining the $Q\cdot Q$-driven behavior of the wavefunctions. This Hamiltonian in its zeroth-order approximation (for parameter $\gamma \to 0$) and for a valence shell goes back to the established Elliott model. We take the coupling constant χ to be proportional to $\hbar\Omega$ and, to leading order, to decrease with the total number of HO excitations, as shown by Rowe [40] based on self-consistent arguments.

As the interaction and the model space are carefully selected to reflect the most relevant physics, the outcome reveals a quite remarkable agreement with the experiment (Fig. 2a). The low-lying energy spectrum and eigenstates for ^{12}C were calculated using the NCSpM with H of Eq. (2) for $\hbar\Omega = 18$ MeV given by the empirical estimate $\approx 41/A^{1/3}$ (e.g., see [1]). The results are shown for $N_{\max} = 20$, which we found sufficient to yield convergence. This $N_{\max}$ model space is further reduced by selecting the most relevant symplectic irreps, namely, five symplectic bandheads and all symplectic multiples thereof up through $N_{\max} = 20$ of total dimensionality of about 10^4. The five bandheads are, for positive-parity states, the spin-zero ($S = 0$) 0$\hbar\Omega$ 0p-0h (0 4), 2$\hbar\Omega$ 2p-2h (6 2), and 4$\hbar\Omega$ 4p-4h (12 0) symplectic bandheads together with the $S = 1$ 0$\hbar\Omega$ 0p-0h (1 2), and for negative-parity states, the 1p-1h (3 3) bandhead. In comparison to experiment (Fig. 2a), the outcome reveals that the lowest 0^+, 2^+, and 4^+ states of the 0p-0h symplectic slices calculated with JISP16 and $V_{mN}^{\rm eff}$ with $\gamma = -1.71 \times 10^{-4}$ closely reproduce the

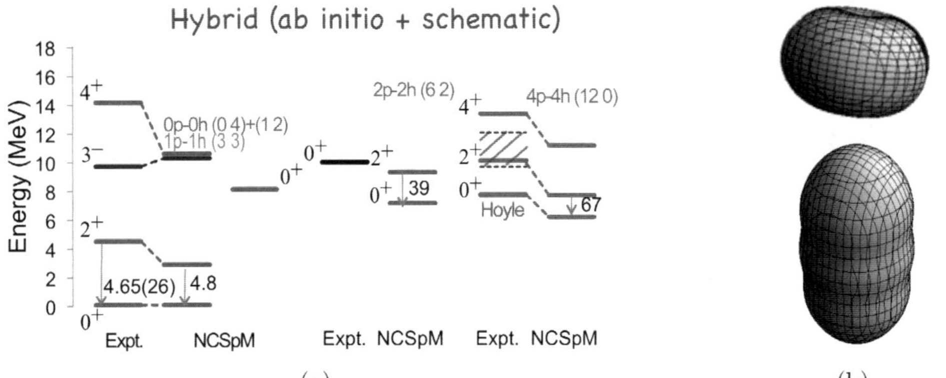

(a) (b)

Figure 2. (a) ^{12}C energy spectrum calculated by the NCSpM using $V^{\rm eff}_{mN}$ with $\gamma = -1.71\times10^{-4}$ and the JISP16 in an $N_{\rm max} = 20$ model space and compared to experiment ("Expt."). Experimental data is from [41], except the latest results for 0^+_3 [31] and the states above the Hoyle state, 4^+ [42] and 2^+ [34] (with a shaded area showing the energy range from [29, 30, 31, 32, 33]). $B(E2)$ transition rates are in W.u. units. (b) NCSpM matter-density profile (with respect to the intrinsic frame) of the ^{12}C ground state (top) and the Hoyle state (bottom) showing the formation of three clusters in the Hoyle state within a no-core shell-model framework.

g.st. rotational band, indicating clear oblate shapes (Fig. 2b, top). In addition, the calculated lowest states of the $4\hbar\Omega$ 4p-4h (12 0) slice are found to lie close to the Hoyle-state rotational band, revealing alpha-clustering and prolate shapes (Fig. 2b, bottom). Furthermore, the lowest 0^+ of the $2\hbar\Omega$ 2p-2h (6 2) slice is found to lie around the 10-MeV 0^+ resonance (third 0^+ state), while the 1p-1h (3 3) slice is used to describe the lowest unnatural-parity 3^- state. The model successfully reproduces other observables for ^{12}C that are informative of the state structure, such as $B(E2)$ transition strengths (Fig. 2a), ground-state matter rms radius, $r^{\rm NCSpM}_{\rm rms} = 2.46$ fm compared to $r^{\rm Expt}_{\rm rms} = 2.43(2)$ fm, and the 2^+_1 electric quadrupole moment, $Q^{\rm NCSpM}_{2^+_1} = +6.2$ $e$fm^2 compared to $Q^{\rm Expt}_{2^+_1} = +6(3)$ $e$fm^2. For the Hoyle state and its 2^+ excitation, theoretical calculations suggest matter rms radii, $r_{\rm rms} = 2.95$ fm and $r_{\rm rms} = 2.97$ fm, respectively.

While the model includes an adjustable parameter, γ in $V^{\rm eff}_{mN}$, this parameter only controls the presence of many-body interactions that cannot be informed by existing realistic NN and NNN interactions. The entire many-body apparatus is fully microscopic and no adjustments are possible. Hence, as γ varies, there is only a small window of possible γ values that, for large enough $N_{\rm max}$, closely reproduces the relative positions of the ^{12}C three lowest 0^+ states, as well as mass rms radii, electric quadrupole moments and $B(E2)$ transition strengths. This many-nucleon long-range interaction ($V^{\rm eff}_{mN}$) is found to be foremost responsible for including alpha-particle correlations and for bringing the Hoyle state down in energy (Fig. 3a). However, using only $V^{\rm eff}_{mN}$ yields a compressed energy spectrum, which is remedied by a mixing of symplectic slices introduced by the bare JISP16. The spin-1 (1 1) component of JISP16 (or a spin-orbit-like term) is found to contribute the most to this effect (Fig. 3b), while the spin-2 tensor-force-like component of JISP16 appears to influence the ^{12}C energy spectrum only slightly (Fig. 3c). This outcome is in a close agreement with the results of Ref. [12], where the mixing of symplectic slices is successfully accounted for by an empirical spin-orbit force.

In addition, NCSpM eigenstates reveal a preponderance of the (0 4) $S = 0$ configuration and also (1 2) $S = 1$ configuration for the ^{12}C *g.st.* rotational band, further confirming an oblate shape. As shown in Fig. 3d for the 4^+ state, NCSpM results are in a close agreement with the *ab*

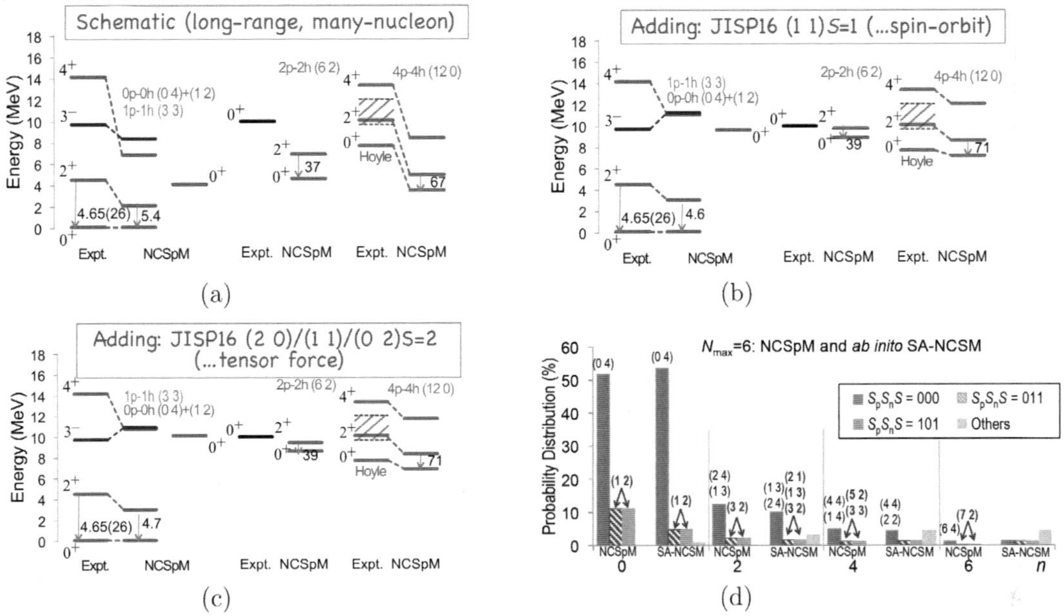

Figure 3. (a)-(c) Same as Fig. 2a, but showing the effect of various interaction components on the ^{12}C energy spectrum: (a) only $V^{\rm eff}_{mN}$; (b) $V^{\rm eff}_{mN} + V^{(1\,1)S=1}_{NN}$ [adding spin-orbit-like $(1\,1)S=1$ components of JISP16]; and (c) $V^{\rm eff}_{mN} + V^{(1\,1)S=1}_{NN} + V^{(2\,0),(1\,1),(0\,2)S=2}_{NN}$ (with the addition of a JISP16 tensor force). (d) Close agreement of probability distributions for the ^{12}C 4^+_1 state (similarly, for the *g.st.* and 2^+_1) as calculated by NCSpM (left) and *ab initio* SA-NCSM with JISP16 (right) for $N_{\rm max}=6$ and $\hbar\Omega = 18$ MeV, shown vs. the n total excitations and with dominant deformation modes, specified by $(\lambda\,\mu)$.

initio SA-NCSM in the reasonable $N_{\rm max}=6$ model space. Furthermore, the outcome shows that among all possible configurations present in the SA-NCSM, only the states of the $(0\,4)$ and then $(1\,2)$ symplectic slices indeed appear dominant in the *g.st.* rotational band. The wavefunctions of the Hoyle-state rotational band can be obtained only in the NCSpM (for $N_{\rm max}=20$) and are found to include shapes of very large prolate deformation, with the largest contribution of $(16\,0)$ (consistent with the prolate shape shown in Fig. 2b).

The outcome of the present analysis is not limited to ^{12}C. The model we find is also applicable to the low-lying states of other *p*-shell nuclei, such as ^{8}Be, as well as *sd*-shell nuclei without any adjustable parameters [37, 43]. In particular, using the same $\gamma = -1.71\times 10^{-4}$ as determined for ^{12}C, we describe selected low-lying states in ^{8}Be in an $N_{\rm max}=24$ model space with only 3 spin-zero $0\hbar\Omega$ $(4\,0)$, $2\hbar\Omega$ $(6\,0)$, and $4\hbar\Omega$ $(8\,0)$ symplectic irreps [37]. Furthermore, we have successfully applied the NCSpM without any adjustable parameters to the ground-state rotational band of heavier nuclei, such as ^{20}O, ^{20,22,24}Ne, ^{20,22}Mg, and ^{24}Si [43]. This suggests that the fully microscopic NCSpM model has indeed captured an important part of the physics that governs the low-energy nuclear dynamics and informs key features of the interaction and nuclear structure primarily responsible for the formation of such simple patterns.

In short, by utilizing symmetries that are found to underpin nuclear dynamics, we offer a novel framework, based on the no-core shell model, to further understand highly deformed states, exemplified by the Hoyle state and its rotational band, as well as unnatural-parity states.

Acknowledgments
We thank the PetaApps Collaboration, in particular, J. P. Vary and P. Maris, as well as D. Rowe and G. Rosensteel for useful discussions. This work was supported by the U.S. NSF (OCI-0904874), the U.S. DOE (DE-SC0005248), and SURA. ACD acknowledges support by the U.S. NSF (grant 1004822) through the REU Site in Dept. of Physics & Astronomy at LSU. We also acknowledge DOE/NERSC, LSU/LONI, and NSF & U. of Illinois/Blue Waters for providing HPC resources.

References

[1] Bohr A and Mottelson B R 1969 *Benjamin, New York* **1**
[2] Elliott J P 1958 *Proc. Roy. Soc. A* **245** 128
[3] Elliott J P 1958 *Proc. Roy. Soc. A* **245** 562
[4] Elliott J P and Harvey M 1962 *Proc. Roy. Soc. A* **272** 557
[5] Hecht K T 1971 *Nucl. Phys. A* **170** 34
[6] Rosensteel G and Rowe D J 1977 *Phys. Rev. Lett.* **38** 10
[7] Rowe D J 1985 *Reports on Progr. in Phys.* **48** 1419
[8] Draayer J, Weeks K and Rosensteel G 1984 *Nucl. Phys.* **A419** 1
[9] Bahri C and Rowe D J 2000 *Nucl. Phys. A* **662** 125
[10] Vargas C, Hirsch J and Draayer J 2001 *Nucl. Phys. A* **690** 409
[11] Dytrych T, Launey K D, Draayer J P, Maris P, Vary J P, Saule E, Catalyurek U, Sosonkina M, Langr D and Caprio M A 2013 *Phys. Rev. Lett.* **111** 252501
[12] Dreyfuss A C, Launey K D, Dytrych T, Draayer J P and Bahri C 2013 *Phys. Lett. B* **727** 163
[13] Machleidt R 2001 *Phys. Rev. C* **63** 024001
[14] Entem D R and Machleidt R 2003 *Phys. Rev. C* **68** 041001
[15] Shirokov A, Vary J, Mazur A and Weber T 2007 *Phys. Lett. B* **644** 33
[16] Navrátil P, Vary J P and Barrett B R 2000 *Phys. Lett.* **84** 5728
[17] Maris P, Shirokov A M and Vary J P 2010 *Phys. Rev. C* **81** 021301(R)
[18] Navrátil P, Quaglioni S, Stetcu I and Barrett B R 2009 *J. Phys. G: Nucl. Part.* **36** 083101
[19] Roth R and Navrátil P 2007 *Phys. Rev. Lett.* **99** 092501
[20] Abe T, Maris P, Otsuka T, Shimizu N, Utsuno Y and Vary J 2012 *Phys. Rev. C* **86** 054301
[21] Pieper S C, Varga K and Wiringa R B 2002 *Phys. Rev. C* **66** 044310
[22] Wloch M, Dean D J, Gour J R, Hjorth-Jensen M, Kowalski K, Papenbrock T and Piecuch P 2005 *Phys. Rev. Lett.* **94** 212501
[23] Tsukiyama K, Bogner S K and Schwenk A 2011 *Phys. Rev. Lett.* **106** 222502
[24] Epelbaum E, Krebs H, Lee D and Meissner U G 2011 *Phys. Rev. Lett.* **106** 192501
[25] Hoyle F 1954 *Astrophys. J. Suppl. Ser.* **1** 121
[26] Chernykh M, Feldmeier H, Neff T, von Neumann-Cosel P and Richter A 2007 *Phys. Rev. Lett.* **98** 032501
[27] Khoa D T, Cuonga D C and Kanada-En'yo Y 2011 *Phys. Letts. B* **695** 469
[28] Fynbo H O U *et al.* 2005 *Nature* **433** 136
[29] Freer M *et al.* 2009 *Phys. Rev. C* **80** 041303
[30] Hyldegaard S *et al.* 2010 *Phys. Rev. C* **81** 024303
[31] Itoh M *et al.* 2011 *Phys. Rev. C* **84** 054308
[32] Zimmerman W R, Destefano N E, Freer M, Gai M and Smit F D 2011 *Phys. Rev. C* **84** 027304
[33] Raduta A R *et al.* 2011 *Phys. Letts. B* **705** 65
[34] Zimmerman W R *et al.* 2013 *Phys. Rev. Lett.* **110** 152502
[35] Marin-Lambarri D, Bijker R, Freer M, Gai M, Kokalova T, Parker D and Wheldon C 2014 *Phys. Rev. Lett.* **113** 012502
[36] Maris P, Vary J P and Shirokov A M 2009 *Phys. Rev. C* **79** 014308 LANL ArXiV: 0808.3420
[37] Launey K D, Dytrych T, Draayer J P, Tobin G K, Ferriss M C, Langr D, Dreyfuss A C, Maris P, Vary J P and Bahri C 2013 *Proceedings of the 5th International Conference on Fission and properties of neutron-rich nuclei, ICFN5, November 4 - 10, 2012, Sanibel Island, Florida, edited by J. H. Hamilton and A. V. Ramayya* (World Scientific, Singapore) p 29
[38] Harvey M 1968 *Adv. Nucl. Phys.* **1** 67
[39] Rosensteel G and Draayer J P 1985 *Nucl. Phys. A* **436** 445
[40] Rowe D J 1967 *Phys. Rev.* **162** 866
[41] Ajzenberg-Selove F and Kelley J 1990 *Nucl. Phys. A* **506** 1
[42] Freer M *et al.* 2011 *Phys. Rev. C* **83** 034314
[43] Tobin G K, Ferriss M C, Launey K D, Dytrych T, Draayer J P and Bahri C 2014 *Phys. Rev. C* **89** 034312

Journal of Physics: Conference Series **569** (2014) 012062 doi:10.1088/1742-6596/569/1/012062

The Hoyle state and its relatives

Thomas Neff[1] **and Hans Feldmeier**[1,2]

[1] GSI Helmholtzzentrum für Schwerionenforschung GmbH and
ExtreMe Matter Institute EMMI, Planckstraße 1, 64291 Darmstadt, Germany
[2] Frankfurt Institute for Advanced Studies, Max-von-Laue Straße 1, 60438 Frankfurt, Germany

E-mail: t.neff@gsi.de

Abstract. The Hoyle state and other resonances in the continuum above the 3 α threshold in ^{12}C are studied in a microscopic cluster model. Whereas the Hoyle state is a very sharp resonance and can be treated reasonably well in bound state approximation, the other higher lying states require a proper treatment of the continuum. The model space consists of an internal region with 3 α particles on a triangular grid and an external region consisting of the ^{8}Be ground state and excited (pseudo)-states of ^{8}Be with an additional α. The microscopic R-matrix method is used to match the many-body wave function to the asymptotic Coulomb behavior of bound states, Gamow states and scattering states. ^{8}Be-α phase shifts are analyzed and resonance properties like radii and transition strengths are investigated.

1. Introduction

It is surprising that the structure of a stable nucleus like ^{12}C is still very much under discussion. The first excited 0^+ state, the famous Hoyle state, has been in the focus of innumerable studies in recent years. This is motivated both by the exotic properties of this state, like a very large extension confirmed by precise electron scattering data [1, 2], and the fact that it provides both a challenge and a benchmark for nuclear structure models. Microscopic cluster models have been able to describe many properties of the Hoyle state [3, 4]. More recently it was shown that the same results could be obtained with the comparatively simple THSR ansatz [5]. Whereas the α cluster structure in these calculations has been assumed from the beginning, AMD [6] and FMD [1] approaches use a Gaussian wave-packet basis and cluster structures appear naturally in a variational procedure. The Hoyle state is also a challenge for textitab initio methods that try to solve the many-body problem exactly for realistic two- and three-body interactions. Within the harmonic oscillator basis it is extremely difficult to describe the asymptotic behavior of three loosely interacting α-particles [7]. This problem can be addressed by the symmetry-adapted NCSM [8]. A new *ab initio* approach, that is not based on a wave function method, is using chiral perturbation theory on a lattice with Monte-Carlo techniques [9].

The Hoyle state is located just above the three α threshold and has a very narrow width of only 8.5 eV. This is not the case for other cluster states that have been investigated in recent years. The properties of the second 2^+ state could be determined unambiguously by excitation with real photons [10, 11]. A second 4^+ state (actually lower in energy than the 4^+ member of the ground state band) was found in [12]. In addition negative parity states were identified [13]. Nevertheless there are still open questions. For example β-decays from ^{12}B and ^{12}N apparently do not populate the second 2^+ state mentioned above but another higher lying 2^+ state [14].

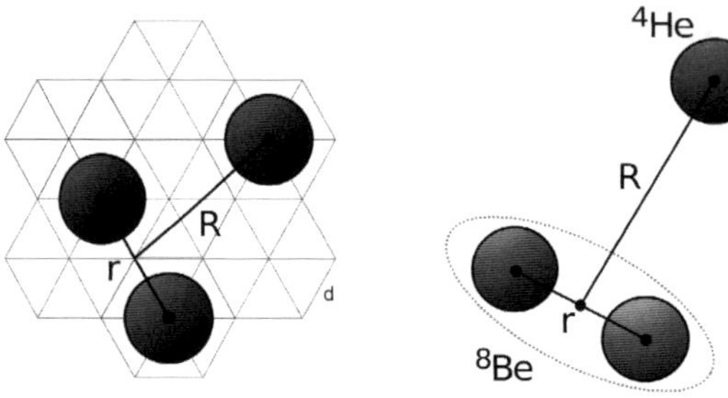

Figure 1. In the internal region three-alpha configurations are generated by putting three alphas on a triangular grid (left). ^{8}Be-α configurations are characterized by the distance R between ^{8}Be eigenstates and the third alpha (right).

From the theoretical point of view these states can be discussed in a meaningful way only if the continuum is treated properly.

Our goal is to perform a microscopic calculation within fermionic molecular dynamics (FMD) [15] including a ^{8}Be+^{4}He continuum in the spirit of our calculation for the ^{3}He$(\alpha,\gamma)^7$Be cross section [16]. As a preparation for the full calculation we perform a simplified calculation within the microscopic cluster model that is presented in this contribution. Compared to earlier studies using a very similar approach [17, 18] a larger basis is used for the ^{8}Be wave functions and a larger number of ^{8}Be pseudo-states is included in the calculation.

2. Microscopic Cluster Model

We use the microscopic cluster model with Brink-type wave functions, where basis states are given as antisymmetrized intrinsic wave functions of three α clusters, projected on parity and angular momentum:

$$|\Psi^{3\alpha}_{JMK\pi}(\mathbf{R}_1, \mathbf{R}_2, \mathbf{R}_3)\rangle = \hat{P}^{\pi}\hat{P}^{J}_{MK}\hat{\mathcal{A}}\left\{|\Psi^{\alpha}(\mathbf{R}_1)\rangle \otimes |\Psi^{\alpha}(\mathbf{R}_2)\rangle \otimes |\Psi^{\alpha}(\mathbf{R}_3)\rangle\right\} . \tag{1}$$

As in our previous work [1] we use the Volkov V2 interaction and the α particle parameters proposed by Kamimura [3, 4] that provide Hoyle state properties in good agreement with experiment. In the internal region an efficient way to obtain a complete basis of three-α configurations is to put the three α's on a triangular grid as shown in Fig. 1. The grid spacing parameter d is chosen as 1.75 fm. The basis states can then be sorted according to the hyperradius $\rho^2 = \frac{1}{2}\mathbf{r}^2 + \frac{2}{3}\mathbf{R}^2$ with $\mathbf{r} = \mathbf{R}_1 - \mathbf{R}_2$ and $\mathbf{R} = \frac{1}{2}(\mathbf{R}_1 + \mathbf{R}_2) - \mathbf{R}_3$.

In the external region ^{8}Be+α configurations are used. The ^{8}Be eigenstates are obtained by diagonalizing α-α configurations projected on angular momentum with distances up to 10 fm:

$$|\Psi^{^8\mathrm{Be}}_{IK}\rangle = \sum_i \hat{P}^{I}_{K0}\hat{\mathcal{A}}\left\{|\Psi^{\alpha}(-\tfrac{r_i}{2}\mathbf{e}_z\rangle \otimes |\Psi^{\alpha}(+\tfrac{r_i}{2}\mathbf{e}_z\rangle\right\} c^I_i . \tag{2}$$

We include the 0^+ ground state, the second 0^+ state, two 2^+ states and a 4^+ state as ^{8}Be configurations. Whereas the ground state is a very narrow resonance (with the Volkov interaction used here it is actually bound by 50 keV) the 2^+ and 4^+ states are very wide resonances. With the exception of the ground state, the eigenstates used here therefore should be considered as pseudo-states that allow us to improve the description of the continuum. In principle the correct asymptotics would require to match to a real three-body Coulomb continuum. However this is not doable in a microscopic approach. Furthermore experiment tells us that the ^{12}C resonances with natural parity predominantly decay through the ^{8}Be ground state.

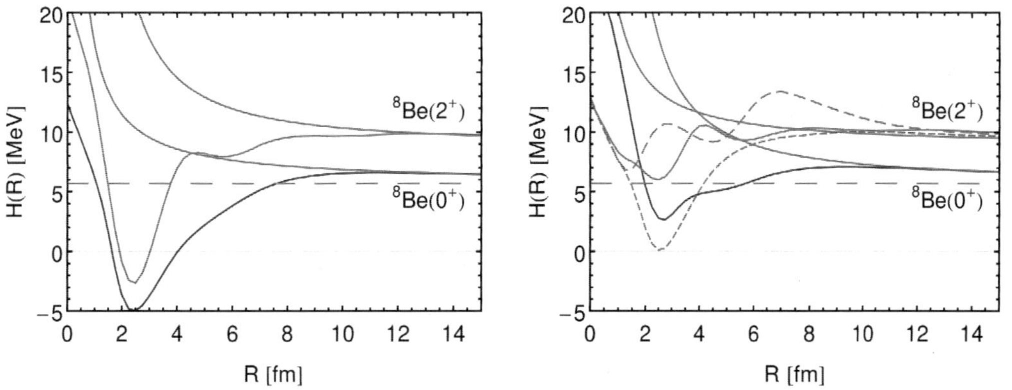

Figure 2. GCM ^{8}Be-α energy surfaces for total spin 0^+ (left) and 2^+ (right). The blue curves correspond to the cluster configurations with the ^{8}Be ground state and the red curves to the different K-projections of the 2^+ state in ^{8}Be. For total spin 0^+ only $K = 0$ is possible, for total spin 2^+ we have $K = 0, 1, 2$. $K = 2$ (red short-dashed) gives the most attraction. The gray curves denote the sum of the threshold energies, the centrifugal energy and the Coulomb energy between the clusters. The gray dashed line indicates the kinetic energy of the relative motion for the GCM basis states due to localization.

The ^{8}Be-α basis states depend on the ^{8}Be eigenstate (with angular momentum I and projection K) and the generator coordinate R_j for the distance between the ^{8}Be and α clusters. The product wave function has to be projected on total angular momentum J, projection M and parity π:

$$|\Psi^{^8\mathrm{Be},\alpha}_{IK;JM\pi}(R_j)\rangle = \hat{P}^\pi \hat{P}^J_{MK} \hat{\mathcal{A}} \left\{ |\Psi^{^8\mathrm{Be}}_{IK}(-\tfrac{R_j}{3}\mathbf{e}_z)\rangle \otimes |\Psi^\alpha(+\tfrac{2R_j}{3}\mathbf{e}_z\rangle \right\} . \tag{3}$$

Evaluating matrix elements with these basis states leads to a five-dimensional integration and therefore a large numerical effort. The GCM energy surfaces shown in Fig. 2 give us some first hints about the ^{12}C structure. From the minima in the energy surfaces we conclude that ^{8}Be-α configurations with a distance of about 2.5 fm should have a large overlap with the members of the ground state band. Furthermore both the 0^+ ground state and the 2^+ state in ^{8}Be appear to be important. At large distances the lowest energy surface always (for natural parity states) corresponds to the ^{8}Be ground state curve. The energy surfaces also tell us that the Coulomb barrier is located at pretty large distances of about 10-12 fm and has a height of about 1.5 MeV. This is consistent with a spatially very extended but also very sharp Hoyle state resonance, that sits just above the threshold. On the other hand we would expect rather large resonance widths for states significantly above the threshold, as the barrier is not very high.

In the GCM basis states the relative motion of the clusters is entangled with the total center of mass motion. To perform the matching to the asymptotic Coulomb solutions the GCM basis states have to be rewritten in terms of RGM basis states [19]:

$$|\Psi^{^8\mathrm{Be},\alpha}_{IK;JM\pi}(R_j)\rangle = \sum_L \left\langle \begin{matrix} I & L \\ K & 0 \end{matrix} \middle| \begin{matrix} J \\ K \end{matrix} \right\rangle \int dr\, r^2 \sqrt{\frac{2L+1}{4\pi}} \Gamma_L(R_j;r)\, |\Phi^{^8\mathrm{Be},\alpha}_{(IL)JM\pi}(r)\rangle \otimes |\Phi^{\mathrm{cm}}\rangle , \tag{4}$$

with the Gaussians localized at R_j projected on orbital angular momentum ($\mu_A = 8/3$):

$$\Gamma_L(R_j;r) = 4\pi \left(\frac{\mu_A}{\pi a}\right)^{3/4} \exp\left(-\mu_A \frac{r^2+R_j^2}{2a}\right) i_L\left(\mu_A \frac{rR_j}{a}\right) . \tag{5}$$

Journal of Physics: Conference Series **569** (2014) 012062 doi:10.1088/1742-6596/569/1/012062

In the RGM basis we no longer have the projection K of the ^{8}Be angular momentum onto the symmetry axis but the relative orbital angular momentum L as a quantum number. The overlap of RGM basis states is given by the RGM norm kernel that becomes orthogonal only at large distances r between the ^{8}Be and α clusters:

$$N_{c,c'}(r,r') = \langle \Phi_c(r) | \Phi_{c'}(r') \rangle \overset{r,r' \to \infty}{\longrightarrow} \delta_{cc'} \frac{\delta(r-r')}{rr'} \tag{6}$$

We use here a short-hand notation $c = \{(IL)JM\pi\}$ for the channel quantum numbers. With the help of the RGM norm kernel we can map the many-body wave function $|\Psi\rangle$ onto the overlap function

$$\psi_c(r) = \int dr' r'^2 \, N_{c,c'}^{-1/2}(r,r') \langle \Phi_{c'}(r') | \Psi \rangle \,, \tag{7}$$

that can be interpreted as the relative wave function of the two clusters. It can be used for matching the many-body wave functions to the asymptotic behavior of two point-like clusters interacting only via Coulomb. To perform this matching we employ the microscopic R-matrix method developed by the Brussels group [20]. The matching is done at the channel radius $a = 16.5\,\text{fm}$ outside the range of the nuclear interaction. For bound states the overlap function is matched to a Whittaker function

$$\psi_c(r) = A_c \frac{1}{r} W_{-\eta_c, L_c+1/2}(2\kappa_c r), \qquad \kappa_c = \sqrt{-2\mu(E-E_c)}\,, \tag{8}$$

where A_c is the asymptotic normalization coefficient and $E - E_c$ are the energies with respect to the corresponding ^{8}Be-α thresholds. For resonances we use Gamow boundary conditions where the overlap function is matched to a purely outgoing Coulomb scattering solution with a complex energy $E = E_R - \frac{i}{2}\Gamma$:

$$\psi_c(r) = A_c \frac{1}{r} O_{L_c}(\eta_c, k_c r), \qquad k_c = \sqrt{2\mu(E-E_c)}\,. \tag{9}$$

Scattering states (incoming channel c_0) are matched to linear combinations of incoming and outgoing Coulomb solutions connected by the scattering matrix S_{c,c_0}:

$$\psi_c(r) = \frac{1}{r} \left\{ \delta_{L_c,L_0} I_{L_c}(\eta_c, k_c r) - S_{c,c_0} O_{L_c}(\eta_c, k_c r) \right\}, \qquad k_c = \sqrt{2\mu(E-E_c)}\,. \tag{10}$$

3. Scattering and Resonances

In Fig. 3-5 we show the phase shifts for total spin 0^+, 2^+, 4^+. The eigenphases δ_α are obtained by diagonalizing the scattering matrix $S = U^{-1}DU$, $D_{\alpha\alpha} = \exp\{2i\delta_\alpha\}$. The diagonal phase shifts δ_c and inelasticity parameters η_c are obtained from the diagonal scattering matrix elements $S_{cc} = \eta_c \exp\{2i\delta_c\}$. To make the plots not too busy we show here the results including only the ^{8}Be 0^+ ground state and the first excited 2^+ state. To calculate the phase shifts the energy is scanned in steps of 50 keV. It is therefore very difficult to resolve very narrow resonances. By employing the Gamow boundary conditions we find an additional 0^+ resonance at an energy of 0.29 MeV and a width of 18 eV corresponding to the Hoyle state, and a 4^+ resonance at 1.17 MeV with a width of 8 eV that is the 4^+ member of ground state band in the cluster model. Analyzing the phase shifts in the 0^+ channel we find a second resonance at 4.11 MeV and a width of 120 keV that shows a strong coupling between ^{8}Be 0^+ and 2^+ channels. In the 2^+ channel we have a first resonance at 1.51 MeV ($\Gamma = 0.32\,\text{MeV}$) that like the Hoyle state decays through the ^{8}Be ground state. There is a second resonance at 4.31 MeV ($\Gamma = 0.14\,\text{MeV}$) that dominantly decays through the ^{8}Be 2^+ state with relative orbital angular momentum $L = 2$. In

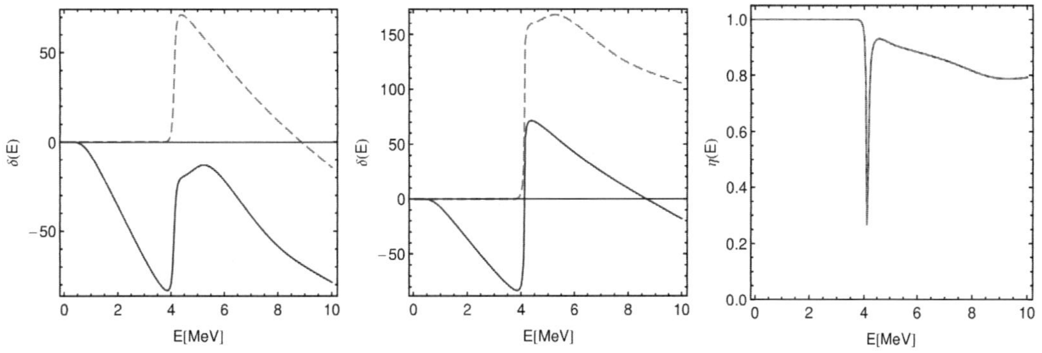

Figure 3. ^{8}Be-α scattering phase shifts for total spin 0^+. The 0^+ ground state and the first excited 2^+ state in ^{8}Be are included in the calculation. On the left the eigenphases, in the middle the diagonal phase shifts and on the right the inelasticities. For the diagonal phase shifts the solid line (blue) indicates the ^{8}Be(0^+) channel and dashed (red) the ^{8}Be(2^+) channel. In case of the eigenphases the channels are mixed but have a dominant component indicated by the colors.

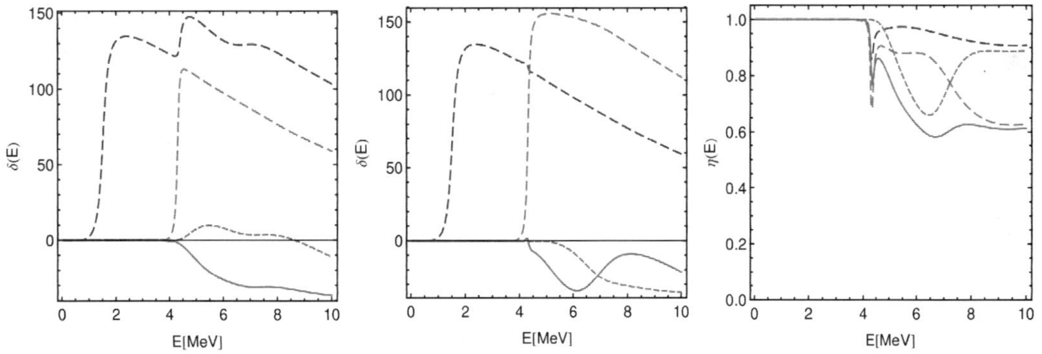

Figure 4. ^{8}Be-α scattering phase shifts for total spin 2^+. There are three ^{8}Be(2^+) channels with relative orbital angular momentum $L = 0$ (solid), $L = 2$ (long dashed) and $L = 4$ (short dashed).

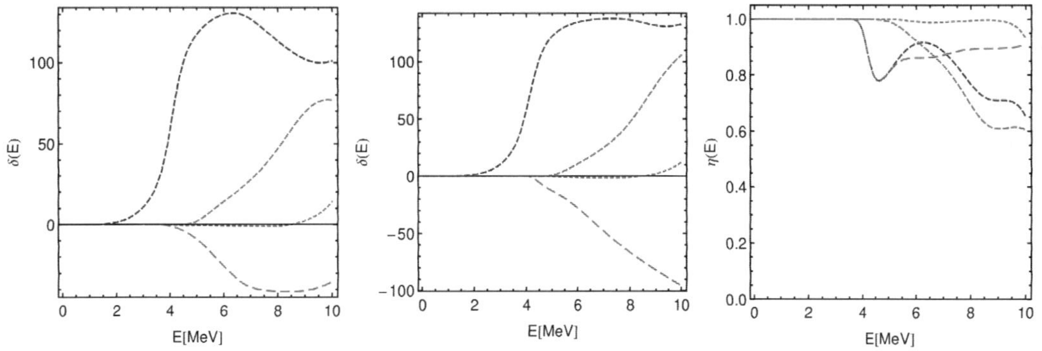

Figure 5. ^{8}Be-α scattering phase shifts for total spin 4^+. There are again three ^{8}Be(2^+) channels with relative orbital angular momentum $L = 2$ (long dashed), $L = 4$ (short dashed) and $L = 6$ (dotted).

Table 1. Observables calculated for increasing model space sizes in bound state approximation and with Gamow states. Energies are given in MeV, radii in fm, $M(E0)$ matrix elements in $e\,\mathrm{fm}^2$, $B(E2)$ strengths in $e^2\,\mathrm{fm}^4$.

	$\rho < 6\,\mathrm{fm}$ $R < 9\,\mathrm{fm}$	$\rho < 6\,\mathrm{fm}$ $R < 12\,\mathrm{fm}$	$\rho < 6\,\mathrm{fm}$ $R < 15\,\mathrm{fm}$	$\rho < 6\,\mathrm{fm}$ Gamow	Experiment
$E(0_1^+)$	-89.64	-89.64	-89.64	-89.64	-92.16
$E^*(2_1^+)$	2.54	2.54	2.54	2.54	4.44
$E^*(0_2^+)$, $\Gamma_\alpha(0_2^+)$	7.82	7.78	7.76	$7.76,\ 3.04 \cdot 10^{-3}$	$7.65,\ 8.5(10) \cdot 10^{-6}$
$E^*(2_2^+)$, $\Gamma_\alpha(2_2^+)$	9.18	9.08	8.93	8.98, 0.46	$10.13(5),\ 2.08^{+0.33}_{-0.26}$
$r_{\mathrm{ch}}(0_1^+)$	2.53	2.53	2.53	2.53	2.47(2)
$r(0_1^+)$	2.39	2.39	2.39	2.39	–
$r(0_2^+)$	3.68	3.78	3.89	4.08 + 0.07i	–
$B(E2, 2_1^+ \to 0_1^+)$	9.12	9.08	9.08	9.08	7.6(4)
$M(E0, 0_1^+ \to 0_2^+)$	6.55	6.40	6.27	6.02 + 0.01i	5.47(9)
$B(E2, 2_2^+ \to 0_1^+)$	2.48	2.09	1.33	2.11 + 1.41i	$1.57^{+0.14}_{-0.11}$

the 4^+ channel we find besides the very narrow resonance belonging to the ground state band a resonance at 4.06 MeV ($\Gamma = 0.98\,\mathrm{MeV}$) that dominantly decays through the ^{8}Be ground state.

The Hoyle state, the 2^+ resonance at 1.51 MeV and the 4^+ resonance at 4.06 MeV might be considered as members of a rotational band built on the ^{8}Be ground state with the third α orbiting around ^{8}Be with relative orbital angular momentum 0, 2 or 4, respectively. We also find higher lying resonances that are built on the ^{8}Be 2^+ state. However the situation gets more complicated when we include channels based on additional ^{8}Be pseudostates. For energies below about 4 MeV we do not see significant effects on the phase shifts. At higher energies, where we find many overlapping resonances, the situation becomes even more complex. This probably indicates that we are dealing here with real three-body decays.

4. Observables

Electromagnetic transitions are powerful tools to test the calculated wave functions. In case of ^{12}C the transition density from the ground state to the Hoyle state could be extracted from electron scattering data very precisely and confirmed the theoretical prediction of a very large radius for the Hoyle state [1, 2]. The theoretical calculations of the transition form factor or the transition density always employed the bound state approximation. This was assumed to be justified, as the Hoyle state is a very sharp resonance. We can now test this assumption using the Gamow states. In Tab. 1 we show the results for various observables for four model spaces. In all four model spaces we include three α configurations on the triangular grid up to a hyperradius ρ of 6 fm. In addition we include ^{8}Be-α configurations with distances R up to 9, 12, and 15 fm without enforcing boundary conditions. These are therefore calculations in a bound state approximation with increasing model space sizes. We finally show results where we use the resonance wave functions obtained as Gamow states. For the evaluation of observables we follow the procedure of Berggren [21, 22]. As expected the results for the binding energy and the radius of the ground state are the same in all model spaces. For the Hoyle state energy we can observe a small change from an excitation energy of 7.82 MeV in the smallest model space to 7.76 MeV in the largest model space. The bound state approximation is also in agreement with the resonance energy obtained for the Gamow state. Including all the ^{8}Be channels we find a resonance width of 3 keV for the Hoyle state. However, a much larger model space dependence is found for the radius of the Hoyle state. It increases from 3.68 fm to 3.89 fm in the bound state approximations and for the Gamow state we obtain a real part of the radius of 4.08 fm and a small imaginary part. The results in bound state approximation are consistent with our previous results [1] ($r(0_2^+) = 3.71\,\text{fm}$) where we used a different set of three α configurations and with the THSR results [5] ($r(0_2^+) = 3.83\,\text{fm}$). This model space dependence is also seen in the monopole matrix element. The result with the Gamow state is now closer to the experimental value extracted from the electron scattering data [2]. An even stronger dependence on the model space can be observed for the second 2^+ state. The resonance width is much larger and even the resonance energy can not be determined reliably in bound state approximation. For the $B(E2)$ transition strength the results in bound state approximation differ by a factor of two. The Gamow state result is in reasonable agreement with the reanalyzed result of Zimmerman *et al.* [10, 11]. The large imaginary part indicates that one has to be careful in disentangling the resonance contribution from background contributions in the experimental data.

5. Summary and Outlook

We have presented an investigation of the ^{12}C continuum above the three α threshold within a microscopic α cluster model. In addition to the Hoyle state we find a second 2^+ and a second 4^+ state that are all built on ^{8}Be-α cluster configurations with ^{8}Be in the ground state. Resonances are described as Gamow states that can be used to calculate electromagnetic transition strengths. Alternatively we could use the scattering states to obtain strength distributions. This might be the better way to compare with experimental data, especially in case of overlapping resonances or significant background contributions. In the future we want to perform a fully microscopic calculation including continuum within the FMD approach with a realistic nucleon-nucleon interaction. This will hopefully provide us with a consistent picture of the ^{12}C structure, including for example also β-transitions and information about T=1 states, that can not be described in an α-cluster model.

References

[1] Chernykh M, Feldmeier H, Neff T, von Neumann-Cosel P and Richter A 2007 *Phys. Rev. Lett.* **98** 032501
[2] Chernykh M, Feldmeier H, Neff T, von Neumann-Cosel P and Richter A 2010 *Phys. Rev. Lett.* **105** 022501
[3] Fukushima Y and Kamimura M 1978 *Proceedings of the International Conference on Nuclear Structure* vol 44 ed Muramori T (Phys. Soc. Jpn. Suppl.) p 225
[4] Kamimura M 1981 *Nucl. Phys.* **A351** 456
[5] Funaki Y, Tohsaki A, Horiuchi H, Schuck P and Röpke G 2003 *Phys. Rev. C* **67** 051306(R)
[6] Kanada-En'yo Y 2007 *Prog. Theor. Phys.* **117** 655
[7] Neff T 2012 *J. Phys.: Conf. Ser.* **403** 012028
[8] Dreyfuss A C, Launey K D, Dytrych T and Draayer J P 2013 *Phys. Lett.* **B727** 511
[9] Epelbaum E, Krebs H, Lee D and Meißner U G 2011 *Phys. Rev. Lett.* **106** 192501
[10] Zimmerman W R, Ahmed M W, Bromberger B, Stave S C, Breskin A, Dangendorf V, Delbar T, Gai M, Henshaw S S, Mueller J M, Sun C, Tittelmeier K, Weller H R and Wu Y K 2013 *Phys. Rev. Lett.* **110** 152502
[11] Weller H R 2013 private communication
[12] Freer M *et al.* 2011 *Phys. Rev. C* **83** 034314
[13] Marín-Lámbarri D J, Bijker R, Freer M, Gai M, Kokalova T, Parker D J and Wheldon C 2014 *Phys. Rev. Lett.* **113** 012502
[14] Hyldegaard S *et al.* 2010 *Phys. Rev. C* **81** 024303
[15] Neff T and Feldmeier H 2008 *Eur. Phys. J Special Topics* **156** 69
[16] Neff T 2011 *Phys. Rev. Lett.* **106** 042502
[17] Descouvemont P and Baye D 1987 *Phys. Rev. C* **36** 54
[18] Arai K 2006 *Phys. Rev. C* **74** 064311
[19] Baye D, Heenen P H and Libert-Heinemann M 1977 *Nucl. Phys.* **A291** 230
[20] Descouvemont P and Baye D 2010 *Rep. Prog. Phys.* **73** 036301
[21] Berggren T 1968 *Nucl. Phys.* **A109** 265
[22] Berggren T 1996 *Phys. Lett.* **B373** 1

Appearance of α-cluster structure in Be isotopes based on Monte Carlo shell model

Tooru Yoshida[1], **Noritaka Shimizu**[1], **Takashi Abe**[2] **and Takaharu Otsuka**[1,2,3]

[1]Center for Nuclear Study, University of Tokyo, Bunkyo, Tokyo 113-0033, Japan
[2]Department of Physics, University of Tokyo, Bunkyo, Tokyo 113-0033, Japan
[3]National Superconducting Cyclotron Laboratory, Michigan State University, East Lansing, Michigan, 48824, USA

E-mail: yoshida@cns.s.u-tokyo.ac.jp

Abstract. Properties of Be isotopes are investigated by using Monte Carlo shell model (MCSM). The structure of ^{10}Be is studied through the intrinsic density. With this density, the appearance of the α-cluster structure and molecular orbits of low-lying states is discussed. Excitation energies for the 2^+ states are compared with experiments and other theoretical calculations.

1. Introduction

In light nuclei, the appearance of the α-cluster structure has been studied from many aspects [1]. For instance, the structure of the developed α cluster has been investigated by using phenomenological α-cluster models near α-cluster thresholds. Recently, *ab initio* calculations have started to reveal the properties of light nuclei and their cluster nature. The analysis of $A = 8$ nuclei based on the Green's function Monte Carlo (GFMC) [2] has been one of the milestones. In the study, the two-α-cluster shape has been found to emerge *a priori* in the density distribution without assuming any cluster configurations. After that, the *ab initio* calculations have been associated with the cluster structure in some cases [3, 4]. For p-shell nuclei, calculations based on the shell model without assuming an inert core have been performed. When the model space is large enough, it is natural to expect that the shell model is able to describe the cluster configuration. The example is the investigation by using the translationally invariant density based on the No-Core Full Configuration (NCFC) [5]. In fact, the deformation and cluster-like shape for Li isotopes have been investigated. Here, the wave functions are not assumed to have any cluster structure *a priori*. As another approach, the Monte Carlo shell model (MCSM) has been developed [6, 7]. The MCSM without assuming an inert core has an advantage in the applicability for the variety of light nuclei [8]. The use of MCSM wave functions is also beneficial for extracting the intrinsic structure. Here, only the wave function before the angular-momentum projection is needed. In this study, we focus on the properties of Be isotopes. In Be isotopes, observables such as $E0$ and $E2$ transition strengths have been studied [9–15]. Recently, the shell model calculations based on NCFC [16] have well reproduced the physical quantities for Be isotopes such as the 2_1^+ energies and $B(E2)$ values. There are also studies of Be isotopes based on MCSM [17]. In this MCSM calculation, the truncation of the model space has been $N_{shell} = 4$, which means that all single-particle orbits below pf shell are taken into account. In their study, the harmonic oscillator parameter has been $\hbar\omega$ =16 MeV, which has been treated empirically. We investigate the properties of the 0^+ states of ^{10}Be with the model space $N_{shell} = 4$. Here, the appearance of the α-cluster structure and molecular orbits have been discussed. When we calculate the 2_1^+ energies of Be isotopes, we increase the size of the model space from $N_{shell} = 4$ to 6. The contents of the proceedings are as follows. In Sect. 2, we show the formulation of MCSM, the numerical setup and the definition of the intrinsic state. In Sect. 3, we discuss the 0_1^+ and 0_2^+ states of ^{10}Be. We also show the theoretical results of the 2_1^+ energies of ^{8,10,12}Be. Finally, we make conclusions and comments for future plans.

2. Formulation

2.1. Calculations of the MCSM wave function

The basis state of the MCSM wave function is described by a Slater determinant, $|\phi_n\rangle$, which is given as

$$|\phi_n\rangle = \prod_{\alpha=1}^{N_p} \sum_{i=1}^{N_{sp}} c_i^\dagger D_{i\alpha}^{(n)} |-\rangle, \tag{1}$$

where, $|-\rangle$ shows a vacuum state and the $c_i^\dagger$ is a creation operator for a single-particle orbit, i. Here, i is specified by t_z, n, l, j (denoting quanta for the isospin, radial node, orbital angular momentum and total angular momentum, respectively). N_p and N_{sp} denote the numbers of nucleons and single-particle orbits, respectively. The resultant wave function is given by a superposition of the N_b Slater determinants as follows,

$$|\Psi(J^\pi)\rangle = \sum_{K=-J}^{J} P_{MK}^{J^\pi} |\Phi_K\rangle, \ \ |\Phi_K\rangle = \sum_{n=1}^{N_b} g_K^n f_n |\phi_n\rangle. \tag{2}$$

Here, $P_{MK}^{J^\pi}$ is the parity and angular-momentum projection operator. This superposition has been obtained iteratively by increasing N_b. The coefficients, g_K^n and f_n, are determined by diagonalizing the Hamiltonian and norm matrices ($\langle\phi_i|HP_{KK'}^{J^\pi}|\phi_j\rangle$ and $\langle\phi_i|P_{KK'}^{J^\pi}|\phi_j\rangle$, respectively). For the n-th step of the iteration, the coefficients, $D_{i\alpha}^{(n)}$, are optimized by the conjugate gradient method to minimize the sum of the energy eigenvalues of the desired states. As for the interaction, we use the JISP16 non-local NN interaction which reproduces two-nucleon scattering data and properties of deuteron and other light nuclei without using three-body interactions explicitly [18]. In the present work, the Coulomb interaction is neglected for simplicity. The contamination of the center of mass motion is factorized approximately by the Lawson's procedure [19], namely adding the Hamiltonian of the center of mass motion, H_{cm}, to the original Hamiltonian, H_0. The Hamiltonian to be diagonalized is given as

$$H = H_0 + \beta\left(H_{cm} - \frac{3}{2}\hbar\omega\right). \tag{3}$$

In the present study, the coefficient is fixed to $\beta\frac{\hbar\omega}{A} = 1$.

2.2. Setup for numerical calculations

The numerical calculation of $P_{M,K}^{J^\pi}$ includes the integration of the Euler angles, $\Omega = (\alpha, \beta, \gamma)$. For the computations of the integral, we need to discretize these angles. We use about 20,000 mesh points for $N_{shell} = 4$ and about 50,000 mesh points for $N_{shell} = 6$. The number of the slater determinants, N_b, is set around 100. For efficient calculations of the matrix elements, OpenMP+MPI hybrid parallelization is applied. The computations were performed on the K computer (RIKEN AICS), FX10 (the University of Tokyo) and T2K Open Supercomputers (the University of Tokyo and University Tsukuba).

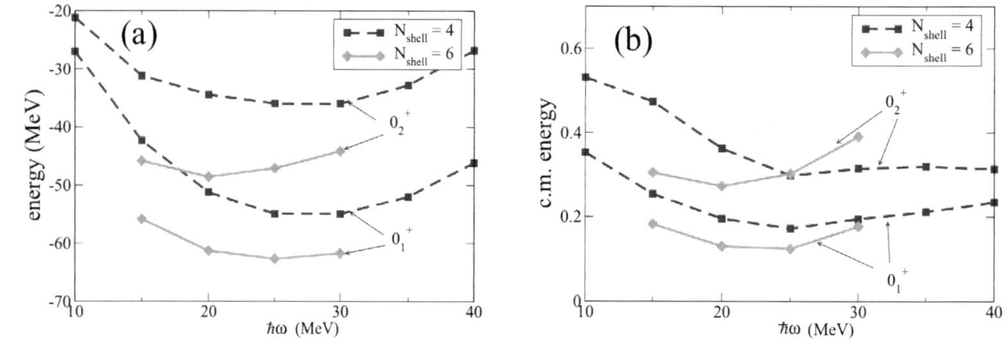

Figure 1. (a) Energies of the 0_1^+ and 0_2^+ of ^{10}Be as functions of $\hbar\omega$. The blue dotted and cyan solid lines show the results with N_{shell} =4 and 6, respectively. (b) Energies of center of mass motion divided by $\hbar\omega/A$ to show the dimensionless values.

2.3. Definition of the intrinsic state

In the J^π-projected wave function, the cluster-like structure is difficult to be visualized. To investigate the wave function in detail, we go back to the wave function before the angular-momentum projection, $|\Phi_K\rangle$, which is a linear combination of the unprojected basis states. However, this state is difficult to be interpreted as an intrinsic state. The reason is that the principal axes of the basis states, $|\phi_n\rangle$, do not have the same direction one another. As a result, the intrinsic structure such as cluster configurations may be hidden. In order to extract the wave function of the intrinsic state, we rotate all basis states so that their diagonalized q-moments are aligned along the z-axis, following the concept of the intrinsic state defined in the GFMC [2]. The formula is already applied on MCSM in the previous study [20]. The intrinsic wave function $|\Phi^{\rm intr}\rangle$ is defined as

$$|\Phi^{\rm intr}\rangle \equiv \sum_n f_n R(\Omega_n)|\phi_n\rangle = \sum_n f_n|\phi_n^R\rangle, \tag{4}$$

with the rotational operator $R(\Omega_n)$ and the transformed basis state $|\phi_n^R\rangle$.

3. Results

In this section, we investigate the properties of ^{8,10,12}Be based on MCSM and compare them with the experiments and other theoretical calculations. We need to confirm which $\hbar\omega$ value to use, since the physical values may strongly depend on it. For this purpose, we focus on the $\hbar\omega$ dependence of 0^+ states of ^{10}Be.

In Fig. 1(a), we depict the energy curve with respect to $\hbar\omega$. Here, we optimize the two lowest 0^+ states. In Fig. 1(b), the expectation values of the Hamiltonian of the center of mass motion are shown. The contamination of the center of mass motion is larger than that for the ground state. Since all values in the figure are small, the approximate factorizations of the center of mass motion are obtained well both for N_{shell} = 4 and 6. The energy minimum of the curve for N_{shell} = 4 is around $\hbar\omega$ = 25 MeV for both 0_1^+ and 0_2^+. We investigate the properties of the two states at $\hbar\omega$ = 25 MeV and N_{shell} =4.

In Fig. 2, the intrinsic densities of the $|\Psi(0_1^+)\rangle$ and $|\Psi(0_2^+)\rangle$ for ^{10}Be are shown. The density distributions of the proton and matter indicate the existence of two-α cluster states, which has been already discussed based on MCSM for the 0_1^+ state [7, 20]. The density of the valence

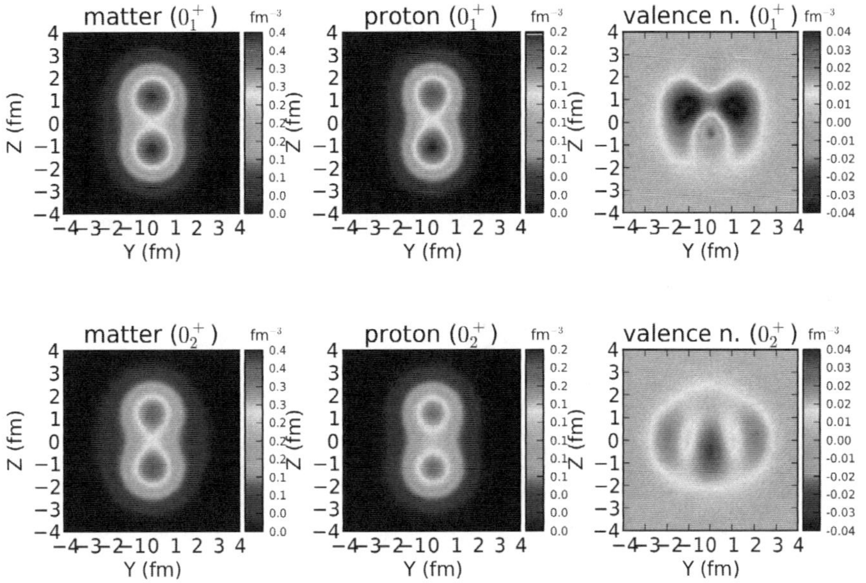

Figure 2. Density distributions of ^{10}Be. The upper (lower) figures show the 0_1^+ (0_2^+) state for the matter (left), proton (middle) and valence neutron (right). The parameters of the MCSM are $N_{shell} = 4$ and $\hbar\omega = 25$ MeV. The number of the basis states is $N_b \sim 100$.

neutron is defined by the difference between the neutron and proton densities. For the 0_1^+ state, the orbits of the valence neutrons show a node at the center of the two-α clusters. The density is located outside the z-axis which is along the two-α clusters. For the 0_2^+ state, the density of the valence neutrons is located mainly at the center of two-α clusters. These valence neutrons mainly occupy the p and sd shells for the 0_1^+ and 0_2^+ states, respectively. These two specific features of the 0_1^+ and 0_2^+ states correspond to the picture of the π and σ orbits in the molecular orbit model [21], respectively. As for the eigenenergy, the model space is enlarged from $N_{shell} = 4$ to $N_{shell} = 6$. For $N_{shell} = 6$, the energy minima for the 0_1^+ and 0_2^+ states are located around $\hbar\omega =$ 25 and 20 MeV, respectively. For simplicity, we fixed $\hbar\omega = 20$ MeV. The excitation energies of the 0_2^+ state are 19.0 and 12.7 MeV for $N_{shell} = 4$ and 6, respectively. These values are larger than the experimental one, 6.1793(7) MeV. However, they are in a reasonable region when we consider that the energy convergence with respect to N_{shell} has some remaining effects.

In Fig. 3, 2_1^+ energies for Be isotopes are shown. In the MCSM calculation, we optimize two states for the 0^+ states and one state for the 2^+ state. Here, the MCSM, NCFC, GFMC and experimental results are shown. As the red open diamonds show, there are discrepancies between the MCSM results before the extrapolation and the other results. The value for ^{8}Be is consistent with NCFC. Even though the discrepancy increases as the neutron number increases, we can still see the tendency to have a kink at neutron number $N = 8$ similar to the other methods. Since the number of the Slater determinants is far below the dimension of the model space defined by the N_{shell}, the calculated eigenenergies give the upper limit of the exact one.

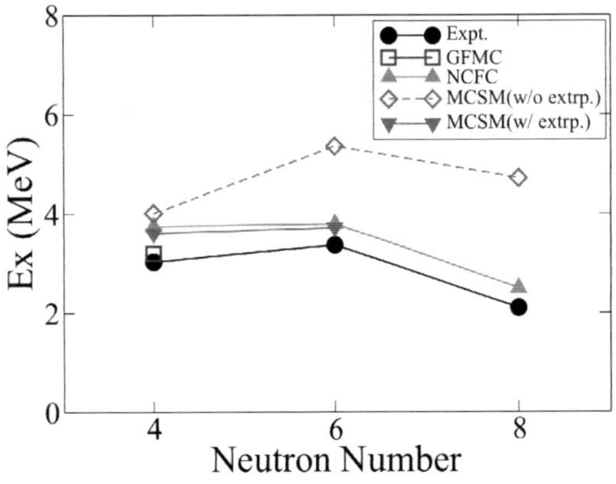

Figure 3. Excitation energies for the 2_1^+ state of Be isotopes for the experiments (black closed circles), GFMC [2] (blue open squares), NCFC [16] (brown triangle), MCSM without the extrapolation (red open diamond) and MCSM with the extrapolation (red triangle). The parameters of the MCSM are $N_{shell} = 6$ and $\hbar\omega = 20$ MeV.

To estimate the exact eigenenergy, we apply the extrapolation method [7, 22, 23]. After the extrapolation, we find a good agreement between the two for ^{8,10}Be (red triangle). We notice that the deviation of the energies before the extrapolations is mainly due to the larger energy variance of the 2_1^+ states than those of the 0_1^+ states. For ^{12}Be, there are still uncertainties in the extrapolation. This is due to the limited number of the basis states for the extrapolation.

4. Conclusions

We investigate the properties of Be isotopes based on no-core MCSM. For the 0^+ states of ^{10}Be with $N_{shell} = 4$, the molecular orbits appear. Their features of the molecular orbits are consistent with those derived from another phenomenological model. For the study of the 2_1^+ state of ^{8,10,12}Be, the model space is set to N_{shell} =6. Even though the number of the basis states in the MCSM before the extrapolation is limited, the tendency of the excitation energies is similar to the results obtained from the NCFC and experiments. However, there is a certain amount of disagreement. The improvement is obtained by the extrapolation method for ^{8,10}Be. To perform the extrapolation for ^{12}Be, we need further investigations in terms of the model space.

5. Acknowledgments

We would like to thank Prof. Pieter Maris, Prof. James P. Vary, and Prof. Yutaka Utsuno for fruitful discussions. We would also like to thank Mr. Yusuke Tsunoda for the discussion of the efficient coding of the intrinsic density. One of the author (T.Y.) would like to thank Prof. Makoto Ito for giving us the information of the cluster development and radii based on the cluster model. This work has been supported by HPCI (hp130024) and the CNS-RIKEN joint project for large-scale nuclear structure calculations.

Journal of Physics: Conference Series **569** (2014) 012063 doi:10.1088/1742-6596/569/1/012063

References

[1] Fujiwara Y, Horiuchi H, Ikeda K, Kamimura M, Katō K, Suzuki Y and Uegaki E 1980 *Prog. Theor. Phys. Suppl.* **68** 29
[2] Wiringa R B, Pieper S C, Carlson J and Pandharipande V R 2000 *Phys. Rev.* C **62** 014001
[3] Epelbaum E, Krebs H, Lee D and Meißner U G 2011 *Phys. Rev. Lett.* **106** 192501
[4] Dytrych T, Launey K D, Draayer J P, Maris P, Vary J P, Saule E, Catalyurek U, Sosonkina M, Langr D, Caprio M A 2014 *et al. Phys. Rev. Lett.* **111** 252501
[5] Cockrell C, Vary J P and Maris P 2012 *Phys. Rev.* C **86** 034325
[6] Otsuka T, Honma M, Mizusaki T, Shimizu N and Utsuno Y 2001 *Prog. Part. Nucl. Phys.* **47** 319
[7] Shimizu N, Abe T, Tsunoda Y, Utsuno Y, Yoshida T, Mizusaki T, Honma M and Otsuka T 2012 *Prog. Theor. Exp. Phys.* **2012** 01A205
[8] Abe T, Maris P, Otsuka T, Shimizu N, Utsuno Y and Vary J P 2012 *Phys. Rev.* C **86** 054301
[9] Datar V M, Chakrabarty D R, Kumar S, Nanai V, Pastore S, Wiringa R B, Behera S P, Chatterjee A, Jenkins D, Lister C J, Mirgule E T, Mitra A, Pillay R G, Ramachandran K, Roberts O J, Rout P C, Shrivastava A and Sugathan P 2013 *Phys. Rev. Lett.* **111**, 062502
[10] Shimoura S, et al. 2007 *Phys. Lett.* B **654** 87
[11] Imai N, et al. 2009 *Phys. Lett.* B **673** 179
[12] Ito M, Itagaki N, Sakurai H and Ikeda K 2008 *Phys. Rev. Lett.* **100** 182502
[13] Kanada-En'yo Y and Horiuchi H 2001 *Prog. Theor. Phys. Supp.* **142** 205
[14] Kanada-En'yo Y and Horiuchi H 2003 *Phys. Rev. C* **68** 014319
[15] Krieger A, Blaum K, Bissell M L, Frömmgen N, Geppert Ch, Hammen M, Kreim K, Kowalska M, Krämer J, Neff T, Neugart R, Neyens G, Nörtershäuser W, Novotny Ch, Sánchez R and Yordanov D T 2012 *Phys. Rev. Lett.* **108** 142501
[16] Maris P 2013 *Journal of Physics: Conf. Ser.* **445** 012035
[17] Liu L, Otsuka T, Shimizu N, Utsuno Y, and Roth R 2012 *Phys. Rev.* C **86** 014302
[18] Shirokov A M, Vary J P, Mazur A I, Zaytsev S A and Weber T A 2005 *Phys. Lett.* B **621** 96
[19] Gloeckner D and Lawson R 1974 *Phys. Lett.* B **53** 313
[20] Yoshida T, Shimizu T, Abe T, Otsuka T 2013 *Few-Body Systems* **54** 1465
[21] Itagaki N and Okabe S 2000 *Phys. Rev.* C **61** 044306
[22] Mizusaki T and Imada M 2002 *Phys. Rev. C* **65** 064319
[23] Shimizu N, Utsuno Y, Mizusaki T, Otsuka T, Abe T and Honma M 2010 *Phys. Rev. C* **82** 061305(R)

Journal of Physics: Conference Series **569** (2014) 012064 doi:10.1088/1742-6596/569/1/012064

Rotational character of the ^{8}Be and ^{12}C spectra investigated through inelastic cross sections via photon emission

E. Garrido[1], A.S. Jensen[2] and D.V. Fedorov[2]

[1]Instituto de Estructura de la Materia, CSIC, Serrano 123, E-28006 Madrid, Spain
[2]Department of Physics and Astronomy, Aarhus University, DK-8000 Aarhus C, Denmark

E-mail: e.garrido@csic.es

Abstract. The electric quadrupole transitions between 0^+, 2^+, and 4^+ states in ^{8}Be and ^{12}C are investigated by discretization of the continuum with a box boundary condition. The γ-emission cross sections and the corresponding transition strengths are computed. The consistency of these transition strengths with the expected behavior for transitions between states in a rotational band is investigated.

1. Introduction

In this work we investigate the spectra of ^{8}Be and ^{12}C, understood as a two-alpha and a three-alpha system, respectively. In particular, we focus on the electric quadrupole γ-emission processes between the states in these two spectra. Experimentally it is well known that in both nuclei, ^{8}Be and ^{12}C, it is possible to identify several sequences of states with angular momentum and parity 0^+, 2^+, and 4^+. This fact strongly suggests that these sequences of states could correspond to states in a rotational band. The goal of this work is to investigate the rotational character of the spectra, and, at the same time, extract as much information as possible about the continuum properties of ^{8}Be and ^{12}C.

We first calculate cross sections defined by specified initial and final state energies. From the cross sections we can obtain the $\mathcal{B}^{(E2)}$ transition strengths. These strengths are used to investigate the rotational character of the spectrum. A rotational band is defined as the sequence of states arising from quantization of the rotational motion of an (almost) frozen deformed structure. This leads to the well-known sequence of excited states with energies following the $J(J+1)$-rule where J is the angular momentum quantum number. However, this behavior of the energy spectrum is not enough. The wave functions for different angular momenta should simultaneously describe the same rotating structure. This condition leads to electromagnetic transition probabilities from one of these states to another given by simple geometric factors depending only on the angular momentum quantum numbers. In this work we investigate to what extent the spectra of ^{8}Be and ^{12}C fulfill this condition for the transition strengths.

We also investigate the dependence on the initial and final state energies and on the chosen $\alpha-\alpha$ interaction. In fact, even if two different two-body potentials reproduce equally well the $\alpha-\alpha$ phase shifts, they can still give rise to two-body wave functions with different short-distance behavior. This different behavior at short distances could produce different cross sections, since the radial matrix element of the electromagnetic operator is a crucial ingredient.

2. Theoretical procedure: Cross sections

At the two-body level, given an initial continuum state with energy E and angular momentum J, the cross section for a γ-transition into some other continuum state with energy E' and angular momentum J' is given by [1]:

$$\frac{d\sigma^{(\lambda)}}{dE'} = \nu \frac{2(2J+1)}{(2J_a+1)(2J_b+1)} \frac{(2\pi)^3(\lambda+1)}{\lambda\,[(2\lambda+1)!!]^2} \frac{1}{k^2} \left(\frac{E_\gamma}{\hbar c}\right)^{2\lambda+1} \left.\frac{d\mathcal{B}^{(\lambda)}}{dEdE'}\right|_{J\to J'}, \tag{1}$$

where J_a and J_b are the angular momenta of the two particles, $E_\gamma = E - E'$ is the photon energy, λ is the multipolarity, ν is the number of identical particles, and $k^2 = 2\mu_{ab}E/\hbar^2$, where μ_{ab} is the reduced mass of the two-body system made of particles a and b.

For three-body systems the expression analogous to the one above takes the form [2]:

$$\frac{d\sigma^{(\lambda)}}{dE'} = \nu! \frac{2(2J+1)}{(2J_a+1)(2J_b+1)(2J_c+1)} \frac{(2\pi)^3(\lambda+1)}{\lambda\,[(2\lambda+1)!!]^2} \frac{32\pi}{\kappa^5} \left(\frac{E_\gamma}{\hbar c}\right)^{2\lambda+1} \left.\frac{d\mathcal{B}^{(\lambda)}}{dEdE'}\right|_{J\to J'}, \tag{2}$$

where again, J_a, J_b, and J_c are the angular momenta of the three particles, and $\kappa^2 = 2mE/\hbar^2$, and m is the normalization mass used to define the Jacobi coordinates.

The double differential transition strength contained in Eqs.(1) and (2) is given by:

$$\left.\frac{d\mathcal{B}^{(\lambda)}}{dEdE'}\right|_{J\to J'} = \frac{1}{2J+1} \left|\langle \Phi_J(E) || \hat{O}_\lambda || \Phi_{J'}(E') \rangle\right|^2 \tag{3}$$

where $\Phi_J(E)$ and $\Phi_{J'}(E')$ are the initial and final continuum two- or three-body wave functions, and $\hat{O}_\lambda$ is the electromagnetic operator with multipolarity λ.

The total cross section for a scattering process with incident energy E is obtained after integration of Eq.(1) or (2) over E'. However, to integrate over all possible final energy values does not make much sense. In fact, experimentally the final energy is restricted to some energy window, $\Delta E'$, which is taken around the energies of interest, typically the resonance energies. In particular, since resonances do not have a well defined energy we have that given an initial energy E we have to consider the transitions into all the states within some energy range around the resonance energy. In this work we choose this energy window as $\Delta E' = E'_R \pm \Gamma'_R$, where E'_R is the energy of the resonance and Γ'_R is the resonance width. Obviously, this choice is a matter of taste, and therefore there is always an ambiguity in the definition of the cross section.

An important point is that the calculations are made on the real energy axis. The wave functions in Eq.(3) are continuum wave functions at a certain energy, and there is no particular treatment of the resonances. Therefore, in order to know the position and the width of the final-state energy windows it is necessary to obtain by some other procedure the energy and width of the resonances. In this work this is made by use of the complex scaling method.

As a final technical remark, let us mention that since we are dealing with continuum states, they should be normalized in the continuum sense, i.e., in terms of a Dirac delta. This can be easily implemented at the two-body level, since in this case the asymptotic behavior of the wave function for two charged particles in the continuum is well known. However, for three charged particles this is not true anymore, and the correct normalization of the continuum three-body wave functions becomes a very delicate task. For this reason, for ^{12}C it is much more convenient to discretize the continuum spectrum, for instance by imposing a box boundary condition. When done, the "discrete" continuum wave functions are normalized in terms of a Kronecker delta, just like bound states. The details about the equivalence of these two methods can be found in Ref.[3], and it is illustrated in Fig.1 for the case of the $4^+ \to 2^+$ transition in ^8Be.

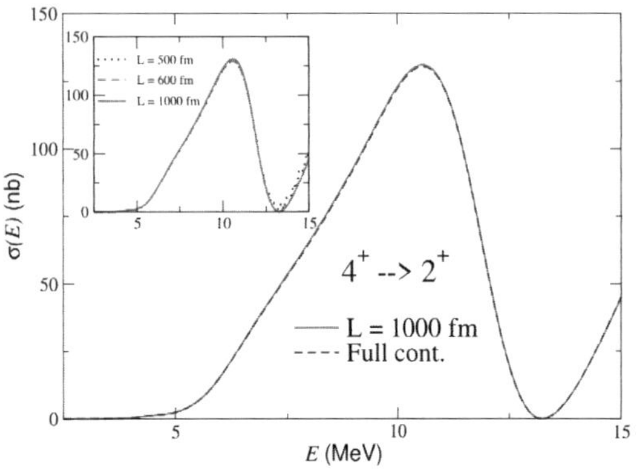

Figure 1. Cross section for the $4^+ \to 2^+$ transition in $\alpha + \alpha$ scattering as a function of the incident energy E. The curves are obtained after integration of Eq.(1) over the final energy E' within the energy window $\Delta E' = 3 \pm 1$ MeV. The big panel shows the results with and without discretization of the continuum (solid and dashed curves, respectively). The discretization has been made in a box with a size of 1000 fm. The inset shows the results obtained for different sizes of the discretization box.

3. Transition strengths

In this section we discuss the $\mathcal{B}^{(E2)}$ transition strengths obtained for ^{8}Be and ^{12}C. The results reported here have been obtained using the Ali-Bodmer $\alpha - \alpha$ potential specified in Ref.[4]. No relevant differences have been found when the Buck potential is used. For both nuclei a previous complex scaling calculation has been performed in order to determine the corresponding resonance energies and widths, which are obtained as poles of the S-matrix. For ^{8}Be a sequence of 0^+, 2^+, 4^+, 6^+, and 8^+ resonances are found [4]. The computed energies and widths of the first three resonances agree with the known experimental values. For ^{12}C three 0^+ states (the first one bound), three 2^+ states (the first one bound), and three 4^+ resonances have been found [5]. These states form three $\{0^+, 2^+, 4^+\}$ bands. The possible rotational character of the first two bands will be investigated. The third band is excluded due to the fact that the energy of the third 0^+ state is experimentally known to be above the third 2^+ state.

Table 1. $\mathcal{B}^{(E2)}$ strengths (in e^2fm^4) for the transitions in ^{8}Be indicated in the first column. The second column shows the results obtained in this work. The third column gives the strengths computed in Ref.[6]. The last two columns are the estimates given by rotational model (see text). The numbers within parenthesis are the ratios in Eq.(6) taking the first reaction as reference.

	This work	Ref.[6]	Rotational model	
			$Z_0 = 3$ fm	$Z_0 = \sqrt{\langle r^2 \rangle}$
$2^+ \to 0^+$	79.1 (1)	71.3 (1)	6.4 (1)	84.0 (1)
$4^+ \to 2^+$	22.1 (0.28)	18.0 (0.25)	9.2 (1.43)	18.1 (0.22)
$6^+ \to 4^+$	10.1 (0.13)	–	10.1 (1.57)	9.1 (0.11)
$8^+ \to 6^+$	13.0 (0.16)	–	10.6 (1.65)	7.6 (0.09)

The differential $E2$-transition strengths, $d\mathcal{B}^{(E2)}/dE$, have been obtained by integrating the cross section in Eq.(1) for ^{8}Be, and in Eq.(2) for ^{12}C, over E', with E' restricted to the final energy window $\Delta E' = E'_R \pm \Gamma'_R$, where E'_R and Γ'_R are the energy and width of the final state resonance. Subsequently, the integrated cross section is divided by all the factors that in Eqs.(1) and (2) multiply the differential transition strength. Finally, the total strength $\mathcal{B}^{(E2)}$ for a transition between some specific initial state into the selected final state is obtained after integration over the initial energy E, where, as in the case of E', the energy E is limited to the corresponding energy window around the initial resonance energy.

Journal of Physics: Conference Series **569** (2014) 012064 doi:10.1088/1742-6596/569/1/012064

For ^{8}Be, the computed transition strengths between the different states are given in the second column of Table 1. The computed values for the $2^+ \to 0^+$ and the $4^+ \to 2^+$ reactions agree with the results given in Ref.[6], where a cluster model was also used. Also, the 22.1 e^2fm^4 obtained for the $4^+ \to 2^+$ transition agrees with the experimental value of 25 ± 8 e^2fm^4 reported in Ref.[7].

For ^{12}C, the computed transition strengths are given in the third column of Table 2. For those reactions involving the second 2^+ state (the lowest 2^+ resonance), two different results are given, which correspond, respectively, to placing the resonance energy at 1.8 MeV and 2.5 MeV. The first energy corresponds to the three-body calculation such that a phenomenological three-body force is used to fit the experimental separation energy of the bound 2^+ state, and in the second case the three-body force is weakened in order to obtain a resonance energy in better agreement with the recent experimental value given in Ref.[8]. In the second column of the table the available experimental data are given.

Table 2. $\mathcal{B}^{(E2)}$ strengths (in e^2fm^4) for the transitions in ^{12}C indicated in the first column. The available experimental data are given in the second column. The third column are the results in this work. For the cases where the second 2^+ state enters the results for two energies of the 2^+ resonance are given. The fourth and fifth columns are the strengths given in Refs.[9] and [10], respectively. The last column is the prediction by the rotational model.

Transition	Exp. [9, 11, 8]	This work		Ref.[9]	Ref.[10]	Rot. model
$2_1^+ \to 0_1^+$	7.6 ± 0.4	10.6		9.16	8.4	18
$4_1^+ \to 2_1^+$		14.5			15.8	25
		E_{2^+}=1.8 MeV	E_{2^+}=2.5 MeV			
$2_2^+ \to 0_2^+$		153	185		102	155
$4_2^+ \to 2_2^+$		0.6	1.7		595	185
$2_1^+ \to 0_2^+$	2.6 ± 0.4	1.1		0.84	5.1	
$4_1^+ \to 2_2^+$		3.6	5.2		7.5	
$2_2^+ \to 0_1^+$	0.73 ± 0.13	2.6	1.8	1.99	0.4	
	1.57 ± 0.14					

Our results agree reasonably well with the cluster-model calculations described in Ref.[9] (fourth column in the table). When compared to the AMD calculations in Ref.[10] significant differences are observed for some of the reactions, specially for the $4_2^+ \to 2_2^+$ transition. Unfortunately, in Ref.[9] this transition was not considered.

For the $2_2^+ \to 0_1^+$ transition the experimental values given in Table 2 are taken from Refs.[11, 8], which are obtained after photo-dissociation of the ^{12}C ground state. The experimental data, whose analysis is obviously still to be clarified, are shown by the square and the circles in Fig.2, respectively. The solid and dashed curves in the figure are the cross sections obtained with the Ali-Bodmer and Buck potentials. As seen in the figure, the shape of the cross section is nicely reproduced. However, in order to get this agreement the computed curves have been divided by a factor of three. This point reflects the fact that a description of the ^{12}C ground state as a three-alpha system is not appropriate, and only a small component of the ground state wave function corresponds to such structure.

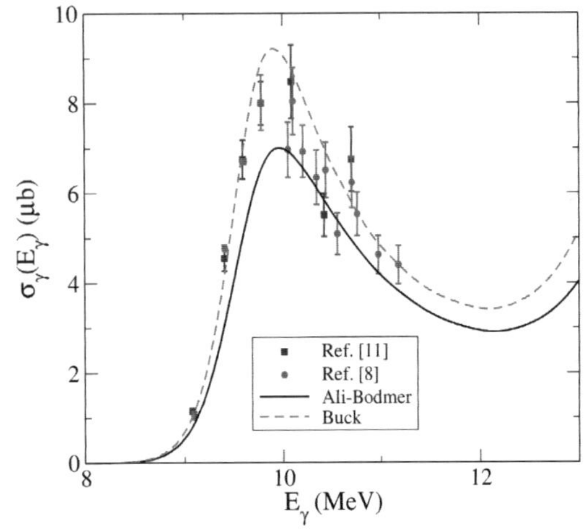

Figure 2. Photo-dissociation cross section for the $^{12}C + \gamma \to \alpha + \alpha + \alpha$ reaction between the 0^+ ground state and the continuum 2^+ states as a function of the photon energy. The computed curves have been divided by a factor of 3. The experimental data are from Refs.[8, 11]. Only the results with the 2^+ resonance at 2.5 MeV are shown.

4. Rotational model

For transitions between states within a rotational band the quadrupole strength is given by

$$\mathcal{B}^{(E2)}(J \to J') = \frac{5}{16\pi} e^2 Q_0^2 \langle J0; 20|J'0\rangle^2, \tag{4}$$

where the rotational band has been assumed to zero projection of the angular momentum over the intrinsic symmetry axis, and where the intrinsic quadrupole moment Q_0 is given by:

$$Q_0 = \frac{4}{3} \langle \sum_i Z_i r_i^2 \rangle \delta, \tag{5}$$

where i runs over all the charged particles (with charge Z_i and center of mass coordinate r_i) and δ is the deformation parameter. From the definition above we have that $Q_0 = 4\delta\langle r^2\rangle/3$ and $Q_0 = 2\delta\langle \rho^2\rangle/3$ for a two-α and a three-α system, respectively, where r is the relative distance between the two particles, and ρ is the hyperradius defined in the three-body system.

The intrinsic quadrupole moment Q_0 is the same for all the states in the band, which is related to the frozen structure of a rotating rigid rotor. Therefore, it is clear that for transitions between states in a rotational band the following ratio between transition strengths holds:

$$\frac{\mathcal{B}^{(E2)}(J \to J')}{\mathcal{B}^{(E2)}(\tilde{J} \to \tilde{J}')} = \frac{\langle J0; 20|J'0\rangle^2}{\langle \tilde{J}0; 20|\tilde{J}'0\rangle^2}. \tag{6}$$

In Ref.[3] it is shown how the average distance between the two alpha-particles in ^{8}Be is far from being constant for the different resonances. In fact, if a constant distance of 3 fm is assumed, the rotational estimate given by Eq.(4), which is given in the fourth column in Table 1, clearly disagrees with our results (second column). In the table we show within parenthesis the results of the ratios in Eq.(6) when the first transition is taken as a reference. Nevertheless, if we compute the intrinsic quadrupole moment in Eq.(5) using the different computed α-α distance for each resonance, and this value is used in Eq.(4), we then get the transition strengths and ratios given in the last column of Table 2. These results agree much better with our calculations. Therefore, the ^{8}Be spectrum has a rotational character provided that the α-α distance is angular momentum dependent, and the principal α-cluster structure is maintained.

About ^{12}C, it is found that the computed value of $\langle \rho^2 \rangle$ is reasonably constant for the 0^+, 2^+, and 4^+ states belonging to each of the two bands under investigation. This is consistent with the expected constant intrinsic quadrupole moment for all the states within a rotational band. Using the computed values of $\langle \rho^2 \rangle$, see Ref.[5], we obtain by means of Eq.(4) the transition strengths shown in the last column of Table 2.

For the two transitions within the first band, $2_1^+ \to 0_1^+$ and $4_1^+ \to 2_1^+$, the values obtained with the rotational picture agree reasonably well with our calculation. The agreement would be even better if we use the known relation between the static and intrinsic quadrupole moments, which permit us to estimate the deformation parameter, which is found to be $\delta^2 \approx 0.55$. Using this value we get that the rotational picture predicts transition strength values of about 10 e^2fm^4 and 14 e^2fm^4 for these two transitions.

For the transitions within the second band, $2_2^+ \to 0_2^+$ and $4_2^+ \to 2_2^+$, the situation is different. For the first of them we can consider that our calculations and the rotational estimate agree. However, our results for the $4_2^+ \to 2_2^+$ transition and the rotational estimate are very far from each other. One could think that a very small deformation parameter could correct the problem, but that would imply to break the reasonable agreement found in the $2_2^+ \to 0_2^+$ case. Simultaneous agreement between our calculations and the rotational picture for the two reactions is not possible.

For completeness in the lower part of Table 2 we show the strengths corresponding to several transitions between states located in different bands.

5. Conclusions

In this work continuum to continuum transitions are used to investigate the properties of the ^{8}Be and ^{12}C spectra. The rotational character of the $\{0^+, 2^+, 4^+, \cdots\}$ bands is investigated.

We have found that the idea of ^{8}Be as a rigid rotor of two α-particles at a fixed distance is questionable. This picture does not agree with the computed transition strengths. However, if the distance between the two particles is allowed to depend on the angular momentum, the agreement is then good.

For ^{12}C the conclusion is that the $E2$-transition strengths between the 0_1^+, 2_1^+, and 4_1^+ states are consistent with the behavior expected for states that belong to a rotational band. For the 0_2^+, 2_2^+, and 4_2^+ states only the strength corresponding to the $2_2^+ \to 0_2^+$ transition can be considered to agree with a rotational behavior, but the 4_2^+ state does not seem to belong to such rotational band, since the strength for the transition into the 2_2^+ state is much smaller than the expected rotational value.

Acknowledgments

This work was partly supported by MINECO (Spain) under contract FIS2011-23565.

[1] Garrido E, Jensen A S and Fedorov D V 2014 *Few-Body Syst.* **55** 101–119
[2] Fowler W A, Caughlan G R and Zimmerman B A 1967 *Annu. Rev. Astro. Astrophys.* **5** 525–570
[3] Garrido E, Jensen A S and Fedorov D V 2013 *Phys. Rev. C* **88** 024001
[4] Garrido E, Jensen A S and Fedorov D V 2012 *Phys. Rev. C* **86** 064608
[5] Álvarez Rodríguez R, Garrido E, Jensen A, Fedorov D and Fynbo H 2007 *Eur. Phys. J A* **31** 303–317
[6] Langanke K 1986 *Phys. Lett. B* **174** 27–31
[7] Datar V M, Kumar S, Chakrabarty D R, Nanal V, Mirgule E T, Mitra A and Oza H H 2005 *Phys. Rev. Lett.* **94** 122502
[8] Zimmerman W R 2013 *Ph.D. Thesis, University of Connecticut Graduate School* URL `http://digitalcommons.uconn.edu/cgi/viewcontent.cgi?article=6439&context=dissertations`
[9] Chernykh M, Feldmeier H, Neff T, von Neumann-Cosel P and Richter A 2007 *Phys. Rev. Lett.* **98** 032501
[10] Cuong D C, Khoa D T and Kanada-En'yo Y 2013 *Phys. Rev. C* **88** 064317
[11] Zimmerman W R *et al* 2013 *Phys. Rev. Lett.* **110** 152502

Journal of Physics: Conference Series **569** (2014) 012065 doi:10.1088/1742-6596/569/1/012065

Role of tensor force in light nuclei with tensor-optimized shell model

T. Myo[1,2], A. Umeya[3], K. Horii[4], H. Toki[2] and K. Ikeda[5]

[1] General Education, Faculty of Engineering, Osaka Institute of Technology, Osaka, Osaka 535-8585, Japan

[2] Research Center for Nuclear Physics (RCNP), Osaka University, Ibaraki, Osaka 567-0047, Japan

[3] Human cience and Common Education, Faculty of Engineering, Nippon Institute of Technology, aitama 345-8501, Japan

[4] Department of Physics, Tokyo Institute of Technology, Meguro 152-8551, Japan

[5] RIKEN Nishina Center, Wako, aitama 351-0198, Japan

E-mail: `myo@ge.oit.ac.jp`

Abstract. We investigate the structures of light nuclei focusing on the role of the tensor force. We describe the tensor and short-range correlations with the tensor optimized shell model (TO M) and the unitary correlation operator method (UCOM), respectively. In the TO M, the 2p2h states are fully optimized to describe the large tensor contribution that brings high momentum components explicitly in the wave function. We use a bare nucleon-nucleon interaction AV8′ and discuss the structures of the He and Li isotopes, such as the excitation energy spectra and the radii. We also investigate the structures of ^{8}Be from the ground-band states to the highly excited states. These states of ^{8}Be are indicated to have different structures of the 2α clustering states and the shell-model like states.

1. Introduction

One of the important issues in nuclear physics is to understand the nuclear structure in connection with the properties of the nucleon-nucleon (NN) interaction. The nucleon-nucleon (NN) interaction has strong tensor forces at long and intermediate distances caused by the pion exchange, and also strong central repulsions at short distance caused by the quark dynamics [1, 2]. It is important to investigate the nuclear structure by treating these characteristics of the NN interaction. The tensor force mainly comes from the pion exchange interaction, which brings the high momentum component in nuclei. The tensor force is of intermediate range and we are able to express the strong tensor correlation in a reasonable shell model space[3]. We name this method as tensor optimized shell model (TOSM), in which the wave function is constructed in terms of the shell-model basis states with full optimization of the two particle-two hole (2p2h) states. The spatial shrinkage of the particle states is essential to obtain convergence of the tensor contribution involving high momentum, and we do not put any truncation to describe the particle states in TOSM. The other is the unitary correlation operator method (UCOM) to treat the short-range correlation coming from the short-range NN repulsion [4]. We shall combine two methods, TOSM and UCOM, to describe nuclei using bare interaction. In this paper, we report our recent applications of TOSM to the He and Li isotopes and discuss their structures focusing on the roles of the tensor force on the energies and configurations. We also discuss the structure of ^{8}Be from the ground-band states to the highly excited states in the aspects of the $\alpha - \alpha$ clustering and the shell-model like states.

2. Tensor Optimized Shell Model (TOSM)

We employ the Hamiltonian having the bare nucleon-nucleon interaction, AV8′[1]. We explain the TOSM wave function Ψ as

$$\Psi = \sum_{k_0} A_{k_0} |\text{0p0h}\rangle_{k_0} + \sum_{k_1} A_{k_1} |\text{1p1h}\rangle_{k_1} + \sum_{k_2} A_{k_2} |\text{2p2h}\rangle_{k_2}. \tag{1}$$

The 1p1h and 2p2h states have the various radial components for particle states (p) which are distinguished by the labels k_0, k_1 and k_2, respectively. The variational coe cients are given as $\{A_k\}$ and are determined by the diagonalization of the Hamiltonian matrix elements. The hole states (h) are described by harmonic oscillator basis states in the TOSM. For the p-shell nuclei, the hole states are defined as the $0s$ and $0p$ shells and the other higher shells are tread as the particle states. The 0p0h states are regarded as the standard shell-model state with A nucleons. In the TOSM, we consider the $0s$+$0p$ space of the hole states and allow up to the two particle excitations from the $0s$ orbit to the $0p$ orbits owing to the presence of the two-body interactions. From each configurations in the 0p0h states, up to two nucleons can be excited to the particle states to make the 1p1h and 2p2h configurations in the TOSM. The orbits of particle states are taken as much as possible until we get the convergence of the solutions such as the energy and the configurations. Therefore, there is no truncation of the particle states included in the 1p1h and 2p2h states in the TOSM.

For the description of the particle states, we employ the aussian basis functions to express single-particle states in the higher shell. We prefer the aussian wave functions over the shell-model states in order to effectively include the necessary high-momentum components due to the tensor force [3, 5]. When we superpose a su cient number of aussian wave functions with various length parameters, the radial components of the particle states can be fully expressed. aussian basis states should be orthogonalized to the hole states and among themselves. This condition is imposed by using the ram-Schmidt orthonormalization. We construct the following orthonormalized single-particle basis function ψ^n_α using a linear combination of aussian bases $\{\phi_\alpha\}$ with length parameter $b_{\alpha,i}$:

$$\psi^n_\alpha(\boldsymbol{r}) = \sum_{i=1}^{N_\alpha} d^n_{\alpha,i}\ \phi_\alpha(\boldsymbol{r}, b_{\alpha,i}), \tag{2}$$

$$\langle \psi^n_\alpha | \psi^{n'}_{\alpha'} \rangle = \delta_{n,n'}\ \delta_{\alpha,\alpha'}, \qquad \text{for } n = 1, \cdots, N_\alpha, \tag{3}$$

where N_α is a number of basis functions for the orbit α, and i is an index to distinguish the bases with a aussian length of $b_{\alpha,i}$. The explicit form of the aussian basis function is written as

$$\phi_\alpha(\boldsymbol{r}, b_{\alpha,i}) = N_l(b_{\alpha,i}) r^l e^{-(r/b_{\alpha,i})^2/2} [Y_l(\hat{\boldsymbol{r}}), \chi^\sigma_{1/2}]_j, \tag{4}$$

where l and j are the orbital and total angular momenta of the basis states, respectively. $N_l(b_{\alpha,i})$ is a normalization factor. The weight coe cients $\{d^n_{\alpha,i}\}$ are determined to satisfy the overlap condition in Eq. (3). Following this procedure, we obtain the new single-particle basis states $\{\psi^n_\alpha\}$ in Eq. (2) used in the TOSM. The particle states in 1p1h and 2p2h states are prepared to specify the basis wave functions, whose amplitudes are determined by the variational principle. In the numerical calculation, we prepare at most 10 aussian basis functions with various range parameters to get a convergence of the energy and Hamiltonian components.

We use short-range UCOM[4] to include the short-range correlation caused by the NN interaction in the TOSM basis states. We employ the form of the correlator of UCOM in the same manner as proposed by Ref. [4] for four channels of spin-isospin pairs. The parametrization of the correlator is explained in our previous papers on He and Li isotopes [6, 7].

For energy variation, we minimize the total energies E by optimizing the amplitudes $\{A_k\}$ in Eq. (1) and the length parameters $\{b_{\alpha,i}\}$ of the aussian bases for all the hole states and the particle states in each J^π state of the particular nucleus. The variational equations are given by

$$\frac{\partial\langle\Psi|H-E|\Psi\rangle}{\partial b_{\alpha,i}} = 0\ , \qquad \frac{\partial\langle\Psi|H-E|\Psi\rangle}{\partial A_{k_j}} = 0 \qquad \text{for } j=0,1,2. \tag{5}$$

3. Results of TOSM+UCOM

3.1. He and Li isotopes

We explain the results of He and Li isotopes, where the partial wave of the particle states is taken as 10 to get a su cient convergence. The excitation energies of He isotopes are shown in Fig. 1, and also shown in Fig. 2 for Li isotopes. For the excitation energies, we see good agreement with the experimental ones, so that we can discuss the structure differences between energy levels. We also predict some energy levels, which has not been observed experimentally yet or not settled to assign the spin of the states, in particular, in the neutron-rich side, ^{7,8}He and ^{8,9}Li. In most of the He and Li isotopes, the ground states have the largest contributions of the tensor force for a particular nucleus [6, 7]. The tensor contributions are dominantly produced by the 2p2h excitations into the particle states from the 0p0h hole states. The corresponding high momentum components induced by the tensor force are confirmed as the enhancement of the kinetic energies from the values of the naive shell model configurations.

It is also found that the resulting level spacing of the Li isotopes is slightly compact than the experimental spectra. For example, in ^{9}Li, the small energy difference between the lowest $3/2^-$ and $1/2^-$ states can be seen. These characteristics are commonly obtained in the FMC calculation within the two-body interaction level [1]. The additional genuine three-body interaction can be one of the components to reproduce the experimental situation.

In the TOSM, the length parameters of the basis states are determined variationally, hence we can discuss the radial properties of nuclei. The matter radii of He and Li isotopes are shown in Fig. 3 and the systematic trend agrees with the experiments. For ^{6}He and ^{8}He, the results are good to explain the enhancement of radius from ^{4}He because of loosely binding extra neutrons in two nuclei, which are however, slightly smaller than the experiments. This difference can be recovered including the continuum effect of extra neutrons. The value of ^{6}Li is also smaller than the experiment, which can be recovered including the asymptotic α+d clustering component in addition to the shell model basis states in the TOSM.

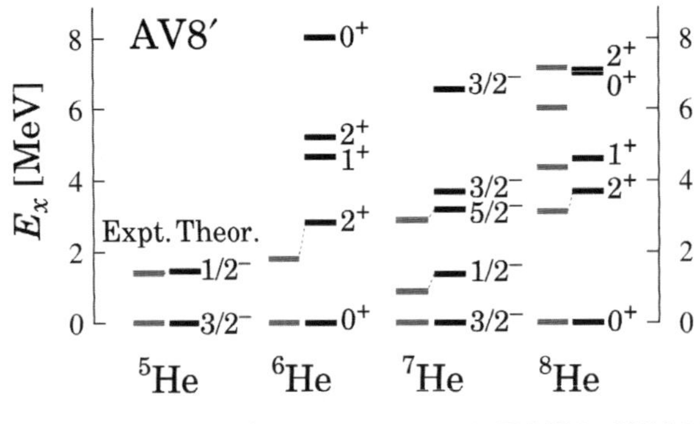

Figure 1. Excitation energies of He isotopes with TOSM+UCOM using AV8′.

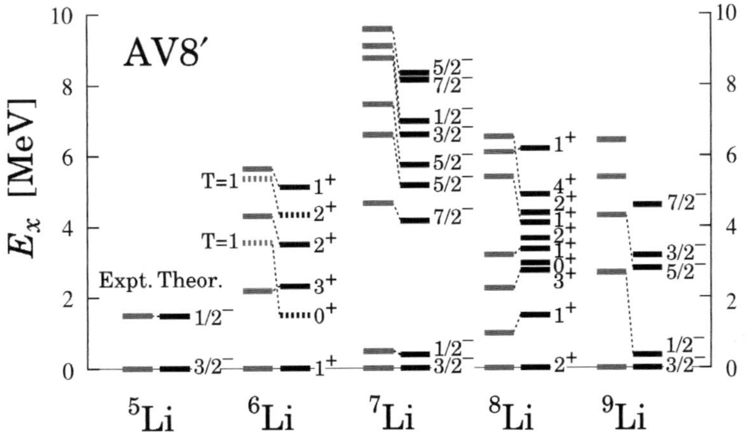

Figure 2. Excitation energies of Li isotopes with TOSM+UCOM using AV8′.

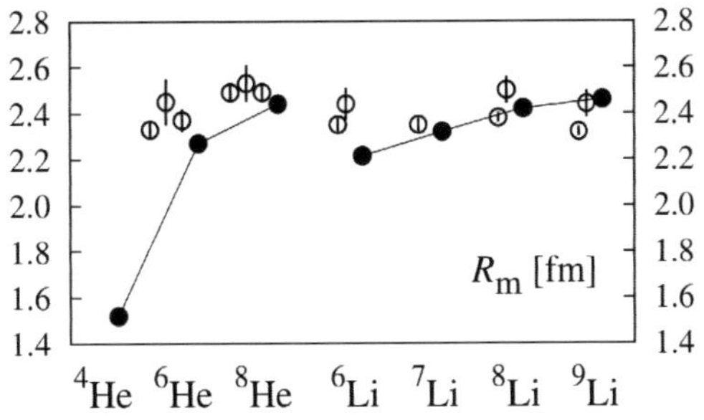

Figure 3. Matter radii of He and Li isotopes in fm (solid circles) with the experiments (open circles)[8, 9, 10].

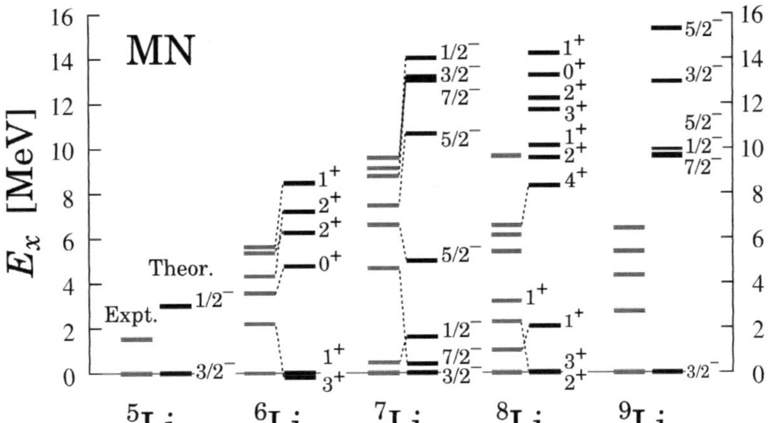

Figure 4. Excitation energies of Li isotopes using the Minnesota interaction (MN) with TOSM.

Table 1. Occupation numbers in each orbit of ^{4}He using AV8′ and Minnesota (MN) interactions.

^{4}He(J^π)	$0s_{1/2}$	$p_{1/2}$	$p_{3/2}$	$1s_{1/2}$	$d_{3/2}$	$d_{5/2}$
AV8′	3.72	0.07	0.05	0.05	0.04	0.02
MN	3.94	0.01	0.03	0.01	0.004	0.006

To see the effect of the tensor force in the TOSM clearly, we show the results with the effective Minnesota (MN) NN interaction, which does not have the tensor force. We choose the u parameter as 0.95 for the central force and use set III of the LS force [11, 12]. The energy spectra of Li isotopes using the MN interaction are shown in Fig. 4. In the MN case, the excitation energies are split in two regions in comparison with the experiments. One of the reasons is the too large splitting energy between $p_{3/2}$ and $p_{1/2}$ components mainly given by the LS interaction. When we adjust the MN interaction such as changing the strength of the LS interaction to fit the LS splitting energy of ^{5}Li, it is still di cult to reproduce the whole trend of the excitation energies of Li isotopes consistently. A large difference between the AV8′ and MN interactions is the tensor force. The comparison of the energy spectra using the two interactions indicates the effect of the tensor force, which has a large impact on the excitation energy spectra of the He and Li isotopes.

We compare the occupation numbers of nucleons in the α particle (^{4}He) between AV8′ and MN interactions in Table 1. In AV8′, it is shown that the $p_{1/2}$ orbit has the largest contribution among the particle states according to the large 2p2h mixing. This selectivity of the $p_{1/2}$ orbit can be explained considering the properties of the tensor operator S_{12} [13, 3, 7], which changes both the orbital angular momentum and spin by two with the opposite direction of their z-components. In the MN case, it is found that the component of the $0s$ orbit is larger than the AV8′ case and the enhancement of the $p_{1/2}$ orbit is not obtained. The $(0s)^4$ configuration in MN is quite dominant in ^{4}He by 96.6 using MN interaction. This comparison means that the tensor force introduces the specific excitations from the s-shell to the p- and sd-shells in ^{4}He. This tensor-driven excitation produces the auli-blocking effect on the excited states of the neutron-rich He and Li isotopes [6, 7].

3.2. ^{8}*Be*

The tensor force contributes to making the α particle strongly. It is known that the tensor contribution in the α particle is generally large [2]. In light nuclei, it is often observed that the α particles are strongly developed as a cluster, e.g., in ^{8}Be and the Hoyle state of ^{12}C [14]. For ^{8}Be, the ground, 2_1^+, and 4_1^+ states are regarded as states consisting of two weakly interacting α particles. On the other hand, the excited states above 4_1^+ can be considered to have shell-like structures, because the α decay is not always favored for the states. In the highly excited states, the T=1 states are degenerated with the T=0 states and are considered as isobaric analog states of ^{8}Li and ^{8}B. These facts of ^{8}Be indicate that the internal structures of the highly excited states are quite different from the ground-band states. In this sense, the ^{8}Be nucleus possesses different features in the ground band and highly excited states. It is theoretically known that a single α particle contains the strong tensor correlation and thus it would be interesting to see how the tensor force affects the variety of structures of ^{8}Be.

In this study, we investigate the different structures of ^{8}Be mentioned above from the viewpoint of the tensor force [16]. For the α clustering, we discuss how well the TOSM describes the two-α components of ^{8}Be. We show the excitation energy spectra of ^{8}Be in Fig. 5. We normalize the energy spectrum to the $2^+(T$=1) state in the highly excited states, because this state is the isobaric analog state of ^{8}Li and the TOSM has nicely described the structures of Li

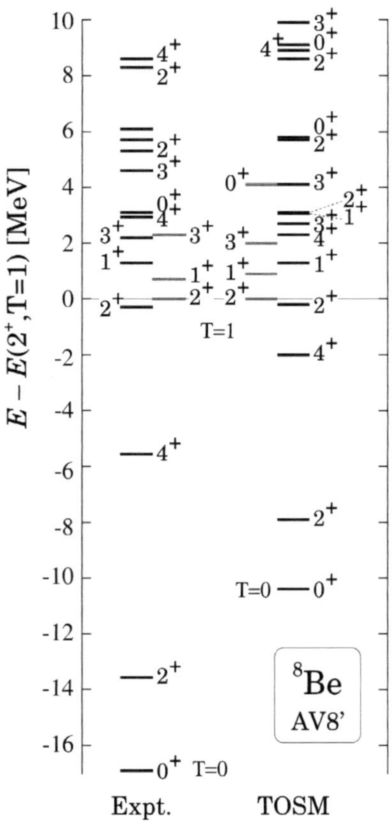

Figure 5. Energy spectrum of ^{8}Be, normalized to the $2^+(T{=}1)$ state.

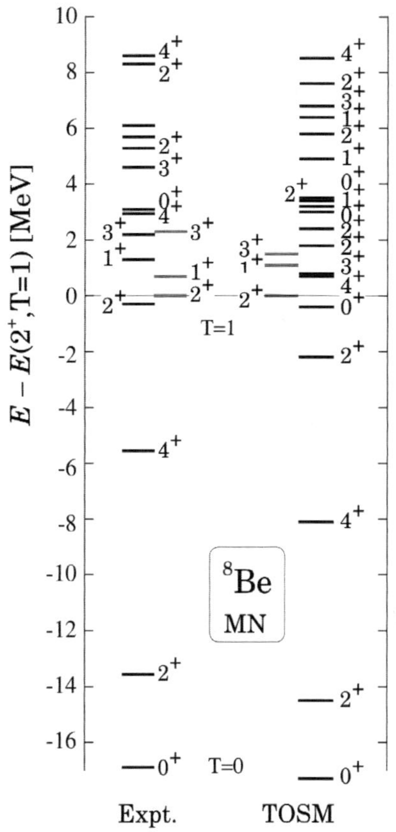

Figure 6. Excitation energy spectrum of ^{8}Be using the Minnesota interaction (MN).

isotopes as shown in Fig. 2 [7]. It is found that there are two groups of states in the spectrum; one is the three ground-band states of 0^+_1, 2^+_1, and 4^+_1, and the other is the highly excited states above 4^+_1. The lowest three states are considered as deformed (clustered) rotational states. The highly excited states include the $T = 1$ states starting from 2^+, the isobaric analog state of the ^{8}Li ground state. In the experiment, the $2^+(T = 1)$ state is almost degenerated with the $2^+(T = 0)$ state in energy and this situation is nicely reproduced in the TOSM.

It is found that the TOSM reproduces fairly well the excitation energy spectrum of ^{8}Be, except for the energy spacing between the ground-band states (0^+, 2^+ and 4^+) and the highly excited states. The resulting small energy spacing is related to the lack of the α clustering component in the grand-band states in the TOSM. In fact, the ^{8}Be ground state possesses almost twice of the values of the α particle for kinetic energies and the central contributions, which can be the signature of two-α clustering of ^{8}Be [16]. From the viewpoint of the shell model, it is generally di cult to express the asymptotic form of the spatially developed α clustering states that contain high shell quanta components. In the TOSM, it is found that specific 2p2h excitations involving high-momentum components are essential to incorporate the tensor correlation in the single α particle [3, 6]. This fact indicates that when two-α particles are established in ^{8}Be, each α particle independently needs the 2p2h components to express the tensor correlation. The other possibility is that many kinds of particle-hole excitations in the shell model bases are also important to describe the formation of the separated two-α clusters in space. In the present TOSM, we take up to 2p2h excitations and this approximation restricts the spatial cluster formation in ^{8}Be, which can require the 4p4h excitations for the α clustering. In addition, the tensor correlations in these α particles might require higher excitations.

In the highly excited states, the calculated spectrum in the TOSM reproduces the experimental level order and the relative energies of each level for both T=0 and T=1 states. The results indicate that the highly excited states of ^{8}Be are regarded as the shell-model like states and the TOSM can treat these states. The tensor contributions in the T=0 states are obtained as stronger than the T=1 states [16], which is consistent to the state dependence of the tensor force.

We compare the energy spectrum of ^{8}Be with those using the effective MN interaction to find the effect of the tensor force. In this case, we reduce the strength of the LS force by 30 to give the same LS splitting energy as ^{5}He obtained using AV8′ [6, 7]. We discuss the excitation energy spectrum using MN shown in Fig. 6, which reproduces the overall excitation energies. We obtain ground-band states and the rotational structure is nicely reproduced. Above 14 MeV excitation energy, many spin states are obtained. For T=0 states, the relative energy between the 4^+_1 and 2^+_2 states is fairly well reproduced, which is different from the AV8′ results shown in Fig. 5. In MN, the effect of the tensor force is renormalized into the model space, hence the 2p2h excitation induced by the tensor force is not necessary to describe ^{4}He. As a result, some amount of two-α clustering component in ^{8}Be can be described in the TOSM in terms of the 2p2h excitations, which could be the reason for the good energy spacing between 4^+_1 and 2^+_2. This situation is different from the case of AV8′ with the tensor force.

For highly excited states, the $T = 0$ states are located lower in energy than the $T = 1$ states by about 2 MeV. This comes from the strong $T = 0$ channel of the central force in MN. In T=0, it is found that the level order is often different from the experiment. From the results of AV8′ and MN, the level order of the higher excited states are rather better described in AV8′ in Fig. 5 than the MN case. This difference is related to the tensor force that is missing in MN explicitly. The tensor force can give the correct state dependence to reproduce the excitation energy spectrum of ^{8}Be.

Acknowledgments

This work was supported by JS S KAKENHI rant Numbers 24740175, 23224004. Numerical calculations were performed on a computer system at RCN , Osaka University.

References

[1] Pieper C and Wiringa R B 2001 *Annu. Rev. Nucl. Part. Sci* **51** 53
[2] Kamada H *et al* 2001 *Phys. Rev.* C **64** 044001
[3] Myo T, Toki H and Ikeda K 2009 *Prog. Theor. Phys.* **121** 511
[4] Feldmeier H, Neff T, Roth R and chnack J 1998 *Nucl. Phys.* A **632** 61
[5] Hiyama E, Kino Y and Kamimura M 2003 *Prog. Part. Nucl. Phys.* **51** 223
[6] Myo T, Umeya A, Toki H and Ikeda K 2011 *Phys. Rev.* C **84** 034315
[7] Myo T, Umeya A, Toki H and Ikeda K 2012 *Phys. Rev.* C **86** 024318
[8] Tanihata I *et al* 1988 *Phys. Lett.* B **206** 592
[9] Dobrovolsky A *et al* 2006 *Nucl. Phys.* A **766** 1
[10] Kiselev O A *et al* 2005 *Eur. Phys. J.* A **25** uppl.1 215
[11] Reichstein I and Tang Y C 1970 *Nucl. Phys.* A **158** 529
[12] Tang Y C, LeMere M and Thompson D R 1978 *Phys. Rep.* **47** 167
[13] Myo T, ugimoto , Katō K, Toki H and Ikeda K 2007 *Prog. Theor. Phys.* **117** 257
[14] Horiuchi H, Ikeda K and Katō K 2012 *Prog. Theor. Phys. Suppl.* **192** 1
[15] Pieper C, Wiringa R B and Carlson J 2004 *Phys. Rev.* C **70** 054325
[16] Myo T, Umeya A, Horii K, Toki H and Ikeda K 2014 *Prog. Theor. Exp. Phys.* **2014** 033D01

Journal of Physics: Conference Series **569** (2014) 012066 doi:10.1088/1742-6596/569/1/012066

Effect of tensor interactions in ^{16}O studied ia (p,d) reaction

H J Ong[1], I Tanihata[1,2], A Tamii[1], T Myo[1,3], K Ogata[1], M Fukuda[4], K Hirota[1], K Ikeda[5], D Ishikawa[1], T Kawabata[6], H Matsubara[1,9], K Matsuta[4], M Mihara[4], T Naito[1], D Nishimura[4,10], Y Ogawa[1], A Ozawa[7], D Y Pang[2], H Sakaguchi[1], K Sekiguchi[5,11], T Suzuki[1], M Taniguchi[8], M Takashina[1,12], H Toki[1], Y Yasuda[1], M Yosoi[1], J Zenihiro[1,5].

[1] Research Center for Nuclear Physics, Osaka University, Ibaraki, Osaka 567-0047, Japan
[2] School of Physics and Nuclear Energy Engineering, Beihang University, Beijing, 100191, China
[3] General Education, Faculty of Engineering, Osaka Institute of Technology, Osaka 535-8585, Japan
[4] Department of Physics, Osaka University, Toyonaka, Osaka 560-0043, Japan
[5] RIKEN Nishina Center, Wako, Saitama 351-0198, Japan
[6] Department of Physics, Kyoto University, Kitashirakawa, Kyoto 606-8502, Japan
[7] Institute of Physics, Tsukuba University, Tsukuba, Ibaraki 305-8571, Japan
[8] Department of Physics, Nara Women's University, Nara 630-8506, Japan

E-mail: onghjin@rcnp.osaka-u.ac.jp

Abstract. The differential cross sections of the ^{16}O(p,d) reaction populating the ground state and several low-lying excited states in ^{15}O were measured using 198-, 295- and 392-MeV proton beams at the Research Center for Nuclear Physics (RCNP), Osaka University, to study the effect of the tensor interactions in ^{16}O. Dividing the cross sections for each excited state by the one for the ground state and comparing the ratios over a wide range of momentum transfer, we found a marked enhancement of the ratio for the positive-parity state(s). The observation is consistent with large components of high-momentum neutrons in the ground-state configurations of ^{16}O due possibly to the tensor interactions.

1. Introduction

Tensor interactions play important roles in atomic nuclei, and are essential to explain the observed quadrupole moment [1] as well as the binding [2] of the deuteron. Besides the deuteron, earlier theoretical [3] studies had also pointed out the importance of the tensor interactions to the binding of ^{3}H and ^{3,4}He; these predictions were later confirmed by experiments [4] which showed the existence of D-wave components. For nuclei heavier than ^{4}He, no direct

[9] Present address: National Institute of Radiological Sciences (NIRS), Chiba 263-8555, Japan
[10] Present address: Department of Physics, Tokyo University of Science, Noda, Chiba 278-8510, Japan
[11] Present address: Department of Physics, Tohoku University, Sendai, 980-8578, Japan
[12] Present address: Graduate School of Medicine, Osaka University, Suita, Osaka 565-0871, Japan
[5] Present address: RIKEN Nishina Center, Wako, Saitama 351-0198, Japan

experimental evidence of the tensor interactions has been reported thus far. Theoretically, Ab-initio calculations [5] have predicted essential importance of the tensor interactions for binding nuclei up to mass number $A = 12$. In addition, extensive studies on the mass data [6] and the subsequent theoretical studies [7] have indicated a possible important role of the tensor interactions in changing the magic numbers and the orders of the single-particle orbitals in neutron-rich nuclei.

Here, we present a brief report of a direct search of the tensor-force effect in ^{16}O [8] using the one-neutron transfer (p,d) reaction. The tensor interactions mix large orbital angular momentum states, giving rise to high momentum components [9] through D-wave mixing in the relative coordinate of two nucleons in finite nuclei. In fact, theoretical calculations [10] have predicted enhanced momentum distributions at around 2 fm^{-1} due to the tensor interactions. In this work, we measured the cross sections of the one-neutron pickup reaction at momentum transfer around 2 fm^{-1} by observing the ground state as well as excited states in ^{15}O. We found strong relative enhancements of the cross section to the positive-parity excited state(s), the $\frac{1}{2}^+$ and/or $\frac{5}{2}^+$ states that could be made by D-wave mixing, at high momentum transfer.

The (p,d) reaction has been applied extensively to study the single particle nature of nuclei. In a single-step neutron pickup reaction on a deuteron target, the momentum transfer, $\vec{P}_{\rm d}$-$\vec{P}_{\rm p}$, is equivalent to the momentum of the picked-up neutron in the target deuteron. Neutron pickup reactions from $A \geq 3$ target nuclei in which deuterons are observed at small scattering angles are expected to occur under the same reaction mechanism, and thus can be used to extract spectroscopic information on the neutrons residing in the target nuclei. Hence, one can probe the neutron momentum distributions by varying the momentum transfer and measuring the picked-up cross sections.

2. Experiment

The experiment was performed at the WS beamline of the RCNP cyclotron facility. Proton beams at $E_{\rm p} = 198$, 295 and 392 MeV were provided by the RCNP ring cyclotron, and bombarded a self-supporting windowless thin ice target [11] placed in a scattering chamber. The deuterons produced in the one-neutron pickup reactions were momentum analyzed by the Grand Raiden spectrometer [12] and detected by two multi-wire drift chambers and two 10-mm thick plastic scintillation detectors placed at the exit focal plane. To cover momentum transfer at around 2 fm^{-1}, we measured the outgoing deuterons at several finite angles from 5° to 25°. For details of the experiment, see Ref. [11].

3. Results and Discussion

Figure 1(a) and (b) show the excitation energy spectra for the reactions at 10° deuteron scattering angle, with proton beams at 392 and 198 MeV. Several peaks corresponding to the ground state and the excited states of the residual nuclei ^{15}O were clearly observed. For reference, the level scheme is shown at the top of the figure. We note that although the 5.183-MeV, $\frac{1}{2}^+$ and the 5.240-MeV, $\frac{5}{2}^+$ states in ^{15}O were not resolved in the present experiment, this should not alter our conclusion. Since the ground $\frac{1}{2}^-$ and the 6.176-MeV excited $\frac{3}{2}^-$ states in ^{15}O can be assumed to be neutron $p_{1/2}$ and $p_{3/2}$ hole states, one expects such states to be relatively strongly populated through direct pickup of a neutron. It is, however, surprising that the positive-parity ($\frac{5}{2}^+$ or $\frac{1}{2}^+$) states are also strongly populated in the ^{16}O(p,d)^{15}O reaction at high energies, which may indicate contributions from the sd-shell. At lower energy, as shown in Fig. 1(c), the cross section to the positive-parity state an order of magnitute smaller than those to the negative-parity states.

In general, the cross sections of all states diminish with increased proton beam energy, i.e. increased momentum transfer, due to the decreases in the amplitudes of the wave functions

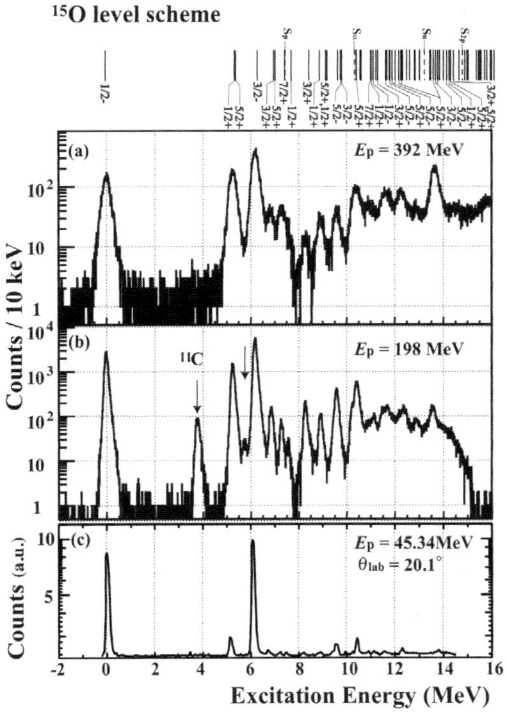

Figure 1. Typical excitation energy spectra for the ^{16}O(p,d)^{15}O reactions at 10° deuteron scattering angle, with proton beams at (a) 392 MeV and (b) 198 MeV. The arrows in (b) indicate components due to the ^{12}C contaminant in the target. For comparison, an excitation energy spectrum, which was replicated from the figure in Ref. [13], for 45.34-MeV proton beam at 20.1° deuteron scattering angle is shown in (c).

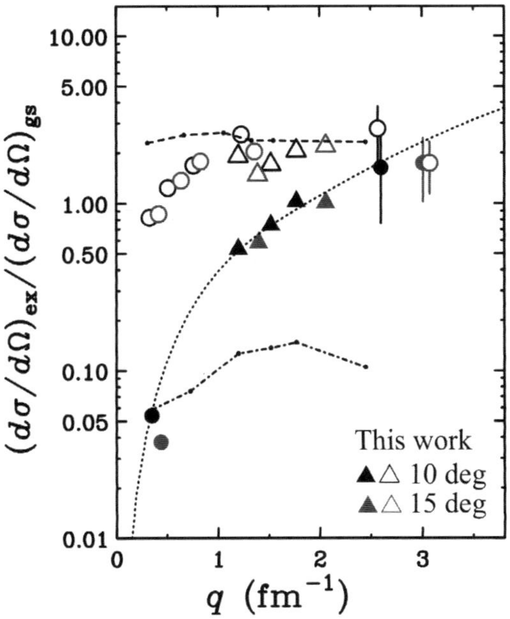

Figure 2. Ratios of the intensities of the $\frac{3}{2}^-$ (open symbols) and the positive-parity ($\frac{5}{2}^+$ and/or $\frac{1}{2}^+$) (filled symbols) excited states to that of the ground $\frac{1}{2}^-$ state as functions of momentum transfer. The black (blue) symbols represent the data taken at 10 (15) deg scattering angle. The dashed (dash-dotted) curve represents the calculated ratios, for 10-deg scattering angle, of the $1p_{3/2}$ ($1d_{5/2}$) and $1p_{1/2}$, obtained by zero-range CDCC-BA with finite-range correction using the Dirac phenomenological potentials. The dotted curve is the ratio of squared momentum wave functions for D- and P-wave components, constructed mainly from Gaussian basis functions with small length parameters that yield large tensor-interaction matrix elements.

at higher momentum. This trend is particularly pronounced for the ground state, which is an evidence of diminishing high-momentum neutrons in the initial ground-state configurations. Notice that the relative intensity of the positive-parity states around 5.2 MeV in ^{15}O increases at very large momentum transfer, which apparently indicates slower decrease or relative increase compared with the negative-parity states.

To examine the relative strength of the excited states, we divided the cross sections for the excited states by that of the ground state. The ratios thus obtained were plotted against the averaged momentum transfer as shown in Fig. 2. The black (blue) symbols represent the data for 10° (15°) scattering angle. The filled symbols represent the ratios (denoted by R_+) for the positive-parity ($\frac{5}{2}^+$ and/or $\frac{1}{2}^+$) states, while the open symbols represent the ratios (denoted by R_-) for the negative-parity, $\frac{3}{2}^-$ state. The filled and open triangles, with statistical errors (< 3), are the data obtained in the present work. The contributions from the ^{12}C contaminant (< 2) as indicated by the arrows in Fig. 1(b) were estimated and subtracted. Other data were taken from the previous measurements with proton energies between 45 and 800 MeV [13,14]. Notice that the ratios for the positive-parity states increase drastically by a factor of over 30 from q transfer 0.3 fm^{-1} to 3.0 fm^{-1}, whereas the ones for the negative-parity states only triple.

We performed theoretical calculations to determine R_- and R_+ at 10° to understand the relative strengths. The dashed (dash-dotted) curve in Fig. 1(c) show the calculated R_- (R_+), obtained with the Continuum-Discretized Coupled-Channel method with Born Approximation to the transition operator $\hat{V}_{\rm tr}$, i.e. CDCC-BA [15] calculation. We made zero-range approximation with finite-range correction to $\hat{V}_{\rm tr}$, and used nucleon-nucleus distorting potentials based on the Dirac phenomenology [16]. In obtaining the calculated R_-, we have adopted the shell-model spectroscopic factors of 1.68 and 3.7 [17] for the $p_{1/2}$ and $p_{3/2}$ states respectively. The calculations are qualitatively consistent with the experimental data at $q \geq 0.8$ fm^{-1}, indicating that the ratios of the $1p_{3/2}$ and $1p_{1/2}$ states can be understood within the present shell-model framework.

The calculations using the shell model that include two-particle two-hole (2p-2h) configuration reproduced the experimental data at proton energy below 45.34 MeV. The reported sum spectroscopic factors were as small as 0.15 and 0.02 for the $1d_{5/2}$ and the $2s_{1/2}$ states respectively [13]. Assuming only the $1d_{5/2}$ orbital and the spectroscopic factor of 0.15, we calculated R_+ as shown in Fig. 1(c). The calculated ratio remains constant from q =1.0 fm^{-1} onwards, and underestimate the experimental data at large momentum transfer by an order of magnitude. This result is expected since it is well known [18] that the conventional shell model does not supply enough high-momentum components. To estimate possible contributions from multi-step processes, we performed a two-step CDCC-BA calculation for the reaction at 200-MeV proton energy populating the $5/2^+$ state, taking into account the inelastic channel via the 3^- state in ^{16}O. The contribution from this process was found to be only 0.5 of the one-step cross section, which is insufficient to explain the discrepancy.

Recently, theoretical calculations using the Tensor-Optimized Shell Model incorporating the Unitary Correlation Operator Method (TOSCOM) have been performed for ^{4}He [19] and ^{9,10,11}Li [20]. In those calculations, the authors included 2p-2h configurations, which contain high-momentum components generated by tensor interactions, in the ground state. Hence, to examine possible contributions from the tensor interactions in the ground-state configurations of ^{16}O, we consider the Gaussian basis wave functions in Ref. [19]. We calculated the squared momentum wave functions for the D-wave and P-wave components. The dotted curve in Fig. 2 was obtained with the amplitudes of squared momentum wave functions relevant for He and Li isotopes [19,20]. The corresponding ratio of the momentum amplitudes for the two P waves (with similar wave functions) is obviously momentum independent and thus is consistent with the observation as well as the CDCC-BA calculation. This marked difference between the D and the P waves indicates that the ratio of momentum amplitudes is the essential ingredient for the observed momentum transfer dependence. Hence, the strong enhancement of the relative cross section for the positive-parity state transition is qualitatively understood by the inclusion of high-momentum wave functions expected with tensor interactions.

4. Conclusion and Future Prospect

We have measured the differential cross sections of the ^{16}O(p,d) reaction populating the ground states as well as several low-lying excited states in ^{15}O at RCNP, Osaka University using the proton beams at 198 MeV, 295 MeV and 392 MeV to search for a direct evidence on the effect of the tensor interactions. By considering the ratio of the cross section for each excited state and the one for the ground state over a wide range of momentum transfer, we observed a marked enhancement in the ratio for the $\frac{1}{2}^+$ and/or $\frac{5}{2}^+$ state(s) in ^{15}O. The result indicates relative increase of high-momentum neutrons in the initial ground-state configuration with neutron(s) in $1s_{1/2}$ and/or $1d_{5/2}$ orbital, and is consistent with the enhancement of the high momentum component expected by a model that includes tensor interactions explicitly.

The present experiment, while offering an intriguing hint on a possible evidence of an effect of the tensor interactions in ^{16}O, exposes needs for more experimental studies. One of the important issues is that other correlations such as the pairing correlations of like nucleons cannot be ruled out by the present experiment, even though they are expected to be smaller than the tensor correlations at around 2 fm^{-1} [10]. Another issue is related to possible effects of reaction mechanisms at finite deuteron scattering angles. To confirm that the observed enhancement is not due to reaction mechanisms, we have performed ^{16}O(p,d) measurements at forward deuteron scattering angle near zero degree at RCNP and the GSI Helmholtz Centre for Heavy Ion Research in Germany. We will also perform (p,dp) and (p,dn) measurements to determine the ratio of the p-n and n-n pairs at momentum transfer region around 2 fm^{-1}.

Acknowledgments

The authors wish to thank the RCNP Accelerator staff for the stable beams. H J Ong and I Tanihata would like to acknowledge the support of Prof. Akihiro Tohsaki (Suzuki) and his spouse. This work was supported by Grant-in-Aid for Scientific Research No. 20244030, 20740163 and 23224008 from Monbukagakusho, Japan.

References

[1] Schwinger J 1939 *Phys. Rev.* **55** 235
Kellog J M B *et al* 1939 *Phys. Rev.* **56** 728
[2] Bethe H A 1940 *Phys. Rev.* **57** 390
Rarita V and Schwinger J 1941 *Phys. Rev.* **59** 436
[3] Gerjuoy E and Schwinger J 1942 *Phys. Rev.* **61** 138
[4] Knutson L D *et al* 1975 *Phys. Rev. Lett.* **35** 1570
Karp B C *et al* 1984 *Phys. Rev. Lett.* **53** 1619
Weller H R *et al* 1984 *Phys. Rev. Lett.* **53** 1325
[5] Pieper S C and Wiringa R B 2001 *Annu. Rev. Nucl. Part. Sci.* **51** 53
[6] Ozawa A *et al* 2000 *Phys. Rev. Lett.* **84** 5493
[7] Otsuka T *et al* 2005 *Phys. Rev. Lett.* **95** 232502
[8] Ong H J *et al* 2013 *Phys. Lett.* B **725** 277
[9] Ikeda K *et al* 2010 *Lecture Note in Physics* (Springer) **818** 165
[10] Neff T and Feldmeier H 2003 *Nucl. Phys.* A **713** 311
Horiuchi W and Suzuki Y 2007 *Phys. Rev.* C **76** 024311
Schiavilla R *et al* 2007 *Phys. Rev. Lett.* **98** 132501
[11] Kawabata T *et al* 2001 *Nucl. Instr. Meth.* A **459** 171
[12] Fujiwara M *et al* 1999 *Nucl. Instr. Meth.* A **422** 484
[13] Snelgrove J L *et al* 1969 *Phys. Rev.* **187** 1246
[14] Roos P G *et al* 1975 *Nucl. Phys.* A **255** 187
Lee J K P *et al* 1968 *Nucl. Phys.* A **106** 357
Abegg R *et al* 1989 *Phys. Rev.* C **39** 65
Smith G R *et al* 1984 *Phys. Rev.* C **30** 593
[15] Yahiro M *et al* 2012 *Prog. Theo. Phys.* Suppl. **196** 87
[16] Hama S *et al* 1990 Phys. Rev. C **41** 2737
Cooper E D *et al* 1993 Phys. Rev. C **47** 297

[17] Brown B A *Lecture Notes in Nuclear Structure Physics* (NSCL, Michigan State University, 2005) (unpublished) page 250
[18] Wall N S and Roos P R 1966 *Phys. Rev.* **150** 811
[19] Myo T *et al* 2007 *Prog. Theo. Phys.* Suppl. **117** 257
[20] Myo T *et al* 2007 *Phys. Rev.* C **76** 024305

Resonant states in ^{13}C and ^{16,17}O at high excitation energy

M R D Rodrigues[1, 9], T Borello-Lewin[1], H Miyake[1], J L M Duarte[1], C L Rodrigues[1], L B Horodynski-Matsushigue[1], G M Ukita[2], F Cappuzzello[3,4], M Cavallaro[4], A Foti[3,5], C Agodi[4], A Cunsolo[4], D Carbone[4], M Bondi[4], M De Napoli[4], B T Roeder[6], R Linares[7] and I Lombardo[8]

[1] Instituto de Física, Universidade de São Paulo, São Paulo, Brasil.
[2] Faculdade de Psicologia, Universidade Santo Amaro, São Paulo, Brasil.
[3]Università di Catania, Dipartimento di Fisica e Astronomia, Catania, Italia.
[4] I.N.F.N., Laboratori Nazionali del Sud, Catania, Italia.
[5] I.N.F.N., Sezione di Catania, Catania, Italia.
[6]Cyclotron Institute, Texas A&M University, College Station, USA.
[7] Instituto de Física , Universidade Federal Fluminense, Niterói, Brasil.
[8] I.N.F.N., Sezione di Napoli, Napoli, Italia.

Abstract. The ^{9}Be(^{6}Li,d)^{13}C and ^{12,13}C(^{6}Li,d)^{16,17}O reactions were measured at the São Paulo Pelletron-Enge-Spectrograph facility at 25.5 MeV incident energy. The nuclear emulsion detection technique was applied. Several narrow resonances were populated up to approximately 17 MeV of excitation energy. An excellent energy resolution was obtained: 40 keV for ^{13}C and 15-30 keV for ^{16}O. The upper limit for the resonance widths were determined. Recently, d-α angular correlations were measured at $\theta_d = 0°$ with incident energy of 25 MeV using the LNS Tandem-MAGNEX Spectrometer facility.

1. Introduction

The cluster phenomenon is important to understand the structure of light nuclei. This phenomenon is expected to be observed at excitation energies close to the cluster decay threshold, according to the Ikeda diagram [1]. The precise nature of cluster formation is still not completely understood. The main purpose of the present work is to study the structure of ^{13}C, ^{14}N and ^{16,17}O nuclei in order to investigate the cluster phenomenon at high excitation energy. In particular, the α cluster had a renewed interest with the interpretation of the Hoyle state as an α condensate state. By consequence, states analogous to the Hoyle state not only for (xα) nuclei but also (xα+n) ones become interesting objects of investigation. The question of whether the cluster has a fairly rigid crystal-like or a gas-like structure is still open to discussion [2-10]. In this context the study of the experimental evolution of the α-cluster phenomenon in (xα) and (xα+n) nuclei through (^{6}Li,d) transfer reactions has been pursued on ^{13}C and ^{16,17}O [11]. In addition, nuclear spectroscopy information of ^{14}N through (^{6}Li,α) reaction was obtained. The measurements performed at the São Paulo Pelletron-Enge-Spectrograph facility, used a ^{6}Li beam of 25.5 MeV employing the nuclear emulsion detection technique. The spectra reveal an unprecedented energy resolution of 15-40 keV, allowing the observation of unresolved states and measurement of resonant states. The absolute cross section and the respective angular distributions

[9] marciadr@if.usp.br.

were determined up to 17 MeV excitation energy. Recently, in order to extract model independent the L transfer values, measurements at $\theta_d = 0°$ of d-α angular correlations [12] were obtained at an incident energy of 25 MeV using the LNS Tandem-MAGNEX Spectrometer facility.

2. Experimental Procedure

A 25.5 MeV ^{6}Li beam from the São Paulo Pelletron Accelerator bombarded uniform ^{9}Be (131 μg/cm^2) and natural C targets (110 and 30 μg/cm^2). The deuterons emerging from (^{6}Li,d) reactions ($\Delta\Omega$ = 0.268(3) msr, $\Delta\Theta_H = 0.38°$) and ($\Delta\Omega$ = 1.237(4) msr, $\Delta\Theta_H = 1.75°$) were momentum analyzed by the magnetic field of the Enge-Spectrograph and detected in emulsion plates (Fuji G6B, 50 μm thick). Spectra of ^{9}Be(^{6}Li,d)^{13}C and ^{12}C(^{6}Li,d)^{16}O associated respectively with seven scattering angles between 3° and 20° and eleven scattering angles between 5° and 32° in the laboratory frame, each one along 50 cm of the focal plane, were measured to investigate resonant states. The plates, after processing, were scanned in strips of 200 μm and a resolution of 40 keV was obtained for the first reaction. For the second reaction, excellent energy resolution of 30 keV and 15 keV was achieved for the thick and thin targets respectively. Due to the joint presence of tracks of deuterons and of elastically scattered ^{6}Li (charge state of two) above approximately 13 MeV of excitation in ^{13}C, the selective reading methodology was applied. The same energy of the beam and methodology were applied to investigate also the ^{13}C(^{6}Li,d)^{17}O reaction, however the data acquisition should be extended to reach the region the 4α + n threshold of interest. Figures 1, 2 and 3 displays portions of the measured position deuteron spectra for the ^{13}C, ^{16}O and ^{17}O respectively for illustration.

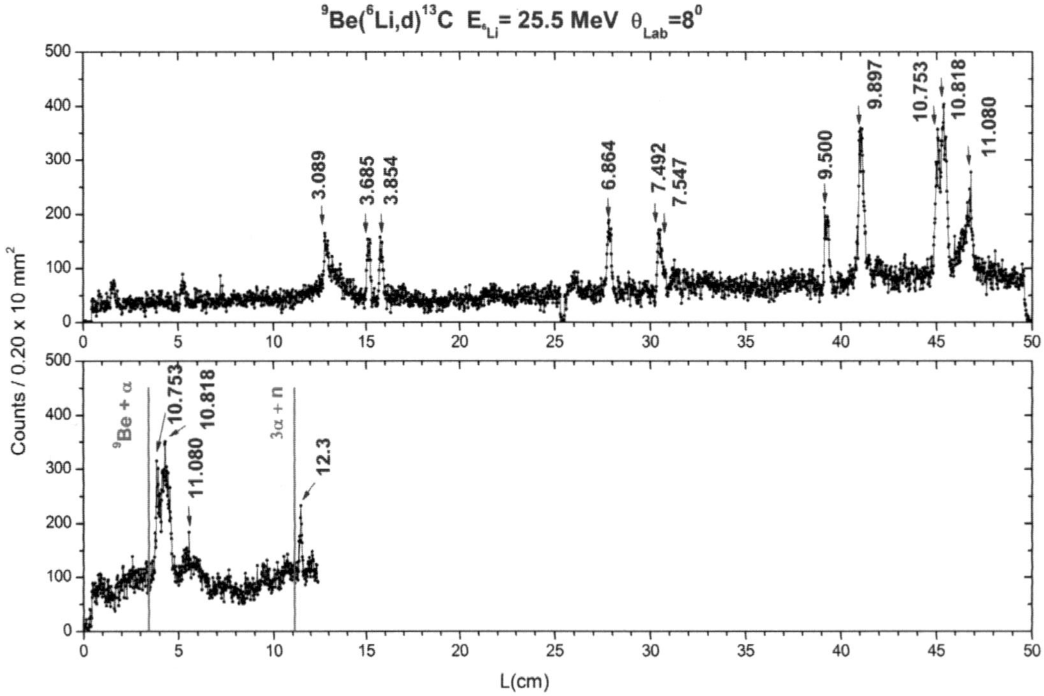

Figure 1. Portion of deuteron spectrum for ^{13}C at $\theta_{Lab} = 8°$. The excitation energies are in MeV and the ^{9}Be+α and 3α+n thresholds are indicated.

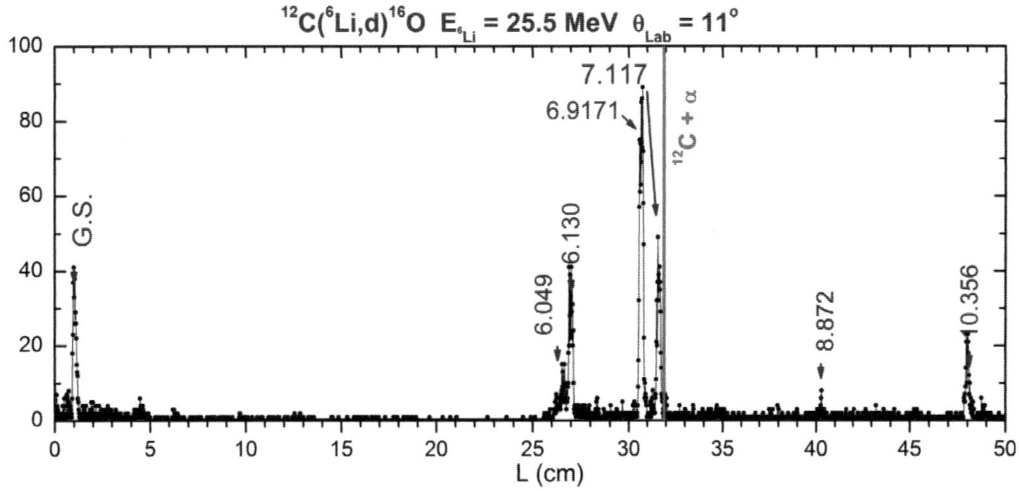

Figure 2. Portion of deuteron spectrum for ^{16}O at $\theta_{Lab} = 11°$. The excitation energies are in MeV and the $^{12}C+\alpha$ threshold is indicated.

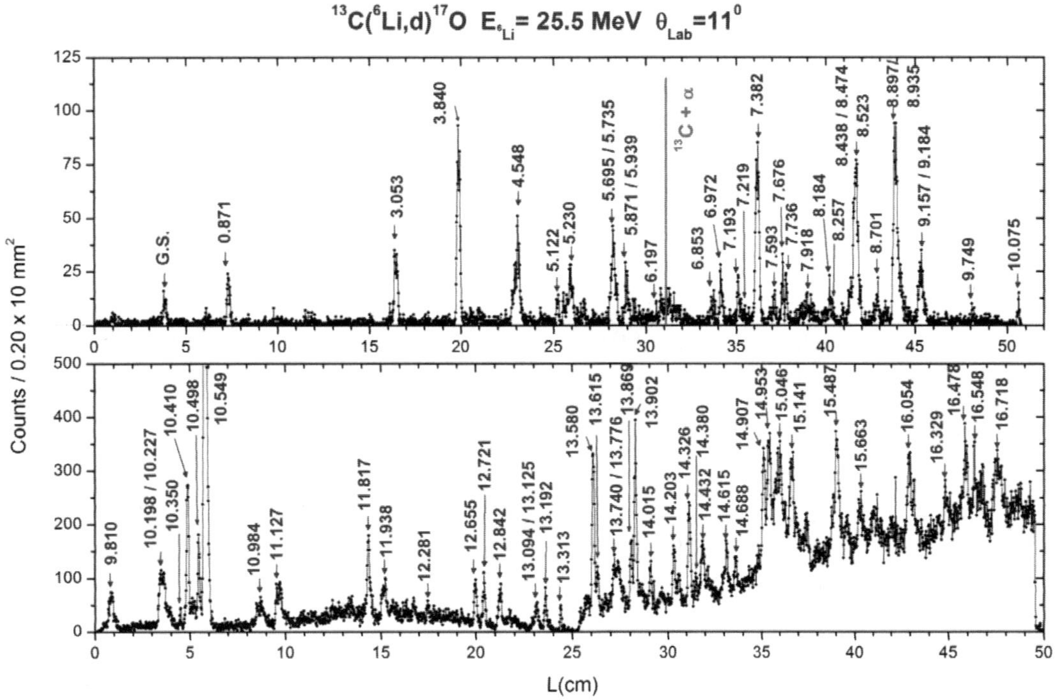

Figure 3. Portion of deuteron spectrum for ^{17}O at $\theta_{Lab} = 11°$. The excitation energies are in MeV and the $^{13}C+\alpha$ threshold is indicated.

The absolute scale of the cross sections was determined by the comparison of the optical model predictions for elastic scattering with measurements in the same target under similar conditions. The relative normalization of the spectra was referred to the beam total charge collected in each run.

In order to complement the data taken in São Paulo measurements of d-α angular correlation were obtained at similar conditions using the LNS Tandem- MAGNEX Spectrometer facility. The experiment was performed with an incident energy of 25.0 MeV, a natural C target with 17 $\mu g/cm^2$ thickness and a solid angle of $\Delta\Omega$ = 50 msr (total angular width $\Delta\theta_H = 12^0$). The α particle were detected using the DINEX detector at the scattering chamber covering an angular range $\Delta\theta_H = 19^0$. Alpha angular distributions were measured from $\theta_{Lab} = 10^0$ to 60^0. The deuterons were detected at zero degree using the Focal Plane Detector (FPD). The trajectory reconstruction technique is applied in order to compensate the aberration due to the large solid angle of the spectrometer. The analysis is in progress.

3. Preliminary analysis and results

Several narrow resonances were populated in ^{13}C and $^{16,17}O$. The alpha width Γ_α and line width $\Gamma_{c.m.}$ can be considered equivalent in the context of one level approximation [11]. In fact $\Gamma_\alpha = \Gamma_{c.m.}$ has been considered often in the literature [13]. In the present analysis the $\Gamma_{c.m.}$ was assumed as an upper limit for the resonance width Γ_α. The upper limit widths obtained are practically equal to the energy experimental resolution.
One step alpha transfer finite-range DWBA calculations using the code DWUCK5, in this preliminary analysis, were performed to describe mainly the shape of the experimental angular distributions associated with natural parity resonant states. The optical model used for the entrance and exit channels took the global parameter sets of Cook [14] and Daehnick et al. [15] respectively. The binding potential of Kubo and Hirata [16] was taken for the α + d description of ^6Li and, although α-resonant, the states under consideration were assumed to be bound by 100 keV in a Woods-Saxon binding potential (r_0 = 1.25 fm, a = 0.65 fm). Relative to the ^9Be core, G [10] values of 4 and 5 were applied respectively for negative and positive parity alpha states. Relative to the ^{12}C core, values 8 and 9 for positive and negative parity respectively were considered. For instance, the experimental angular distributions associated with the α resonances near the ^9Be + α and 3α+ n thresholds are presented in comparison with DWBA predictions in figure 4 [17]. For the resonance at 10.753 MeV, the DWBA angular distribution, calculated considering the wave function predicted by a simplified description of ^9Be + α system [18], is also shown.

4. Conclusions

Several narrow resonances were populated in ^{13}C and $^{16,17}O$ up to approximately 17 MeV of excitation energy. An unprecedented energy resolution was obtained: 40 keV for ^{13}C and 15-30 keV for ^{16}O. The upper limit for the resonance widths obtained is near the energy resolution. The present work is in progress and further analysis is undergoing.

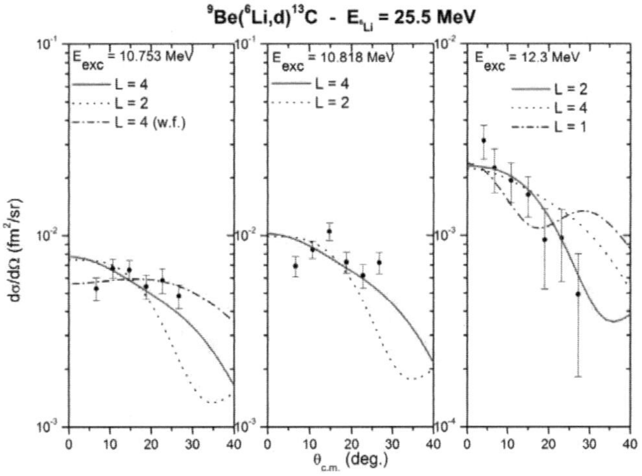

Figure 4. Experimental angular distributions associated with the α resonances near the ^{9}Be + α and 3α + n thresholds and DWBA predictions for different angular momentum transferred (L). The prediction labelled with w.f. considers the wave function obtained by a simplified description of ^{9}Be + α system [18].

Acknowledgments
This work was partially supported by the Brazilian funding agencies FAPESP.

References

[1] Ikeda K, Takigawa N, and Horiuchi H 1968 *Prog. Theor. Phys. Suppl.* E **68**, 464
[2] Funaki Y, Tohsaki A, Horiuchi H, Schuck P, and Röpke G 2005 *Eur. Phys. J.* A **24**, 321
[3] Itoh M et al., *Phys. Rev.* C **84** 2011 054308
[4] Ebran J P, Khan E, Niksic T, and Vretenar D 2012 *NATURE* **487**, 341
[5] Kawabata T, Varenna 2009, *CERN Proc.* 2010-01, 95
[6] Haigh P J, et al. 2010 J. Phys. G: *Nucl. Part. Phys.* **37** 035103
[7] Yamada T and Funaki Y, 2008 *Int. J. Mod. Phys.* E **17** 2101
[8] Yoshida T, Itagaki N, and Otsuka T, 2009 *Phys. Rev.* C **79**, 034308 1
[9] Ohkubo S and Hirabayashi Y, 2010 *Phys. Lett.* B **684**, 127
[10] Kanada-En'yo Y, 2014 *Phys. Rev.* C **89**, 024302
[11] Rodrigues M R D et al., 2014 *Phys. Rev.* C **89** 024306
[12] Cunsolo A et al. 1980 *Phys. Rev.* C **21,** 2345
[13] Becchetti F., Flynn E, Hanson D and Sunier J 1978 *Nucl. Phys.* A **305**, 293
[14] Cook J, *Nucl. Phys.* A 1982 **388,** 153
[15] Daehnick W W, Childs J D, and Vrcelj Z 1980 *Phys. Rev.* C **21**, 2253
[16] Kubo K I and Hirata M, 1972 *Nucl. Phys.* A **187**, 186
[17] Borello-Lewin T et al., 2011 *Int. J. Mod. Phys.* E **20**, 1018
[18] Souza M A and Miyake H, 2007 Proceedings of the XXX RTFNB, Brazil, 46

Study of Nuclear Structure of ^{13}C and ^{20}Ne by Low Energy Nuclear Reactions

I Lombardo[1,2], L Campajola[1,2], D Dell'Aquila[3,4], M La Commara[1,2], A Ordine[2], E Rosato[1,2], G Spadaccini[1,2] and M Vigilante[1,2]

[1] Dipartimento di Fisica, Università di Napoli Federico II, Via Cinthia, 80126, Napoli, Italy
[2] Istituto Nazionale di Fisica Nucleare - ezione di Napoli, Via Cinthia, 80126, Napoli, Italy
[3] Dipartimento di Fisica ed Astronomia, Università di Catania, Via . ofia, 95125, Catania, Italy
[4] Istituto Nazionale di Fisica Nucleare - Laboratori Nazionali del ud, Via . ofia, 95125, Catania, Italy

E-mail: ivlombardo@na.infn.it

Abstract. We report some recent experimental results on the spectroscopy of ^{13}C and ^{20}Ne nuclei by means of low energy nuclear reactions carried out with high resolution electrostatic accelerators. In the case of ^{13}C we investigated the possible existence of α-cluster states above the α emission threshold by means of low energy elastic resonant scattering α+^{9}Be in direct kinematics. Excitation functions show the presence of various resonances that have been reproduced by R-matrix fit. We studied also the structure of ^{20}Ne by means of the ^{19}F(p,α_0) reaction at sub-barrier energies. The spectroscopy of ^{20}Ne excited states in the region $E_x \approx$ 13.5-14.0 MeV can be probed by analyzing experimental angular distributions and excitation functions. This reaction plays an important role also in the CNOF cycle and is an important ingredient to describe hydrogen-induced destruction of fluorine in massive stars. For this reason we investigated the trend of S-factor, that has been compared with results previously reported in the literature.

1. Introduction

Low Energy Nuclear Reactions between light ions are an important tool for carefully probing the structure of light nuclei [1]. In particular, it is possible to deduce information about the possible existence of α-cluster states in light nuclei, a subject of large interest in Nuclear hysics [1, 2]. To this aim, different type of nuclear reactions can be used. For example, direct reactions as (^{6}Li,d) and (^{7}Li,t) at some tens of MeV, leading to α-transfer processes on the target nucleus, have been often used to probe the existence of α-structures in light nuclei [3, 4]. Other experimental ways to probe the α-structure of light nuclei are the study of resonant elastic scattering (in direct or inverse kinematic) of ^{4}He nuclei on various targets [5, 6, 7] or, in general, the study of nuclear reactions leading to the formation of a given compound nucleus that subsequently de-excites by emitting α-particles. In the first part of this paper we describe some investigations on the structure of the ^{13}C nucleus via the resonant elastic scattering of α particles on ^{9}Be in direct kinematics [8, 9]. In this way it is possible to explore an excitation energy region in ^{13}C where the existence of rotational bands built on molecular states has been suggested [10]. In the second part of the paper we discuss on the spectroscopy of ^{20}Ne excited states by means

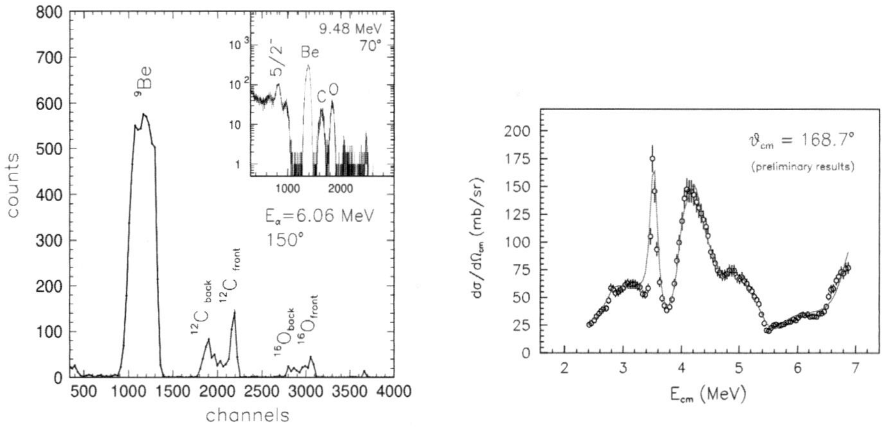

Figure 1. (Left panel) Experimental spectra of $\alpha+^9$Be scattering at 150° and 70° (insert), at 6.06 and 9.48 MeV bombarding energies. (Right panel) Elastic scattering excitation function obtained at θ_{cm}=168.7° (open circles) and preliminary R-matrix fit to the experimental data (red line).

of the ^{19}F(p,α_0) reaction at sub-Coulomb energies [11]. In this case the analysis of excitation functions and angular distributions allows to refine the spectroscopy of ^{20}Ne in the excitation energy range $E_x \approx$ 13.5-14.0 MeV, where uncertain J^π assignment are still persisting in the literature. Furthermore we extracted the S-factor in the energy region $E_{cm} \simeq$ 0.57-1.0 MeV, where two different trends are reported in the literature. Our work allows to shed light on this discrepancy and points out the need for further measurements at lower energies. ood quality data at low energies are important because this reaction is involved in the CNOF cycle during the hydrogen-burning phase in stars [12] and represents an important channel of fluorine destruction in hydrogen-rich stellar environments [13, 14]. Both the experiments have been carried out at the TTT-3 tandem in Naples [15] and will be briefly described in the following paragraphs.

2. Structure of ^{13}C via $\alpha+^9$Be elastic scattering

Between the various non-self-conjugated nuclei, the study of ^{13}C excited states above the α emission threshold probably constitutes one of the most interesting physical cases. In fact, the structure of some excited states could be described in term of an excitation of the Hoyle state in a ^{12}C* core coupled with a covalent neutron [16]. Information on the structure (included the radius and the moment of inertia) of these states could be useful to shed light on the structure of the ^{12}C nucleus. In this framework the possible presence of molecular-like states in highly-lying excited states of ^{13}C has been suggested in [10] starting from the available data reported in the literature [17]. This and other theoretical works stimulated experimentalists to refine the spectroscopy of ^{13}C near and above the α threshold.

A powerful way to perform these investigations is represented by the analysis of resonances in $\alpha+^9$Be elastic scattering excitation functions at backward angles. Unfortunately, very few experimental data have been reported in the literature. An old paper by oss et al [5] reported on the elastic scattering in direct kinematics in a broad angular range (50-160°) but in a limited bombarding energy domain (E_{lab} ≈2.0-5.0 MeV), while in a more recent paper Freer et al [6] reported an excitation function obtained with the inverse kinematic thick target method at around 180° in the energy domain E_{cm} ≈2.0-7.0 MeV. In both cases, information on the J^π of various states have been deduced by R-matrix fit of experimental data. Anyway, some uncertainties on J^π assignments are still persisting, thus stimulating a new experiment covering a broader energy and angular domain. We performed this experiment in direct kinematics at the TTT-3 tandem accelerator in Naples [15]; details on this experiment can be found in [8, 9]. We checked the absolute scale of experimental cross sections by performing both thin and thick target experiments [8]. An example of spectra obtained at backward angles is reported in the left panel of Figure 1. The presence of elastically and inelastically scattered α particles from ^{9}Be and contaminants constituting the target is recognized by kinematics and energy loss considerations.

An example of excitation function of the $\alpha+^9$Be elastic scattering at θ_{cm}=168.7° is reported in the right panel of Figure 1 (open circles). The red solid line indicates the result of a preliminary R-matrix fit to the experimental data, obtained by using the *minRmatrix* code [18, 19]. The effect of target thickness has been taken into account in performing the fit procedure. These results are still preliminary and will be extended in the future to other excitation functions obtained at different angles. Nevertheless, we can extract some interesting information from this preliminary analysis. First of all, in the energy domain here explored, it seems that no 1/2 states are excited, in agreement with data by [6]. Therefore the excitation energy of a possible state obtained by coupling in s-wave a neutron with a ^{12}C core excited to the Hoyle state should be lower than the minimal excitation energy probed in this experiment, and this is in agreement with recent AMD calculations [16]. Another interesting point concerns the resonant structure at E_{cm} ≈3.5 MeV. Our preliminary results pointed out the presence of two close-lying states with J^π=5 2^-,7 2^+ with quite low values of dimensionless reduced widths for the α_0 channel (of the order of some percent of the Wigner limit). These preliminary results are in good agreement with the findings of [5] and seems to rule out the presence of a 9 2^- state at this excitation energy, needed to support the existence of the molecular negative-parity rotational band in ^{13}C. Further investigations on the present experimental data are currently going on.

3. Investigation of the ^{19}F(p,α_0) reaction at low energies

The investigation of low energy ^{19}F(p,α_0) and ^{19}F(p,α_π) reactions has been used in the past to probe the structure of ^{20}Ne, which is a self-conjugated nucleus and can show pronounced α-cluster structures [20]. To this aim experimental efforts have been performed [21, 22] to evidence the existence of quartet excitations in this nucleus [23, 24]. Unfortunately, ambiguities in the spectroscopy of some excited states are still persisting, especially in the excitation energy region around 13.3-14.0 MeV [25]. Beyond the spectroscopic motivation, the ^{19}F(p,α) reaction plays a significative role in Nuclear Astrophysics because it closes the CNOF cycle in the hydrogen-burning phase of massive stars [12] and can represent an important channel of fluorine destruction in hydrogen-rich stellar environments [13, 14]. Despite of this interest, very few (and often contrasting) data-sets of the S-factor at $E_{cm} < 1$ MeV have been reported in the literature, thus preventing an accurate reaction rate determination at astrophysical energies [26]. Taking into account both these aspects, we performed a new experiment at the TTT-3 tandem accelerator in Naples, aimed at measuring angular distributions and excitation functions of the ^{19}F(p,α_0) reaction in the bombarding energy domain E_p ≈0.6-1.0 MeV. The details of this experiment are widely discussed in [11]. In the following sections we report some experimental findings related to this experiment.

An array of 12 silicon detectors covering a large angular range has been used as detection system. Thin aluminium foils preceded silicon detectors in order to stop the high flux of scattered particles; at variance, they allow the passing through of high energy α-particles emitted in the ^{19}F(p,α_0) reaction (Q=8.114 MeV). The obtained experimental spectra are practically background-free. Due to the low bombarding energy domain (E_p ≈0.6-1 MeV), only s, p and d partial waves are expected to mainly contribute; for this reason, angular distributions have been

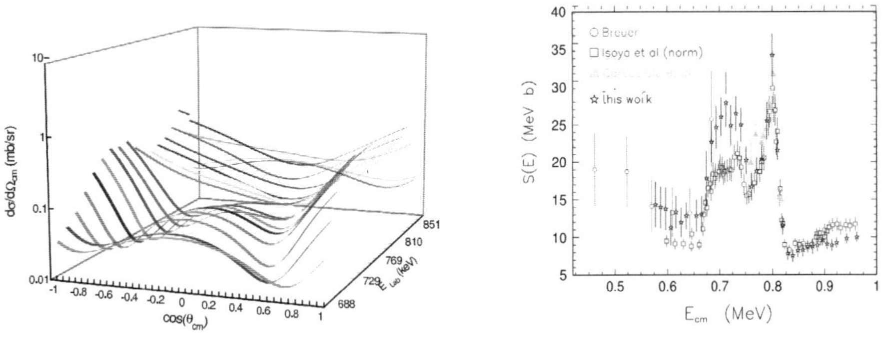

Figure 2. (Left panel) Best fits of ^{19}F(p,α_0) experimental angular distributions in terms of Legendre polynomials, as a function of the bombarding energy. (Right panel) ^{19}F(p,α_0) S-factor extracted in the present experiment [11] and compared with literature data [21, 27, 28] as reported in NACRE [26].

fitted with Legendre polynomials up to 4th degree [21, 27, 28]. The distributions of coe cients of the polynomial expansions around resonance energies allow to estimate the relative orbital angular momentum in the outgoing channel and therefore the J^π of excited states [21, 27, 28]. The evolution of angular distributions as a function of bombarding energy is displayed in Figure 2 (left panel). A continuous shape evolution is evident, pointing out the excitation of various states with different J^π. The analysis of angular distributions has been largely discussed in Ref. [11]; in this way we obtained J^π assignment of various states for which contrasting values were reported in the literature [25]. For example, the E_p=842 keV (E^*=13.642 MeV) state has been ambiguously reported to be 0^+ [21, 29] or alternatively 2^+ [28]. From the analysis of the present data in the E_p=810-860 keV interval we deduce a 0^+ assignment for the 842 keV state. Moreover, a contribution from a 1^- state at $E_p \approx$ 825 keV partially merged into the previous one has been seen [11]. Further details on this analysis can be found in Ref. [11].

As discussed at the beginning of this paragraph, the ^{19}F(p,α) reaction plays an important role also in Nuclear Astrophysics, with the ^{19}F(p,α_0) channel being dominant at $T < 0.1$ K [30]. The recommended values of the ^{19}F(p,α_0) S-factor at low energies are reported in NACRE [26]. They are taken from Breuer (E_{cm}=461-684 keV [27]), Isoya et al (E_{cm}=598-1385 keV [28]), Caracciolo et al (E_{cm}=760-817 keV [21]) and Cuzzocrea et al (E_{cm}=1476-2544 keV, [22]). In more recent times, indirect data obtained with the Trojan Horse Method pointed out the role played by the low-energy resonance at E_{cm} =113 keV [14].

We can recognize two different behaviors of experimental data at $E_{cm} < 0.7$ MeV, where the data of Refs. [27] and [28] differ by a factor of ≈ 40 . This discrepancy affects the non-resonant extrapolation made in [26], and therefore the reaction rate determination at stellar energies. In the experiment here discussed, we extracted the S-factor from the experimental differential cross sections in the range E_{cm}=577-982 keV [11]. The results are shown in the right panel of Figure 2 together with data reported in NACRE [26]. A reasonable agreement (within the errors) between present results and Refs. [21, 28] is seen in the energy region E_{cm} ≈800 keV. Furthermore, the S-factor smoothly increases with decreasing energy at $E_{cm} < 650$ keV, in good agreement with results by [27]. This behavior could be attributed to the excitation of two broad states $(0^+, 1^-)$ at E^* =13.224 MeV [27]. Moreover, our data and Breuer ones [27] in the $E_{cm} \approx$ 550-650 keV interval are about 40 larger than the Isoya et al ones [28]; consequently, extrapolations mainly based on this data set can underestimate the low energy non-resonant S-factor values. These findings stimulate to explore the low energy region of the ^{19}F(p,α_0) S-factor with new direct investigations. To this aim, a new experiment has been recently performed at Laboratori

Nazionali di Legnaro, exploring the ^{19}F(p,α_0) S-factor down to $E_{cm} \approx= 180$ keV. Data analysis is currently going on.

4. Conclusions

In this paper we reported some recent results about the structure of ^{13}C and ^{20}Ne nuclei, investigated by means of resonant elastic scattering (α+^{9}Be) and low energy nuclear reactions (^{19}F(p,α_0)) respectively. On the side of ^{13}C structure, the appearance of anomalies in the differential cross sections is associated to the excitation of various states in ^{13}C above the α threshold. R-matrix analysis of experimental data is currently in progress. Concerning the ^{19}F(p,α_0) reaction, from the analysis of angular distributions we estimated the J^π of various excited states of the compound nucleus ^{20}Ne, trying to solve some ambiguities still persisting in the literature. The trend of the S-factor has been also explored down to $E_{cm} \approx$0.55 MeV and has been compared with results reported in the literature, pointing out the need to increase the non-resonant contribution with respect to the currently adopted extrapolations. In this framework a new experiment, aimed at exploring the S-factor down to $E_{cm} \approx$0.18 MeV, has been recently performed and the analysis is currently ongoing.

We are strongly indebted to rof. . Rogachev for the use of *minRmatrix* code. We acknowledge also rof. E. erillo for useful discussions.

References

[1] von Oertzen W, Freer M and Kanada En yo Y 2006, *Phys. Rep.* 432 43
[2] von Oertzen W 1996, *Zeit. Phys.* A 354 37
[3] Rodrigues MRD et al 2014, *Phys. Rev.* C 89 024306, and *this conference proceedings.*
[4] Wheldon C et al 2012, *Phys. Rev.* C 86 044328
[5] Goss JD et al 1973, *Phys. Rev.* C 7 1837
[6] Freer M et al 2011, *Phys. Rev.* C 84 034317
[7] Zadro M et al 2007, *Nucl. Instrum. Methods* B 259 836
[8] Lombardo I et al 2013, *Nucl. Instrum. Methods* B 302 19
[9] Lombardo I et al 2013, *J. Phys. Conf. Ser.* 436 012021
[10] Milin M and von Oertzen W 2002, *Eur. Phys. Jour.* A 14 295
[11] Lombardo I et al 2013, *J. Phys. G: Nucl. Phys.* 40 125102
[12] Rolfs C and Rodney W 1988, *Cauldrons in the Cosmos*, Chicago Univ. Press
[13] Lucatello et al 2011, *Astr. Jour.* 729 40
[14] La Cognata M et al 2011, *Astr. Jour. Lett.* 739 L54
[15] Campajola L et al 1987, *Nucl. Instrum. Methods* B 29 129
[16] Chiba Y and Kimura M 2014, *this conference proceedings.*
[17] Ajzenberg- elove F 1991, *Nucl. Phys.* A 523 1
[18] Rogachev G et al 2014, *this conference proceedings.*
[19] Rogachev G et al 2007, *Phys. Rev.* C 75 014603
[20] Beck C (ed.) 2010, *Lect. Not. Phys.* 818, Vol. 1
[21] Caracciolo R et al 1974, *Lett. Nuovo Cim.* 11 33
[22] Cuzzocrea P et al 1980, *Lett. Nuovo Cim.* 28 515
[23] Arima A, Gillet V and Ginocchio J 1970, *Phys. Rev. Lett.* 25 1043
[24] Middleton R, Garret JD and Fortune HT 1971, *Phys. Rev. Lett.* 27 950
[25] Tilley DR et al 1998, *Nucl. Phys.* 636 249
[26] Angulo C et al 1999, *Nucl. Phys.* 656 3
[27] Breuer G 1959, *Zeit. Phys.* 154 339
[28] Isoya A, Ohmura H and Momota T 1958, *Nucl. Phys.* 7 116
[29] Webb T , Hagedorn F B, Fowler W A and Lauritsen C C 1955, *Phys. Rev.* 99 138
[30] pyrou K et al 2000, *Eur. Phys. Jour.* 7 79

Molecular Structures and monopole excitations in light neutron-excess systems

M. Ito[1,2] and K. Ikeda[2,3]

[1]Department of Pure and Applied Physics, Kansai University, 3-3-35 Yamatecho, uita, Osaka 564-8680, Japan.

[2]Research Center for Nuclear Physics (RCNP), Osaka University, Mihogaoka, 10-1, uita 567-0047, Japan.

[3]RIKEN Nishina Center for Accelarator based cience, RIKEN, 2-1 Hirosawa, Wako, aitama 351-0198, Japan.

E-mail: itomk@kansai-u.ac.jp

Abstract. The generalized two-center cluster model (GTCM), which can treat covalent, ionic and atomic configurations in general systems with two inert cores plus valence nucleons, is formulated on the basis of the microscopic cluster model. In this model, the covalent configurations constructed by the molecular orbital (MO) method and the atomic (or ionic) configuration obtained by the valence bonding (VB) method can be handled in a consistent manner. GTCM is applied to the light neutron-rich system, ^{12}Be = α + α + 4N. The energy spectra are characterized in terms of the excitation degrees of freedom, such as the the excitation of two α relative motion and the neutrons single particle excitation. A large enhancement in a monopole transition from a ground MO state to the excited states is found, which seems to be consistent to a recent observation. The systematics of the 0^+ states in even Be isotopes over a wide range of the excitation energy is also discussed.

1. Introduction

In the last two decades, developments of experiments with secondary RI beam have extensively advanced the studies on light neutron-rich nuclei. In particular, much efforts have been devoted to the investigation of molecular structure in Be isotopes. The Be isotopes can be considered as typical examples of two-center superdeformed systems which build on an α + α rotor of ^{8}Be. Theoretically, molecular orbital (MO), such as the π^- and σ^+ orbitals associated with the covalent bonding in molecular systems, have been successful in understanding the low-lying states of Be isotopes as well as B isotopes [1, 2, 3].

There is a long history in the MO model, and the MO picture has been extended to various systems beyond the Be and B isotopes [4, 5]. The MO model can describe many kinds of characteristic properties of Be and B isotopes, but they are mainly limited to the analysis on low-lying bound states, and theoretical studies on the highly excited states above the particle-decay threshold is still open area. In contrast to the situation of theoretical studies, recent experiments on ^{12}Be revealed the existence of many resonant states [6, 7], which strongly decay into ^{6}He$_{g.s.}$+^{6}He$_{g.s.}$ and α+^{8}He$_{g.s.}$. These resonances are expected to have the structures similar to the ionic or valence bonding (VB) structure in molecular physics, in which electrons are trapped in one of ions and generate the exchange force due to the anti-symmetrization effect.

In the present study, we investigate the exotic molecular structures appearing in an unbound region of ^{12}Be and even Be isotopes. In order to do a unified study of the ionic VB structures in a continuum and the low-lying MO structures, the traditional MO model [1, 2] must be extended because the MO structures do not satisfy the asymptotic boundary condition of the scattering states in a continuum, which are described by the ionic VB structures. Therefore, the MO and VB structures are combined to each other for the study of various molecular structures, appearing from bound states to continuum states.

For this purpose, we have formulated the generalized two-center cluster model (TCM) [8]. In this model, the covalent MO configuration can be smoothly connected to the atomic or ionic VB configuration, in which valence neutrons are localized around one of the α cores. Furthermore, it becomes possible to describe both the formation of the covalent MO structures and the reaction process induced by the collision of the ionic VB state. In ^{12}Be for instance, this model can handle the formation of an atomic structure of ^{6}He+^{6}He, ionic structures of α+^{8}He and ^{5}He+^{7}He, and covalent MO structures in a consistent manner. In this report, we investigate the exotic molecular structures based on two α cores plus excess neutrons in ^{12}Be, which appears in the continuum energy region. In addition, we also discuss the monopole excitation in ^{12}Be. The monopole excitation is considered to be a candidate of a sign to identify the molecular structure [9, 10]. Finally, the application of TCM is extended to even Be isotopes, and the systematic analysis of even Be isotopes are presented.

2. Theoretical Framework

The detailed explanation of TCM has already been published in Refs. [8, 9], and we briefly explain its formulation in the following. In this model, the basis function of Be $= \alpha + \alpha +$ N is given by the $\{\Phi_{\mathbf{m}}^{J^\pi K}(S)\}$, where,

$$\Phi_{\mathbf{m}}^{J^\pi K}(S) = \hat{P}_K^{J^\pi} \mathcal{A} \left\{ \psi_L(\alpha)\psi_R(\alpha) \prod_{j=1}^{X} \varphi_j(m_j) \right\}_S . \quad (1)$$

The α-cluster $\psi_n(\alpha)$ (n=L, R) is expressed by the (0s)^4 configuration of the harmonic oscillator (HO) centered at the left(L)- or right(R)- side with the relative distance S. The single-particle wave function for the valence neutrons localized around one of the α clusters is given by an atomic orbital (AO) $\varphi(p_k, i, \tau)$, and 0p-orbitals p_k ($k = x, y, z$) around i ($= L$ or R) with the spin τ ($= \uparrow$ or $\downarrow$).

Here, $\{m_j\}$ are indices of AO (p_k, i, τ) and $\mathbf{m}$ represents a set of AOs for the valence neutrons, $\mathbf{m}$=($m_1 \sim m_X$). The intrinsic basis functions with the full anti-symmetrization $\mathcal{A}$ are projected to the eigenstate of the total spin J, its intrinsic angular projection K, and the total parity π by the projection operator $\hat{P}_K^{J^\pi}$.

The total wave function is finally given by taking the superposition over S, $\mathbf{m}$ and K as

$$\hat{\Psi}_\nu^{J^\pi} = \int dS \sum_{\mathbf{m},K} C_{\mathbf{m}K}^\nu(S)\, \Phi_{\mathbf{m}}^{J^\pi K}(S) . \quad (2)$$

The coe cients for the ν-th eigenstate, $C_{\mathbf{m}K}^\nu(S)$, are determined by solving a coupled-channel CM (enerator Coordinate Method) equation [11]. The present calculation is restricted to the J=K=0 case; however, we include all the possible AO configurations for the excess neutrons within this approximation. Therefore, the MO structures, where each valence neutron rotates around two centers simultaneously, and the VB structures with the ionic configuration of the excess neutrons, can be handled in a consistent manner [8, 9].

As for the nucleon-nucleon interaction, we use the Volkov No.2 [12] and the 3RS [13] for the central and spin-orbit parts, respectively. The parameters in the interactions and the size parameter of HO are the same as those applied in Refs. [8, 9], which successfully reproduce the properties of ^{10,12}Be. The adopted parameter set reasonably reproduces the threshold energies of ^{x}He+^{y}He in respective Be isotopes.

The unified treatment of the formation of the intrinsic states, which have the MO or VB structures, and their coupling to the scattering continuum can be achieved by the combined method of TCM [8, 9] and the closed state method [14, 15]. In this method, the total wave function is given by the following linear combination

$$\Psi^{J^\pi(+)} \quad = \quad \sum_\beta \varphi_\beta \, \chi_\beta^{(+)} \; + \; \sum_\nu b_\nu \hat{\Psi}_\nu^{J^\pi} \; . \tag{3}$$

The first term stands for the "open channels" labeled by β, on which the scattering boundary condition is explicitly imposed [8, 9]. Here, φ_β and $\chi_\beta^{(+)}$ denote the internal and relative wave-functions of the two scattering nuclei, and the whole nucleons included in the scattering nuclei are completely anti-symmetrized. In the calculation of the $^{12}\mathrm{Be} = \alpha + \alpha + 4N$ system, for instance, we consider three rearrangement channels, such as α+$^8\mathrm{He}_{g.s.}$, $^6\mathrm{He}_{g.s.}$+$^6\mathrm{He}_{g.s.}$ and $^5\mathrm{He}_{g.s.}$+$^7\mathrm{He}_{g.s.}$, as open channels.

The second term stands for the "intrinsic states" confined within the interaction region, which is a linear combination of the solutions in Eq. (2). The solutions in Eq. (2) are simply calculated by the diagonalization procedure. Therefore, the basis of the intrinsic states, $\hat{\Psi}_\nu^{J^\pi}$ labeled by an eigenvalue number ν, is damped in the asymptotic region. The second term describes the closed compound states formed before decaying into the binary open channels.

3. Results

3.1. Classification of the levels according to the excitation degrees of freedom

The energy levels of ^{12}Be are calculated under the appropriate boundary conditions; the bound state boundary condition and the scattering boundary condition. The internal MO states, $(\pi_{3/2}^-)^2(\sigma_{1/2}^+)^2$ in 0_1^+ and $(\pi_{3/2}^-)^2(\pi_{1/2}^-)^2$ in 0_2^+, appear in the bound region. In addition to bound MO states, we newly identified the four resonant states with a various structures: α+$^8\mathrm{He}_{g.s.}$ in 0_3^+, $^5\mathrm{He}_{g.s.}$+$^7\mathrm{He}_{g.s.}$ in 0_6^+, $^6\mathrm{He}_{g.s.}$+$^6\mathrm{He}_{g.s.}$ in 0_4^+, $\tilde{\varphi}(L)\tilde{\varphi}(R)(\sigma_{1/2}^+)^2$ in 0_5^+.

The excitation schemes of these six 0^+ states are explained in Ref. [8], and they are classified in Fig. 1. We briefly summarized the excitation schemes as follows: (1) Molecular orbital (MO) excitation: The excited states with a different type of the covalent MO structures corresponds to the MO excitation from the ground states. For example, the excitation of $0_1^+ \to 0_2^+$ can be interpreted in terms of the MO excitation with $(\sigma_{1/2}^+)^2 \to (\pi_{1/2}^-)^2$. In the similar manner, the 0_2^+ and 0_5^+ states are characterized by the MO excitation from the ground 0_1^+ state. These excitations involve the pair rearrangements of the neutrons: $(\sigma_{1/2}^+)^2 \to (\pi_{1/2}^-)^2$ for 0_2^+ and $(\pi_{3/2}^-)^2 \to \tilde{\varphi}(L)\tilde{\varphi}(R)$ for 0_5^+, in which $\tilde{\varphi}(i)$ denotes the atomic orbit localized at the center i $(=L, R)$. The MO excitations involves the change in the mean distance of two α cores.

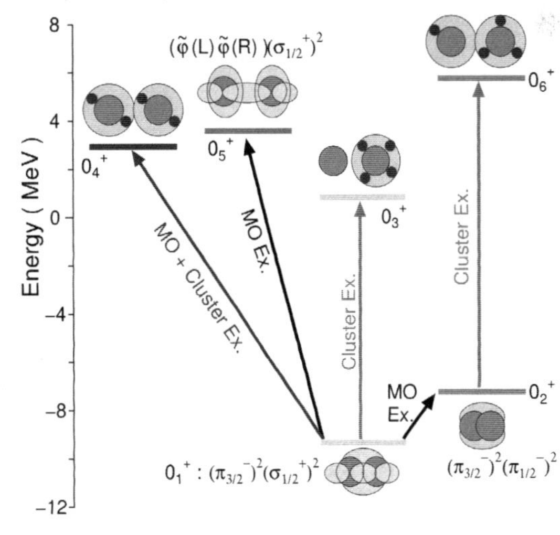

Figure 1. Energy spectra classified by the excitation degrees of freedom in ^{12}Be (J^π=0^+). See text for details.

(2) Cluster excitation: The 0_3^+ state has a dominant amplitude of α + ^{8}He$_{g.s.}$, and this state is realized as a result of the excitation of the $\alpha - \alpha$ relative wave function from the 0_1^+ state. Specifically, the 0_3^+ state corresponds to the higher nodal state of the 0_1^+ state. In a similar classification, the 0_6^+ state with ^{5}He$_{g.s.}$ + ^{7}He$_{g.s.}$ corresponds to the excited state of the two α relative motion from the 0_2^+ state.

(3) Double excitation: The main component of 0_4^+ is a linear combination of the MO configuration of $(\pi_{3/2}^-)^2(\pi_{3/2}^+)^2$ and the VB configuration of ^{6}He$_{g.s.}$+^{6}He$_{g.s.}$. The mixture of these two configurations is generated by the double excitations in the $\alpha - \alpha$ relative motion and the excitation of the MO motion of the excess neutrons.

The ^{12}Be nucleus is constructed by adding four excess neutrons to the unbound two α cores of ^{8}Be. This means that, in ^{12}Be = α + α + $4N$, the ground state contains two degrees of freedom by nature: the $\alpha - \alpha$ relative motion and the single particle motion of the four excess neutrons. All the excited states are realized by activating these two degrees of freedom although these degrees of freedom are not strongly activated in the ground state.

3.2. Monopole transition in ^{12}Be

It is very interesting to consider the dynamical excitation of the static structures summarized in Fig. 1 in realistic experiments. We consider the monopole excitation, which is a direct transition from the ground 0_1^+ to the excited 0^+ states. The monopole transition is expected to be an important key to identify the excitation degrees of freedom [8, 10]. The monopole operators for the isoscalar transition ($\hat{\mathcal{M}}(IS)$) and isovector transition ($\hat{\mathcal{M}}(IV)$) are given by

$$\hat{\mathcal{M}}(IS) = \sum_{i=1}^{A} \xi_i^2, \quad \hat{\mathcal{M}}(IV) = \sum_{i=1}^{A} \xi_i^2 \tau_{3i} \ , \tag{4}$$

where ξ_i is a coordinate of i-th nucleon measured from the center of mass in the total system. Both the operators in Eqs. (4) are composed of the squared radius operators for composite nucleons and hence, monopole excitations are sensitive to the (transition) density distribution of a nucleus; specifically, the isoscalar operator strongly sensitive to a total density distribution of a nucleus, while the isovector operator strongly responds to a difference of the neutron and proton distributions.

Here we focus on the case of ^{12}Be and discuss the ratio of the monopole matrix elements, such as

$$\left|\tilde{M}(T)\right| \equiv \left|\frac{M(T)}{M^{s.p.}}\right|, \quad T = IS, IV \ . \tag{5}$$

Here $M(T)$ denotes the monopole matrix elements defined by

$$M(T) = < 0_\nu^+ | \ \hat{\mathcal{M}}(T) \ | \ 0_1^+ > \quad (\nu = 2 \sim 6) \tag{6}$$

with $T = IS$ or IV, while $M^{s.p.}$ represents the single particle limit of the monopole strength [9].

In table 1, the ratio of the monopole strength for the isoscalar (IS) and isovector (IV) are shown. All the isoscalar strength ($|\tilde{M}(IS)|$) are comparable to or a few times larger than the single particle strength. In a recent experiment [7], the IS monopole matrix element for the 0_3^+ state, corresponding to the band head of α + ^{8}He$_{g.s.}$, has been measured [7]. The observed $|\tilde{M}(IS)|$ strength is about 2.08, which is consistent to the TCM calculation, $|\tilde{M}(IS)| = 3.53$, shown in table 1. The isovector ($|\tilde{M}(IV)|$) strength is weaker than the isoscalar strength, but all the magnitudes are still comparable to the single-particle limit value of $M^{s.p.}$. Thus, the monopole strength comparable to the single particle strength appears at $E_x \leq 20$ MeV in the present system.

State	$\vert\tilde{M}(IS)\vert$	$\vert\tilde{M}(IV)\vert$
0_2^+	2.59	0.98
0_3^+	3.53	0.75
0_4^+	0.92	0.69
0_5^+	1.48	0.90
0_6^+	1.76	0.57

Table 1. The ratios of the various monopole strength and the single particle strength in the 1p-shell. Individual strength $\tilde{M}$ are defined by $\tilde{M}(T) \equiv M(T)/M^{s.p.}, T = IS, IV$.

The results shown in table 1 are in marked contrast to a naive mean field picture. In a naive mean field picture, a $2\hbar\omega$ ($\sim$ 35 MeV) jump is needed for monopole excitations because a monopole excitation of a single particle wave function involves one higher nodal excitation. The present calculations predict that the monopole strength comparable to the mean-field model appears in much lower energy region than the region expected in a naive mean-field picture.

3.3. Systematics in even Be isotopes

The systematics of the 0^+ energy levels in even Be isotopes is shown in Fig. 2. The systems becomes the bound states (discrete levels) and continuum states (continuous reaction probabilities) below and above the lowest He decay threshold energy, respectively. We can observe the formation of intruder 0^+ states along the levels connected by the dashed lines. In ^{8}Be, a well developed α + α state exists just above the α threshold. This unbound two-α state becomes a bound 0_2^+ state at $E_x \sim 6$ MeV in ^{10}Be with two extra neutrons. The extra neutrons form the MO configuration of $(\sigma^+)^2$ around two α cores. The 0_2^+ state in ^{10}Be becomes the ground 0_1^+ state in ^{12}Be, which also has the MO configuration of $(\pi^+)^2(\sigma^+)^2$, and the systematics of ^{8,10,12}Be nicely reproduce the experimental observation. The energy of the intruder 0^+ state becomes high as a system goes to the neutron drip-line nuclei, ^{14,16}Be.

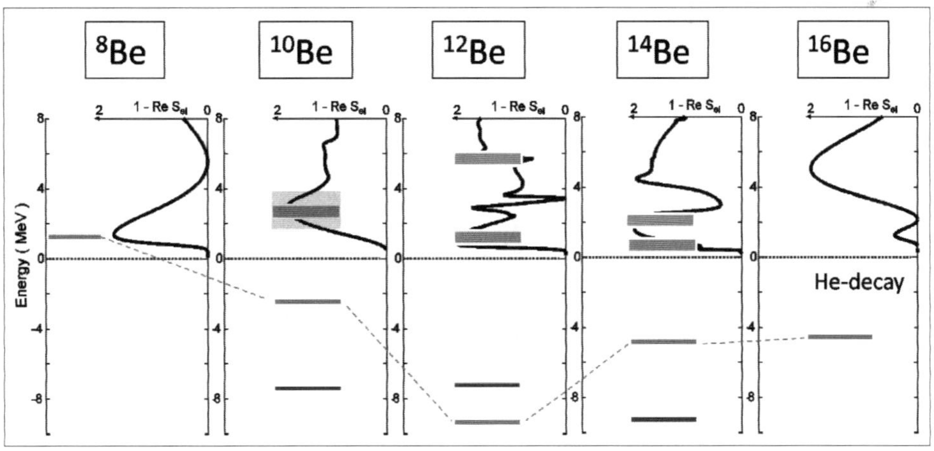

Figure 2. Systematics of the $J^\pi = 0^+$ energy levels for even Be isotopes. The lowest He-decay thresholds are taken to be zero energy point. The solid curves above the threshold represent the total reaction probabilities for the central collision. See text for details.

Above the He-decay threshold (dotted lines at zero energy), the several resonances are identified, and some of them are plotted by the solid lines in Fig. 2. The resonant levels mainly have the VB structure of ^{x}He + ^{y}He, such as ^{10}Be = α + ^{6}He, ^{12}Be = α + ^{8}He and ^{14}Be = ^{6}He + ^{8}He. In ^{10,12,14}Be, the resonant VB levels are realized as the excitation of the $\alpha - \alpha$ relative motions from the bound 0_1^+ and 0_2^+ states. On the contrary, the 0_2^+ state corresponds to the excited states of the neutrons' MO motion from the ground 0_1^+ state. Therefore, the excited states in Be isotopes are generated by the excitations of two degrees of freedom: the $\alpha - \alpha$ relative motion and neutrons' MO motion. This results means that a ground states of Be isotopes have a hybrid structure of two α-cores and extra neutrons, and various excited states are generated by the excitation of these two degrees of freedom.

4. Summary

In summary, we have studied the exotic molecular structure in ^{12}Be and even Be isotopes from bound states to a continuum. To achieve the unified studies of the molecular orbit (MO) structure in a bound region and the valence bonding (VB) structure in a continuum, we have applied the generalized two-center cluster model (TCM) [8, 9]. The energy levels are calculated under the appropriate boundary conditions. Various molecular structures appear in the excited states, and these molecular structures are classified in terms of the excitation degrees of freedom; the 2α excitations and the excitation of excess neutrons.

The isoscalar and isovector monopole strengths are investigated. We have found that all the energy levels have the monopole strengths comparable with or larger than the respective single-particle strength. As a consequence, in the ^{12}Be nucleus, strong monopole transitions can be observed in the lower energy region than the region expected from a naive mean-field picture. The magnitude of the monopole strength is consistent with recent observations [7].

The application of TCM has been extended to even Be isotopes, and the energy levels of the 0^+ state are shown over a wide range of the excitation energy. In the bound region below the lowest He decay threshold, the covalent MO structures appears, while the ionic VB structures, such as ^{x}He+^{y}He, are realized in the unbound region. These two kinds of configurations are generated by the two α excitation and excitation of the excess neutrons. Therefore, the coupling of the α clusters and the excess neutrons is essential in the formation of the excited states.

The α cluster structure is not peculiar in Be isotopes but generally appear in the excited states in light $N = Z$ systems, such as ^{12}C = 3α, ^{16}O = α + ^{12}C and ^{20}Ne = α + ^{16}O, for example. Therefore, such the wide variety induced by the α cluster and the excess neutrons is expected to be general in the light nuclear systems, and the systematic studies is interesting. The application of TCM to other cluster systems is now under progress.

References

[1] Itagaki N and Okabe 2000 Phys. Rev. C **61** 044306 (and references therein.)
[2] Furutani H, Kanada H, Kaneko T, Nagata , Nishioka H, Okabe , aito , akuda T, eya M 1980 uppl. Prog. Theor. Phys. **68** 193; eya M, Kohno M, and Nagata 1981 Prog. Theor. Phys. **65** 204
[3] von Oertzen W 1996 Z. Phys. A **354** 37
[4] von Oertzen W 1997 Z. Phys. A **357** 355
[5] Kimura M 2007 Phys. Rev. C**75** 034312
[6] A. aito et al. 2010 Mod. Phys. Lett. A **25** 1858 (and references therein.)
[7] Yang Z H, et al. 2014 Phys. Rev. Lett. **112** 162501 (and references therein.)
[8] Ito M 2012 Phys. Rev. C **85**, 044308
[9] Ito M 2011 Phys. Rev. C **83** 043319
[10] Yamada T, Funaki Y, Horiuchi H, Ikeda K, and Thosaki A 2008 Prog. Theor. Phys. **120** 1139
[11] Horiuchi H 1977 uppl. Prog. Theor. Phys. **62** 90
[12] Volkov A B 1965 Nucl. Phys. **74** 33
[13] Yamaguchi N, Kasahara T, Nagata and Akaishi Y 1979 Prog. Theor. Phys. **62** 1018
[14] Kamimura M 1977 Prog. Theor. Phys. uppl. **62** 236
[15] Hamaguchi K, Hatsuda T, Kamimura M, Kino Y and Yanagida T T 2007 Phys. Lett. B **650** 268

Journal of Physics: Conference Series **569** (2014) 012070 doi:10.1088/1742-6596/569/1/012070

A solution to the old puzzle of 0^+ resonances above the 0_2^+ Hoyle state in ^{12}C

— *A new analysis with complex-scaled 3α OCM* —

S Ohtsubo[1], **Y Fukushima**[1], **M Kamimura**[2] **and E Hiyama**[2]

[1] Department of Applied Physics, Fukuoka University, Fukuoka 814-0180, Japan
[2] RIKEN Nishina Center, RIKEN, Wako 351-0198, Japan

E-mail: mkamimura@riken.jp

Abstract. We have developed the complex-scaling method (CSM) by using the complex-range (or oscillating) Gaussian basis functions that are suited for describing highly oscillating few-body wave functions. The eigenvalue distribution of the complex scaled Hamiltonian becomes much more precise and the maximum scaling angle becomes drastically larger than those given by the use of real-range Gaussians. Owing to this advantage, we were able to isolate the S-matrix pole of the new broad 0_3^+ resonance from the 3α continuum. This confirms the Kurokawa-Kato's prediction (2005) of the new 0_3^+ resonance, which is considered to correspond to the newly observed 0_3^+ resonance (E_x = 9.04 MeV,Γ = 1.45 MeV) by Itoh *et al.* (2013). As a result the long-standing puzzle for the 0^+ and 2^+ resonances above the 0_2^+ Hoyle state in ^{12}C was solved. In this paper, the negative parity resonances with $J = 1^-, 2^-, 3^-, 4^-$ and 5^- are newly calculated.

1. Introduction

There was a long-standing puzzle, as shown in Fig. 1, for the 0^+ and 2^+ resonances above the 0_2^+ Hoyle state in ^{12}C; observation [1] and the typical 3α-cluster-model calculations are inconsistent. Recently, this puzzle has been almost solved by the new cluster-model calculation by Kurokawa and Kato [2] and the new experiment by Itoh *et al.* [3] as shown in Fig. 2.

The authors of Ref. [2] predicted a new broad 0_3^+ resonances at $E_{\rm x}$ = 8.95 MeV with Γ = 1.48 MeV using the complex-scaling method (CSM) (see Ref. [4] for a review) within the 3α orthogonality-condition model (OCM). However, the prediction is not very satisfactory from the following point: Their CSM calculation was performed at the scaling angle of 16° that is the largest angle available in their calculation using the real-range Gaussian basis. Since the angle is not enough for separating the low-lying broad 0_3^+ resonance from the continuum eigenvalues, they made an extrapolation by the method of analytic continuation of the coupling constant (ACCC) [5] to derive the complex energy of the 0_3^+ resonance (cf. Fig. 6 below).

In order to overcome this difficulty and confirm the prediction in Ref. [2], we proposed [6] to use the complex-range Gaussian basis functions [7, 8],

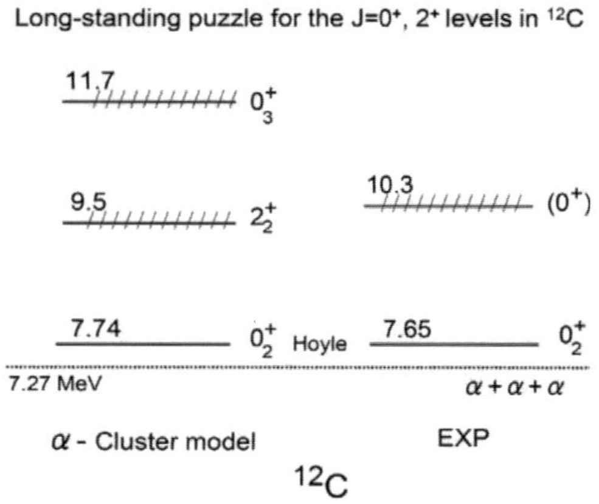

Figure 1. The 0^+ and 2^+ resonances in ^{12}C above the 0_2^+ Hoyle state that were inconsistent between the experiments [1] and the typical cluster-model calculations.

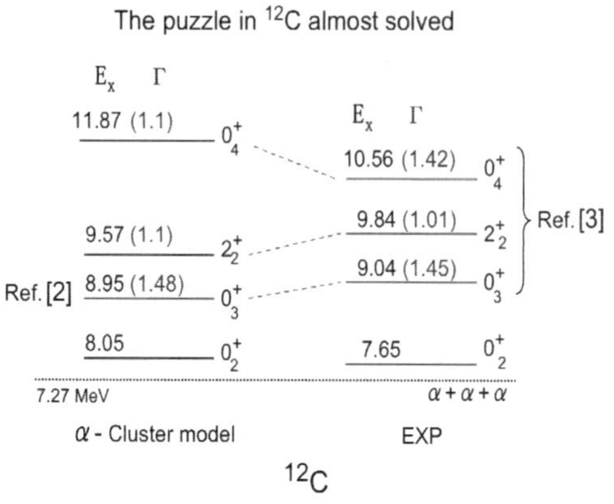

Figure 2. The new cluster-model calculation for the 0_3^+ resonance [2] and the new experiment for the 0_3^+, 2_2^+ and 0_4^+ resonances [3]. All the calculated results are cited from Ref. [2]. The energy E_x and the width Γ are in MeV.

$$r^l\, e^{-(1\pm i\omega)(r/r_n)^2} Y_{lm}(\hat{\mathbf{r}}), \qquad \text{or equivalently} \qquad r^l\, e^{-(r/r_n)^2} \{ {\cos\omega(r/r_n)^2 \atop \sin\omega(r/r_n)^2} \} Y_{lm}(\hat{\mathbf{r}}) \tag{1}$$

with r_n in a geometric progression, in the calculation of three-body resonances with CSM in which use is often made of the real-range Gaussian basis functions [9, 7, 4],

$$r^l\, e^{-(r/r_n)^2} Y_{lm}(\hat{\mathbf{r}}). \tag{2}$$

Though the latter set is suitable for describing the short-distance structure and the asymptotic decaying behavior of few-body systems [7], the former basis set is more powerful when describing the resonant and non-resonant continuum states in CSM with highly oscillating amplitude at large scaling angles θ.

We applied the new basis functions (both for the two Jacobi coordinates, $\mathbf{r}$ and $\mathbf{R}$) to the 3α OCM-CSM calculation [6] of the 3α resonances with $J = 0^+, 2^+$ and 4^+ in ^{12}C (see Ref. [6] for details). In this paper, we newly calculate the negative parity resonances with $J = 1^-, 2^-, 3^-, 4^-$ and 5^-. We took the same model and the Hamiltonian as used in the 3α OCM-CSM calculation in Ref. [2].

2. The broad new 0_3^+ resonance

Figure 3, taken from Ref. [2], shows the 0^+ eigenvalue distribution of the complex scaled Hamiltonian calculated with the real-range Gaussian basis functions. The scaling angle $\theta = 16°$ was the maximum angle available in the calculation. The new broad 0_3^+ state was predicted at $E_{\rm res} = 1.66 - i\,0.74$ MeV ($E_{\rm x} = 8.95 - i\,0.74$ MeV) but this complex energy is not isolated from the continuum states in Fig. 3 at $\theta = 16°$. The energy was derived by an extrapolation based on the method of analytic continuation of the coupling constant [5] (cf. Fig. 6 below).

Figure 4 illustrates the result of our CSM calculation [6] for the $J = 0^+$ states at $\theta = 16°$ and $26°$. The distribution of the S-matrix becomes very much improved and precise. However, even at $\theta = 26°$, the 0_3^+ resonance is not isolated from the continumm states.

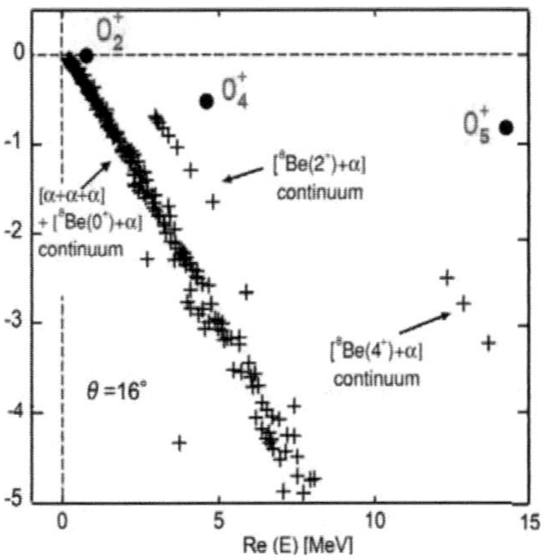

Figure 3. The 0^+ eigenvalue distribution of the complex scaled Hamiltonian for the 3α system obtained by Kurokawa and Katō [2] using the real-range Gaussian basis functions. The scaling angle is $\theta = 16°$. The 0_3^+ state was predicted at $E = 1.66 - i\,0.74$ MeV (see the text).

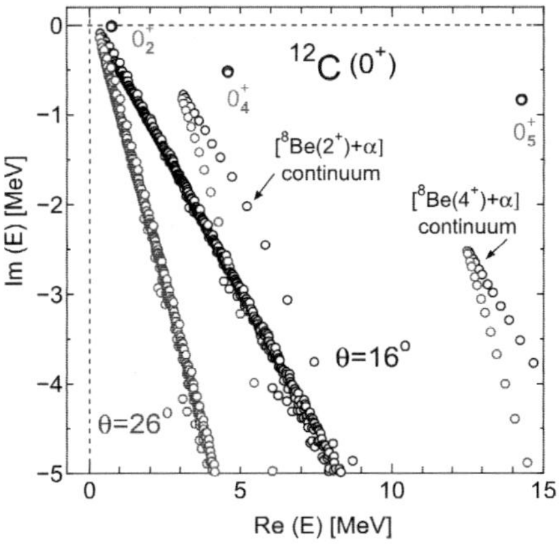

Figure 4. The 0^+ eigenvalue distribution of the complex scaled Hamiltonian for the 3α system with the use of the complex-range Gaussian basis functions [6]. The scaling angles are $\theta = 16°$ (black) and 26° (blue). This figure is to be compared with Fig. 3.

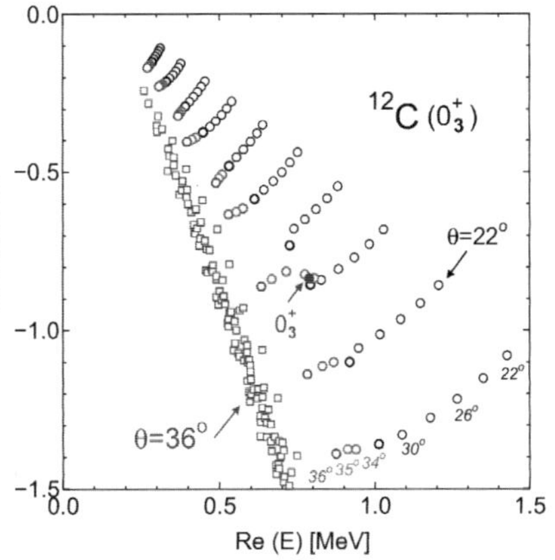

Figure 5. The 0^+ eigenvalue distribution of the complex scaled Hamiltonian, where the angle θ is varied from 22° to 36° [6]. The 0_3^+ resonance appears, as the closed blue circle (36°), at $E = 0.79 - i\,0.84$ MeV. Only for $\theta = 36°$ both the continua of $[\alpha + \alpha + \alpha]$ (open blue boxes) and $[^8\text{Be}(0^+) + \alpha]$ (open blue circles) are illustrated, but the former is omitted for $\theta < 36°$ for clarity of the figure.

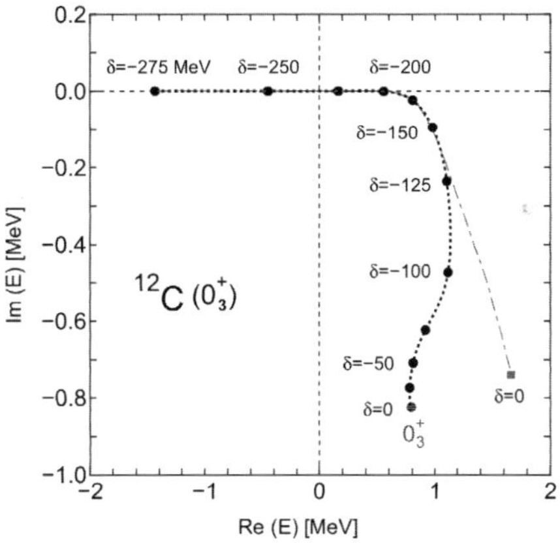

Figure 6. Trajectory of the 0_3^+ state obtained by changing the strength parameter δ of the auxiliary three-body potential [6]. The blue closed circle at $\delta = 0$ corresponds to the 0_3^+ resonance in Fig. 5. The green box denotes the 0_3^+ state predicted by Kurokawa and Katō [2] on the basis of the extrapolation (the dash-dotted green curve) using the ACCC+CSM.

Journal of Physics: Conference Series **569** (2014) 012070 doi:10.1088/1742-6596/569/1/012070

We illustrate, in Fig. 5, the 0^+ distribution of complex eigenvalues. Only for $\theta = 36°$ both the $[\alpha+\alpha+\alpha]$ continuum (open blue boxes) and the $[^8\mathrm{Be}(0^+)+\alpha]$ continuum (open blue circles) are given, but the former is omitted for $\theta < 36°$ to avoid complexity of the figure. We observe a converged resonance pole at $E_{\mathrm{res}} = 0.79 - i\,0.84$ MeV as shown by the closed blue circle. We identify it as the third 0^+ state that was predicted in Ref. [2]. The position and width of the resonance, however, differ slightly from the result of Ref. [2], $E_{\mathrm{res}} = 1.66 - i\,0.74$ MeV.

We explain the reason of this difference with Fig. 6 which is to be compared with Fig. 2 of Ref. [2]. Fig. 6 illustrates the trajectory of the 0_3^+ state on the complex energy plane, which was obtained by changing the strength parameter δ of the auxiliary three-body potential $V_{\mathrm{aux.}} = \delta \exp[-\mu(r_1^2 + r_2^2 + r_3^2)]$ with $\mu = 0.15$ fm^{-2} (Eq. (4) of Ref. [2]) that is added to the 3α OCM Hamiltonian. The closed blue circle for $\delta = 0$ in Fig. 6 is the same as that for the 0_3^+ resonance in Fig. 5. On the other hand, in Ref. [2], the direct CSM calculation of the 0_3^+ resonance was not possible when the auxiliary 3α potential is less attractive than $\delta = -120$ MeV. The green box that indicates the 0_3^+ state of Ref. [2] was therefore estimated by the extrapolation (the dash-dotted green curve) using the ACCC+CSM. We thus understand that the difference in the resonance-pole position between the two calculations comes from the error of the extrapolation by the ACCC.

We concluded [6] that we confirmed the prediction by Kurokawa-Katō [2] about the appearance of a new 0_3^+ broad resonance which is located slightly above the Hoyle state (0_2^+).

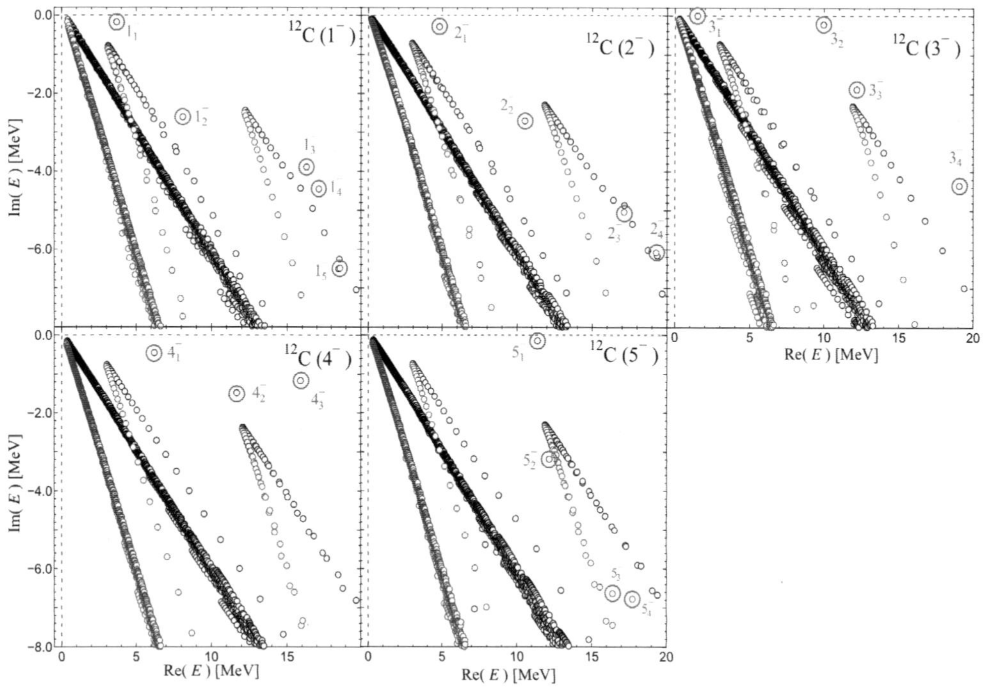

Figure 7. The eigenvalue distribution of the complex scaled Hamiltonian for $J = 1^-, 2^-, 3^-, 4^-$ and 5^- states of the 3α system with the use of the complex-range Gaussian basis functions [6]. The scaling angles are $\theta = 16°$ (black) and $26°$ (blue). The resonace poles are stable against the change of the scaling angle, and their energies and widths are summarized in Table 1.

3. Negative-parity states

In this paper, we newly calculated the negative parity resonances with $J = 1^-, 2^-, 3^-, 4^-$ and 5^- using the same model and interactions as those in Refs. [2, 6]. Figure 7 shows the eigenvalue distribution of the complex scaled Hamiltonian for the negative-parity states at the scaling angles $\theta = 16°$ (black) and $26°$ (blue).

The resonance poles are stable against the change of the scaling angle, and their energies and widths are summarized in Table 1 together with the calculated result by Kurokawa and Kato [2] and the experimental data [1]. Owing to the advantage of using the complex-range Gaussian basis functions, the very broad resonances that were not given in Ref. [2] are obtained. The experimental data for $J = 1_1^-, 2_1^-, 3_1^-$ and 4_1^- are reproduced well. Further observation of the negative-parity states are expected to examine our prediction in Table 1.

Table 1. Summary of the negative parity resonaces in ^{12}C by the present work together with the result by Ref. [2] and the experimental data [1]. All quantities are given in MeV.

^{12}C	present work			Ref. [2]			Experimental data [1]		
J^π	E_x	E_r	Γ	E_x	E_r	Γ	E_x	E_r	Γ
3_1^-	8.79	1.49	2.1×10^{-3}	8.80	1.51	2.0×10^{-3}	9.641(5)	2.366	3.4×10^{-2}
1_1^-	10.98	3.68	0.35	10.94	3.65	0.30	10.844(16)	3.569	0.315(25)
2_1^-	12.10	4.80	0.57	11.97	4.68	0.42	11.828(16)	4.553	0.260(25)
4_1^-	13.51	6.21	0.45	12.45	5.16	0.12	13.352(17)	6.077	0.375(40)
1_2^-	15.40	8.10	5.20						
3_2^-	17.25	9.95	0.45	18.29	11.0	0.5			
2_2^-	17.80	10.50	5.40	16.60	9.31	4.65			
5_1^-	18.66	11.36	0.29	18.69	11.4	0.3			
4_2^-	18.97	11.67	2.93						
5_2^-	19.47	12.17	6.37						
3_3^-	19.49	12.19	3.67						
4_3^-	23.15	15.85	2.34						
1_3^-	23.58	16.28	7.79						
5_3^-	23.70	16.40	13.27						

References

[1] Ajzenberg-Selobe F 1990 *Nucl. Phys.* A **506** 1
[2] Kurokawa C and Katō K 2005 *Phys. Rev.* C **71** 021301(R); 2007 *Nucl. Phys.* A **792** 87
[3] Itoh M *et al.* 2011 *Phys. Rev.* C **84** 054308
[4] Aoyama S, Myo T, Katō K and Ikeda K 2006 *Prog. Theor. Phys.* **116** 1
[5] Kukulin V I and Krasnopol'sky V M 1977 *J. Phys.* A **10** L33
[6] Ohtsubo S Fukushima Y Kamimura M and Hiyama E 2013 *Prog. Theor. Exp. Phys.* **2013** 073D02
[7] Hiyama E Kino Y and Kamimura K 2003 *Prog. Part. Nucl. Phys.* **51** 223
[8] Hiyama E 2012 *Prog. Theor. Exp. Phys.* **2012** 01A204
[9] Kamimura M 1988 *Phys. Rev.* A **38** 621

A newly observed state in ^{12}C – characterisation and consequences

D J Marín-Lámbarri[1], M Freer[1], Tz Kokalova[1], C Wheldon[1], D J Parker[1]

[1]School of Physics and Astronomy, University of Birmingham, Birmingham, B15 2TT, UK.

E-mail: `djm179@bham.ac.uk`

Abstract. Measurements of the ^{12}C(^{4}He,^{4}He)^{12}C* reaction at 40 MeV were performed using a charged-particle detector array composed of four double-sided silicon strip detectors. Resonances in ^{12}C were reconstructed with the requirement that the α-decay proceeded first via the ^{8}Be ground state. Of primary significance was the population of the 13.3 and 22.4 MeV resonances, the latter, observed for the first time in this measurement. By using the angular correlation technique the spins of the resonances, have been established as J^π=4$^+$ and J^π=5$^-$ respectively.

1. Introduction

Despite all the knowledge acquired over several decades, there are still important details that remain unexplained regarding the nature of the ^{12}C nucleus. Its second excited state at 7.654 MeV (0$^+$) was first proposed 61 years ago by Fred Hoyle in order to explain the abundance of ^{12}C in the universe [1]. Carbon-12 is of great importance for the creation of the elements key to organic life, as well as being the primary step in the synthesis of heavier nuclei in stars. The structure of the Hoyle state is a significant open question. It is known to be composed of three α-particles, but the precise arrangement remains unclear. *Ab initio* calculations suggest the state is configured as a bent-arm [2], in which the three α-particles are located at the vertices of an open triangle. The measurement reported here was motivated by the idea of measuring and characterising the 13.3 MeV excited state in ^{12}C which had previously been observed by members of the Charissa collaboration [3, 4], and which is thought to be a candidate member of the Hoyle-state band. Recent investigations show the existence of a 2$^+$ resonance, which could be a collective rotation or vibration of the Hoyle-state [5, 6]. The 13.3 MeV resonance may be one of these collective resonances. Ultimately, by making evermore precise measurements, the data may be sufficiently discriminating to refine current nuclear models. Some results from the present study have been published in [7].

2. Experimental Details

The measurements were performed at the University of Birmingham's MC40 cyclotron with a ^{4}He beam at an energy of 40 MeV incident on a 100 μgcm^{-2} ^{12}C target. The reaction of interest was ^{12}C(^{4}He,^{4}He+^{4}He+^{4}He)^{4}He which corresponds to the inelastic excitation of ^{12}C above the three α-decay threshold at 7.275 MeV. A bespoke charged-particle detector array of four double-sided silicon strip detectors (DSSSDs) was used. Each detector is subdivided by 16 vertical and 16 horizontal strips on the front and back faces respectively giving a 5 $\times$ 5 cm^2 surface with

Journal of Physics: Conference Series **569** (2014) 012071 doi:10.1088/1742-6596/569/1/012071

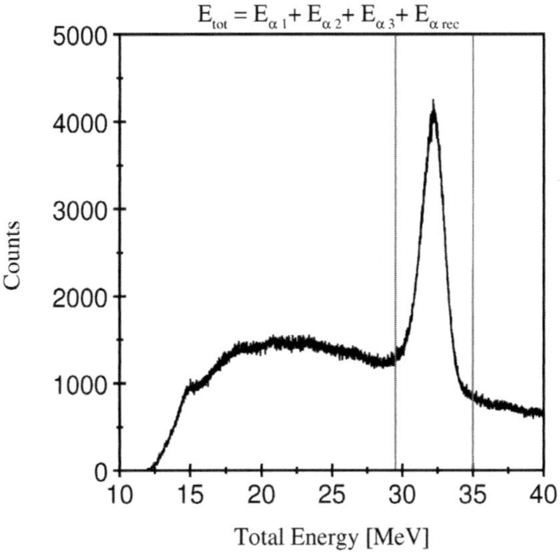

Figure 1. Total energy spectrum. The region between the two red lines was selected, as this corresponds to events proceeding from the $^{12}C(^4He,^4He+^4He+^4He)^4He$ reaction (Q=−7.275 MeV).

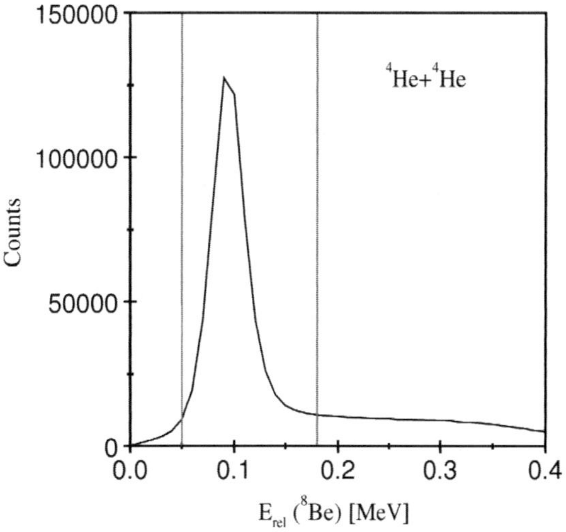

Figure 2. Peak associated with the decay from the ^{8}Be ground state at 92 keV. Data between the two red vertical lines were selected for further analysis.

a thickness of 500 μm. The master trigger was such that at least three of the four final-state particles were recorded via detectors placed at distances 13.0, 11.0, 11.0, 13.0 cm from the target at in-plane angles 62.0°, 32.0°, −32.0°, −62.0°, respectively (positive and negative values indicate opposite sides of the beam axis). The array covered an angular range of θ_{lab} = 20° to 75°. The detectors were calibrated with a triple α-particle source.

3. Analysis and Results

The DSSSDs do not have explicit particle identification so event selection was achieved through the full kinematic reconstruction allowing the energy and scattering angle of each particle to be determined and, therefore, the momentum. By conservation of momentum and energy the fourth, undetected α-particle was reconstructed. Using the information from all the products and the beam energy it was then possible to create the total energy spectrum for the reaction, as shown in figure 1. This allowed events from the reaction of interest to be selected, with Q= −7.275 MeV. A second filter, which gives an additional level of selectivity, required events originating only from the decay of the ^{8}Be nucleus in its ground state, reconstructed from pairs of detected α-particles for events within the peak of figure 1. The centroid of the peak of figure 2, at a relative energy, E_{rel} = 92 keV, clearly identifies the decay of the ^{8}Be ground state.

To visualise the event energy landscape, a Dalitz plot was generated as shown in figure 3. Excitation energies in ^{12}C were calculated for the case in which three α-particles from the decay from ^{12}C were detected (horizontal axis). For the vertical axis the excitation energies were calculated for events in which one of the ^{12}C break-up α-particles was undetected, *i.e.* the reconstructed particle; the scattered beam particle being one of the three hits. It is possible to observe features running parallel to the horizontal and the vertical axes, as well as diagonal ridges (with a gradient equal to -1). The horizontal and vertical features correspond to resonances in ^{12}C and the diagonal lines to states in ^{8}Be (from right to left the ground state is the first diagonal

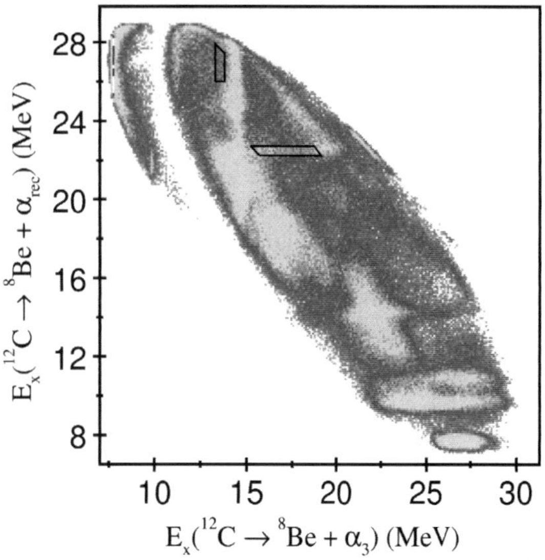

Figure 3. Dalitz plot of the $^{12}C(^4He,^4He)^{12}C^*$ reaction measured at a beam energy of 40 MeV. The solid boxes shows the region selected for the angular correlation analysis.

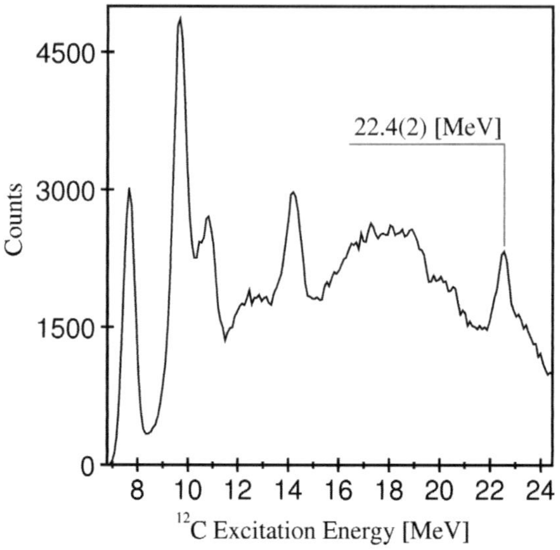

Figure 4. Projection onto the vertical axis of the Dalitz plot of figure 3. Different known resonances are shown as well as the newly observed state at a 22.4(2) MeV.

line). The projection of the Dalitz plot is shown in figure 4, in which excited states at 7.654 (0^+), 9.641 (3^-), 10.844 (1^-), 14.083 (4^+) MeV are observed. Additionally, the presence of a resonance at 22.4(2) MeV is observed for the first time.

3.1. Angular Correlations

With the aim of gaining an insight into the nature of the newly observed state in ^{12}C the distribution pattern of the final products (α-particles) in the decay has been analysed. This analysis method is formally known as angular correlations and is model independent when the initial and final particles all have spin zero, as here. A more detailed explanation of this method can be found in [8]. For this correlation analysis it is necessary to define the centre-of-mass emission angle θ^* of the ^{12}C nucleus with respect to the beam axis, and the emission angle Ψ of the first α-particle from the decay of the resonance in ^{12}C, also with respect with the beam axis, and within the ^{12}C centre-of-mass frame. Using the Dalitz plot of figure 3 it was possible to select a specific excitation energy to be examined. The event selection windows are shown by the solid rectangular boxes in figure 3, for the 13.3 and 22.4 MeV states, determining both the excitation energy and the origin of the particles detected, and therefore, enabling the angular correlation plots for the selected data to be produced. This procedure was first applied to well known resonances, 7.654 (0^+), 9.641 (3^-), 10.844 (1^-) and 14.083 (4^+) MeV, with the aim of demonstrating the effectiveness of the procedure. Subsequently, this method was applied to the 13.3 MeV and the newly observed 22.4 MeV states. The angular correlation plots for the 13.3 MeV and the newly observed resonance are shown in figure 5, demonstrating the key features in which the structure in the θ^*-Ψ plane are dependent on

$$W(\theta^*, \Psi) \propto | P_J[\cos(\Psi + \Delta\Psi)] |^2, \tag{1}$$

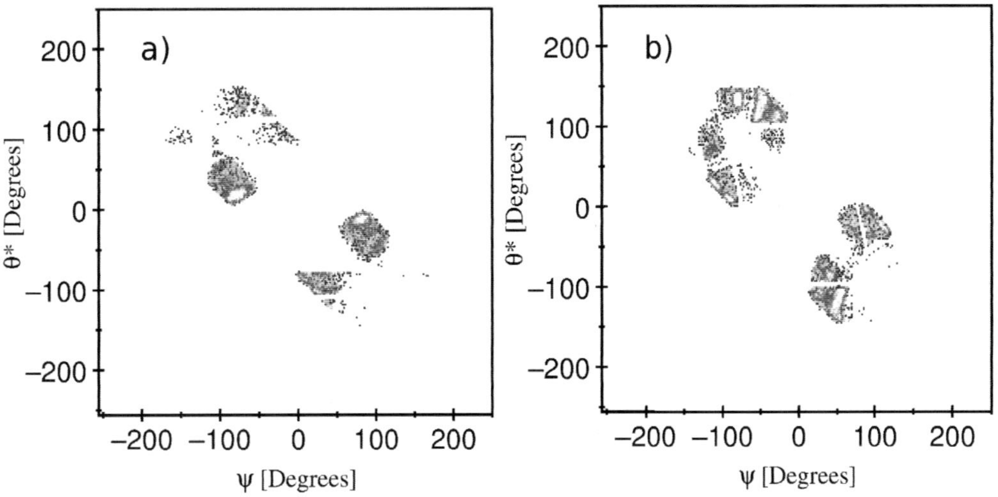

Figure 5. Angular correlation plot a) for the 13.3 MeV resonance and b) for the 22.4 MeV resonance in ^{12}C. The gradient of the angular correlation features increases as the spin increases.

to first order. A small change in Ψ, $\Delta\Psi$, is related to a small change in θ^*, $\Delta\theta^*$, through

$$\frac{\Delta\theta^*}{\Delta\Psi} = \frac{J}{l_{gr} - J}, \tag{2}$$

where l_{gr} is the grazing angular momentum J is the spin of the state and P_J are the Legendre polynomial of order J. Within the θ^*-Ψ plane, the formation of sloped features will appear with a specific gradient $\Delta\theta^*/\Delta\Psi$, given by equation 2 for which the periodicity reveals the spin of the resonance under analysis [8]. For the case of a spin-zero state the gradient will be zero, increasing as the spin increases, as shown in figure 5. In order to analyse the periodicity, the angular correlation plot is projected onto the Ψ axis at a specific angle $\Delta\theta^*/\Delta\Psi$, determined by the spin of the state. For the present reaction, the grazing angular momentum was estimated to be 9.8 $\hbar$ using r_0=1.2 fm, for which the optimum angles for the projection of the data for the 9.641 (3^-), 10.844 (1^-) and 14.083 (4^+) were found to be $\Delta\theta^*/\Delta\Psi$= 6.5°, 25.3° and 39.7° respectively. The known states all exhibit the correct periodicity for the published spin value. From figure 6 it is possible to observe that the 13.3 MeV data appear to best match the periodicity of the Legendre polynomial J=4 and for the case of figure 7 (22.4 MeV) the periodicity corresponds to order J=5. The associated angles are $\Delta\theta^*/\Delta\Psi$= 39.7° and 60.0° for 13.3 MeV and 22.4 MeV respectively. Recent measurements using ^{3}He projectiles observed a peak at 22.2(3) MeV, confirming this state [9].

4. Conclusions

The 13.3 MeV state must have natural parity as the decay involves only spin-zero nuclei, yielding a spin and parity of $J^\pi = 4^+$. Similarly for the 22.4 MeV level, the spin and parity are found to be J^π=5^-. The observation of a 5^- resonance at 22.4 MeV is strong evidence for triangular $\mathcal{D}_{3h}$ symmetry [10], based on the pattern of excited states 0^+, 2^+, 3^-, $4^\pm$ and 5^-, predicted by the algebraic cluster model [10], for an equilateral triangle structure. As a result, this symmetry is now established in nuclear physics [7]. The model also predicts additional states such as the 4^- state, degenerate with the 4^+ level; a strong signature of $\mathcal{D}_{3h}$ symmetry [3, 10]. The significance of this result is that the algebraic model also predicts the structure of the Hoyle

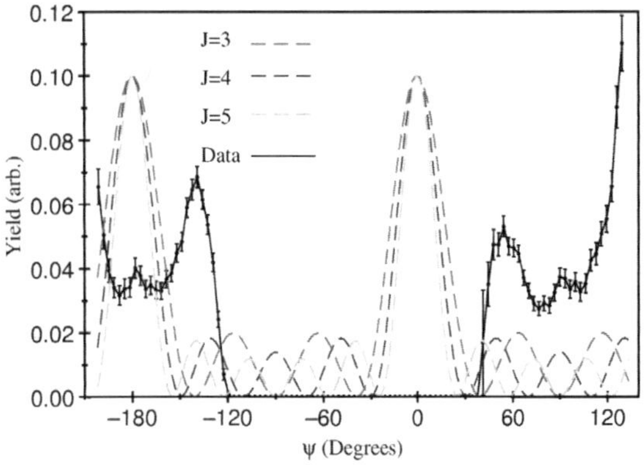

Figure 6. Projection of the angular correlation data for the 13.3 MeV resonance onto the Ψ axis at an angle of 39.7° (black line). The Legendre polynomials of order J=3, 4 and 5 are plotted for comparison (red, blue and green respectively).

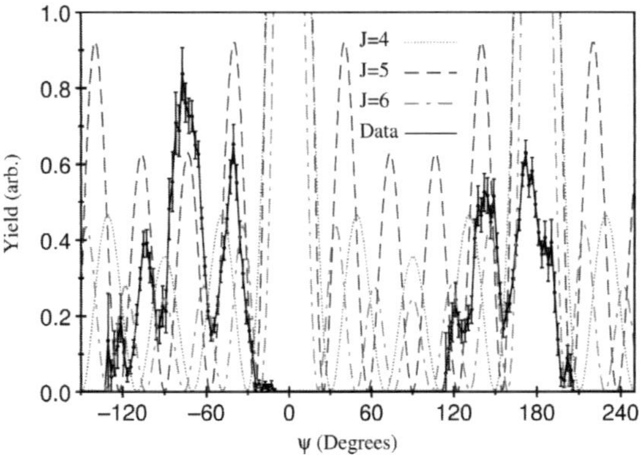

Figure 7. Projection of the angular correlation data for the 22.4 MeV resonance onto the Ψ axis at an angle of 60.0° (black line). The Legendre polynomials of order J=4, 5 and 6 are plotted for comparison (red, blue and magenta respectively), generated from [7].

state to correspond to an expanded equilateral triangle, the excitations of which are calculated to be 0^+, 2^+, 3^-, $4^\pm$, 5^-. The expanded structure means that the energy range is more compact, placing the calculated 5^- Hoyle resonance at around 17 MeV. Such concrete predictions offer a tangible challenge for future experiments.

5. Summary

A measurement of the $^{12}C(^{4}He,^{4}He)^{12}C^{*}$ reaction establishes evidence for a newly observed resonance at 22.4(2) MeV for which an angular correlation analysis was performed. The spin and parity of the state was found to be J^π=5^- confirming the observation of a triangular oblate spinning top with $\mathcal{D}_{3h}$ symmetry for the first time in nuclear physics. The spin and parity of the 13.3 MeV Hoyle-band candidate has been measured unambiguously as $J^\pi = 4^+$.

This work is supported in part by the University of Birmingham and in part from CONACyT, Mexico.

References

[1] Hoyle F 1954 *Astro. Phys. J. (Suppl.)* **1**
[2] Epelbaum E *et al.* 2011 *Phys. Rev. Lett.* **106**
[3] Freer M *et al.* 2011 *Phys. Rev. C* **83**
[4] Kirsebom O *et al.* 2010 *Phys. Rev. C* **81**
[5] Freer M *et al.* 2009 *Phys. Rev. C* **80**
[6] Hyldegaard S *et al.* 2010 *Phys. Rev. C* **81**
[7] Marín-Lámbarri D J *et al.* 2014 *Phys. Rev. Lett.* **113**
[8] Freer M 1996 *Nucl. Instr. and Meth. A* **383** 463–472
[9] Wheldon C *et al.* 2014 *Phys. Rev. C* **90**
[10] Bijker R *et al.* 2000 *Phys. Rev. C* **61**

Journal of Physics: Conference Series **569** (2014) 012072 doi:10.1088/1742-6596/569/1/012072

A resonance problem on the low-lying resonant state in the ^{9}Be system

Myagmarjav Odsuren[1], **Kiyoshi Katō**[2], **Yuma Kikuchi**[3], **Masayuki Aikawa**[2] **and Takayuki Myo**[4,5]

[1]Nuclear Research Center, National University of Mongolia, Ulaanbaatar 210646, Mongolia
[2]Nuclear Reaction Data Centre, Faculty of Science, Hokkaido University, Sapporo 060-0810, Japan
[3]RIKEN Nishina Center, Wako, 351-0198, Japan
[4]General Education, Faculty of Engineering, Osaka Institute of Technology, Osaka 535-8585, Japan
[5]Research Centre for Nuclear Physics (RCNP), Osaka University, Ibaraki 567-0047, Japan

E-mail: odsurenn@gmail.com

Abstract. The photo-disintegration cross section of ^{9}Be is investigated in a framework of the $\alpha + \alpha + n$ three-body cluster model. Much interest is concentrated on the nature of the $1/2^+$ state of ^{9}Be located just above the three-body threshold. The existence of this $1/2^+$ resonance is a long-standing problem and has been a topic of quite interest in relation to the breakup mechanism of ^{9}Be into the ^{8}Be(0^+)+n and $(\alpha + \alpha + n)$ channels. The newly measured breakup cross sections of the $1/2^+$ state of Ref. [1] is inconsistent with the old experimental values [2, 3]. The purpose of this study is to clarify the properties of the $1/2^+$ state of ^{9}Be with the possibility of a genuine three-body $(\alpha + \alpha + n)$ resonance or the two-body (^{8}Be(0^+)+n) virtual state.

1. Introduction

The neutron capture (n,γ) and its inverse (γ,n) reactions are the highly interesting topics in current nuclear physics, because these reactions play important roles in nuclear reactor physics and astronuclear physics. So far, the resonant structure [4, 5] and a virtual characteristic state [6] of ^{9}Be($1/2^+$) have been studied based on the various theoretical approaches and photo-dissociation experiments. However, a complete understanding on the nature of ^{9}Be($1/2^+$) state has not yet been obtained. It is desired to perform a more comprehensive study based on an $\alpha+\alpha+n$ three-body calculations which can treat the bound ground state and also the unbound resonant states of ^{9}Be in an unified framework. In this work, we apply the complex scaling method (CSM) [7] to an $\alpha + \alpha + n$ three-cluster model for understanding the nuclear structure and (γ,n) reactions for low-lying states in ^{9}Be.

For the purpose of this work we treat the unbound nature of the $1/2^+$ state of ^{9}Be and investigate the E1 transition between $1/2^+$ and $3/2^-$ states, which contributes to the (γ,n) reaction dominantly. We examine the resonance formation in the (γ,n) reaction to see whether a direct three-body breakup or a two-step process through the ^{8}Be+n intermediate resonant state is dominant.

Journal of Physics: Conference Series **569** (2014) 012072 doi:10.1088/1742-6596/569/1/012072

2. Three-body model and complex scaling method

The Hamiltonian for the relative motion of the $\alpha + \alpha + n$ three-body system for ^{9}Be is given as

$$\hat{H} = \sum_{i=1}^{3} t_i - T_{c.m} + \sum_{i=1}^{2} V_{\alpha n}(\xi_i) + V_{\alpha\alpha} + V_3 + V_{PF}. \tag{1}$$

The Hamiltonian consists of the kinetic energy operators of the first and second terms, the two-body potential terms of subsystems for $\alpha+\alpha$ and $\alpha+n$, the three-body $\alpha+\alpha+n$ potential term and the last term that projects out the Pauli forbidden (PF) state from the relative motion. The details are given in Ref. [8]. The KKNN potential [9] and the folding potential of an effective nuclear and the Coulomb forces are applied to $V_{\alpha n}$ and $V_{\alpha\alpha}$, respectively. Here ξ_i is the relative coordinate between two clusters (α and n). The three-body (three-cluster) potential V_3 is explicitly given by the following one-range Gaussian form with the strength v_{3b}:

$$V_3(r_1,\ r_2) = \mathrm{v}_{3b} \exp\left(-\mu\rho^2\right), \tag{2}$$

where ρ is the hyper-radius of the $\alpha + \alpha + n$ system.

The lowest threshold of ^{9}Be is a three-body breakup into $\alpha + \alpha + n$. Only the ground state of ^{9}Be is bound, so ^{9}Be is a Borromean system. We investigate the properties of the unbound states of ^{9}Be using the CSM. The CSM has been applied to many kinds of nuclear systems, particularly as a useful method to describe many-body resonance states together with bound states [7, 8, 10]. The complex-scaled Schrödinger equation is expressed using the complex-scaled Hamiltonian $H(\theta)$ as

$$H(\theta)\Psi(\theta) = E\Psi(\theta). \tag{3}$$

The relative coordinates $\vec{r}_j$ of the three-body system $\alpha + \alpha + n$ are transformed as $\vec{r}_j \to \vec{r}_j e^{i\theta}$ ($j = 1, 2$) with a real parameter θ. We can approximate the eigen function $\Psi^\nu_J(\theta)$ for the state ν with the total spin J in Eq. (3) using the following basis expansion:

$$\Psi^\nu_J(\theta) = \sum_{\beta=1}^{N} c^\nu_\beta(\theta)\psi^J_\beta, \tag{4}$$

where ψ^J_β is the basis function for three-body states of ^{9}Be and given as

$$\psi^J_\beta = \left[\left[\phi^{i_\beta}_{l_\beta}(\vec{r}^{\,c}_1) \otimes \phi^{j_\beta}_{\lambda_\beta}(\vec{r}^{\,c}_2)\right]_{L_\beta} \otimes \chi^\sigma_{1/2}\right]_J. \tag{5}$$

The index β represents a set of $\{c, i, l, j, \lambda, L\}$ where c =1 and 2 specifies a channel of $(\alpha+\alpha)+n$ and $(\alpha + n) + \alpha$, respectively. The orbital angular momenta l and λ are corresponding to the relative motion for $\vec{r}_1$ and $\vec{r}_2$, respectively, in the channel c. Furthermore, L is the total orbital angular momentum. The indices i and j are to distinguish the radial basis functions. We employ the Gaussian basis functions to describe the radial component for each partial wave:

$$\phi^k_\ell(\vec{r}) = N^k_\ell r^\ell e^{-\frac{a_k}{2}r^2} Y_\ell(\hat{r}), \qquad a_k = a_0\eta^{k-1}, \tag{6}$$

where the details are explained in Ref. [11].

The cross section $\sigma_{\gamma n}$ of the photo-disintegration

$$^9\mathrm{Be}(3/2^-) + \gamma \to n + \alpha + \alpha. \tag{7}$$

is calculated in terms of the multi-pole response can be expressed as the following form

$$\sigma^\gamma_{E\lambda}(E_\gamma) = \frac{(2\pi)^2(\lambda+1)}{\lambda[(2\lambda+1)!!]^2}\left(\frac{E_\gamma}{\hbar c}\right)^{2\lambda-1}\frac{dB(E\lambda, E_\gamma)}{dE_\gamma}. \tag{8}$$

In this work, we consider the $E1$ transitions from the $3/2^-$ ground state to the unbound $1/2^+$ state in ^{9}Be using the complex-scaled Green's function [7, 8], which contribute to the $^9\mathrm{Be}(\gamma,n)$ ^{8}Be cross section as

$$\sigma^{\gamma}_{E1}(E_\gamma) = \frac{16\pi^3}{9\hbar c} E_\gamma \frac{dB(E1, E_\gamma)}{dE_\gamma}. \tag{9}$$

3. Results and Discussion

The $3/2^-$ ground state of ^{9}Be is calculated at -2.16 MeV from the $\alpha+\alpha+n$ threshold with no three-body potential (v_{3b}=0) [12]. The two-cluster potentials $V_{\alpha n}$ and $V_{\alpha\alpha}$ are fixed so as to reproduce the experimental data of the corresponding sub-systems. Here, we introduce a repulsive three-body potential (v_{3b}=6.57 MeV, $\mu = 0.1$ fm^{-2}) to fit the observed binding energy (-1.574 MeV) of ^{9}Be. The Hamiltonian Eq. (1) reproduces all the threshold energies of $\alpha+\alpha+n$, ^{8}Be+n and ^{5}He+α in the ^{9}Be nucleus. Using this Hamiltonian, we do not obtain the $1/2^+$ resonance of ^{9}Be. This result is the same as that obtained in Refs. [6, 12]. To clarify the resonance nature of the $1/2^+$ state, we carry out three-body calculations of the $1/2^+$ state by adding the three-body potential. With a strong attractive three-body potential with a negative value of v_{3b}, we can obtain the $1/2^+$ state as a bound state as shown in Fig. 1. From calculations for various values of v_{3b}, we find that the $1/2^+$ state becomes bound with the value of v_{3b} below -26 MeV.

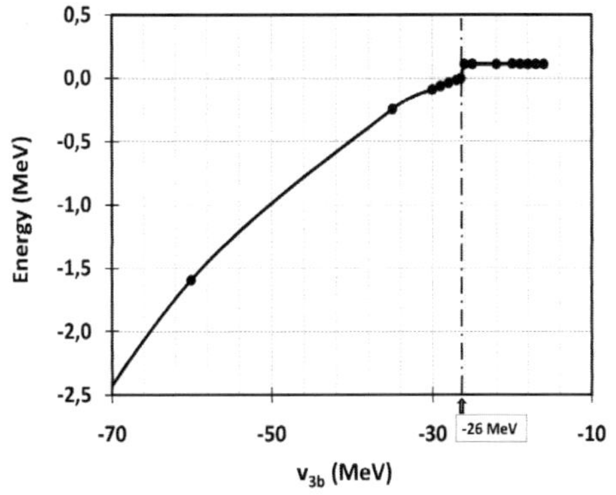

Figure 1. Energies of the $1/2^+$ state measured from the α+α+n threshold, as a function of the strength three-body potential v_{3b}.

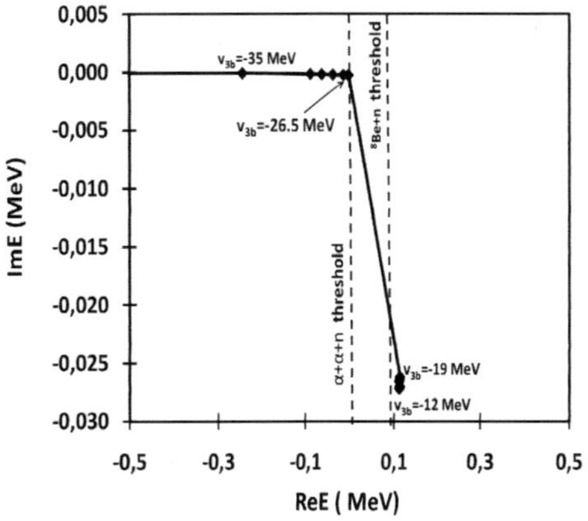

Figure 2. The eigenvalue trajectory of the $1/2^+$ state in the complex-energy plane.

Increasing the value of v_{3b} from -26 MeV gradually, we search for the $1/2^+$ resonance solutions in the CSM, whose eigenvalue trajectory is shown in Fig. 2. We see a jump from the $\alpha+\alpha+n$ threshold energy to a complex value, the real part of which corresponds to a energy just above the ^{8}Be+n threshold. This discontinuity of the trajectory may be understood by considering the properties of solutions in the CSM. In the CSM, resonance poles existing below the 2θ-line of continuum states starting from the lowest threshold of $\alpha+\alpha+n$ cannot be obtained as isolated eigenvalues. Because of the analyticity condition for the present complex scaled Hamiltonian, the θ value must be smaller than 45° [7]. For virtual states (whose energy eigenvalue is located at the negative real axis of the second Riemann surface), the angle θ has to be larger than 90°. When a virtual state exists in the ^{8}Be+n channel, a jump is expected in the resonance trajectory as discussed in Ref. [13]. The present case is not a two-body system

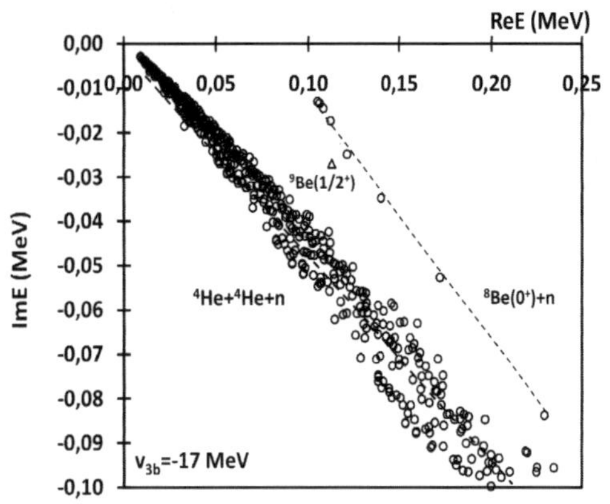

Figure 3. The $1/2^+$ energy eigenvalue distribution in the complex-scaled $\alpha + \alpha + n$ model.

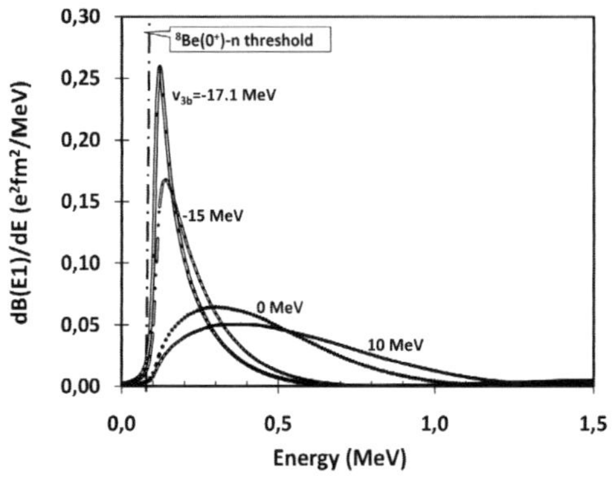

Figure 4. The $E1$ transition strength for ^{9}Be$(3/2^- \rightarrow 1/2^+)$ as a function of energy E, changing the strength v_{3b} of the three-body potential.

but the $\alpha + \alpha + n$ three-body one. It is difficult to trace the jump behavior of the pole solutions in the CSM.

Figure 3 displays an example of the eigenvalue distribution of ^{9}Be$(1/2^+)$ for v_{3b}=-17 MeV with $\theta = 15°$. Besides the triangle corresponding to the resonance, all solutions for continuum states described by open circles lie on two straight lines starting from positions of the $\alpha + \alpha + n$ three- and ^{8}Be$(0^+) + n$ two-body thresholds.

Using the solutions of $1/2^+$ and ground states, we calculate the $E1$ transition strength to investigate the effect of the $1/2^+$ resonance of ^{9}Be. In Fig. 4, the results are shown as a function of the energy measured from the $\alpha + \alpha + n$ threshold for several values of the three-body strength v_{3b}. We applied attractive ($v_{3b} = -15$ and -17.1 MeV) or repulsive ($v_{3b} = 0$ and 10 MeV) three-body potentials. The $E1$ transition strength for $v_{3b} = -17.1$ MeV rises sharply from the ^{8}Be$(0^+) + n$ two-body threshold. The strength reaches a maximum of 0.26 e^2fm^2/MeV

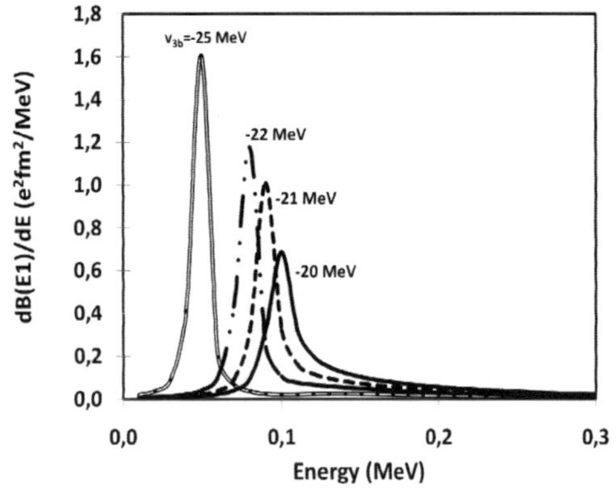

Figure 5. The $E1$ transition strength of ^{9}Be$(3/2^- \rightarrow 1/2^+)$ calculated for $v_{3b} = -20, -21, -22, -25$ MeV.

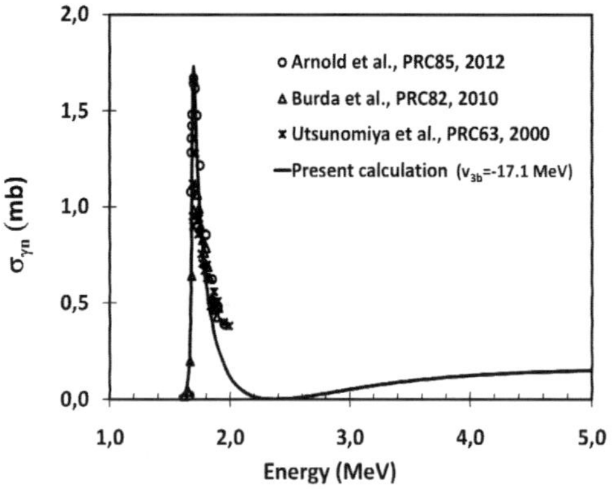

Figure 6. The ^{9}Be$(\gamma, n)^8$Be cross section $\sigma_{\gamma n}$ as a function of excitation energy E. The experimental data are taken from Refs. [1, 2, 3].

at $E = 0.11$ MeV, and decreases to 0.01 $e^2\text{fm}^2/\text{MeV}$ at $E = 0.5$ MeV. As v_{3b} increases, the peaks of the $E1$ transition strength decrease gradually and move into higher energies. In the cases of v_{3b}=0 and 10 MeV, the peaks of the $E1$ strength are broad and their energy positions are distant from the $^8\text{Be}(0^+) + n$ threshold. It is noted that the $E1$ transition strength is very small in the energy region below the $^8\text{Be}(0^+) + n$ two-body threshold.

The $E1$ transition strengths calculated for $v_{3b} = -25 \sim -20$ MeV are shown in Fig. 5. They have strengths in the energy region between $\alpha+\alpha+n$ and the $^8\text{Be}(0^+)+n$ thresholds. Each $E1$ transition strength seems to be expressed by the Breit-Wigner form. In the case of $v_{3b} < -25$ MeV, where the $1/2^+$ state becomes a bound state, the $E1$ transition strengths have not a characteristic structure around $\alpha + \alpha + n$ and $^8\text{Be}(0^+) + n$ threshold energies.

Comparing the calculated photo-disintegration cross section of the $1/2^+$ state with the recent new experiment [1], we find that $v_{3b} = -17.1$ MeV gives a good agreement as shown in Fig. 6. Using this v_{3b}, we obtain $1/2^+$ as a three-body resonance. This result indicates that the experimental cross section can be explained in terms of the $1/2^+$ resonance of ^{9}Be.

4. Summary

In this work, we investigate the properties of the $1/2^+$ state of ^{9}Be varying values of three-body potential. Without the three-body potential, we cannot obtain the $1/2^+$ resonance of ^{9}Be, which is consistent to the results in Refs. [6, 12]. Using the attractive three-body potential, we can obtain the $1/2^+$ resonance, which is responsible to explain the $E1$ transition strength of ^{9}Be with an enhancement near the $\alpha + \alpha + n$ threshold energy. The calculated photo-disintegration cross section is in good agreement with new experimental data [1] using the resonance solution of $^9\text{Be}(1/2^+)$ with an appropriate strength of the three-body potential.

Acknowledgements

The numerical calculations were done with use of computing system of the Nuclear Research Centre, National University of Mongolia and Theoretical Nuclear Physics Laboratory, Hokkaido University. Two of the authors, M.O. and K. K., are thankful for the support from a Grant-in Aid Scientific Research (No. 25400241) of the Ministry of Education, Science and Culture, Japan.

References

[1] Arnold C W, Clegg T B, Iliadis C, Karwowski H J, Rich G C, Tompkins J R and Howell C. R 2012 *Phys. Rev.* C **85** 044605
[2] Burda O, von Neumsnn-Cosel P, Richter A, Forssén C and van Overstraeten Brown B A 2010 *Phys. Rev.* C **82** 015808
[3] Utsunomiya H, Yonezawa Y, Akimune H, Yamagata T, Ohta M, Fujishiro M, Toyokawa H and Ohgaki H 2000 *Phys. Rev.* C **63** 018801
[4] Efros V D, von Neumann-Cosel P and Richter A 2014 *Phys. Rev.* C **89** 027301
[5] Gariddo E, Fedorov D V and Jensen A S 2010 *Phys. Lett.* B **684** 132
[6] Arai K, Descouvemont P, Baye D and Catford W N 2003 *Phys. Rev.* C **68** 014310
[7] Aoyama S, Myo T, Katō K and Ikeda K 2006 *Prog. Theor. Phys.* **116** 1
[8] Myo T, Katō K, Aoyama S and Ikeda K 2001 *Phys. Rev.* C **63** 054313
[9] Kanada H, Kaneko T, Nagata S and Nomoto M 1979 *Prog. Theor. Phys.* **61** 1327
[10] Odsuren M, Katō K, Aikawa M and Myo T 2014 *Phys. Rev.* C **89** 034322
[11] Hiyama E, Kino Y and Kamimura M 2003 *Prog. Part. Nucl. Phys.* **51** 223
[12] Kato M, Master Thesis in Hokkaido University (March, 2012) and private communication.
[13] Masui H, Aoyama S, Myo T Katō K, and Ikeda K 2000 *Nucl. Phys.* A **673** 207

Journal of Physics: Conference Series **569** (2014) 012073 doi:10.1088/1742-6596/569/1/012073

Complete kinematics study of the p + ^{11}B → ^{12}C reaction

K L Laursen, H O U Fynbo, A M Howard and O S Kirsebom

Department of Physics and Astronomy, Aarhus University, 8000 Aarhus C, Denmark

E-mail: klj06@phys.au.dk

Abstract. New data and preliminary results from an experimental study of the p + ^{11}B reaction are presented. Using proton energies in the range of 167 - 170 keV the 2^+ resonance in ^{12}C at an excitation energy of 16.11 MeV was populated. Detecting the emitted α-particles in full kinematics allows us to study ^{12}C resonances and their properties. In addition to the 3-α break-up of the 16.11 MeV resonance we observe γ-transitions to lower lying resonances. Transitions to the 3^- state at 9.64 MeV, the 1^- state at 10.84 MeV and the 1^+ state at 12.71 MeV are clearly seen. The transitions to the 1^- state has not been observed previously. In addition we see decays to structures of natural parity at excitation energies around 11 - 13 MeV. The results illustrate that the indirect detection of γ-decays is an effective technique for studying the low lying resonance spectrum of ^{12}C.

1. Introduction

The nature of the second 0^+ state in ^{12}C, known as the Hoyle state, is one of the most eminent tests of our understanding of structure in light nuclei. Since its prediction and first observation much work has been focussed on describing the structure of this state. Morinaga gave a structural interpretation in 1956 as a linear chain of α-particles. Advanced no-core shell model calculations fail to reproduce the excitation energy of the Hoyle state (see e.g. Ref. [1]) whereas its properties are well described in cluster models. Even in fermionic molecular dynamical calculations, which are able to describe both cluster and shell model like configurations, the Hoyle state has cluster structure (see e.g. [2]). Indeed the Hoyle state is believed to have a triple α cluster structure. The detailed structure, such as whether it is in a triangular or linear form or even some other arrangement, is however still a subject of much debate.

A test of how the three α-particles arrange themselves in the Hoyle state is to measure the moment of inertia of the rotational band build upon it, if it exists. Therefore much effort has been directed on observing the first 2^+ excitation (2_2^+) and recently there have been significant experimental progress in this area. Evidence for a 2_2^+ excitation in the region of interest has emerged from experiments using inelastic alpha scattering [3], inelastic proton scattering [4], an intense γ-ray beam [5] and β-decay [6][1]. Work still needs to be done on establishing more precisely its properties and searching for new broad states will continue to be important in order to reveal information concerning the detailed structure of the Hoyle State [7].

This contribution presents new results on low lying ^{12}C resonances populated by γ-decays.

[1] Probably not the same 2^+ state as found in the other experiments.

2. Experiment: p + ^{11}B

We have used the p + ^{11}B reaction to populate the (J^π,T) = (2^+, 1) resonance at 16.11 MeV. This resonance predominantly breaks up into three α-particles. Measuring these will allow information about the resonance and the break up mechanism to be extracted. It is known as reported in Ref. [8] that this state decays by emitting γ rays. This contribution deals with γ-decays to unbound resonances leaving results on direct 3-α decay for later publications.

2.1. Method

Motivated by the need for a more detailed experimental understanding of the ^{12}C resonance spectrum our idea is to address low lying resonances in ^{12}C by using γ-decays from the 16.11 MeV resonance (see Figure 1). This approach introduces selectivity between states of different spin and parity due to the electromagnetic transition rules. Transitions to the broad 0^+ state around 10 MeV and the 3^- state at 9.64 MeV can complicate the observation of 2^+ states as was the case in earlier experiments. In the present case (see Table 1) isovector M1 transitions to 2^+ states will be favoured over isovector E2 transitions to 0+ states as there is no expected collective enhancement of isovector E2 transitions. E1 transitions can populate the 3^- state, but since this state is quite narrow it is likely to pose less of a problem than the broad 0^+. Choosing the 0^+ resonance at 17.76 MeV (E_{proton} = 2.0 MeV) could be more advantageous with no transitions to 0^+ states and negligible 3^- γ-population. See the contribution from O. S. Kirsebom where this case is studied.

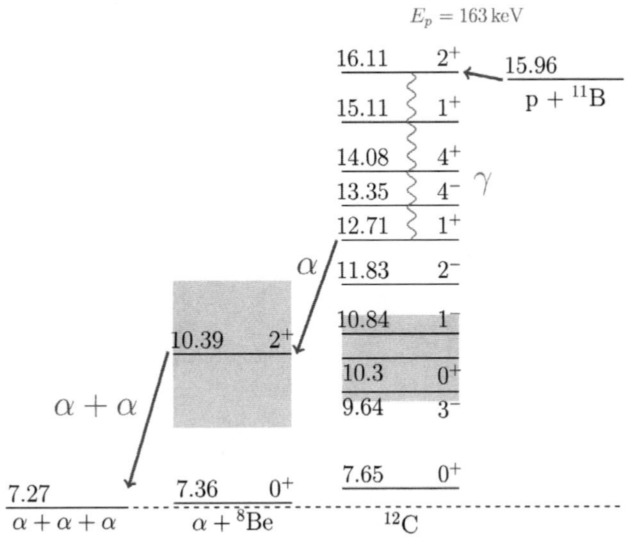

Figure 1. Level scheme showing low energy ^{12}C resonances. A specific decay sequence leading to the 3-α continuum is shown in the figure.

The small γ-branching (of the order 10^{-4} to unbound resonances) and the fact that most states in the 3-α continuum are broad makes this experimental approach quite difficult. Certainly it would be an immense challenge when doing conventional γ-spectroscopy. As this contribution shows we are able to identify new transitions to at least one broad state by detecting the γ's indirectly.

Instead of detecting the decay γ's directly we detect the γ-delayed α-particles in full kinematics. This means that the energies and directions are detected for two or all the three

α's. This gives full kinematical information about the break up. The results presented here are based on multiplicity 3 events only. The ^{12}C excitation energy from where the triple α break-up originated is calculated in the $\mathrm{p} + {}^{11}\mathrm{B}$ center of mass frame (CM) as

$$E_{\mathrm{exc.}}^{^{12}\mathrm{C}} = \sum_{i=1}^{3} E_{\alpha} + 7274.8\,\mathrm{keV}, \tag{1}$$

where 7274.8 keV is the 3-α threshold energy. Equation (1) works exactly for the 3-α break-up of the 16.11 MeV level and due to the small recoil energy following a γ-decay it gives the energy of γ-populated levels within 1 keV.

Table 1. Possible spin and parities for states to be reached by E1, M1 and E2 transitions from a 2^+ state.

El/Ml	16.11 MeV, 2^+
E1	$1^-, 2^-, 3^-$
M1	$1^+, 2^+, 3^+$
E2	$0^+, 1^+, 2^+, 3^+, 4^+$

2.2. Setup

The experiment was carried out at the *400 keV Van de Graaff* accelerator at Aarhus University, Denmark. Our detection setup consisted of two double sided silicon strips detectors. Both are 60 μm thick with front deadlayer thicknesses of 200 nm and 700 nm. The typical energy resolution in a given detector strip was around 20 keV (1σ). The detectors typically covered CM polar angles in the range of 55° - 160° and 30° - 120° respectively. The total solid angle coverage was $\sim$ 18%. For each recorded data event both ADC (energy) and TDC (time) information was stored.

The experiment was carried out with proton energies in the range from 167 keV to 170 keV with beam currents in the order of 1 - 2 nA. The target foils consisted of natural boron (typically 10 μg/cm^2) with a backing layer of carbon (4 μg/cm^2). The beam spot size at the target was in the order of 4 mm^2. Calibration of the detectors were done with a ^{228}Th source. Calibration runs was carried out many times during the period of the experiment (4 months). The multiplicity-3 data presented here was acquired over a period of 250 hours.

2.3. Data and analysis

The single α-particle energy spectrum recorded in one of the detectors is shown in Figure 2. The sharp peak near 6 MeV is due to the first emitted α when the break up proceeds through the ^{8}Be ground state. The broad distribution is both due to the secondary emitted α-particles in the ^{8}Be ground state channel and the α-particles emitted when going through the broad ($\Gamma = 3$ MeV) first excited state of ^{8}Be.

In Figure 3 the reconstructed ^{12}C excitation energy spectrum is shown. It is calculated by use of Eq. (1) using all triple coincidence energies. We mainly observe the decay of the 16.11 MeV state. At lower energies we also see triple events which could be candidates for γ-decays.

In order to remove the background events we need to impose several cuts to the data. The source of the background is mainly from random coincidence events containing two α-particles from the same physical break-up with and a third detector-hit caused by either noise, protons from the beam or a different α-particle. The most important cuts we apply are

- Time cut by use of TDC data. Removes more than 95% of random coincidences.

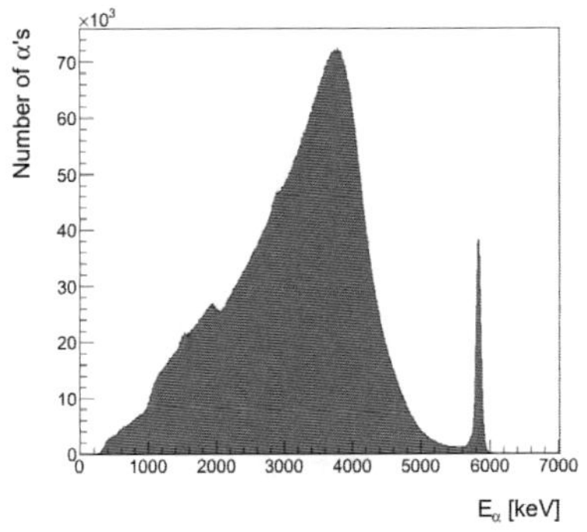

Figure 2. Single α energy spectrum in detector 1 for a subset of the total data collected.

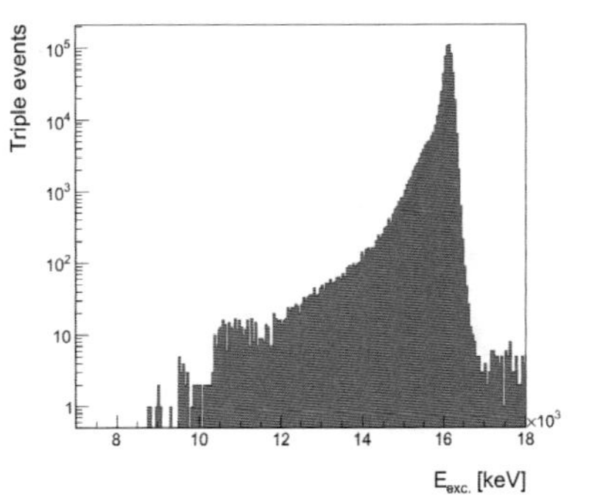

Figure 3. Reconstructed ^{12}C excitation energy where no cuts have been applied to the data.

- Momentum cut. In the CM frame the total momentum ($\sum_{i=1}^{3} \mathrm{P}_\alpha$) and its components ($\sum_{i=1}^{3} \mathrm{P}_{x,y,z,\alpha}$) should equal zero.
- Different cuts on the angles of the α-particles (sum of the relative angles should give 360° and the three momentum vectors should all lie in the same plane)

Figure 4 shows in the top part how the total momentum for triple-α events that survive the time cut are distributed as function of excitation energy. Clear background tails are visible stretching into the region of high sum momentum. Besides from the maximum intensity region, resulting from the 3-α break-up of the 16.11 MeV level, we notice interesting events at low sum momentum. The dotted red line shows a typical cut level applied in the analysis.

3. Results

The final excitation energy spectrum, after applying all cuts, is seen in the bottom part of Figure 5. We now clearly see evidence for γ-decays giving rise to distinct structures. We identify them as the 3^- state at 9.64 MeV, the 1^- state at 10.84 MeV and the 1^+ state at 12.71 MeV. The fact that we see transitions to the broad 1^- state, which are seen for the first time in this work, illustrates the applicability of this technique for studying the low energy excitation spectrum of ^{12}C.

With our detection technique we can discriminate the break-up decays proceeding through the ^{8}Be ground state. Those decays are represented by the hatched blue histogram in Figure 5, whereas the red histogram is due to decays taking a different route. Since parity conservation forbids break up via ^{8}Be (gs) for unnatural parity states the 12.71 MeV state cannot decay by this route. This is consistent with our observation. We see events in the region around 11.6 - 11.8 MeV. It is unclear if these are from the 2^- state at 11.83 MeV.

The top part of Figure 5 illustrates how each of the three α-particle energies are distributed as function of excitation energy. A useful feature in this plot is that the energy of the first emitted α will be on the diagonal line (slope of 2/3) when it goes to the ^{8}Be (gs). Looking at the events between 11 MeV and 13 MeV this indicates decays to states of natural parity. This is of high interest as there are no known natural parity states in this region. The observation of e.g. a 3^- state, which could be likely through E1 transitions, would be important for different models pinning down the exact structure of the Hoyle state (see e.g. the discussion in [7]). Additional data on this reaction is clearly necessary in order to analyze this in more detail. A future experiment is already planned on this reaction with an improved detection setup consisting of

4 detectors.

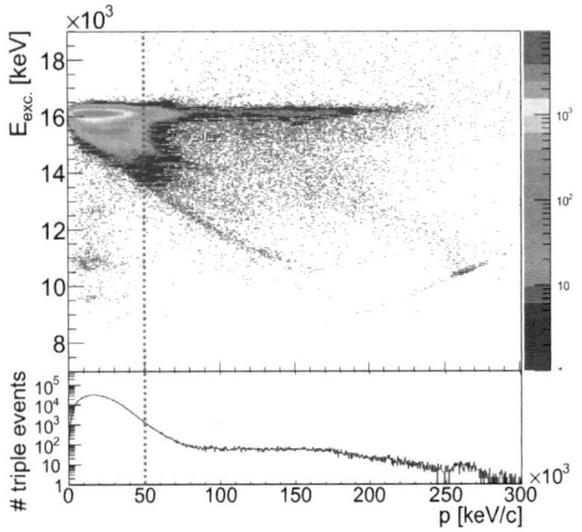

Figure 4. Top part: Excitation energy plotted against the total momentum of the three recorded *particles*. Bottom part: projection of top figure on the momentum axis. The red dotted line illustrates the cut on the total momentum.

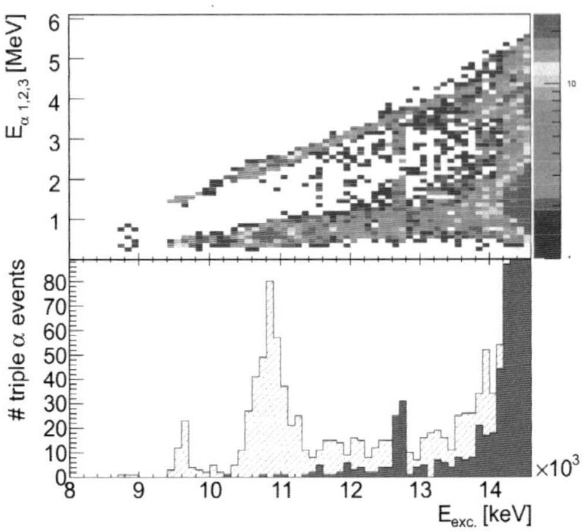

Figure 5. Top part: α-particle energies vs. excitation energy with all cuts applied. Bottom part: Projection on first axis. The hatched blue histogram correspond to decays through the ^{8}Be ground state while the red one corresponds to decays not going through that channel.

4. Summary

Through indirect detection of γ-decays from the 16.11 MeV 2^{+} state in ^{12}C we investigate the low energy resonance spectrum of this nucleus. Through this method, where the three emitted α-particles are detected in full kinematics, we observe clear evidence for decays to three known states. One of these has never been seen populated before from the 16.11 MeV level. Left as an open question and motivation for future studies we observe transitions to natural parity states in a region where no such states are known.

Acknowledgments

We acknowledge financial support from the European Research Council under ERC starting grant LOBENA, No. 307447.

References

[1] Roth R, *et al.* 2011 *Phys. Rev. Lett.* **107** 072501
[2] Chernykh M, *et al.* 2007 *Phys. Rev. Lett.* **98** 032501
[3] Itoh M, *et al.* 2011 *Phys. Rev.* C **84** 054308
[4] Freer M, *et al.* 2009 *Phys. Rev.* C **80** 041303
[5] Zimmerman W R, *et al.* 2013 *Phys. Rev. Lett.* **110** 152502
[6] Hyldegaard S, *et al.* 2010 *Phys. Rev.* C **81** 024303
[7] Marín-Lámbarri D J, *et al.* 2014 *Phys. Rev. Lett.* **113** 012502
[8] Adelberger E G, *et al.* 1977 *Phys. Rev.* C **15** 484

Journal of Physics: Conference Series **569** (2014) 012074 doi:10.1088/1742-6596/569/1/012074

Microscopic cluster model of $\alpha + \alpha$ bremsstrahlung following a Siegert approach

J Dohet-Eraly[1,2] **and D Baye**[2]

[1] TRIUMF, 4004 Wesbrook Mall, Vancouver, BC V6T 2A3, Canada.

[2] Physique Quantique and Physique Nucléaire Théorique et Physique Mathématique, C.P. 229, Université libre de Bruxelles (ULB), B-1050 Brussels, Belgium.

E-mail: jdoheter@triumf.ca

Abstract. We have recently developed a microscopic cluster model of nucleus-nucleus bremsstrahlung, which takes implicitly account of a part of the meson-exchange current effects, by applying an extension of the Siegert theorem valid for arbitrary photon energy. This model is applied to the $\alpha+\alpha$ bremsstrahlung. The differences between the bremsstrahlung cross sections obtained in the non-Siegert and Siegert approaches are displayed and discussed. The theoretical bremsstrahlung cross sections are compared with experimental measurements. Whereas the agreement with quite old experimental data is fair, a rather important discrepancy between the theoretical results and the more recent experimental data is noticed. Some paths to investigate this disagreement are proposed.

1. Introduction

The $\alpha + \alpha$ bremsstrahlung is an $\alpha + \alpha$ collision during which a part of the kinetic energy between the α particles is converted to a photon. In other words, this is a radiative transition between $\alpha+\alpha$ scattering states corresponding to different energies. This specific nucleus-nucleus bremsstrahlung is particularly interesting to be studied because on the one hand, the $\alpha + \alpha$ scattering states are well described by relatively simple cluster models and on the other hand, some experimental $\alpha + \alpha$ bremsstrahlung cross sections are available [1–4] making possible a comparison between theory and experiment.

While most of previous nucleus-nucleus bremsstrahlung studies neglect the effects of meson-exchange currents, we have recently proposed [5, 6] to include them partially in an implicit way by using an extended version of the Siegert theorem [7]. Following this approach, a microscopic cluster model of bremsstrahlung has been developed and applied to the $\alpha+\alpha$ [5] and $\alpha+N$ [6] bremsstrahlung reactions. Recently, new measurements of $\alpha+\alpha$ bremsstrahlung cross sections have been performed for four beam energies [4]. We apply the model developed in [5] at these colliding energies and compare the theoretical and experimental results each other.

2. The $\alpha+\alpha$ bremsstrahlung cross sections

In the center-of-mass (c.m.) frame, two α particles collide with initial relative wave vector in the z direction and relative energy E_i. After emission of a photon with energy $E_\gamma = \hbar c k_\gamma$, the nuclear system has relative wave vector $\mathbf{k}_f$ in direction $\Omega_f = (\theta_f, \varphi_f)$ and energy E_f given by

$$E_f = E_i - E_\gamma, \tag{1}$$

up to small recoil corrections.

The multipole matrix elements are defined by

$$u^{\sigma}_{\lambda\mu}(\Omega_f) = 2\pi^{1/2}\alpha^{\sigma}_{\lambda}\sum_{l_i l_f}(l_i\lambda 0\mu|l_f\mu)(2l_f+1)^{-1/2}Y^{\mu}_{l_f}(\Omega_f)e^{2i(\sigma_{l_f}+\delta_{l_f})}\langle\psi^f_{l_f}||\mathcal{M}^{\sigma}_{\lambda}||\psi^i_{l_i}\rangle \tag{2}$$

where σ_{l_f} and δ_{l_f} are the Coulomb and quasinuclear phase shifts, $\mathcal{M}^{\sigma}_{\lambda\mu}$ are the electromagnetic transition multipole operators, λ is the order of the multipole, μ is its component, $\sigma = 0$ or E corresponds to an electric multipole and $\sigma = 1$ or M corresponds to a magnetic multipole, $\alpha^{\sigma}_{\lambda}$ is given by

$$\alpha^{\sigma}_{\lambda} = -\frac{\sqrt{2\pi(\lambda+1)}i^{\lambda+\sigma}k^{\lambda}_{\gamma}}{\sqrt{\lambda(2\lambda+1)}(2\lambda-1)!!}, \tag{3}$$

the reduced matrix elements are defined by the following convention

$$\langle\psi^f_{l_f m_f}|\mathcal{M}^{\sigma}_{\lambda\mu}|\psi^i_{l_i m_i}\rangle = (l_i\lambda m_i\mu|l_f m_f)\langle\psi^f_{l_f}||\mathcal{M}^{\sigma}_{\lambda}||\psi^i_{l_i}\rangle, \tag{4}$$

and $\psi^c_{l_c m_c}$ are the partial waves of the unit-flux $\alpha+\alpha$ scattering wave function at colliding energy E_c with $c = i$ or f. The form of the partial waves $\psi^c_{l_c m_c}$ in the microscopic cluster approach is discussed in section 4.

From the multipole matrix elements, the bremsstrahlung cross sections $d\sigma/dE_\gamma$ can be evaluated,

$$\frac{d\sigma}{dE_\gamma} = \frac{E_\gamma k_f^2}{2\pi^2\hbar^3 c}\sum_{\sigma\lambda\mu}\int_0^{\pi}(2\lambda+1)^{-1}|u^{\sigma}_{\lambda\mu}(\theta_f,0)|^2\sin\theta_f d\theta_f. \tag{5}$$

Explicit formulas giving the angular differential bremsstrahlung cross sections from the multipole matrix elements $u^{\sigma}_{\lambda\mu}$ can be found in [8]. Since the $E2$ transitions are dominant for the $\alpha+\alpha$ bremsstrahlung, the sums in (5) are restricted to $\sigma = E$ and $\lambda = 2$, here. The Siegert and non-Siegert forms of the electric transition multipole operators are defined in the next section.

In the experiments from [3,4], some $\alpha+\alpha$ bremsstrahlung cross sections, integrated over the angles and some range of photon energies, are measured. They are defined from $d\sigma/dE_\gamma$ by

$$\sigma = \int_{E_{\min}}^{E_{\max}}\frac{d\sigma}{dE_\gamma}dE_\gamma, \tag{6}$$

where $[E_{\min}, E_{\max}]$ is the energy range of the detected photons.

3. Electric transition multipole operators

In the non-Siegert approach, the electric transition multipole operators are defined from the intrinsic nuclear current density $\boldsymbol{J}$ by

$$\mathcal{M}^{E}_{\lambda\mu} = \sqrt{\frac{\lambda}{\lambda+1}}\frac{(2\lambda+1)!!}{k^{\lambda}_{\gamma}c}\int\boldsymbol{J}\cdot\boldsymbol{A}^{E}_{\lambda\mu}d\boldsymbol{r}, \tag{7}$$

where $\boldsymbol{A}^{E}_{\lambda\mu}$ is the electric multipole defined, in the Coulomb gauge, as [9]

$$\boldsymbol{A}^{E}_{\lambda\mu}(\boldsymbol{r}) = \frac{i}{k_\gamma\sqrt{\lambda(\lambda+1)}}\left(k^2_\gamma\boldsymbol{r} + \boldsymbol{\nabla}\frac{\partial}{\partial r}r\right)\phi_{\lambda\mu}(k_\gamma\boldsymbol{r}) \tag{8}$$

with

$$\phi_{\lambda\mu}(k\boldsymbol{r}) = j_\lambda(kr)Y_{\lambda\mu}(\Omega) \tag{9}$$

and (r,Ω) are the spherical coordinates of $\boldsymbol{r}$.

In the Siegert approach, the electric multipole is divided into a gradient term and a rest,

$$\boldsymbol{A}^E_{\lambda\mu}(\boldsymbol{r}) = \frac{i}{k_\gamma}\sqrt{\frac{\lambda}{\lambda+1}}\boldsymbol{\nabla}\phi_{\lambda\mu}(k_\gamma\boldsymbol{r}) + \boldsymbol{A}'^E_{\lambda\mu}(\boldsymbol{r}). \tag{10}$$

The gradient term dominates at low photon energy. The Siegert electric transition multipole operators are defined by [5]

$$\mathcal{M}^{E(\mathrm{S})}_{\lambda\mu} = \frac{(2\lambda+1)!!}{k_\gamma^\lambda}\int\left[\phi_{\lambda\mu}(k_\gamma\boldsymbol{r})\rho(\boldsymbol{r}) + \frac{1}{c}\sqrt{\frac{\lambda}{\lambda+1}}\boldsymbol{J}\cdot\boldsymbol{A}'^E_{\lambda\mu}\right]d\boldsymbol{r}, \tag{11}$$

where $\rho(\boldsymbol{r})$ is the charge density. The label (S) is added to recall that the extended Siegert theorem is used. At low photon energy, since the density-dependent term is dominant, the main effects of meson-exchange currents are implicitly included in the Siegert approach. The explicit forms of the electric transition multipole operators, which are considered here, can be found in [5].

If consistent current and charge densities are considered and the exact scattering wave functions are used, the non-Siegert operators (7) and the Siegert operators (11) lead exactly to the same results. However, in microscopic cluster approaches, simplified versions of the current and charge densities and of the wave functions are considered to keep the complexity of the models reasonable. In this case, differences between Siegert and non-Siegert results occur and give some hint about the accuracy of these approximations.

4. Microscopic cluster approach

The microscopic description of the $\alpha+\alpha$ system relies on the internal 8-body Schrödinger equation

$$H\Psi = E_T\Psi, \tag{12}$$

where H is the internal Hamiltonian, Ψ is the internal wave function, and E_T is the total energy of the system in the c.m. frame. The internal Hamiltonian H is given by

$$H = \sum_{i=1}^{8}\frac{p_i^2}{2m_N} + \sum_{i>j=1}^{8} v_{ij} - T_{\mathrm{c.m.}}, \tag{13}$$

where $\boldsymbol{p}_i$ is the momentum of nucleon i, m_N is the nucleon mass, v_{ij} is the two-body potential describing the interaction between nucleons i and j, and $T_{\mathrm{c.m.}}$ is the center-of-mass (c.m.) kinetic energy. The nucleon-nucleon potential v_{ij} is divided into two parts: a nuclear part v_{ij}^N and a Coulomb part v_{ij}^C. For the nuclear part, a central potential with spin-isospin dependence, namely the Minnesota potential [10], is considered. This potential leads to a good description of the $\alpha+\alpha$ elastic phase shifts by the microscopic cluster approach.

The internal wave function Ψ is expanded in partial waves ψ_{lm} which are in turn expanded as a sum of Generator Coordinate Method (GCM) basis functions $\Phi_{lm}(R_n)$,

$$\psi_{lm} = \sum_n f_n^l\Phi_{lm}(R_n), \tag{14}$$

where R_n are discrete values of the so-called generator coordinate. The GCM functions are given by

$$\Phi_{lm}(R) = \mathcal{A}\phi_\alpha\phi_\alpha\int Y_l^m(\Omega_R)\Gamma(\boldsymbol{\rho}-\boldsymbol{R})d\Omega_R, \tag{15}$$

Journal of Physics: Conference Series **569** (2014) 012074 doi:10.1088/1742-6596/569/1/012074

where ϕ_α is the internal α wave function, $\boldsymbol{\rho}$ is the relative coordinate between the centers of mass of the two α particles, (R, Ω_R) are the spherical coordinates of $\boldsymbol{R}$, and $\Gamma(\boldsymbol{\rho} - \boldsymbol{R})$ is a Gaussian centered at $\boldsymbol{R}$. The internal α wave function ϕ_α is obtained by removing a c.m. factor from the Slater determinant describing the ground state of the α cluster within the harmonic oscillator shell model. The same oscillator parameter is considered for the cluster wave function ϕ_α and the relative wave function Γ. The GCM functions, defined by (15), have the advantage that they can be written, after multiplication by a c.m. factor, as integrals of Slater determinants, which makes rather easy the calculation of the matrix elements between these functions. However, the GCM functions are not able to reproduce the correct asymptotic behavior of a scattering wave function. This problem is solved by the Microscopic R-matrix Method (MRM) [11, 12]: the configuration space is divided in two regions at the channel radius a: an internal region ($\rho < a$) where the partial wave functions are described by the GCM, i.e. by (14), and an external region ($\rho > a$) where the antisymmetrization between nucleons is reduced to the symmetrization between the α particles and the $\alpha\alpha$ interaction is considered purely Coulombic. The wave functions are evaluated by solving the Bloch-Schrödinger equation, derived from (12), on the internal region and by requiring the continuity of the wave function at the channel radius a. They are insensitive to the value of a if a is chosen large enough.

5. Results

In [5], some differential $\alpha + \alpha$ bremsstrahlung cross sections as well as some integrated ones are calculated within the microscopic cluster model and compared with experimental data [1–3]. All partial waves are included up to $l_i, l_f = 8$. Coulomb effects from higher partial waves are approximately included following the approach developed in [8]. The theoretical and experimental bremsstrahlung cross sections are noted to be in good agreement. However, these experimental data are few and quite inaccurate. Recently, new measurements of integrated $\alpha + \alpha$ bremsstrahlung cross sections have been performed for colliding energy $E_i = 9.22$, 10.90, 12.04, and 14.20 MeV [4]. The experimental error bars associated with these measurements are significantly smaller than the ones associated to the previous experimental data from [4]. In figure 1, the integrated bremsstrahlung cross sections σ, defined by (6), are calculated in the microscopic cluster approach for these four colliding energies and compared with these recent experimental data. Needless to say, the photon-energy ranges that are considered in the calculations are the same as the experimental ones [13]. The number and the values of the generator coordinates, the value of the channel radius, and other computational details can be found in [5].

Except at $E_i = 12.04$ MeV, a large discrepancy is noted between the calculated and measured bremsstrahlung cross sections. There are several possible explanations to this discrepancy but further investigations are required to verify or reject them. On the theoretical side, the use of cluster basis functions (15) could be responsible for a part of the discrepancy between theory and experiment. However, the gap between the Siegert and non-Siegert bremsstrahlung cross sections, which is an indicator of the validity of the cluster approximation, is significantly smaller than the gap between the calculated and measured bremsstrahlung cross sections. It is possible that the Minnesota potential, which is designed to reproduce the elastic $\alpha + \alpha$ phase shifts, is not able to reproduce the $\alpha + \alpha$ bremsstrahlung cross sections. The development of *ab initio* methods of nucleus-nucleus bremsstrahlung would be useful to determine if more realistic potentials lead to $\alpha + \alpha$ bremsstrahlung cross sections in better agreement with the experimental data from [4]. However, even the description of the $\alpha + \alpha$ elastic scattering is not yet reachable by *ab initio* approaches. Finally, more experimental data for a larger range of colliding energies and for smaller photon-energy ranges would be interesting to understand the gap between experiment and theory at some colliding energies and the good agreement at some other ones.

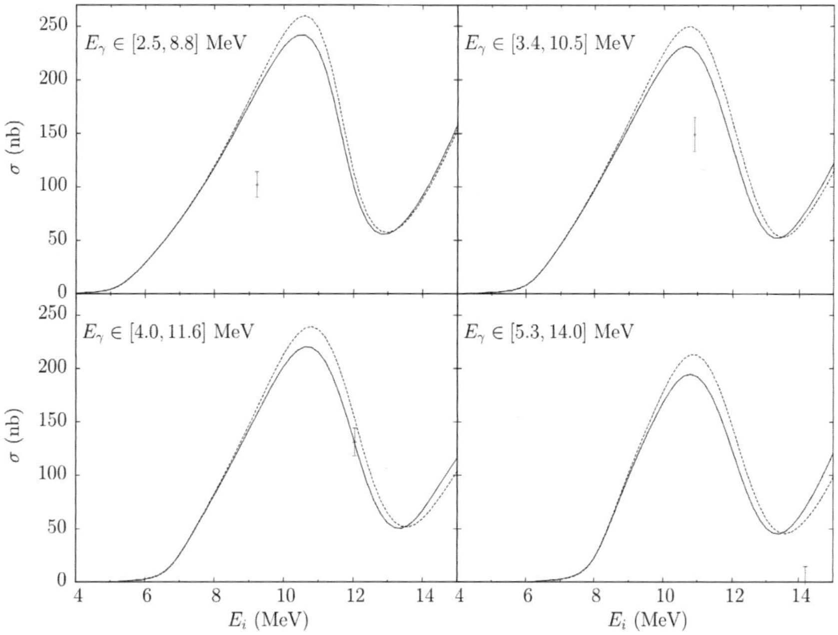

Figure 1. The $\alpha + \alpha$ integrated bremsstrahlung cross sections σ in the Siegert (full lines) and non-Siegert (dashed lines) approaches for different ranges of the photon energy E_γ. Experimental data from [4].

Acknowledgments

We thank D. R. Chakrabarty for the communication of some experimental details. This text presents research results of the interuniversity attraction pole programme P7/12 initiated by the Belgian-state Federal Services for Scientific, Technical and Cultural Affairs. A part of this work was done with the support of the F.R.S.-FNRS. TRIUMF receives funding via a contribution through the National Research Council Canada.

References

[1] Peyer U, Hall J, Müller R, Suter M and Wölfli W 1972 *Phys. Lett.* **41B** 151
[2] Frois B, Birchall J, Lamontagne C R, von Moellendorff U, Roy R and Slobodrian R J 1973 *Phys. Rev.* C **8** 2132
[3] Datar V M, Kumar S, Chakrabarty D R, Nanal V, Mirgule E T, Mitra A and Oza H H 2005 *Phys. Rev. Lett.* **94** 122502
[4] Datar V M, Chakrabarty D R, Kumar S, Nanal V, Pastore S, Wiringa R B, Behera S P, Chatterjee A, Jenkins D, Lister C J, Mirgule E T, Mitra A, Pillay R G, Ramachandran K, Roberts O J, Rout P C, Shrivastava A and Sugathan P 2013 *Phys. Rev. Lett.* **111** 062502
[5] Dohet-Eraly J and Baye D 2013 *Phys. Rev.* C **88** 024602
[6] Dohet-Eraly J 2014 *Phys. Rev.* C **89** 024617
[7] Schmitt K M, Wilhelm P, Arenhövel H, Cambi A, Mosconi B and Ricci P 1990 *Phys. Rev.* C **41** 841
[8] Baye D, Sauwens C, Descouvemont P and Keller S 1991 *Nucl. Phys.* A **529** 467
[9] Messiah A 1962 *Mécanique Quantique* vol 2 (Paris: Dunod)
[10] Thompson D R, LeMere M and Tang Y C 1977 *Nucl. Phys.* A **286** 53
[11] Baye D, Heenen P H and Libert-Heinemann M 1977 *Nucl. Phys.* A **291** 230
[12] Descouvemont P and Baye D 2010 *Rep. Prog. Phys.* **73** 036301
[13] Chakrabarty D R Private communication

Journal of Physics: Conference Series **569** (2014) 012075 doi:10.1088/1742-6596/569/1/012075

Transient effects in highly-excited fissioning systems

Y. Ayyad[1,7], J. Benlliure[1], J.L. Rodríguez-Sánchez[1], A. Bacquias[2], A. Boudard[3], E. Casarejos[1], T. Enqvist[4], M. Fernandez[1], V. Henzl[2], V. Henzlova[2], B. Jurado[5], A. Kelić-Heil[2], T. Kurtukian[5], S. Lukić[2], P. Nadtochy[2], D. Pérez-Loureiro[1], R. Pleskač[2], F. Farget[6], M. V. Ricciardi[2], K.-H. Schmidt[2], C. Schmitt[6] and Son Nguyen Ngoc[2]

[1] Universidade de Santiago de Compostela, Santiago de Compostela, Spain
[2] GSI, Darmstadt, Germany
[3] CEA-Saclay/IRFU, Gif-sur-Ivette, France
[4] Oulu southern Institute and Department of Phyiscs, University of Oulu, Finland
[5] Universite Bordeaux I, CNRS/IN2 P3, CENBG, BP 120, Gradignan, France
[6] GANIL, BP 55027, Caen Cedex 05, France

E-mail: `yassid.ayyad@rcnp.osaka-u.ac.jp`

Abstract.
In this work we report the proton- and deuteron-induced fission of ^{208}Pb at 500A MeV in inverse kinematics. We obtained two observables that allow us to investigate dynamical effects in the fission process: partial fission cross sections and the width of the fission fragment charge distribution as a function of the atomic number of the fissioning system. Results are compared to nuclear reaction model calculations in order to describe the evolution of the system from ground to saddle.

1. Introduction

Spallation reactions at high kinetic energies are a promising tool to investigate dissipative and dynamical effects because thermalized pre-fragments are produced with small distortion, low angular momentum and high excitation energy [1]. In spallation reactions, the interaction of a fast light projectile with the target nucleus can be described as binary nucleon-nucleon collisions through an intra-nuclear cascade. During the process the target gains thermal excitation energy and angular momentum proportionally to particle-holes excitations that take place in the initial distribution of the target. Subsequently, the excited system enters into pre-equilibrium emission leading to a thermalized pre-fragment through emission of fast nucleons. This thermalized system enters then into a second statistical de-excitation stage where the equilibrated pre-fragment dissipates its excitation energy by light charged particle and cluster evaporation, γ-ray emission, fission and multi-fragmentation [2].

Fission is usually described as a statistical process where the fission probability only depends on the density of states above the barrier, according to Bohr-Wheeler model [3]. However, this

[7] Present address: Research Center for Nuclear Physics, Osaka University, Japan.

model fails in reproducing pre-scission neutron multiplicities [4]. Presently, it is generally accepted that fission is a diffusion process whose evolution is governed by a reduced dissipation coefficient [5]. Dissipation arises from the excitation energy exchange rate between intrinsic and collective degrees of freedom of the fissioning system. The collective modes need a certain time to explore the potential-energy landscape. As consequence, the fission width is highly suppressed at the onset of the process, requiring a certain transient time (delay time) to establish a quasi stationary flux over the barrier [6]. The fission delay allows the system to cool down by particle evaporation. Under the proper conditions stated before, the lifetime of the system is comparable to such fission delay time and dynamical effects manifest.

In this work we investigated dynamical effects on fission by means of the spallation reactions p+^{208}Pb and d+^{208}Pb at 500A MeV in inverse kinematics. By using a dedicated setup for fission measurements [7] we determined the atomic number of the fissioning system. We deduced two observables with clear signatures of nuclear dissipation: partial fission cross sections and width of the fission fragment charge distribution σ_z as a function of the atomic number of the fissioning system. The former yields information about the spallation stage of the process and the evolution from ground to saddle. σ_z is strongly correlated to the temperature at saddle and thus gives us information about the fission delay and the dissipation [8, 9, 10]. By using two different targets we investigate the temperature-dependence of nuclear dissipation. In addition, the spherical configuration of ^{208}Pb makes it suitable to investigate such dynamical effects since the process starts at ground deformation. Finally, we compare our results with calculations performed with two different intra-nuclear cascade codes, INCL [11] and ISABEL [12], coupled to ABLA [13] statistical de-excitation code. In ABLA code, the fission width can be calculated according to the Bhor-Wheeler statistical model, the time-independent picture of Kramers and a time-dependent fission width obtained from Fokker-Planck equation. More details of these codes applied to spallation reactions can be found in Refs. [14, 15].

2. Experimental details and results

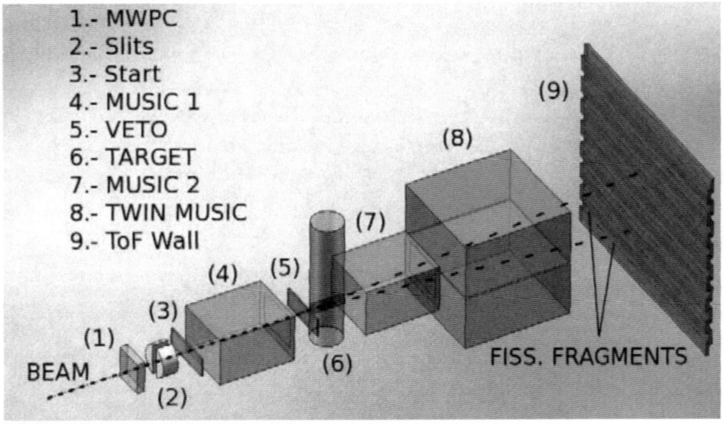

Figure 1. Sketch of the experimental setup used for the present work.

The experiment was conducted at GSI (Darmstadt - Germany) using a dedicated setup (see Fig. 1). Comprehensive details can be found in Ref. [16]. A ^{208}Pb beam, accelerated to 500A MeV, impinged on a cell with thin titanium foils containing liquid hydrogen and deuterium.

Due to the kinematics, fission fragments are emitted in forward direction with high kinetic energies. Two multi-sample ionization chambers (MUSIC) surrounded the target to identify the products of the reaction and nuclei produced in other layers of matter placed upstream of the target. A double ionization chamber (Twin MUSIC) was placed downstream of the second MUSIC to detect both fission fragments in coincidence. By measuring the energy loss of the fission fragments in the Twin MUSIC detector we identified their atomic number and we reconstructed the charge of the fissioning system Z_1+Z_2 by summing each fission fragment atomic number, taking into account that the proton evaporation probability is negligible.

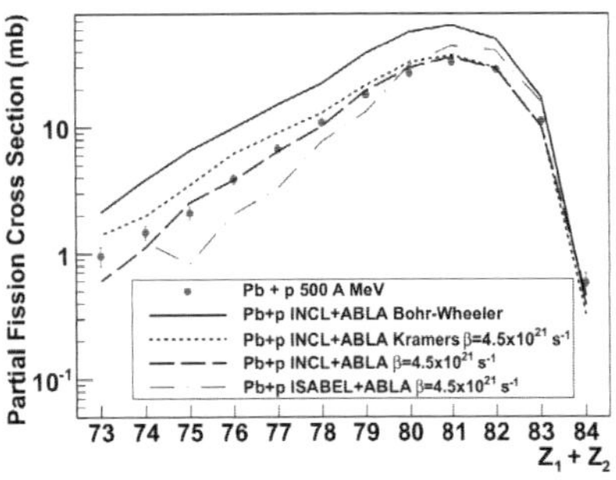

Figure 2. Partial fission cross sections for the reaction p+^{208}Pb.

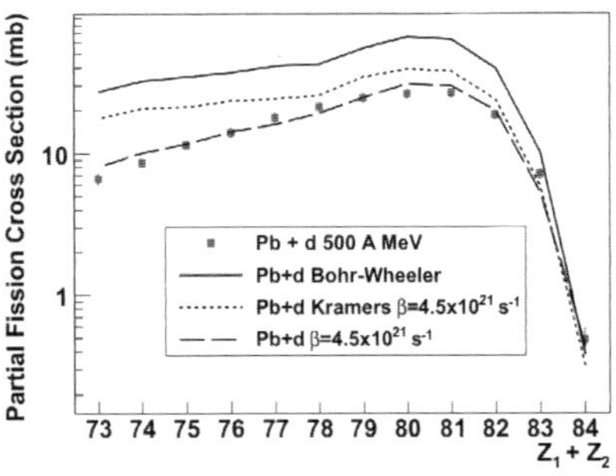

Figure 3. Partial fission cross sections for the reaction d+^{208}Pb.

Partial fission cross sections for the reactions p+^{208}Pb and d+^{208}Pb were obtained by normalizing the number of fission events for each Z_1+Z_2 value to the beam intensity and the target thickness (yield). The contribution of the titanium foils of the target was subtracted after measuring the yield with empty target. Corrections were applied due to the beam attenuation in the target and secondary reactions of the fission fragments. The geometrical acceptance of the setup for each fission fragment was evaluated by means of a monte-carlo simulation. As can be seen in Figs. 2 (p+^{208}Pb) and 3 (d+^{208}Pb), partial fission cross sections for the case of reactions with deuteron target are larger for fissioning systems with lower values of Z_1+Z_2 due to the larger kinetic energy of the reaction. We also compared the partial fission cross sections with calculations performed with INCL+ABLA for reactions with proton and deuteron and ISABEL+ABLA for the case of the reaction with proton. One can see that the shape of the partial fission cross section curve depends strongly on the entrance channel defined in the intra-nuclear cascade because two codes provide rather different results. On the other hand, the magnitude of the curve is sensitive to the de-excitation stage. Calculations with INCL+ABLA according to Bohr-Wheeler transition-state (solid lines) and Kramers (dotted lines) models clearly over predict the experimental values. By considering a dynamical fission delay with a reduced dissipation strength of $\beta = 4.5 \times 10^{21}$ s^{-1} (dashed line), calculations agree fairly well with the experimental data. Therefore, partial fission cross sections yield information about the evolution of the system from ground to saddle.

We also obtained the width of the fission fragment charge distribution σ_z for the reactions p+^{208}Pb and d+^{208}Pb (Fig. 4). It can be clearly seen that for both reactions we obtain very similar values for the entire Z_1+Z_2 range studied here. This fact indicates that the observable does not depend on the entrance channel but on the temperature and excitation energy at the

Journal of Physics: Conference Series **569** (2014) 012075 doi:10.1088/1742-6596/569/1/012075

saddle point for each Z_1+Z_2. We also performed calculations with INCL+ABLA (see Fig. 4) according to the statistical model (solid line) and Kramers time-independent fission width (dash dotted line). Both calculations over predict the experimental results in the Z_1+Z_2 range where the excitation energy is larger and dynamical effects manifest. However, calculations considering a fission delay induced by a dissipation with a strength of $\beta = 4.5 \times 10^{21}$ s^{-1} (dashed and dotted lines for proton and deuteron, respectively) are in very good agreement with the data. Dynamical calculations show that the reduction of the temperature at saddle (which also reduces σ_z) is because of the fission delay instead of the reduction of the asymptotic value of the fission width proposed by Kramers. Indeed, the nucleus cools down by particle evaporation because the time needed to reach the saddle point is longer due to the fission delay.

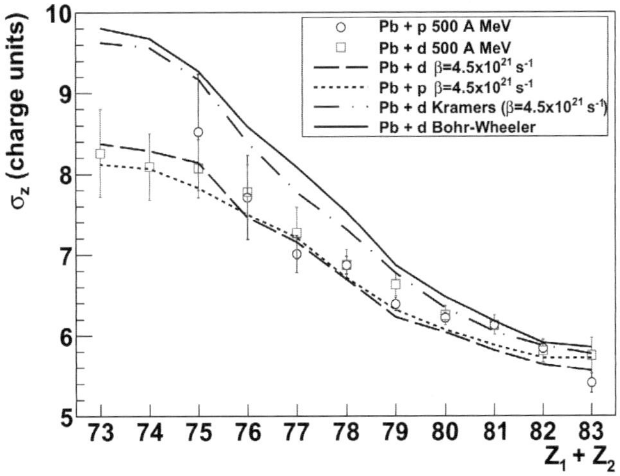

Figure 4. Width of the fission fragment charge distribution as a function of the atomic number of the fissioning nuclei for both reactions investigated in this work.

3. Conclusions

We have investigated transient and dissipative effects in proton- and deuteron-induced fission of ^{208}Pb at 500A MeV by using a dedicated setup for fission experiments in inverse kinematics. We have obtained the partial fission cross sections and the width of the fission fragment charge distribution σ_z both as a function of the atomic number of the fissioning system. Partial fission cross sections yield information about the evolution of the system from ground to saddle, while σ_z gives information about the temperature of the system at the saddle point. This temperature does not depend on the entrance channel of the reaction for each system with a defined atomic number Z_1+Z_2. Finally, by performing calculations with intra-nuclear cascade code INCL coupled to ABLA de-excitation code we show that a fission delay with a dissipation strength of $\beta = 4.5 \times 10^{21}$ s^{-1} is needed to explain the experimental data.

References

[1] Bhatt K H, Grangé P and Hiller B 1986 *Phys. Rev.* C **33** 954

[2] Bohr N 1936 *Nature,* **137** 344
[3] Bohr N and Wheeler J A 1939 *Phys. Rev.* **56** 426
[4] Gavron A, Beene J R, Ferguson R L, Obenshain F E, Plasil F, Young G R, Petitt G A, Jaaskelainen M, Sarantites D G and Maguire C F 1981 *Phys. Rev. Lett* **47** 1255
[5] Kramers H A 1940 *Physika* VII **4** 284
[6] Grangé P, Jun-Qing L and Weidenmüller H A 1983 *Phys. Rev.* C **27** 2063
[7] Schmidt K H et al 2000 *Nucl. Phys.* A **665** 221
[8] Benlliure J et al 2002 *Nucl. Phys.* A **700** 69
[9] Jurado B, Schmitt C, Schmidt K H, Benlliure J, Enqvist T, Junghans A R, Kelic A and Rejmund F 2004 *Phys. Rev. Lett.* **93** 072501
[10] Schmitt C, Schmidt K H, Kelić A, Heinz A, Jurado B and Nadtochy P N 2010 *Phys. Rev.* C **81** 064602
[11] Boudard A, Cugnon J, David J C, Leray S and Mancusi D 2013 *Phys. Rev.* C **66** 014606
[12] Yariv Y and Fraenkel Z 1979 *Phys. Rev.* C **20** 2227
[13] Kelic A, Ricciardi M V and Schmidt K H 2008 *Proc. of Joint ICTP-IAEA Advanced Workshop on Model Codes for Spallation Reactions, ICTP Trieste, Italy, 4-8 February*
[14] Benlliure J et al. 2006 *Phys. Rev.* C **74** 014609
[15] Ayyad Y et al. 2014 *Phys. Rev.* C **89** 054610
[16] Schmidt K H et al. 2013 *Phys. Rev.* C **87** 034601

Journal of Physics: Conference Series **569** (2014) 012076 doi:10.1088/1742-6596/569/1/012076

Tensor force and delta excitation for the structure of light nuclei

K. Horii[1], T. Myo[2,3], and H. Toki[3]

[1] Department of Physics, Tokyo Institute of Technology, Meguro 152-8551, Japan
[2] Faculty of Engineering, Osaka Institute of Technology, Osaka 535-8585, Japan
[3] Research Center for Nuclear Physics (RCNP), Osaka University, Osaka 567-0047, Japan

E-mail: horii@rcnp.osaka-u.ac.jp, myo@ge.oit.ac.jp, toki@rcnp.osaka-u.ac.jp

Abstract. We treat explicitly $\Delta(1232)$ isobar degrees of freedom using a bare nucleon-nucleon interaction for few-body systems, where Δ excitations can be the origin of the three-body force via the pion exchange. We adopt the Argonne two-body potential including Δ, named as AV28 potential, and study the role of Δ explicitly in two-body and three-body systems. It was found that the additional Δ states generate strong tensor correlations caused by the transitions between N and Δ states, and change tensor matrix elements largely from the results with only nucleons. We studied the effects of three-body force in the triton and obtained 0.8 MeV attraction due to the intermediate Δ excitation. Due to the lack of the total binding energy for the triton in the delta model, we further studied carefully the effects of the delta excitation in various two body channels and compared with the nucleon only model in the AV14 potential. We modified slightly the AV28 potential in the singlet S channel so that we could reproduce the triton binding energy due to the appropriate amount of the three-body force effects.

1. Introduction

For a quantitative description of nuclei, it is important to include three-body force [1]. The contributions of the three-body force turned out to be 0.85 MeV for the triton and 4.26 MeV for ^{4}He in the GFMC simulations with the AV18/UI Hamiltonian. The origin of the three body force is considered to come from the delta intermediate excitation and the relativistic effect [1]. Particularly for the delta intermediate excitation one delta is excited by two nucleon interaction and the excited delta decays back to a nucleon by interacting with another nucleon, which is named as the Fujita-Miyazawa mechanism [2]. Usually this three-body process is introduced in the calculation of nuclear many-body systems as a phenomenological three-body force expressed in the nucleon space [1, 3].

In the literature, there were a few attempts to include delta degrees of freedom explicitly for the discussion of the three-body force due to the delta excitation. Sauer and his collaborators studied the effect of single delta excitation with the knowledge of the pion exchange based on the Bonn potential [4]. Pikclesimer and his collaborators studied the three body system using the Argonne AV28 potential, where the excitation of one and two delta's was included for the construction of the nucleon-nucleon potential [5]. They used the Faddeev equation to solve for the triton using both the single delta and double delta amplitudes. Their results were quite reasonable as compared with the three body contribution of the Argonne group. The Los Alamos group, however, found that the net binding energy for the triton turned out to be

Journal of Physics: Conference Series **569** (2014) 012076 doi:10.1088/1742-6596/569/1/012076

7.29 MeV short of the experimental binding energy of 8.48 MeV. Hence, it was concluded that the delta excitation model was not able to explain the triton binding energy.

Recently, there was a renewed interest on the nucleon-nucleon interaction in the singlet S channel. In the neutron-neutron quasi-free scattering cross section in the neutron-deuteron break-up reaction process, Witala and Glockle showed the the quasi-free scattering cross section was underestimated, when they used the standard two-body nucleon-nucleon interaction in the Faddeev calculation [6]. They suggested that the neutron-neutron interaction in the singlet S channel should be increased by about 10 so that the quasi-free cross section was reproduced. It is therefore very important to revisit the effect of the delta excitation in order to understand the vast amount of the three-body scattering data accumulated until now [7]. The increase of the neutron-neutron attraction in the singlet S channel by 10 might remove the discrepancy found in the calculation of the Los Alamos group for the triton.

The inclusion of the delta isobar excitation needs a large amount of model space. Recently, we have developed the tensor optimized few-body model (TOFM) for few body systems in order to take into account the strong tensor force in the nucleon-nucleon interaction. In the TOFM, we include only a single Y_2 component in the few body wave function, which has a direct connection to the major Y_0 configuration by the tensor interaction [8]. The TOFM is able to truncate the model space largely and the computational effort becomes much lighter than the rigorous method [9]. Hence, we should be able to reduce the computational effort for few body calculations in the case of explicit introduction of the delta degrees of freedom. Particularly, when we like to calculate heavier systems, the success of TOFM in few body systems encourage the use of the tensor optimized shell model (TOSM) including the delta degrees of freedom. We are then able to analyze the wave functions and the dynamics easier than other sophisticated methods.

2. Three body system in the tensor optimized few body model

We discuss here the structure of the triton including Δ degrees of freedom. We calculate the ground state energy for the triton using the few-body method [8]. Our model space allows single and double Δ excitations, and the wave function is described as,

$$|\Psi\rangle = |\Psi_{NNN}\rangle + |\Psi_{N\Delta N}\rangle + |\Psi_{\Delta\Delta N}\rangle \ . \tag{1}$$

Using the Jacobi coordinate, we treat the strong short-range repulsion and the strong tensor correlation. We consider all the nucleons excite to the Δ states. Hence, we include all the possible anti-symmetrization. In this study, we perform variational calculations to obtain the triton state by using Stochastic Variational Method which is developed by the Niigata group [9].

The wave function is expressed in terms of the basis function of the correlated Gaussian basis, $\Psi_{JM} = \sum_k c_k \psi_k$, where ψ_k is given by

$$\psi_k = \mathcal{A}[\psi_L(x, A_k, u_k)\chi_s]_{JM}\eta_t \ . \tag{2}$$

The spin wave function is $\chi_s = [[\chi_1\chi_2]_{s_{12}}\chi_3]_s$ with χ_1, χ_2, χ_3 are the spin wave functions of nucleon or Δ. The isospin wave function is $\eta_t = [[\eta_1\eta_2]_{t_{12}}\eta_3]_t$, where η_1, η_2, η_3 are the isospin wave functions of nucleon or Δ. As for the spatial part of ψ_L, the basis functions are described by using the Correlated Gaussian basis with the global vector $\psi_L = e^{-\frac{1}{2}\tilde{x}Ax}|\tilde{u}x|^L Y_L(\hat{ux})$. In the present calculation, we include also $[Y_1 \times Y_1]_1$ components so that we can compare with the rigorous calculations of Niigata group [9].

3. AV28 potential and two-body systems

3.1. Two nucleon interaction: AV28

Wiringa *et al.* constructed phenomenological two-body NN interactions with and without $\Delta(1232)$ degrees of freedom, Argonne v14 (AV14) and v28 (AV28) models, respectively [10]. We

Table 1. Various quantities of the deuteron with the AV14 and AV28 potentials.

^{2}H	Energy	Kinetic	Central	Tensor	LS	L^2	$(LS)^2$
AV14	−2.2	19.1	−1.9	−18.8	0.4	3.1	−4.0
AV28	−2.2	23.9	8.8	−35.3	0.8	3.5	−4.1

review briefly the content of the NN interactions of the AV28 potential together with the AV14 potential. These potentials have three radial components: long-range one-pion exchange part v_π, and phenomenological intermediate-range and short-range parts v_I and v_S:

$$\mathrm{V}_{ij}(r_{ij}) = \sum_p [\mathrm{v}_\pi^p(r_{ij}) + \mathrm{v}_I^p(r_{ij}) + \mathrm{v}_S^p(r_{ij})]\mathcal{O}_{ij}^p \ , \tag{3}$$

where operators $\mathcal{O}^p$ represent operators of spin, isospin, tensor and other forms. For the AV14 potential, it is described by the sum of 14 operators ($p = 1, ..14$). On the other hand, the AV28 potential includes the delta excitation, and contains 14 components in the NN channel and have 6 operators in the $N\Delta$ channel and 8 operators in the $\Delta\Delta$ channels to represent all the possible processes due to $\pi N\Delta$ and $\pi\Delta\Delta$ vertices.

The $NN - NN$ interaction is expressed by the first 14 operators with the same form as the AV14 potential. The AV28 potential including the transition between N and Δ appearing in $NN - N\Delta, NN - \Delta\Delta, N\Delta - \Delta\Delta, N\Delta - \Delta N$ processes are assumed to be made only through the pion exchange (v_π) along with the corresponding spin-isospin and tensor-isospin operators. The ratio of coupling constants are taken as $f^2_{\pi N\Delta}/f^2_{\pi NN}$=4 and $f^2_{\pi\Delta\Delta}/f^2_{\pi NN}$=1/25 [10]. The diagonal $N\Delta - N\Delta$ and $\Delta\Delta - \Delta\Delta$ processes include not only the pion-exchange part but also the short-range repulsion (v_I and v_S) . These parameters in the short-range and intermediate-range potentials for $N\Delta$ and $\Delta\Delta$ channels are taken as the same values as those in the NN channel. Thus, the total number of parameters does not change for the AV28 potential compared with the AV14 potential. All the phenomenological parameters should be determined to reproduce the experimental data of NN scattering phase shifts and deuteron properties for both AV14 and AV28 potentials.

3.2. Triplet-even (^3E) channel

In order to understand the roles of Δ in nuclei, we start with the deuteron including explicitly Δ degrees of freedom. In Ref. [10], some deuteron properties have been presented such as the binding energy and the probabilities of the wave function components. The wave function of the deuteron consisting of NN and $\Delta\Delta$ states with the isospin $T = 0$ channel is given as, $|\Psi\rangle = |\Psi_{NN}\rangle + |\Psi_{\Delta\Delta}\rangle$. The NN wave function is written by the $^3S_1$ and $^3D_1$ channels and $\Delta\Delta$ state is written in four channels: $^3S_1$, $^3D_1$, $^7D_1$ and $^7G_1$. We present the energy and various Hamiltonian matrix elements using the AV28 potential [10] in Table 1 and also in Fig. 1. We also compare the results with those of the AV14 potential [10]. Both results provide the same binding energies of 2.2 MeV, but the matrix elements are largely different. Especially the tensor force component becomes large with the AV28 potential as −35.3 MeV in comparison with the value −18.8 MeV of the AV14 potential. The enhancement of tensor force contribution increases the kinetic energy, where the contributions of the mass difference between NN and $\Delta\Delta$ is obtained as 3.1 MeV. This mass contribution is much smaller than the original value of the mass difference of about 600 MeV, which is due to the small probabilities of about 0.5 for the total $\Delta\Delta$ state. Most of the kinetic and central energy components come from the NN wave functions. However as for the tensor energy components, even the small mixing of Δ generates the large tensor matrix elements in the deuteron. We find that the additional Δ comprises about −10 MeV of

Table 2. Various energy components for a two-body system in the ^1E channel with the AV28 and AV14 potentials using the radius constraint. The upper part shows various energy components in unit of MeV, and the lower one for the probabilities of the wave function in .

^{1}E	Energy	Kinetic	Central	Tensor
AV14	1.6	8.3	−6.7	0.0
AV28	1.5	14.0	2.2	−14.7
P [ ]	^{1}S(NN)	^{1}S(ΔΔ)	^{5}S(ΔΔ)	^{5}D($N\Delta$)
AV28	98.86	0.02	0.13	0.99

the tensor matrix elements, and the transitions between $^3S_1$ of NN state and $^7D_1$ of $\Delta\Delta$ state play the dominant role. The probability of the ^7D-state is 0.42 , which is particularly larger than other $\Delta\Delta$ states. In comparison to the same D-state between ^7D with spin 3 and ^3D with the total spin 1, the tensor correlations prefer the state which has spin alignment. The central force components provide the repulsive effect to cancel out the large attraction of the tensor force, which is the opposite sign to the result of the AV14 potential.

3.3. Singlet-even (^1E) channel

In two-body system, $T = 1$ channels are constructed by not only $\Delta\Delta$ state but also $N\Delta$ state. Specifically, it is important to investigate the role of $N\Delta$ state with $T = 1$, because single Δ excitation components in three-body system would contribute to the three-body process such as the Fujita-Miyazawa model. In order to examine the properties of $N\Delta$ state, we discuss another two-body system of the singlet-even (^1E) channel. The wave function of ^1E channel with the total spin J=0 consists of NN, $N\Delta$ and $\Delta\Delta$ states, $|\Psi\rangle = |\Psi_{NN}\rangle + |\Psi_{N\Delta}\rangle + |\Psi_{\Delta\Delta}\rangle$, where the components of each wave function are $^1S_0$ state for the NN channel, $^5D_0$ state for the NΔ channel and $^1S_0$ and $^5D_0$ states for the $\Delta\Delta$ channel.

Since the ^1E channel is unbound, we use a method to deal with unbound systems as bound states by using a Hamiltonian with radius constraint. It enables us to find out the energies and the matrix elements for the unbound two-body systems. The Hamiltonian with the constraint for the radius is given as, $\tilde{H} = H + \lambda\hat{r}^2$, where $\hat{r}^2$ is the operator of the radius, and we can obtain the energy by subtracting the effect of the radius operator. We constrain the two-body system with the same radius as the deuteron 2.0 fm and show the results in Table 2.

As a result we obtain a large tensor energy of −14.7 MeV, where the total energy is 1.5 MeV with the kinetic energy 14.0 MeV and central energy 2.2 MeV. It is very important to point out that the energy gained by adding the delta component is about 15 MeV, while the probability is only about 1 . The NN wave function consisting of only S-wave does not make the tensor correlations by itself, hence this large tensor components come from the transitions between N and Δ. Particularly we estimate the couplings between $^1S_0$ of NN state and $^5D_0$ of $N\Delta$ state are dominant in this tensor component, −13.2 MeV.

4. The three body force in the triton

We show here the results for the triton together with the deuteron with both the AV28 and AV14 potentials in Fig. 1. It is very interesting to see that the results of AV28 are largely different from those of AV14 for both the two and three nucleon systems. As for the binding energy of the triton, the result with the AV28 potential was less than that of the AV14 potential. We calculated the triton system by dropping the three body process and found the effects of three-body force due to the Δ excitations as 0.8 MeV attraction. Hence, we concluded that the Δ excitations gave a reasonable amount of the three-body effect. This means that the two body interaction terms of AV28 gave a smaller binding energy as compared to AV14 by about 1 MeV. We have checked the two body systems where the AV28 potential gave almost the same amount of attraction as compared to the AV14 potential. Hence, we concluded that the attraction in the singlet S channel in AV28 was decreased due to the blocking effect in the three body system. We multiplied then the central interaction in the singlet S channel by 0.93 and calculated the triton binding energy with the result of -7.8 MeV as compared with the value of -7.7 MeV for the case of the AV14 potential. In this case, the energy of the singlet S channel becomes 0.5 MeV as compared with the value of 1.6 MeV as shown in Table 2.

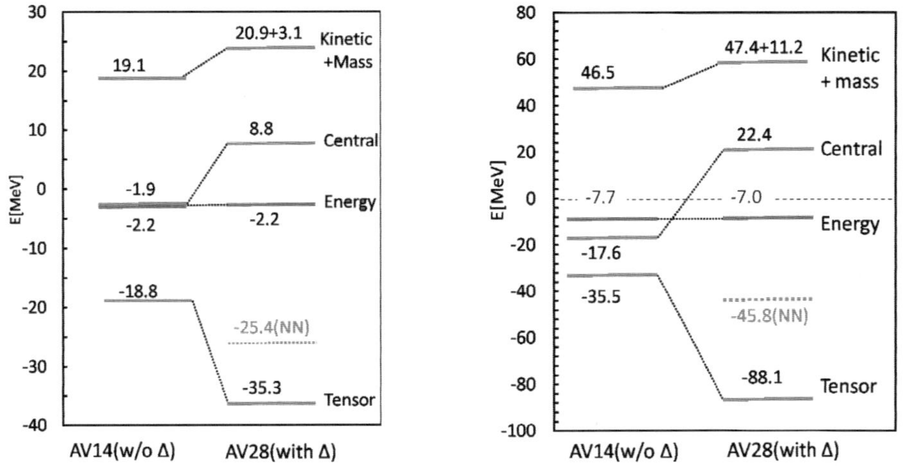

Figure 1. Calculated results for the binding energies and various matrix elements in the deuteron and the triton with AV28 and AV14.

5. Discussion and Summery

We have studied the two body and three body systems using the AV28 and AV14 potentials. We have found that the three body force effect in fact came from the delta excitation by the right amount of 0.8 MeV. We have found at the same time that the attractive effect in the singlet S channel in two body interaction was decreased by about 1 MeV in the triton system. This amount corresponds to the necessary strength for the reproduction of the neutron-neutron quasi-elastic process in the pd reactions [6]. It is very interesting to see that the two body interaction changes in the triton due to the Pauli blocking effect for the case of the AV28 potential.

References

[1] Pieper S C and Wiringa R B 2001 *Annu. Rev. Nucl. Part. Sci.* **51** 53
[2] Fujita J and Miyazawa H 1957 *Prog. Theor. Phys.* **17** 360
[3] Pudliner B S, Pandharipande V R, Carlson J and Wiringa R B 1995 *Phys. Rev.* **C74** 4396
[4] Deltuva A, Fonseca A C and Sauer P U 2009 *Mod. Phys. Lett.* **A24** 785
[5] Pikclesimer A, Rice R A and Branderburg R 1992 *Phys. Rev.* **C45** 2624
[6] Witala H and Gloekel W 2011 *Phys. Rev.* **C83** 034004
[7] Sekiguchi K *et al.* 2014 *Phys. Rev.* **C89** 064007
[8] Horii K, Toki H, Myo T and Ikeda K 2011 *Prog. Theo. Phys.* **127** 1109
[9] Suzuki Y, Horiuchi W, Orabi M and Arai K 2008 *Few-Body System* **42** 33
[10] Wiringa R B, Smith R A and Ainsworth T L 1984 *Phys. Rev.* **C29** 1207

Journal of Physics: Conference Series **569** (2014) 012077 doi:10.1088/1742-6596/569/1/012077

$K^- d \to \pi\Sigma n$ reactions and structure of the $\Lambda(1405)$

S Ohnishi[1,2], Y Ikeda[2], T Hyodo[3], E Hiyama[2], W Weise[4,5]

[1] Department of Physics, Tokyo Institute of Technology, Tokyo 152-8551, Japan
[2] RIKEN Nishina Center, Wako, aitama 351-0198, Japan
[3] Yukawa Institute for Theoretical Physics, Kyoto University, Kyoto 606-8502, Japan
[4] ECT*, Villa Tambosi, I-38123 Villazzano (Trento), Italy
[5] Physik Department, Technische Universität München, D-85747 Garching, Germany

E-mail: s_ohnishi@riken.jp

Abstract. We report on the first results of a full three-body calculation of the $\bar{K}NN$-πYN amplitude for the $K^- d \to \pi\Sigma n$ reaction, and examine how the $\Lambda(1405)$ resonance manifests itself in the neutron energy distributions of $K^- d \to \pi\Sigma n$ reactions. The amplitudes are computed using the $\bar{K}NN$-πYN coupled-channels Alt-Grassberger- andhas (AG) equations. Two types of models are considered for the two-body meson-baryon interactions: an energy-independent interaction and an energy-dependent one, both derived from the leading order chiral U(3) Lagrangian. These two models have different off-shell properties that cause correspondingly different behaviors in the three-body system. As a remarkable result of this investigation, it is found that the neutron energy spectrum, reflecting the $\Lambda(1405)$ mass distribution and width, depends quite sensitively on the (energy-dependent or energy-independent) model used. Hence accurate measurements of the $\pi\Sigma$ mass distribution have the potential to discriminate between possible mechanisms at work in the formation of the $\Lambda(1405)$.

1. Introduction

Understanding the structure of the $\Lambda(1405)$ with spin-parity $J^\pi = 1/2^-$ and strangeness $S = -1$ is a long-standing issue in hadron physics. The mass of the $\Lambda(1405)$ is slightly less than the $\bar{K}N$ threshold energy. The $\Lambda(1405)$ can be considered as a $\bar{K}N$ quasi-bound state embedded in the $\pi\Sigma$ continuum [1, 2]. uided by this picture, $\bar{K}N$ interactions which reproduce the mass of $\Lambda(1405)$ and two-body scattering data have been constructed phenomenologically [3, 4]. On the other hand, $\bar{K}N$ interactions have been studied for a long time based on chiral SU(3) dynamics [5, 6, 7]. Between the phenomenological and chiral SU(3) $\bar{K}N$ interactions, subthreshold $\bar{K}N$ amplitudes are quite different [8]. The phenomenological model describes $\Lambda(1405)$ as a single pole of the scattering amplitude around 1405 MeV. The $\bar{K}N$ amplitude from the interaction based on chiral SU(3) dynamics has two poles, one of which located not at 1405 MeV but around 1420 MeV [9, 10]. The differences in the pole structure come from the different off-shell behavior, especially as a consequence of the energy-dependence of the $\bar{K}N$ interaction. The $\bar{K}N$ interaction based on chiral SU(3) dynamics is energy-dependent, and its attraction becomes weaker as one moves below the $\bar{K}N$ threshold energy. Hence the (upper) pole of the $\bar{K}N$ amplitude shows up around 1420 MeV. On the other hand, the phenomenological $\bar{K}N$ interaction is energy-independent and strongly attractive so that the pole shows up around 1405 MeV. These differences are enhanced in the so-called few-body kaonic nuclei, such as the strange dibaryon resonance under discussion in the $\bar{K}NN$-πYN

coupled system [11, 12, 13, 14, 15, 16, 17, 18, 19, 20, 21]. How a possible signature of this strange dibaryon resonance shows up in the resonance production reaction is also of interest as it reflects the two-body dynamics of the $\Lambda(1405)$ [22].

One of the possible kaon-induced processes forming the $\Lambda(1405)$ is $K^-d \to \Lambda(1405)\,n$. The signature of the $\Lambda(1405)$ was observed in an old bubble-chamber experiment that measured the $\pi\Sigma$ invariant mass distribution in the $K^-d \to \pi^+\Sigma^-n$ reaction [23]. A new experiment is planned at J- ARC [24]. Theoretical investigations of the $K^-d \to \pi\Sigma n$ reaction have previously been performed in simplified models assuming a two-step process [25, 26, 27, 28].

In this contribution we examine how the $\Lambda(1405)$ resonance shows up in the $K^-d \to \pi\Sigma n$ reaction by making use of the approach based on the coupled-channels Alt- rassberger-Sandhas (A S) equations developed in Refs. [19, 20, 21, 22]. This is the first calculation of this process which incorporates the full three-body dynamics.

2. Three-body Scattering Equations

Throughout this paper, we assume that the three-body processes take place via separable two-body interactions, which have the following form in the two-body center-of-mass (CM) frame,

$$V_{\alpha\beta}(\vec{q}_\alpha, \vec{q}_\beta; E) = g_\alpha(\vec{q}_\alpha)\lambda_{\alpha\beta}(E)g_\beta(\vec{q}_\beta)\ , \tag{1}$$

where $\vec{q}_\alpha$ $[g_\alpha(\vec{q}_\alpha)]$ is the relative momentum [form factor] of the two-body channel α; E is the total energy of the two-body system. With this assumption the amplitudes for the quasi-two-body scattering of an "isobar" and a spectator particle, $X_{ij}(\vec{p}_i, \vec{p}_j; W)$, are then obtained by solving the A S equations [29, 30],

$$\begin{aligned} X_{ij}(\vec{p}_i, \vec{p}_j, W) &= (1-\delta_{ij})Z_{ij}(\vec{p}_i, \vec{p}_j, W) \\ &+ \sum_{n\neq i}\int d\vec{p}_n Z_{in}(\vec{p}_i, \vec{p}_n, W)\tau_n\left(W - E_n(\vec{p}_n)\right)X_{nj}(\vec{p}_n, \vec{p}_j, W)\ . \end{aligned} \tag{2}$$

Here the subscripts i, j, n specify the reaction channels; W and $\vec{p}_i$ are the total scattering energy and the relative momentum of channel i in the three-body CM frame, respectively; $Z_{ij}(\vec{p}_i, \vec{p}_j; W)$ and $\tau_i\left(W - E_i(\vec{p}_i)\right)$ are the one-particle exchange potential and the two-body propagator.

With the quasi-two-body amplitudes, the scattering amplitudes for the break-up process $d + \bar{K} \to \pi + \Sigma + N$ are obtained as

$$\begin{aligned} T_{\pi\Sigma N\text{-}\bar{K}}\ (\vec{q}_N, \vec{p}_N, \vec{p}_{\bar{K}}, W) &= g_{Y_\pi}(\vec{q}_N)\tau_{Y_\pi Y_K}\left(W - E_N(\vec{p}_N)\right)X_{Y_K}\ (\vec{p}_N, \vec{p}_{\bar{K}}, W) \\ &+ g_{Y_\pi}(\vec{q}_N)\tau_{Y_\pi Y_\pi}\left(W - E_N(\vec{p}_N)\right)X_{Y_\pi}\ (\vec{p}_N, \vec{p}_{\bar{K}}, W) \\ &+ g_{N^*}(\vec{q}_\Sigma)\tau_{N^*N^*}\left(W - E_\Sigma(\vec{p}_\Sigma)\right)X_{N^*}\ (\vec{p}_\Sigma, \vec{p}_{\bar{K}}, W) \\ &+ g_{\ y}(\vec{q}_\pi)\tau_{\ y\ y}\left(W - E_\pi(\vec{p}_\pi)\right)X_{\ y}\ (\vec{p}_\pi, \vec{p}_{\bar{K}}, W)\ , \end{aligned} \tag{3}$$

where $X_{Y_K}\ (\vec{p}_N, \vec{p}_{\bar{K}}, W)$ is the quasi-two-body amplitude anti-symmetrized for two nucleons; the subscripts denote the isobars. The notations for the isobars are $Y_K = \bar{K}N$, $Y_\pi = \pi Y$, $d = NN$, $N^* = \pi N$ and $d_y = YN$, respectively.

In this contribution we employ the first two terms of Eq. (3) as a first step. These terms emerge directly from the $\Lambda(1405)$ in the final state interaction; they are the dominant parts of the full T-matrix. Using this T-matrix, the differential cross section of the break-up process $d + \bar{K} \to \pi + \Sigma + N$ is calculated as:

$$\begin{aligned} \frac{d\sigma}{dE_n} &= (2\pi)^4 \frac{E\ E_{\bar{K}}}{Wp_{\bar{K}}} \frac{m_N m_\Sigma m_\pi}{m_N + m_\Sigma + m_\pi} \\ &\times \int d\ \ _{p_N} d\ \ _{q_N} p_N q_N \sum_{\bar{i}f} |< N\Sigma\pi|T(W)|d\bar{K} >|^2\ , \end{aligned} \tag{4}$$

Table 1. Cutoff parameters of $\bar{K}N$-πY interaction.

	$\Lambda^{I=0}_{\bar{K}N}$(MeV)	$\Lambda^{I=0}_{\pi\Sigma}$(MeV)	$\Lambda^{I=1}_{\bar{K}N}$(MeV)	$\Lambda^{I=1}_{\pi\Sigma}$(MeV)	$\Lambda^{I=1}_{\pi\Lambda}$(MeV)
E-dep	1000	700	725	725	725
E-indep	1000	700	920	960	640

where E_n is the neutron energy in the center-of-mass frame of $\pi\Sigma$ defined by

$$E_n = m_N + \frac{p_N^2}{2\eta_N} \ . \tag{5}$$

3. Models of Two-body Interaction

We use two-body s-wave meson-baryon interactions obtained from the leading order chiral Lagrangian,

$$L_{\ \ T} = \frac{i}{8F_\pi^2} Tr(\bar{\psi}_B\gamma^\mu[[\phi,\partial_\mu\phi],\psi_B]). \tag{6}$$

Here, we examine two interaction models, both of which are derived from the above Lagrangian but have different off-shell behavior: one is the energy dependent model (E-dep),

$$V_{\alpha\beta}(q',q;E) = \ - \ \lambda_{\alpha\beta}\frac{1}{32\pi^2F_\pi^2}\frac{2E-M_\alpha-M_\beta}{\sqrt{m_\alpha m_\beta}}\left(\frac{\Lambda_\alpha^2}{q'^2+\Lambda_\alpha^2}\right)^2\left(\frac{\Lambda_\beta^2}{q^2+\Lambda_\beta^2}\right)^2, \tag{7}$$

while the other is the energy independent model (E-indep),

$$V_{\alpha\beta}(q',q) = \ - \ \lambda_{\alpha\beta}\frac{1}{32\pi^2F_\pi^2}\frac{m_\alpha+m_\beta}{\sqrt{m_\alpha m_\beta}}\left(\frac{\Lambda_\alpha^2}{q'^2+\Lambda_\alpha^2}\right)^2\left(\frac{\Lambda_\beta^2}{q^2+\Lambda_\beta^2}\right)^2. \tag{8}$$

Here, m_α (M_α) is the meson (baryon) mass; F_π is the pion decay constant; $\lambda_{\alpha\beta}$ are determined by the flavor SU(3) structure of the chiral Lagrangian.

In the derivation of these potentials we have assumed the so-called "on-shell factorization" [6] for Eq. (7) and $q, q' \ll M_\alpha$ for Eq. (8). The cutoff parameters Λ are determined by fitting experimental data as shown in Table 1.

In the E-dep model, the $\bar{K}N$ amplitudes have two poles for $l = I = 0$ in the $\bar{K}N$-physical and $\pi\Sigma$-unphysical sheets, corresponding to those derived from the chiral unitary model [10]. On the other hand, the E-indep model has a single pole that corresponds to $\Lambda(1405)$. It is interesting to examine how this difference of the two-body interaction models appears in the neutron energy spectrum of the $K^-d \to \pi\Sigma n$ reaction.

4. Results and Discussion

In Fig.1, we present the differential cross section of $K^-d \to \pi\Sigma n$ [Eq. (4)] computed using the E-dep (a) and E-indep (b) models, respectively. We investigate the cross section for initial kaon momentum $p^{lab}_{K^-} = 1000$ MeV in accordance with the planned J- ARC experiment [24]. Here, we decompose the isospin basis states into charge basis states using Clebsch— ordan coe cients: the solid curve represents the $K^- + d \to \pi^+ + \Sigma^- + n$; the dashed curve refers to the $K^- + d \to \pi^- + \Sigma^+ + n$; the dotted curve represents the $K^- + d \to \pi^0 + \Sigma^0 + n$ reaction, respectively.

Journal of Physics: Conference Series **569** (2014) 012077 doi:10.1088/1742-6596/569/1/012077

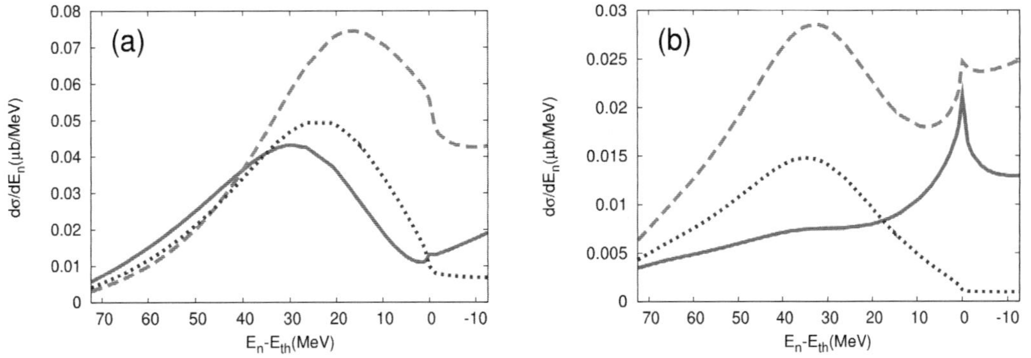

Figure 1. Differential cross section $d\sigma/dE_n$ for $K^-+d \to \pi+\Sigma+n$. The initial kaon momentum is set to $p^{lab}_{K^-} = 1000$ MeV. anel (a): the E-dep model; anel (b) the E-indep model. Solid curves: $\pi^+\Sigma^-n$; dashed curves: $\pi^-\Sigma^+n$; dotted curves: $\pi^0\Sigma^0n$ in the final state, respectively.

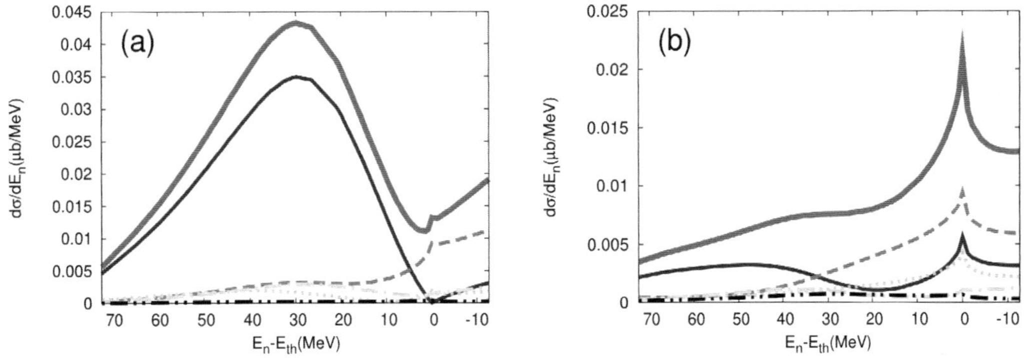

Figure 2. Contribution of each partial wave component to the differential cross section $d\sigma/dE_n$ for $d+K^- \to \pi^+ + \Sigma^- + n$. anel (a): the E-dep model; anel (b) the E-indep model. The thick solid curve represents the summation of total orbital angular momentum $L = 0$ to 14; The thin solid curve represents $L = 0$ only; The dashed curve represents $L = 1$ only; The dotted curve represents $L = 2$ only; The dashed-dotted curve represents $L = 3$ only; The dashed-two-dotted curve represents $L = 4$ only, respectively. The initial kaon momentum is set to $p^{lab}_{K^-} = 1000$ MeV.

We subtract the neutron energy E_{th} at which the amplitudes have the $\bar{K}N$ threshold cusp from the neutron energy E_n, i.e. $\bar{K}N$ threshold cusp shows up on the differential cross section at $E_n - E_{th} = 0$. Well defined maxima are found at $E_n \sim$ 17-30 MeV for the E-dep model and a peak or bumps at $E_n \sim$ 32-38 MeV for the E-indep model, depending in the charge combination of $\pi\Sigma$ in the final state. These peak and bump structures appear about 5 MeV higher in energy than the calculated binding energy of the $\Lambda(1405)$ ($E_B \sim 13$ MeV for the E-dep model and $E_B \sim 28$ MeV for the E-indep model). The magnitude of the differential cross section for the E-dep model is twice larger than that for the E-indep model, and the interference patterns with backgrounds are quite different between these two models. This clear difference in the differential cross section, arising from the dependence on the two-body interaction models, suggests that the $K^-d \to \pi\Sigma n$ reaction can indeed provide useful information on the $\bar{K}N$-πY system.

Finally, we examine the contribution of each partial wave component for total orbital angular momentum L to the differential cross section (Fig. 2). We conclude that the s-wave component is dominant in the low-energy region, but around the $\bar{K}N$ threshold higher partial waves such as the p-wave component become important.

In summary, we have calculated the differential cross sections (4) for $K^- + d \to \pi + \Sigma + n$ reactions. We have found peak and bump structures in the neutron energy spectrum, and therefore it is possible to observe the signal of the $\Lambda(1405)$ resonance in the physical cross sections. We have also shown that the $K^-d \to \pi\Sigma n$ reactions are useful for judging existing dynamical models of $\bar{K}N$-$\pi\Sigma$ coupled systems with $\Lambda(1405)$. Further improvements of the present model to account for the neglected contributions in Eq. (3) and relativistic corrections are under investigation.

Acknowledgments
The simulation has been performed on a supercomputer (NEC S 8R) at the Research Center for Nuclear hysics, Osaka University. This work was partly supported by RIKEN Junior Research Associate rogram, by RIKEN iTHES roject and by JS S KAKENHI rants Nos. 25800170, 24740152 and 23224006.

References
[1] Dalitz R and Tuan 1959 *Phys. Rev. Lett.* **2** 425–428
[2] Dalitz R and Tuan 1960 *Annals Phys.* **10** 307–351
[3] Akaishi Y and Yamazaki T 2002 *Phys. Rev.* **C65** 044005
[4] hevchenko N 2012 *Phys. Rev.* **C85** 034001 (*Preprint* 1103.4974)
[5] Kaiser N, iegel P and Weise W 1995 *Nucl. Phys.* **A594** 325–345 (*Preprint* nucl-th/9505043)
[6] Oset E and Ramos A 1998 *Nucl. Phys.* **A635** 99–120 (*Preprint* nucl-th/9711022)
[7] Hyodo T and Jido D 2012 *Prog. Part. Nucl. Phys.* **67** 55–98 (*Preprint* 1104.4474)
[8] Hyodo T and Weise W 2008 *Phys. Rev.* **C77** 035204 (*Preprint* 0712.1613)
[9] Oller J and Meissner U G 2001 *Phys. Lett.* **B500** 263–272 (*Preprint* hep-ph/0011146)
[10] Jido D, Oller J, Oset E, Ramos A and Meissner U 2003 *Nucl. Phys.* **A725** 181–200 (*Preprint* nucl-th/0303062)
[11] Yamazaki T and Akaishi Y 2002 *Phys. Lett.* **B535** 70–76
[12] Yamazaki T and Akaishi Y 2007 *Phys. Rev.* **C76** 045201 (*Preprint* 0709.0630)
[13] Dote A, Hyodo T and Weise W 2008 *Nucl. Phys.* **A804** 197–206 (*Preprint* 0802.0238)
[14] Dote A, Hyodo T and Weise W 2009 *Phys. Rev.* **C79** 014003 (*Preprint* 0806.4917)
[15] Wycech and Green A 2009 *Phys. Rev.* **C79** 014001 (*Preprint* 0808.3329)
[16] Barnea N, Gal A and Liverts E 2012 *Phys. Lett.* **B712** 132–137 (*Preprint* 1203.5234)
[17] hevchenko N, Gal A and Mares J 2007 *Phys. Rev. Lett.* **98** 082301 (*Preprint* nucl-th/0610022)
[18] hevchenko N, Gal A, Mares J and Revai J 2007 *Phys. Rev.* **C76** 044004 (*Preprint* 0706.4393)
[19] Ikeda Y and ato T 2007 *Phys. Rev.* **C76** 035203 (*Preprint* 0704.1978)
[20] Ikeda Y and ato T 2009 *Phys. Rev.* **C79** 035201 (*Preprint* 0809.1285)
[21] Ikeda Y, Kamano H and ato T 2010 *Prog. Theor. Phys.* **124** 533–539 (*Preprint* 1004.4877)
[22] Ohnishi , Ikeda Y, Kamano H and ato T 2013 *Phys. Rev.* **C88** 025204 (*Preprint* 1302.2301)
[23] Braun O, Grimm H, Hepp V, trobele H, Thol C *et al.* 1977 *Nucl. Phys.* **B129** 1
[24] Noumi H *et al.* 2009 J-PARC proposal E31 URL http://j-parc.jp/researcher/Hadron/en/pac_0907/pdf/Noumi.pdf
[25] Jido D, Oset E and ekihara T 2009 *Eur. Phys. J.* **A42** 257–268 (*Preprint* 0904.3410)
[26] Miyagawa K and Haidenbauer J 2012 *Phys. Rev.* **C85** 065201 (*Preprint* 1202.4272)
[27] Jido D, Oset E and ekihara T 2013 *Eur. Phys. J.* **A49** 95 (*Preprint* 1207.5350)
[28] Yamagata- ekihara J, ekihara T and Jido D 2013 *PTEP* **2013** 043D02 (*Preprint* 1210.6108)
[29] Alt E, Grassberger P and andhas W 1967 *Nucl. Phys.* **B2** 167–180
[30] Amado R D 1963 *Phys. Rev.* **132** 485–494

Journal of Physics: Conference Series **569** (2014) 012078 doi:10.1088/1742-6596/569/1/012078

Search for $^{6}_{\Lambda}$H hypernucleus by the (π^-, K^+) reaction at J-PARC

Hitoshi Sugimura for J-PARC E10 Collaboration

Advanced Science Research Center, Japan Atomic Energy Agency, Tokai, Iabaraki, 319-1195, Japan

Abstract. We have carried out an experiment to produce the neutron-rich hypernucleus $^{6}_{\Lambda}$H by the (π^-, K^+) reaction on ^{6}Li target at the pion beam momentum of 1.2 GeV/c (J-PARC E10). In order to calibrate the scale of the missing-mass or of the Λ binding energy of the hypernucleus, we also measured the $^{12}C(\pi^+, K^+)^{12}_{\Lambda}C$, $p(\pi^-, K^+)\Sigma^-$ and $p(\pi^+, K^+)\Sigma^+$ reactions. The experiment was performed at the K1.8 beam line of J-PARC Hadron Experimental Facility. The overall collected data sample corresponds to an integrated beam intensity of 1.4×10^{12} pions.

1. Introduction

One of the most important topics in strangeness nuclear physics is the study of neutron-rich Λ hypernuclei [1, 2]. It is expected that a Λ hyperon plays a glue-like role in nuclei beyond the neutron drip-line. The knowledge of the behavior of the hyperon in a neutron-excess environment significantly impacts on our understanding of the neutron stars, because the addition of hyperons softens the Equation of State of matter at the core [3]. Akaishi and collaborators [4, 5] suggested that the coherent ΛN-ΣN coupling can resolve the over binding problem in $^{5}_{\Lambda}$He [6] and that such a coupling is enhanced in a neutron-excess environment.

2. Spectroscopy of Λ hypernuclei

One of the important subjects in studying Λ hypernuclei in the past was the precise measurements of the level structures of the Λ hypernuclei, which made possible to discuss the underlying hyperon-nucleon strong interaction. The similarity of the Λ hyperon with the nucleon is one of the key properties which brings the rich spectra of Λ hypernuclei. Another important property is the additional binding energy due to the Λ hyperon. Then we expect the hypernuclear chart is even richer than the ordinary nuclear chart.

On the other hand, we have identified only a small fraction of hypernuclei in the hypernuclear chart. One reason of the limited observation is that we mainly used the (K^-, π^-) and the (π^+, K^+) reactions to produce Λ hypernuclei. Figure 1(b) shows the hypernuclei which have ever produced in the A$\leq$15 region: purple colored ones were directly produced via the (K^-, π^-) and the (π^+, K^+) reactions on stable nuclear targets while green colored ones were observed as hyperfragments in the nuclear emulsion experiments. The chart looks already compatible with that of the ordinary nuclei (Fig. 1(a)), but the information on the hyperfragments from the nuclear emulsion experiments was quite limited.

To explore wider area of the hypernuclear chart in further detail, we need new spectroscopic tools. If we exploit charge exchange reactions, we can directly produce many neutron-rich Λ hypernuclei as shown in Fig. 1(c). Hypernuclei in the light blue boxes are produced by the single charge-exchange reactions and the ones in yellow boxes are produced by the double charge-exchange reactions (DCX), such as the (K^-, π^+) and (π^-, K^+) reactions.

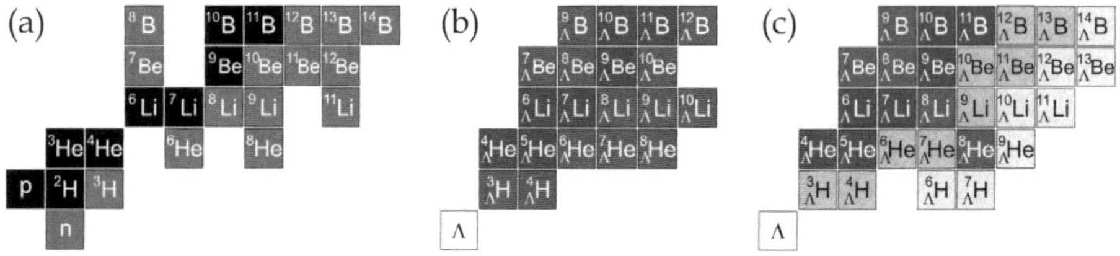

Figure 1. (a) chart of light ordinary nuclei. (b) chart of Λ hypernuclei ever produced in the A≤15 region. Purple boxes correspond to directly produced hypernuclei via the (K^-, π^-) and (π^+, K^+) reactions on stable nuclear targets and green boxes correspond to hypernuclei observed as hyperfragments in the emulsion experiments. (c) blue boxes show hypernuclei to be produced by single charge-exchange reactions and yellow boxes correspond to hypernuclei to be produce via the (K^-, π^+) and (π^-, K^+) reactions on stable targets.

3. Study of double charge-exchange reactions

3.1. $(K^-_{stopped}, \pi^+)$ *reaction*

A pilot experiment aiming to produce Λ hypernuclei was performed at KEK-PS by using the $(K^-_{stopped}, \pi^+)$ reaction [7]. In that experiment, only upper limits were obtained for the production rate of neutron-rich Λ hypernuclei (${}^{6}_{\Lambda}$He, ${}^{12}_{\Lambda}$Be and ${}^{16}_{\Lambda}$C) due to tiny branching ratios to the DCX channel and a huge background from the in-flight hyperon decays, $\Sigma^+ \rightarrow n\pi^+$. An improved study has been carried out for the ${}^{6}_{\Lambda}$H, ${}^{7}_{\Lambda}$H, ${}^{9}_{\Lambda}$He and ${}^{12}_{\Lambda}$Be hypernuclei by the FINUDA Collaboration at Frascati-DAΦNE [8, 9]: they claimed the observation of 3 events of ${}^{6}_{\Lambda}$H hypernuclei and their mesonic weak decays [10].

3.2. (π^-, K^+) *reaction*

Another promising DCX reaction to produce the neutron-rich Λ hypernuclei is the (π^-, K^+) reaction. An attempt to produce the neutron-rich Λ hypernucleus ${}^{10}_{\Lambda}$Li was carried out at KEK-PS by exploiting the (π^-, K^+) reaction at 1.05 and 1.2 GeV/c pion beam momenta (KEK-E521 experiment) [11]. In that experiment, clear signal events were observed in the Λ bound region in the missing mass spectrum of the ^{10}B(π^-, K^+) reaction. The production cross section of the ${}^{10}_{\Lambda}$Li hypernucleus was estimated to be very small (~10 nb/sr), roughly 10^{-3} of that of the (π^+, K^+) reaction (typically 10 μb/sr). Compared with the $(K^-_{stopped}, \pi^+)$ reaction, the (π^-, K^+) reaction is almost background free in the Λ bound region.

4. J-PARC E10 experiment

The J-PARC E10 experiment was proposed to produce the neutron-rich Λ-hypernucleus ${}^{6}_{\Lambda}$H by using the ^{6}Li(π^-, K^+) reaction at 1.2 GeV/c and to study its structure. The experiment was performed at the K1.8 beam line of the J-PARC Hadron Experimental Facility. The K1.8 beam line spectrometer and the Superconducting Kaon Spectrometer (SKS) were used. Fig. 2 shows the K1.8 beam line and the SKS spectrometers.

The K1.8 beam line spectrometer consists of a gas Čerenkov counter (GC), a scintillating fiber tracker (BFT), QQDQQ magnets, two drift chambers (BC3 and BC4) and a timing plastic scintillation hodoscope (BH2).

The SKS spectrometer consists of a scintillating fiber tracker (SFT), three drift chambers (SDC2, SDC3 and SDC4), a superconducting magnet, a timing plastic scintillator hodoscope (TOF) and two threshold type Čerenkov counters (LAC and LC).

An enriched ^{6}Li target (95.54%) of 3.5 g/cm^2 in thickness, 70 mm in width and 40 mm in height was used. To reconstruct the reaction vertex point precisely and to find the right combination of upstream and downstream tracks for multi-track events, silicon strip detectors (SSD) were installed just upstream of the target.

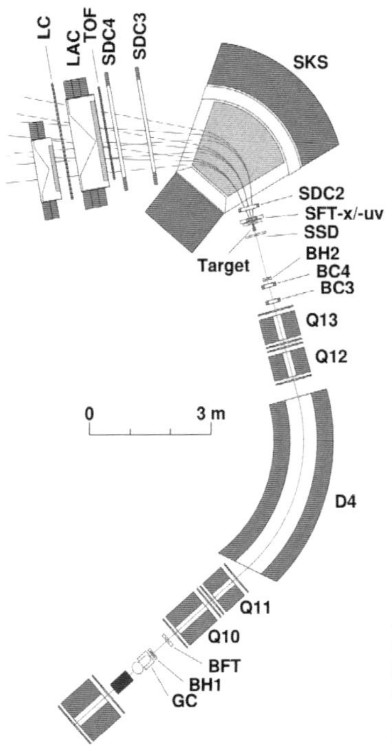

Figure 2. Schematic view of the K1.8 beam line spectrometer (from GC to BH2), the target area (SSD and Target) and the SKS spectrometer (from SFT toe LC). See the text for more details.

5. Results

Table 1 shows a run summary of the J-PARC E10 experiment. We started the calibration run of $^{12}_{\Lambda}$C production at a beam momentum of 1.2 GeV/c to evaluate the missing-mass resolution and to confirm that all the detectors were working properly. Next, we made the measurement of the Σ^- and Σ^+ production reactions from hydrogen on a $(CH_2)_n$ target at a beam momentum of 1.37 GeV/c. By using this way, we could calibrate momenta measured by the beam line spectrometer and SKS with a systematic error of 1.34 MeV/c. The π beam intensity at the target position was typically 4.1×10^6/spill (2.0 s spill length) for the $^{12}C(\pi^+, K^+)^{12}_{\Lambda}C$ and the $p(\pi^+, K^+)\Sigma^+$ reactions and 1.3×10^7/spill for the $p(\pi^-, K^+)\Sigma^-$ and the $^6Li(\pi^-, K^+)^6_{\Lambda}H$ reactions. We also performed the beam-through runs at four momentum settings, 0.8, 0.9, 1.0, and 1.2 GeV/c, with and without the ^{6}Li target to know the amount of energy straggling and the energy loss in the target; in addition we evaluated the difference between the momentum values measured by the K1.8 beam line and the SKS spectrometers.

Table 1. Run Summary of the J-PARC E10 experiment

reaction	momentum (GeV/c)	intensity (/spill)	time (hour)
$^{12}C(\pi^+, K^+)^{12}_{\Lambda}C$	1.2	4.1×10^6	24
$p(\pi^-, K^+)\Sigma^-$	1.37	1.3×10^7	4
$p(\pi^+, K^+)\Sigma^+$	1.37	3.5×10^6	1
beam-through	0.8, 0.9, 1.0, 1.2	$\sim 10^4$	2
$^6Li(\pi^-, K^+)^6_{\Lambda}H$	1.2	$1.2\text{–}1.4 \times 10^7$	276

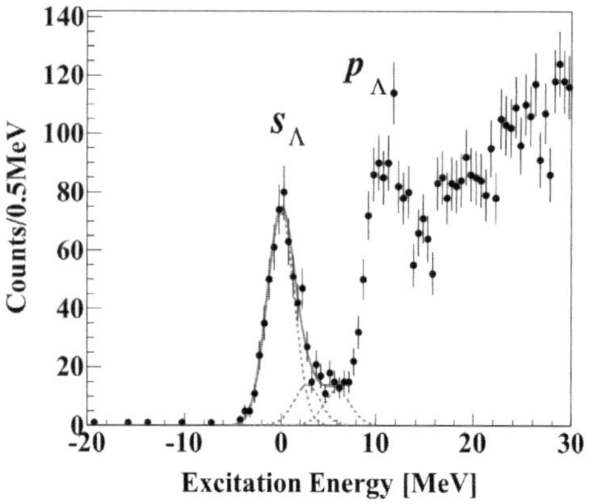

Figure 3. Excitation energy spectrum of the $^{12}C(\pi^+, K^+)^{12}_{\Lambda}C$ reaction at the beam momentum of 1.2 GeV/c. The ground (s_Λ) and excited (p_Λ) states are clearly observed. The missing-mass resolution is estimated by fitting the ground and known excited states. The dashed curves show the best fit Gaussian functions for these states and the solid curve is the sum.

The same target thickness, 3.5 g/cm^2, and the same beam momentum, 1.2 GeV/c, as those in the KEK-E521 experiment were employed in the measurement of the $^6Li(\pi^-, K^+)$ reaction. As the final result, the number of pions delivered to the experiment reached 1.4×10^{12} taking into account the DAQ efficiency.

Figure 3 shows the excitation energy spectrum of the (π^+, K^+) reaction on the graphite target. The ground (s_Λ) and excited (p_Λ) states of the $^{12}_{\Lambda}C$ hypernucleus are clearly observed. The ground state region is fitted with three Gaussian functions corresponding to the ground and to known excited states at 2.833 and 6.050 MeV[12]. A same width of the Gaussian function is used for the three states. The missing-mass resolution is estimated to be 3.2 MeV (FWHM).

Figure 4 shows the missing-mass spectrum of the $^6Li(\pi^-, K^+)$ reaction. The vertical axis shows the double differential cross section in the laboratory frame averaged over the scattering angle from 2° to 14°, $d^2\overline{\sigma}_{\rm lab}/d\Omega/dM$ in a unit of nb/sr/(MeV/c^2). The scattering angle region has small ambiguity in spectrometer acceptance. The continuum of the unbound Λ formation reaction and the component of the Σ^- quasi-free production reaction are observed in the missing-mass regions of 5810–5880 MeV/c^2 and above 5880 MeV/c^2, respectively. A magnified view in the missing-mass range of 5795–5830 MeV/c^2 is shown in the inset. Around the $^4_{\Lambda}H$+$2n$ particle decay threshold indicated by the arrow (5801.7 MeV/c^2), no significant peak structure is observed.

There were 3 events in bound region with missing-mass window of 5.4 MeV which range is 2σ of the mass resolution of $^{12}_{\Lambda}C$. An upper limit was estimated by using the Poisson statistics. We employ upper limit of 6.68 events at a 90% confidence level coming from the background free hypothesis as a conservative estimation. As shown in Fig. 4, the differential cross sections were 0.18 nb/sr for 1 event. By using this value, the upper limit of the differential cross section averaged in the scattering angle from 2° to 14° is estimated to be 1.2 nb/sr at a 90% confidence level.

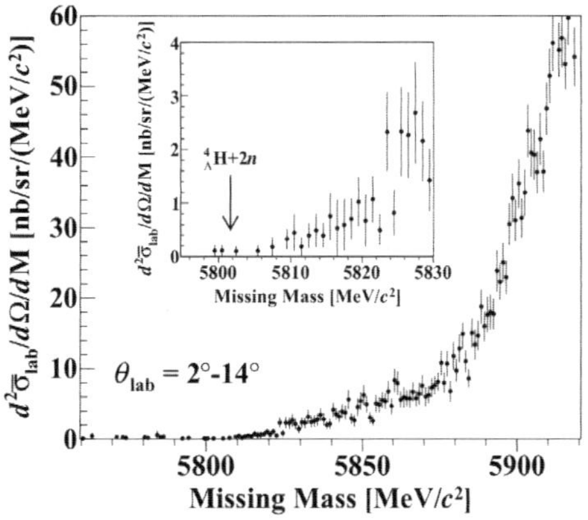

Figure 4. Missing-mass spectrum of the ^{6}Li(π^-, K^+) reaction at 1.2 GeV/c. The ordinate shows the double differential cross section averaged over the angular range from 2° to 14°. A magnified view around the Λ bound region is shown in the inset. The arrow labeled as $^4_\Lambda$H+2n shows the particle decay threshold (5801.7 MeV/c^2).

6. Discussion

In our spectrum, neither significant peak structure nor a large yield is observed around the $^4_\Lambda$H+2n particle decay threshold. The $^6_\Lambda$H hypernucleus is believed to have the $^4_\Lambda$H+2n structure dominantly and have the 0^+ ground and the 1^+ excited states which are analogous to the 0^+ and 1^+ spin-doublets in $^4_\Lambda$H. Our reaction should favor the 1^+ state direct population at foward angles, because the dominant spin non-flip amplitude of the (π^-, K^+) reaction would not change the spin of $^6_\Lambda$H with the same spin of ^{6}Li. As far as the FINUDA result is concerned, the observed $^6_\Lambda$H candidate events were interpreted as the primary population of the excited 1^+ state by the ($K^-_{stopped}$, π^+) reaction followed by the γ-ray transistion to the ground 0^+ state because the direct population of the 0^+ state should be suppressed due to the small spin-flip amplitude in the ($K^-_{stopped}$, π^+) reaction. The FINUDA results indicate that the excited 1^+ state, whose excitation energy is considered to be about 1 MeV, should be particle bound, otherwise the γ-ray transistion to the 0^+ ground state should be impossible. If the 1^+ state was bound and the production cross section was comparable with that for $^{10}_\Lambda$Li of 11.3 nb/sr, more than 60 events should be observed as a peak in the Λ bound region. Therefore, our observation is in conflict with the FINUDA observation.

The FINUDA collaboration also discussed another scenario for the spectrum of $^6_\Lambda$H in which two out of the three candidate events came from the population of the spin-tripret states, 1^+, 2^+, and 3^+, at around 3 MeV excitation [13]. With this scenario, it is interesting to compare our upper limit, 1.2 nb/sr, with quantitative theoretical estimations of the production cross sections because the cross sections are sensitive to the binding energies and the wave-functions of the low-lying states.

References

[1] R.H. Dalitz and R. Levi Setti, Nuovo Cimento **30**, 489 (1963).
[2] L. Majling, Nucl. Phys. A **585**, 211c (1995).
[3] M. Baldo, G.F. Burgio, H.-J.Schulze, Phys. Rev. C **61**, 055801 (2000).
[4] Khin Swe Myint, T. Harada, S. Shinmura, Y. Akaishi, Few-Body Syst. Suppl.**12**, 383 (2000).
[5] Y. Akaishi, T. Harada, S. Shinmura, Khin Swe Myint, Phys. Rev. Lett. **84**, 3539 (2000).
[6] R.H. Dalitz, R.C. Herndon, Y.C. Tang, Nucl. Phys. B **47**, 109 (1972).
[7] K. Kubota *et al*, Nucl. Phys. A **602**, 327 (1996).
[8] M. Agnello *et al*, Nucl. Phys. A **752**, 139c (2005).
[9] M. Agnello *et al*, Phys. Lett. B **640**, 145 (2006).
[10] M. Agnello *et al*, Phys. Rev. Lett. **108**, 042501 (2012).
[11] P.K. Saha *et al*, Phys. Rev. Lett. **94**, 052502 (2005).
[12] K. Hosomi *et al*, Nucl. Phys. A **914** 184 (2013).

[13] M. Agnello *et al*, Nucl. Phys. A **881** 269 (2012).

Possibility of a new neutral hypernucleus $_{\Lambda\Lambda}^{4}\text{n} = (n, n, \Lambda, \Lambda)$[1]

Jean-Marc Richard[1], **Qian Wang**[2], **and Qiang Zhao**[3]

[1]Université de Lyon, Institut de Physique Nucléaire, UCBL–IN2P3-CNRS, 4, rue Enrico Fermi, Villeurbanne, France
[2]Institut für Kernphysik, Institute for Advanced Simulation, and Jülich Center for Hadron Physics, D-52425 Jülich, Germany
[3]Institute of High Energy Physics and Theoretical Physics Center for Science Facilities, Chinese Academy of Sciences, Beijing 100049, China

E-mail: j-m.richard@ipnl.in2p3.fr, q.wang@fz-juelich.de, zhaoq@ihep.ac.cn

Abstract. The status of the light hypernuclei is reviewed, and discussed with models based either on the Nijmegen-RIKEN baryon–baryon interaction or on recent studies using chiral effective field theory. The latter suggests a significantly shorter range for the $\Lambda\Lambda$ interaction, and this favours the formation of a Borromean state made of two neutrons and two hyperons. Various corrections are discussed, in particular the coupling of $N\Lambda$ to $N\Sigma$, or of $\Lambda\Lambda$ to $N\Xi$, and the role of tensor forces. The new nucleus $_{\Lambda\Lambda}^{4}\text{n} = (n, n, \Lambda, \Lambda)$ could be produced in various reactions, in particular deuteron–deuteron scattering with the simultaneous production of two charged kaons, for which an estimate of the cross section is provided.

1. Introduction

The physics of light hypernuclei has been the subject of many papers, and there is a renewed interest in recent months, see, for instance, [1–3] and references therein. In this contribution, we wish to stress the possibility of Borromean states such as the neutral $T = (n, n, \Lambda, \Lambda)$ with strangeness -2, which is more likely bound using the recent interaction models inspired by chiral effective field theory (CEFT) than in more conventional models based on meson exchange and SU(3) flavour symmetry.

After a brief reminder on the allowed window for Borromean binding, we shall discuss the interaction models, present our results on light hypernuclei, and then suggest some reaction mechanisms to produce light hypernuclei.

2. Borromean binding

By a kind of miracle, ^{6}He, considered as (α, n, n) is stable, though neither $^5\text{He} = (\alpha, n)$ nor (n, n) are bound. In quantum mechanics, this can be understood even without invoking 3-body forces. Binding two bosons, for instance, with a short-range potential $g\,V(r)$, where $V(r)$ is attractive and $g > 0$, requires a minimal strength, say $g > g_2$. There is also a threshold coupling $g > g_3$ for binding three bosons with the same interaction, $g \sum_{i<j} V(r_{ij})$, but the interesting

[1] Talk given by JMR at SOTANCP3, 3rd International Workshop on State of the Art in Nuclear Cluster Physics, Yokohama, Japan, May 26-30, 2014

Journal of Physics: Conference Series **569** (2014) 012079 doi:10.1088/1742-6596/569/1/012079

observation is that $g_3 < g_2$! Hence in the window $g \in (g_3, g_2)$, one gets 3-body binding without 2-body binding. This is implicit in the pioneering paper by Thomas [4], who observed a large ratio of 3-body to 2-body binding energies for $g \gtrsim g_2$, and deduced a limit $m_\pi \lesssim 200\,\mathrm{MeV}$ on the range of nuclear forces before the discovery of the pion. The word "Borromean" became popular in the context of halo nuclei [5] and is now of current use.

Borromean binding is allowed in a limited range of couplings. For n bosons, one can demonstrate [6] that the threshold coupling g_n obey

$$(n+1)\, g_{n+1} \geq n\, g_n \;. \tag{1}$$

A system of unequal masses (m, m, M) involves two couplings g_{mm} and g_{mM}, which can be normalised so that $g = 1$ corresponds to the coupling threshold for the (m, m) and (m, M) pairs, respectively, one can draw a map of the $\{g_{mm}, g_{mM}\}$ region for which Borromean binding is not forbidden [6]. One gets typically one third of the square $\{g_{mm} < 1,\ g_{mM} < 1$ allowed, so it is conceivable to get (n, n, Λ) bound in some models where neither (n, n) nor (n, Λ) are stable.

In the case of a 4-body system with two different masses, i.e. (m, m, M, M), there are three different couplings, g_{mm}, g_{MM} and g_{mM}. Again, they can be normalised such that $g_{ij} = 1$ which is the coupling threshold for the (m_i, m_j) pair. In [6] and our recent study [7], the best frontier of the allowed region for binding inside the cube $\{g_{mm} < 1,\ g_{MM} < 1,\ g_{mM} < 1\}$ is drawn. A large fraction of the cube $\{g_{ij} \leq 1\}$ can be eligible for Borromean binding.

This preliminary study provides some hopes for Borromean binding of ${}^{4}_{\Lambda\Lambda}\mathrm{n} = (n, n, \Lambda, \Lambda)$. Atomic physics gives us some further encouragement to search for Borromean 4-body systems. With a pure Coulomb interaction, the system of four unit charges (m^+, m^-, M^+, M^-) remains bound against dissociation into two neutral atoms in the region of mass ratio $M/m \simeq 2$ (or, of course, $m/M \simeq 2$), although none of 3-body subsystems such as (m^+, M^+, M^-) or (m^+, m^-, M^+) are stable [8]. A group of Como [9] has studied how the hydrogen atom, ions and molecules evolve when the Coulomb interaction becomes screened: there is a domain of the screening parameter for which the hydrogen molecule remains stable while the hydrogen atom is unbound.

The inequality (1), and the allowed domains in the case of unequal masses, are derived from a Hall–Post type of decomposition of the n-body Hamiltonian in terms of $n-1$-body or 2-body sub-systems, with proper removal of the centre-of-mass motion [6]. Hence, they are saturated in the case of a potential with an external barrier and an inner part that coincides with a harmonic oscillator. The chances of getting Borromean binding decrease if the shape of the potential evolves towards a monotonic potential, and further decrease if the potential acquires an inner repulsive core of increasing size and strength [10].

In Refs. [7, 11], another point of view was adopted, not based on the shape of potential, but on its low-energy scattering properties. It was shown that for a given scattering length (negative so that the 2-boson system is attractive without bound state), the Borromean binding in the three-boson system depends dramatically on the effective range r_{eff}. Moreover, the curve giving the 3-body energy E_3 vs. r_{eff} is almost universal, i.e., independent of the shape of the potential.

Thus, for our purpose, it is extremely important to use toy potentials that fit accurately not only the scattering lengths but also the effective ranges.

3. The 3- and 4-body problem

Our study is devoted to light hypernuclei with mass number $A \leq 4$. The 2-body problem has been solved by standard techniques to fix the parameters of some simple potentials as to reproduce the deuteron binding energy, the scattering length and effective range of ΛN and $\Lambda\Lambda$ provided by the Nijmegen group and the Bonn-Jülich EFT group.

The 3-body problem has been solved in perimetric coordinates $x_1 = |\vec{r}_2 - \vec{r}_3|$, etc., using the variational wave function

$$\Psi = \sum_i \alpha_i \exp[-a_i\, x_1 - b_i\, x_2 - c_i\, x_3]\ , \tag{2}$$

that is widely used in atomic physics. The permutation symmetry, if any, can be checked *a posteriori* or imposed *a priori* assigning the same coefficient α_i to several terms. To avoid instabilities in the variational optimisation, it is convenient to choose the inverse ranges a_i, b_i and c_i in a geometric series, and thus to minimise only two parameters, given that the α_i and the variational energy are computed from a simple eigenvalue equation. This is somehow similar to the method used by Hiyama et al. [12] for the Gaussian expansion.

As a check, we also solve the 3-body problem by the method of Gaussian expansion in the stochastic variant of Suzuki and Varga, using their computer code [13, 14]. This latter method was also applied to the 4-body systems.

4. Results

Once again, our preliminary study is based on a kind of "model of model", namely a simple potential that reproduce the low-energy properties of a more complicated model of the baryon-baryon interaction. A similar approach was used in Refs. [15, 16]. As a check, we reproduce the results of this latter group, but we note that the models they used, borrowed from the former reference, differ from ours.

More precisely, we adopted two sets of parameters, given in Table 4, the first one, labelled ESC08, corresponding to the Nijmegen-RIKEN group [17], the second one, labelled CEFT, to Refs. [18, 19]. We also use standard values for np in both spin $s = 0$ and $s = 1$ states (not shown). The most remarkable feature is the much shorter effective range for $\Lambda\Lambda$ in the new

Table 1. Low-energy scattering parameters in the two models (in fm)

	ESC08		CEFT	
Pair	a	r_{reff}	a	r_{eff}
nn	−16·51	2·85	−18·9	2·75
$(n\Lambda)_{s=0}$	−2·7	2·97	−2·9	2·65
$(n\Lambda)_{s=1}$	−1·65	3·63	−1·51	2·64
$\Lambda\Lambda$	−0·88	4·34	−1·54	0·31

CEFT model, as compared to ESC08.

The results are the following:

$\mathbf{A = 2}$: The deuteron $d = (n, p)$ is of course well reproduced once the low-energy parameters of np are fitted. There are also other bound states with strangeness $S = -1$ or -2.

$\mathbf{A = 3}$: The $S = -1$ system (n, p, Λ) is bound in both models for isospin $I = 0$ and spin $s = 1/2$, below the $d + \Lambda$ threshold. With CEFT, the partner with $I = 0$ and $s = 3/2$ and the one with $I = 1$ and $s = 1/2$ are nearly bound. We do not find any binding for (n, n, Λ) nor for (n, Λ, Λ), but we have not yet introduced any mixing with configurations involving Σ or Ξ.

$\mathbf{A = 4}$: We agree with the conclusions of Ref. [16] regarding ${}^{\;4}_{\Lambda\Lambda}\text{H} = (n, p, \Lambda, \Lambda)$. Our main concern is ${}^{\;4}_{\Lambda\Lambda}\text{n} = (n, n, \Lambda, \Lambda)$: this state is found unbound with ESC08, in agreement with [20]. However, ${}^{\;4}_{\Lambda\Lambda}\text{n}$ is at the edge of binding with CEFT. For instance, if we simulate the low-energy parameters of the CEFT column in Table 4 by an exponential potential $V(r) = -g\, \exp(\mu\, r)$ with g and μ properly tuned for each baryon-baryon pair in each spin state, we obtain a binding of about 1 MeV for ${}_{\Lambda\Lambda}^{4}\mathrm{n}$ using about 80 terms (plus the ones deduced by permutation) in the Gaussian expansion scheme [13, 14].

Several improvements might enforce or spoil the binding. In particular it would be desirable to push further the CEFT calculations to test the stability in the 2-body sector and access to the 3-body forces. It is well-known that in the strangeness $S = 0$ sector, 3-body forces are crucial to reproduce the properties of light nuclei.

Another concern is the neglect of tensor forces, which, in our approach, are replaced by a reinforcement of the effective central interaction. The validity of this approach was discussed many years ago and summarised in the textbook by Blatt and Weisskopf [21]. It is argued that in the deuteron, the tensor forces couple D-states in the continuum to a system located at -2 MeV, while in the triton, they couple the same continuum to a system lying at -8 MeV. This explains while models with pure central forces simulate too much tensor interaction beyond the deuteron, and in particular, overbind the triton. In the case of, e.g., the $N\Lambda$ interaction, the zero-energy where the scattering length and the effective range are fixed almost coincides with the energy of the very weakly bound states such as (n, n, Λ, Λ), and the tensor forces playing a comparable role in 2-body and few-body systems, can be mimicked more safely within the same central potential.

A related important issue is the role of the couplings $\Lambda N \leftrightarrow \Sigma N$ and $\Lambda\Lambda \leftrightarrow \Xi N$. Part of it can be absorbed into an effective diagonal potential in the ΛN and $\Lambda\Lambda$ channels. However, as stressed in the literature [1–3, 22, 23], the ΣN and ΞN channels sometimes open new isospin couplings within the hypernuclei, and this will lower the energy.

The shape and the strength of the $\Lambda\Lambda$ is intimately related to the existence of the H dibaryon. The chromomagnetic binding of $H = (u, u, d, d, s, s)$ below the $\Lambda\Lambda$ threshold was proposed by Jaffe [24], with, however, some unjustified assumptions about the SU(3) flavour symmetry and the short-range correlations in the 6-body system [25]. Many experiments have searched for the H, without success. It remains that the chromomagnetic interaction at the quark level induces a short-range attraction between two $\Lambda\Lambda$, and it is interesting that this property is recovered in the CEFT approach. If the $\Lambda\Lambda$ system itself is not bound, the attraction might be sufficient to bind ${}_{\Lambda\Lambda}^{4}\mathrm{n}$.

5. Production of ${}_{\Lambda\Lambda}^{4}\mathrm{n}$

Several reactions have been used to produce double hypernuclei. One possibility is to use a Ξ^- exotic atom as initial state [26]. Another way is to consist of (K^-, K^+) inelastic scattering on a light nucleus, with a recoiling system of strangeness $S = -2$ [27]. There is also the possibility of inclusive production in electron or hadron scattering.

In the case of the tetrabaryon $T = {}_{\Lambda\Lambda}^{4}\mathrm{n}$, the exclusive reaction

$$d + d \to T + K^+ + K^+ \ , \tag{3}$$

where d is the deuteron, seems particularly promising [7]. The leading mechanism is a virtual π^0 exchanged between the two protons. This means twice the reaction $\pi^0 + p \to K^+\Lambda$, whose cross-section is known. One can thus estimate rather safely the order of magnitude of the reaction (3), given the very low background. Our estimate [7] gives a cross-section of the order of $1 - 2$ nb for an energy $\sqrt{s} \sim 5.5$ GeV. Thus the reaction (3) appears as an interesting tool to search for new hypernuclei.

6. Outlook

The recent studies of the hyperon-nucleon and hyperon-hyperon interaction indicates a significantly shorter range, as compared to the conventional approaches. This modifies the spectroscopy of light hypernuclei, and conversely, the study of light hypernuclei can probe the new models of baryon–baryon interaction.

We presented here the first results of a study restricted to simple pairwise potentials that reproduce the low-energy properties of each baryon-baryon pair. In the case where the potentials are tuned to reproduced the low-energy scattering parameters of Chiral Effective Field Theory, there are indications that the tetrabaryon ${}^{4}_{\Lambda\Lambda}\mathrm{n} = (n, n, \Lambda, \Lambda)$ might be bound. If so, this would be a fully Borromean system whose 2-body as well as 3-body sub-systems are unbound. The role of 3-body forces and the changes induced by the coupling of ΛN to ΣN and of $\Lambda\Lambda$ to ΞN will be the subject of further investigations.

Acknowledgments
We thank the organisers for this very stimulating meeting SOTANCP3, providing the opportunity to discuss with many experts.The participation by JMR was made possible by the French-Japanese project CNRS-JPS #169725. We are also grateful to Prs. V. Belyaev and A. Gal for useful comments.

This work was done within the China–France program FCPPL, and is also supported, in part, by the National Natural Science Foundation of China (Grant No. 11035006), the Chinese Academy of Sciences (KJCX3-SYW-N2), and DFG and NSFC funds to the Sino-German CRC 110.

References

[1] Hiyama E, Ohnishi S, Gibson B and Rijken T A 2014 *Phys.Rev. C* **89** 061302
[2] Garcilazo H and Valcarce A 2014 *Phys. Rev. C* **89**(5) 057001
[3] Gal A and Garcilazo H 2014 *Phys.Lett.* **B736** 93–97 (*Preprint* 1404.5855)
[4] Thomas L 1935 *Phys.Rev.* **47** 903–909
[5] Zhukov M, Danilin B, Fedorov D, Bang J, Thompson I *et al.* *Phys.Rept.* **231** 151–199
[6] Richard J M and Fleck S 1994 *Phys.Rev.Lett.* **73** 1464–1467
[7] Richard J M, Wang Q and Zhao Q 2014 (*Preprint* 1404.3473)
[8] Richard J M 2003 *Phys. Rev. A* **67**(3) 034702
[9] Bertini L, Mella M, Bressanini D and Morosi G 2004 *Phys. Rev. A* **69**(4) 042504
[10] Moszkowski S, Fleck S, Krikeb A, Theussl L, Richard J M *et al.* 2000 *Phys.Rev.* **A62** 032504
[11] Thøgersen M, Fedorov D V and Jensen A S 2008 *Phys. Rev. A* **78**(2) 020501
[12] Hiyama E, Kino Y and Kamimura M 2003 *Prog.Part.Nucl.Phys.* **51** 223–307
[13] Varga K and Suzuki Y 1996 *Phys.Rev.* **A53** 1907–1910
[14] Varga K and Suzuki Y 1997 *Comput.Phys.Commun.* **106** 157–168
[15] Filikhin I and Gal A 2002 *Phys.Rev.Lett.* **89** 172502
[16] Nemura H, Akaishi Y and Myint K S 2003 *Phys.Rev.* **C67** 051001
[17] Rijken T, Nagels M and Yamamoto Y 2013 *Few Body Syst.* **54** 801–806
[18] Polinder H, Haidenbauer J and Meissner U G 2007 *Phys.Lett.* **B653** 29–37
[19] Haidenbauer J, Petschauer S, Kaiser N, Meissner U G, Nogga A *et al.* 2013 *Nucl.Phys.* **A915** 24–58
[20] Lekala M, Rampho G, Adam R, Sofianos S and Belyaev V 2014 *Phys.Atom.Nucl.* **77** 472–478
[21] Blatt J M and Weisskopf V F 1979 *Theoretical nuclear physics* (New York, NY: Springer)
[22] Gibson B and Lehman D 1977 *Phys.Rev.* **C16** 1679
[23] Garcilazo H, Valcarce A and Carames T 2014 *J.Phys.* **G41** 095103
[24] Jaffe R L 1977 *Phys.Rev.Lett.* **38** 195–198
[25] Oka M, Shimizu K and Yazaki K 1983 *Phys.Lett.* **B130** 365
[26] Zhu D, Dover C, Gal A and May M 1991 *Phys.Rev.Lett.* **67** 2268–2271
[27] Yamamoto Y, Wakai M, Motoba T and Fukuda T 1997 *Nucl.Phys.* **A625** 107–142

Journal of Physics: Conference Series **569** (2014) 012080 doi:10.1088/1742-6596/569/1/012080

Search for the K^-pp bound state via the ^{3}He(K^-, n) reaction at 1 GeV/c

T. Hashimoto[1], S. Ajimura[2], G. Beer[3], H. Bhang[4], M. Bragadireanu[5], . Buehler[6], L. Busso[7,8], M. Cargnelli[6], S. Choi[4], C. Curceanu[9], S. Enomoto[2], D. Faso[7,8], H. Fujioka[11], Y. Fujiwara[13], T. Fukuda[12], C. Guaraldo[9], R. S. Hayano[13], T. Hiraiwa[2], M. Iio[14], M. Iliescu[9], K. Inoue[10], Y. Ishiguro[11], T. Ishikawa[13], S. Ishimoto[14], T. Ishiwatari[6], K. Itahashi[1], M. Iwai[14], M. Iwasaki[15,1,], Y. Kato[1], S. Kawasaki[10], . Kienle[16], H. Kou[15], Y. Ma[1], J. Marton[6], Y. Matsuda[17], Y. Mizoi[12], O. Morra[7], T. Nagae[11], H. Noumi[2], H. Ohnishi[1], S. Okada[1], H. Outa[1], K. iscicchia[9], M. oli Lener[9], A. Romero Vidal[9], Y. Sada[11], A. Sakaguchi[10], F. Sakuma[1], M. Sato[1], A. Scordo[9], M. Sekimoto[14], H. Shi[13], D. Sirghi[9,5], F. Sirghi[9,5], K. Suzuki[6], S. Suzuki[14], T. Suzuki[13], K. Tanida[4], H. Tatsuno[9], M. Tokuda[15], D. Tomono[1], A. Toyoda[14], K. Tsukada[18], O. Vazquez Doce[9,19], E. Widmann[6], B. K. Wuenschek[6], T. Yamaga[10], T. Yamazaki[13,1], H. Yim[20], Q. Zhang[1], and J. Zmeskal[6]
(J- ARC E15 Collaboration)

[1]RIKEN Nishina Center, RIKEN, Wako, 351-0198, Japan
[2]Research Center for Nuclear Physics (RCNP), Osaka University, Osaka, 567-0047, Japan
[3]Department of Physics and Astronomy, University of Victoria, Victoria BC V8W 3P6, Canada
[4]Department of Physics, eoul National University, eoul, 151-742, outh Korea
[5]National Institute of Physics and Nuclear Engineering - IFIN HH, Romania
[6] tefan-Meyer-Institut für subatomare Physik, A-1090 Vienna, Austria
[7]INFN ezione di Torino, Torino, Italy
[8]Dipartimento di Fisica Generale, Universita di Torino, Torino, Italy
[9]Laboratori Nazionali di Frascati dell INFN, I-00044 Frascati, Italy
[10]Department of Physics, Osaka University, Osaka, 560-0043, Japan
[11]Department of Physics, Kyoto University, Kyoto, 606-8502, Japan
[12]Laboratory of Physics, Osaka Electro-Communication University, Osaka, 572-8530, Japan
[13]Department of Physics, The University of Tokyo, Tokyo, 113-0033, Japan
[14]High Energy Accelerator Research Organization (KEK), Tsukuba, 305-0801, Japan
[15]Department of Physics, Tokyo Institute of Technology, Tokyo, 152-8551, Japan
[16]Technische Universität München, D-85748, Garching, Germany
[17]Graduate chool of Arts and ciences, The University of Tokyo, Tokyo, 153-8902, Japan
[18]Department of Physics, Tohoku University, endai, 980-8578, Japan
[19]Excellence Cluster Universe, Technische Universität München, D-85748, Garching, Germany
[20]Korea Institute of Radiological and Medical ciences (KIRAM), eoul, 139-706, outh Korea

E-mail: `tadashi.hashimoto@riken.jp`

Abstract. The J-PARC E15 experiment searches for the simplest kaonic nuclear bound state, K^-pp, via the ^{3}He(K^-, n) reaction at the K1.8BR beam-line in the J-PARC hadron experimental facility. We performed the first data-taking in May, 2013, with 5×10^9 incident kaons on the ^{3}He target. The preliminary results are reported focusing on the forward-neutron spectrum, which shows a clear peak structure composed of the quasi-elastic K^-"n" $\rightarrow K^-n$ and the charge-exchange K^-"p" $\rightarrow \bar{K}^0 n$ reactions as expected.

1. Introduction

The $\bar{K}N$ interaction is known to be strongly attractive in isospin 0 channels by the measurements of kaonic hydrogen x-rays[1] and low-energy $\bar{K}N$ scattering[2]. Consequently, a possible existence of kaonic nuclear bound states was predicted more than 10 years ago[3]. However, the existence of such exotic states has not been established even for the lightest kaonic nucleus, so-called K^-pp[4]. New experimental data with various reaction are definitely important to clarify the nature of kaonic nuclei together with existing data in stopped-K^-[5], pp-collision[6], and γ-induced[7] reactions.

The J- ARC E15 experiment searches for K^-pp bound state via the ^{3}He(K^-,n) reaction at 1 eV c[8]. This reaction channel is one of the simplest reaction among elementally-$\bar{K}$ induced reactions. We expect a good signal-to-noise ratio in the region of interest since background neutrons from non-mesonic two-nucleon absorptions and hyperon decays are expected to be substantially suppressed and kinematically separated. In addition, by using a liquid ^{3}He target and a large acceptance detector surrounding it, the experiment is designed to detect the expected decay, $K^-pp \rightarrow \Lambda p \rightarrow \pi^- pp$, to fully reconstruct the reaction kinematics.

The first data-taking was performed in May 2013. About $5{\times}10^9$ kaons were irradiated on the helium-3 target during 88-hour machine time. In this paper, we describe our spectrometer system, its performance and a preliminary result of the forward neutron spectrum.

2. Experimental apparatus

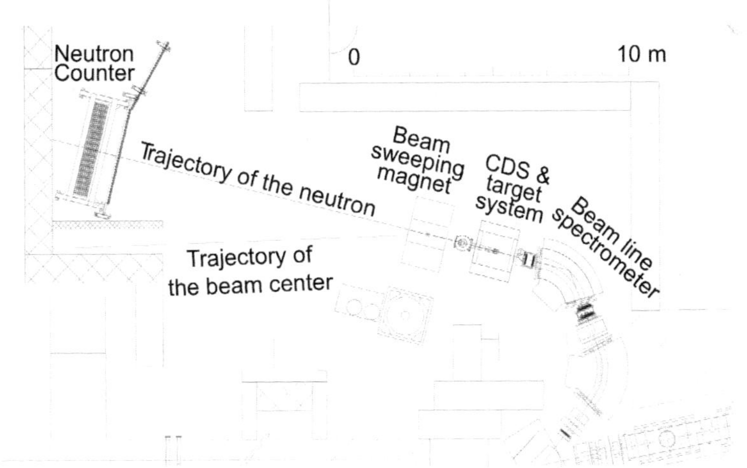

Figure 1. Schematic view of the J- ARC K1.8BR spectrometer. Taken from Ref. [9].

A dedicated spectrometer was designed and constructed at the secondary beam-line K1.8BR in the J- ARC hadron experimental facility. Here, we briefly describe the spectrometer system shown in Fig. 1. More detailed information can be found elsewhere[9].

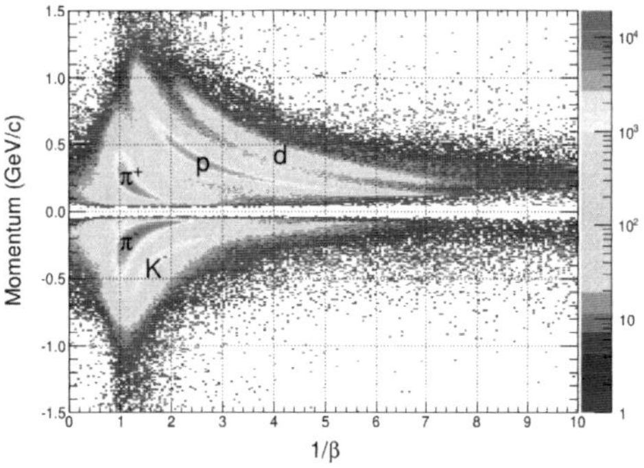

Figure 2. erformance of particle identification by the CDS. 1 β was measured by the TOF from the reaction point to CDH, and the momentum was obtained by the curvature of the CDC track. Taken from Ref. [10].

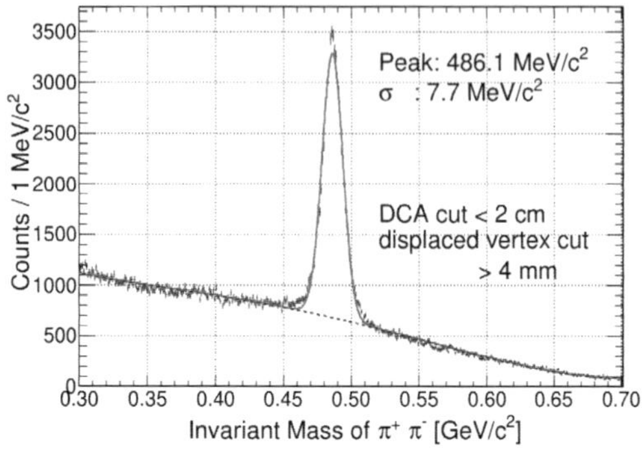

Figure 3. Invariant mass distribution of $\pi^+\pi^-$ pairs reconstructed by the CDS. The spectrum is fitted with a aussian and a background curve. The centroid and resolution of K_s^0 are well reproduced by a detailed detector simulation. Taken from Ref. [10].

The K1.8BR beam line delivers secondary charged beams with momenta up to 1.2 eV c, which are purified with an electrostatic separator. At 1 eV c, a kaon is identified with an aerogel Cherenkov counter, and its momentum is analyzed by a beam-line spectrometer with a momentum resolution of 2.2 MeV c (σ). A typical K^- yield at an accelerator power of 24 kW was 1.5×10^5 per spill with a K^-/π^- ratio of 0.45.

A liquid helium-3 target was placed at the center of a cylindrical detector system (CDS), which detects charged particles emitted in a reaction at the target. The main components of the CDS are a 15-layer cylindrical drift chamber (CDC) and a 36-segmented cylindrical detector hodoscope (CDH). The momentum of a charged particle is analyzed by the CDC and a particle identification is done together with the timing information provided by the CDH, as shown in Fig. 2. The CDH was also used as a trigger counter with a polar-angle acceptance from 54 to 126 degrees. Figure 3 shows the invariant-mass distribution of $\pi^+\pi^-$ pairs detected by the CDS, where the $K_s^0 \to \pi^+\pi^-$ decay was successfully reconstructed. The CDC track also determines the reaction vertex point together with a beam-line drift chamber placed just upstream the target. The vertex information is essential to select reactions on helium-3 and to determine the flight-length for a forward-neutral particle.

One of the most unique feature of our spectrometer system is the high-resolution large-acceptance neutron TOF counter array (NC). The NC locates at a distance of ~15 m from the target position and covers a solid angle of ~20 msr at 0 degree with 112-segment plastic scintillators. The detection e ciency for a 1 eV c neutron is estimated to be ~30 . Figure 4 shows a $1/\beta$ spectrum of forward neutral particles, where charged particles are removed by two types of charged-veto counter arrays and a beam-sweeping magnet. The γ-ray peak is clearly separated from the neutron continuum, which indicates a good accidental-background suppression. The TOF resolution of the system was evaluated to be ~160 ps (σ) by using the γ peak. It corresponds to ~10 MeV c (σ) momentum resolution in the region of interest (1.2–1.4 eV c). The uncertainty of the neutron momentum dominantly contribute to the missing-mass resolution of ~10 MeV c^2 (σ) at around the K^-pp binding threshold.

Journal of Physics: Conference Series **569** (2014) 012080 doi:10.1088/1742-6596/569/1/012080

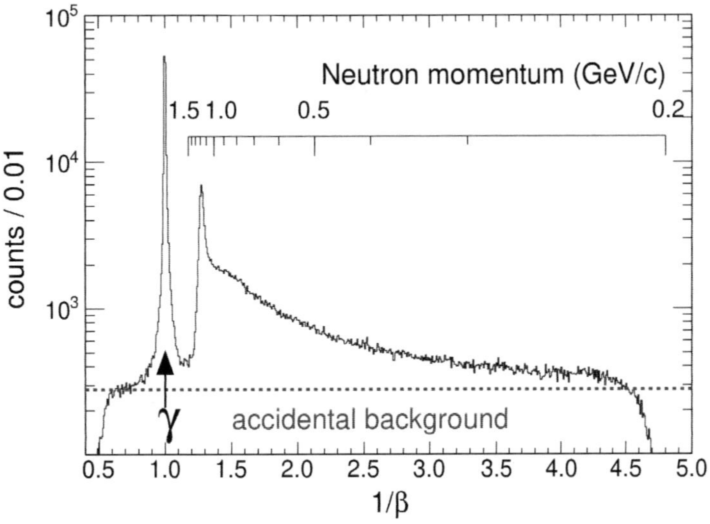

Figure 4. 1 β spectrum of forward-neutral particles, obtained by TOF analysis using a time-zero counter (T0) and the NC. The dotted line shows an accidental background contribution estimated from the left shoulder of the γ peak. Taken from Ref. [10].

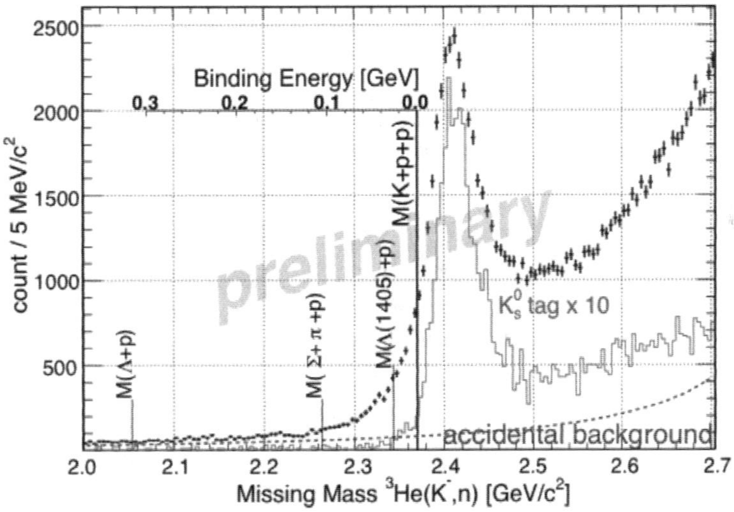

Figure 5. Missing mass spectrum of neutrons at a forward angle. At least one charged track in CDS is requested to reconstruct the reaction vertex. A spectrum with K_s^0 tag by the CDS is overlaid with a scale factor of 10. The dotted line shows the accidental background obtained in Fig. 4. The K^-pp binding threshold is indicated by a solid line. Taken from Ref. [10].

3. reliminary results of the first data-taking

Figure 5 shows the $^3\mathrm{He}(K^-, n)X$ missing-mass spectrum obtained under a *semi*-inclusive condition. Here, "semi-inclusive" means that we required at least one charged track in the CDS to reconstruct the reaction vertex. Thus, the obtained spectrum is somehow distorted by the limited acceptance of the CDS. The most prominent structure in the spectrum is a peak structure around 2.4 eV c^2. The peak is a so-called quasi-free peak attributed to the quasi-elastic K^-"n" $\rightarrow K^-n$ and the charge-exchange K^-"p" $\rightarrow \bar{K}^0n$ reactions. It is confirmed by requesting K^0_s reconstructed in the CDS as superimposed in Fig. 5. The K^0_s-tagged spectrum also shows that the missing-mass scale is correctly tuned at a 10 MeV c^2 level, and there is a tail-structure in the K^-pp bound region which cannot be explained by the Fermi motion. To further investigate the bound-region structure, a few background sources were studied. The most trivial one is the accidental background evaluated from the left shoulder of the γ peak in the $1/\beta$ spectrum. An analytical conversion to the missing-mass spectrum is plotted in Fig. 5. Other background sources were also evaluated including neutrons and secondary γ-rays from know elementary processes, and contaminations from the target cell. These contribution explains observed yields in the deep-binding region, while we need other non-ordinary processes to explain the yield of the tail just below the K^-pp binding threshold.

4. Summary

We successfully conducted the first data-taking of the J- ARC E15 experiment, which aimed to search for the K^-pp bound state via the ^{3}He$(K^-, n)X$ reaction at 1 eV c. The semi-inclusive neutron spectrum at forward angle, obtained with 5×10^9 incident kaons, shows a clear peak structure composed of the quasi-elastic K^-"n" $\rightarrow K^-n$ and the charge-exchange K^-"p" $\rightarrow \bar{K}^0n$ reactions as expected. The analysis on the forward neutron will be finalized soon. Analyses of semi-inclusive ^{3}He(K^-, p) and exclusive ^{3}He$(K^-, \Lambda pn)$ channels are also in progress.

Reference

[1] M. Iwasaki et al., Phys. Rev. Lett. **78** (1997) 3067;
G. Beer et al., Phys. Rev. Lett. **94** (2005) 212302;
M. Bazzi et al., Phys. Lett. B **704** (2011) 113.
[2] A. D. Martin, Nucl. Phys. B **179** (1981) 33.
[3] Y. Akaishi and T. Yamazaki, Phys. Rev. **C65** (2002) 044005.
[4] T. Yamazaki and Y. Akaishi, Phys. Lett. **B535** (2002) 70.
[5] M. Agnello *et al.*, Phys. Rev. Lett. **94** (2005) 212303.
[6] T. Yamazaki *et al.*, Phys. Rev. Lett. **104** (2010) 132502.
[7] A. O. Tokiyasu et al., Phys. Lett. B **728** (2014) 616.
[8] M. Iwasaki *et al.*, J-PARC E15 proposal, `http://j-parc.jp/NuclPart/pac_0606/pdf/p15-Iwasaki.pdf`.
[9] K. Agari *et al.*, Prog. Theor. Exp. Phys. (2012) 02B011.
[10] T. Hashimoto *et al.*, EPJ Web of Conferences **66**, 09008 (2014)

First experimental evidence for heavy hyperhydrogen ${}^{6}_{\Lambda}$H by the FINUDA experiment at DAΦNE

Alessandro Feliciello

Istituto Nazionale di Fisica Nucleare, Sezione di Torino, Via P. Giuria 1, I-10125 Torino (Italy)

E-mail: Alessandro.Feliciello@to.infn.it

Abstract. The FINUDA experiment, carried out at the Frascati ϕ-factory DAΦNE, allowed to find the first experimental evidence for the existence of the neutron-rich hypernucleus ${}^{6}_{\Lambda}$H by measuring in coincidence (π^+,π^-) pairs from the $K^-_{\rm stop}$ + ^{6}Li $\rightarrow$ ${}^{6}_{\Lambda}$H + π^+ production reaction, followed by the ${}^{6}_{\Lambda}$H $\rightarrow$ ^{6}He + π^- mesonic weak decay. The production rate of ${}^{6}_{\Lambda}$H undergoing such a two-body π^- decay turned out to be $(2.9 \pm 2.0) \times 10^{-6}/K^-_{\rm stop}$. A binding energy value $B_\Lambda({}^{6}_{\Lambda}\mathrm{H}) = (4.0 \pm 1.1)$ MeV was evaluated with respect to the (^{5}H + Λ) threshold by combining the information from both production and decay processes. A systematic difference of (0.98 ± 0.74) MeV observed between B_Λ values when derived separately from ${}^{6}_{\Lambda}$H formation and from its decay was tentatively assigned to the ${}^{6}_{\Lambda}$H $0^+_{g.s.} \rightarrow 1^+$ excitation.

1. Introduction

A neutron-rich Λ-hypernucleus is a bound strange[1] nuclear system with an unbalanced number of constituent neutrons or, in other words, with an unusually high neutron to proton ratio [1]. The search for such systems is legitimated by the experimental observation of the so-called glue-like role that the Λ hyperon manifests when it is embedded in a nucleus [2]. This property is essentially due to the attractive nature of the $\Lambda\mathcal{N}$ interaction and to the fact that the Pauli principle is not effective for a single strange particle inside a nucleus. Unstable nuclear systems, like for instance ^{8}Be, become then particle stable thanks to the critical contribution to the overall binding energy carried by the Λ. This effect is particularly evident for nuclei close to the proton- and neutron-drip lines, thus potentially expanding the boundary of the nuclear stability valley.

The motivations for trying to observe and for studying neutron-rich Λ-hypernuclei span over a wide spectrum. The discovery of these exotic systems could open very interesting scenarios as far as the investigation of few-body nuclear systems and of $\Lambda\mathcal{N}$ interactions (both strong and weak) in a low density nuclear medium are concerned. This environment is then best suited to put in evidence the effects of the three-body forces involving hyperons ($\Lambda\mathcal{N}\mathcal{N}$) which are believed to reduce the stiffness of the nuclear matter equation of state (EoS) in neutron stars [3, 4]. In this respect, a precise knowledge of the light neutron-rich Λ-hypernuclei energy level structure could imply far-reaching consequences on strange dense stellar matter properties. Indeed, by constraining the $\Lambda\mathcal{N}$ potential it could be possible to make more reliable predictions about the internal structure of neutron stars and in particular about their mass maximum value [5]. However, the recent measurement of two neutron star masses with values of the order of two solar masses [6, 7] seriously questions the real presence of hyperons in the neutron star core

[1] i.e. with an explicit strangeness content

composition [8]. A possible way to face with this "hyperon crisis" is an extension of the three-body forces definition, by including the additional $YY\mathcal{N}$ and YYY channels as well. This way it would be possible to provide enough repulsive strength at high density stellar matter, thus reducing the softening of the EoS and then justifying the observation of very massive neutron stars.

When focusing the attention on the strict hypernuclear physics field, such studies offer the possibility to evaluate the strength of the coherent $\Lambda\mathcal{N} - \Sigma\mathcal{N}$ mixing and to check whether such mechanism is able to generate a sizeable attractive three-body $\Lambda\mathcal{N}\mathcal{N}$ force, making the strange nuclear system extremely deeply bound [9, 10, 11, 12, 13, 14].

The existence and the observability of neutron-rich Λ-hypernuclei were first discussed by Dalitz and Levi Setti [1]. Afterwards, Majling endorsed this hypothesis and provided even more stringent predictions about their production rates [15].

Experimental evidences were actually found for ${}^{6}_{\Lambda}$He, ${}^{7}_{\Lambda}$Be, ${}^{8}_{\Lambda}$He, ${}^{9}_{\Lambda}$Be and ${}^{10}_{\Lambda}$B hypernuclei in emulsion experiments [17]. However no unstable-core hydrogen Λ-hypernuclei were observed until very recently. ${}^{6}_{\Lambda}$H is the simplest neutron-rich hypernucleus and, at the same time, it is the bound strange nuclear system with the largest possible neutral baryon excess. Actually it would be $(N + Y)/Z = 5$, with $Y = 1$ for a Λ hyperon, or $N/Z = 4$, larger than the maximal value in light nuclei $N/Z = 3$ for ^{8}He [16]. In this case the stabilized core nucleus is ^{5}H.

The accelerator-aided neutron-rich Λ-hypernucleus production can be achieved by exploiting the two-body double charge-exchange reaction

$$K^- + {}^{A}Z \rightarrow {}^{A}_{\Lambda}(Z-2) + \pi^+, \tag{1}$$

induced on nuclear targets by K^- meson both in flight and at rest, or

$$\pi^- + {}^{A}Z \rightarrow {}^{A}_{\Lambda}(Z-2) + K^+ \tag{2}$$

with π^- mesons in flight ($p_{\pi^-} > 0.89$ GeV/c).

Both these reactions can be described as two-step processes involving two different protons of the same nucleus, in which they are sequentially transformed into a Λ particle and a neutron (or vice versa), leading to a final bound strange nuclear system. Another possible mechanism could be a one-step double charge exchange $m_i^- p \rightarrow \Sigma^- m_f^+$ (where m_i^- stands for the initial π^-/K^- meson and m_f^+ for the final K^+/π^+ one) feeding the Σ component coherently admixed into the final Λ hypernuclear state. The two-step processes (1) and (2) are expected to occur at a rate two orders of magnitude smaller [18] than the ones for "normal" Λ-hypernuclei production via the corresponding one-step two-body reactions (K^-,π^-) and (π^+,K^+).

The first experimental attempt to produce neutron-rich hypernuclei was carried out by exploiting the reaction (1) at KEK [19]. Upper limits were obtained for the production of ${}^{9}_{\Lambda}$He, ${}^{12}_{\Lambda}$Be and ${}^{16}_{\Lambda}$C (on ^{9}Be, ^{12}C and ^{16}O targets respectively), all lying in the (0.6 – 2.0) $\times 10^{-4}/K^-_{\text{stop}}$ range. A second KEK experiment (E521), reported the production of ${}^{10}_{\Lambda}$Li in the (π^-, K^+) reaction on a ^{10}B target using a 1.2 GeV/c π^- beam, with a cross section of (11.3 ± 1.9) nb/sr, integrated over the whole Λ bound region [20].

2. The FINUDA experiment

2.1. The apparatus and its performances

FINUDA was a complex magnetic spectrometer installed on the DAΦNE e^+e^- collider, the Italian $\phi(1020)$-factory put in operation at the INFN National Laboratories of Frascati, near Rome. It was carefully designed and specially dedicated to an extensive program of hypernuclear physics, mainly focussed on spectroscopy of light Λ-hypernuclei and on systematic study of their decay modes. The apparatus was carefully modeled and optimized by means of a sophisticated

Table 1. Upper limits at a 90% C.L. for neutron-rich Λ-hypernuclei production rate R_{π^+} per stopped K^- for the (K^-,π^+) reaction. (From [25]).

target	Λ-hypernucleus	R_{π^+} [$\times 10^{-5}/K^-_{\rm stop}$]
^{6}Li	$^6_\Lambda$H	$\leqslant 2.5 \pm 0.4_{\rm stat}\ ^{+0.4}_{-0.1}{}_{\rm syst}$
^{7}Li	$^7_\Lambda$H	$\leqslant 4.5 \pm 0.9_{\rm stat}\ ^{+0.4}_{-0.1}{}_{\rm syst}$
^{12}C	$^{12}_\Lambda$Be	$\leqslant 2.0 \pm 0.4_{\rm stat}\ ^{+0.3}_{-0.1}{}_{\rm syst}$

simulation code, based on the GEANT3 software package [21]. This phase was driven by the constraints imposed by the DAΦNE accelerator peculiar footprint and by the requirement of fully exploiting both the K^- source topology and the K^- "beam" characteristics. The final outcome was a detector characterized by a complex architecture, whose description can be found in [22].

However, it is important to mention here at least the main spectrometer performances, relevant to the present analysis. For a ~250 MeV/c π^+ the tracker momentum resolution, measured by looking at the peak due to monochromatic 236.5 MeV/c μ^+ from $K_{\mu 2}$ decay, was $\sigma_p = (1.1 \pm 0.1)$ MeV/c [23]; this figure translated in a resolution on the kinetic energy $\sigma_T = 0.96$ MeV. The precision on the absolute momentum calibration was better than 0.12 MeV/c in the case of ^{6}Li targets, giving a systematic deviation on the kinetic energy $\sigma_{Tsyst}(\pi^+) = 0.1$ MeV. For a ~130 MeV/c π^- the momentum resolution and the absolute energy calibration were evaluated by measuring the peak due to monochromatic 132.8 MeV/c π^- coming from the two-body mesonic weak decay of $^4_\Lambda$H hyperfragments produced with a formation probability of ~10^{-3}–10^{-2} per stopped K^- [24]; in this case a resolution $\sigma_p = (1.2 \pm 0.1)$ MeV/c, giving $\sigma_T = 0.84$ MeV, and a precision of 0.2 MeV/c, giving a systematic deviation on the kinetic energy $\sigma_{Tsyst}(\pi^-) = 0.14$ MeV, were achieved.

2.2. The $^6_\Lambda$H search strategies

The FINUDA Collaboration pursued the objective of producing and of detecting $^6_\Lambda$H, $^7_\Lambda$H and $^{12}_\Lambda$Be since the very beginning of its activity, by exploiting the unrivaled characteristics of the kaon source available at DAΦNE jointly to the excellent capabilities of the apparatus.

The strategy adopted in analyzing the first limited data sample ($\mathcal{L}_{\rm int} \sim 220$ pb^{-1}) consisted in searching for discrete peaks in the region of interest in the momentum spectrum of π^+ emitted following the K^- capture at rest on ^{6}Li, ^{7}Li and ^{12}C targets. Because of the overwhelming background due to concurrent K^- induced processes, it was possible to provide only upper limits at a 90% C.L., listed in table 1. In the case of $^{12}_\Lambda$Be, the FINUDA determination lowered by a factor ~3 the aforementioned KEK results [19].

A second data taking campaign allowed to increase the statistics collected with ^{6}Li targets by a factor 5 ($\mathcal{L}_{\rm int} \sim 1156$ pb^{-1}). However even in this case it was not possible to observe in the inclusive π^+ momentum spectrum clear peaks that could be unambiguously attributed to the two-body, double charge exchange reaction:

$$K^-_{\rm stop} + {}^6{\rm Li} \rightarrow {}^6_\Lambda{\rm H} + \pi^+ \qquad (p_{\pi^+} \approx 252 \text{ MeV/c} \quad \text{when} \quad B_\Lambda \approx 5 \text{ MeV}). \tag{3}$$

Then, by exploiting the larger number of reconstructed events, a new analysis strategy was adopted with the aim of reducing the physical background. It consisted in looking at the π^+ momentum distribution by requiring in coincidence the π^- coming from the $^6_\Lambda$H mesonic decay:

$${}^6_\Lambda{\rm H} \rightarrow {}^6{\rm He} + \pi^- \qquad (130 \lesssim p_{\pi^-} \lesssim 140 \text{ MeV/c}). \tag{4}$$

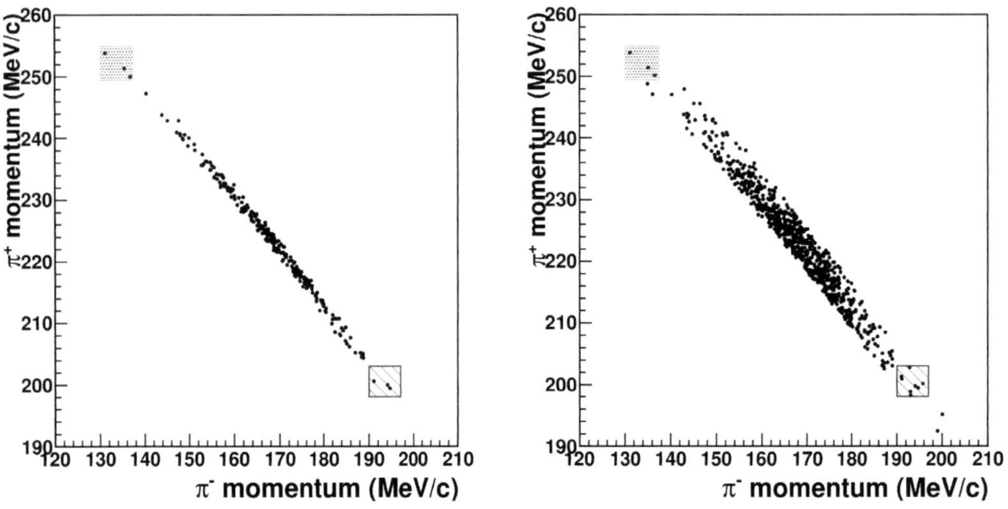

Figure 1. π^+ momentum vs. π^- momentum plot for ^{6}Li target events with 202 MeV $\leqslant T_{\rm tot} \leqslant$ 204 MeV (left panel) and with 200 MeV $\leqslant T_{\rm tot} \leqslant$ 206 MeV (right panel). Red hatched areas correspond to event subsets with 250 MeV/c $\leqslant p_{\pi^+} \leqslant$ 255 MeV/c and 130 MeV/c $\leqslant p_{\pi^-} \leqslant$ 137 MeV/c, respectively. Blue hatched areas are the symmetric event subsets. (From [27]).

In order to identify the events due to the $^6_\Lambda$H possible formation in a bound state, energy conservation was imposed to both reactions (3) and (4). Momentum conservation is automatically satisfied because both formation and decay processes occur at rest: (3) by definition and (4) because $^6_\Lambda$H stopping time in the target is indeed shorter than its lifetime. It was then realized that, while both π^- and π^+ momenta depend on the energy level in which the Λ-hypernucleus is formed, the sum of their kinetic energies ($T_{\rm tot} = T_{\pi^+} + T_{\pi^-}$) is, to a first approximation, independent of it and equal to 203.0 MeV [27]. By taking into account the tracker energy resolution, events with $T_{\rm tot} = (203 \pm 1)$ MeV and satisfying additional cuts on π^+ and π^- momentum values ($250 \leqslant p_{\pi^+} \leqslant 255$ MeV/c and $130 \leqslant p_{\pi^-} \leqslant 137$ MeV/c) were selected. These latter limits are equivalent to constrain the $B_\Lambda(^6_\Lambda\rm H)$ value in the plausible range from 2 to 6 MeV, that is from the (Λ + ^{3}H + 2n) threshold (about 2 MeV in the $^6_\Lambda$H continuum) down to a $^6_\Lambda$H bound somewhat deeper than predicted in [9, 10, 11, 12, 13, 14].

Three events, out of the 27×10^6 ones due to K^- stopped in the two ^{6}Li targets, survived to the filtering procedure [26], described in detail in [27]. Figure 1 shows graphically the effect of the selection cuts. Alternative choices for $T_{\rm tot}$ interval width (2 – 6 MeV) and central value (202 – 204 MeV) and for $p_{\pi^\pm}$ accepted ranges (5 – 10 and 8 – 15 MeV/c, respectively) with fixed limits at 250 and 137 MeV/c, respectively, to exclude the unbound region, do not affect the population of the selected region (red areas in figure 1). For instance, no additional candidate events appear in the red areas upon extending the $T_{\rm tot}$ window from 202 – 204 MeV to 200 – 206 as it is possible to see in the right panel of figure 1. On the contrary, a similar stability is not observed in the blue hatched areas, where on the top of selected events with the first set of cuts, six additional events appear. The $^6_\Lambda$H mass values evaluated for each event both from production (3) and decay (4) processes are listed in table 2; the quoted errors are evaluated directly from the tracker momentum resolution for π^+ and π^- recalled in Sec. 2.1. Such nuclear

Table 2. Pion momenta $p_{\pi^\pm}$ and mass values for the three ${}^{6}_{\Lambda}$H candidate events. Masses are evaluated both from production (3) and decay (4) reactions. In the last two columns the mass mean value and the difference between production and decay masses are reported. (From [27]).

p_{π^+} [MeV/c]	p_{π^-} [MeV/c]	M(${}^{6}_{\Lambda}$H) prod. [MeV]	M(${}^{6}_{\Lambda}$H) decay [MeV]	M(${}^{6}_{\Lambda}$H) mean [MeV]	ΔM(${}^{6}_{\Lambda}$H) [MeV]
251.3 ± 1.1	135.1 ± 1.2	5802.33 ± 0.96	5801.41 ± 0.84	5801.87 ± 0.96	0.92 ± 1.28
250.1 ± 1.1	136.9 ± 1.2	5803.45 ± 0.96	5802.73 ± 0.84	5803.09 ± 0.96	0.72 ± 1.28
253.8 ± 1.1	131.2 ± 1.2	5799.97 ± 0.96	5798.66 ± 0.84	5799.32 ± 0.96	1.31 ± 1.28

mass values yield a mean value $M({}^{6}_{\Lambda}\mathrm{H}) = (5801.4 \pm 1.1)$ MeV where the error is essentially due to the spread of the average mass values for the three events. The inferred ${}^{6}_{\Lambda}$H binding energies $B_\Lambda = (4.0 \pm 1.1)$ MeV, with respect to the $(\Lambda + {}^{5}\mathrm{H})$ threshold, and $B_\Lambda = (0.3 \pm 1.1)$ MeV, with respect to the lowest particle stable threshold (${}^{4}_{\Lambda}$H + 2n), are in excellent agreement with the predictions of both [1] and [15]. On the contrary the first value is significantly lower than the outcome of the calculations described in [9, 10, 11, 12, 13, 14], where the coherent $\Lambda\mathcal{N}$-$\Sigma\mathcal{N}$ mixing effect is invoked in order to justify a considerable higher binding energy for ${}^{6}_{\Lambda}$H. Then the present result seems to support the hypothesis of a reduced effectiveness of such a mechanism, in line with the output of some recent shell-model calculations [28].

The mass values obtained from production process (3) turned out to be systematically higher than those inferred from decay one (4) by (0.98 ± 0.74) MeV. A possible physical origin for such a difference, discussed in [26] and [27], could be related to the ${}^{6}_{\Lambda}$H excitation spectrum: the production could occur preferentially to an ~1 MeV 1^+ excited state, then decaying through a fast M1 γ-transition to the 0^+ ground state of ${}^{6}_{\Lambda}$H, which in turn undergoes the mesonic weak decay.

In order to provide a correct evaluation of the ${}^{6}_{\Lambda}$H production rate, a careful study of the possible background sources was performed. A complete simulation of K^-_{stop} absorption reactions, both on single and on correlated few-nucleons clusters, leading to the formation and the decay of Λ and Σ hyperons was then carried out. The full description of this task can be found in [27]; here it is enough to focus on just one chain of reactions likely to produce (π^+,π^-) pairs in coincidence overlapping with those selected to identify ${}^{6}_{\Lambda}$H formation (3) and decay (4) processes, namely Σ^+ production

$$K^-_{\mathrm{stop}} + {}^{6}\mathrm{Li} \rightarrow \Sigma^+ + {}^{4}\mathrm{He} + n + \pi^- \qquad (p_{\pi^-} \leqslant 190\ \mathrm{MeV/c}), \tag{5}$$

followed by in-flight Σ^+ decay

$$\Sigma^+ \rightarrow n + \pi^+ \qquad (p_{\pi^+} \leqslant 282\ \mathrm{MeV/c}). \tag{6}$$

The Σ^+ production was treated in the quasi-free approach, according to the outcome of the previous FINUDA study of the $A(K^-_{\mathrm{stop}}, \pi^\pm\Sigma^\mp)A'$ reaction on p-shell nuclei [29]; in the signal region a small contribution of (0.16 ± 0.07) expected background events was estimated. Another possible source of physical background, is the ${}^{4}_{\Lambda}$H hyperfragment production and its subsequent two-body decay. This contamination is strongly suppressed by the cut on T_{tot}, then the expected number of background events turned out to be as small as (0.04 ± 0.01). All other reaction chains that could produce (π^+,π^-) pairs in coincidence, falling within the above described regions of interest, were filtered thanks to the applied selection cuts. As far as the instrumental backgrounds are concerned, they could mainly result from fake tracks, included in the data sample because of wrong momentum reconstruction. In order to quantify such a contribution, events originated in the other targets used in the same runs (^{7}Li, ^{9}Be, ^{13}C, ^{16}O) were processed through the same analysis procedure. Only one event, coming from the ^{9}Be target, was filtered. Accordingly, the estimated number of fake events from the ^{6}Li target was (0.27 ± 0.27). The total number of background events was then (0.43 ± 0.28). Thus, by using Poisson statistics, it was possible to conclude that the three $^6_\Lambda$H-assigned events do not arise from background at a 99% confidence level, the statistical significance of the result being S = 3.9. By taking into account the above considerations, the apparatus reconstruction efficiency for $\pi^\pm$, the target purity and the cut effectiveness the product $R_{\pi^+}\cdot$ BR(π^-), where R_{π^+} is the $^6_\Lambda$H production rate per $K^-_{\rm stop}$ in reaction (3) and BR(π^-) is the unknown branching ratio for the two-body mesonic π^- decay (4), turned out to be:

$$R_{\pi^+}(^6_\Lambda\mathrm{H})\cdot BR(\pi^-) = (2.9 \pm 2.0)\times 10^{-6}/K^-_{\rm stop}. \quad (7)$$

Finally, by assuming BR(π^-) = 49%, in analogy to the experimental result for $^4_\Lambda\mathrm{H}\rightarrow {}^4\mathrm{He}+\pi^-$ decay [24], $R_{\pi^+} = (5.9 \pm 4.0)\times 10^{-6}/K^-_{\rm stop}$, a value fully consistent with the previous published result [25] and two to three orders of magnitude smaller than summed Λ-bound production rates R_{π^-} of "normal" light Λ-hypernuclei in the ($K^-_{\rm stop}$,π^-) reaction [23].

3. Discussion

The FINUDA claim is now triggering a lively debate on the $^6_\Lambda$H existence and its structure both in the theoretical physicist community and in the experimental physicist one. Very recently, Hiyama *et al.* computed the $^6_\Lambda$H binding energy by exploiting a four-body cluster-model [30]. According to these calculations, which are able to reproduce the ^{5}H properties, the Λ separation energy amounts to B_Λ = 2.47 MeV only. This means that both 0^+ and 1^+ states are unbound, in contrast to the FINUDA experimental evidence. Gal and Millener performed a shell-model calculation in order to predict in turn the binding energy for the system made by a $^4_\Lambda$H core plus a pair of neutrons [31]. In this case the obtained binding energy value was B_Λ = (3.83 ± 0.08 ± 0.22) MeV, thus leaving room for an alternative interpretation of the FINUDA observation. They suggested that the 1^+ excited state could be unbound but characterized by a particle decay width extremely small and comparable with the one of the aforementioned M1 γ-transition to the 0^+ ground state. Such a hypothesis could be justified on the basis of a kinematical and dynamical suppression of the two neutron emission from the 1^+ state.

From the experimental point of view, it is clear that the $^6_\Lambda$H discovery must be confirmed in an independent way and with much higher statistical significance. In this perspective, the E10 experiment was proposed and approved to run on the K1.8 beam line at the Japan Proton Accelerator Research Complex (J-PARC). It has been specifically designed to search for neutron-rich Λ-hypernuclei, namely $^6_\Lambda$H and $^9_\Lambda$He, and it could be considered the natural extension of the already mentioned KEK E521 one [20]. Actually, the neutron-rich Λ-hypernucleus production reaction exploited is the (π^-,K^+) at 1.2 GeV/c. The E10 Collaboration took data in December 2012 and January 2013 with a ^{6}Li target. The careful analysis of the measured missing mass spectrum found no significant indication for $^6_\Lambda$H existence, neither below nor above the ($^4_\Lambda$H + 2n) particle decay threshold [32]. On the basis of three observed events in the region of interest, an upper limit for the production cross section of a bound $^6_\Lambda$H hypernucleus was estimated to be 1.2 nb/sr at a 90% C.L., a value one order of magnitude lower that the one reported in the analogous KEK measurement on ^{10}B target [20].

4. Conclusions

The FINUDA Collaboration presented the first experimental evidence for heavy hyper-hydrogen $^6_\Lambda$H, based on the observation of three events, shown to be free from any instrumental and/or physical backgrounds. The derived ${}^{6}_{\Lambda}$H binding energy value drastically downsizes the strength of the coherent $\Lambda\mathcal{N} - \Sigma\mathcal{N}$ mixing effect predicted for neutron-rich strange matter [9, 10, 11, 12, 13, 14], together with the conjectured $0^+ - 1^+$ doublet splitting.

However, the recent search for ${}^{6}_{\Lambda}$H performed by the E10 Collaboration at J-PARC did not confirm this observation [32].

Further measurements are then required in order to confirm whether ${}^{6}_{\Lambda}$H really exists. An important effort should be put in place on theoretical ground as well, to restore order among the different calculations and, possibly, to reconcile experimental evidence and theoretical prediction.

Acknowledgments
This work has been partially supported by the Networking Activity SPHERE (WP6), part of the "Study of Strongly Interacting Matter" Integrated Activity (HadronPhysics3) of the VII European Union Framework Program (FP7), Grant Agreement Number 283286.
The support received from the Japanese Grant-In-Aid for Scientific Research on Innovative Areas "Nuclear Matter in Neutron Stars Investigated by Experiments and Astronomical Observations" is acknowledged as well.

References

[1] Dalitz R H and Levi Setti R 1963 *Nuovo Cimento* **30** 489–501
[2] Tanida K *et al.* 2001 *Phys. Rev. Lett.* **86** 1982–1985
[3] Schaffner-Bielich J 2008 *Nucl. Phys.* A **804** 309–321
[4] Schaffner-Bielich J 2010 *Nucl. Phys.* A **835** 279–286
[5] Vidaña I *et al.* 2011 *Europhys. Lett.* **94** 11002
[6] Demorest P B *et al.* 2010 *Nature* **467** 1081–1083
[7] Klähn T *et al.* 2013 *Phys. Rev.* D **88** 085001
[8] Lattimer J M and Steiner A W 2014 *The Astrophysical Journal* **784** 123
[9] Akaishi Y and Yamazaki T 1999 *Physics and detectors for DAΦNE* Frascati Physics Series **XVI**, ed S Bianco *et al.* November 16–19, 1999, Frascati (RM), Italy (SIS-LNF: Frascati (RM)) pp 59–74
[10] Myint K S, Harada T, Shinmura S and Akaishi Y 2000 *Few Body Syst. Suppl.* **12** 383–386
[11] Myint K S and Akaishi Y 2002 *Progr. Theor. Phys. Suppl.* **146** 599–600
[12] Shinmura S, Myint K S, Harada T and Akaishi Y 2002 *J. Phys. G* **28** L1–L7
[13] Akaishi Y and Myint K S 2007 *New facet of three nucleon force - 50 years of Fujita Miyazawa three nucleon force* AIP Conf. Proc. **1011**, ed H Sakai *et al.* October 29–31, 2007, Tokyo, Japan (AIP Publishing: Melville) pp 277–286
[14] Akaishi Y 2010 *Prog. Theor. Phys. Suppl.* **186** 378–383
[15] Majling L 1995 *Nucl. Phys.* A **585** 211c–214c
[16] Tilley D R *et al.* 2004 *Nucl. Phys.* A **745** 155–362
[17] Jurić M *et al.* 1973 *Nucl. Phys.* B **52** 1–30
[18] Chrien R E, Dover C B and Gal A 1992 *Czech. J. Phys.* **42** 1089–1113
[19] Kubota K *et al.* 1996 *Nucl. Phys.* A **602** 327–333
[20] Saha P K *et al.* 2005 *Phys. Rev. Lett.* **94** 052502
[21] Brun R, Bruyant F, Maire M, McPherson A C and Zanarini P 1987 European Organization for Nuclear Research, Report No. CERN-DD-EE-84-1
[22] Feliciello A 2013 *Mod. Phys. Lett.* A **28** 1330029
[23] Agnello M *et al.* 2011 *Phys. Lett.* B **698** 219–225
[24] Tamura H *et al.* 1989 *Phys. Rev.* C **40** 479–482
[25] Agnello M *et al.* 2006 *Phys. Lett.* B **640** 145–149
[26] Agnello M *et al.* 2012 *Phys. Rev. Lett.* **108** 042501
[27] Agnello M *et al.* 2012 *Nucl. Phys.* A **881** 269–287
[28] Millener D J 2007 *Lect. Notes Phys.* **724** 31–79
[29] Agnello M *et al.* 2011 *Phys. Lett.* B **704** 474–480
[30] Hiyama E, Ohnishi S, Kamimura M and Yamamoto Y 2013 *Nucl. Phys.* A **908** 29–39
[31] Gal A and Millener J D 2013 *Phys. Lett.* B **725** 445–450
[32] Sugimura H *et al.* 2014 *Phys. Lett.* B **729** 39–44

Journal of Physics: Conference Series **569** (2014) 012082 doi:10.1088/1742-6596/569/1/012082

Progress on the study of double-Λ hypernuclei

Kazuma Nakazawa, Shinji Kinbara, Akihiro Mishina, Myint Kyaw Soe, Khin Than Tint, Junya Yoshida

Physics Department, Gifu University, Gifu 501-1193, Japan

E-mail: nakazawa@gifu-u.ac.jp

Abstract. With use of nuclear emulsion, double-hypernuclei have been studied for aiming to understand Λ-Λ interaction via at-rest-capture of Ξ^- hyperon originated by quasi-free (K^-, K^+) reactions, where the beam K^- momentum was 1.66 GeV c. Under the process for studying them, elementary feature for not only quasi-free reaction but also production of nuclei with single and double unit(s) of strangeness have been roughly understood, even the statistics were not large. o far we have succeeded to analyze 5 events of double-hypernuclei which introduced binding energies of two Λ hyperons on several nuclei. However its knowledge is very limited. We have started a new experiment, J-PARC E07, to give us rich information in $S = -2$ world. An event which may give us a new information about Ξ-nucleus interaction shall be introduced.

1. Introduction

Transfer reaction of two units of strangeness is quite interesting, because the reaction provides us information of double-strangeness systems such as H-dibaryon, hypernucleus and double-Λ hypernucleus. Conversely say, that transfer reaction would be the uniquely available source of information on study of Λ-Λ and -N interction.

The first observation of a double-Λ hypernucleus was performed among expected four events of $^-$ hyperon capture at rest in nuclear emulsion[1]. The capture reaction of $^-$ hyperon is more reliable to produce double-Λ hypernuclei than directly production via (K^-, K^+) reaction due to small Q-value of at most 28 MeV, *i.e.* $^- + p \rightarrow \Lambda + \Lambda + 28$ MeV. When we detected typical sequential decay topologies of light double-Λ hypernuclei, kinematical analysis has been made for measurement of the mass of the nuclei under the energy and momentum conservation. The energies and momentum of decay daughter particles can be obtained by using the relation between track range and energy in the emulsion. The relation was calibrated with ranges of particles, which would be unique kinetic energies of 8.784 MeV and 5.367 MeV for α particles from ^{212} o and ^{228}Th, respectively, and 4.259 MeV for μ^+ meson from stopped π^+ meson.

In the E176 experiment at KEK- S, since the emulsion was used not only as a detector of production and decay of double-Λ hypernucleus but also for the (K^-, K^+) reaction target, we have studied elementary processes for (K^-, K^+) reaction regardless of whether quasi-free reaction, even the statistics were not so much. If we select events with K^+ momentum higher than 1.0 eV c, those events would present quasi-free production of $^-$ hyperons very well. Sticking probabilities of single and double strangeness are discussed for light (C, N, O) and heavy (Ag, Br, I) elements in nuclear emulsion. Those probabilities are strongly related to theoretical calculation of formation and transition of $S = -2$ nuclear systems [2]. In the E373 experiment, (K^-, K^+) reactions were taken place in a diamond block located upstream the emulsion stack to get more events with $^-$ hyperon stopping. Unfortunately, we suspended to

Journal of Physics: Conference Series **569** (2014) 012082 doi:10.1088/1742-6596/569/1/012082

study such sticking probability in the E373 experiment at KEK- S, because we need to develop some technology to identify $^-$ stopping events except for clear events with double strangeness. However we have succeeded to detect several double-Λ hypernuclei, especially NA ARA event presented unique information on Λ-Λ interaction[3, 4, 5, 6].

In the coming E07 experiment at J- ARC, to search for nuclear mass dependence of binding energy of two Λ hyperons in nuclei, $B_{\Lambda\Lambda}$, nearly one hundred double-Λ hypernuclei can be detected by using fully automated system for tracking of $^-$ hyperons. By the development of "*Overall-scanning*" method[7] to pick up characteristic decay topologies with three vertices of double-Λ hypernuclei, more than one thousand of the nuclei are expected to be observed.

In this report, past obtained knowledge about double strangeness reaction and systems will be summarized, briefly, and it will be introduced for a new event of uniquely identified twin hypernuclei which was detected in test operation of *Overall-scanning* method.

2. resent knowledge with (K^-, K^+) reaction

2.1. strangeness transfer via (K^-, K^+) reaction

The setup around target emulsion was seen for the E176 experiment at KEK- S in Figure 1. In the experiment, the emulsion was used for the target of (K^-, K^+) reaction and for the detector of double-hypernuclei. The K^- π^- ratio in beam particles of 1.66 eV c momentum was typically 1 3 with the intensity of 3×10^3 K^- spill. Tracks of K^+ mesons reconstructed by silicon microstrip detectors (VSSD) were scanned at Changeable emulsion sheet, and if they were detected, we followed up them to the (K^-, K^+) reaction vertices. If some track(s) were emitted from the vertex, we followed them to their end points and checked the production and decay of hypernucleus.

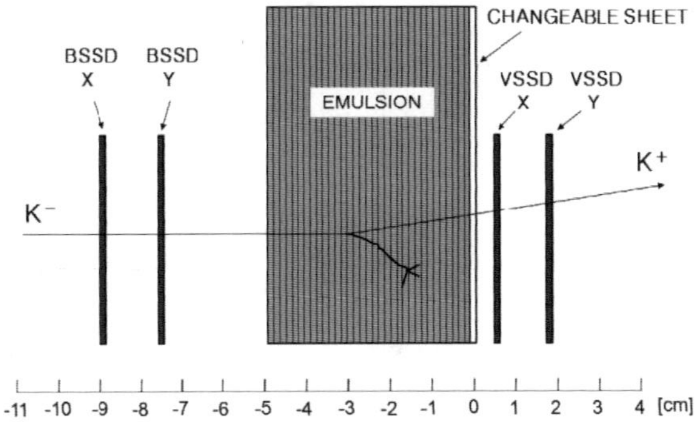

Figure 1. (K^-, K^+) reactions were recorded in the emulsion stack. The K^+ mesons were identified by the spectrometer system with USHIWAKA magnet. Detected K^+ mesons were located at the changeable emulsion sheet which recorded 10 track density of the emulsion stack. The detected K^+ mesons were followed up to the reaction vertex.

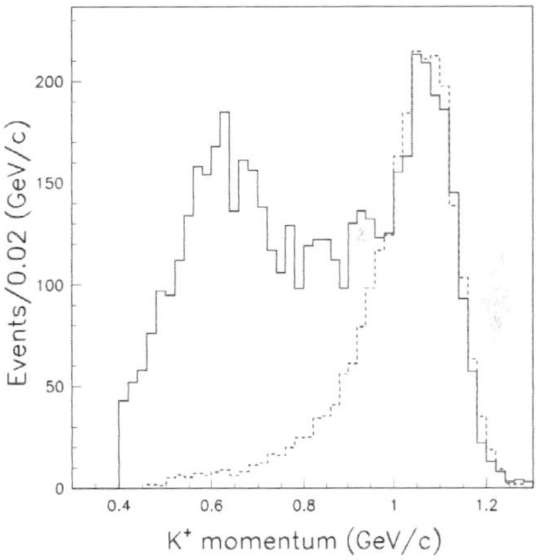

Figure 2. K^+ momentum spectra obtained by the experiment (solid line) and by simulation with fermi-motion (dotted line).

In Figure 2, K^+ momentum spectra are presented with the result of a simulation beased on quasi-free 'p'(K^-, K^+) $^-$ reaction. In the higher momentum region ($p_{K^+} \geq 1.0$ eV c), the (K^-, K^+) reaction is considered to be in quasi-free process.

We have located 796 (K^-, K^+) events in the region of $p_{K^+} \geq 1.0$ eV c. After rejection of events without the production of $^-$ hyperon such as $^{*-}$ production, $^0\pi^-$ production, two step process, and so on, 761 events were studied for interaction cross section $^-$ hyperon and nucleon, $\sigma_{\Xi N}$. Since $^-$ emission rate from target nuclei in the emulsion was 73.9 $^{+4.5}_{-4.6}$ ($\pm$ 5 from non quasi-free process), we obtained mean-free-path of $^-$ hyperon in nucleus and $\sigma_{\Xi N}$ to be 4.7 $^{+1.5}_{-1.0}$ fm and 12.7 $^{+3.5}_{-3.1}$ mb, respectively. In 18 events, hyperfragment was emitted from (K^-, K^+) reaction vertex [8].

Regarding non quasi-free reaction ($p_{K^+} < 1.0$ eV c), 901 events have been located in the emulsion. Single-Λ and double-Λ hypernuclei were produced in 42 and 3 events, respectively, at (K^-, K^+) reaction point. roduction rate, 4.6 ± 0.7 , of single-Λ hypernuclei was in good agreement with old data [9]. It was found that the 0.3 ± 0.2 would be the production rate for double-Λ hypernuclei in (K^-, K^+) reaction with non quasi-free process.

2.2. strangeness transfer via $^-$ capture at rest

We have located 77.6 ± 5.1 events with at rest capture of $^-$ hyperon in E176. Among them, 42 of the events were occurred in capture by light elements, ^{12}C, ^{14}N and ^{16}O, in the emulsion nuclei. On the other hand, capture reaction by heavy elements, Ag and Br, was 58 . One event of double-Λ hypernucleus and three events of twin single-Λ hypernuclei provided us lower limits of sticking probabilities for two and at least one Λ hyperon(s) to be 4.8 and 48.4 , respectively, at 90 confidence level for light elements. Regarding heavy elements, 1.7 and 35.7 were for sticking probabilities of two and at least one Λ hyperon(s) at 90 confidence level, respectively[5].

Table 1. Double-Λ hypernuclei in the world.

Event	$_{\Lambda\Lambda}^{A}$	$^-$ captured	$B_{\Lambda\Lambda} - B_{\Xi^-}$ (MeV)	$\Delta B_{\Lambda\Lambda} - B_{\Xi^-}$ (MeV)	Level	$B_{\Lambda\Lambda}$ (MeV)	$\Delta B_{\Lambda\Lambda}$ (MeV)
NA ARA	$^{6}_{\Lambda\Lambda}$He	^{12}C	$B_{\Lambda\Lambda} = 6.79 + 0.91B_{\Xi^-}(\pm 0.16)$		3*D*	6.91	0.67
			$\Delta B_{\Lambda\Lambda} = 0.55 + 0.91B_{\Xi^-}(\pm 0.17)$				
MIKA E	$^{6}_{\Lambda\Lambda}$He	^{12}C	9.88 ± 1.71	3.64 ± 1.71	3*D*	10.01	3.77
	$^{11}_{\Lambda\Lambda}$Be	^{14}N	21.95 ± 2.67	3.73 ± 2.71	3*D*	22.12	3.94
DEMACHI-ANA I	$^{10}_{\Lambda\Lambda}$Be*	^{12}C	11.77 ± 0.13	-1.65 ± 0.15	3*D*	11.90	-1.52
					$Ex(^{10}_{\Lambda\Lambda}\mathrm{Be}^*) \simeq 2.8$ MeV		
HIDA	$^{11}_{\Lambda\Lambda}$Be	^{16}O	20.60 ± 1.27	2.38 ± 1.34	3*D*	20.83	2.61
	$^{12}_{\Lambda\Lambda}$Be	^{14}N	22.31 ± 1.21	–	3*D*	22.48	–
E176	$^{13}_{\Lambda\Lambda}$B	^{14}N	–	–	3*D*	23.3 ± 0.7	0.6 ± 0.8
Danysz	$^{10}_{\Lambda\Lambda}$Be	–	Reconstructed by decay daughters			14.7 ± 0.4	1.3 ± 0.4

In the E373 experiment at KEK- S, we have expected to observe nearly 10 double-Λ hypernuclei. We have changed the (K^-, K^+) reaction target and SSDs to a diamond block (2 × 2 × 3 cm^3) upstream of the emulsion and fiber bundle detectors, respectively, for getting much events for $^-$ hyperon capture at rest. We followed $^-$ hyperon candidate tracks reconstructed by fiber bundle detector. The expected number of $^-$ stopping events was nearly 10^3. Although we did not get the correct number of $^-$ stopping events due to impossibility of checking (K^-, K^+) reaction vertices in the diamond target, 4 events from 7 samples of double-Λ hypernucleus were succeeded to be analyzed. Among them, the NA ARA event was the event uniquely identified as the production and decay of a $^{6}_{\Lambda\Lambda}$He nucleus. Assuming $^-$ hyperon capture at atomic 3*D* state in the NA ARA event, two Λ binding energy, $B_{\Lambda\Lambda}$, and Λ-Λ interaction energy, $\Delta B_{\Lambda\Lambda}$, obtained by $2B_{\Lambda}(^{5}_{\Lambda}\mathrm{He}) - B_{\Lambda\Lambda}(^{6}_{\Lambda\Lambda}\mathrm{He})$ to be 6.91 ± 0.16 MeV and 0.67 ± 0.17 MeV, respectively, where B_{Λ} is a Λ binding energy of single-hypernucleus. Under the consistency of $\Delta B_{\Lambda\Lambda}$ of the NA ARA event, other three and two events by E176[3, 5] and Danysz *et al.* [1] were summarized as listed in Table 1. Since the error of the atomic 3*D* state

Journal of Physics: Conference Series **569** (2014) 012082 doi:10.1088/1742-6596/569/1/012082

is quite small, the errors of $B_{\Lambda\Lambda}$ and $\Delta B_{\Lambda\Lambda}$ are same as ones of $B_{\Lambda\Lambda} - B_{\Xi^-}$ and $\Delta B_{\Lambda\Lambda} - B_{\Xi^-}$, respectively. The difference, 2.8 MeV, of $B_{\Lambda\Lambda}$ between the events of DEMACHI ANA I and Danysz *et al.* is considered as the excitation energy of ${}^{10}_{\Lambda\Lambda}$Be nucleus.

3. The E07 experiment at J- ARC

In the E07 experiment, we will study nuclear mass dependence of $B_{\Lambda\Lambda}$, precisely, by detecting many double-Λ hypernuclei. Its statistics is expected to be ten times more than the previous E373 experiment by following $^-$ hyperons detected by SSDs. We also measure the energies of -ray emitted from atomic levels to know the energy shift by the strong interaction of $^-$ hyperon capture on nucleus by using e detector array, for the first time. Since the $^-$ stopping events are clearly detected in the emulsion, the measurement of -rays will be performed in almost background free for several thousands' events of $^-$ stopping.

The tracking e ciency of $^-$ hyperons is about 10 of 'p'(K^-, K^+) $^-$ events, therefore nearly 10^3 double-Λ hypernuclei stay in the emulsion. To perform scanning for the full volume of the emulsion and detect typical topologies of three vertices of double-Λ hypernuclei, we have been developing "*Overall-scanning*" method in two steps[7] which is listed in Table 2.

In the 1st step, we drive the microscope stage without stopping by taking images in a half side of the emulsion plate by CCD camera. It is necessary for 1.5 seconds to take images and 1.5 seconds to store image data. After that, we apply graphic processing method to image data. raphics process consists of routines by contrast maximization, background homogenization with aussian filter, track sharpness with matrix filter, binarization, track edge detection, thinning, Hough transformation and vertex detection.

Table 2. Specifications for two steps of *Overall-scanning* method.

	the 1st step	the 2^{n} step
Objective lens	× 50 (NA. 0.9)	× 20 (NA. 0.35)
Camera	C_HR300	H C20
Frame rate (Hz)	100	800
Device	CCD	CMOS
pixel	512 × 440	2048 × 358
Area (μm^2)	130 × 110	1,140 × 200
Stage	non-stop	non-stop
driving	Motor	iezo
Capture rate (Hz)	0.3	5

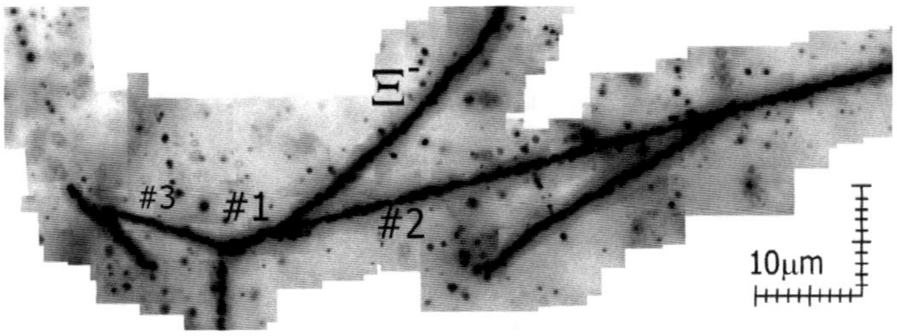

Figure 3. A newly found event of twin hypernuclei.

During test operations of the 1st step applying E373 emulsion plate, we have detected a typical event showing twin hypernuclei from 7.9×10^6 images as shown in Figure 3. A $^-$ hyperon was brought to rest after $\sim$ 8 mm flying from the top of the emulsion and produced two single-Λ hypernuclei (tracks #1 and #2). Two charged particles were emitted from the end point of track #1. Among them, track #3 is typical shape of '*hammer track*' of the decay of ^{8}Be$^*(2^+)$ nucleus to two α particles. We will measure the binding energy of $^-$ hyperon and the analysis is on-going.

We obtained quite high speed with $\times$ 300 than the system of the 1st step by iezo driving of the microscope head. This speed assures the detection of $\sim 10^3$ double-Λ hypernuclei in the E07 emulsion plates within a few years.

4. Concluding remarks

We have performed two experiments, E176 and E373 at KEK- S, to study doubly strangeness systems for these twenty years.

In the E176 experiment, we have studied not only quasi-free (K^-, K^+) reaction but also (K^-, K^+) reaction itself in the range of (0.4 eV c $< p_{K^+} <$ 1.2 eV c) with separated K^- beams (p_{K^-}= 1.66 eV c) at KEK- S. From the quasi-free reaction (1.0 eV c $\leq p_{K^+}$), mean-free-path of $^-$ hyperon in nucleus was obtained to be 4.7 $^{+1.5}_{-1.0}$ fm. By this result, we presented the interaction cross section $\sigma_{\Xi N}$ = 12.7 $^{+3.5}_{-3.1}$ mb. Hyperfragments were produced at 18 reaction vertices among 796 reactions. Regarding non quasi-free reaction ($p_{K^+} <$ 1.0 eV c), 901 reaction vertices were located in the emulsion. roduction rates of double-Λ and single-Λ hypernuclei for non quasi-free reaction were obtained to be 0.3 $\pm$ 0.2 and 4.6 $\pm$ 0.7 , respectively. At rest captures of $^-$ hyperon were seen in 77.6 $\pm$ 5.1 events which were caused in 58 and 42 by heavy (Ag, Br) and light (C, N, O) elements, respectively, in the emulsion. Sticking provabilities of two Λ hyperons or at least one Λ hyperon(s) in light nuclei would be 4.8 or 48.4 at 90 confidence level. In heavy nuclei, such probvabilities of two or at least one Λ hyperon(s) was 1.7 or 35.7 (90 CL.).

Double-Λ hypernuclei in several nuclear species were detected. The NA ARA event was uniquely assigned to be the production and decay of $^{6}_{\Lambda\Lambda}$He double-Λ hypernucleus and presented $B_{\Lambda\Lambda}$ and $\Delta B_{\Lambda\Lambda}$ for $6.79 + 0.91 B_{\Xi^-} (\pm 0.16) MeV$ and $0.55 + 0.91 B_{\Xi^-} (\pm 0.17) MeV$, respectively, as a fuction of B_{Ξ^-}. If we took the atomic $3D$ capture of $^-$ hyperon into account, $B_{\Lambda\Lambda}$ and $\Delta B_{\Lambda\Lambda}$ were obtained to be 6.91 $\pm$ 0.16 MeV and 0.67 $\pm$ 0.17 MeV, respectively. By applying result to other nuclei, the excitation energy was obtained of $^{10}_{\Lambda\Lambda}$Be as 2.8 MeV.

To study nuclear mass dependence of $B_{\Lambda\Lambda}$ with several tens' double-Λ hypernuclei, the E07 experiment at J- ARC was started. *Overall-scanning* method is now developing to search for double-Λ hypernuclei with very high speed in full volume of the emulsion. In the test operations of the method, a twin hypernulcear event was detected which will be useful for study -Nucleus interaction.

Acknowledgment

The authors would like to express our thanks to members of the E176, E373 and E07 experiment. This work was supported by the Endo Seijiro Science Foundation in 2007, by JS S rants No. 14340069 and No. 23224006, by ME T rants No. 08239103, No. 15001001 and No. 24105002 (Innovative Area 2404).

References

[1] M. Danysz, et al., Nucl. Phys. **49**, 121 (1963); R. H. Dalitz, et al., Proc. R. oc. London, **426**, 1 (1963); D. H. Davis, et al., Nucl. Phys. A **754**, 3c (2005).
[2] Y. Yamamoto, et al., Prog. Theor. Phys. uppl. No. 117, 281 (1994).
[3] . Aoki, et al., Prog. Theor. Phys. **85**, 1287 (1991).
[4] H. Takahashi, et al., Phys. Rev. Lett. **87**, 212502 (2001).
[5] . Aoki et al., Nucl. Phys. A **828**, 191 (2009).
[6] K. Nakazawa and H. Takahashi, Prog. Theor. Phys. uppl. **185**, 335 (2010); J. K. Ahn et al., Phys. Rev. C **88**, 014003 (2013).
[7] J. Yoshida, K. Nakazawa and K. Umehara, JP Conf. Proc. **1**, 013070 (2014);to be submitted to NIM.
[8] . Aoki et al., Nucl. Phys. A **644**, 365 (1998).
[9] Fletcher et al., Phys. Lett. **3** 280 (1964).

Structure of *p-sd* Shell Λ Hypernuclei Studied with AMD

Masahiro Isaka

RIKEN Nishina Center, RIKEN, Wako, Saitama 351-0198, Japan

E-mail: masahiro.isaka@riken.jp

Abstract. The structures of the ground and low-lying states of $^{21}_{\Lambda}$Ne and $^{12}_{\Lambda}$Be hypernuclei are studied with the antisymmetrized molecular dynamics. In these hypernuclei, first, the difference of the binding energy on structures is discussed. Particularly, in $^{12}_{\Lambda}$Be, it is found that the parity-inverted ground state of ^{11}Be is reverted by adding a particle. Furthermore, in $^{21}_{\Lambda}$Ne, it is discussed that the reduction of nuclear radii is larger in the α + ^{16}O + band than the ground band and the difference appears as the difference of the $B(E2)$ reduction.

1. Introduction

We focus on the ground and low-lying states of $^{21}_{\Lambda}$Ne and $^{12}_{\Lambda}$Be and discuss the difference of the binding energy as well as structure changes by a particle based on the antisymmetrized molecular dynamics (AMD). In *sd*-shell nuclei and neutron-rich nuclei, it is discussed that the many kinds of (cluster) structures coexist in the ground and low-lying energy regions. For example, ^{20}Ne is a typical *sd*-shell nucleus having the various structures within the small excitation energy [1, 2, 3]. The ground band ($K^{\pi} = 0^{+}_{1}$) and the $K^{\pi} = 0^{-}_{1}$ band built on the 1^{-}_{1} state at 5.79 MeV constitute the parity doublet associated with α+^{16}O clustering. It is known that the $K^{\pi} = 0^{-}_{1}$ band has the pronounced α+^{16}O clustering, while the ground band has the mixing of deformed shell structure and α + ^{16}O cluster structure. Therefore, it is of interest to investigate how a particle affects and modifies their structures. In the neutron-rich nucleus ^{11}Be, the ground state is $1/2^{+}$, which is inconsistent with the ordinary shell-model picture where the seven neutrons may have a $1/2^{-}$ state as the ground state. One of the main reasons for the parity inversion is molecular-orbit structure of ^{11}Be [4]. In the $1/2^{+}$ state, one of the extra neutrons occupies the σ orbit around the 2α cluster structure and enhance the 2α clustering. The deformation of the $1/2^{-}$ state is smaller than that of the $1/2^{+}$ state because the extra neutrons occupy the π orbit and reduce the clustering. By adding a hyperon to such nuclei, it is expected that the binding energy is different depending on the deformations and the level structure will be changed.

In this article, we focus on the low-lying structure of $^{21}_{\Lambda}$Ne and $^{12}_{\Lambda}$Be hypernuclei with the coexistence of various (exotic) cluster structures. Especially, first, we focus on the difference of the binding energy (B_{Λ}) depending on the structure. In neutron-rich $^{12}_{\Lambda}$Be, the difference of B_{Λ} may affect the parity-inverted ground states of ^{11}Be. Furthermore, in $^{21}_{\Lambda}$Ne, it is expected that the particle modifies the α + ^{16}O cluster structure in the $K^{\pi} = 0^{-}_{1}$ band and the structure change is different from the $K^{\pi} = 0^{+}_{1}$ band. To study such phenomena, we have applied the AMD model extended to hypernuclei (HyperAMD) to these hypernuclei[5]. Combined with the generator coordinate method (GCM), it is possible to predict the low-lying level structure and structure changes of hypernuclei[6].

2. Theoretical Framework

In this study, we have applied the HyperAMD [5, 6]. The Hamiltonian used in this study is given as,

$$\hat{H} = \hat{T}_N + \hat{T}_\Lambda - \hat{T}_g + \hat{V}_{NN} + \hat{V}_{\Lambda N} + \hat{V}_C \tag{1}$$

where $\hat{T}_N$, $\hat{T}_\Lambda$ and $\hat{T}_g$ are the kinetic energies of nucleons, hyperon and the center-of-mass motion. We used the Gogny D1S interaction as an effective nucleon-nucleon interaction V_{NN}, which has been successfully applied to the stable and unstable nuclei. The YNG interactions have been employed as -nucleon interaction $V_{\Lambda N}$ [7, 8]. Since the YNG interactions depend on the nuclear density through the Fermi momentum k_F, we adopted $k_F = 1.17\ \mathrm{fm}^{-1}$ and $k_F = 0.97\ \mathrm{fm}^{-1}$ for ${}^{21}_{\Lambda}\mathrm{Ne}$ and ${}^{12}_{\Lambda}\mathrm{Be}$, respectively. The Coulomb interaction is approximated by the sum of seven Gaussians.

The variational wave function of a single hypernucleus is described by the parity-projected wave function,

$$\Psi^\pi = \hat{P}^\pi\{\Psi_N \otimes \varphi_\Lambda\},\ \Psi_N = \frac{1}{\sqrt{A!}}\det\{\phi_i(r_j)\}, \tag{2}$$

$$\phi_i \propto e^{-\sum \nu_\sigma (r_\sigma - Z_{i\sigma})^2}(u_i\chi_\uparrow + v_i\chi_\downarrow)\eta_i \tag{3}$$

$$\varphi_\Lambda \propto \sum_{m=1}^{M} c_m(a_m\chi_\uparrow + b_m\chi_\downarrow)e^{-\sum \nu_\sigma (r_\sigma - z_{m\sigma})^2}, \tag{4}$$

where ϕ_i is the nucleon single-particle wave packet consisting of spatial, spin and isospin η_i parts. The single particle wave function of , φ_Λ is represented by a superposition of Gaussian wave packets. The variational parameters $\vec{Z}_i$, $\vec{z}_m$, ν_σ, u_i, v_i, a_m, b_m, and c_m are set to minimize the total energy under the constraint on the matter quadrupole deformation β.

After the variation, we perform the angular momentum projection and the generator coordinate method (GCM) calculations [9],

$$\Psi^{J\pi}_{MK}(\beta) = \frac{2J+1}{8\pi^2}\int d\Omega D^{J*}_{MK}(\Omega)R(\Omega)\Psi^\pi(\beta), \quad \Psi^{J\pi}_n = \sum_\beta \sum_{K=-J}^{J} c_{nK}\Psi^{J\pi}_{MK}(\beta), \tag{5}$$

where the wave functions with differing values of K and β are superposed. The coefficients c_{npK} are determined by solving the Griffin-Hill-Wheeler equation.

To analyze each state obtained after the GCM calculation, we calculate the overlap between the $\Psi^{J\pi}_n$ and $\Psi^{J\pi}_{MK}(\beta)$, so called GCM overlap, defined as,

$$O^{J\pi}(\beta) = |\langle\Psi^{J\pi}_n|\Psi^{J\pi}_{MK}(\beta)\rangle|^2. \tag{6}$$

It is noted that the GCM overlap $O^{J\pi}(\beta)$ is a function of β. In this study, we regard the deformation of each state as the β which gives the largest $O^{J\pi}(\beta)$ in each state.

Finally, we introduce the energy gain of the J^π state in hypernuclei from the core state j^π to investigate the binding energy, as,

$$B_\Lambda = E({}^{A}Z(j^\pi)) - E({}^{A+1}_{\Lambda}Z(J^\pi)), \tag{7}$$

where $E({}^{A}Z(j^\pi))$ and $E({}^{A+1}_{\Lambda}Z(J^\pi))$ are the calculated energies of the j^π and J^π states, respectively.

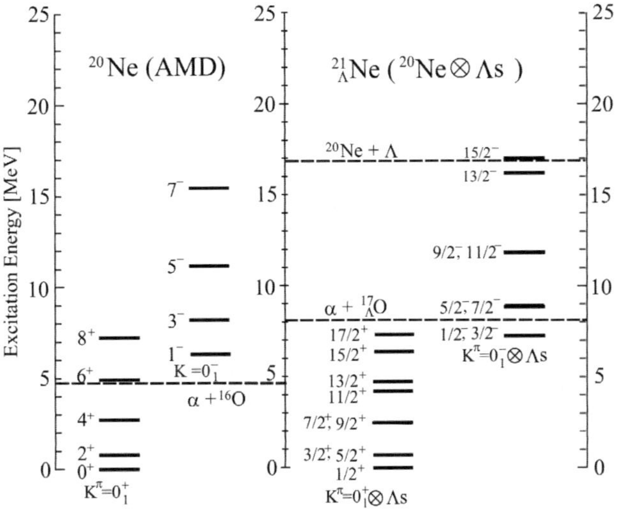

Figure 1. Calculated excitation spectra of the $K^\pi = 0_1^+$ and $K^\pi = 0_1^-$ bands in ^{20}Ne and the corresponding bands in $^{21}_{\Lambda}$Ne.

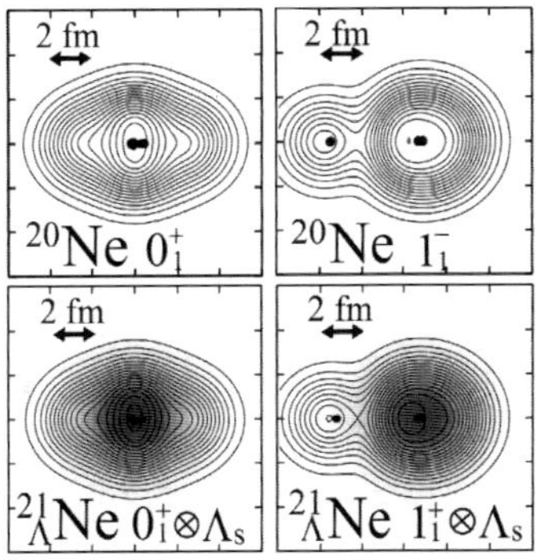

Figure 2. Solid lines show the matter density distributions for the band head states 0_1^+ and 1_1^- in ^{20}Ne and the corresponding states in $^{21}_{\Lambda}$Ne. Color plots show the density distribution of .

Table 1. Calculated energy E, excitation energy E_x, and binding energy B_Λ for the band head states of the $0_1^+ \otimes \;_s$ and $0_1^- \otimes \;_s$ bands are listed in MeV. The definition of B_Λ is given by the equation(7). For the comparison, theoretical and experimental E and E_x values of the 0_1^+ and 1_1^- states in ^{20}Ne are also presented.

	Band	E(MeV)	E_x(MeV)	B_Λ(MeV)
$^{21}_{\Lambda}$Ne(HyperAMD)	$(K^\pi = 0_1^+) \otimes \;_s$	-176.5	0.0	16.9
$^{21}_{\Lambda}$Ne(HyperAMD)	$(K^\pi = 0_1^-) \otimes \;_s$	-169.2	7.3	15.9
^{20}Ne(AMD)	$K^\pi = 0_1^+$	-159.6	0.0	
^{20}Ne(AMD)	$K^\pi = 0_1^-$	-153.3	6.3	
^{20}Ne(Exp.)	$K^\pi = 0_1^+$	-160.6	0.0	
^{20}Ne(Exp.)	$K^\pi = 0_1^-$	-154.9	5.8	

3. Results and Discussions

3.1. Ground and α + ^{16}O + bands in $^{21}_{\Lambda}$Ne

By performing the GCM calculations, it is found that many excited states with positive parity appear in $^{21}_{\Lambda}$Ne. Among them, we obtain the two rotational bands of $^{21}_{\Lambda}$Ne, which correspond to the $K^\pi = 0_1^+$ and $K^\pi = 0_1^-$ bands of ^{20}Ne. Therefore, we denote these bands in $^{21}_{\Lambda}$Ne as $K^\pi = 0_1^+ \otimes$ and $K^\pi = 0_1^- \otimes$, respectively. The resulting excitation spectra of these bands are shown in figure 1.

Let us compare B_Λ, defined by the equation(7), of the band head states of the $K^\pi = 0_1^+ \otimes \;_s$ and $K^\pi = 0_1^- \otimes \;_s$ bands. As mentioned above, the mean-field like and α + ^{16}O structures are mixed in the $K^\pi = 0_1^+$ band, while the $K^\pi = 0_1^-$ band has a pronounced α+^{16}O cluster structure. In table 1, the total energy E, excitation energy E_x and binding energy B_Λ are listed. It shows that the B_Λ of the $K^\pi = 0_1^+ \otimes \;_s$ (ground) state is larger than that of the $K^\pi = 0_1^- \otimes \;_s$ band head state. In figure 2, the density distributions of the band head states of $K^\pi = 0_1^+ \otimes \ _s$ and $K^\pi = 0_1^- \otimes \ _s$ bands of $^{21}_{\Lambda}$Ne are presented. It is clearly seen that the hyperon coupled to the $K^\pi = 0_1^-$ band head state of ^{20}Ne is localized around the ^{16}O cluster. This is because the single-particle potential of $\ _s$ is not parity symmetric due to the α+^{16}O clustering and is deeper around ^{16}O cluster. Indeed, B_Λ of $^{17}_{\Lambda}$O may be more than 12 MeV, while that of $^{5}_{\Lambda}$He is about 3 MeV. On the other hand, the hyperon coupled to the $K^\pi = 0_1^+$ band head state locates at the center of ^{20}Ne nucleus and interacts with all nucleons. Therefore, the $\ _s$ in the the $K^\pi = 0_1^+ \otimes \ _s$ band head state is more deeply bound than that in the $K^\pi = 0_1^- \otimes \ _s$ band head state.

Since the particle in the $K^\pi = 0_1^- \otimes \ _s$ band is located around the ^{16}O cluster, it is not an eigenstate of parity of single particle state of $\ _s$. Thus, the p-orbit component of the hyperon should contribute to the $K^\pi = 0_1^- \otimes \ _s$ state. Such mixed nature was called "parity coupling"[10] for $^{21}_{\Lambda}$Ne or "inter-shell coupling"[11]. It was argued that parity coupling could occur because the energy difference between the positive and negative parity states in the core nucleus is similar to that between the s- and p-orbits of the hyperon [10]. However, in the present result, the parity coupling is due to the asymmetry of α+^{16}O clustering. To see the contribution of the particle in p orbit, as discussed in the reference [10], we also analyze the GCM overlap, defined by the equation (6), of the $1/2^-$ state which is the band head state of the $K^\pi = 0_1^- \otimes \ _s$. The details are discussed in the reference [6].

Table 2. RMS radii (fm) for the $K^\pi = 0_1^+$ band in ^{20}Ne and the corresponding band in $^{21}_{\Lambda}$Ne. $\Delta r_{\rm rms}$ is defined as a subtraction of RMS radius: $\Delta r_{\rm rms} = r_{\rm rms}(^{21}_{\Lambda}\text{Ne}) - r_{\rm rms}(^{20}\text{Ne})$ for each state.

^{20}Ne	$r_{\rm rms}$	$^{21}_{\Lambda}$Ne	$r_{\rm rms}$	$\Delta r_{\rm rms}$
0^+	2.97	$1/2^+$	2.92	-0.05
2^+	2.96	$3/2^+$	2.91	-0.05
		$5/2^+$	2.91	-0.05
4^+	2.93	$7/2^+$	2.87	-0.06
		$9/2^+$	2.88	-0.04
6^+	2.87	$11/2^+$	2.81	-0.06
8^+	2.82	$15/2^+$	2.77	-0.04

Table 3. Same as table2 but for the $K^\pi = 0^-$ band in ^{20}Ne and the corresponding band in $^{21}_{\Lambda}$Ne.

^{20}Ne	$r_{\rm rms}$	$^{21}_{\Lambda}$Ne	$r_{\rm rms}$	$\Delta r_{\rm rms}$
1^-	3.27	$1/2^-$	3.15	-0.11
		$3/2^-$	3.15	-0.11
3^-	3.24	$5/2^-$	3.13	-0.11
		$7/2^-$	3.14	-0.10
5^-	3.23	$9/2^-$	3.11	-0.12
		$11/2^-$	3.11	-0.13
7^-	3.23	$13/2^-$	3.06	-0.17

Next, we discuss the difference in the reduction of the nuclear radii by adding a particle, so called "shrinkage effect", between the $K^\pi = 0_1^+ \otimes \ _s$ and $K^\pi = 0_1^- \otimes \ _s$ bands in $^{21}_{\Lambda}$Ne. In tables

2-3, the nuclear RMS radii for the $K^\pi = 0_1^+$ and $K^\pi = 0_1^-$ bands of ^{20}Ne and the corresponding bands with $_s$ are listed. The RMS radii for the $K^\pi = 0_1^-$ band change more than those of the $K^\pi = 0_1^+$. This is due to the difference in the clustering of these bands. Since the $K^\pi = 0_1^-$ band has well developed α + ^{16}O cluster structure, hyperon reduces the inter-cluster distance.

The difference of the shrinkage effect in $^{21}_{\Lambda}$Ne affect the reduction of the intra-band $B(E2)$ values in $K^\pi = 0_1^+ \otimes \ _s$ and $K^\pi = 0_1^- \otimes \ _s$ bands. To compare $B(E2)$ values of $^{21}_{\Lambda}$Ne with those of ^{20}Ne, we corrected them under the assumption that a hyperon occupies the s-orbit for each hypernuclear state in the $K^\pi = 0_1^+ \otimes \ _s$ and $K^\pi = 0_1^- \otimes \ _s$ bands [6]. Both the corrected and uncorrected $B(E2)$ values for the $K^\pi = 0_1^+ \otimes \ _s$ and $K^\pi = 0_1^- \otimes \ _s$ bands are presented in tables 4-5.

Tables 4-5 shows that a particle causes the $B(E2)$ reduction in the $K^\pi = 0_1^- \otimes \ _s$ and the $K^\pi = 0_1^+ \otimes \ _s$ bands. In the reference [10], the $B(E2)$ reductions predicted by Yamada et al. are more than 20 for those bands. However, in the present study, the corrected $B(E2)$ values for the $0_1^+ \otimes \ _s$ band reduce less than 20 , while those for the $K^\pi = 0_1^+ \otimes \ _s$ band is almost about 20 , as shown in tables 4-5. We consider the difference in the $B(E2)$ reduction mainly comes from the difference in the reduction of RMS radii between these two bands, as discussed above.

Table 4. Intra-band $B(E2)$ values in e^2 fm^4 for the $K^\pi = 0_1^+$ in ^{20}Ne and the corresponding band in $^{21}_{\Lambda}$Ne. $B(E2)$ and $cB(E2)$ represent the uncorrected and corrected $B(E2)$ values, respectively. The correction of the $B(E2)$ values is explained in the reference [6].

$K^\pi = 0_1^+$	$B(E2)$	$0_1^+ \otimes \ _s$	$B(E2)$	$cB(E2)$	( )
$2^+ \to 0^+$	72.2	$3/2^+ \to 1/2^+$	63.7	63.7	-11.8
		$5/2^+ \to 1/2^+$	63.9	63.9	-11.5
$4^+ \to 2^+$	86.9	$7/2^+ \to 3/2^+$	64.3	71.4	-17.8
		$9/2^+ \to 5/2^+$	75.7	75.7	-13.0
$6^+ \to 4^+$	55.1	$11/2^+ \to 7/2^+$	40.3	41.9	-23.9
		$13/2^+ \to 9/2^+$	48.0	48.0	-12.9
$8^+ \to 6^+$	17.0	$15/2^+ \to 11/2^+$	15.9	16.2	-4.6
		$17/2^+ \to 13/2^+$	17.1	17.1	0.8

Table 5. Same as table4 but for the $K^\pi = 0^-$ band in ^{20}Ne and the corresponding band in $^{21}_{\Lambda}$Ne.

$K^\pi = 0_1^-$	$B(E2)$	$0_1^- \otimes \ _s$	$B(E2)$	$cB(E2)$	( )
$3^- \to 1^-$	221.2	$5/2^- \to 1/2^-$	139.2	179.0	-19.1
		$7/2^- \to 3/2^-$	178.5	178.5	-19.3
$5^- \to 3^-$	249.3	$9/2^- \to 5/2^-$	184.2	195.4	-21.6
		$11/2^- \to 7/2^-$	189.3	189.3	-24.1
$7^- \to 5^-$	240.3	$13/2^- \to 9/2^-$	164.3	166.7	-30.6

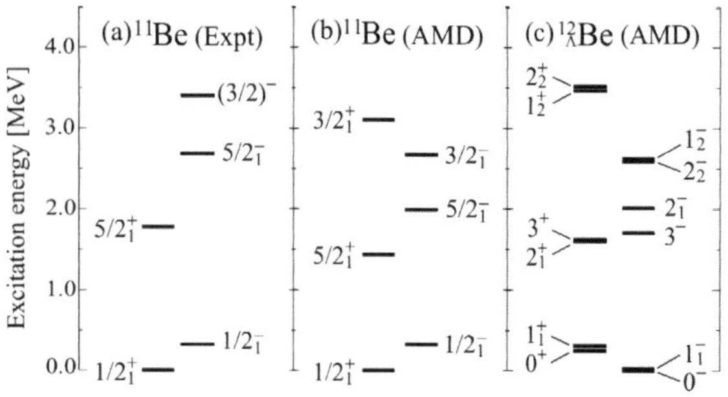

Figure 3. (a) Observed excitation spectra of ^{11}Be. (b),(c) Calculated spectra of ^{11}Be and $^{12}_{\Lambda}$Be.

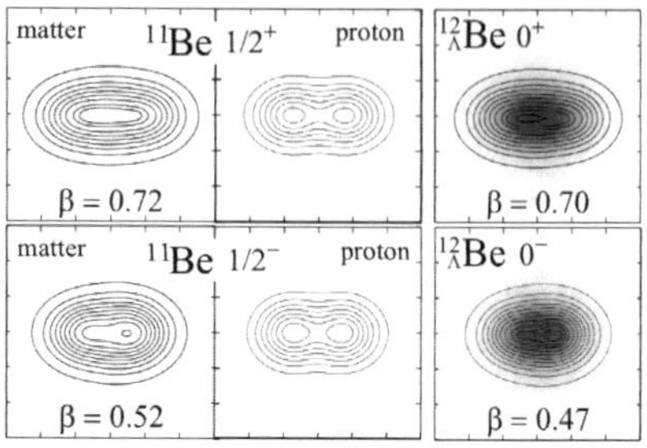

Figure 4. Solid lines show the matter and proton density distributions of the $1/2^+$ and $1/2^-$ states of ^{11}Be and the corresponding states in $^{12}_{\Lambda}$Be. Color plots show the density distribution of the hyperon.

Table 6. uadrupole deformation β, calculated energy E and excitation energy E_x for the $1/2^+_1$ and $1/2^-_1$ states in ^{11}Be and the corresponding 0^+ and 0^- states in $^{12}_{\Lambda}$Be. binding energy B_Λ is also listed for $^{12}_{\Lambda}$Be.

	J^π	β	E(MeV)	E_x(MeV)	B_Λ(MeV)
$^{12}_{\Lambda}$Be(HyperAMD)	0^+_1	0.70	-74.44	0.25	9.67
$^{12}_{\Lambda}$Be(HyperAMD)	0^-_1	0.47	-74.69	0.00	10.24
^{11}Be(AMD)	$1/2^+_1$	0.72	-64.77	0.00	
^{11}Be(AMD)	$1/2^-_1$	0.52	-64.45	0.32	
^{11}Be(Exp.)	$1/2^+_1$		-65.48	0.00	
^{11}Be(Exp.)	$1/2^-_1$		-65.16	0.32	

3.2. Parity reversion of the $^{12}_{\Lambda}$Be *ground state*

First, let us discuss the structure of the core nucleus ^{11}Be. As shown in figure 3(a), the ^{11}Be ground state has positive parity, and the order of the $p_{3/2}$ and sd shells looks inverted, which is called "parity inversion" [12, 13]. In the present study, the AMD calculation successfully describes the parity inversion of the ground state in ^{11}Be as shown in figure 3(b).

The low-lying states of Be isotopes are known to have a 2α cluster core and valence neutrons occupying the molecular orbits around the core, which are called π and σ orbits [14]. The formation of the 2α cluster core in each state is confirmed by the proton density shown in figure 4. In the ground state $1/2^+$, two of the valence neutrons occupy the π orbit, and the third valence neutron occupies the σ orbit. In terms of the spherical shell model, a neutron is promoted to sd shell across the $N = 8$ shell gap (breakdown of magic number $N = 8$). On the contrary, in the first excited state $1/2^-$, all valence neutrons occupy the π orbit or p shell, which corresponds to the normal shell order. As we can see from figure 4 and table 6, the ground state has a more pronounced 2α clustering and a larger quadrupole deformation β than the first excited state.

figure 3(c) shows the level structure of $^{12}_{\Lambda}$Be with a hyperon in s orbit. It is found that the ground state parity of $^{12}_{\Lambda}$Be becomes negative. Namely, the parity reversion of the $^{12}_{\Lambda}$Be ground state will occur by adding a particle. As shown in table 6, the B_Λ for the 0^-_1 state is larger than the 0^+_1 state by about 500 keV. The difference in B_Λ mainly comes from the N potential

Journal of Physics: Conference Series **569** (2014) 012083 doi:10.1088/1742-6596/569/1/012083

energy $V_{\Lambda N}$, and it originates in the difference of the nuclear deformation. It is expected that the difference of the deformation between the $1/2^+$ and $1/2^-$ states can be confirmed by observing the parity-reverted ground state of ${}^{12}_{\Lambda}$Be.

4. Summary

In summary, we applied an extend version of AMD to hypernuclei to ${}^{21}_{\Lambda}$Ne and ${}^{12}_{\Lambda}$Be. We discussed the difference of the binding energy (B_Λ) depending on the structure of the core states. In ${}^{21}_{\Lambda}$Ne, it is discussed that B_Λ is different between the ground and $K^\pi = 0_1^- \otimes \;_s$ band, reflecting the difference of their structure. In ${}^{12}_{\Lambda}$Be, it is found that the parity reversion in the ground state will occur by a particle due to the difference of B_Λ, originated in the difference of deformations between the $1/2^+$ and $1/2^-$ states in ^{11}Be. Furthermore, in ${}^{21}_{\Lambda}$Ne, intra-band $B(E2)$ reduction in the $K^\pi = 0_1^- \otimes \;_s$ band is larger than that in the $K^\pi = 0_1^+ \otimes \;_s$ band. This is mainly due to the reduction of the inter-cluster distance between α and ^{16}O clusters in the $K^\pi = 0_1^-$ band.

References

[1] Ikeda K, Marumori T, Tamagaki R and Tanaka H 1972 *Prog. Theor. Phys. Suppl.* **52** 1
[2] Ikeda K, Marumori T, Horiuchi H and Saito S 1980 *Prog. Theor. Phys. Suppl.* **68** 1
[3] Kimura M 2004 *Phys. Rev.* C **69** 044319
[4] Kanada-En'yo Y and Horiuchi H 2002 *Phys. Rev.* C **66** 024305
[5] Isaka M, Kimura M, Dote A and Ohnishi A 2011 *Phys. Rev.* C **83** 044323
[6] Isaka M, Kimura M, Dote A and Ohnishi A 2011 *Phys. Rev.* C **83** 054304
[7] Yamamoto Y, Motoba T, Himeno H, Ikeda K and Nagata S 1994 *Prog. Theor. Phys. Suppl.* **117** 361
[8] Hiyama E, Kamimura M, Motoba T, Yamada T and Yamamoto Y 1997 *Prog. Theor. Phys.* **97** 881
[9] Hill D L and Wheeler J A 1953 *Phys. Rev.* **89** 1102
[10] Yamada T, Ikeda K, Bandō H and Motoba T 1984 *Prog. Theor. Phys.* **71** 985
[11] Motoba T 1998 *Nucl. Phys.* A **629** 398c
[12] Wilkinson D H and Alburger D E 1959 *Phys. Rev.* **113** 563
[13] Talmi U and Unna I 1960 *Phys. Rev. Lett.* **4**, 469
[14] von Oertzen W, Freer M and Kanada-En'yo Y 2006 *Phys. Rep.* **432** 43

Journal of Physics: Conference Series **569** (2014) 012084 doi:10.1088/1742-6596/569/1/012084

Essential $\bar{K}$ cluster "K^-pp" studied with a coupled-channel Complex Scaling Method + Feshbach method

A. Dote[1,2], T. Inoue[3] and T. Myo[4,5]

[1] KEK Theory Center, Institute of Particle and Nuclear tudies (IPN), High Energy Accelerator Research Organization (KEK), 1-1 Oho, Tsukuba, Ibaraki, 305-0801, Japan
[2]J-PARC Branch, KEK Theory Center, IPN , KEK, 203-1, hirakata, Tokai, Ibaraki, 319-1106, Japan
[3] Nihon University, College of Bioresource ciences, Fujisawa 252-0880, Japan
[4] General Education, Faculty of Engineering, Osaka Institute of Technology, Osaka 535-8585, Japan
[5] Research Center for Nuclear Physics (RCNP), Osaka University, Ibaraki 567-0047, Japan

E-mail: `dote@post.kek.jp`

Abstract. We have investigated an essential $\bar{K}$ cluster "K^-pp", which is the simplest system of kaonic nuclei, with a coupled-channel Complex caling Method (ccC M). Combining the ccC M with Feshbach method we can handle a coupled-channel problem effectively as a single-channel problem. As a result of a study of the K^-pp with the ccC M+Feshbach method using an energy-dependent chiral-theory based potential, it is found that the K^-pp is shallowly bound with the binding energy of around 20-35 MeV. The mesonic decay width is rather dependent on the interaction parameters and ansatz; the half decay width is ranging from 20 to 65 MeV.

1. Introduction

In strange nuclear physics and hadron physics, kaonic nuclei (nuclear system with anti-kaons, $\bar{K} = K^-, \bar{K}^0$) have been a hot topic since the formation of dense state are interestingly expected due to strong $\bar{K}N$ attraction. To consider such an interesting system, let us start with an excited hyperon $\Lambda(1405)$ ($J^\pi = 1/2^-$, $I = 0$). The early work with a quark model succeeded the reproduction of most spectra of P-wave baryons, except for the $\Lambda(1405)$ whose mass was resulted to be larger than the observed value [1]. Since the $\Lambda(1405)$ exists below the $\bar{K}N$ threshold by just 30 MeV, it is expected to be a $\bar{K}N$ quasi-bound state. Actually, chiral unitary models, which are based on a meson-baryon picture, have succeeded to explain various properties of the $\Lambda(1405)$ [2]. Thus, we believe that the $\Lambda(1405)$ is a $\bar{K}N$ quasi-bound state, rather than a genuine three quark state. Based on this picture, Akaishi and amazaki proposed a phenomenological $\bar{K}N$ potential, which is nowadays called A potential [3]. A potential is quite attractive especially in the isospin-zero channel. As a result of a fully-microscopic calculation of kaonic nuclei with antisymmetrized molecular dynamics (AMD) method using A potential [4], it is found that a single K^- meson can be deeply bound in various light nuclei with 100 MeV binding energy. The AMD study shows exotic properties of kaonic nuclei. Due to the strong $\bar{K}N$ attraction, nucleons are attracted by a K^- meson to form dense nuclear state; the average density amounts

Table 1. Summary of theoretical studies of K^-pp. “$B(K^-pp)$” and “Γ_M” are the binding energy and mesonic decay width of K^-pp, respectively. They are in unit of MeV. “Method” and “ otential” mean a method and a $\bar{K}N$ potential employed in each calculation, respectively. In variational calculations two kinds of basis function are used: aussian function “(auss)” and Hyperspherical Harmonics “(H.H.)”. In these calculations, phenomenological potentials (“ heno.”) and chiral-theory based potentials (“Chiral”) are examined. In the latter potentials, there are energy-dependent (“(E-dep.)”) and energy-independent (“(E-indep.)”) versions. “Non-rel.” and “Semi-rel.” indicate non-relativistic and semi-relativistic kinematics, respectively.

	Dote-Hyodo -Weise [6]	Akaishi - amazaki [7]	Barnea- al -Liverts [8]	Ikeda-Sato [9]	Shevchenko- al -Mares [10]
$B(K^-pp)$	20 ± 3	47	16	$60 \sim 95$	$50 \sim 70$
Γ_M	$40 \sim 70$	61	41	$45 \sim 80$	$90 \sim 110$
Method	Variational (auss)	Variational (auss)	Variational (H.H.)	Faddeev-A S	Faddeev-A S
otential	Chiral (E-dep.)	heno.	Chiral (E-dep.)	Chiral (E-indep.)	heno.
Kinematics	Non-rel.	Non-rel.	Non-rel.	Semi-rel.	Non-rel.

to several times higher than the normal nuclear density. In addition even three protons can be bound by a single K^- meson with a funny structure. Thus, the kaonic nuclei are expected to be a exotic system which could be in particular a doorway to dense nuclear matter.

To clarify exotic properties, which are expected, involved with kaonic nuclei, we have investigated the most simplest kaonic nucleus “K^-pp” before considering complicated many-body systems. The K^-pp seems to be an *essential cluster* of kaonic nuclei, while the $\Lambda(1405)$ (= a quasi-bound state of K^-p) can be regarded as a building block of kaonic nuclei. Since the K^-pp is only a three-body system composed of two protons and a single K^- meson, it has been investigated with various ways as summarized in Table 1. The binding energy and decay width are rather dependent on approaches and employed potentials. However, all theoretical studies show that the K^-pp can be bound with less than 100 MeV binding energy. Since there is $\pi\Sigma N$ threshold at 103 MeV below $\bar{K}NN$ threshold, these calculations indicate that the K^-pp should be a resonant state between the two thresholds. In addition, we know the fact that the $\bar{K}N$ is strongly coupled with the $\pi\Sigma$ through the study of the $\Lambda(1405)$ [2]. Thus, we consider that *1. Resonance* and *2. Coupled-channel problem* are key ingredients in the theoretical study of the K^-pp. We, here, employ a coupled-channel Complex Scaling Method (ccCSM) since this approach can simultaneously treat these two ingredients. It should be noted that the complex scaling method has greatly succeeded in the study of resonances of unstable nuclei [11]. In our previous study [12], we applied the ccCSM to the two-body system of $\bar{K}N$-πY. (Y means Λ and Σ hyperons.) Since that study shows that the ccCSM is a useful tool also for the study of hadronic system, we now tackle the three-body system of K^-pp.

2. Methodology

In this section, we explain our method to investigate the K^-pp resonance shortly. At first, we remark that the K^-pp is considered as a coupled-channel system of $\bar{K}NN$-$\pi\Sigma N$-$\pi\Lambda N$ with $J^\pi = 0^-$ and $I = 1/2$ in theoretical studies.

Basically, we follow the usual prescription of the complex scaling method [11] to calculate

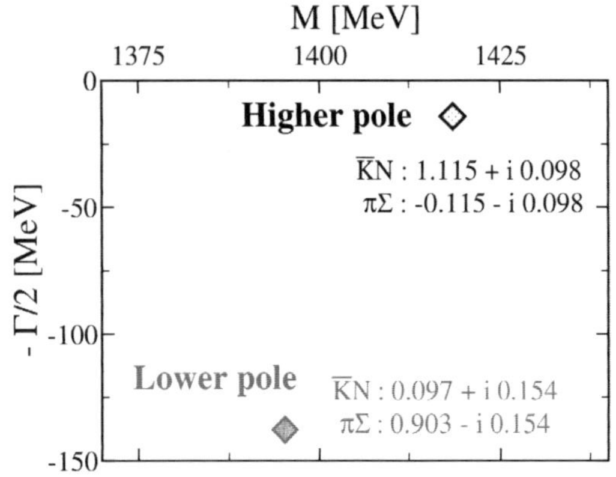

Figure 1. Double pole of $\Lambda(1405)$ obtained with the ccCSM. NRv2 potential with $f_\pi = 110$ MeV is used. The complex values of "$\bar{K}N$" and "$\pi\Sigma$" are norms of $\bar{K}N$ and $\pi\Sigma$ components in the pole states, respectively.

complex eigenvalues of the three-body system of K^-pp. The Hamiltonian for the K^-pp is complex-scaled, with a complex-scaling operator $U(\theta)$ the coordinate and the conjugate momentum is transformed as $\boldsymbol{r} \to \boldsymbol{r}e^{i\theta}$ and $\boldsymbol{p} \to \boldsymbol{p}e^{-i\theta}$, respectively. Diagonalizing the complex-scaled Hamiltonian with a basis function, complex eigenvalues are obtained. Among those eigenvalues, the eigenvalues which are independent of the scaling angle θ are those of resonant states. In our study we employ the correlated aussian function [13] as a basis function. We give detailed explanation on important points of our study as below.

2.1. A chiral-theory based $\bar{K}N$ potential and treatment of its energy dependence

In our previous work [12], we proposed a chiral-theory based $\bar{K}N$-πY potential which is a local potential with a single-range aussian form factor in the coordinate space so that we can easily handle it with aussian basis function:

$$V_{ij}(r) = -\frac{C_{ij}}{8f_\pi^2}(\omega_i + \omega_j) \times [\text{flux factor}] \times (\ \text{aussian form factor}), \tag{1}$$

where ω_i indicates the meson energy in the channel i and C_{ij} is the Clebsch- ordan coe cient in the SU(3) algebra. In the present study we use one of non-relativistic versions of our potential, which is denoted as NRv2 potential. Detail of the NRv2 potential is given in Eq. (8) in Ref. [12]. Note that the pion decay constant f_π in Eq. (1) is treated as a parameter in our potential.

Here, we mention the recent progress of our study. According to many studies based on the chiral SU(3) theory, there appear two poles in s-wave and $I = 0$ channel, which corresponds to the $\Lambda(1405)$, when we use the energy-dependent chiral potential [2]. However, such a *double-pole structure* of the $\Lambda(1405)$ was not clearly observed in our previous work [12], though the potential was constructed based on the chiral SU(3) theory. Recently, we have successfully confirmed the double-pole structure in our potential by using the improved aussian basis function [14], as shown in Fig. 1. Analyzing the norm of $\bar{K}N$ and $\pi\Sigma$ components in each pole state, we have found that the higher-pole state is dominated by the $\bar{K}N$ component since the $\bar{K}N$ norm is a large magnitude. On the other hand, in the lower pole the $\pi\Sigma$ component is found to be a major component.

By the way, as mentioned above our $\bar{K}N$ potential has an energy dependence. We need to take into account the self-consistency for $\bar{K}N$ energy when we treat bound and resonant states. But, the definition of the $\bar{K}N$ energy in the K^-pp is non-trivial because the two-body $\bar{K}N$ is a subsystem of the three-body $\bar{K}NN$ system. We define the $\bar{K}N$ energy (E_{KN}) in such a way as proposed in an earlier study [6]. In that study the E_{KN} is defined in two ways by considering extreme two pictures; The antikaon is regarded as a field (Ansatz 1) or it is considered as a particle (Ansatz 2). Details of calculation of the E_{KN} is given in Eqs. (20) and (21) in Ref. [6]. Also in the present study, we examine these two ansatz. We remark that this self-consistency is realized for the *complex* $\bar{K}N$ energy since the complex energy of the resonance pole is directly treated in the current work, while the real $\bar{K}N$ energy is considered in in the earlier work with a variational approach [6].

2.2. Essence of the ccCSM+Feshbach method

The K^-pp system is a coupled-channel system of $\bar{K}NN$, $\pi\Sigma N$ and $\pi\Lambda N$. We treat such a multi-channel problem as a single-channel problem with help of Feshbach method [15]. In Feshbach method, we set a model space (P space) and outer space of the model space (Q space). Then, the Schrödiner equation is given as a coupled-channel equation of wave functions for P and Q spaces. Eliminating the Q-space wave function, we can construct a Schrödiner equation only for the P-space wave function Φ_P as $[T_P + U_P^{Eff}(E)]\Phi_P = E\Phi_P$. ($T_P$ is a kinetic-energy operator for the P space.) Here, the effective potential for P space, $U_P^{Eff}(E)$, is formally given as

$$U_P^{eff}(E) \;=\; V_{PP} \,+\, V_{PQ}\, G_Q(E)\, V_{QP} \quad \text{with} \quad G_Q(E) \;=\; \frac{1}{E - H_{QQ}}, \tag{2}$$

where $G_Q(E)$ is the reen function for the Q space and $\{V_{XY}\}$ indicates the coupled-channel potential for the P and Q spaces with (X, Y)=P or Q.

Certainly, we can eliminate the Q-space component with the Feshbach method. However, the problem is how to express the $G_Q(E)$ in actual calculations. We overcome this matter with an interesting nature of the complex scaling method. The closure relation is proven to hold also in the CSM, including resonant states explicitly in addition to bound and non-resonant continuum states. (*Extended Closure Relation*, ECR [11]). The ECR is known to be well described approximately with a set of finite number of the eigenstates $\{|\chi_n^\theta\rangle\}$ which are obtained by the diagonalization of a complex-scaled Hamiltonian H^θ with a aussian basis function; $\sum_n |\chi_n^\theta\rangle\langle\tilde{\chi}_n^\theta| \simeq 1$ [16].

With help of the ECR, the complex-scaled reen function $G_Q^\theta(E) = (E - H_{QQ}^\theta)^{-1}$ is represented as

$$G_Q^\theta(E) \;=\; \sum_n \frac{|\chi_{Q,n}^\theta\rangle\langle\tilde{\chi}_{Q,n}^\theta|}{E - \epsilon_n^\theta} \quad \text{with} \quad H_{QQ}^\theta\,|\chi_{Q,n}^\theta\rangle \;=\; \epsilon_n^\theta\,|\chi_{Q,n}^\theta\rangle, \tag{3}$$

where the energy eigenvalue ϵ_n^θ and eigenstate $|\chi_{Q,n}^\theta\rangle$ are obtained by the diagonalization of the complex-scaled Hamiltonian H_{QQ}^θ with aussian basis functions. Since the original reen function for the Q space is obtained from $G_Q^\theta(E)$ as $G_Q(E) \;=\; U^{-1}(\theta)\, G_Q^\theta(E)\, U(\theta)$, the effective potential for the P space is

$$U_P^{eff}(E) \;=\; V_{PP} \,+\, V_{PQ}\, G_Q(E)\, U^{-1}(\theta)\, G_Q^\theta(E)\, U(\theta)\; V_{QP}. \tag{4}$$

Here, the $G_Q^\theta(E)$ is given as Eq. (3) and is represented with aussian functions. Therefore, since the effective potential $U_P^{eff}(E)$ is also composed of aussian functions, it can be used as usual in the complex scaling method with aussian basis function.

We apply this technique to the calculation of the K^-pp system. Setting the $\bar{K}N$ channel as P space and the $\pi\Sigma$ and $\pi\Lambda$ channels as Q space, we construct an effective $\bar{K}N$ potential $U_{\bar{K}N}^{eff}(E)$. In other words, the πY channels are eliminated at the step of two-body calculation. The $U_{\bar{K}N}^{eff}(E)$ plugged in the three-body Hamiltonian of the $\bar{K}NN$, we solve the single-channel problem of the $\bar{K}NN$ system with the complex scaling method.

Journal of Physics: Conference Series **569** (2014) 012084 doi:10.1088/1742-6596/569/1/012084

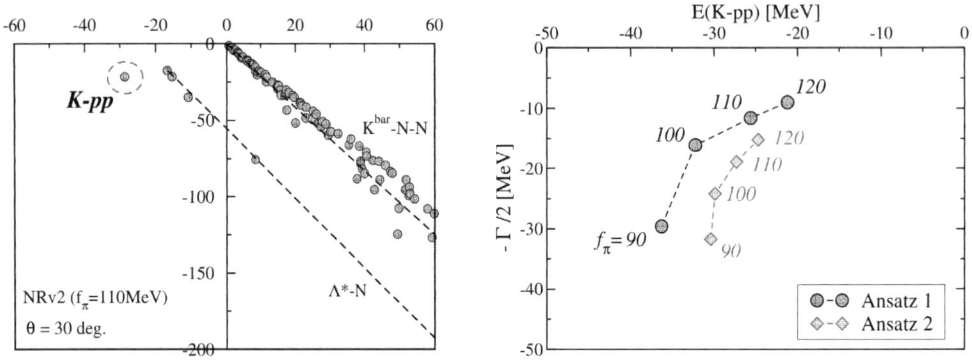

Figure 2. (Left) Complex-energy eigenvalue distribution when $\bar{K}N$ energy for the effective potential is fixed to the Λ^* energy. (Right) ole position of K^-pp obtained by self-consistent calculation, where f_π is varied from 90 MeV to 120 MeV and two ansatz for $\bar{K}N$ energy are examined.

3. Result

We show our result of the K^-pp calculated with the ccCSM+Feshbach method. Here, we employ a version of our potential, NRv2, in which the f_π value is set to be 110 MeV as a typical case.

At first, we consider the case where the $\bar{K}N$ energy in the energy-dependent $\bar{K}N$ potential is fixed to that for the $\Lambda(1405)$. In other words, the $\bar{K}N$ energy is not self-consistent in the K^-pp, but it is self-consistent for the isolated two-body system of $\bar{K}N$-$\pi\Sigma$ which forms the $\Lambda(1405)$ resonance. The obtained complex-energy eigenvalue distribution is shown in the left panel of Fig. 2. In the figure the origin of the real energy axis corresponds to the $\bar{K}NN$ three-body threshold. It is known that in the complex scaling method the continuum states appear on so-called 2θ line ($\tan^{-1}(\mathrm{Im}\,E/\mathrm{Re}\,E) = -2\theta$) when the scaling angle is θ [11]. Therefore, the eigenvalues along the 2θ line running from the origin indicate the continuum states of $\bar{K}NN$ three-body system. There are eigenvalues on the other line. The starting point of this line is $(-16.7, -17.5)$ MeV which is almost identical to the complex energy of the Λ^* [12]. (Λ^* denotes the higher pole of the $\Lambda(1405)$.) So, the eigenvalues on the second line indicate the Λ^*N two-body continuum states. We can see a point which is isolated from the two lines mentioned above, as marked with a red circle in the figure. This point means the K^-pp resonance. In this case where the $\bar{K}N$ energy is fixed at that of the Λ^*, the pole energy of the K^-pp resonance is found to be $(-B(K^-pp), -\Gamma_M/2) = (-28.6, -21.6)$ MeV.

Next, we take into account the self-consistency for the $\bar{K}N$ energy in the three-body system K^-pp, following the two ansatz as explained in the section 2.1. Searching a self-consistent solution, we have succeeded to find such a solution of the K^-pp resonance. The resonance energy is obtained to be $(-B(K^-pp), -\Gamma_M/2) = (-25.6, -11.6)$ MeV for Ansatz 1 and $(-27.3, -18.9)$ MeV for Ansatz 2. The binding energy $B(K^-pp)$ is not so dependent on the ansatz, whereas the decay width Γ_M strongly depends on it. Compared with the earlier work of the variational calculation with a chiral potential [6], the present calculation gives slightly deeper binding and narrower width.

We examine several f_π values, since the f_π value is a parameter of our $\bar{K}N$ potential. We have found the pole of the K^-pp resonance when the f_π value is varied from 90 MeV to 120 MeV, as depicted in the right panel of Fig. 2. As a result, the binding energy is found to be small, 20-35 MeV. The decay width is obtained to spread widely, 20-65 MeV. We note on the dependence of the ansatz. Both ansatz give similar binding energy. However, they give rather different decay width; Ansatz 1 tends to give small decay width, compared to Ansatz 2.

4. Summary and future plans

We have investigated the essential $\bar{K}$ cluster, K^-pp, with a coupled-channel Complex Scaling Method combined with Feshbach method. In the method, we reduce a coupled-channel problem to a single-channel problem in order to handle the system more easily, by utilizing the extended closure relation in the complex scaling method. Applying this method to the K^-pp, we have confirmed that the method works well to find the resonant pole clearly on the complex energy plane. Using a chiral SU(3)-based $\bar{K}N$ potential with a local aussian form, we have found that the binding energy and the mesonic decay width of K^-pp are roughly 20-35 MeV and 20-65 MeV, respectively. We consider that the K^-pp is shallowly bound as suggested by the early studies with variational methods using an energy-dependent chiral potential.

On the experimental side, there are several reports related to the K^-pp. According to the reports of Refs. [17, 18], if the observed state is a K^-pp state, the binding energy is more than 100 MeV. However, the situation is still controversial [19]. Recently, two experimental results on the K^-pp have been reported from J- ARC [20, 21]. Further analysis of these results will help to reveal the detailed nature of the K^-pp, compared with theoretical studies.

In the present study, we have incorporated the $\pi\Sigma$ and $\pi\Lambda$ components into the effective $\bar{K}N$ potential. In our future, we will consider the $\bar{K}NN$-$\pi\Sigma N$-$\pi\Lambda N$ coupled-channel problem without any channel elimination. By the explicit treatment of $\pi\Sigma$ and $\pi\Lambda$ channels we will obtain more accurate result on the K^-pp. In addition, the role of the πYN three-body dynamics is expected to be clarified as well as that of the $\bar{K}NN$ dynamics.

Acknowledgments

This work was developed through several activities (meetings and workshops) at the J- ARC Branch of the KEK Theory Center. This work is supported by JS S KAKENHI rant Number 25400286 and partially by rant Number 24105008.

References

[1] Isgur N and Karl G, 1978 *Phys. Rev.* D **18** 4187
[2] Hyodo T and Jido D 2012 *Prog. Part. Nucl. Phys.* **67** 55
[3] Akaishi Y and Yamazaki T 2002 *Phys. Rev.* C **65** 044005; Yamazaki T and Akaishi Y 2002 *Phys. Lett.* B **535** 70
[4] Doté A, Horiuchi H, Akaishi Y and Yamazaki T 2004 *Phys. Lett.* B **590** 51; 2004 *Phys. Rev.* C **70** 044313
[5] Gal A 2010 *Prog. Theor. Phys. Suppl.* **186** 270
[6] Doté A, Hyodo T and Weise W 2008 *Nucl. Phys.* A **804** 197; 2009 *Phys. Rev.* C **79** 014003
[7] Yamazaki T and Akaishi Y 2007 *Phys. Rev.* C **76** 045201
[8] Barnea N, Gal A and Liverts E Z 2012 *Phys. Lett.* B **712** 132
[9] Ikeda Y and ato T 2007 *Phys. Rev.* C **76** 035203
[10] hevchenko N V, Gal A, Mares J and Révai J 2007 *Phys. Rev.* C **76** 044004
[11] Aoyama , Myo T, Kato K and Ikeda K 2006 *Prog. Theor. Phys.* **116** 1
[12] Doté A, Inoue T and Myo T 2013 *Nucl. Phys.* A **912** 66
[13] uzuki Y and Varga K 1998 *Stochastic Variational Approach to Quantum-Mechanical Few-Body Problems (Lecture Notes in Physics Monographs)* (pringer)
[14] Doté A and Myo T *Preprint* ar iv:1406.1540 [nucl-th], J-PARC-TH-0037, to be published in *NPA*
[15] Feshbach H 1958 *Ann. Phys.* **5** 357; 1962 *Ann. Phys.* **19** 287
[16] Myo T, Ohnishi A and Kato K, 1998 *Prog. Theor. Phys.* **99** 801
[17] Agnello M *et al.* (FINUDA collaboration) 2005 *Phys. Rev. Lett.* **94** 212303
[18] Yamazaki T *et al.* (DI TO collaboration) 2010 *Phys. Rev. Lett.* **104** 132502
[19] Magas V K, Oset E, Ramos A and Toki H 2006 *Phys. Rev.* C **74** 025206
[20] akuma F *et al.* (J-PARC E15 Collaboration) 2014 *Acta Phys. Polon.* B **45** 767
[21] Ichikawa Y *et al.* (J-PARC E27 Collaboration) 2013 *Few-Body Syst.* **54** 1191; 2014 *PoS Hadron2013* 181; *Preprint* ar iv:1407.3051 [nucl-ex]

Journal of Physics: Conference Series **569** (2014) 012085 doi:10.1088/1742-6596/569/1/012085

Spectroscopy of charmed baryons at the J-PARC high-momentum beam line

K. Shirotori[1], T. Ishikawa[2], Y. Miyachi[3], Y. Morino[4], T. Nakano[1], M. Naruki[5], H. Noumi[1], K. Ozawa[6], Y. Sugaya[1], K. Tanida[7] and T. Yamaga[1]

[1] Research Center for Nuclear Physics (RCNP), Osaka University, Osaka 567-0047, Japan
[2] Research Center for Electron Photon cience (ELPH), Tohoku University, Miyagi 982-0826, Japan
[3] Physics Department, Yamagata University, Yamagata 990-8560, Japan
[4] RIKEN, aitama 351-0198, Japan
[5] Department of Physics, Kyoto University, Kyoto 606-8502, Japan
[6] Institute of Particle and Nuclear tudies (IPN), High Energy Accelerator Research Organization (KEK), Tsukuba 305-0801, Japan
[7] Department of Physics and Astronomy, eoul National University, eoul 151-747, Korea

E-mail: sirotori@rcnp.osaka-u.ac.jp

Abstract.
The heavy quark baryon spectroscopy is a key way to understand what the building block of hadrons is. The diquark correlation which is expectedly an essential degree of freedom to describe the hadron structure can be investigated from the spectroscopy of charmed baryons. An experiment to observe and investigate the charmed baryons was proposed at the J-PARC high-momentum beam line. The missing mass spectroscopy experiment via the $\pi^- p \rightarrow Y_c^{*+} D^{*-}$ reaction at 20 GeV c will be performed for the systematic measurement of the excitation energy, the production rates and the decay products of charmed baryons. In addition, the properties of the strange baryons can also be measured systematically by using the same experimental setup. From the systematic study of both charmed baryons and hyperons, the diquark correlation can be revealed which is expectedly an essential degree of freedom to describe the hadron structure.

1. Introduction

One of fundamental questions in the hadron physics is how hadrons are formed. The constituent quark model well describes the properties of hadrons with constituent quarks as an effective degree of freedom. The hadron properties, such as hadron mass and magnetic moments of ground state baryons, are well reproduced by the model. However, the constituent quark model doesn't have enough description for the excited states. There are many undiscovered states which are called as missing resonances. Recently, exotic hadrons, such as Θ^+ [1] pentaquark and X, Y, Z mesons [2], have been reported. The existence of those exotic states suggests that hadrons have much rich internal structure which is beyond the naive quark model. Therefore, investigations of proper effective degrees of freedom to describe hadrons are essential to understand hadron structure.

In order to answer this question, it is necessary to understand the interaction between quarks in hadrons. The quark-quark correlation, namely diquark correlation, is expected to be an

Journal of Physics: Conference Series **569** (2014) 012085 doi:10.1088/1742-6596/569/1/012085

essential degree of freedom to describe hadrons. In the case of light baryons, 3 diquark pairs are correlated each other in an equal weight so that the extraction of diquark correlation is unclear. The magnitude of the color-spin interaction between quarks is proportional to the inverse of the quark mass. When one quark in a baryon is replaced to a heavy quark, one expects that the correlation of two light quarks is stronger than that of the other pairs. As a result, a diquark correlation is expected to be isolated and developed in the baryon structure. Charmed baryons which have one heavy charm quark can provide unique opportunities to study diquark correlation.

2. Diquark correlation in charmed baryons

The nature of the diquark correlation can appear in level structure, decay branching ratios and production rates of charmed baryons. Two excitation modes, called isotope shift, emerge to the level structure. The ρ and λ modes are a rotation of the diquark and an orbital excitation between the diquark and the other heavy quark, respectively. Those excited states caused by the diquark correlation appear in the excitation spectrum. It could not be observed in light quark baryons due to the same dynamics of three diquark pairs. From the study of the excitation spectrum of charmed baryons, we can access to the ρ and λ excitation modes which are strongly related to the internal structure of the charmed baryons.

One of the approaches to reveal those excitation modes is to measure decay properties. In the case of ρ mode, the rotating diquark decays by emitting a pion. Then, the charmed baryon goes to a light-meson and a charmed baryon pair. On the other hand, in the λ mode case, the charmed baryon decays to a charmed meson and a light nucleon pair. Difference of the decay branching ratios is expected for the decay property which provides us information of the excitation modes. In addition, it is found that the production cross section is strongly related to the excitation modes. Charmed baryons are produced via the hadronic reaction, the $\pi^- \, p \rightarrow Y_c^{*+} \, D^{*-}$ reaction. For the production process, the D^* exchange can be taken into account. The generated charm quark from u quark is attached to a spectator diquark pair from the initial state in the proton wave function. Then, a charmed baryon forms from a diquark and a charm quark with an orbital excitation. In this reaction process, the production cross section can be described by the overlapping of wave functions between initial and final states. The production rate has the relation to the spin isospin configuration of charmed baryons and momentum transfer. As a result, the production cross section has strong dependence on the excitation modes of charmed baryons [3]. From the D^* exchange reaction process, the forward production can produce the λ-mode excitation because only the orbital motion between the diquark and the charm quark can be excited from the single-step reaction. In the calculation, the interaction between the diquark and the charm quark is neglected so that the ρ-mode excitation cannot be produced. The ρ-mode excitation is expected to be produced by the double-step process which includes the interaction between the diquark and the charm quark. From the measurement of the $\pi^- \, p \rightarrow Y_c^{*+} \, D^{*-}$ reaction, we can observe the charmed baryon states with the different yield according to its internal structure. The production mechanism can gives us unique information of the diquark correlation of charmed baryons.

The purposes of experiment are to observe and investigate the excited states of the charmed baryons [4]. The goal is to reveal the diquark correlation which is expected to be an essential degree of freedom to describe hadrons. The excitation energy and width are measured by the missing mass spectroscopy method by which all the excited states can be observed independent of any dependence of the final state. The measurement of the production rate which strongly related to the spin isospin configuration of charmed baryons can provide the unique information of the diquark correlation. In addition, the analysis of the decay particles and its decay chain which decays to the known states, the absolute decay branching ratio and the spin parity of the excited states could also be determined. From the systematic measurement of the excitation

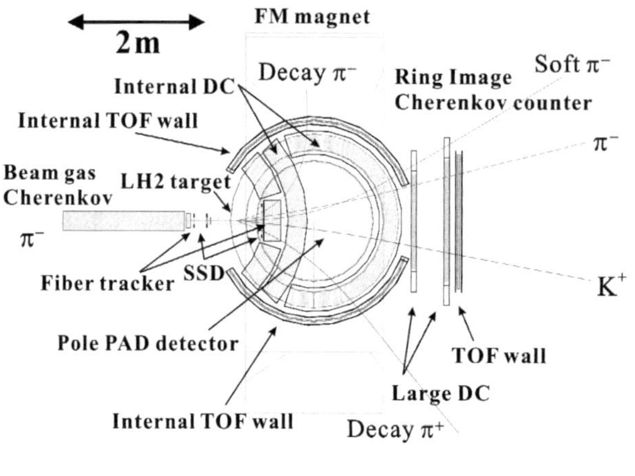

Figure 1. The schematic view of the charmed baryon spectrometer.

energy, the production cross section and the decay properties, the charmed baryon states could completely be measured.

3. Experiment at the J- ARC High-Momentum Beam Line

The experiment will be performed at the high-momentum beam line which is being constructed until the beginning of 2016 in the J- ARC hadron experimental facility [5]. The high-momentum beam line will be upgraded to unseparated secondary beams with the high-intensity of more than 10^7 spill and the momentum of up to 20 eV c. The momentum resolution of the secondary beams is expected to be less than 0.1 by using the momentum dispersive optical method. The high-momentum secondary beams with both high-intensity and high-resolution can be delivered to the experimental area.

For the charmed baryon production, the $\pi^- \, p \to Y_c^{*+} \, D^{*-}$ reaction is used with the beam momentum of 20 eV c. The decay chain of the D^{*-} meson, $D^{*-} \to \bar{D}^0 \, \pi^-$ (branching ratio of 67.7) and $\bar{D}^0 \to K^+ \, \pi^-$ (branching ratio of 3.88), is detected for measuring missing masses of produced charmed baryons. The decay products of K^+ and π^- of 2–16 eV c from $\bar{D}^0$ and π^- of 0.5–1.7 eV c from D^{*-} are detected by a spectrometer. The decay measurement can also be performed by detecting decay products from the produced charmed baryon, Y_c^{*+}, such as the $Y_c^{*+} \to \Sigma_c^{*++,0} \, \pi^{-,+}$ and $Y_c^{*+} \to p \, D^0$ channels. The recoil momentum of Y_c^{*+} is measured by the missing mass method so that the mass of the decay products ($\Sigma_c^{*++,0}$ or D^0) can be obtained as a missing mass by only detecting the emitted pion and proton with the momentum of 0.2–1.5 eV c.

Figure 1 shows a conceptual design of the charmed baryon spectrometer [6]. In the case of the fixed target experiment with the high-momentum beam, all the generated particles, not only the scattered high-momentum particles from the D^{*-} decay but also the decay products from the produced Y_c^{*+}, are scattered to the forward direction. Therefore, the dipole magnet system which commonly measures both the particles from the D^{*-} decay for the missing mass method and the decay products from the produced Y_c^{*+} for the decay measurement is used for the charmed baryon spectrometer. Since the production cross section of charmed baryons is estimated to be 10^{-4} smaller than that of the strangeness production (10–100 μb of the $\pi^- \, p \to Y^* \, K^*$ reaction) [3], the charmed baryon production cross section of 1 nb was assumed for the experimental design. To obtain the production yield, the beam intensity of 6×10^7 pulse is used. The high-rate detectors such as scintillating fiber trackers will be installed for the

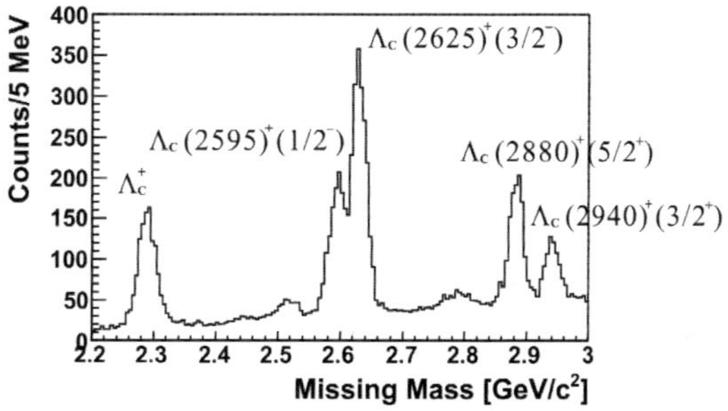

Figure 2. The expected missing mass spectra simulated with the known charmed baryon states. The production rate from the theoretical calculation [3] in the case of $\sigma(\Lambda_c^+)$ =1 nb are used for the simulation. The spin parity of $\Lambda_c(2940)$ has not been determined yet. In this simulation, it is assumed to be $J_P = 3/2^+$.

measurements of the beam momentum and profiles at focal plane and at the upstream of the experimental target, respectively. For beam particle identification, the high-rate Ring-Image Čerenkov counter will be used. The D^{*-} meson is measured by the forward detector system, the scintillating fiber trackers at the downstream of the target, and the drift chambers and the Ring-Image Čerenkov counter for detecting the high-momentum K^+ and π^- from $\bar{D}^0$ at the exit of the magnet. For measuring the slow π^- from D^{*-} at the forward region and the decay particles from Y_c^* with the wide angular coverage, it is necessary to detect the particles inside of the magnet. The tracking detectors which have the large angular acceptance are installed at the downstream of the target. The horizontal and vertical directions can be covered by using the detectors installed around the magnet pole and the pad-type detector on the magnet pole face are installed, respectively.

The acceptance for detecting the D^{*-} decay was estimated to be 50−60 for the excitation states up to 1 eV by assuming the angular distribution from the dominance of the t-channel production process. The momentum resolution of 0.2 was obtained at 5 eV c so that the invariant mass resolution for reconstructing the $\bar{D}^0$ and D^{*-} are estimated to be 5.5 MeV and 0.6 MeV, respectively. The missing mass resolution of excited states above 2.8 eV c^2 are estimated to be ~10 MeV. For the decay measurement, both the polar and azimuthal angles are completely covered more than $\cos\theta_{CM} = -0.9 \sim -0.5$ for the $\Lambda_c(2940)^+ \rightarrow \Sigma_c(2455)^{++,0}\ \pi^{-,+}$ decay modes.

Figure 2 shows an expected missing mass spectrum simulated by the realistic experimental conditions. The main strangeness production background processes were simulated by the JAM code [7], while the THIA code [8] was also used for the comparison. The most effective method of the background reduction is the D^* tagging. Both the mass region of the $\bar{D}^0$ mass and the Q-value corresponded to the $\bar{D}^{*-}$ decay are selected. By using the D^* tagging, the background reduction of 2×10^6 can be achieved from the JAM simulation result. The sensitivity of the production cross section was found to be 0.1−0.2 nb in the missing mass region of 2.2 − 3.4 eV c^2.

For simulating the missing mass spectrum, the relative production rates calculated by Ref. [3] are input. In the expected spectrum, the strong dependence of the production rate can be observed. One of the subjects is to measure the production rate of the λ-mode doublet states, such as $\Lambda_c(2595)(\frac{1}{2}^-)$ and $\Lambda_c(2625)(\frac{3}{2}^-)$ and excited states with $\Lambda_c(2880)(\frac{5}{2}^+)$ and $\Lambda_c(2940)(\frac{3}{2}^+)$ which are enhanced in the missing mass spectrum. By taken into account the λ-mode picture,

Journal of Physics: Conference Series **569** (2014) 012085 doi:10.1088/1742-6596/569/1/012085

those λ-mode doublet states have production ratios with $R(\frac{3/2^-}{1/2^-}) \sim 2$ and $R(\frac{5/2^+}{3/2^+}) \sim \frac{3}{2}$, respectively [3]. The measurement of the production rates of charmed baryons is expected to give a strong evidence of the existence of the excitation modes from the diquark correlation.

4. Systematic study with strange sector

For further understand the excitation mode, the systematic studies of both charm and strange sector are an another key way. From the studies of both sectors, the heavy quark mass dependence of the λ and ρ excitation mode can be investigated. By changing the mass of heavy quark, the level splitting of the excitation states by the spin-spin interaction are changed according to the heavy quark mass. In addition, in the strange sector, the $\lambda/\rho-$mixing due to the smaller mass of the strange quark has to be taken into account. Excited states of hyperons can be produced via the $\pi^-\ p \to Y^*\ K^*$ reaction. The t-channel $\pi^-\ p \to Y^*\ K^*$ reaction favors to produce the λ-mode. However, the $\rho-$mode states can be produced with a larger fraction compared with the charm sector from a larger fraction of the $\lambda/\rho-$mode mixing. The measurements of hyperons are inseparable for determining the $\lambda/\rho-$mode mixing ratio from the experimental data.

From the measurement of the strange sector, the production rate and the decay branching ratios of known excited states can precisely be determined due to the huge statistics from the much larger production cross section of strangeness. The $\lambda/\rho-$mixing can be determined from the strange sector. In the charm sector, the widths of excitation states are narrower than those of hyperons so that the separation of the excitation states can be much clearer than that of strange sector. In addition, the production rate is found to have strong dependence of the excitation state. The clear distinction of the λ and ρ excitation is expected in the charm sector. However, the excited states of not only the charmed baryons but also the hyperons are not completely understood because of the lack of the experimental data. For the J- ARC experiment, the systematic measurements of both charm and strange sectors are also essential to understand the internal structure of hadrons.

5. Summary

The charmed baryon spectroscopy give us a chance to understand the hadron structure. For the properties of the charmed baryon, two light quarks are isolated and form the diquark correlation due to the small color-magnetic interaction between the charm and lighter quarks. The excitation modes from the diquark correlation are reflected to the structure of the excited spectrum and the decay modes of excited states. In addition, the production cross section via the hadronic reaction gives us unique information of the charmed baryon structure. For investigating charmed baryons, we proposed a spectroscopy experiment to observe and investigate excited states of charmed baryons at the J- ARC high-momentum beam line. The systematic measurement of the excitation energy, the production rate and the decay products of charmed baryons will be performed by the spectroscopy experiment. From the systematic study of both charmed baryons and hyperons, the diquark correlation can be revealed which is expectedly an essential degree of freedom to describe the hadron structure.

[1] T. Nakano *et al.*, 2003 Phys. Rev. Lett. **91** 012002.
[2] . K. Choi *et al.*, 2003 Phys. Rev. Lett. **91** 262001.
[3] . H. Kim, A. Hosaka, H. C. Kim, H. Noumi, K. hirotori, 2014 Prog. Theor. Exp. Phys., in press.
[4] H. Noumi *et al.*, 2012 J-PARC E50 Proposal.
[5] K. H. Tanaka *et al.*, 2010 Nucl. Phys. A **835** 81.
[6] K. hirotori *et al.*, 2013 Po (Hadron 2013) 130.
[7] Y. Nara *et al.*, 2000 Phys. Rev. C **61** 024901.
[8] T. jöstrand, . Mrenna and P. kands, 2006 JHEP05 026, 2008 Comput. Phys. Comm. **178** 852.

Fragmentation in nuclear reaction and its relation to EOS

Akira Ono

Department of Physics, Tohoku University, endai 980-8578, Japan

E-mail: ono@nucl.phys.tohoku.ac.jp

Abstract. Heavy-ion collisions in the incident energy region from several ten to several hundred MeV nucleon are closely related to the properties of nuclear matter under various conditions of the density, the temperature and the neutron-proton asymmetry. The emission of a large number of clusters in collisions implies strong cluster correlations in nuclear matter that is expanding and breaking up into many fragments. The antisymmetrized molecular dynamics approach has been extended in order to describe the cluster emissions properly. Calculations show that the formation of α clusters affects strongly the observables, such as the ^{3}H/^{3}He spectrum ratio, that are usually considered as probes of the density dependence of nuclear symmetry energy.

1. Heavy-ion collisions, nuclear matter EOS, and cluster correlations

Heavy-ion collisions go through various conditions of nuclear matter depending on the incident energy and the impact parameter. In central collisions at several ten to several hundred MeV nucleon, the initially compressed system expands toward low densities where many intermediate mass fragments are formed, which is closely related to the instability of uniform matter and or the liquid-gas phase transition. It has been expected that one can explore the nuclear matter equation of state (EOS) in a wide density region from low ($\sim \frac{1}{10}\rho_0$) to high ($\sim 2\rho_0$) by carefully studying the heavy-ion collision observables, where ρ_0 is the nuclear saturation density. The knowledge of the EOS is also important for astrophysical problems such as the properties of neutron stars and the explosion mechanism of supernovae. The temperature of the order of 1-10 MeV achieved in heavy-ion collisions is similar to the situation of supernovae, while the information from heavy-ion collisions is still useful for neutron star problems as far as one can take care of the temperature dependence.

Recently much effort of experimental and theoretical studies has been made in order to get some information of the EOS of asymmetric nuclear matter by varying the neutron to proton ratio of the reaction system. The energy of nuclear matter per nucleon can be written as $E/A = E_0(\rho) + S(\rho)\delta^2 + \cdots$, where $\delta = (\rho_n - \rho_p)/\rho$ represents the neutron-proton asymmetry. If the EOS of symmetric nuclear matter is known, the symmetry energy $S(\rho)$, which is a function of the density ρ, is the key quantity to characterize the EOS of asymmetric nuclear matter. Constraints on $S(\rho)$ obtained so far from nuclear physics information are summarized in Refs. [1, 2]. The masses of neutron-rich nuclei place a stringent constraint on the value of $S(\rho_1)$ at $\rho_1 \approx \frac{2}{3}\rho_0$, while the E1 strength distribution (such as the electric dipole polarizabilty) is sensitive to the parameter $L = 3\rho_0(dS/d\rho)_{\rho_0}$ which characterizes the density dependence of the symmetry energy. Analyses of heavy-ion observables have also placed some constraints in the plane of L and $S(\rho_0)$. The representative is that from the isospin diffusion analysis of the exchange of neutrons and protons between the projectilelike and targetlike parts in semiperipheral collisions of nuclei with different neutron-to-proton ratios [3]. It seems that these constraints are consistent with the ones from nuclear structure information. However, these relatively reliable observables are constraining $S(\rho)$ only at subsaturation densities $\frac{1}{3}\rho_0 \lesssim \rho \lesssim \rho_0$. It is important to obtain some information at higher densities $\rho \sim 2\rho_0$ in near future from experiments at RIBF RIKEN and future facilities.

Nuclear matter information at much lower densities $\rho \sim \frac{1}{10}\rho_0$ have been extracted by a Texas A&M group by analyzing the yields of clusters emitted from the intermediate velocity source in heavy-ion collisions [4, 5, 6]. Their essential assumption is that the clusters with the same velocity (at the surface of the source) were emitted simultaneously from an equilibrated source which is expanding. They have extracted the evolution of the temperature as a function of the density, the symmetry energy at low densities, the binding energies of clusters in medium, and so on. Some results are compared with the calculation that explicitly takes into account the clusters [7, 8]. These studies clearly show that cluster correlations play essential roles in low-density nuclear matter EOS. It should be noted that, due to the existence of cluster correlations, the low-density EOS exhibits nontrivial dependence on ρ, δ and T that cannot be simply represented by $S(\rho)$.

Isospin diffusion in collisions at around 50 MeV nucloen is known to be sensitive to the symmetry energy at higher subsaturation densities $\frac{1}{3}\rho_0 \lesssim \rho \lesssim \rho_0$ [3]. An interesting and important question is whether cluster correlations are important in the isospin diffusion reactions. Coupland et al. [9] compared two BUU calculations with and without cluster correlations, and found that the fragmentation dynamics and the degree of isospin diffusion depend on the existence of cluster correlations. In this approach with clusters [10], collision terms couple the BUU equations for the phase-space distribution functions of different species of clusters with $A \leq 3$. This example shows that the study of nuclear matter properties should also treat clusters carefully to obtain a robust constraint on EOS.

Collisions at higher incident energies of several hundred MeV nucleon are generally suitable to get information of high density EOS since the system is compressed up to a high density ($\sim 2\rho_0$) in the early stage of a reaction. Calculations have shown that the neutron-to-proton ratio in the high density part of the compressed system is sensitive to the symmetry energy $S(\rho)$ at high densities in neutron-rich systems. The yield ratio of charged pions π^-/π^+ has been proposed as a probe for this high density effect, since many of the pions originate from this high density part where Δ resonances are produced by two-nucleon collisions [11, 12]. Currently, different transport model calculations resulted in contradicting conclusions on the high density symmetry energy [13, 14], based on the same existing FO I data [15]. Theoretical work is required to resolve the origin of this contradiction, while experimental data for collisions with various neutron-to-proton ratios will be useful to clarify the situation. The study of pions should be done together with the study of other observables, such as the neutron and proton flows, to obtain a consistent picture. It should also be remarked that, even in these high energy collisions, a major part of nucleons are bound in clusters at the end of the collisions [16], and therefore cluster formation may influence the global expansion dynamics and the emissions of particles.

Thus the formation and or existence of clusters seem to be essential not only for themselves but also for the study of nuclear matter EOS in many situations. It may be required that transport models should be able to describe clusters in order to obtain reliable conclusions on EOS from heavy-ion collisions. In the following, we will review the basic idea to treat cluster correlations in the antisymmetrized molecular dynamics (AMD) approach. The composition of clusters is investigated in the calculation of heavy-ion collisions, which suggests that the α cluster formation should be carefully studied in order to get the symmetry energy information

Journal of Physics: Conference Series **569** (2014) 012086 doi:10.1088/1742-6596/569/1/012086

from typical observables such as the ^{3}H/^{3}He ratio.

2. Cluster correlations in AMD

2.1. Basic version of AMD

AMD [17] solves the time evolution of many-nucleon system starting with two boosted nuclei with a given impact parameter and a suitable distance between them. To describe the state at each time, AMD employs a single Slater determinant of aussian wave packets for A nucleons

$$\langle \mathbf{r}_1 \ldots \mathbf{r}_A | \Phi(Z) \rangle \propto \det_{ij} \Big[\exp\Big\{ -\nu(\mathbf{r}_i - \mathbf{Z}_j/\sqrt{\nu})^2 \Big\} \chi_{\alpha_j}(i) \Big], \tag{1}$$

where χ_{α_i} are the spin-isospin states with $\alpha_i = p\uparrow, p\downarrow, n\uparrow$, or $n\downarrow$. Thus the many-body state $|\Phi(Z)\rangle$ is parametrized by a set of complex variables $Z = \{\mathbf{Z}_i\}_{i=1,\ldots,A}$. The width parameter $\nu = (2.5\ \mathrm{fm})^{-2}$ is treated as a constant parameter common to all the wave packets. The time evolution of the wave packet parameters Z is determined, up to the extensions described later, by applying the time-dependent variational principle from which the equation of motion for Z is obtained, $\frac{}{t}\mathbf{Z}_i = \{\mathbf{Z}_i, \mathcal{H}\}_{\mathrm{PB}}$ with suitably defined oisson brackets. The Hamiltonian $\mathcal{H}$ is the expectation value of the Hamiltonian operator with an additional correction for the spurious zero-point energies of the center-of-mass motions of fragments [17]. The effective interactions for the mean-field calculations, such as the ogny force and the Skyrme force, have been usually employed. The calculated results presented in this paper were obtained with the Skyrme SLy4 force [18] unless otherwise mentioned. The equation of motion can be interpreted intuitively as representing the motion of individual wave packets in the mean-field potential.

In addition to the mean-field effect, the two-nucleon collisions play important roles. This effect is treated as a stochastic transition form an AMD state to one of the possible other AMD states,

$$W_{i\to f} = \frac{2\pi}{\hbar} |\langle \Psi_f | V | \Psi_i \rangle|^2 \delta(E_f - E_i). \tag{2}$$

When the nucleons 1 and 2 collide in the initial state $|\Psi_i\rangle$ that has corresponding physical coordinates $W_i = \{\mathbf{W}_1, \ldots, \mathbf{W}_A\}$ [17], the set of final states $\{|\Psi_f\rangle\}$ is usually constructed by changing only the two physical coordinates $W_f = \{\mathbf{W}'_1, \mathbf{W}'_2, \mathbf{W}_3, \ldots, \mathbf{W}_A\}$, under the constraints of the momentum and energy conservation and the auli principle. The transition probability is given based on the differential cross section $(d\sigma/d\ )_{\mathrm{NN}}$ of the two-nucleon scattering, which depends on the scattering energy and may be modified in nuclear medium. In the calculations presented here, we employ the in-medium density-dependent total cross sections by Li and Machleidt [19] for the scattering energies $E > 35$ MeV in the nucleon-nucleon center-of-mass system, while the cross sections at $E = 35$ MeV are employed at lower energies. Angular distributions are assumed to be similar to the experimental data in the free space.

This basic version of AMD, however, cannot well describe fragmentation reactions in most cases except for very light systems. For example [20], in central e + Sn collisions at 50 MeV nucleon, the calculation often produces two heavy fragments which are remnants of initial nuclei but do not exist in experimental data. Too many protons and too few α particles are emitted compared with the data.

2.2. Consideration on cluster correlations

In low density nuclear matter, clusters such as deuterons and α particles can exist as if they have a bound intrinsic state as shown by some calculations [10, 8, 7]. The existence of a bound cluster depends on the density and the temperature of the medium, and the momentum of the cluster relative to the medium.

Figure 1 shows the internal binding energy of an α cluster that is placed in a ^{124}Sn nucleus. The center-of-mass of the AMD wave function for the ^{124}Sn bound state is placed at the origin

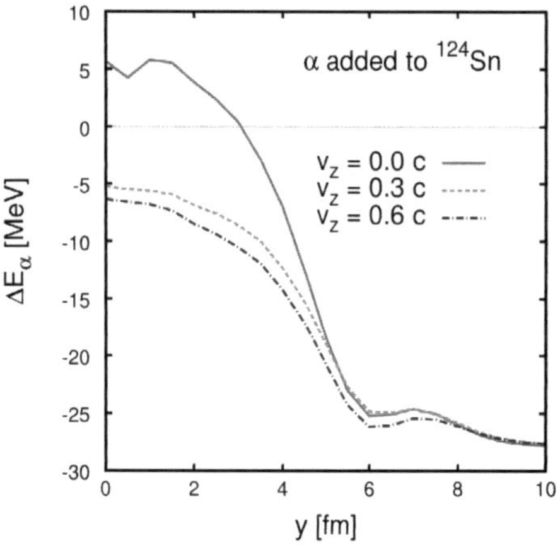

Figure 1. Internal binding energy of an α cluster placed in the ground state ^{124}Sn nucleus as a function of the position $(0, y, 0)$ and the velocity $(0, 0, v_z)$ of the cluster relative to the center of the nucleus. See the text for the definition of the binding energy.

in the phase space, and an α cluster, that is represented by four aussian wave packets with different spin and isospin states, is placed at $(0, y, 0)$ and $(0, 0, v_z)$ in the coordinate and velocity spaces, respectively. The total wave function is always antisymmetrized. The energy of this state E_α is evaluated relative to the energy of the ^{124}Sn ground state. Similarly single-nucleon energies ($E_{p\uparrow}$, $E_{p\downarrow}$, $E_{n\uparrow}$ and $E_{n\downarrow}$) are evaluated by adding a nucleon at $(0, y, 0)$ and $(0, 0, v_z)$ to ^{124}Sn. The α binding energy B_α is defined by $-B_\alpha = \Delta E_\alpha = E_\alpha - (E_{p\uparrow} + E_{p\downarrow} + E_{n\uparrow} + E_{n\downarrow})$ as a function of y and v_z. The calculated result shows that the α cluster at rest ($v_z = 0$) has a positive binding energy ($\Delta E_\alpha < 0$) at the surface of the nucleus and outside it, while the α binding energy is always positive ($\Delta E_\alpha < 0$) if a su cient momentum is given as in the cases of $v_z = 0.3c$ and $0.6c$ in the figure.

The above result suggests that the AMD wave function is suitable for the description of cluster correlations. However, this is not su cient for the proper description of emergence of cluster correlations in nuclear reactions. The probability of formation of bound clusters is dominantly influenced by the phase space or the density of states. When a bound state exists, the correct density of states for the internal state of the subsystem contains the bound state contribution as $D(E) = \delta(E - \Delta E_\alpha) + D_{\text{continuum}}(E)$. On the other hand, when the AMD equation of motion $\frac{}{t}\mathbf{Z}_i = \{\mathbf{Z}_i, \mathcal{H}\}_{\text{PB}}$ is solved (with the usual two-nucleon collisions included), the system tends to follow the classical phase space $\tilde{D}(E)$. If the cluster binding energy is correctly described by AMD, then $\tilde{D}(E) = 0$ for $E \leq \Delta E_\alpha$ and it starts at $E = \Delta E_\alpha$ continuously. The probability of finding a cluster is related to the bound phase-space volume $\int_{\Delta E_\alpha}^{0} \tilde{D}(E)dE$ which is likely much smaller than the correct quantum-mechanical value $\int_{\Delta E_\alpha}^{0} D(E)dE = 1$. This is probably the reason why the usual AMD calculations underestimate the yield of clusters.

2.3. AMD with cluster correlations

uided by the above consideration, we try to recover the correct quantum-mechanical probability of cluster formation by choosing the set of the final states $\{|\Psi_f\rangle\}$ of each two-nucleon collision in such a way that the configuration of bound cluster formation is explicitly included in the set [20]. We assume here that clusters with $A = 2, 3$ and 4 should be treated explicitly.

To obtain the transition probabilities to clusterized states, we employ an approximation similar to Ref. [10]. As one of possible final states $\{|\Psi_f\rangle\}$ for a collision of two nucleons N_1 and N_2 with the initial relative velocity v_{NN}, let us consider a case that N_1 (N_2) forms a cluster C_1 (C_2) with another nucleon B_1 (B_2) in the final state. The partial differential cross section to this final channel is given by

$$v_{\mathrm{NN}} d\sigma(\mathrm{N_1B_1N_2B_2} \to \mathrm{C_1C_2}) = \frac{2\pi}{\hbar} |\langle\varphi_1'|\varphi_1^{+\mathbf{q}}\rangle|^2 \, |\langle\varphi_2'|\varphi_2^{-\mathbf{q}}\rangle|^2 \, |M|^2 \, \delta(\mathcal{H} - E) \frac{p_{\mathrm{rel}}^2 dp_{\mathrm{rel}} d}{(2\pi\hbar)^3}, \quad (3)$$

where M is the matrix element for the two-nucleon scattering to the final state with the relative momentum p_{rel} and the scattering angle in the two-nucleon center-of-mass system. The overlap matrix $\langle\varphi_1'|\varphi_1^{+\mathbf{q}}\rangle$ is taken between $|\varphi_1^{+\mathbf{q}}\rangle = e^{i\mathbf{q}\cdot\mathbf{r}_1}|\varphi_1\rangle$ and $|\varphi_1'\rangle$, where $|\varphi_1\rangle$ and $|\varphi_1'\rangle$ are the initial and final states of the $\mathrm{N_1} + \mathrm{B_1}$ system, respectively, and the operator $e^{i\mathbf{q}\cdot\mathbf{r}_1}$ gives the momentum transfer to the nucleon $\mathrm{N_1}$. The clusterized states $|\varphi_1'\rangle$ and $|\varphi_2'\rangle$ are approximated by the simple harmonic oscillator $(0s)^n$ configuration with the oscillator constant associated with the wave packet width ν in AMD, so that any final state of the collision is represented by an AMD wave function. By identifying the two-nucleon matrix element M with that of the usual two-nucleon scattering, the cluster-forming cross section can be expressed by using the two-nucleon collision cross section that is an input to the calculation.

The actual situation of a two-nucleon collision requires more considerations because there are many possible ways of forming a cluster with different nucleons (B's) in the system for each of the scattered nucleons $\mathrm{N_1}$ and $\mathrm{N_2}$. It is important to take care of the non-orthogonality of the final states corresponding different ways of cluster formation.

The procedure is repeated for the cluster formation with nucleons (B's) with other spin and isospin directions until formation of an α cluster is considered. The particle N should be regarded as a cluster, instead of a nucleon, if a (sub)cluster has been already formed in previous steps of the repetition.

Many of light nuclei (Li, Be *etc.*) have only one or a few bound states which may be regarded as bound states of internal clusters. Then the same issue exists about the mismatch of classical and quantum-mechanical density of states. Therefore, for a better description, inter-cluster correlation is introduced as a stochastic process of bonding clusters. The relative momentum between clusters is replaced by zero if moderately separated clusters ($R_{\mathrm{rel}} < 5$ fm) are moving away from each other with a small relative momentum ($\mathbf{R}_{\mathrm{rel}} \cdot {}_{\mathrm{rel}} < 0$ and $P_{\mathrm{rel}}^2/2\mu < 8$ MeV).

3. Cluster emission in fragmentation reactions

This version of the AMD approach with cluster correlations has been shown to be able to describe the important features in multifragmentation reactions at $E \gtrsim 50$ MeV nucleon reasonably well, including the yields of protons, clusters and heavier nuclei [20]. In fact, the effect of cluster correlations is so strong that it changes not only the yields of emitted clusters but also the whole dynamics of expansion and fragmentation of the system.

We choose here a typical central $^{124}\mathrm{Sn} + {}^{124}\mathrm{Sn}$ collisions at 50 MeV nucleon. Such a neutron-rich system is suitable to see the effect of the density dependence of the symmetry energy or the parameter $L = 3\rho_0(dS/d\rho)_{\rho_0}$. In the early stage of collisions, the difference of the calculated neutron and proton densities in the high density region shows a very clear effect of the density dependence of the symmetry energy. In the late stage of the reaction, the calculation clearly shows that the symmetry energy influences the degree of fractionation, i.e., how much the gas part is neutron rich compared with the liquid part of the system. Fragment nuclei with $Z \gtrsim 3$ are usually regarded as belonging to the liquid part, while the gas part consists of emitted nucleons and light clusters. The effects in some final observables are shown in Ref. [21]. Direct information of the liquid part may be obtained from the isotope distribution of fragment nuclei. The width of the isotope distribution has been found to be sensitive to the symmetry energy parameter L, as well as the mean value of the distribution. However, in order to obtain a robust constraint on the L parameter from the comparison with experimental data, it is necessary to better understand the decay of primary fragments. On the other hand, direct information of the gas part should be observed in the yields and spectra of nucleons and light clusters. In fact,

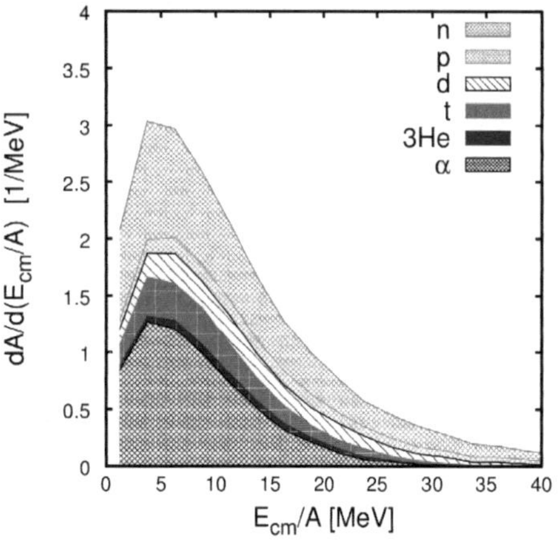

Figure 2. Calculated distributions of the kinetic energy per nucleon $E_{\rm cm}/A$ for the nucleons and clusters emitted in transverse directions $70^\circ < \theta_{\rm cm} < 110^\circ$ in central ^{124}Sn + ^{124}Sn collisions at 50 MeV nucleon. Each colored area shows the number of the specific cluster multiplied by the cluster mass number.

the dependence of the calculated ^{3}H and ^{3}He spectra on the L parameter can be understood as a consequence of fractionation that is sensitive to the symmetry energy at low densities [21].

In order to better understand the composition of the gas part, Fig. 2 shows the distributions of the kinetic energy per nucleon $E_{\rm cm}/A$ for the nucleons and clusters emitted in transverse directions $70^\circ < \theta_{\rm cm} < 110^\circ$. Each colored area shows the number of the specific cluster multiplied by the cluster mass number A, so that the figure shows the decomposition of nucleons into clusters for each emission velocity. Except for the high velocity part, the α cluster formation is quite important since about half of gas nucleons are bound in α particles. The gas part is generally much more neutron-rich than the total system as a consequence of fractionation. For the lowest velocity part $0 < E_{\rm cm} < 5$ MeV, the neutron-to-proton ratio is $N_{\rm gas}/Z_{\rm gas} = 2.07$ for this calculated case. As the α particles contain equal numbers of protons and neutrons, the gas part becomes more neutron-rich if the α particles are excluded. In fact, the neutron-to-proton ratio for $A \le 3$ particles is $(N_{\rm gas} - 2M_\alpha)/(Z_{\rm gas} - 2M_\alpha) = 3.92$ in this case. This is the reason of the large n p and ^{3}H/^{3}He ratios. In the calculated result with the SLy4 interaction ($L = 46$ MeV), the ^{3}H/^{3}He is about 2 for large velocity ($E_{\rm cm}/A \sim 20$ MeV) while it becomes much larger (about 5-6) for low energies $0 < E_{\rm cm}/A < 5$ MeV. This behavior is qualitatively similar to the experimental data from MSU [22] but the ratio observed in experiment is larger than the calculated ratio at low energies. However, it should be kept in mind that the ^{3}H/^{3}He ratio is influenced not only by the density dependence of the symmetry energy but also by the formation of α particles as mentioned above. In fact, a 10 change of α multiplicity for the low energy part will result in a 15 change of the ^{3}H/^{3}He ratio, if the total numbers of gas protons and neutrons do not change. Therefore, a careful comparison of α particle spectra between theory and experiment is quite important.

4. Summary

Clusters are important for the study of heavy-ion collisions to explore EOS, because formation and existence of light clusters influence very much the global reaction dynamics and the bulk nuclear matter properties. AMD has been extended to include cluster correlations in the final states of two-nucleon collisions. Some cluster observables, such as ^{3}H/^{3}He ratio, are sensitive to the density dependence of symmetry energy. It is important to describe precisely the α -particle formation for the study of nuclear symmetry energy.

Acknowledgments
This work was supported by JS S KAKENHI rant Numbers 21540253 and 24105008. The numerical calculations were partly carried out on SR16000 at IT in Kyoto University.

References

[1] Tsang M B *et al* 2012 *Phys. Rev.* C **86** 015803
[2] Horowitz C J, Brown E F, Kim Y, Lynch W G, Michaels R, Ono A, Piekarewicz J, Tsang M B and Wolter H H 2014 *J. Phys. G: Nucl. Part. Phys.* **41** 093001
[3] Tsang M B, Zhang Y, Danielewicz P, Famiano M, Li Z, Lynch W G and teiner A W 2009 *Phys. Rev. Lett.* **102** 122701
[4] Qin L *et al* 2012 *Phys. Rev. Lett.* **108** 172701
[5] Wada R *et al* 2012 *Phys. Rev.* C **85** 064618
[6] Hagel K *et al* 2012 *Phys. Rev. Lett.* **108** 062702
[7] Typel , Röpke G, Klähn T, Blaschke D and Wolter H H 2010 *Phys. Rev.* C **81** 015803
[8] Röpke G 2011 *Nucl. Phys.* A **867** 66
[9] Coupland D D , Lynch W G, Tsang M B, Danielewicz P and Zhang Y 2011 *Phys. Rev.* C **84** 054603
[10] Danielewicz P and Beartsch G F 1991 *Nucl. Phys.* A **533** 712
[11] Li B A, Yong G C and Zuo W 2005 *Phys. Rev* C **71** 014608
[12] Ferini G, Gaitanos T, Colonna M, Di Toro M and Wolter H H 2006 *Phys. Rev. Lett.* **97** 202301
[13] iao Z, Li B A, Chen L W, Yong G C and Zhang M 2007 *Phys. Rev. Lett.* **102** 062502
[14] Feng Z Q and Jin G M 2010 *Phys. Lett.* B **683** 140
[15] Reisdorf W *et al* 2007 *Nucl. Phys.* A **781** 459
[16] Reisdorf W *et al* 1997 *Nucl. Phys.* A **612** 493
[17] Ono A, Horiuchi H, Maruyama T and Ohnishi A 1992 *Prog. Theor. Phys.* **87** 1185
[18] Chabanat E, Bonche P, Haensel P, Meyer J and chaeffer R 1998 *Nucl. Phys.* A **635** 231
Chabanat E, Bonche P, Haensel P, Meyer J and chaeffer R 1998 *Nucl. Phys.* A **643** 441(E)
[19] Li G Q and Machleidt R 1993 *Phys. Rev.* C **48** 1702
Li G Q and Machleidt R 1994 *Phys. Rev.* C **49** 566
[20] Ono A 2013 *J. Phys.: Conf. Ser.* **420** 012103
[21] Ono A 2013 *J. Phys.: Conf. Ser.* **436** 012068
[22] Liu T *et al.* 2012 *Phys. Rev.* C **86** 024605

Journal of Physics: Conference Series **569** (2014) 012087 doi:10.1088/1742-6596/569/1/012087

Spin-*M1* and *E1* responses of nuclei probed by proton inelastic scattering

Atsushi Tamii[1] and Hiroaki Matsubara[1,2]

[1] Research Center for Nuclear Physics (RCNP), Osaka University, 10-1 Mihogaoka, Ibaraki 567-0047, Japan

E-mail: tamii@rcnp.osaka-u.ac.jp

Abstract. We pick up two studies on the nuclear responses from the recent experiments of high-resolution proton inelastic scattering at the Research Center for Nuclear Physics, Osaka University; 1) study of the nuclear symmetry and the neutron skin thickness by the measurement of energy electric dipole (*E1*) response of ^{208}Pb, and 2) study of the tensor correlation in the ground state by the measurement of the spin-*M1* responses of even-even self-conjugate nuclei in the *sd*-shell nuclei.

1. Electric dipole response of ^{208}Pb and the symmetry energy

The electric dipole (*E1*) response is one of the fundamental responses of atomic nuclei to the electromagnetic external field. The electric giant dipole resonances (GDR) was the first observed [1] collective motions in nuclei after prediction by Migdal [2]. GDR is described by a picture, given by Goldhaber and Teller [3], as a relative dipole oscillation between the protons and the neutrons. The *E1* strength above and below the neutron separation energy (S_n) were studied intensively by (γ, xn) and nuclear resonance fluorescence (NSR) measurements, respectively. However, the strength around the neutron threshold could not been studied well. Recently *E1* strength concentration around the neutron threshold has been found in neutron rich heavy nuclei. The strength is called the low-energy dipole resonance, or often pygmy dipole resonance (PDR), and is attracting a lot of experimental and theoretical studies. In several theoretical models, the PDR is visualized as a collective oscillation of the neutron skin against the core. Thus the PDR naturally carries information on the neutron skin thickness and isospin dependence of nuclear equation of state, although the structure of the PDR is still under much discussion.

The electric dipole polarizability is another promising quantity to study the neutron skin thickness and the isospin dependence of nuclear equation of state, especially for the slope parameter of the symmetry energy term [4, 5]. The electric dipole polarizability is defined as an inversely energy weighted sum-rule of the *E1* reduced transition probability $B(E1)$. Thus the determination of the complete $E1$ strength distribution is required including the region of neutron separation energy. In addition concentration of magnetic dipole ($M1$) excitation also exists in the same excitation energy region [6]. Hence the decomposition of the $E1$ and $M1$ strengths is crucial for the study.

[2] Present address: National Institute of Radiological Sciences (NIRS), Chiba 263-8555, Japan

We have measured proton inelastic scattering from ^{208}Pb at very forward angles including zero degrees at the Research Center for Nuclear Physics, Osaka University for studying the $E1$ strength distributions in ^{208}Pb. By applying missing mass spectroscopy without detecting decaying particles or photons, the $E1$ strengths could be determined from low (5 MeV) to high (20 MeV) excitation energies regardless of the neutron threshold. The forward scattering angle is essential to enhance the $E1$ Coulomb excitation. The spin-$M1$ excitation strengths were separated out from $E1$ by using polarization transfer coefficients or by angular distributions at angles close to zero degrees. The details of the experimental method and the analysis can be found in Refs. [7, 8, 9].

The overall $B(E1)$ distribution determined by the measurement is shown in Fig. 1. The bump centered at ~13 MeV corresponds to the GDR and the strength concentration around $7-9$ MeV corresponds to the pygmy dipole resonance (PDR). The $B(E1)$ strength distribution of ^{208}Pb has been determined from 5 to 20 MeV which fully covers the PDR and GDR regions, as well as the region of the neutron separation energy. The obtained $B(E1)$ strength distribution was integrated with the inverse weight of the excitation energy and yielded a dipole polarizability up to 20 MeV. By including the gamma absorption data above 20 MeV [11], the dipole polarizability of ^{208}Pb up to 130 MeV has been determined as $\alpha_D = 20.1 \pm 0.6$ fm^3 [8].

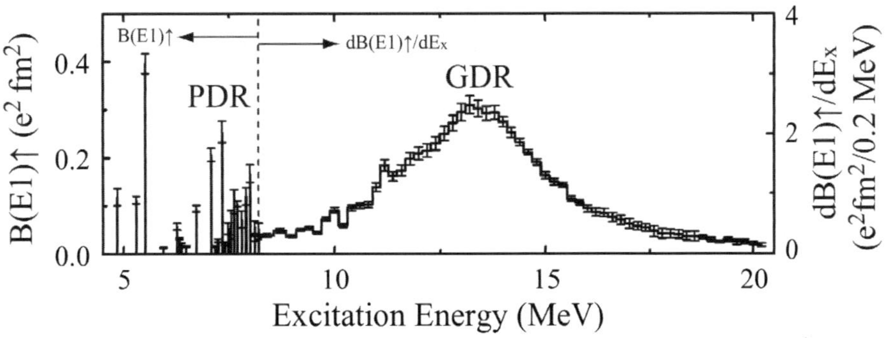

Figure 1. $B(E1)$ distribution in ^{208}Pb deduced from the proton inelastic scattering measurement at forward angles. The bump centered at ~13 MeV corresponds to the giant dipole resonance, and the strength concentration at around 7-9 MeV to the pygmy dipole resonance. The figure has been adapted from a figure in Ref. [10].

A linear correlation between the slope parameter (L) of the symmetry energy and the neutron skin thickness of ^{208}Pb is predicted by self-consistent mean field model calculations with various sets of interaction parameters [12, 13]. Here, the neutron skin thickness Δr_{np} is defined as the difference of the root-mean-square radii of the neutrons and the protons. Thus, experimental data on the neutron skin thickness may provide constraints on the slope parameter.

Roca-Maza *et al.* [5] have found that $\alpha_D J$ has good correlation with Δr_{np}. The correlation could be naturally explained by a macroscopic droplet model. Adopting the measured α_D and $J = 31 \pm (2)_{\rm est}$ MeV as a realistic range of values for the symmetry energy [14, 15], they extracted the constraint on the neutron skin thickness of ^{208}Pb as [5]

$$\Delta r_{np} = 0.165 \pm (0.009)_{\rm expt} \pm (0.013)_{\rm theor} \pm (0.021)_{\rm est}\ {\rm fm}, \tag{1}$$

where "est" uncertainty is associated with the estimates on J.

Roca-Maza *et al.* have also shown a strong correlation between $\alpha_D J$ and L [5]. By adapting the theoretical correlation we have extracted a constraint band in the J-L plane of the symmetry energy from the measured dipole polarizability data. We have taken the quadratic sum of the 99.9% theoretical confidence level band from the work by Roca-Maza *et al.* [5] and the one-sigma experimental uncertainty of α_D [8]. The result is plotted in Fig. 2 for the one-sigma uncertainty in comparison with various constraints from other works. These include heavy ion collisions (HIC), pygmy dipole resonance (PDR), isobaric analog states (IAS), nuclear mass formula with finite range droplet model (FRDM), analysis of neutron star observation data (n-star). A throughout discussion of these methods and the corresponding references are given in Ref. [14]. Additionally, constraints from a chiral effective field theory (χEFT) calculation including $3N$ forces [16] and a quantum Monte-Carlo (QMC) calculation [17] are shown together. This work was performed by the RCNP-E282 collaboration.

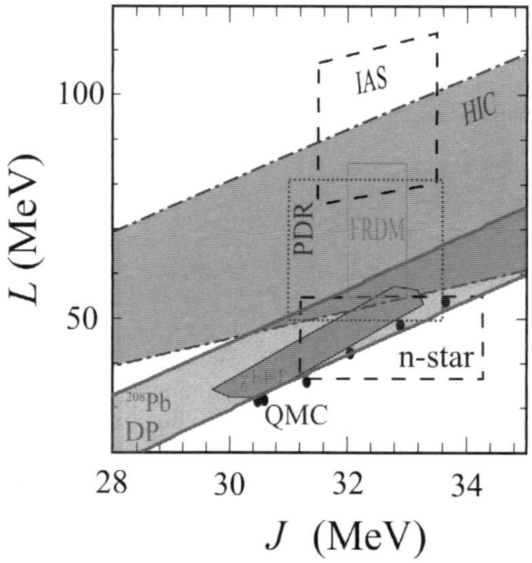

Figure 2. Constraints on the symmetry energy parameters J and L from various methods. Besides the present work based on the ^{208}Pb dipole polarizability (DP), the results in this figure are taken from Ref. [14] except the ones labeled (χEFT) from Ref. [16] and QMC (solid circles) from Ref. [17]. The figure has been adapted from a figure in Ref. [10].

2. Isoscalar and Isovector spin-$M1$ responses and the tensor correlation in the ground state

The tensor interaction between two nucleons is playing an essential role to bound atomic nuclei. In fact, a deuteron is the only bound two nucleon system owing to the tensor interaction. The tensor interaction gives rise to a correlation between nucleons in nucleus, *tensor correlation.* The tensor correlation is quite important to study the properties and structures of nuclei.

Experimental evidence on the tensor correlation in nuclei is quite limited. Due to the short-range nature of the tensor interaction, a nucleon wave-function component with high-momentum is induced especially for a pair of a proton (p) and a neutron (n). The high-momentum component has been probed by electron scattering [18], (p, d) reaction [19], or by $(p, 2p)$ reaction [20]. In this article, we show a completely different experimental approach to probe the tensor correlation in nuclear ground states. We focus on the correlation between spins of protons and neutrons in nuclei, which could be a clear signature of the tensor correlation.

We define a total proton (neutron) spin operator $\vec{S}_p(n)$ as

$$\vec{S}_{p(n)} = \sum_{i}^{Z(N)} \vec{s}_{p(n),i}, \tag{2}$$

where $\vec{s}_{p(n),i}$ is the spin operator of the ith proton (neutron). We define a p-n spin-correlation operator as $\vec{S}_p \cdot \vec{S}_n$, and a p-n spin-correlation function as an expectation value of the operator for the ground state, $< \vec{S}_p \cdot \vec{S}_n >$. The p-n spin-correlation function is considered to take a positive value for the tensor-correlated components, while it vanishes for the case without tensor-correlation.

The following equations can be derived by applying the completeness relation,

$$< (\vec{S}_p - \vec{S}_n)^2 > \; = \; \frac{1}{4}\sum |M(\vec{\sigma}\tau_z)|^2 \tag{3}$$

$$< (\vec{S}_p + \vec{S}_n)^2 > \; = \; \frac{1}{4}\sum |M(\vec{\sigma})|^2, \tag{4}$$

where $M(\vec{\sigma}\tau_z)$ and $M(\vec{\sigma})$ are isovector and isoscalar spin-matrix elements, respectively, and the sum is taken for the matrix elements between the ground state and all the excited states. The p-n spin-correlation function can be written by the spin-matrix elements as

$$\begin{aligned} < \vec{S}_p \cdot \vec{S}_n > \; &= \; \frac{1}{4}\{< (\vec{S}_p + \vec{S}_n)^2 - (\vec{S}_p - \vec{S}_n)^2 >\} \\ &= \; \frac{1}{16}\{\sum |M(\vec{\sigma})|^2 - \sum |M(\sigma\vec{\tau}_z)|^2\}. \end{aligned} \tag{5}$$

In the case of even-even self-conjugate (N=Z) nuclei in the sd-shell, the ground state spin-parity and isospin is 0^+ and 0, respectively, without exception. Thus the first sum of Eq. (5) will be naturally taken for isoscalar (T=0) 1^+ excited states and the second sum for isovector (T=1) 1^+ excited states, since all the other matrix elements vanish, assuming the isospin as a good quantum number.

We have measured the the isoscalar and isovector 1^+ excitations of even-even self-conjugate nuclei in the sd-shell by high-resolution proton inelastic-scattering measurements to obtain the spin-M1 matrix elements [7, 21, 22]. The sum of the spin-M1 strengths have been taken up to 16 MeV excitation energy. The details of the analysis can be found in Ref. [23].

Although still preliminary, the experimental results showed that the p-n spin correlation function takes positive values around 0.1 for the measured nuclei of ^{24}Mg, ^{28}Si, ^{32}S, and ^{36}Ar, showing a positive signature of the tensor correlation in the ground state. In contrast a shell-model calculation with USD-A or USD-B effective interactions [24] with bare spin-operators predicts the p-n spin-correlation function of close to 0. The predictions do not improve even by using effective spin-operators. It is intuitive to compare the result with the predictions for the case of ^{4}He nucleus, since high-precision calculations with realistic nucleon-nucleon interactions are available for ^{4}He. A global vector method has been applied [25]. The calculated p-n spin correlation function with the AV8', G3RS and Minesota potentials are 0.135, 0.109 and -0.020, respectively. The predictions of the former two potentials, which incorporates the tensor interaction, are positive and close to 0.1 , while the prediction of the last potential, which does not incorporates the tensor interaction, is close to zero. Further analysis is still in progress. This work was performed by the RCNP-E299 collaboration.

Acknowledgments

We are indebted to the RCNP accelerator staff and operators for providing the excellent beams. We appreciate the discussions with W. Horiuchi. The above works were supported in part by JSPS with Grant Numbers of 20302804 and 25105509.

Journal of Physics: Conference Series **569** (2014) 012087 doi:10.1088/1742-6596/569/1/012087

References

[1] Baldwin G C, Klaiber G S 1947 *Phys. Rev.* **71** 3
[2] Migdal A B 1944 *J. Phys. Acad. Sci. USSR* **8** 331
[3] Goldhaber M and Teller E 1948 *Phys. Rev.* **74** 1046
[4] Reinhard P-G, Nazarewicz W 2010 *Phys. Rev.* C **81** 051303(R)
[5] Roca-Maza X, Brenna M, Colò G *et al.* 2013 *Phys. Rev.* C **88** 024316
[6] Heyde K, von Neumann-Cosel P, Richter A 2010 *Rev. Mod. Phys.* **82** 2365
[7] Tamii A *et al.* 2009 *Nucl. Instrum. Meth. in Phys. Res.* A **605** 326 (2009)
[8] Tamii A *et al.* 2011 *Phys. Rev. Lett.* **107** 062502
[9] Poltoratska I *et al.* 2012 *Phys. Rev.* C **85** 041304(R)
[10] Tamii A, von Neumann-Cosel P, Poltoratska I 2014 *Euro. Phys. J.* A **50** 28
[11] Schelhaas K P *et al.* 1988 *Nucl. Phys.* A **489** 189
[12] Roca-Maza X, Centelles M, Viñas X, Warda M 2011 *Phys. Rev. Lett.* **106** 252501
[13] Piekarewicz J *et al.* 2012 *Phys. Rev.* C **85** 041302(R)
[14] Tsang M B *et al.* 2012 *Phys. Rev.* C **86** 015803
[15] Lattimer J M, Lim Y 2013 *Astrophys. J.* **771** 51
[16] Hebeler K, Lattimer J M, Pethick C J, Schwenk A *Phys. Rev. Lett.* **105** 161102
[17] Gandolfi A, Carlson J, Reddy S, Steiner A W, Wiringa R B 2014 *Euro. Phys. J.* A **50** 10
[18] Subedi R *et al.* 2008 *Science* **320** 1476
[19] Ong H J *et al.* 2013 *Phys. Lett. B* **725** 277
[20] Miki K *et al.* 2013 *Few-Body Systems* **54** 1607
[21] Matsubara H *et al.* 2009 *Nucl. Instrum. and Meth. in Phys. Res.* B **267** 3682
[22] Matsubara H *et al.* 2012 *Nucl. Instrum. and Meth. in Phys. Res.* A **678** 122
[23] Matsubara H 2010 *doctoral dissertation*, Osaka University; Matsubara H *et al.* to be submitted elsewhere.
[24] Brown B A, Richter W A 2006 *Phys. Rev.* C **74** 034315
[25] Horiuchi W *private communication*; Horiuchi W, Suzuki Y 2013 *Phys. Rev.* C **87** 034001 and the references teherein.

Cluster correlations in dilute matter and equation of state

Stefan Typel

GSI Helmholtzzentrum für Schwerionenforschung GmbH, Planckstraße 1, D-64291 Darmstadt, Germany

E-mail: s.typel@gsi.de

Abstract. Cluster correlations are an important feature in strongly interacting matter at subsaturation densities. They modify the chemical composition of the system and the thermodynamic properties, which are encoded in the equation of state. The formation and dissolution of clusters can be described in a generalized relativistic density functional approach. Clusters are included as explicit degrees of freedom with medium-dependent properties. They are considered as quasiparticles with scalar and vector self-energies that represent the effective in-medium interaction. Essential ingredients in the self-energies are rearrangement contributions, which guarantee the thermodynamic consistency of the model, and mass shifts of composite particles, which take into account the effects of the Pauli exclusion principle. They cause the dissolution of clusters at high densities. The occurence of α-particle clusters on the surface of heavy nuclei can be considered by adapting the generalized relativistic density functional approach to nuclear structure calculations. A systematic variation of the α-particle abundance and neutron skin thickness with the neutron excess of the nucleus and the strength of the isovector contribution to the effective interaction is observed. The surface α clustering affects the correlation of the neutron skin thickness of heavy nuclei with the density dependence of the symmetry energy, which is relevant for calculations of neutron star structure.

1. Introduction

Dilute matter, i.e. matter with densities smaller than nuclear saturation density $n_{\text{sat}} \approx 0.15 \text{ fm}^{-3}$, can be found at various places in nature, e.g. in astrophysical environments such as the crust of neutron stars or expanding matter in core-collapse supernovae. It can be studied experimentally in the laboratory in heavy-ion collisions or surface-dominated reactions of atomic nuclei.

All the above mentioned systems are very different. They cover microscopic to macroscopic length scales under cold to hot thermal conditions in static equilibrium or dynamical evolution. Nevertheless, they are all interacting many-body systems where correlations are essential. Assuming equilibrium conditions, their thermodynamic properties and chemical composition can be described with appropriate equations of state (EoS). One important aspect is the occurence of clustering at low densities due to the action of competing attractive and repulsive interactions between the constituents. Low-density matter is found on the surface of heavy nuclei where the nucleon densities drop steeply with increasing distance from the center. Four-nucleon correlations, in particular, can appear as α particles, which have to be preformed in order to facilitate the radioactive decay of unstable heavy nuclei.

Two specific systems have to be distinguished in the discussion. On the one hand, only strongly interacting particles are considered in *nuclear matter*, where the electromagnetic interaction is not taken into account. In isospin asymmetric systems with baryon number densities n_b below $n_{\rm sat}$ at not too high temperatures T, a 'non-congruent' liquid-gas phase transition is observed with the coexistence of a more isospin asymmetric low-density phase and a more isospin symmetric high-density phase [1]. On the other hand, both the strong and electromagnetic interactions have to be considered in *stellar matter* with hadrons and leptons as constituent particles under the specific condition of charge neutrality. Under these circumstances, new particle species (nuclei) or 'pasta' phases are formed, e.g. in the crust of neutron stars. Coulomb correlations are responsible for the occurence of lattice structures and the gas-solid phase transition at low temperatures.

Information on correlations in interacting many-body systems is contained in spectral functions, which show a complicated structure in general. The description is simplified by using a quasiparticle approximation with self-energies that take into account a part of the correlations by an in-medium change of the particle properties. With this approach the strength of residual correlations is reduced. The quasiparticle concept is very successful in nuclear physics. E.g. phenomenological mean-field models of Skyrme, Gogny or relativistic type [2] use nucleonic quasiparticles. Pairing correlations are efficiently treated by introducing quasiparticles with the help of a Bogoliubov transformation. In the extreme limit, an exact diagonalisation of a interacting many-body Hamiltonian leads to a system of independent quasiparticles, which are many-body states in this case, without residual correlations. In dilute matter at densities much below $n_{\rm sat}$ clusters/nuclei appear as new degrees of freedom and the properties of the system are described by the model-independent virial equation of state (VEoS) [3], which can be used as a benchmark for model calculations. The considerations above initiated the development of a generalized relativistic density functional (gRDF) for nuclear and stellar matter with the correct limits and explicit cluster of freedom [4, 5, 6].

2. Generalized relativistic density functional

A thermodynamic consistent formulation of the gRDF with an extended set of constituents is most simply obtained by using a grand canonical approach. The grand canonical potential density $\omega(T, \{\mu_i\})$ depends on the temperature T and the chemical potentials μ_i of all particles, which are related by the condition of chemical equilibrium. It is obtained by extending well-known relativistic mean-field models with density dependent meson-nucleon couplings [7]. In addition to nucleons, composite particles, i.e. clusters or nuclei, are introduced as explicit degrees of freedom that change their properties inside the medium. E.g. their mass or binding energy can shift and internal states can be excited thermally. Nucleon-nucleon scattering correlations are considered, too, since they are essential to reproduce the correct low-density limit given by the VEoS. In a relativistic approach, particles and antiparticles can be naturally incorporated. So-called rearrangement contributions guarantee the thermodynamic consistency of the approach. The parameters of this phenomenological model are obtained from a fit to selected properties of finite nuclei [4]. In nuclear matter the condensation of bosons (e.g. deuterons or α particles) can be discussed. In stellar matter the effect of Coulomb correlations with the charge neutrality condition and the transition to the crystal phase are of interest.

The list of constituent particles in the full gRDF model comprises baryons (nucleons and hyperons at densities above $n_{\rm sat}$), four light nuclei (^{2}H, ^{3}H, ^{3}He, ^{4}He) and 16745 heavy nuclei $^{A_i}Z_i$ with mass numbers $A_i > 4$, neutron numbers $N_i \leq 184$ and proton numbers $Z_i \leq 184$. Experimental binding energies of the nuclei are taken from the 2012 Atomic mass evaluation [8] when available. For all other nuclei between the neutron and protons driplines (determined without the Coulomb contribution to the energy) the predictions of the DZ10 model [9] are used. Nucleon-nucleon scattering correlations are represented by effective resonances in the continuum [5]. In stellar matter, leptons and photons are added without any problems. The dispersion relation of all interacting quasiparticles i depends on scalar potentials S_i and vector potentials V_i that represent the effective in-medium interaction. For pure nucleonic matter they can be parametrized by suitable functions [10]. In general, scalar and vector potentials receive contributions from Lorentz scalar mesons (σ, δ, σ_*, ...) and Lorentz vector mesons (ω, ρ, ϕ, ...), respectively. These are treated as classical fields in the model. The coupled field equations are derived in the standard way by variation. They have to be solved self-consistently for given temperature, baryon number density and baryonic charge fraction. The parametrization DD2 [4] is used for the density dependence of the nucleon-meson couplings. This effective interaction is characterized by very reasonable nuclear matter parameters (saturation density $n_{\rm sat} = 0.149$ fm^{-3}, energy per nucleon at saturation $E/A|_{\rm sat} = -16.02$ MeV, incompressibility $K = 242.7$ MeV, symmetry energy $J = 31.67$ MeV, symmetry energy slope parameter $L = 55.04$ MeV), which are compatible with most of current constraints. The neutron matter EoS lies nicely inside the error bounds given by ab-initio N^{3}LO calculations in chiral effective field theory (χEFT) [11, 12], see figure 1. Note that effects of clustering or phase transitions on the symmetric matter EoS [6] are not taken into account in the figure.

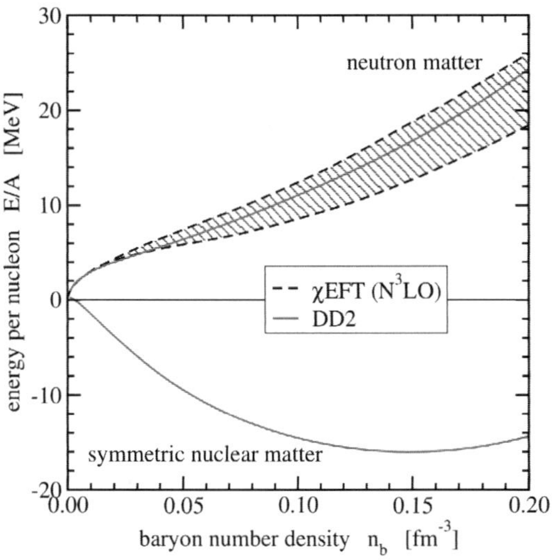

Figure 1. Zero-temperature equation of state for symmetric nuclear matter and pure neutron matter with the DD2 parametrization [4] in the gRDF model with only nucleons as degrees as freedom and without clustering effects or liquid-gas phase transition. The hatched area indicated the bounds for the neutron matter equation of state from chiral effective field theory (χEFT) calculations in references [11, 12].

In addition to the standard mesonic terms, the vector potential V_i contains "rearrangement" contributions and an effective electromagnetic contribution that represents Coulomb correlations, which even appear in uniform systems. The scalar potential S_i includes the medium dependent mass shifts Δm_i of composite particles. It allows to describe their formation and dissolution with changing density and temperature. There are two contributions to the mass shift Δm_i. The strong shift originates mainly from the Pauli exclusion principle. It blocks states in the medium that can no longer participate in the formation of cluster correlations. Thus the binding energies are reduced and the Mott effect, i.e. the dissolution of clusters, is observed with increasing density. This microscopic mechanism replaces the traditional geometric excluded-volume mechnism in order to suppress the cluster formation at high densities. The electromagnetic contribution to the mass shift appears only in stellar matter. It increases the binding energies of clusters because the almost uniformly distributed electrons lead to a screening of the Coulomb field. The mass shifts of light clusters can be calculated by solving in-medium few-body Schrödinger-like equations with realistic nucleon-nucleon potentials. A simple parametrization of the extracted mass shift is used in the gRDF calculations [4]. For heavy nuclei fully self-consistent spherical Wigner-Seitz cell calculations with the gRDF for nucleons and electrons are performed in an extended Thomas-Fermi approximation [13]. From a comparison with calculations of uniform matter, the increase of the binding energy due to cluster formation

Journal of Physics: Conference Series **569** (2014) 012088 doi:10.1088/1742-6596/569/1/012088

can be determined. Presently, only a simplified parametrization of the thus determined mass shifts is employed. The description will be improved with more systematic calculations in the future covering the whole chart of nuclei for various temperatures and densities of the medium.

The modification of the in-medium properties of composite particles has important consequences for the chemical composition of matter. These effects can be explored by studying the mass number fractions $X_i = A_i n_i / n_b$ with $n_b = \sum_i A_i n_i$. For given temperature T and proton fraction Y_p, only two-body correlations are relevant at low n_b and the chemical composition is dominated by nucleons and a few deuteron-like correlations. With increasing density the fraction X_d increases and three-, four- and many-nucleon cluster become more and more significant. Close to the saturation density n_{sat}, the clusters dissolve and nucleonic matter remains at higher densities, see, e.g., the right panel of figure 10 in reference [13]. In figure 2, the mass fractions of nucleons, light and heavy clusters, and electrons are depicted as a function of the temperature T at given baryon number density $n_b = 0.001$ fm^{-3} and total proton fraction Y_p from a preliminary calculation in the gRDF approach. At high temperatures the matter is mainly composed of nucleons, electrons and very few deuterons. The abundancies of light clusters increase when the system is cooled down. At temperatures below approx. 3 MeV, light clusters disappear and heavy clusters dominate the chemical composition. The corresponding evolution of the average mass and charge numbers of the heavy clusters is shown in figure 3. Their strong increase with decreasing temperature is obvious. At very low temperatures, a phase transition from the gas to the solid crystal phase is expected. According to Monte Carlo simulations in the so-called one-component plasma (OCP) model this transition occurs at a plasma parameter $\Gamma_{\text{heavy}} = \langle Z \rangle^{5/3}_{\text{heavy}} e^2/(a_e T) \approx 175$ containing the characteristic length scale $a_e = [(3/(4\pi n_e))]^{1/3}$ that depends on the electron density n_e, see e.g. reference [14]. The vertical dashed line in figure 3 indicates the corresponding temperature for solidification or melting. All the effects of cluster formation and phase transitions have to be implemented in EoS models to provide a reliable input to astrophysical simulations.

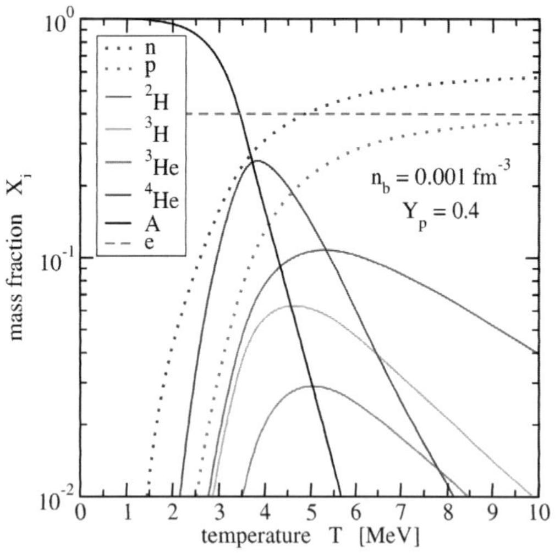

Figure 2. Temperature dependence of the mass fractions X_i for nucleons (n,p), light and heavy clusters (A) and electrons (e) in stellar matter at baryon number density $n_b = 0.001$ fm^{-3} and proton fraction $Y_p = 0.4$.

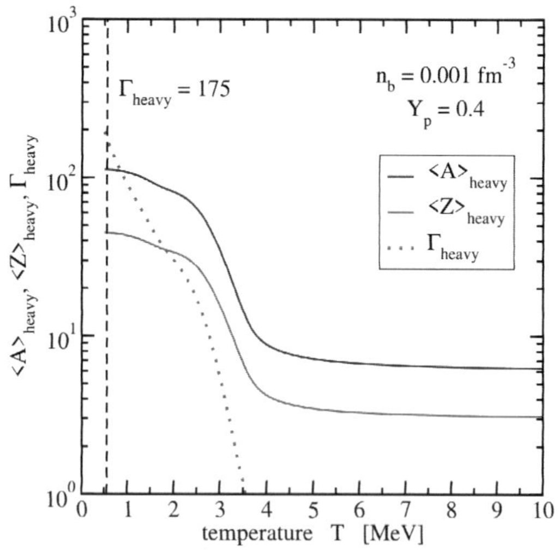

Figure 3. Temperature dependence of the average mass and charge number of heavy clusters and of the plasma parameter Γ_{heavy} in stellar matter at baryon number density $n_b = 0.001$ fm^{-3} and proton fraction $Y_p = 0.4$.

3. Symmetry energy and neutron skins

The energy of nuclear matter at given baryon density n_b depends strongly on the isospin asymmetry of the system, i.e. the proton-to-neutron ratio. This effect can be quantified with the help of the density dependent symmetry energy $E_s(n_b)$ of nuclear matter. It measures the difference of the energy per baryon in pure neutron matter and symmetric nuclear matter. There is a well-known correlation between the neutron skin thickness of heavy nuclei, i.e. the difference between the neutron and proton root-mean-square radii, and the stiffness of the neutron matter EoS [15, 16] or, equivalently, the slope parameter $L = 3n_{\rm sat}\, dE_s(n_b)/dn_b|_{n_b=n_{\rm sat}}$ of the symmetry energy [17], see, e.g., figure 2 in reference [18]. There have been many attempts in recent years to determine experimentally the symmetry energy at saturation $J = E_s(n_{\rm sat})$ and the slope parameter L, see, e.g., references [18, 19]. One approach, PREX@JLab [20], uses parity violating electron scattering on ^{208}Pb that is mainly sensitive to the neutron distribution in nuclei. Combining the results with charge distribution measurements in ordinary electron scattering, the neutron skin thickness $r_{\rm skin}$ can be determined. The $r_{\rm skin} \leftrightarrow L$ correlation can then be used to extract the slope coefficient L of the symmetry energy. However, this correlation is based only on relativistic and non-relativistic mean-field calculations of nuclei that do not take into account the effect of clustering beyond pairing correlations. Thus the question arises how strongly will the $r_{\rm skin} \leftrightarrow L$ correlation be affected by clustering in the neutron skin of nuclei.

Finite temperature gRDF calculations in spherical Wigner-Seitz cells using an extended relativistic Thomas-Fermi (RTF) approximation can describe the formation of heavy nuclei inside dilute stellar matter [13]. They show an increased probability of finding light clusters on the surface of nuclei when these additional degrees of freedom are considered, see figure 11 in reference [13]. In order to apply the gRDF approach to nuclei at zero temperature in the vacuum, the model has to be adapted to this particular situation. When T approaches zero, only α particles remain as relevant clusters. Since they are bosons, a Thomas-Fermi approximation cannot be used for them but their density distribution has to be obtained from a ground-state wave function, which is calculated in the WKB approximation self-consistently with the nucleon distributions. The original DD2 parametrization [4] had to be modified slightly in the Thomas-Fermi calculation in order to improve the description of heavy nuclei because the parameter set was determined by fits to properties of nuclei in the relativistic Hartree (RH) approximation but not in TF approximation. This was achieved by rescaling the σ meson mass and coupling without affecting nuclear matter parameters.

The neutron skin thickness of Sn nuclei with mass numbers from $A = 107$ to $A = 133$ is depicted in figure 4. Without α-particle correlations on the surface there is an almost linear increase of $r_{\rm skin}$ with the mass number A. Taking α-cluster formation into account modifies the mass number dependence of $r_{\rm skin}$ clearly. The largest effect is observed at intermediate mass numbers where the occurrence of α particles reduces the neutron skin thickness considerably since the asymmetry of the density distributions of neutrons and protons is reduced. For very neutron-rich nuclei, almost no α clusters can form in the highly neutron-rich low-density matter on the surface and the reduction of the neutron skin thickness is insignificant. For nuclei with more or less equal numbers of protons and neutrons, no neutron skin develops even if α particles can form abundantly on the nuclear surface. The density distribution of α particles always peaks close to the nuclear matter radius, see figure 1 in reference [21], since α clusters cannot penetrate into the bulk nucleon distributions. The absolute effective number of α particles is always less than one half in the present gRDF model calculations as depicted in figure 4 of reference [21].

In order to study the change of the neutron skin thickness of ^{208}Pb with the slope parameter L including surface α-particle clustering, the isovector part of the effective interaction, which is represented by the ρ meson exchange, was systematically varied. The density dependent ρ meson coupling $\Gamma_\rho(n_b) = \Gamma_\rho(n_{\rm ref}) \exp[-a_\rho(n_b/n_{\rm ref} - 1)]$ depends on two parameters: $\Gamma_\rho(n_{\rm ref})$ at the reference density $n_{\rm ref} = n_{\rm sat}$ and a_ρ. These two parameters were determined by fits to

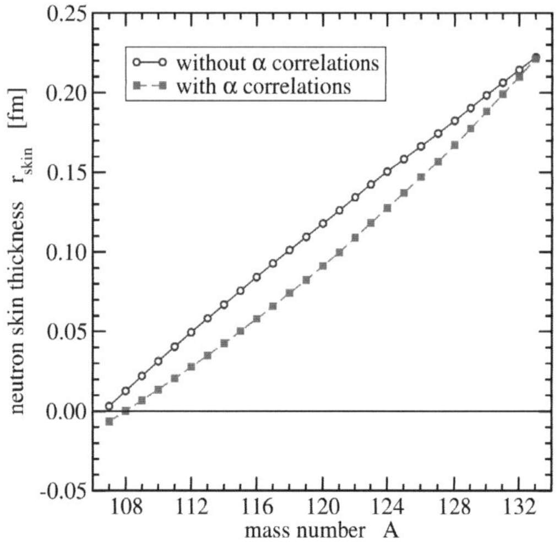

Figure 4. Neutron skin thickness of Sn nuclei as a function of the mass number A in gRDF model calculations with and without α-particle correlations on the nuclear surface.

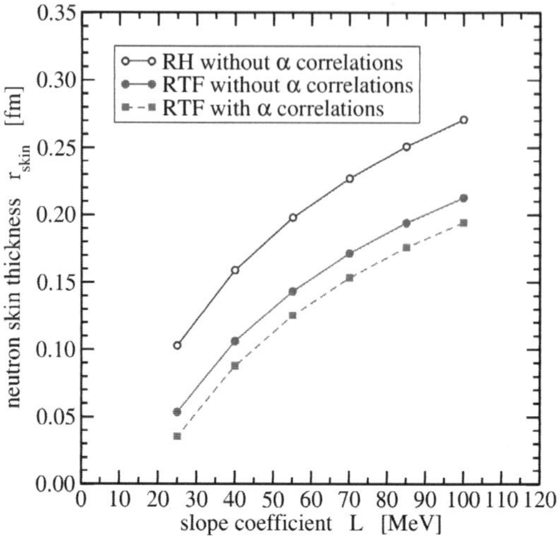

Figure 5. Neutron skin thickness of ^{208}Pb as a function of the slope parameter L in RH and RTF calculations with and without α-particle correlations.

properties of finite nuclei as for the original DD2 parametrization, however demanding specific values of the slope parameter L between 25 MeV and 100 MeV. Actual values for $\Gamma_\rho(n_{\text{ref}})$, a_ρ and the symmetry energy J at saturation are given in table 1 of reference [21]. The correlation of the neutron skin thickness of ^{208}Pb with L is shown in figure 5 for three different calculations. The full blue line with circles represents the results of the original relativistic Hartree calculation without α-particle correlations for different parametrizations of the isovector interaction. When L rises from 25 MeV to 100 MeV, the neutron skin thickness increases from about 0.10 fm to approx. 0.27 fm. The increase is not linear as in the usual representations of the $r_{\text{skin}} \leftrightarrow L$ correlation because only the parameters of the ρ-meson coupling were adjusted in the fit and not all coupling parameters. Changing from the RH to the RTF calculations causes an overall reduction of the neutron skin thickness, however the trend is very similar as in the RH calculation. When α-particle correlations on the surface are included, a further reduction of r_{skin} in the order of 0.02 fm is found. This is a sizeable fraction of the neutron skin thickness if the slope parameter L and correspondingly r_{skin} is already small. The α-clustering effect leads a shift of the $r_{\text{skin}} \leftrightarrow L$ correlation that should be taken into account at least as a systematic error when L is determined from experimentally measured neutron-skin thicknesses of heavy nuclei.

4. Conclusions

Thermodynamic properties and the chemical composition of nuclear and stellar matter can be described properly only if many-body correlations are taken into account with suitable methods. In a generalized relativistic density functional approach cluster correlations are included as explicit degrees of freedom. All constituents in dense matter are considered as quasiparticles with medium-dependent properties, which are represented by scalar and vector self-energies. The effective in-medium interaction is modeled by the exchange of mesons with density-dependent couplings, which are well constrained by properties of finite nuclei. Composite particles receive additional contributions to their self-energies from explicit mass shifts. They incorporate the Pauli exclusion principle and lead to the dissolution of clusters at high densities. In order to reproduce the correct low-density limit, which is given by the virial equation of state, nucleon-

nucleon scattering correlation have to be included in the model. A proper treatment of Coulomb correlations is required to describe the phase transition to a crystal at low temperatures. The thermodynamic consistency of the approach is ensured through the appearance of rearrangement contributions.

The gRDF approach can be applied to the construction of equation of state tables for astrophysical simulations such as core-collapse supernovae or for the description of static properties of neutron stars. A substantial variation of the chemical composition is observed, in particular in dilute matter at subsaturation densities, as a function of temperature, baryon density and isospin asymmetry.

With appropriate adjustments, the gRDF can also be used to describe the properties of atomic nuclei in vacuum at zero temperature. Since atomic nuclei are surrounded by very dilute nucleonic matter it is expected that cluster correlations can affect their surface properties. In first calculations employing simple approximations, it is found that α-particle correlations appear on the surface of heavy nuclei. The actual amount of α clusters depends strongly on the neutron excess of the nucleus. It modifies the neutron thickness as compared to conventional mean-field calculations without explicit four-nucleon correlations. This effect pertains also to the correlation between the neutron skin thickness of heavy nuclei and the slope coefficient of the nuclear symmetry energy, which in turn is related to the stiffness of the neutron matter equation of state.

The prediction of α-cluster correlations on the surface of heavy nuclei can be tested experimentally with quasifree (p,pα) reactions. Such investigations are planned for the future at RCNP Osaka [22].

Acknowledgments

This work was supported by the Helmholtz Association (HGF) through the Nuclear Astrophysics Virtual Institute (VH-VI-417). The author gratefully acknowledges discussions and collaboration with S Antić, T Aumann, D Blaschke, T Klähn, G Poghosyan, G Röpke, T Uesaka, M Voskresenskaya and H H Wolter. He thanks T Yamada and RIKEN for financial support.

References

[1] Müller H and Serot B D 1995 *Phys. Rev.* C **52** 2072
[2] Bender M, Heenen P-H and Reinhard P-G 2003 *Rev. Mod. Phys.* **75** 121
[3] Horowitz C J and Schwenk A 2006 *Nucl. Phys.* A **776** 55
[4] Typel S, Röpke G, Klähn T and Wolter H H 2010 *Phys. Rev.* C **81** 015803
[5] Voskresenskaya M D and Typel S 2012 *Nucl. Phys.* A **887** 42
[6] Typel S, Wolter H H, Röpke G and Blaschke D 2014 *Eur. Phys. J.* A **50** 17
[7] Typel S and Wolter H H 1999 *Nucl. Phys.* A **656** 331
[8] Wang M *et al* 2012 *Chinese Physics* C **36** 1603
[9] Duflo J P and Zuker A P 1995 *Phys. Rev.* C **52** R23
[10] Röpke G, Bastian N-U, Blaschke D, Klähn T and Typel S 2013 *Nucl. Phys.* A **897** 70
[11] Tews I, Krüger T, Hebeler K and Schwenk A 2013 *Phys. Rev. Lett.* **110** 032504
[12] Krüger T, Tews I, Hebeler K and Schwenk A 2013 *Phys. Rev.* C **88** 025802
[13] Typel S 2013 *AIP Conf. Proc.* **1520** 68
[14] Potekhin A Y and Chabrier G 2000 *Phys. Rev.* E **62** 8554
[15] Brown B A 2000 *Phys. Rev. Lett.* **85** 5296
[16] Typel S and Brown B A 2001 *Phys. Rev.* C **64** 027302
[17] Furnstahl R 2002 *Nucl. Phys.* A **706** 85
[18] Viñas X, Centelles M, Roca-Maza X and Warda M 2014 *Eur. Phys. J.* A **50** 27
[19] Lattimer J M and Lim Y 2013 *Astrophys. J.* **771** 51
[20] Horowitz C J *et al* 2012 *Phys. Rev.* C **85** 032501
[21] Typel S 2014 *Phys. Rev.* C **89** 064321
[22] Aumann T and Uesaka T 2014 *private communication*

Journal of Physics: Conference Series **569** (2014) 012089 doi:10.1088/1742-6596/569/1/012089

Search for electromagnetic transitions between molecular states : results from the radiative capture channel

Sandrine Courtin

Institut Pluridisciplinaire Hubert Curien, UMR 7178, Université de Strasbourg / CNRS-IN2P3. 23, rue du Loess, 67037 Strasbourg, France

E-mail: sandrine.courtin@iphc.cnrs.fr

Abstract. This paper discusses the characterization of the cluster nature of resonances in $^{12}C+^{12}C$ and $^{12}C+^{16}O$ by measuring their γ decay. Results from resonant radiative capture measurements in these systems are presented. Spin attributions as well as specificities of the decay pattern are described. For the $^{12}C+^{12}C$ system, identification of γ transitions within the molecular band are discussed as well as a new method to measure the eventual persistence of the resonant behaviour down to astrophysical energies.

1. Introduction

Resonances are well established phenomena in light heavy-ion collisions. They have been observed in the collisions of identical bosons like $^{12}C+^{12}C$, $^{14}C+^{14}C$, $^{16}O+^{16}O$, $^{24}Mg+^{24}Mg$ and $^{28}Si+^{28}Si$ and also in some asymmetric systems like $^{12}C+^{16}O$, at energies ranging from the Coulomb barrier to a few MeV per nucleon. They have been shown to be related to the number of open channels in the reaction [1] and in some cases been interpreted as the result of molecular configurations of the compound system. Despite the wealth of experimental data and theoretical descriptions, the clear origin of these resonances is still a debated question. From an experimental point of view, clustering can be related to properties of the deformed coumpound system or specificities of its decay and fragmentation. In this contribution, we will focus on the simple assumption that if resonances are associated with molecular, very deformed configurations, they should show hints of deformation in their electromagnetic decay. In that sense, cluster states could be grouped into bands with large moment of inertia and connected by very collective transitions. In this paper, we will focus on heavy-ion systems where the resonances are the most pronounced, i.e. $^{12}C+^{12}C$ and, to a lesser extent $^{12}C+^{16}O$, and discuss the experimental study of the electromagnetic decay of resonances of possible molecular nature.

- For lower spin members of the molecular bands in these systems (0^+ to 6^+), radiative capture experiments will be described;
- For higher spin members of the molecular band (8^+ - 10^+), perspectives to measure the gamma transitions between molecular states will be presented;

- As future perspectives, projects to measure deep sub-barrier fusion cross-section in the $^{12}C+^{12}C$ system, towards astrophysical energies will be briefly discussed.

2. Clusters in $^{12}C+^{12}C$ and $^{12}C+^{16}O$: results from the radiative capture experiments

The $^{12}C(^{12}C,\gamma)^{24}Mg$ and $^{12}C(^{16}O,\gamma)^{28}Si$ radiative capture reactions have been performed at the Triumf laboratory (Vancouver) using the Dragon spectrometer and the associated BGO γ array [2, 3, 4], on resonances previously observed by Sandorfi et al. in the 80^{ies} in radiative capture experiments [5]. In these pioneering experiments, excitation functions were measured using a single large NaI γ-detector. Due to the piling-up of low energy γ-rays coming from the much more intense fusion-evaporation channels, the detection was limited to high energy γ-rays feeding low-lying states in ^{24}Mg and ^{28}Si with excitation energies less than $\sim$4.5 MeV. The measured capture cross sections to the first members of the g.s. bands show a series of resonances in both cases. After this radiative capture campaign, Sandorfi et al. improved their experimental set-up by using a Wien filter at $\sim 3°$ for the $^{12}C(^{16}O,\gamma)^{28}Si$ reaction to select the compound nucleus. They could thus measure the feeding of states up to $\sim$9.5 MeV and found, in particular a first indication of the feeding of intermediate states at $\sim$7 MeV in ^{28}Si from a direct γ transition from the resonance.

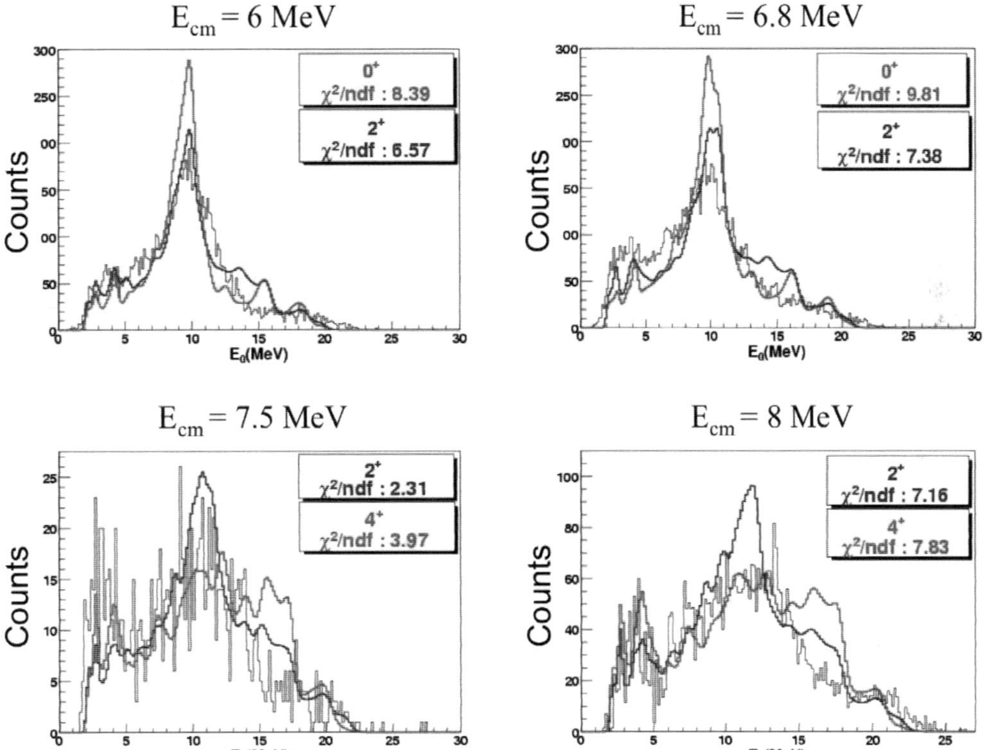

Figure 1. . Radiative capture spectra of the highest energy gamma-ray detected together with numerical simulations for different resonances characteristics.

Our experimental campaigns at Triumf have been performed at energies around and below the Coulomb barrier for both $^{12}C+^{12}C$ and $^{12}C+^{16}O$ systems, where resonances have been found to be narrow ($\Gamma \sim 200$ KeV) and where the number of open channels is rather small.

The Dragon spectrometer was used to select the coumpound nuclei at 0° with an incident beam rejection factor as good as 10^{12}. This type of heavy-ion radiative capture experiments is a real *tour de force* since the process is dominated by fusion-evaporation which is 5 orders of magnitude more important than the radiative capture channel. The coincident γ-rays were detected in an array of 30 BGO scintillators arranged in a compact configuration. This has allowed to measure the complete γ decay scheme from the resonances and the relative decay strengths of all branches in both cases. Surprisingly enough, for both systems, the feeding of intermediate doorway states has been identified around E_γ = 10 and 11 MeV for $^{12}C+^{12}C$ and $^{12}C+^{16}O$ respectively. Monte Carlo GEANT calculations have been performed to simulate the response function of the experimental setup to different entrance channel conditions, i.e. energy and spin of the resonance. Details of these simulations can be found in [3, 4] and we will recall the main steps here. The acceptance of the spectrometer and its transmission is indeed depending on the γ decay path. The complete detection setup has been described in a GEANT code. Properties of the recoiling nuclei have been taken into account such as their mass, recoiling energy and known level scheme, i.e. the full decay path and angular distribution of E_0. The complete geometry of the DRAGON spectrometer as well as its optical elements and the BGO detectors have been taken into account in the calculated spectra. Total number of counts in the simulated γ spectra have been normalized to the total number of counts in the experimental spectra. Our experimental spectra were subsequently compared with simulated ones with different entrance-channel conditions (spins) to infer on the spin of the resonance and the feeding of specific states of the compound nucleus.

- For the $^{12}C+^{12}C$ experiment, five energies have been used: four on resonance at $E_{c.m.}$=6.0, 6.8, 7.5 and 8 MeV, and one off resonance at 6.4 MeV. For the first time in the study of the capture process, the relative strength of all the decay branches could be extracted. We found an important new decay branch to ^{24}Mg states just above the α-threshold (9.3 MeV) which represents ~50 % of the total γ decay width.
  We were able to assign spin 0^+ and 2^+ for the resonances at 6.0 and 6.8 MeV and spin 4^+ or 2^+ would be compatible with our data for the resonances at 7.5 and at 8 MeV (see Figure 1). Enhanced feeding of deformed prolate states in ^{24}Mg directly from the resonance was found for the three highest explored energy. At $E_{c.m.}$=6.0 MeV, namely at the Coulomb barrier, surprisingly enhanced feeding of 1^+ T=1 state around 11 MeV directly from the entrance resonant state was identified. Our results are summarized in Table 1.

Table 1. Radiative capture experimental campaign. Summary of the results for $^{12}C+^{12}C$.

Energy (MeV)	8.0	7.5	6.8	6
Spin/parity	4^+-2^+	4^+-2^+	0^+-2^+	2^+
Observations	Feeding of deformed states			Enhanced feeding of 1^+ T = 1

- For the $^{12}C+^{16}O$ experiments, five energies have been used: four on resonance at $E_{c.m.}$=6.6, 7.2, 8.5 and 9 MeV and one off resonance at 8.8 MeV. Experimental procedure and analysis is similar as for the $^{12}C+^{12}C$ system. Our spin attributions are the following : above the

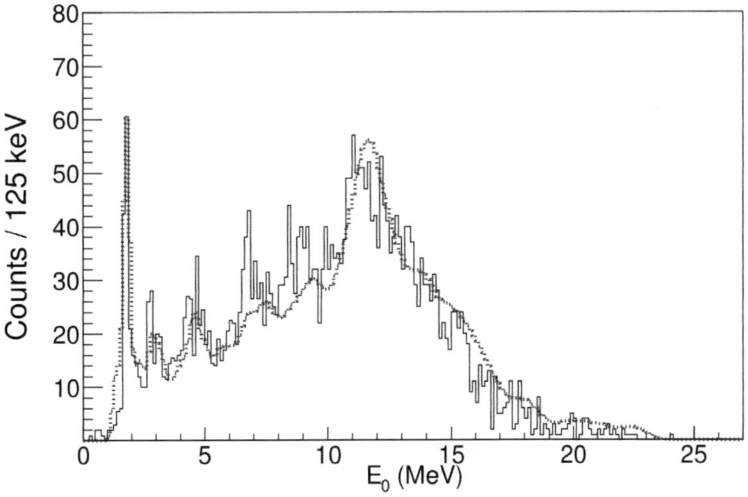

Figure 2. Reaction $^{12}C(^{16}O,\gamma)^{28}Si$, $E_{c.m.}$ = 6.6 MeV: highest energy γ-ray spectrum compared with numerical simulations for a 2^+ entrance resonance and enhanced feeding of $J^\pi = 1^+$, T = 1 states around 11 MeV in ^{28}Si (dashed blue). In the region below 8 MeV, some of the lines may come from the fusion-evaporation reaction leading to ^{28}Si and produced from a possible ^{13}C contamination of the target.

Coulomb barrier ($E_{c.m.}$= 7.2, 8.5 and 9 MeV), spins 6^+ and 5^- are found, in agreement with previous fusion data [6] and below the barrier, a spin 2^+ is found for the resonance. Enhanced feeding of deformed states such at the 3^- state at at 6.88 MeV, from the resonance.This state is the bandhead of the $K^\pi = 3^-$ ^{28}Si octupole band which mainly decays to the g.s. with a strong E3 transition. At the lowest explored energy, we have also observed the enhanced feeding of 1^+ T = 1. The B(M1) values extracted from our experimental data for the transition from the resonance to the T = 1 states are around 1.25 W.u., among the strongest measured in this mass region [7]. Fig. 2 shows the highest energy γ-ray spectrum compared with numerical simulations for a 2^+ entrance resonance and enhanced feeding of $J^\pi = 1^+$, T = 1 states. Experimental results are summarized in Table 2.

Table 2. Radiative capture experimental campaign. Summary of the results for $^{12}C+^{16}O$.

Energy (MeV)	9.0	8.8	8.5	6.6
Spin/parity	6^+	5^-	5^--6^+	2^+
Observations	Feeding of deformed states Agreement with fusion data			Enhanced feeding of 1^+ T = 1 B(M1) $\sim$ 1.25 W.u.

The $^{12}C(^{16}O,\gamma)^{28}Si$ reaction has been performed in the past at higher energies, to study the isospin mixing in the ^{28}Si excitation energy region around 35 MeV. The reaction mechanism was discussed in terms of populating the giant dipole resonance (GDR) built on excited states [11]. It might be that in our studies, the giant quadrupole resonance (GQR) is excited, built on the ^{24}Mg prolate ground state for ^{12}C+^{12}C [12] and on the prolate excited band in the case of ^{12}C+^{16}O.

Both experimental campaings described in this paper used the Triumf BGO array for the measurement of the resonance decay. This allowed high efficiency measurements. The radiative capture studies in these two systems would greatly benefit from better resolution experiments to unambiguously identify the resonance decay scenarios, i.e. the specific states fed and single spins of the resonances. This could be achieved with the use of new generation γ detector arrays based on scintillators such as $LaBr_3$. Such a project, called PARIS [13] has started some years ago and we plan to study radiative capture in the ^{12}C+^{12}C and ^{12}C+^{16}O systems using this array in the coming years.

3. Clusters in ^{12}C+^{12}C : transitions between molecular resonances and future investigations at very low energies

The "smoking gun" in favour of the existence of ^{12}C-^{12}C cluster configurations in ^{24}Mg would certainly be the measurement of electromagnetic transitions between the resonances. This has recently been reported by Datar et al. for a transition from 4^+ to 2^+ resonances in the light ^{8}Be cluster nucleus in the radiative capture experiment ^{4}He+^{4}He [14]. This kind of signature is essentially missing for heavier clusters like ^{12}C-^{12}C. To our knoweledge, there is only one pioneering study which was reported by F. Haas et al. for ^{12}C+^{12}C. They reported the measurement of a handful of events for the gamma transition between the 10^+ and 8^+ resonant cluster state in ^{24}Mg using the Château de Cristal scintillator array for γ detection and two position sensitive detectors for fragment detection installed at the Orsay Tandem (France) [15].
Advances in detector technologies for particle and fragment detection as well as for γ detection with significantly better resolution and efficiency at high energy, will allow to study the γ emission between cluster resonances and should provide unique experimental validations of the cluster models. In that sense, the PARIS calorimeter [13] could be used for γ detection and could be associated to a good coverage fragment detection system.

The ^{12}C+^{12}C reaction is known to be of major importance in the C burning during late stages of the evolution of massive stars. Low energy resonances in the ^{12}C+^{12}C system have thus received large experimental attention. Methods used to measure the low energy ^{12}C+^{12}C fusion cross sections were based on the detection of α or protons from the ^{20}Ne+α and ^{23}Na+p exit channels [16, 17, 18] or the detection of γ-rays of the residues [19, 20]. More recently a rather strong resonance was reported at $E_{c.m.}$ = 2.14 MeV by T. Spillane et al. [21] close to the upper border of the Gamow window (E_{Gamow} = 1.5 MeV at T = 10^7 K). All these low energy fusion experiments suffer from large backgrounds from hydrogen or deuterium contaminations of the target, proton and deuteron recoil particles. This can lead to deviations among the different experimental results. Recently, a new method was proposed by C. L. Jiang et al. to use a particle-γ coincidence technique to reduce the background. First promising results have been obtained using the Gammasphere array at Argonne for γ detection and 3 double sided silicon strip detectors covering $\sim$ 20 % of 4π [22].
This new technique is certainly very interesting to future determine if the cluster resonances persist in the astrophysical energy region, which could have important consequences on the carbon burning in stars.

Acknowledgments
The author acknoweledges support to attend this workshop from the University of Strasbourg Institute for Advanced Study and the France-Japan International Associated Laboratory agreement LIA-FJNSP.

References

[1] Haas F and Abe Y 1981, *Phys. Rev Lett* **46** 1667
[2] Jenkins D G et al. 2007, *Phys.Rev.* **C76** 044310
[3] Lebhertz D et al. 2012, *Phys. Rev.* **C 85** 034333
[4] Goasduff A et al. 2014, *Phys. Rev.* **C 89** 014305
[5] Sandorfi A M 1984, *Treatise on Heavy-Ion Science*, Vol.2, p.52, Ed. Allan Bromley, Plenum Press, New-York, and references therein
[6] Cujec B and Barnes C A 1976, *Nucl. Phys.* **A 266** 461
[7] Endt P M, 1998, *Nucl. Phys.* **A 633**
[8] Harakeh M N et al. 1986, *Phys. Lett.* **B 176** 297
[9] Cujec B 1993, *Nucl. Phys.* **A 552** 267
[10] Youngblood D H, Lui Y W and Clark H L 2007, *Phys. Rev.* **C 76** 027304, and references therein
[11] Harakeh M N et al. 1986, *Phys. Lett.* **B 176** 297
[12] Cujec B 1993, *Nucl. Phys.* **A 552** 267
[13] Maj A et al. 2009, *Acta Phys. Polonica* **B 40** 565
[14] Datar V M et al. 2013, *Phys. Rev Lett* **111** 062502
[15] Haas F at al. 1997, Il Nuovo Cimento **110** 989
[16] Patterson L J, Winkler H and Zaidins C S 1969, *Astrophys. J.* **157** 367
[17] Kovar D G et al. 1979, *Phys. Rev.* **C 20** 1305
[18] Becker et al. 1981, *Z. Phys.* **A 303** 305
[19] Satkoviak et al. 1982 *Phys. Rev.* **C 26** 2027
[20] High M D and Cujec B 1977, *Nucl. Phys.* **A 282** 181
[21] Spillane T et al. 2007, *Phys. Rev Lett* **98** 122501
[22] Jiang C L et al. 2013, *Journ. of Phys. CS* **420** 012076

Journal of Physics: Conference Series **569** (2014) 012090 doi:10.1088/1742-6596/569/1/012090

Clustering and superdeformation in ^{28}Si

David Jenkins

Department of Physics, University of York, York, YO10 5DD, UK
University of Strasbourg Institute of Advanced Study, Strasbourg, France

E-mail: david.jenkins@york.ac.uk

Abstract. Clustering in nuclei is traditionally explored through reaction studies but observation of electromagnetic transitions can be of high value in establishing, for example, that highly-excited states with candidate cluster structure do indeed form rotational sequences. A topical example is given of the identification of a candidate super deformed band in ^{28}Si where super deformation in this nucleus has been described as originating from ^{24}Mg+α clustering.

1. Introduction

Alpha-clustering is often suggested in various alpha-conjugate light nuclei on the basis of the location of excited states which appear to form rotational sequences from their energy spacing. Confirmation of such assignments and a deeper understanding of the clustering phenomenon can only come, however, from observation of electromagnetic transitions connecting these states, or indeed electromagnetic transitions between these cluster states and excited states of standard shell model character. Unfortunately, there are very few cases where this has been done in practice. The best example, perhaps, and where clustering is observed even in the ground state is the case of ^{8}Be. Datar *et al.* carried out a brute-force determination of the $4^+ \rightarrow 2^+$ transition strength in ^{8}Be an experiment at the Tata Institute for Fundamental Research in Mumbai [1]. In this paper, we describe a recent example where the observation of electromagnetic transitions was studied as a means of probing clustering behaviour in nuclei, specifically, in the context of super deformed bands in light nuclei.

2. Candidate superdeformed band in ^{28}Si

Superdeformed (SD) states in nuclei were first reported in rare-earth isotopes like ^{152}Dy [2], and were later found to exist in several mass regions, including those with A$\sim$150, A$\sim$130, and A$\sim$190 [3, 4]. The identification of these weakly-populated, highly-excited structures came about through a step-change in technology with the advent of highly-segmented, high-resolution gamma-ray detector arrays. These same techniques led to the discovery around ten years ago, of SD bands in the light, alpha-conjugate nuclei, ^{36}Ar [5] and ^{40}Ca [6]. These form fascinating examples of superdeformation since complementary descriptions can be found in terms of particle-hole excitations in the shell model [7, 8], and α-clustering configurations within various cluster models [9, 10, 11]. Key theoretical questions center on whether the clustering is a real feature of the system, or whether it simply corresponds to the appearances but is not a true physical description. In addition, a major question is how such clustered configurations evolve into deformed ones. It is particularly important to locate SD bands in lighter, alpha-conjugate nuclei such as ^{32}S and ^{28}Si for which long-standing theoretical predictions exist and which continue to attract the interest of theory. Recent examples of theory initiatives in this area include AMD calculations for ^{28}Si [12] and ^{32}S [13], and macroscopic-microscopic calculations for both nuclei [14]. In all cases, it is predicted that the SD bands in ^{28}Si and ^{32}S should lie at high excitation energy; i.e., with bandheads around 10 MeV. This has two consequences in terms of the challenge in identifying such states experimentally: firstly, phase space favours high-energy, out-of-band transitions compared to low-energy, in-band ones despite the strong collective character of the latter. Secondly, the bandhead lies on or above the particle-decay threshold meaning that there is competition with particle emission.

Recently, Taniguchi *et al.* [12] carried out an extensive study of collective structures in ^{28}Si using the AMD model. They explore clustering degrees of freedom of the type: ^{24}Mg+α and ^{12}C+^{16}O. These studies reveal a rich diversity of rotational behaviors. An SD band is identified in the AMD calculations [12] with a strong ^{24}Mg+α configuration as well as some ^{12}C+^{16}O component. Such a cluster configuration for the SD minimum is supported by recent macroscopic-microscopic potential-energy surface calculations for ^{28}Si [14] as well as by Nilsson model calculations. The AMD calculations [12] suggest that the SD band should have a moment of inertia $\mathcal{J}^{(1)} \approx 6\ \hbar^2/MeV$, related to the large associated deformation, $\beta_2 \approx 0.8$. It is difficult to identify experimental counterparts for the predicted SD states. Taniguchi *et al.* [12] compare their predictions for the SD band in ^{28}Si with the properties of a so-called "excited prolate" band identified in the early 1980s by Kubono *et al.* [15] using the ^{12}C(^{20}Ne,α)^{28}Si reaction. The experimental assignment of this "excited prolate" band rests on peaks in a charged-particle spectrum, and many of the associated states do not have well-established spin/parity. As shown by Taniguchi *et al.* [12], the states identifed by Kubono *et al.* [15] do not form a smooth sequence characteristic of a rotational band even when making plausible allowance for mixing, and the suggested moments of inertia are higher than the calculated values. Moreover, γ-ray transitions between these states are not observed, and, consequently, transition strengths are unknown. Without the observation of in-band transitions, assigning candidate rotational bands is difficult and potentially ambiguous, although such an approach has been a common procedure in the past for "cluster" bands in light nuclei.

The fact that both recent AMD [12] and other [14, 16] calculations suggest that the SD band in ^{28}Si should have a strong ^{24}Mg+α component, raises the question as to whether the ^{24}Mg(α,γ) radiative capture reaction might prove to be a favoured one to selectively populate SD states in ^{28}Si. Such a possibility was not considered by Taniguchi *et al.* [12], but a detailed review of the literature suggests, in fact, that plausible candidates for SD states may already exist. In a series of articles, Brenneisen *et al.* [17] collate data from studies they carried out with the ^{27}Al(p,γ)^{28}Si and ^{24}Mg(α,γ)^{28}Si reactions. Of the large number of states identified in this systematic study, a number stand out as having unusual characteristics. In particular, a 6^+ state at 12.86 MeV is identified which is populated in the (α,γ), but not in the (p,γ) reaction. This 12.86-MeV level has decay branches to a number of states including a 4^+ state at 10.945 MeV, via a 1.921-MeV transition. The observation of a relatively intense, low-energy E2 transition, in competition with high-energy γ rays, immediately suggests that it must have a large transition strength. Brenneisen *et al.* [17] infer that $(2I+1)\Gamma_\gamma > 0.37$ eV for the 12.86-MeV state which means that the transition to the 10.945-MeV level has an associated B(E2) value exceeding 25 Wu [17]. The unusual character of the 10.94-MeV and 12.86-MeV states becomes clear in conjunction with other work such as the ^{12}C(^{20}Ne,α)^{28}Si reaction studied by Kubono *et al.* [15]. In particular, the 10.94-MeV state is the most strongly populated level below 12 MeV (see Figure 1 (a) of Ref. [15]), and it is populated with more than ten times the cross-section of the 4^+ levels in the prolate and oblate ground-state bands. A ^{24}Mg(^{6}Li,d) reaction by Tanabe *et al.* [18] also shows a remarkably similar spectra of states with selective population. Again, the 10.94-MeV level is the most strongly populated one below 12 MeV, exceeding the cross-section to the other 4^+ levels by a similar factor. These observations taken together would suggest that the 10.94-MeV state has a dominant ^{24}Mg+α configuration. Indeed, it is interesting to consider this in the light of studies of the ^{32}S(^{12}C,α) reaction by Middleton *et al.* [19], where the 0^+ state attributed to the 4p-4h configuration is excited ten times more strongly than the level associated with the 0p-0h configuration. The 8p-8h level is excited 1.5 times more strongly than the 4p-4h one. Indeed, the state most strongly excited in this reaction is at 7.98 MeV in ^{40}Ca which has later been shown to correspond to the 6^+ member of the SD band based on the 8p-8h configuration [6].

A state at 12.8 MeV is strongly excited in both the ^{12}C(^{20}Ne,α)^{28}Si [15] and ^{24}Mg(^{6}Li,d) reactions [18]. Analysis of angular correlations in the ^{12}C(^{20}Ne,α)^{28}Si reaction provides a firm assignment of 6^+ to a 12.8-MeV state [15]. This level is also shown to have a direct proton branch to the $5/2^+$ ground state of ^{27}Al [15], implying L = 4 decay and, hence, there must be an associated $g_{9/2}$ component. This is estimated by Kubono *et al.* [20] as corresponding to a spectroscopic factor for the $g_{9/2}$ component of S = 0.3. This result is reinforced by a parallel ^{24}Mg(α,t) study by Kubono *et al.* [15, 20] which also indicated a sizeable $g_{9/2}$ component in the 12.82-MeV state. In this scenario, a consistent picture emerges where the candidate intruder states discussed by Brenneisen *et al.* [17] appear with unusual selectivity in the ^{12}C(^{20}Ne,α)^{28}Si reaction, and with the suggestion of strong deformation, in the case of the 12.86-MeV state.

3. Gammasphere study

Supporting information on the presence of candidate super deformed states in ^{28}Si was obtained from analysis of a data-set related to a γ-ray spectroscopy study where ^{28}Si was one of the main channels [21]. The original objective of the experiment was, in fact, the study of mirror symmetry in ^{31}S and ^{31}P, for which results were published some years ago [22]. Excited states in ^{28}Si were populated via the ^{12}C(^{20}Ne,α) reaction using a 32-MeV,^{20}Ne beam from the ATLAS accelerator at Argonne National Laboratory. A self-supporting ^{12}C target of 90 μg/cm^2 was bombarded with a 40 pnA ^{20}Ne beam for a period of two days. The resulting γ decays were detected by Gammasphere, an array of 100 Compton-suppressed germanium detectors [23]. The array was operated in stand-alone mode with a trigger condition of two or more coincident γ rays. Since evaporated alpha particles were not detected, γ rays associated with ^{28}Si were strongly Doppler-broadened, but the use of high-fold coincidence data still permitted a level scheme for ^{28}Si to be produced from the analysis of a γ-γ matrix and a γ-γ-γ cube. The analysis confirms the location and decay branching of the candidate states in the intruder band identified by Brenneisen *et al.* [17]. Figure 1 provides a simplified level for ^{28}Si and identifies the candidate SD states. A hierarchy of transition strengths is observed, with a clear preference for decay from the SD candidate states to the prolate (ND) states as opposed to the oblate ground state band.

4. Study with K600 spectrometer

Following the identification of candidate SD states in ^{28}Si, a further study to follow-up this interpretation was carried out at iThemba laboratory in May 2013. In this experiment, the ^{12}C(^{20}Ne,α)^{28}Si reaction originally employed by Kubono *et al.* [15] was used to produce high spin states in ^{28}Si. The K600 spectrometer at iThemba was used to detect the reaction alpha particles at zero degrees as a means of selecting the ^{28}Si states of interest. The particle decay of highly-excited states was examined with a pair of annular silicon detectors around the target position, while a very large sodium iodide detector and germanium detectors were used to detect the gamma-ray decay of these states (see Fig. 2). The angular distribution of alpha particles for breakup of ^{28}Si states to ^{24}Mg$_{(g.s)}$ + α may be fitted with Legendre polynomials to determine their spin.

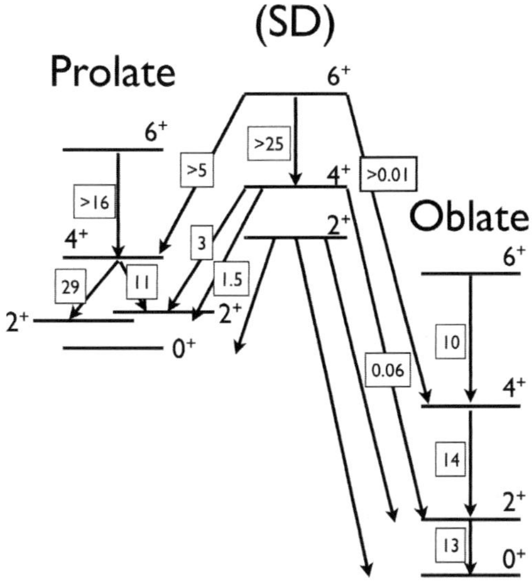

Figure 1. Simplified level scheme for ^{28}Si showing only the oblate ground state band, the prolate band, and the candidate SD band. Transition strengths in Wu are presented for in-band and inter-band transitions originating from the SD band.

5. Future work and conclusions

Several key pieces of information remain outstanding regarding the interpretation of the candidate SD band in ^{28}Si. Firstly, a candidate 0^+ bandhead is not presently known. Secondly, the collectivity associated with these states has not been fully demonstrated. At the present time, there is a lower limit on the B(E2) strength of the $6^+ \rightarrow 4^+$ transition, which albeit a large value is significantly smaller than predicted transitions strengths. At the time of writing, a proposal is being presented to the PAC at RCNP Osaka to carry out a study of the ^{28}Si(α,α') reaction at 400 MeV using the Grand Raiden spectrometer. The CAGRA germanium detector array will be available around the target position and this may be used to study details of the γ decay of excited states produced in the reaction. The state-by-state selection should make the γ spectra very simple, easy to interpret and sensitive to weak γ branches unlike the Gammasphere study where coincident γ gating was required. This study should be well suited to locating a candidate for the 0^+ bandhead of the SD band but also, in particular, it should be possible to search for the $4^+ \rightarrow 2^+$ transition in the SD band. Since the lifetime of the corresponding 4^+ state is known, and the γ branchings measured will be absolute, it should be possible to obtain a B(E2) value for the transition of interest.

In conclusion, candidate SD states in ^{28}Si have been suggested following a synthesis of data on excited states in ^{28}Si from the literature. Analysis of a γ-ray spectroscopy study at Gammasphere supports these conclusions. Recent work with the K600 spectrometer at iThemba and future studies intended for the Grand Raiden spectrometer and CAGRA γ-ray array should place the discussion of the SD band in ^{28}Si on a firm experimental footing.

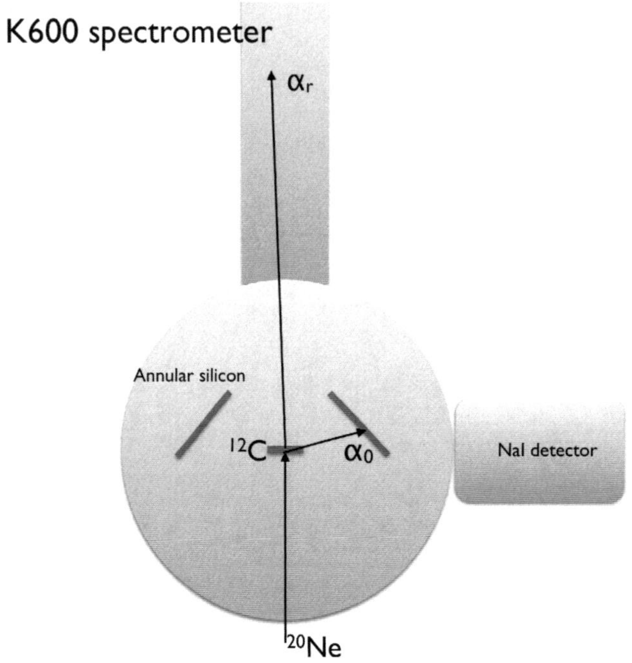

Figure 2. Apparatus used for studies at iThemba

[1] Datar V M *et al.* 2005 Phys. Rev. Lett. **94** 122502
[2] Nyako B *et al.* 1984 Phys. Rev. Lett. **52** 507
[3] Nolan P J and Twin P J 1988 Ann. Rev. Nucl. Part. Sci. 38 533
[4] Janssens R V F and Khoo T L 1991 Ann. Rev. Nucl. Part. Sci. 41 321
[5] Svensson C E *et al.* 2000 Phys. Rev. Lett. **85** 2693
[6] Ideguchi E *et al.* 2001 Phys. Rev. Lett. **87** 222501
[7] Poves A, Caurier E, Nowacki F and Zuker A 2004 Eur. Phys. J. A **20** 119
[8] Caurier E, Menendez J, Nowacki F and Poves A 2007 Phys. Rev. C **75** 054317
[9] Taniguchi Y, Kimura M, Kanada-En'yo Y and Horiuchi H 2007 Phys. Rev. C **76** 044317
[10] Sakuda T and Okhubo S 2004 Nucl. Phys. A **744** 77
[11] Cseh J, Algora A, Darai J, and Hess P O 2004 Phys. Rev. C **70** 034311
[12] Taniguchi Y, Kanada-En'yo Y and Kimura M 2009 Phys. Rev. C **80** 044316
[13] Kimura M and Horiuchi H 2004 Phys. Rev. C **69** 051304
[14] Ichikawa T, Kanada-En'yo Y and Möller P 2011 Phys. Rev. C **83** 054319
[15] Kubono S *et al.* 1986 Nucl. Phys. **A457** 461
[16] Darai J, Cseh J and Jenkins D G 2012 Phys. Rev. C **86** 064309
[17] Brenneisen J *et al.* 1995 Z. Phys. A **352**, 149; ibid. p.279; ibid. p.403.
[18] Tanabe T *et al.* 1983 Nucl. Phys. **A 399** 241
[19] Middleton R, Garrett J D and Fortune H T 1972 Phys. Lett. B **39** 339
[20] Kubono S *et al.* 1986 Phys. Rev. C. **33**, 1524
[21] Jenkins D G *et al.* 2012 Phys. Rev. C **86** 064308
[22] Jenkins D G *et al.* 2005 Phys. Rev. C **72** 031303
[23] Lee I Y 1990 Nucl. Phys. **A520** 641c

Journal of Physics: Conference Series **569** (2014) 012091 doi:10.1088/1742-6596/569/1/012091

$^{12}C + ^{12}C$ resonances by a molecular model

E Uegaki[1] and Y Abe[2]

[1]Graduate School of Engineering and Resource Science, Akita University, Akita 010-8502, Japan

[2]Research Center for Nuclear Physics, Osaka University, Ibaraki 567-0047, Japan

E-mail: uegaki@phys.akita-u.ac.jp

Abstract. Resonances observed in the $^{12}C + ^{12}C$ collisions are studied with a molecular model. At high spins $J = 12$–16, a stable dinuclear configuration is found to be an equator-equator touching one as is obtained in the $^{28}Si + ^{28}Si$ system. With K-quantum number being specified as rotation-vibration model, normal modes have been solved around the equilibrium, firstly. Furthermore, Coriolis coupling has been investigated by diagonalization among low-lying normal-mode states with $K = 0$ and K-excitation. It is found that the Coriolis coupling introduces about 30% of K-mixing into the $K = 0$ molecular ground states of $J = 12$–16. The analyses of the angular momentum coupling show up a dominance of the lowest L in the inelastic channel of the molecular ground state newly obtained. Thus alignments appear strongly in the inelastic channel, which is in agreement with the explanation by the band crossing model. Discussion is given on "why disalignments appear in the $^{28}Si + ^{28}Si$ system, in contrast to the above results".

1. Introduction

The first discovery of the sub-barrier resonances in the $^{12}C + ^{12}C$ system is the starting point of the heavy-ion resonances [1]. Well above the Coulomb barrier, series of resonances have been found with high spins of $J = 12$–16 [2, 3]. Coupled channel calculations have been applied with the elastic and inelastic channels such as 2^+ excitation of ^{12}C nuclei, which are expected to play an important role in the resonances. Band Crossing Model, based on the double resonance mechanism, has successfully explained resonance mechanism and resonance states with the aligned configurations of the orbital angular momentum and the spins of the excited states of ^{12}C [4]. On the other hand, for high-spin resonances observed in $^{24}Mg + ^{24}Mg$ and $^{28}Si + ^{28}Si$, we have developed a new molecular model [5, 6, 7]. In particular, for $^{28}Si + ^{28}Si$, the molecular model naturally explains peculiar disalignments observed between the orbital angular momentum and the spins of ^{28}Si fragments. The resonances in both $^{12}C + ^{12}C$ and $^{28}Si + ^{28}Si$ systems are thought to have dinuclear configurations of oblate constituent nuclei, but they have completely different structures, i.e., aligned in the former while disaligned in the latter, concerning angular momentum coupling. The contrasting features are observed experimentally and well explained by the two models, respectively. Thus, it is keenly interesting to know how they can be understood consistently. In other words, the question is how or in what conditions angular momentum coupling is determined in the resonance states. For the purpose, we take up $^{12}C + ^{12}C$ system with the new molecular model, and show how the aligned configuration arises in the resonances, not the disaligned one.

2. Dinuclear molecular structure of ^{12}C – ^{12}C

2.1. Dynamical property and Normal modes

Assuming a constant deformation and axial symmetry of the constituent nuclei for simplicity, we have seven degrees of freedom $(q_i) = (\theta_1, \theta_2, \theta_3, R, \alpha, \beta_1, \beta_2)$, as illustrated in fig. 1. The relative vector of two ^{12}C nuclei (R, θ_2, θ_1) defines molecular z'-axis. The orientations of the symmetry axes of the ^{12}C nuclei are considered as internal degrees of freedom, and are described with Euler angles (α_i, β_i) referring to the molecular axes. With α_1 and α_2, $\theta_3 = (\alpha_1 + \alpha_2)/2$ of the total K-rotation around z'-axis and $\alpha = (\alpha_1 - \alpha_2)/2$ of twisting motion are defined.

The kinetic energy is given classically in terms of angular velocities, and then the quantization is done to describe the rotation of the whole system with the total angular momentum operator $\hat{J}'$ and the internal motions referring to the molecular axes. The nucleus-nucleus interaction is given with folding potential using the nucleon-nucleon interaction DDM3Y. Expecting a stable configuration, the interaction potential is described with geometrical configurations specified by the internal degrees $(R, \alpha, \beta_1, \beta_2)$. Consistently with the coordinate system, at first we introduce a rotation-vibration type wave function specified with K-quantum number as basis one, $\Psi_\lambda \sim D^J_{MK}(\theta_i)\chi_K(R, \alpha, \beta_1, \beta_2)$, where χ_K describes the internal motions. Correspondingly with specified J and K, we define a multidimensional centrifugal potential

$$T'_{\text{rot}}(J, K; R, \alpha, \beta_1, \beta_2) = \frac{\hbar^2}{2}\Big[\frac{1}{\mu R^2}\Big\{J(J+1) - \frac{3}{2}K^2 + \frac{1}{2}\cos 2\alpha \cot\beta_1 \cot\beta_2 (K^2 - 1)\Big\} + \Big(\frac{1}{I} + \frac{1}{\mu R^2}\Big)\Big(\frac{K^2 - 1}{4\sin^2\beta_1} + \frac{K^2 - 1}{4\sin^2\beta_2} - \frac{1}{2}\Big)\Big], \quad (1)$$

where I denotes the moment of inertia of the constituent nuclei ^{12}C, the value of which is estimated from the excitation energy of the 2^+_1 state of ^{12}C. In order to know dynamical aspects of multi-dimensional internal motion, we calculate the effective potential for given J and K,

$$V_{JK}(R, \alpha, \beta_1, \beta_2) = V_{\text{interaction}}(R, \alpha, \beta_1, \beta_2) + T'_{\text{rot}}(J, K; R, \alpha, \beta_1, \beta_2). \quad (2)$$

In fig. 2, a multidimensional energy surface is displayed. A local minimum is obtained around $\beta = 90°$ and $R = 5.6$ fm, which is called as *equator-equator* (E-E) configuration. The barrier height is 5.7 MeV located at $R \sim 7.5$ fm.

Couplings among various molecular configurations are taken into account by the method of normal mode around the equilibrium at $\beta_i = 90°$, which gives rise to the molecular modes of

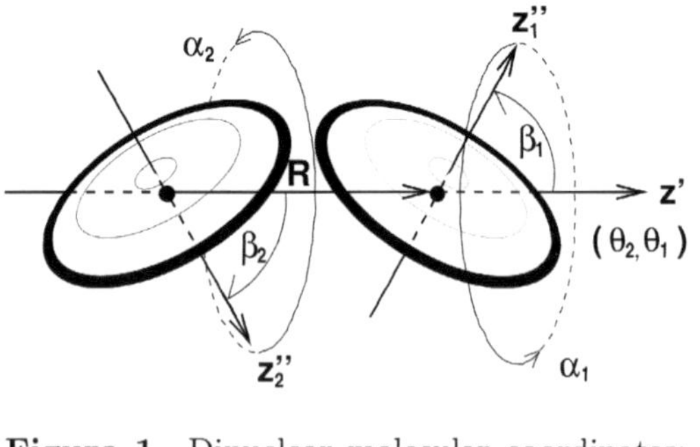

Figure 1. Dinuclear molecular coordinates; seven degrees of freedom of ^{12}C+^{12}C (oblate-oblate system) are described with $(q_i) = (\theta_1, \theta_2, \theta_3, R, \alpha, \beta_1, \beta_2)$.

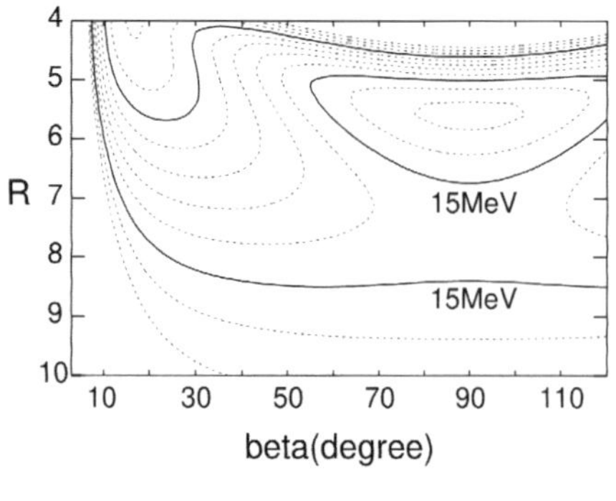

Figure 2. Contour map of the effective potential energy V_{JK} with $J = 14$ and $K = 0$ for the $R - \beta(\beta_1 = \beta_2)$ degrees at $\alpha = \pi/2$. Contours are displayed by 2 MeV.

excitation. Combining the kinetic energy operator and the expanded effective potential, the total hamiltonian is written as

$$H = H_0 + T'_C + (higher\ order), \tag{3}$$

where H_0 and T'_C denote the normal-mode hamiltonian and the Coriolis coupling, respectively. H_0 is separated into each internal degree of freedom, which means the internal degrees of freedom $(R, \alpha, \beta_1, \beta_2)$ are the normal modes themselves, as follows,

$$H_0 = H(R) + H(\alpha) + H(\Delta\beta_1) + H(\Delta\beta_2) + H_{\rm coupl}(\alpha, \beta_1, \beta_2) \tag{4}$$

where $\Delta\beta_i$ denote $\beta_i - \pi/2$. Additional $H_{\rm coupl}$ includes coupling among β_i and α such as $\beta_1\beta_2 \cos 2\alpha$ as well as corrections for confinement of β_i on weak K-dependence. Thus we use basis wave functions, $\Psi_{JMK}(\theta_i; R, \alpha, \beta_1, \beta_2) \sim D^J_{MK}(\theta_i) f_n(R)\phi_\nu(\alpha)\varphi_{n_1}(\beta_1)\varphi_{n_2}(\beta_2)$, where f_n, φ_{n_1} and φ_{n_2} are H.O. functions, while ϕ_ν is a rotational function $e^{i\nu\alpha}$ or $\cos \nu\alpha$, since the confinement in α-motion is very weak. The energy of the basis state is given as

$$\begin{aligned} E_{JK}(n, \nu, n_1, n_2) &= E_0(R_e) + \{J(J+1) - K^2 - 1\}\hbar^2/2\mu R_e^2 + \{K^2 - 2\}\hbar^2/4I \\ &\quad + (n + 1/2)\hbar\omega_R + (n_1 + n_2 + 1)\hbar\omega_\beta + \nu^2\hbar^2/4I, \end{aligned} \tag{5}$$

where $\hbar\omega_\beta \sim 5$ MeV. $H_{\rm coupl}$ and coupling by $H(\alpha)$ have been diagonalized. Since H_0 is almost separable, the eigenenergy of H_0 is very close to eq. (5) except for the following case: $H_{\rm coupl}$ couples strongly between the basis states with $(n_1, n_2, \nu) = (2, 0, 0)$, $(1, 1, 2)$, in which butterfly and anti-butterfly modes appear as likely described with $\beta_\pm = (\Delta\beta_1 \pm \Delta\beta_2)/\sqrt{2}$ [6, 7]. Note that a selection rule exists as $K \pm \nu = even$ in general. In addition, we have rules $(-1)^{n_1} = (-1)^{(K+\nu)/2}$ and $(-1)^{n_2} = (-1)^{(K-\nu)/2}$. In fig. 3, molecular normal modes of $^{12}C + ^{12}C$ with spin 14 are displayed, classified with quantum numbers as $K = 0, 1, 2$. The first excited state of $K = 0$ is twisting excitation with $\nu \sim 4$ of α-rotation, and the next levels are of the butterfly modes. The lowest $K = 1$ state starts with β excitation of $(n_1, n_2, \nu) = (1, 0, 1)$ due to the selection rule.

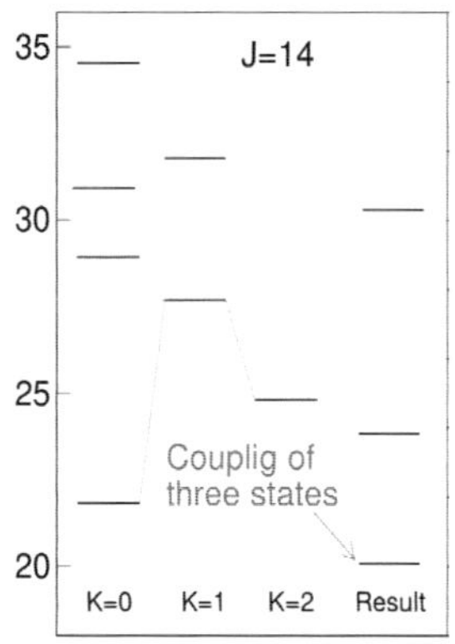

Figure 3. Energy spectra of molecular normal modes of $^{12}C + ^{12}C$ for $J = 14$. From the left, levels with $K = 0, 1$ and 2 are displayed. Diagonalization of the Coriolis coupling with 3 low lying states of $K = 0, 1, 2$ gives a new spectrum in the r.h.s.

2.2. Coriolis coupling

Generally in the rotating system, we have Coriolis coupling, which is denoted as T'_C in eq. (3). It is obtained as many terms described with differential operators of (θ_i) and those of the internal variables $(\alpha, \beta_1, \beta_2)$, but we can make them into a simple form such as a product $-\boldsymbol{J}' \cdot (\boldsymbol{S}'_1 + \boldsymbol{S}'_2)/\mu R^2$ between the total angular momentum operator $\boldsymbol{J}'$ and spins of the constituent nuclei $\boldsymbol{S}'_i$ which refer to the molecular axes. We consider it as residual coupling, which causes K-mixing with $\Delta K = 1$. In $^{12}C + ^{12}C$, in which alignments are expected to be important, we have to take this coupling into account, as it favors alignments. Actually, in the molecular ground state, $(K = 1, \nu = 1)$ mixed component appears to bring strong alignments in the single 2^+ channel.

Since we specify K-quantum number at the first stage, the angular momentum coupling has its own characteristics of each specified K. In fig. 4, a configuration of $^{12}C + ^{12}C$ is displayed. Spin vectors are illustrated as an example, upward ones (black) of which are aligned with the orbital angular momentum $\boldsymbol{L}$, while downward ones (red) are anti-aligned with $\boldsymbol{L}$. The downward spins originate from the upward ones by π-rotation around z'-axis, followed with the phase $\exp(iK\theta_3)$. Thus characteristics of the angular momentum coupling are inferred: 1) amplitudes $A_{J\pm S}$ are almost in the same magnitudes, 2) relative phase between A_{J-S} and A_{J+S} is $(-1)^K$, which is confirmed later in table 1.

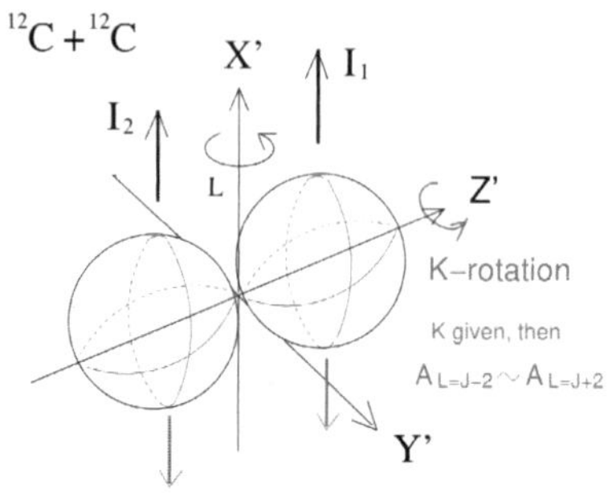

Figure 4. A configuration of $^{12}C + ^{12}C$ is displayed. Upward spin vectors (black) are aligned with $\boldsymbol{L}$, and downward ones (red) are anti-aligned with $\boldsymbol{L}$, both of which are included in the states with specified K.

Usually the spins $\boldsymbol{S}_i'$ are described with differential operators of Euler angles β_i. It is noted here that β_i can describe not only rotational motions but also vibrational motions of the poles of the constituent nuclei, depending on the potential. In our molecular model, with respect to confinements around the equilibrium configurations, the normal modes of β_i-motions are vibrational. For the H.O. states, the differential operators of β_i are written as $-\partial/\partial\beta_i = -\partial/\partial\Delta\beta_i = \sqrt{m\omega_\beta/2\hbar}(b_i^* - b_i)$, and $\Delta\beta_i = \sqrt{\hbar/2m\omega_\beta}(b_i^* + b_i)$, respectively, where b_i^* and b_i denote the creation and annihilation operators of bosons for β_i, respectively. Note that m is defined as $(1/I + 1/\mu R_e^2)^{-1}$, and $\hbar/m\omega_\beta$ is nondimensional. The Coriolis coupling operator $\boldsymbol{J}'(\boldsymbol{S}_1' + \boldsymbol{S}_2')$ connect different K states, namely $\boldsymbol{J}'$ gives lowering and raising operators of K, $J_\pm' = J_1' \pm iJ_2'$, respectively. Thus the Coriolis coupling is written with those operators and we can calculate its matrix elements among K-states.

Low lying three states of $K = 0, 1$ and 2 are taken for basis states, and the Coriolis coupling between them are solved. In fig. 3, energy levels before the diagonalization are connected by red lines and the resulting levels are displayed in the r.h.s for $J = 14$. The molecular ground state of $J = 14$ is obtained with energy gain 1.8 MeV from the $K = 0$ state, in which mixed probability with $K \neq 0$ is 30%. In fig. 5(a), the energy spectrum of basis states for $J = 12 - 16$ are displayed, and in (b), the spectrum obtained after K-mixing are displayed, classified into resultant three bands. For comparison, experimental resonance levels with spin assignments [3] are indicated in the r.h.s. Level energies of the first band are in good agreement with the experiment. Note that for the states of the second and third bands to be reliable solutions, they need more extended basis states than three low-lying basis ones for the diagonalization, as the energies of the states of those excited bands are close to those of the other basis states.

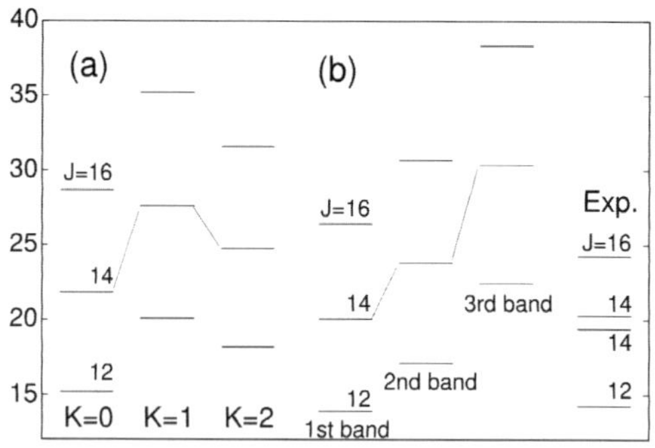

Figure 5. (a) Each energy spectrum of $J = 12 - 16$ for the low lying $K = 0, 1, 2$ states, (b) Whole spectrum after K-mixing are displayed, where thin lines are an eye-guide for $J = 14$. Experimental resonance levels with spin assignments [3] are given.

Journal of Physics: Conference Series **569** (2014) 012091 doi:10.1088/1742-6596/569/1/012091

3. Analyses of angular momentum coupling

We examine the angular momentum coupling of the molecular wave functions, before and after K-mixing, by calculating overlapping between those functions and the channel wave functions. We define a generalized channel wave function, with a given orbital angular momentum l for the relative motion as $\mathcal{Y}_{(I_1I_2)Il;JM} = \sum_{M_I,m}(IlM_Im|JM)\mathcal{S}_{12}[\psi_{(I_1I_2)IM_I}Y_{lm}(\theta_2,\theta_1)]$, where $\psi_{(I_1I_2)IM_I}$ denotes a channel wave function composed with D-functions, i.e., rotational functions for the constituent nuclei with spins (I_1, I_2) and a specified channel spin I. $\mathcal{S}_{12}$ denotes the symmetry operator between two identical constituent nuclei. The overlapping integral is defined as follows, to obtain the distribution on $[(I_1, I_2), L]$,

$$A_{JK;(I_1I_2)L} = \int d\Omega \int \Big[\sum_I \mathcal{Y}^*_{(I_1I_2)IL;JM}\Big] \Psi'_{JMK}(\theta_i;\alpha,\beta_1,\beta_2)\sqrt{\sin\beta_1\sin\beta_2}d\alpha d\beta_1 d\beta_2, \quad (6)$$

where Ψ'_{JMK} is defined as $\Psi_{JMK}/f_n(R)$, dropping the radial wave function out, from the eigenstate of H_0. Ω denotes the solid angle of (θ_i). As we can transform the generalized channel wave function into a kind of molecular wave function, with the rotational function of the whole system $D^J_{MK}(\theta_i)$ in a series of K, we easily calculate overlapping integrals [8]. Note that $\sqrt{\sin\beta_1\sin\beta_2}$ of the integral volume elements in eq. (6) is due to the definition of the volume element $dV = dRd\alpha d\beta_1 d\beta_2$ for the model wave function with vibrational description.

In table 1, the amplitudes $A_{JK;(I_1I_2)L}$ are shown for each K-state of the low-lying series of $J = 14$ and for the resultant molecular ground state with K-mixing by the Coriolis coupling. Only single 2^+ channel is examined, because the amplitudes of the mutual one are small for the low lying K-states adopted. In each K-state, we see almost the same magnitudes of the amplitudes for $L = 12$ and $L = 16$, as expected by the consideration with fig. 5, and thus the states with specified K have not spin alignments. However, after K-mixing by the Coriolis coupling, the components of $L = 14$ and 16 in the single 2^+ channel extremely diminish as given in the rightmost column, and we obtain strong spin alignments.

Table 1. Analyses of alignments by the distribution of the amplitudes $A_{JK;(I_1I_2)L}$ in the elastic and single 2^+ channels of the states of $J = 14$. $A_{JK;(I_1I_2)L}$ of three states, $K = 0, 1$ and 2 are indicated in the left, and the total amplitudes of the mixed results with the coefficients (the values in the parentheses) obtained by the Coriolis diagonalization are given in the r.h.s.

	$K = 0(0.836)$	$K = 1(-0.465)$	$K = 2(-0.289)$	total(K-mixed state)
elastic channel	0.86	0	0	0.72
single 2^+ channel				
$L = 12$	-0.29	0.65	0.36	-0.65
$L = 14$	0.24	-.06	0.80	-1.$\times 10^{-3}$
$L = 16$	-0.31	-0.65	0.30	-0.04

4. Discussion — *Different characteristics in disalignments of $^{28}Si + {}^{28}Si$* —

The resonances in ^{28}Si + ^{28}Si system are expected to be similar to ^{12}C + ^{12}C, with dinuclear configurations of oblate constituent nuclei, but those systems have completely different structures. Namely, disalignments in ^{28}Si + ^{28}Si are observed experimentally and well explained by the molecular model [8]. Thus, it is very interesting to know how they can be understood consistently, i.e., to know what determines the angular momentum coupling of the resonance states.

As explained in [7], we have found an equilibrium configuration like two pancakes sitting on the plane, as is displayed in fig. 6, the shape of which is axially asymmetric as a whole. With tight binding between two ^{28}Si, the whole compound system is expected to behave like *an axial-asymmetric rotator*, which rotates wobblingly on the plane with a high spin. In the high-spin limit ($|K|/J \sim 0$), we obtain analytical wobbling (K-mixing) solutions of the asymmetric rotator as $F_n(K) = H_n(K/b)\exp(-K^2/2b^2)$ with the Hermite polynomial H_n, where K is taken to be $K = even$. Since the intrinsic deformation axes of ^{28}Si are almost orthogonal to the plane, the spin vectors are in the plane, which is the origin of disalignments.

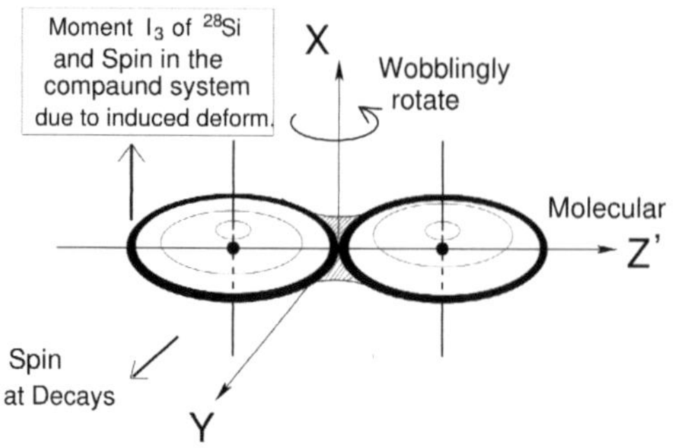

Figure 6. A configuration of ^{28}Si + ^{28}Si is displayed. Spin vectors in the wobbling motions and those at decays are illustrated respectively, as an example, the latter of which are in the reaction plane and are disaligned with $\boldsymbol{L}$.

In this context, we have assumed that the axis of the largest moment is along X-axis in fig. 6, due to the density distribution of the whole system of fig. 6. Namely, in the quantum mechanical understanding, we have assumed new moments of inertia along X-axis, which are expected to come out with induced deformations of the constituent ^{28}Si nuclei. Of course, at decays, the induced deformations vanish and ^{28}Si nuclei return to be normal ones. Note that in the *sticking limit*, the spins $\boldsymbol{S}'_i$ of ^{28}Si nuclei along X-axis are frozen and are constrained to be proportional to $\boldsymbol{J}'$, and thus the the energy of the Coriolis coupling turns to ${J'_x}^2$, as a part of rotator energy. This is the reason why the system turns to be like an axial-asymmetric rotator [7]. Thus the necessary condition is enough binding of the constituent nuclei to be a compound system like an axial-asymmetric rotator.

5. Conclusions

Molecular model has been applied to ^{12}C + ^{12}C system, where the double resonance mechanism is known to be successful. With respect to the Coriolis coupling, K-mixing is calculated. Alignments between the orbital angular momentum and the fragment spin in the single 2^+ channel have been obtained, which is qualitatively in agreement with band crossing model.

In conclusion, description of the molecular model includes various angular momentum couplings, both alignments and disalignments. The answer for "which coupling scheme appears?" depends on strength of nucleus-nucleus interactions. Actually, in the heavier system such as ^{28}Si + ^{28}Si, the interaction between the constituent nuclei is stronger than ^{12}C + ^{12}C case, which is exhibited by the deep potential energy surface around the equilibrium. Moreover, the heavier constituent nuclei are expected to deform easily from their original deformation, to form a total compound system. Thus we can naturally understand the mechanism for the contrasting features of angular momentum couplings between the two systems.

References

[1] Almqvist E, Bromley D A and Kuehner J A 1960 *Phys. Rev. Lett.* **4** 515
[2] Cormier T M *et al* 1978 *Phys. Rev. Lett.* **40** 924
[3] Szanto de Toledo A *et al* 1984 *Phys. Rev.* C **30** 1706
[4] Abe Y, Kondō Y and Matsuse T 1980 *Prog. Theor. Phys. Suppl. No.* **68** 303
[5] Uegaki E and Abe Y 1993 *Prog. Theor. Phys.* **90** 615
[6] Uegaki E and Abe Y 1994 *Phys. Lett.* B **340** 143
[7] Uegaki E and Abe Y 2012 *Prog. Theor. Phys.* **127** 831
[8] Uegaki E and Abe Y 2012 *Prog. Theor. Phys.* **127** 877

A summary of the SOTANCP3 workshop

Daniel Baye

Physique Quantique, and Physique Nucléaire Théorique et Physique Mathématique, C.P. 229
Université Libre de Bruxelles (ULB), B-1050 Brussels, Belgium

E-mail: dbaye@ulb.ac.be

Abstract. This text summarizes the talks presented at the 3rd International Workshop on State of the Art in Nuclear Cluster Physics (SOTANCP3) held in Yokohama (Japan) from 26 to 30 May 2014. Some personal opinions are also expressed on two much debated topics: the ^{12}C spectrum in the continuum and conflicting interpretations of cluster wave functions.

1. Introduction

Summarizing the talks presented at a conference is of course an impossible task, for several reasons. Physical reasons are the limited time allowed to a summary and the existence of parallel sessions which would require some kind of quantum nonlocality for the person in charge of this summary. Psychological reasons are bias introduced by personal tastes and, above all, a limited field of expertise (specially the limited experimental knowledge of a theorist).

Anyway, I tried to find a reasonable first approximation. I thus start with some statistics that reveal some properties of the field. Then comes an attempt to answer the question "What is a cluster?" based on the notions of cluster encountered in the various talks. I also give my opinion on two topics which led to hot debates and which are of special interest for me. A brief summary follows.

Because of space limitations, talks are referred to just by the name of the presenter.

2. Some statistics

The SOTANCP3 workshop has attracted more than 130 physicists to Yokohama. About 40% of them came from abroad.

In 16 plenary sessions, 56 invited talks were presented. Among them, 60% covered theoretical aspects and 40% covered experimental aspects. As the organizers rightly decided to have a maximum of oral presentations, parallel (or clusterized?) sessions were organized. The three parallel sessions allowed 35 oral contributions, 80% concerning theory and 20% experiment. Among a dozen posters, about 90% concerned theory.

These statistics lead to questioning the notion of cluster. Is it more a theoretical concept than an experimental fact? The answers from the participants are summarized in the next sections.

3. What is a cluster?, or What is the nuclear cluster physics?

3.1. A definition

At the fifth cluster conference held in Kyoto in 1988, Ikeda proposed a definition of a cluster [1]. A cluster is a "*spatially localized substructure composed of strongly correlated nucleons*". It is characterized by "*internally strong and externally weak interactions*". Moreover, "*α clustering is the most prominent*". A well-developed cluster structure appears in general near thresholds. This last point is exemplified by the Ikeda diagram [1].

In fact, the present workshop shows that several concepts of "cluster" consistent with this definition coexist: (i) microscopic clusters (where all nucleons are described and the Pauli antisymmetrization is performed), (ii) clusters as structureless component particles, and (iii) clusters as emitted nuclei. So maybe should one rephrase the question "What is a cluster?" as "What is the nuclear cluster physics?".

3.2. Several cluster concepts

The cluster concept is different when treated by theory or by experiment.

In theoretical nuclear cluster physics, a cluster is a pre-assumed substructure in bound states but also in the description of collisions. These pre-assumed clusters describe nuclear properties while simplifying the theoretical treatment and interpretation. As already mentioned different types of models may involve microscopic clusters (§4.1), clusters treated as particles (§4.2) or emitted clusters arising from some fission-like process (§4.3).

A significant part of theoretical nuclear cluster physics concerns the development of efficient calculation techniques (§4.4), which may also be useful in other fields of quantum physics.

An important, rather recent, topic is the explanation of the cluster existence from first principles. There is an increasing number of works aiming at showing the presence of clusters in models where no clusters are assumed to exist at the start (§4.5).

In experimental nuclear cluster physics, the aim is a search for evidence and properties of cluster structures in nuclei. The main topic is α clustering (§5.1) but other clustering aspects appear in the study of nuclear molecules and halo nuclei (§5.2). Cluster radioactivity and fission are examples of the emitted-cluster aspect (§5.2). The cluster concept is also useful in the study of hypernuclei or nuclear properties useful for astrophysics (§5.3).

4. Theoretical nuclear cluster physics

4.1. Microscopic clusters

Microscopic clusters are a 77-year old idea. They were introduced by Wheeler [2] who presented a molecular picture where "*the constituent neutrons and protons are divided into various groups (such as alpha-particles)*". These *groups* are our clusters. Wheeler also introduced an antisymmetrized wave function based on α clusters which is now known as the resonating-group wave function.

The microscopic α clusters are in general made of two protons and two neutrons in a $(0s)^4$ translation-invariant harmonic-oscillator state. Horiuchi has insisted on the notion of duality: several apparently different interpretations may coexist (cluster model, shell model, ...). A number of talks have been devoted to the Tohsaki-Horiuchi-Schuck-Röpke (THSR) basis function and its generalizations. Funaki, Horiuchi, and Schuck have presented condensate interpretations of cluster states and suggested that the THSR wave functions imply a non-localized clustering. Suhara has shown that the ^{12}C and ^{16}O wave functions can have a large overlap with 3α and 4α linear chain states. Zhou has generalized this basis for the description of the positive- and negative-parity bands of ^{20}Ne.

An important topic, the resonating-group method (RGM), was not covered in plenary talks but its results served as a reference for several tests of the THSR functions. In parallel sessions, Tchuvil'sky discussed the $\alpha(^3\text{He},\gamma)^7$Be radiative capture with the algebraic RGM and Dohet-Eraly used the RGM to study the $\alpha(\alpha,\alpha\gamma)\alpha$ bremsstrahlung reaction.

Another traditional model, the two-centre microscopic (Brink) model in a generalized form was used by Ito to study coupled configurations of Be isotopes and the corresponding monopole strengths.

The tensor force is the enemy of the simple $(0s)^4$ α cluster which is too simple to take its effect into account. Its introduction requires a generalization. An important progress is reached with generalized microscopic clusters allowing the efficient use of realistic forces. Generalized cluster internal functions are obtained from variational calculations. The cluster structure describes bound/scattering states in the spirit of the RGM, with possible additional configurations at short distances. Navrátil combined the no-core shell model and resonating-group method (NCSM-RGM) to study with realistic forces two-body collisions and the three-body continuum of ^{6}He. Neff described the ^{12}C bound states and the α+^{8}Be scattering by a coupled-channel cluster model.

4.2. Clusters as structureless component particles

Many cluster-model studies treat clusters as particles (i.e. non-microscopic clusters). This requires using effective forces between the clusters.

Again the most convenient cluster is the α particle. This model was used to describe the $3\alpha \rightarrow {}^{12}\mathrm{C}+\gamma$ reaction. How can this 3-body radiative-capture problem with the additional complication of Coulomb forces be treated at very low temperatures? There is a controversy which started with a CDCC calculation of Ogata *et al* [3, 4] (see §6.1). To clarify this 3α problem, Ishikawa solved the Faddeev equation while Yabana performed a direct calculation of the capture rate with the imaginary-time method without reference to cross sections. In both cases, they find a 3α rate in qualitative agreement with the rate in the 1999 NACRE compilation [5], i.e. without the suggested enhancement at very low temperatures.

Photon emissions can reveal the cluster structure. Garrido related the molecular structure to transitions in the continuum of ^{8}Be and ^{12}C.

Heavier clusters such as ^{12}C and ^{16}O can also be useful. Uegaki studied alignment effects of oblate nuclei in ^{12}C+^{12}C resonances. W. Horiuchi reached convergence in a 5-body ^{12}C+p+p+n+n model of ^{16}O. A particular case is the core cluster appearing in 3-body models of halo nuclei. Kikuchi studied the breakup of core+n+n nuclei (^{6}He, ^{11}Li) with complex scaling. Yarmukhamedov discussed the asymptotic normalization factors for core+n+n systems with an application to ^{6}He.

Hypernuclei can also be studied with a similar model where a Λ particle is added. Hiyama presented a t+n+n+Λ model disfavoring the existence ${}^{6}_{\Lambda}$H and an α+n+n+Λ model of ${}^{7}_{\Lambda}$He.

4.3. Clusters as emitted particles

The name "cluster" also applies to final particles, in other words to observable nuclei emitted in some process.

The simplest example are elastic scattering and breakup. Capel discussed the breakup of halo nuclei with CDCC and eikonal methods. Matsumoto used CDCC for the 3-body breakup of ^{6}He on ^{12}C and the one- and two-neutron stripping in the ^{6}He+^{208}Pb collision.

Cluster formation can be studied in different models. With an extended version of antisymmetrized molecular dynamics (AMD), Ono showed the importance of cluster effects in fragmentation and their relation to the symmetry energy in the equation of state (EOS). With a mean-field approach, Robledo related the octupole degree of freedom with cluster emission and discussed nuclei with exotic shapes.

The fine structure in α decay was studied by Ren with a multichannel cluster model for deformed nuclei. Von Oertzen analyzed potential energy surfaces for collinear ternary fission.

4.4. Techniques related to cluster structures

A much used technique is the complex-scaling method (CSM). Lazauskas applied it to the elastic scattering and breakup of 3-body systems as well as to 4-nucleon scattering. Doté combined a coupled-channel CSM with Feshbach projection to the K$^-$pp system. Using the CSM with a complex Gaussian basis, Kamimura described a successful search for the elusive 0_3^+ state predicted in ^{12}C. Katō studied the bound states and resonances in $A = 5$-8 nuclei with the CSM using Green's function. The CSM is an efficient tool which was also used in several other talks.

How to extract information from collisions was discussed by Khoa. He obtained transition strengths in ^{12}C from the inelastic α+^{12}C scattering.

Cluster aspects of the EOS and of phases of matter were also treated. Khan introduced a localization parameter for the transition between quantum-liquid and crystal states. Typel discussed a relativistic density functional with light and heavy clusters and its relation with the EOS.

4.5. Cluster existence from first principles

The appearance of non pre-assumed clusters was obtained for some time with AMD [6]. With this method, Kimura displayed the variety of cluster structures encountered in ^{24}Mg. Isaka showed that there is no parity inversion in the spectrum of the $^{12}_{\Lambda}$Be hypernucleus, unlike in the corresponding ^{11}Be (halo) nucleus.

The nuclear lattice effective-field theory was applied by Epelbaum in a lattice study of ^{4}He, ^{8}Be, ^{12}C, ^{16}O.

Surprisingly, the shell model can also reveal cluster structures. Myo showed the role of the tensor interaction with the tensor-optimized shell model applied to He, Li, Be isotopes. Launey presented a symmetry-guided reduction of the no-core shell-model space and its application to ^{8}Be, ^{12}C, ... Yoshida employed Monte Carlo shell model wave functions to derive intrinsic densities of Be isotopes, displaying clustering.

5. Experimental nuclear cluster physics

5.1. Alpha clustering

In bound states, clustering is not an observable but rather an interpretation. One can thus only rely on indirect experimental evidence. In collisions, it is customary to consider as "clusters" the incoming-channel nuclei or some reaction products. The indirect experimental constraints for the cluster interpretation can come in increasing order of severity from: (i) searches for bands or for missing states in assumed bands, (ii) determinations of α widths and transition probabilities, (iii) measurements of multiplicities and correlation functions of the emitted particles.

Many experiments concentrate on α cluster structures. A well-known but efficient tool is resonant α scattering. Rogachev used it to study ^{10}Be and ^{18}O and asymptotic normalization coefficients (ANC) in oxygen isotopes. Yamaguchi presented results on resonant α scattering on ^{7}Li/^{7}Be and ^{10}Be. Lombardo studied the α resonant scattering on ^{9}Be.

Inelastic α scattering also provides useful information. Itoh studied near-threshold resonances from inelastic α scattering on ^{12}C and ^{16}O. Kokalova and Marin-Lambarri discussed a new excited state of ^{12}C. Kawabata measured monopole strengths of ^{12}C, ^{16}O, ..., ^{40}Ca.

Another approach consists in studying α decay products in some reaction. Gai discussed the Hoyle 2^+ excitation and proposed a D_{3h} symmetry of the ^{12}C spectrum. Kirsebom detected three α particles produced in p+^{11}B reactions. Akimune measured high α multiplicities in the α+^{36}Ar inelastic scattering.

Other probes can also be used to test clustering. Ong presented a (p,d) study of tensor-force effects. Rodrigues studied α clustering with (^{6}Li,d) transfer reactions.

5.2. Nuclear molecules

Nuclear molecules are weakly bound systems involving α or heavier clusters. The clusters can be nuclei in the entrance channel. Courtin studied ^{12}C+^{12}C and ^{12}C+^{16}O molecular resonances by radiative capture. Jenkins excited superdeformed states with the ^{12}C(^{20}Ne,α)^{28}Si reaction.

Luong showed in subbarrier reactions with ^{6,7}Li, ^{9}Be that nucleon transfer is important and reduces fusion.

Heavy clusters also appear in the output channel. Kohley discussed the two-neutron decays of unbound nuclei with a first measurement of the lifetime of the ^{26}O nucleus. Kondo discussed the spectroscopy of the unbound ^{25}O and ^{26}O nuclei. Andreyev presented results on the β-delayed multimodal (i.e. symmetric and asymmetric) fission of some nuclei.

5.3. Hypernuclei and astrophysics

Experiments on hypernuclei were also discussed. Feliciello presented evidence for the $^{6}_{\Lambda}$H hyperhydrogen (evidence not supported by a J-PARC experiment and by theory). Nakazawa explained the search for double Λ hypernuclei with automated emulsion scanning.

Clusters are also useful for nuclear astrophysics. In addition to the many talks on the ^{12}C spectrum already quoted, Tumino presented applications of the Trojan-horse method, which involves a spectator cluster.

6. Comments on two hot topics

6.1. The ^{12}C spectrum

Why is there such an excitement for the spectrum of this "well-known", abundant, stable nucleus? One third of the plenary talks discussed its unbound spectrum! My opinion is that it is the simplest case where resonances display a surprising variety of shapes while remaining accessible to both accurate experiment and theory. The (relatively) small number of nucleons allows a variety of precise model or *ab initio* studies. The fact that the nucleus is stable and the low density of the continuum allow detailed experiments that should be much more difficult for heavier or unstable nuclei.

The structure and interpretation of this continuum remain nevertheless quite difficult for experiment (overlaps of resonances, large widths) and for theory (unusual structures for which models must be adapted). Of particular interest are the Hoyle resonances and the triple-α rate. There has been a long experimental hunt for the 2_2^+ "Hoyle-band" resonance. This resonance was predicted long ago by various authors [7, 8, 9] and this prediction was so strong that, despite the lack of experimental confirmation, its effect was included in the NACRE compilation [5]. Another strange state is the 0_3^+ resonance predicted by Kurokawa and Katō [10] and confirmed by Kamimura in this workshop [11]. The study of further 0^+ states, other J^π resonances and the general structure of the continuum spectrum of ^{12}C are of great interest.

Let me come back to the problems of the triple-α rate for astrophysics. The role of the Hoyle state is well known. At high temperatures, other resonances in addition to 2_2^+ might affect the rate. At low temperatures, a hot debate about a possible strong rate enhancement with respect to the values in the NACRE compilation was initiated by the CDCC calculation of Ref. [3]. I think that this controversy is to a large extent solved by the calculations of Ishikawa [12] and Yabana [13] presented at this conference. In addition, Yabana proposed a plausible explanation for the qualitative validity of the simple description in NACRE.

The problem is however not completely settled. It would be interesting to understand the origin of the discrepancies in other approaches. An unpleasant aspect of comparing references is that the various calculations of the 3α rate usually differ by the $\alpha\alpha$ effective potential. A comparison of models with a benchmark potential (usable in the different techniques) is crucial for closing the debate. In the longer term, an *ab initio* calculation will be of great interest.

6.2. Localized clustering vs nonlocalized clustering, or THSR vs Brink

The THSR basis function provides an amazingly good description of several states of some nuclei with a *single* (though rather complicated) basis function. It has recently been generalized to

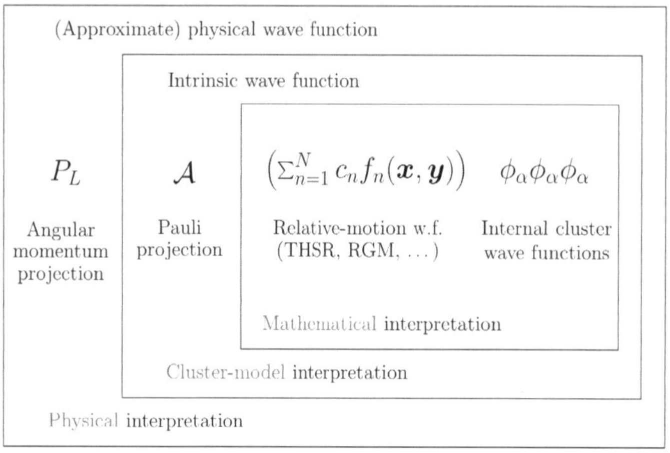

Figure 1. Schematic structure of a microscopic wave function: the 3α example.

other angular momenta and to negative parity. Its validity is established by an almost perfect overlap with accurate RGM wave functions!

So, should we forget the RGM-GCM description with Brink basis functions? The generator-coordinate method (GCM) makes use of linear combinations of angular-momentum and parity projected Slater determinants. More basis states are necessary in this approach. Nevertheless it is very flexible and applicable to most light nuclei. Its results are not improved by the THSR basis which only provides a more elegant presentation. Moreover, its extension to the continuum and to reactions is natural.

I think that while the elegant mathematical simplicity of THSR is a striking discovery, the THSR basis is not competitive (yet?) with the flexibility of Brink expansions in the RGM-GCM.

There are various apparently conflicting visions of the cluster structure. The THSR wave function has led to the suggestion of an interpretation where the clusters are non localized, in contradiction with the definition in §3.1. In order to clarify this question, let me summarize in Fig. 1 the structure of a cluster wave function. To fix ideas, this figure presents a microscopic 3α cluster description of ^{12}C. In this physical cluster-model wave function, one encounters from right to left three translation-invariant internal wave functions ϕ_α of the α particles, a relative wave function depending on two Jacobi coordinates $\boldsymbol{x}$ and $\boldsymbol{y}$ expanded in some basis of N functions f_n, the antisymmetrization projector $\mathcal{A}$ and a projector P_L on angular momentum and parity. The relative wave function $\sum_n c_n f_n(\boldsymbol{x},\boldsymbol{y})$ can be of Brink or THSR or other types. It can thus receive various mathematical interpretations but these are not physically relevant. In the THSR case, the number N of basis states can be very small in some cases. After antisymmetrization, one obtains an intrinsic wave function which can receive a cluster-model interpretation but is not yet physical. Only when good quantum numbers are recovered after projection does one obtain physical, though approximate, wave functions in the laboratory frame.

In the laboratory frame, there is no clear indication of clustering in the properties of 0^+ states obtained with the cluster model. Here one can talk of non-localized clustering. In the intrinsic frame, both the Brink and THSR expansions lead to a similar structure since they have an almost perfect overlap. Intrinsic states may then display a localized clustering in both approaches.

This is exemplified in Figs. 15 and 16 of the *ab initio* ^8Be calculation of Ref. [14]. The 0^+ ground-state density is spherical, displaying no clustering, but the intrinsic density clearly displays two α clusters. More strikingly, the 4^+ density is elongated because of the centrifugal effect and thus different, but the corresponding intrinsic density is very similar to the 0^+ one. More recent *ab initio* calculations show similar patterns. Epelbaum obtained cluster structures on an intrinsic lattice. Yoshida also obtained clusters by transforming from the laboratory-frame shell model to an intrinsic frame.

A number of interpretations concern the relative wave function $\sum_n c_n f_n(\boldsymbol{x}, \boldsymbol{y})$ and are mathematical interpretations: α condensate, particles in a container, localized clusters in the GCM, nonlocalized clusters in THSR functions, gas-like states, ... The cluster-model interpretation is based on intrinsic densities in the intrinsic frame: linear chain, shell-model-like, more or less localized clusters. This approach is the first step in many models: RGM-GCM, THSR, AMD, FMD, lattice EFT, ... Physical interpretations require a projection on good quantum numbers in the laboratory frame where observable properties can be compared with experiment. Such wave functions are directly obtained in *ab initio* calculations: Green-function Monte Carlo, shell models, no-core shell model, ... In this frame, clusters may seem to be absent, or be said nonlocalized or in a gas state.

An example of the difference between the intrinsic and mathematical interpretations can be seen in the THSR 3α description of Funaki. Before antisymmetrization, the clusters do not seem to be localized in the density $|\sum_n c_n f_n(\boldsymbol{x}, \boldsymbol{y})|^2$. After antisymmetrization but before angular-momentum projection, the clusters are localized in the intrinsic THSR density. Another aspect has been illustrated by Zhou: the THSR and RGM intrinsic densities of $\alpha+^{16}O$ look very similar and the clustering into α and ^{16}O can also be said localized.

7. Summary of the summary

Clusters are a useful simplifying approximation in (hyper)nuclear physics. There is a variety of cluster concepts and applications, which is a richness. The theoretical interpretation of a microscopic cluster wave function may be misleading if it is done before antisymmetrization. As the cluster structure is hidden in the laboratory frame, it must be analyzed in the intrinsic frame. Cluster ideas are also very useful in scattering, reactions, decays and many other aspects of nuclear physics, up to now mostly for light nuclei. Significant progresses have recently been made in *ab initio* explanations of the occurrence of a cluster substructure.

The cluster structure is not an observable. Experimental evidence can thus only be indirect. Evidence for a cluster structure requires a variety of experimental information. A new generation of experiments appeared which will allow making more stringent tests of the models.

Nuclear cluster physics has many facets and a lot of exciting work remains to be done. Many thanks to Yamada-san and his team for having perfectly organized this interesting workshop.

Acknowledgments

This text presents research results of the interuniversity attraction pole programme P7/12 initiated by the Belgian-state Federal Services for Scientific, Technical and Cultural Affairs.

References

[1] Ikeda K 1989 *J. Phys. Soc. Jpn.* **58** 277
[2] Wheeler J A 1937 *Phys. Rev.* **52** 1083
[3] Ogata K, Kan M and Kamimura M 2009 *Prog. Theor. Phys.* **122** 1055
[4] Nguyen N B, Nunes F M and Thompson I J 2013 *Phys. Rev.* C **87** 054615
[5] Angulo C et al 1999 *Nucl. Phys.* A **656** 3
[6] Kanada-En'yo Y, Horiuchi H and Ono A 1995 *Phys. Rev.* C **52** 628
[7] Kamimura M 1981 *Nucl. Phys.* A **351** 456
[8] Uegaki E, Okabe S, Abe Y and Tanaka H 1977 *Prog. Theor. Phys.* **57** 1262
[9] Descouvemont P and Baye D 1987 *Phys. Rev.* C **36** 54
[10] Kurokawa C and Katō K 2007 *Nucl. Phys.* A **792** 82
[11] Ohtsubo S-I, Fukushima Y, Kamimura M and Hiyama E 2013 *Prog. Theor. Exp. Phys.* **2013** 073D02
[12] Ishikawa S 2013 *Phys. Rev.* C **87** 055804
[13] Akahori T, Funaki Y and Yabana K 2014 ArXiv:1401.4930v1
[14] Wiringa R B, Pieper S C, Carlson J and Pandharipande V R 2000 *Phys. Rev.* C **62** 014001